THE PYRIMIDINES

SUPPLEMENT I

D. J. Brown
The Australian National University
Canberra

WITH A CHAPTER BY

R. F. Evans
The University of Queensland
Brisbane

AND AN ESSAY BY

T. J. Batterham
The Australian National University
Canberra

1970
WILEY-INTERSCI

a division of

JOHN WILEY & SONS · NEW YORK · LONDON ·
SYDNEY · TORONTO

10 9 8 7 6 5 4 3 2 1

Library of Congress Catalogue Card Number: 59-13038

SBN 471 38116 0

Printed in the United States of America

THE PYRIMIDINES

SUPPLEMENT I

This is the sixteenth volume in the series

THE CHEMISTRY OF HETEROCYCLIC COMPOUNDS

THE CHEMISTRY OF HETEROCYCLIC COMPOUNDS

A SERIES OF MONOGRAPHS

ARNOLD WEISSBERGER and EDWARD C. TAYLOR

Consulting Editors

To
ADRIEN ALBERT

in honour of his significant and continuing
contributions to heterocyclic chemistry

The Chemistry of Heterocyclic Compounds

The chemistry of heterocyclic compounds is one of the most complex branches of organic chemistry. It is equally interesting for its theoretical implications, for the diversity of its synthetic procedures, and for the physiological and industrial significance of heterocyclic compounds.

A field of such importance and intrinsic difficulty should be made as readily accessible as possible, and the lack of a modern detailed and comprehensive presentation of heterocyclic chemistry is therefore keenly felt. It is the intention of the present series to fill this gap by expert presentations of the various branches of heterocyclic chemistry. The subdivisions have been designed to cover the field in its entirety by monographs which reflect the importance and the interrelations of the various compounds, and accommodate the specific interests of the authors.

In order to continue to make heterocyclic chemistry "as readily accessible as possible", new editions are planned for those areas where the respective volumes in the first edition have become obsolete by overwhelming progress. If, however, the changes are not too great so that the first editions can be brought up-to-date by supplementary volumes, supplements to the respective volumes will be published in the first edition.

Research Laboratories
Eastman Kodak Company
Rochester, New York

ARNOLD WEISSBERGER

Princeton University
Princeton, New Jersey

EDWARD C. TAYLOR

The Chemistry of Heterocyclic Compounds

Preface

Although published in 1962, *The Pyrimidines* completely covered the literature only until the end of 1957. The phenomenal advances in pyrimidine chemistry since that year have necessitated the present "Supplement I," which completes the reviewing process until the end of 1967 and covers some important aspects of the 1968 literature. As before, emphasis is placed on practical rather than on theoretical aspects of the subject.

Like any *Ergänzungswerk* volume of *Beilstein*, the present supplement is in no sense a new edition of *The Pyrimidines*, and it must be used in conjunction with the original volume. To facilitate such use each chapter, section, or table heading in the supplement is followed by a reference to the corresponding part of the original volume in the form of a page number preceded by the conventional *H* for *Hauptwerk*; the headings for sections and tables without counterparts in the original volume are followed by *New* in parentheses. The letter *H* after an individual entry in a table indicates that earlier data for the same pyrimidine will be found in the original volume. Access may be gained to the literature and melting point of any simple pyrimidine described to mid-1967 by consulting the appropriate appendix table both in the original volume and in the supplement; the scope and conventions of such tables have been defined (*H* 501 et seq.). To reduce costs and the time-lag in publication the appendix tables (paginated T 1 et seq.) and the single list of references (paginated R 1 et seq.) are reproduced photographically from the typescript; early references (1–2169) are listed only in the original volume (*H* 624). Throughout the supplement the original nomenclature (*H* 3) and general presentation (*H* 501) are retained except that formulas are indicated now by boldface Arabic instead of Roman Numerals.

The origin of recent papers on pyrimidine chemistry is indicated below. Some interesting differences are evident in comparing the list with that given previously (*H* IX): although the United States still occupies first place, its contribution and, more particularly, that of Germany have decreased in favour of those from Russia and Eastern Europe; the British and Japanese percentages remain almost unchanged.

United States of America	31.4%
British Commonwealth	20.2%
Germany (East and West)	13.6%

Japan	8.7%
Russia	6.9%
Eastern Europe	6.0%
France and Switzerland	4.1%
Netherlands and Belgium	2.5%
Austria	2.1%
Italy and Spain	2.0%
Scandinavia	0.9%
China	0.3%
Others including Israel	1.2%

The tasks involved in preparing this supplement have been shared by many kind friends. Dr R. F. Evans and Dr T. J. Batterham willingly put their expertise at my disposal by writing the chapter on reduced pyrimidines and the section on nuclear magnetic resonance spectra, respectively. Professor Adrien Albert made innumerable valuable suggestions; his unfailing encouragement ultimately proved decisive in my completing this work. Dr D. D. Perrin, Dr E. Spinner, Dr W. L. F. Armarego. Dr G. B. Barlin, and Dr J. E. Fildes provided welcome expert advice; Dr T.-C. Lee, Dr H. Yamamoto, and Mr T. Sugimoto translated Chinese or Japanese papers; Mrs D. McLeod solved problems in the library; Mr B. T. England, Mr B. W. Arantz, Mr D. A. Maguire, Mrs P. J. English, and Mrs H. E. Jones assisted in various ways; and Mrs S. M. Schenk cheerfully performed the miracle of producing neat typescript and perfect camera copy from an appalling manuscript. To all these good people, and to my wife and family for their kindly forbearance and practical help, I offer my sincere thanks.

D. J. Brown

The Australian National University, Canberra
September, 1969

Contents

Tables Integrated with the Text

The Appendix

Systematic Tables of Simple Pyrimidines

THE PYRIMIDINES

SUPPLEMENT I

CHAPTER I

Introduction to the Pyrimidines (*H* 1)

Some of the important advances in pyrimidine chemistry during recent years are mentioned very briefly in this Chapter. It is therefore supplementary to Section 4 of the original Chapter (*H* 9) and retains the framework of that general summary. Advances in synthetic methods are deferred to Chs. II and III.

Although no major general reviews of pyrimidine chemistry have appeared recently, the place of pyrimidine in the broad context of heterocyclic chemistry has been defined nicely by A. Albert,*[3774] and in differing ways by several others.[3775–3778]

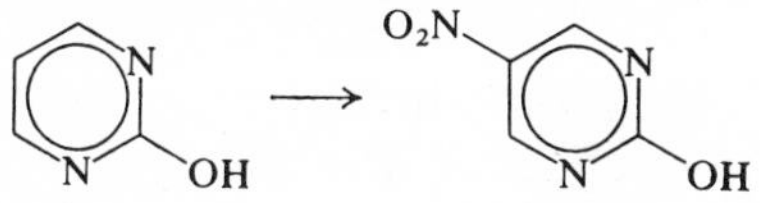

(1) → (2)

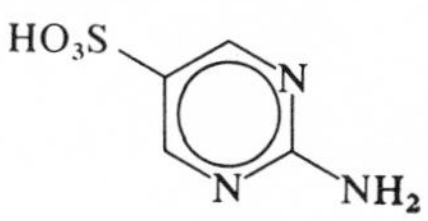

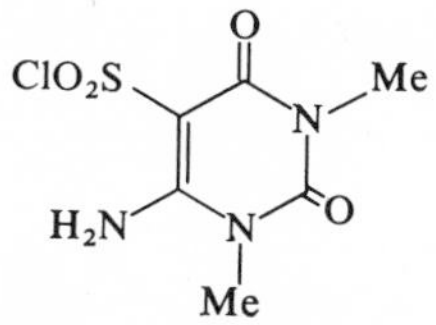

(3) (4) (5)

* Adrien Albert was born in Sydney, Australia, in 1907. He took his first degree at Sydney University followed by the Ph.D. (Medicine) and later the D.Sc. of London University. He was a Member of the Staff at Sydney University for 10 years from 1938, and during much of this time also acted as Advisor to the Australian Army on Medical Chemistry. In 1947 he joined the Wellcome Research Institute in London, but in 1949 became Professor of Medical Chemistry in the Australian National University, a position he still occupies. He was elected a Fellow of the Australian Academy of Science in 1958.

In addition to a great many original papers in *J. Chem. Soc.*, *Biochem. J.*, and *Brit. J. Exper. Path.*, Albert has written several books: *The Acridines*, *Selective Toxicity*, *Heterocyclic Chemistry*, and *Ionization Constants* (with E. P. Serjeant). His passionate devotion to heterocyclic chemistry and its place in medicine is combined with an abiding interest in music, travel, photography, and Australian flora.

The dedication of this book is a small tribute to a teacher, colleague, and friend.

A well documented account of the biological activities of pyrimidines has appeared,[2553] and references to more specialized reviews will be found in appropriate sections of this book.

4. General Summary of Pyrimidine Chemistry (*H* 9)

A. Electrophilic Substitution (*H* 10)

(1) *Nitration and Nitrosation* (*H* 10)

(Ch. V, Sects. 1 and 2)

The presence of at least two electron-releasing groups in pyrimidine is now known to be unnecessary for a successful 5-nitration. Thus 2-hydroxy- (**1**) and 2-amino-pyrimidine both yielded 2-hydroxy-5-nitropyrimidine (**2**) by nitration under very vigorous conditions, and 1,2-dihydro-1-methyl-2-oxopyrimidine gave its 5-nitro derivative similarly.[2431, 3483] In addition, 2,4-diaminopyrimidine has now been nitrated,[2432] and a variety of pyrimidines bearing chloro, alkoxy, alkylthio, and other sensitive groups have been coaxed to yield 5-nitro derivatives.

The oxidation of nitroso- to nitro-pyrimidines,[2445] the recognition of the powerful activating effect of a nitroso group on aminolysis of an alkylthio group,[2473] and the direct use of amino-nitrosopyrimidines in purine syntheses,[2485, 2486, 2497–2500] have stimulated interest in nitrosation processes.

(3) *Halogenation* (*H* 11)

(Ch. VI, Sect. 2.A)

Direct 5-fluorination has been achieved now: 2,4,6-trifluoropyrimidine and silver difluoride in hot triperfluorobutylamine gave tetrafluoropyrimidine (**3**).[2618] Direct 5-iodination, always a difficult task, has been facilitated by using *N*-iodosuccinimide.[2623, 2629]

(4) *Sulphonation* (*H* 11)

(Ch. VIII, Sect. 4.A)

The direct sulphonation of 2-aminopyrimidine to give 2-amino-5-sulphopyrimidine (**4**) has been improved greatly using chlorosulphonic acid;[3232] in some cases, the same reagent has given the sulphonyl chloride, e.g., 4-amino-5-chlorosulphonyl-1,3-dimethyluracil (**5**).[2364]

(6) *Other Electrophilic Attacks* (*H* 11)

Direct introduction of an amide group into the 5-position of 4-amino-2,6-dihydroxypyrimidine resulted from fusion with urea;[3352] although the product (**6**) was first described[3228] as the isomeric 4-ureido derivative [Ch. XI, Sect. 3.A(4)]. The Reimer–Tiemann reaction has been used to make 5-formyluracil (**7**, R = H),[3193] and Vilsmeier reagents (e.g.,

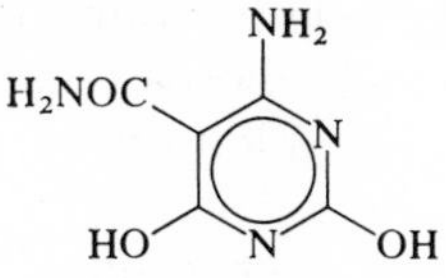

(**6**)

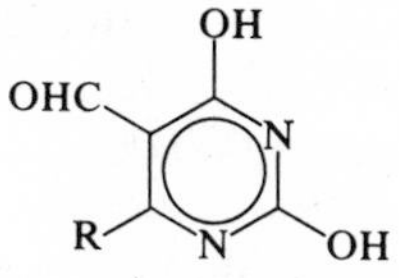

(**7**)

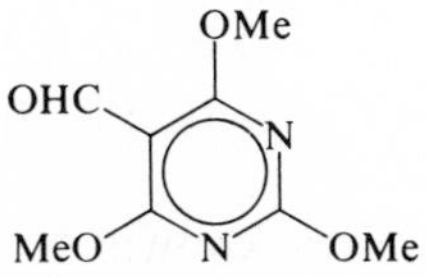

(**8**)

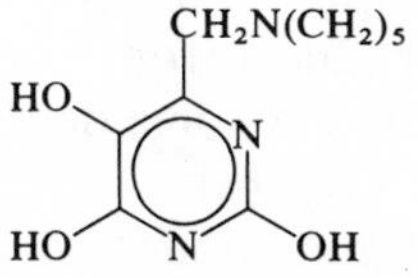

(**9**)

dimethylformamide/phosphoryl chloride) have been applied [Ch. XI, Sect. 5.A(4)] to several alkoxy- and amino-pyrimidines to give, e.g., 5-formyl-2,4,6-trimethoxypyrimidine (**8**);[2815] the latter reagents applied to 6-aminouracil (**7**, R = NH_2) gave 4-amino-2,6-dichloro-5-formylpyrimidine, and occasionally primary-amino groups have become involved too.[2593] The Mannich reaction (Ch. IV, Sect. 2.C) has been used for introducing some quite complicated 5-substituents to give, e.g., 5-bis-(β-chloroethyl)aminomethyluracil.[2658, 2831] The reaction also has been used now to introduce a 6-substituent into suitable pyrimidines: e.g., isobarbituric acid gave its 6-piperidinomethyl derivative (**9**).[3476]

C. Nucleophilic Metatheses (*H* 12)

(1) *Replacement of Halogens* (*H* 13)

(Ch. VI, Sects. 5 and 6)

Despite an increase in the use of alkoxy-, alkylthio-, alkylsulphinyl, and alkylsulphonyl-pyrimidines, halogenopyrimidines remain the most used intermediates for metathesis in the series.

(a) *By Amino or Substituted-amino Groups* (*H* 13). In connexion with the development of a simple method to predict optimum conditions for aminolysis,[2573] some interesting facts emerged: the rate was unaffected by lengthening the alkyl chain of a primary amine or by β- or γ-branching, but one or two α-substituents decreased the rate to 5% and 0.1%, respectively; a secondary dialkylamine was roughly equivalent to a single α-branched primary amine; a 4-chloropyrimidine was usually more reactive than its 2-isomer; the effect of substituents in the pyrimidine ring was indicated in the relative rates for 2-chloro-4,6-dimethyl- (1), 2-chloro- (10), 5-bromo-2-chloro- (200), and 2-chloro-5-nitro-pyrimidine (3,000,000); and the addition of copper had no effect on the rate.[2668]

The use of aq. alcoholic amines has been claimed[2674] to facilitate preferential aminolysis of the 4-position in 2,4-dichloropyrimidines to give, for example, 4-benzylamino-2-chloro-5-methylpyrimidine (**10**). The various products from aminolysis of 2,4,6-trihalogenopyrimidines have been studied in some detail but the picture is still rather confused [Ch. VI, Sect. 5.B(1.b)].

Many 5-bromopyrimidines are now known to undergo normal aminolysis, albeit under vigorous conditions. However, although 4-amino-5-bromo-2,6-dihydroxypyrimidine (**11**) behaved normally with piperidine, with benzylamine it gave only 4-benzylamino-2,6-dihydroxypyrimidine (**12**).[2802] Treatment of 5-chloro-2-methylpyrimidine with sodium amide in liquid ammonia gave 4-amino-2-methylpyrimidine, probably by addition of ammonia to an intermediate 'pyrimidyne' (**13**).[2804] The same reagent produced other fascinating reactions: e.g., 4-chloro-2-phenylpyrimidine gave 2-methyl-4-phenyl-1,3,5-triazine[2554] (see Ch. VI, Sect. 5.I).

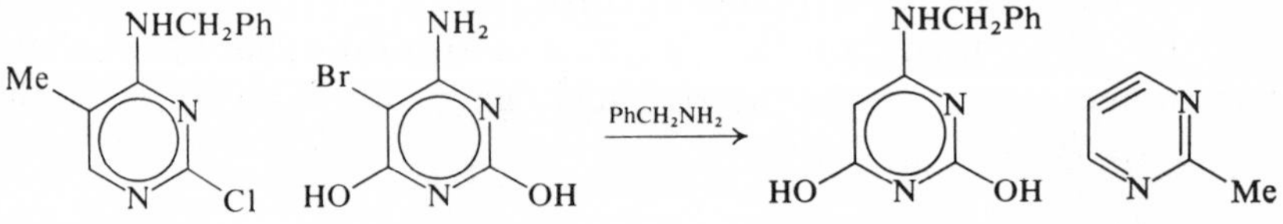

(**10**) (**11**) (**12**) (**13**)

(b) *By Alkoxy Groups* (*H* 13). A considerable selectivity has been achieved recently in the reaction of alkoxides with di- and tri-halogenopyrimidines. For example, 2,4-dichloro- gave 2-chloro-4-methoxy-6-methylpyrimidine (**14**; 90%) under carefully controlled conditions.[2699] The kinetics of the reactions of 2-, 4-, and 5-chloropyrimidine with *p*-nitrophenoxide ion have been measured and compared with those for other chlorodiazines.[3195] A few examples of the reaction of substituted 5-halogenopyrimidines with alkoxides are known now.[2236, 2807–2810]

(c) *By Hydroxy Groups* (*H* 14). The long-standing inhibition of the direct hydrolysis of chloro- to hydroxy-pyrimidines no longer appears to operate. A great many examples are known (Ch. VI, Sect. 5.E) and even preferential hydrolyses are well represented in the formation of 4-chloro-6-hydroxy-5-nitropyrimidine (**15**; by sodium carbonate at 40°) and such like.[2240]

(d) *By Alkylthio and Arylthio Groups* (*H* 14). The treatment of chloropyrimidines with thiophenates is an important route to arylthiopyrimidines which (unlike alkylthiopyrimidines) cannot be made from the corresponding mercaptopyrimidine. 2-Phenylthiopyrimidine (**16**) and many other such thio-ethers were made in this way,[2619] and a surprising number of alkylthiopyrimidines have been made recently *via* the chloropyrimidines (Ch. VI, Sect. 5.F); even some of the familiar carboxymethylthio derivatives have been made using sodium thioglycollate.[2522] Replacement of a 5-bromo substituent by an alkyl- or aryl-thio group is now well documented (Ch. VI, Sect. 6.B).

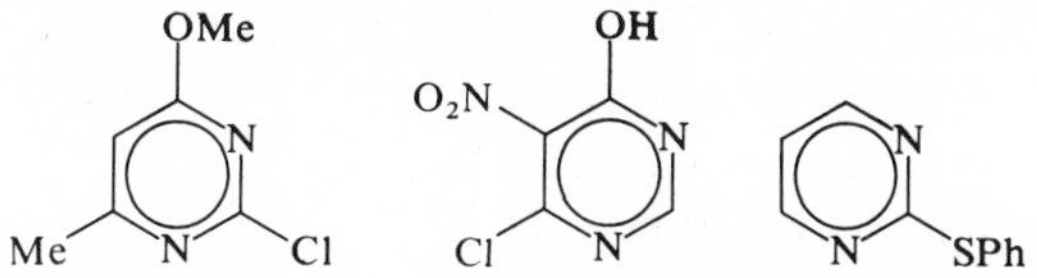

OMe, N, Me, N, Cl **(14)** OH, O_2N, N, Cl, N **(15)** N, N, SPh **(16)**

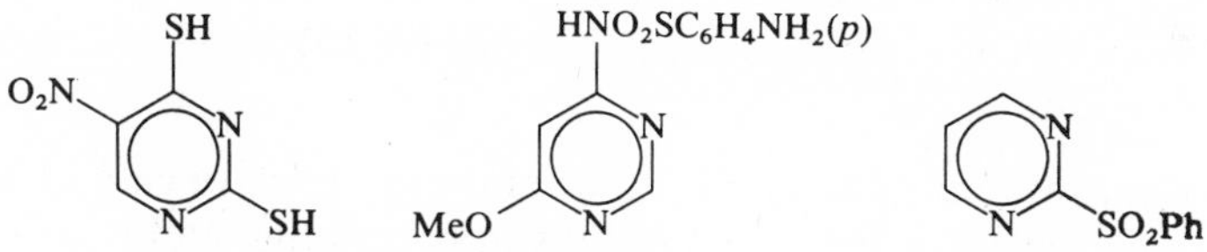

SH, O_2N, N, N, SH **(17)** $HNO_2SC_6H_4NH_2(p)$, N, MeO, N **(18)** N, N, SO_2Ph **(19)**

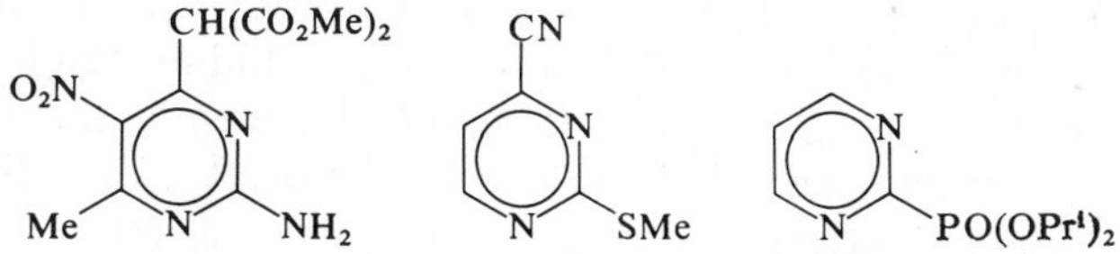

$CH(CO_2Me)_2$, O_2N, N, Me, N, NH_2 **(20)** CN, N, N, SMe **(21)** N, N, $PO(OPr^i)_2$ **(22)**

(e) *By Mercapto Groups* (*H* 14). The conversion of chloro- into mercapto-pyrimidines continues to be used widely. When sodium hydrogen sulphide is used as the reagent, nitro and other susceptible groups are usually reduced at the same time, but occasionally they have survived as in the formation of 2,4-dimercapto-5-nitropyrimidine[2590] (**17**) or 2-amino-4-butylamino-5-*p*-chlorophenylazo-6-mercaptopyrimidine.[2761] The route *via* a thiouronium salt has been used less of recent years.

(f) *By Sulpho, Thiocyanato, and Related Groups* (*H* 15). The replacements of a chloro by sulpho or thiocyanato substituent have been rather neglected.[2475, 2783] However a number of benzenesulphonamidopyrimidines, e.g., (**18**), have been made from chloropyrimidines using sodio sulphanilamide and related reagents,[2682] and sodium benzenesulphinate gave 2-phenylsulphonylpyrimidine (**19**) and related sulphones.[2619, 2771]

(g) *By Other Groups* (*New*). The formation of a C—C bond by treating a chloropyrimidine with the sodio derivative of a reagent having an activated methylene group has been used to make, e.g., 2-amino-4-dimethoxycarbonylmethyl-6-methyl-5-nitropyrimidine (**20**).[2790] The replacement of a 2- or 4-halogeno by a cyano group is almost unknown but 4-iodo- has been converted into 4-cyano-2-methylthiopyrimidine (**21**) by cuprous cyanide in pyridine;[2608] the same reagent in quinoline has been used more widely to introduce 5-cyano groups, for example in making 5-cyano- from 5-bromopyrimidine.[2607] The formation of a C—P bond was successful when, e.g., 2-chloropyrimidine was treated with tri-isopropyl phosphite to give 2-di-isopropylphosphinylpyrimidine (**22**).[2182]

(2) *Replacement of Alkoxy, Alkylthio, Alkylsulphinyl, and Alkylsulphonyl Groups* (*H* 15)

Much recent quantitative information on the relative ease of nucleophilic replacement of the above groups in pyrimidine and related series has been summarized.[2668] Previous impressions (*H* 16) clearly were misleading in part as indicated by the relative activities towards aminolysis given in Table Ia.[2668, 2672] Thus the pyrimidine sulphoxides and sulphones are rather more reactive than the corresponding chloropyrimidines; the ethers and sulphides are comparatively unreactive; and all may be vastly activated by an appropriate electron-withdrawing substituent. In this connexion, the preferential aminolysis[2749] of the chloro substituent in 4-chloro-6-methylsulphonylpyrimidine (**23**) [Ch. VI, Sect. 5.B(6)] was due to stronger electron-withdrawal by the sulphone grouping than by the chloro substituent. Alkylsulphonyl groups in the 2- or 4-position have been replaced directly by cyano, azido, sulphanilamido, and other groups (Ch. VIII, Sect. 5.B), but when

TABLE Ia. Approximate Comparative Figures for the Reactivity of 2-Substituted-pyrimidines Towards Aminolysis (*New*)

Pyrimidine	Relative reactivity
2-Methylthio-	20
2-Methoxy-	80
2-Chloro-	3,000,000
2-Methylsulphinyl-	7,000,000
2-Phenylsulphonyl-	12,000,000
2-Methylsulphonyl-	15,000,000
4,6-Dimethyl-2-methylthio-	1
2-Methoxy-4,6-dimethyl	60
2-Chloro-4,6-dimethyl-	300,000
5-Bromo-2-methylthio-	80
5-Bromo-2-methoxy-	400
5-Bromo-2-chloro-	60,000,000
2-Methylthio-5-nitro-	4,000,000
2-Methoxy-5-nitro-	100,000,000
2-Chloro-5-nitro-	10,000,000,000,000

amylaminolysis of 5-methylsulphonylpyrimidine was attempted, ring fission occurred to give 1,3-diamylimino-2-methylsulphonylpropane (**24**).

During aminolysis of methoxypyrimidines, rearrangement into the corresponding *N*-methylated oxopyrimidine, e.g., (**25**), occurred in part. This rearrangement and the related Hilbert–Johnson reaction have been explored recently in some detail [Ch. XI, Sect. 1.A(4)].

D. Other Metatheses (*H* 16)

(1) *Hydroxy- to Chloro-pyrimidines* (*H* 16)

(Ch. VI, Sect. 1)

Phosphoryl chloride, with or without a tertiary base, remains almost the sole reagent for converting hydroxy- into chloro-pyrimidines. However, chloromethylene dimethyl ammonium chloride (a crystalline reagent made[2848, 2849] from dimethylformamide with phosgene, thionyl chloride, or phosphorus pentachloride) has proved useful, especially when highly acidic conditions must be avoided;[2845–2847] phenyl phosphonic dichloride ($PhPOCl_2$) is a powerful high-boiling reagent;[2593] and even thionyl chloride has occasionally proved effective, e.g., to make 4-chloro-5-ethoxycarbonyl-2-methylthiopyrimidine (**26**) from its 4-hydroxy analogue.[2100, 2599]

(2) *Hydroxy- to Mercapto-pyrimidines* (*H* 16)

(Ch. VII, Sect. 6.B)

It has become quite evident that pyridine (or a homologue) is the best solvent in which to thiate hydroxypyrimidines with phosphorus

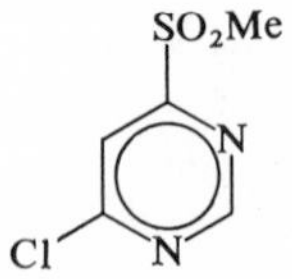

SO_2Me Cl N N **(23)**

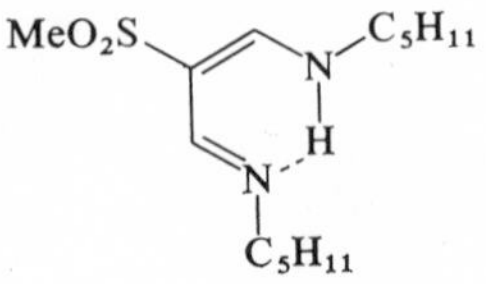

MeO_2S N C_5H_{11} H N C_5H_{11} **(24)**

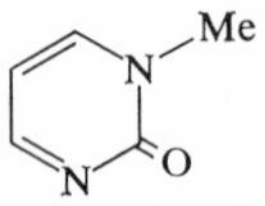

N Me N O **(25)**

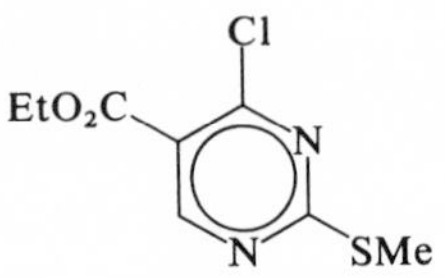

Cl EtO_2C N N SMe **(26)**

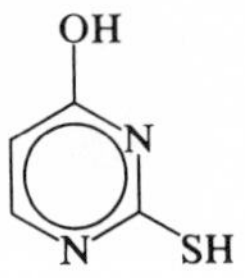

OH N N SH **(27)**

pentasulphide. Thus 2-thiouracil (**27**) gave 2,4-dimercaptopyrimidine (56%) in 15 min. using α-picoline as solvent but 8 hr. was needed in xylene.[356, 2915] The use of pyridine also eliminated the difficulties (*H* 17) formerly associated with thiation of amino-hydroxypyrimidines.

(3) *Mercapto- to Hydroxy-pyrimidines* (*H* 17)

(Ch. VIII, Sect. 1.F)

Of the several known indirect routes for this metathesis, that *via* a sulphone or sulphoxide has been developed into a practical procedure.

(4) *Mercapto- to Amino-pyrimidines* (*H* 17)

[Ch. VIII, Sect. 1.D(5)]

Mercaptopyrimidines are best converted into alkylthio- or alkylsulphonyl-pyrimidines prior to aminolysis. However the direct preferential aminolyses of 4-mercapto group in 2,4-dimercaptopyrimidines continued to be useful; 2,4-diselenylpyrimidine (**28**) behaved similarly to give 4-amino-2-selenylpyrimidine (**29**) and such like.[2779] The rare aminolysis of a 2-mercapto group furnished 5-hexyl-2-hydrazino-4-hydroxy-6-methylpyrimidine.[2225]

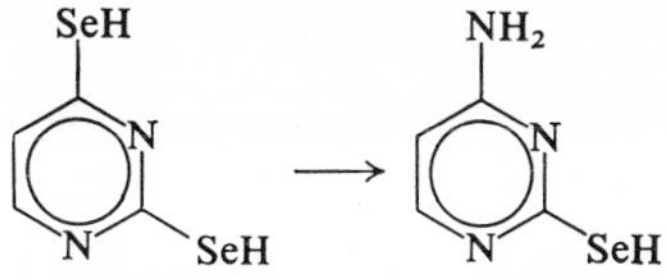

(28) (29)

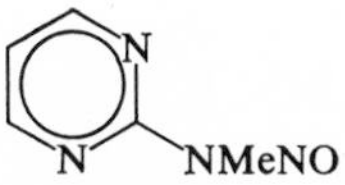

(30) 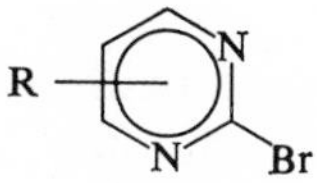 (31)

(5) *Amino- to Hydroxy-pyrimidines* (*H* 17)

(Ch. VII, Sects. 1.D and 2)

The three methods (acidic hydrolysis, alkaline hydrolysis, nitrous acid) have all been used widely. A kinetic study of the alkaline hydrolysis of some simple amino- and methylamino-pyrimidines indicated that 2-amino groups are a little more easily hydrolysed than 4-amino groups, and that the process is facilitated through electron-withdrawal by a 5-bromo substituent and retarded through electron-release by additional methyl groups.[2993] Another semi-quantitative study with more complicated derivatives confirmed these findings.[2992] The nitrous acid route is inapplicable to secondary-amino groups which give *N*-nitroso-derivatives, e.g., (**30**) (Ch. V, Sect. 2.D).

(6) *Amino- to Halogeno-pyrimidines* (*H* 18)

(Ch. VI, Sects. 1.C and 1.D)

Simple bromopyrimidines have now been made by treating the corresponding aminopyrimidines with nitrous acid in the presence of an excess of bromide ion; yields of 2-bromopyrimidine (**31**, R = H) and its derivatives are rather poor but can be improved (if the 5-position is occupied to prevent bromination) by first making the perbromide of the aminopyrimidine.[2604–2606]

(7) *Replacement of Ammonio Groups* (*H* 18)

(Ch. IX, Sect. 8.G)

Quaternary compounds, e.g., (**32**),[2557] have been prepared in increasing numbers because they are good intermediates, especially when used in acetamide solution. From such compounds were made, for example,

4-cyano-2-methylthiopyrimidine (KCN),[2954] 2-ethylthio-5-fluoro-4-sulphanilylaminopyrimidine (**33**; sodium sulphanilamide),[2209] and 4-azido(or 4-fluoro)-2,6-dimethoxypyrimidine (NaN_3 or KF, respectively).[2611] ω-Ammonio groups are also effective as leaving groups.

(8) *Interchange of Halogen Substituents* (*New*)

(Ch. VI, Sect. 1.D–F)

Although phosphorus tribromide has been used to convert 2-chloro- (**34**) into 2-bromo-pyrimidine (**35**) satisfactorily,[2603] it is probable that the reaction is an equilibrium. When the bromo compound was a solid (as above) it could be purified by recrystallization, but liquid bromopyrimidines made in this way proved very difficult to purify from their chloro precursors.[3780]

The 2- or 4-iodopyrimidines, e.g., (**36**), have been made only from appropriate chloro(or bromo)pyrimidines using hydriodic acid,[2608] potassium iodide in dimethylformamide,[2607] sodium iodide in glacial acetic acid,[2599] or sodium iodide and hydriodic acid in acetone.[2609] Fluoropyrimidines have been made from their chloro analogues by vigorous treatment with potassium or silver fluoride or with sulphur tetrafluoride.[2157, 2613] 5-Halogenopyrimidines have not been recorded as undergoing halogen interchange reactions.

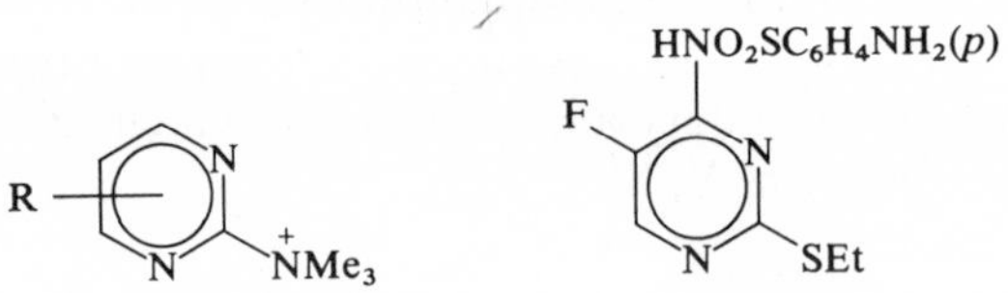

R; N; N; $\overset{+}{N}Me_3$

(**32**)

$HNO_2SC_6H_4NH_2(p)$; F; N; N; SEt

(**33**)

N; N; Cl → N; N; Br → N; N; I

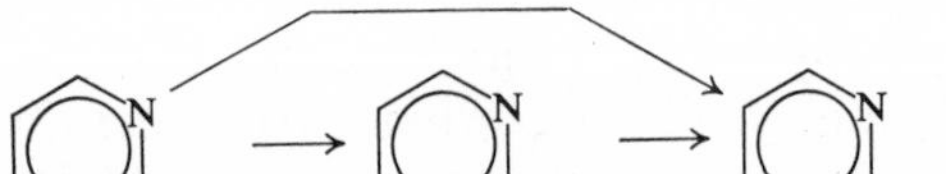

(**34**) (**35**) (**36**)

E. Addition Reactions (*H* 18)

(1) *Quaternization; Dimroth Rearrangement* (*H* 18)

[Ch. X, Sects. 1.A(3) and 2]

Quaternization of a pyrimidine brings about a considerable activation of attached leaving groups [Ch. X, Sect. 1.A(4)] but no systematic study has been made. The quaternization of hydroxypyrimidines to give *N*-alkylated oxopyrimidines has been widely used and a beginning has been made with a systemic study.[2762] In contrast, the quaternization of primary(or secondary)-aminopyrimidines to give *N*-1 or *N*-3-methylated iminopyrimidines and the subsequent Dimroth rearrangement of these imines have been studied in detail during the last few years and summarized recently.[2855] Thus 1,2-dihydro-2-imino-1-methylpyrimidine (**37**) underwent rearrangement ($t_{1/2}$ = 114 min.) in aqueous solution at pH 14 to give 2-methylaminopyrimidine (**39**) *via* an open chain intermediate (**38**).[2627] The mechanism was proven by isotopic tagging[2154, 2986] and the rates of rearrangement for such imines were very sensitive to the electronic nature of substituents present.[2855]

(2) *Formation of* N-*Oxides* (*H* 19)

(Ch. X, Sect. 4)

The peroxide oxidation of 4-methylpyrimidine gave a separable mixture of the *N*-oxides (**40** and **41**), now confirmed in structure. When 2- or 4-hydroxy(or amino)pyrimidine is converted into its *N*-oxide, e.g., (**42**), this is potentially tautomeric with (and may well exist as) the *N*-hydroxy derivative (**43**). The orientation of several *N*-oxides have been found by unambiguous primary syntheses.

(4) *Addition of Water; Photodimers* (*H* 20)

Spontaneous and complete covalent hydration has been observed in the cations of 5-nitro-, 5-methylsulphonyl-, and 5-methylsulphinyl-pyrimidine, which exist in the form (**44**; R = NO_2, SO_2Me, or SOMe);[2688, 3240] some evidence has been advanced for the partial hydration of 2-hydroxypyrimidine and related compounds.[3459]

Other protic solvents behaved similarly: for example 2-hydroxy-5-nitropyrimidine and acetone gave 4-acetonyl-3,4-dihydro-2-hydroxy-5-nitropyrimidine which in alkali gave *p*-nitrophenol![3483]

The well-known photohydration of uracil and its derivatives at the 5,6-bond has continued to attract interest; e.g., the hydrate (**45**) has been synthesised unambiguously.[3112] Even greater interest has attended studies of the photodimers of thymine and related compounds which are currently thought to be implicated in radiation damage of tissues. Thus ultra-violet irradiation of a frozen aqueous solution of thymine gave *cis-syn*-dimer (**46**);[3760] from 1,3-dimethylthymine, all four isomeric dimers have been isolated and identified[3460] (Ch. VII, Sect. 8.A,B).

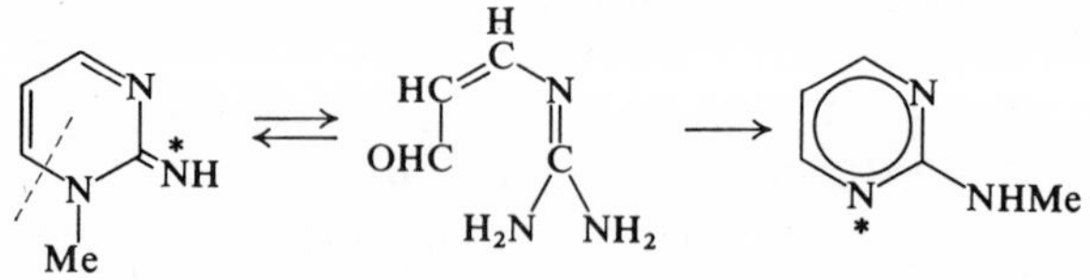

(37) (38) (39)

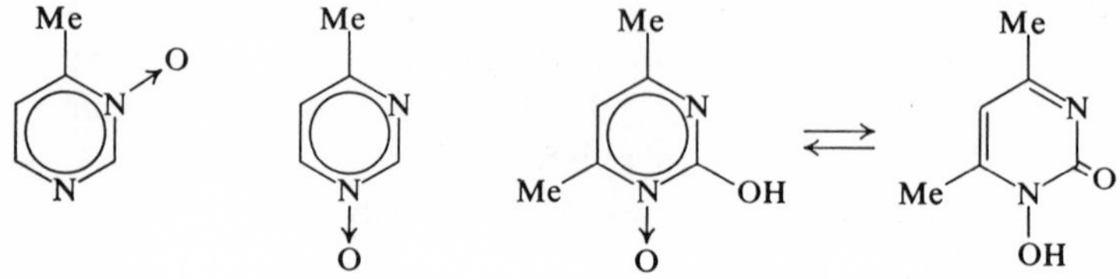

(40) (41) (42) (43)

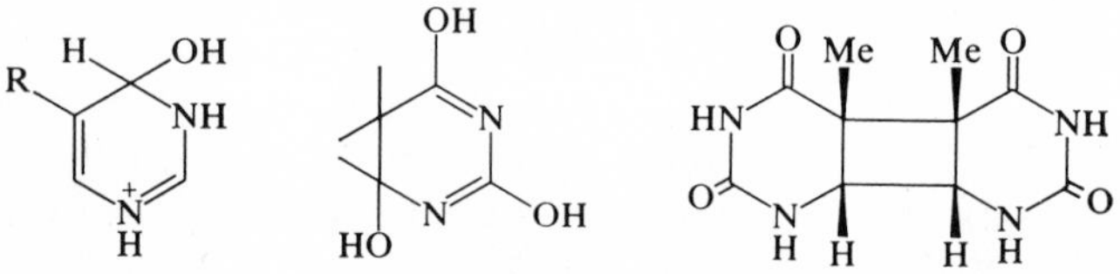

(44) (45) (46)

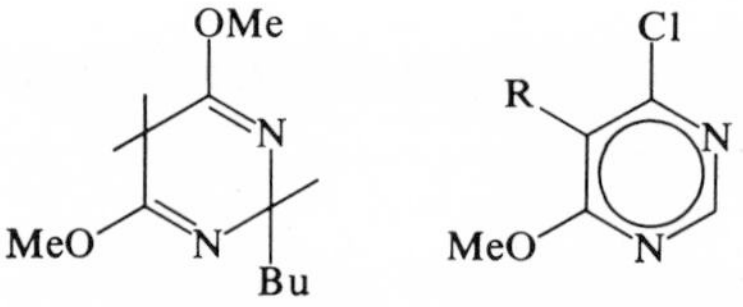

(47) (48)

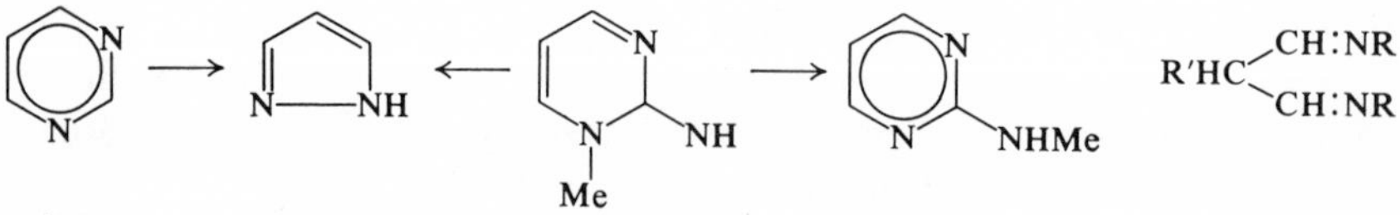

(49) (50) (51) (52) (53)

(5) *Addition of Metal Alkyls* (*H* 20)

(Ch. XII, Sect. 1.I)

t-Butyl magnesium chloride underwent normal 1,6-addition to pyrimidine to give eventually 4-t-butyl-3,4-dihydropyrimidine,[3551] but lithium butyl underwent 2,5-addition to 4,6-dimethoxypyrimidine to give eventually its 2-butyl-2,5-dihydro derivative (**47**).[2519] In contrast, butyl-lithium reacted with 5-bromo-4-chloro-6-methoxypyrimidine (**48**, R = Br) to give lithium derivative (**48**, R = Li) which reacted with carbon dioxide to furnish the carboxylic acid (**48**, R = CO_2H).[2519]

(6) *Addition of Amines* (*New*)

Pyrimidines undergo several interesting reactions, in each of which the first step is addition of an amine. For example, pyrimidine (**49**) and hydrazine gave (Ch. IV, Sect. 1.C) a good yield of pyrazole (**50**);[3246] the same product was obtained (Ch. X, Sect. 2.B) by treatment of 1,2-dihydro-2-imino-1-methylpyrimidine (**51**) with hydrazine, whereas butylamine gave 1,3-bisbutyliminopropane (**53**; R = Bu, R′ = H), dibutylamine caused normal Dimroth rearrangement to 2-methylaminopyrimidine (**52**), and tributylamine did not react because it lacked a mobile proton;[3240] 5-methylsulphonylpyrimidine reacted (Ch. VIII, Sect. 5.B) with amylamine to give 1,3-diamylimino-2-methylsulphonylpropane (**53**; R = C_5H_{11}, R′ = SO_2Me);[3240] and 4-methoxy- or 4-hydrazino-5-nitropyrimidine with an excess of hydrazine (Ch. VII, Sect. 7.C) gave 3-amino-5-nitropyrazole.[2858]

F. Oxidative Reactions (*H* 20)

[Ch. V, Sect. 1.A(3); Ch. VI, Sect. 5.A]

The oxidation of nitrosopyrimidines by hydrogen peroxide in trifluoroacetic acid has furnished a new and useful synthesis of nitropyrimidines such as 4,6-diamino-5-nitro-2-phenylpyrimidine (**54**).[2445]

Several simple nitropyrimidines have been made by oxidative removal of attached hydrazino groups, a process of potential use in the presence of other reduction-labile but oxidation-stable groups. The process, carried out with silver oxide or silver acetate in protic or aprotic media, gave 5-nitropyrimidine (**56**) from its 4,6-dihydrazino derivative (**55**) and has been used also in analogous cases.[2562, 2688, 2858]

G. Reductive Reactions (*H* 21)

(1) *Nuclear Reduction* (*H* 21)

(Ch. XII, Sects. 1.H, 2.D, and 3)

A welcome start has been made to study systematically the nuclear reduction of pyrimidines, especially using palladium as catalyst.[2162, 2685, 2686, 3625] Thus a variety of pyrimidines were reduced to the 1,4,5,6-tetrahydro stage in acidic media, but 5-amino- and 5-hydroxy-pyrimidine formed only dihydro derivatives. The use of a platinum or rhodium catalyst sometimes led to hydrogenolysis of an amino substituent: for example, cytosine (**57**) gave 1,4,5,6-tetrahydro-2-hydroxypyrimidine (**58**).[3627]

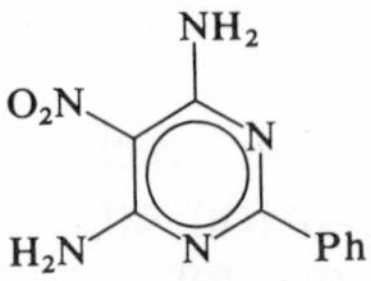

(**54**) 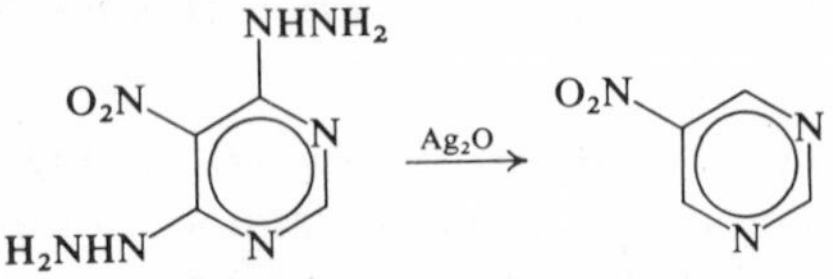 (**55**) (**56**)

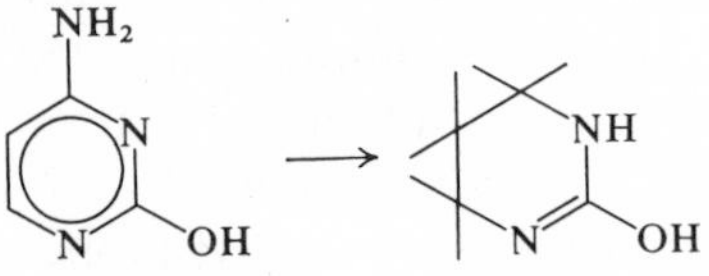

(**57**) (**58**)

A little more has been learned also about reduction with metal hydrides: it appears that sodium borohydride seldom reduces the pyrimidine ring but lithium aluminium hydride usually produces a di- or tetra-hydro derivative according to the substituents present.

(2) *Reductive Removal of Groups* (*H* 22)

No new general methods of reductive removal of groups have emerged. However nickel boride in aqueous suspension under pressure did give 4,5-diaminopyrimidine from its 2-mercapto derivative,[3457] albeit in poorer yield than when Raney nickel was used [Ch. VIII, Sect. 1.D(1)]; 2-, 4-, or even 5-dehalogenation by either zinc dust or hydrogenation has been extended to give a rather wide spectrum of products (Ch. VI, Sects. 5.A and 6.B); and the necessity of making specifically deuterated pyrimidines for p.m.r. and mass-spectral studies has led to the use of palladium-catalysed replacements of Cl by D to give, for example, 4,5-diamino-2-deuteropyrimidine (**60**) from its 2-chloro analogue (**59**).[2835]

(3) *Reductive Modification of Groups* (*H* 23)

The reductions of nitro-, nitroso-, and arylazo- to amino-pyrimidines have continued to be reactions of great importance (Ch. V). Some less well known processes for modification of groups have been explored recently. Preferential reductive alkylation of the 5-group in 4,5-di-

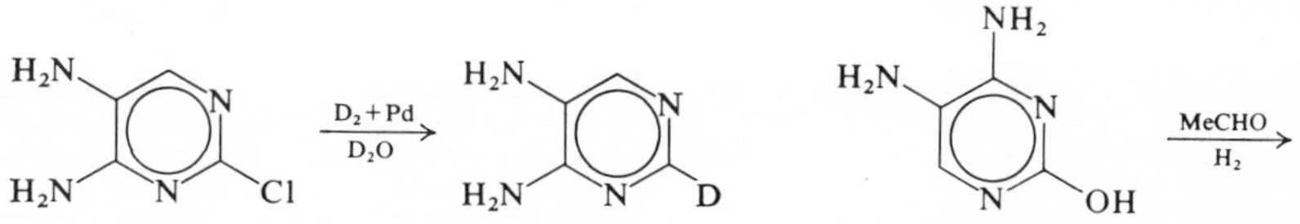

(59) (60) (61) 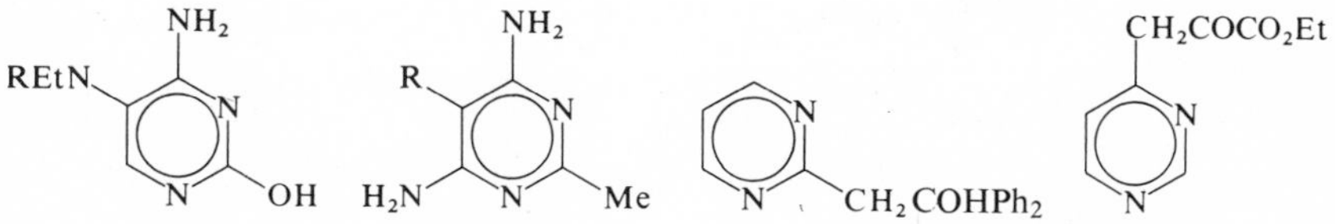 (62) (63) (64) (65)

aminopyrimidines has been done quite widely (Ch. IX, Sect. 2) by hydrogenation in the presence of an appropriate aldehyde: for example, 4,5-diamino- (**61**) gave 4-amino-5-ethylamino(or diethylamino)-6-hydroxypyrimidine (**62**, R = H or Et) according to the relative amount of acetaldehyde used.[3475] The difficult reduction of a cyano to an aminomethyl group (Ch. IX, Sect. 3.A), has been well exemplified in the formation of 4,6-diamino-5-aminomethyl-2-methylpyrimidine (**63**, R = CH_2NH_2) from the nitrile (**63**, R = CN) using a specially active nickel catalyst.[2681] The formation of ω-hydroxyalkylpyrimidines from aldehydes has been extended to ketones (Ch. VII, Sect. 3.F): 2-phenacylpyrimidine and phenylmagnesium bromide gave 2-β-hydroxy-$\beta\beta$-diphenylethylpyrimidine (**64**);[3008] and borohydride reduction of 4-ethoxyoxalylmethylpyrimidine (**65**) gave mainly 4-$\beta\gamma$-dihydroxypropylpyrimidine.[3009]

H. The Modification of Substituents (H 24)

(1) *Amino Groups* (H 24)

(Ch. IX)

Although transamination was not unknown in the pyrimidine series, this process has been used widely only in recent years. For example, 4-amino-2,6-dimethylpyrimidine hydrochloride (**66**) and a slight excess of alkylamine at 170° gave the 4-butylamino (**67**, R = Bu) and other homologues (**67**) in good yield.[2153] The process seems to be more effective with 4- than with 2-aminopyrimidines. The related process of

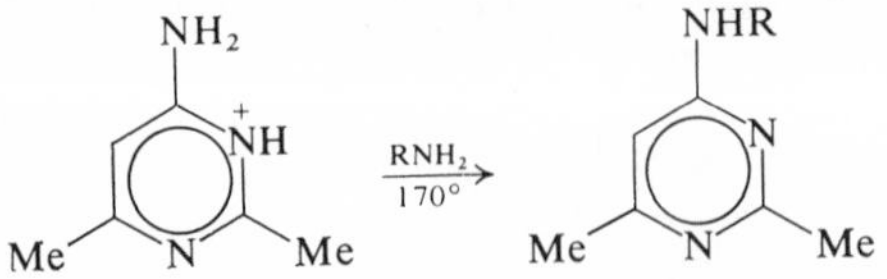

(66) $\xrightarrow[170°]{RNH_2}$ (67) 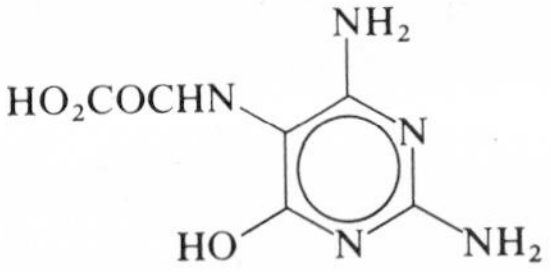 (68)

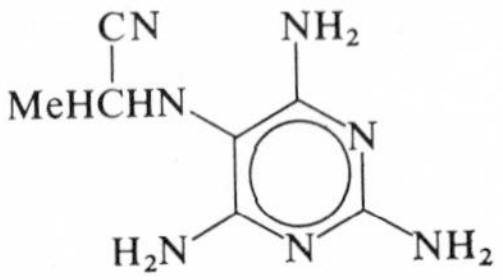

(69) 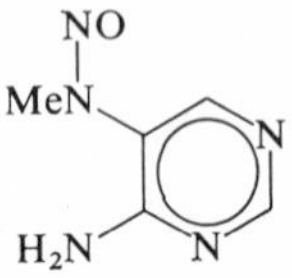 (70)

transacylation has been exemplified in treatment of 2,4-diamino-5-formamido-6-hydroxypyrimidine with aqueous oxalic acid to give the 5-hydroxyoxalylamino analogue (**68**).[3302] Direct α-cyanoalkylation of 5-amino groups has been done by treating the amine in acetic acid first with sodium cyanide and then with an aldehyde: thus tetra-aminopyrimidine and acetaldehyde gave 2,4,6-triamino-5-α-cyano-ethylaminopyrimidine (**69**) which underwent oxidative cyclization to 2,4,7-triamino-6-methylpteridine.[3332] The first nitrosoamine derived from an alkylaminopyrimidine was 4-amino-5-methylnitrosoamino-pyrimidine (**70**), made by the action of isoamyl nitrite on 4-amino-5-methylaminopyrimidine.[2491]

(2) *Hydroxy and Alkoxy Groups* (*H* 25)

(Ch. VII; Ch. X)

Uracil has been monomethylated for the first time by treating its sodium salt with methyl iodide in a medium of low polarity: 1-methyl-uracil was obtained in reasonable yield.[3381] A great many other *N*-alkylations of hydroxypyrimidines have been recorded recently but little headway has been made in understanding the factors controlling orientation.

The Hilbert–Johnson reaction of methoxypyrimidines, e.g., (**71**) with an alkyl iodide to give an *N*-alkylated oxopyrimidine (**72**) has been studied extensively and reviewed.[3400] The less-known thermal re-arrangement of alkoxypyrimidines to similar *N*-alkylated oxopyrimi-dines has been studied also of recent years: the rearrangement of 2-methoxypyrimidine (**71**) to 1,2-dihydro-1-methyl-2-oxopyrimidine (**72**, R = Me) was intermolecular, but followed first-order kinetics; it was accelerated by electron-withdrawing substituents, slowed by electron-releasing substituents, slowed on replacing the methoxy group by a larger alkoxy group, and catalysed by organic bases.[2511, 2630, 2697, 3412]

(4) *Alkyl and Related Groups* (*H* 26)

(Ch. IV, Sect. 2.C)

The nitrosation of a 2- or 4-methyl group has been used to produce the oxime of the corresponding aldehyde and thence by dehydration, the nitrile: an example is the sequence **(73)** → **(75)**;[2592, 2954] nitrosation of the hydroxypyrimidine **(73)** was thought previously (*H* 148) to give the 5-nitroso derivative.

The deuterium exchange of *C*-methyl groups in simple methylpyrimidines proved very slow but a relatively rapid acid-base catalysed exchange took place in the *C*-methyl groups of compounds like 1,2-dihydro-2-imino-1,4,6-trimethylpyrimidine and its 2-oxo analogue.[2861]

Two interesting prototropic changes of a 5-prop-2′-ynyl group (to an allenyl or a prop-1′-ynyl group according to conditions) have been recorded in the pyrimidine series [see footnote to Ch. VI, Sect. 5.B(1a)].[2700]

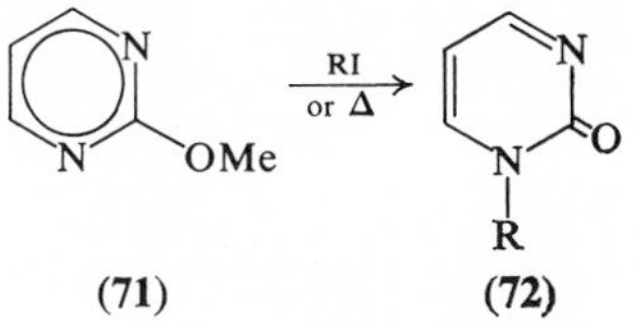

N OMe —RI or Δ→ N O N R

(71) **(72)**

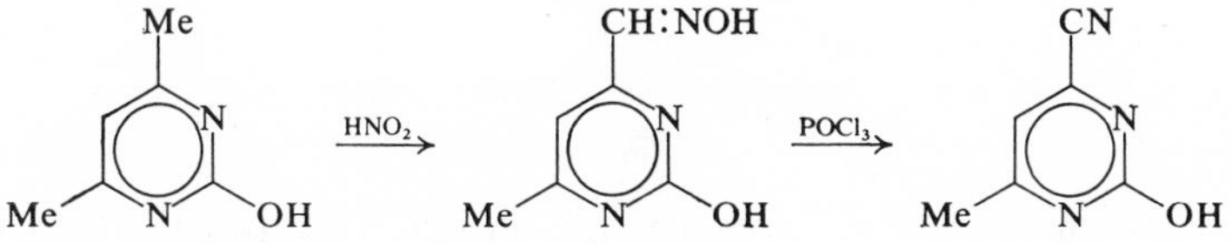

Me N Me N OH —HNO_2→ CH:NOH N Me N OH —$POCl_3$→ CN N Me N OH

(73) **(74)** **(75)**

(5) *Carboxy and Related Groups* (*New*)

(Ch. XI)

Most of the transformations within the carboxy family of substituents are ignored here on account of their ubiquity. However, mention must be made of several processes which have been used more of late: the formation of acid chlorides and their transformation into amides; the direct formation of the ester (**76**, R = H) from the corresponding amide by methanolic hydrogen chloride;[3481] the transalkylation of the ester (**76**, R = OMe) to corresponding benzyloxycarbonylmethyl derivative by benzyl alcohol in the presence of aluminium and mercuric chloride;[3481] the first isolation of a simple pyrimidine iminoether: 2-cyano- (**77**) gave 2-*C*-ethoxy-*C*-iminomethyl- (**78**, R = OEt) and thence 2-amidino-pyrimidine (**78**, R = NH_2);[2806] the conversion of an aldehyde into the corresponding nitrile *via* the oxime,[2611] or directly into a 'pyrimidoin', by potassium cyanide;[2301] and the dehydrogenation of aldoximes to give the first pyrimidine nitrile oxides, e.g., 2,4,6-trimethoxypyrimidine-5-nitrile oxide (**79**).[2815, 3518]

5. Physical Properties of Pyrimidines (*H* 26)

Numerous ionization constants, ultra-violet spectra, and infra-red spectra for pyrimidines continue to be recorded and used. However, since no new principles have emerged of recent years, only a supple-

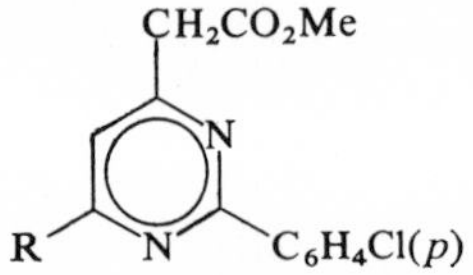

CH_2CO_2Me R N N $C_6H_4Cl(p)$

(76)

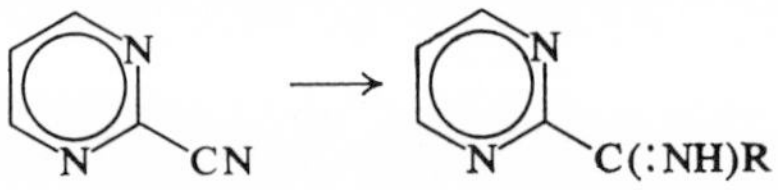

N N CN → N N C(:NH)R

(77) **(78)**

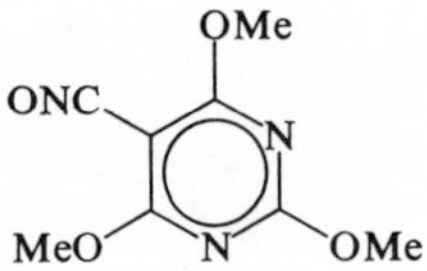

OMe ONC N MeO N OMe

(79)

mentary list of pK_a values appears in Chapter XIII, but the ionization of hydropyrimidines is discussed briefly (Ch. XII, Sect. 5.A). In contrast, the last ten years has seen the first harvest of nuclear magnetic resonance and mass spectral results relevant to pyrimidines.

G. Nuclear Magnetic Resonance Spectra (*New*)

This aspect of pyrimidines is treated for the first time in an essay by T. J. Batterham (Ch. XIII, Sect. 3) but data relevant to hydropyrimidines is contained in the appropriate chapter by R. F. Evans (Ch. XII, Sects. 5.C and D). Apart from their immense value in structural elucidation, these data have been particularly useful in problems associated with tautomeric groups in pyrimidines and with the conformation of hydropyrimidines.

H. Mass Spectra (*New*)

The number of mass spectral data on pyrimidines[2428, 2629, 3272] is still insufficient to make a coherent review, but some examples of their use in hydropyrimidines are discussed (Ch. XII, Sect. 5.E).

CHAPTER II

The Principal Synthetic Method

1. General Scope (*H* 31)

The Principal Synthesis continues to be a prolific source of diverse pyrimidines. Every one of its ten divisions has been extended in scope, not only by numerous new examples analogous to those already known, but by the use of new reagents to produce both new and old types of pyrimidine. Thus, for example, 4(or 6)-alkoxypyrimidines have been made by the Principal Synthesis for the first time by using a three-carbon fragment, e.g., (**25**), having a diethoxymethylene part; again, if in preparing 1,3-dicyclohexylbarbituric acid the conventional disubstituted urea is replaced by *NN*-dicyclohexylcarbodiimide, the yield is increased from 5 to 65%.

In addition, several errors in structure and unexpected ring systems have come to light. For example, the so-called 2,4,6-trimethyl-5-phenylazopyrimidine, prepared from phenylazoacetylacetone and acetamidine, turned out to be an isomeric pyridine; and β-ethoxy-α-methoxymethylenepropionate behaved, not as the expected aldehydo-ester, but as an aldehydo-ether, giving with urea a dihydropyrimidine, with thiourea a 1,3-thiazine, and with *NN'*-dimethylthiourea a mixture of both systems!

Improved methods of preparation have made formamidine (as its acetate,[2170, 2434] hydrochloride,[2170, 2957] hydrobromide,[2957] hydriodide[2957]) and acetamidine (as acetate[2170]) readily available for Principal Syntheses.

2. Use of β-Dialdehydes (*H* 32)

The condensation of malondialdehyde, as 1,1,3,3-tetraethoxypropane (**1**) or a related acetal, with guanidine has been explored under a variety of conditions. Yields of 2-aminopyrimidine (**2**) vary from 21% (in ethanolic sodium ethoxide) to 90% (in ethanolic hydrogen chloride).[2171] Aqueous sulphuric acid, or hydrogen chloride in acetic acid or chloroform, were less satisfactory acidic media (yields 50–70%). Guanidine carbonate (only) may also be used without solvent; by heating it with the acetal in the presence of ammonium acetate, up to 90% of 2-aminopyrimidine results.[2172] The acetal reacts satisfactorily with *p*-acetamidobenzenesulphonylguanidine to yield 2-*p*-acetamidobenzenesulphonamidopyrimidine, which on alkaline hydrolysis gives sulphadiazine.[2171]

1,2-Dihydro-1-methyl-2-oxopyrimidine and its 1-alkyl homologues have been made from the same acetal with the appropriate *N*-alkylurea in ethanolic hydrochloric acid;[2511, 2630] the similar formation of 1,2-dihydro-1-methyl-2-thiopyrimidine using *N*-methylthiourea has been described now in detail.[2173] *β*-*N*-Methylanilinoacrolein (also equivalent to malondialdehyde) reacts with thiourea alone, or better in the presence of alkoxide, to give 2-mercaptopyrimidine.[2512]

A dipyrimidinyl sulphide and two analogous disulphides have been made by the Principal Synthesis using a dialdehyde equivalent. Thus di(*β*-diethylamino-*α*-formylvinyl) sulphide and the corresponding disulphide (**3**) react with guanidine to give di(2-aminopyrimidin-5-yl) sulphide and disulphide, respectively; di(2-phenylpyrimidin-5-yl) disulphide may be made similarly using benzamidine.[2304]

Nitromalondialdehyde has been condensed with 2(or 3)-amidinopyridine to give 5-nitro-2-2′(or 3′)-pyridylpyrimidine,[2603] but a similar reaction with 2-amidino-4,6-dimethylpyrimidine failed to yield a dipyrimidinyl.[2603]

Sodio nitromalondialdehyde has also been condensed with *NN*-dimethylguanidine in aqueous piperidine to give 2-dimethylamino-5-nitropyrimidine in 35% yield, but the same pyrimidine is probably better made by methylation of 2-amino-5-nitropyrimidine with sodium

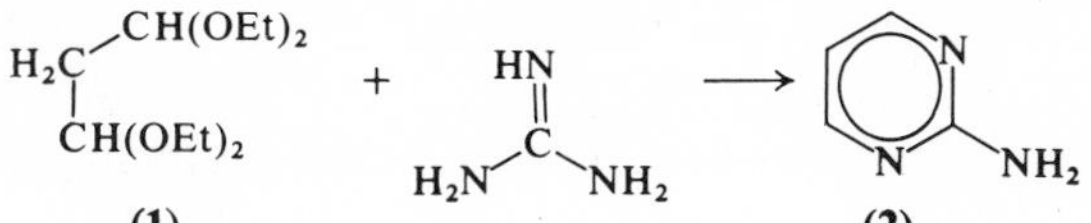

$H_2C(CH(OEt)_2)_2$ (**1**) + $HN{=}C(NH_2)_2$ ⟶ 2-aminopyrimidine (**2**)

$OHC(Et_2NHC{=})C{-}S{-}S{-}C({=}CHNEt_2)CHO$

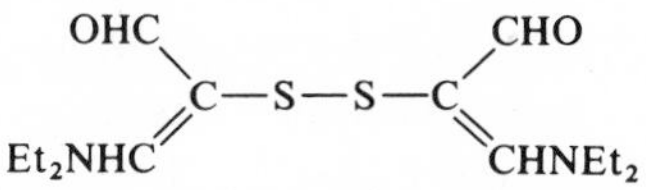

(**3**)

hydride/methyl iodide (85% yield).[3519] The latter method has been used to make 2-di(trideuteriomethyl)amino-5-nitropyrimidine.[3519]

3. Use of β-Aldehydo Ketones (*H* 34)

The formylacetone derivatives, 3-diethoxymethylbutan-2-one (**4**; R = Me) and 3-diethoxymethyl-4-methylpentan-2-one (**4**; R = Pr^i) react under alkaline conditions with *S*-methylthiourea to give 4,5-dimethyl-2-methylthiopyrimidine (**5**; R = Me) and its 5-isopropyl homologue (**5**; R = Pr^i), respectively.[2174]

This group also contains examples of the formation of 5-ethoxycarbonylpyrimidines. Thus diethyl ethoxymethyleneoxalacetate and acetamidine yield 4,5-diethoxycarbonyl-2-methylpyrimidine,[2175] while ethyl α-ethoxymethylenebenzoylacetate and guanidine carbonate yield 2-amino-5-ethoxycarbonyl-6-phenylpyrimidine.[2176] Similarly, ethyl ethoxymethyleneacetoacetate (**6**) reacts with benzamidine and other arylamidines (but not acetamidine) to give 5-ethoxycarbonyl-4-methyl-2-phenyl (or other aryl) pyrimidine (**7**);[2177] with *S*-methyl- and *S*-ethyl-thiourea it gives, respectively, 5-ethoxycarbonyl-4-methyl-2-methylthiopyrimidine (**5**; R = EtO_2C) and its ethylthio homologue.[2174] That ethyl ethoxymethyleneacetoacetate should react above as an aldehydo ketone rather than as an aldehydo ester or keto ester would not be surprising but for previous claims to the contrary, e.g., using *N*-alkyl derivatives of urea[101] and thiourea[274] (see *H* 43, 48). The structural veracity of some of the pyrimidine esters above were incidentally proven[2177] by their hydrolysis to known carboxylic acids.

Attention has been drawn[2178] to benzylideneaminoguanidine (**8**) which condenses[527] with formylacetone to yield 2-hydrazino-4-methylpyrimidine, presumably *via* its benzylidene derivative; this appears to be the first hydrazinopyrimidine to be made by a Principal Synthesis.

Unusual intermediates are exemplified[2298] in 1-acetyl-2-dimethylaminoethylene (which may be made[2299] *in situ* from acetone and bisdimethylamino-methoxymethane[2300]) and 1-benzoyl-2-dimethylaminoethylene. These react with guanidine to give 2-amino-4-methyl- and 2-amino-4-phenyl-pyrimidine respectively.[2299] Two other pyrimidines have been made similarly,[2298] and essentially the same reaction occurs when 1-diethoxyacetyl-2-dimethylaminoethylene reacts with formamidine, acetamidine, benzamidine, urea, guanidine, or *S*-methylthiourea to give 4-diethoxymethylpyrimidine, its 2-methyl, 2-phenyl, 2-hydroxy, 2-amino, or 2-methylthio derivative, respectively.[2301] The preparation of 5-bromo-4-carboxy-2-phenylpyrimidine from bromomucic acid and benzamidine[270] has been modernized.[2603]

Several 4,4′-bipyrimidinyls have been made directly by condensing 1,4-bisethoxymethylenebutan-2,3-dione with two molecules of an amidine or *S*-alkylthiourea.[2390] Thus formamidine gave 4-pyrimidin-4′-ylpyrimidine, acetamidine gave 2-methyl-4-2′-methylpyrimidin-4′-ylpyrimidine, benzamidine the corresponding 2,2′-diphenyl homologue, and *S*-methylthiourea the 2,2′-bismethylthio-analogue.[2390]

1,1-Diethoxybutan-3-one (**4**; R = H) has been shown to react with *N*-methylurea to give only one product,[2602] proven by an unambiguous synthesis[2630] and an associated p.m.r. study[2861] to be 1,2-dihydro-1,6-dimethyl-2-oxopyrimidine; similarly, *N*-ethylurea gives only 1-ethyl-1,2-dihydro-6-methyl-2-oxopyrimidine.[2630]

$MeCOCHRCH(OEt)_2$

(4)

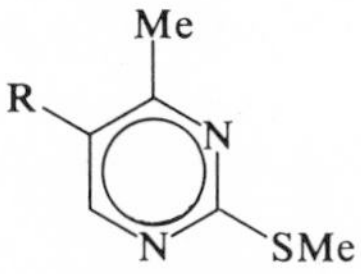

(5)

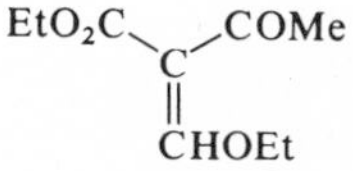

$EtO_2C\text{—}C(\text{=CHOEt})\text{—}COMe$

(6)

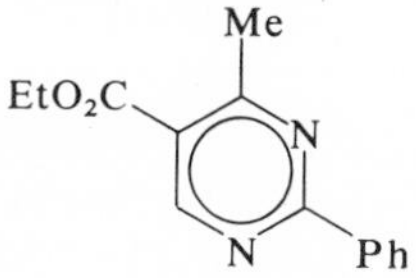

(7)

PhCH:NNHC(:NH)NH_2

(8)

4. Use of β-Diketones (*H* 36)

The condensation of acetamidine with phenylazoacetylacetone has been claimed[2179] to yield 2,4,6-trimethyl-5-phenylazopyrimidine (**9**) as a major product. However, after careful fractionation of the reaction products, it appears[2180] that the readily isolated product is the isomeric 2-amino-4,6-dimethyl-5-phenylazopyridine (**9a**), and that the pyrimidine is formed only in minute yield. The pyridine was identified by reduction and acetylation to 2,5-bisacetamido-4,6-dimethylpyridine which was unambiguously synthesized.[2180] 5-Nitro-2-furamidine condenses with appropriate diketones to yield 4,6-dimethyl-2,5′-nitro-2′-furylpyrimidine (**11**) and the corresponding 4,6-bistrifluoromethyl-, 4-methyl-6-trifluoromethyl-, 4-2′-furyl-6-trifluoromethyl-, and 4-2′-thienyl-6-trifluoromethyl-analogues.[2181] Acetamidine and benzamidine react with (diethoxyacetyl)acetone to give 4-diethoxymethyl-2,6-dimethyl- and 4-diethoxymethyl-6-methyl-2-phenyl-pyrimidine, respectively.[2301] 4,6-Dimethyl-2-(3,4,5-trimethoxyphenyl)pyrimidine is made directly.[2513]

2-Hydroxy-4,6-dimethylpyrimidine may be made by a modified process from acetylacetone and urea in aqueous alcoholic hydrogen chloride.[2182] Similarly, with urea, benzoylacetone gives 2-hydroxy-4-methyl-6-phenylpyrimidine,[2592] 2-ethoxalylcyclohexanone yields 4-ethoxycarbonyl-2-hydroxy-5,6-tetramethylenepyrimidine (**12**),[2183] and methyl acylpyruvates give 2-hydroxy-4-methoxycarbonyl-6-methylpyrimidine and its 6-alkyl homologues.[2184] The same pyruvates with *S*-methylthiourea give 4-methoxycarbonyl-6-methyl-2-methylthiopyrimidine and its 6-alkyl homologues;[2184] and methyl acetylpyruvate with acetamidine gives 4-carboxy-2,6-dimethylpyrimidine, presumably *via* the ester.[2184] The preparation of 2-mercapto-4-methyl-6-4′-pyridylpyrimidine from thiourea and 4-acetoacetylpyridine is better done by heating the dry reactants at 160° than by refluxing in alcoholic hydrogen chloride.[2185] *S*-Methylthiourea condenses with 2-trifluoroacetylcyclopentanone to yield 2-methylthio-4-trifluoromethyl-5,6-trimethylenepyrimidine,[2235] and with (diethoxyacetyl)acetone to give 4-diethoxymethyl-6-methyl-2-methylthiopyrimidine.[2301] 2-Benzylthio-4,6-dimethylpyrimidine is made in 75% yield by simply heating thiourea, benzyl chloride, and acetylacetone in dimethylformamide for 1 hr.[2873]

Guanidine carbonate also reacts successfully with diketones by mere heating at 120–150°. Thus 1-acetoacetylnaphthalene yields 2-amino-4-methyl-6-1′-naphthylpyrimidine;[2186] 2-acetoacetylfuran yields 2-amino-4-2′-furyl-6-methylpyrimidine;[2186] and appropriate diketones lead to 2-amino-4-methyl-6-2′-thienylpyrimidine[2186, 2187] (and its 4-alkyl homologues[2187]), 2-amino-4-methyl-6-2′-pyridylpyrimidine,[2186] 2-amino-4-methyl-6-3′-pyridylpyrimidine[2186, 2187] (and homologues[2187]), and 2-amino-4-methyl-6-4′-pyridylpyrimidine.[2185, 2186] Finally, benzamidoguanidine condenses with acetylacetone to give 2-benzoylhydrazino-4,6-dimethylpyrimidine,[2178, 2188] which undergoes acid hydrolysis to 2-hydrazino-4,6-dimethylpyrimidine, identified by unambiguous synthesis from its 2-chloro analogue.[2188]

N-Methoxyurea condenses with acetylacetone in ethanolic hydrogen chloride to give 1,2-dihydro-1-methoxy-4,6-dimethyl-2-oxopyrimidine, which can be hydrolysed by hydrobromic acid to the 1-hydroxy analogue (an *N*-oxide tautomer); other alkoxyureas behave similarly.[2869]

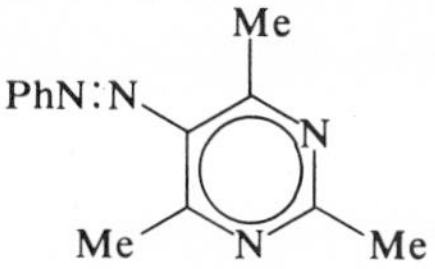

(9) 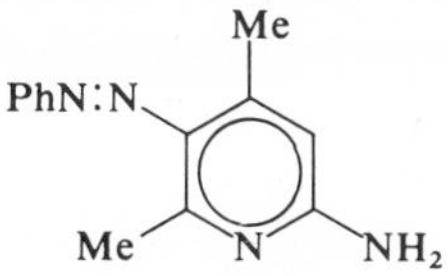 (9a) 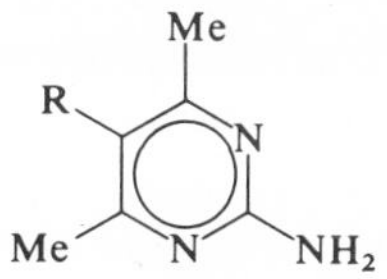 (10)

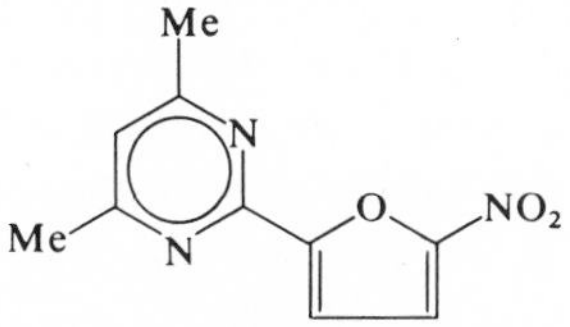

(11) 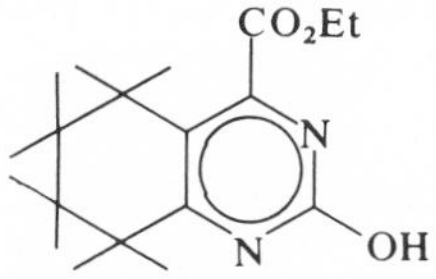 (12)

The formation of pyrimidines with unsaturated groups in the 5-position is exemplified in the condensation under acidic conditions of allyl- or prop-2′-ynyl-acetylacetone with urea to give 5-allyl-2-hydroxy-4,6-dimethyl- and 2-hydroxy-4,6-dimethyl-5-prop-2′-ynyl-pyrimidine (64, 65%; *cf.* 5-propyl analogue, 46%).[2700] The allylic intermediate condensed normally with guanidine carbonate yielding 5-allyl-2-amino-4,6-dimethylpyrimidine (**10**, R = $CH_2CH{:}CH_2$), but the acetylenic intermediate gave a separable mixture of 5-allenyl-2-amino-4,6-dimethylpyrimidine (**10**, R = $CH{:}C{:}CH_2$), 2-amino-4,6-dimethyl-5-prop-1′-ynylpyrimidine (**10**, R = C⋮CMe), and 2-amino-4,6-dimethyl-5-prop-2′-ynylpyrimidine (**10**, R = $CH_2C{⋮}CH$), due to prototropic change under the alkaline conditions pertaining.[2700]

5. Use of β-Aldehydo Esters (*H* 38)

Condensations of amidines with aldehydo esters were poorly represented until recently. Examples now known, include β-ethoxy-α-methoxymethylenepropionate with formamidine (and four other amidines) giving 5-ethoxymethyl-4-hydroxypyrimidine and appropriate 2-alkyl derivatives;[2189, 2190] ethyl β-ethoxy-α-formylpropionate (another form of the last mentioned intermediate) and fluoroacetamidine giving 5-ethoxymethyl-2-fluoromethyl-4-hydroxypyrimidine;[2116] ethyl fluoroformylacetate and formamidine giving 5-fluoro-4-hydroxypyrimidine;[2191] ethyl fluoro(or chloro)formylacetate and acetamidine giving the 5-halogeno-4-hydroxy-2-methylpyrimidine;[2191] ethyl formylacetate and hydroxy- or trifluoro-acetamidine giving 4-hydroxy-2-hydroxymethylpyrimidine[2192] and 4-hydroxy-2-trifluoromethylpyrimidine,[2193] respectively; ethyl allylformylacetate and acetamidine or trifluoroacetamidine giving 5-allyl-4-hydroxy-2-methyl(or trifluoromethyl)pyrimidine;[2194] and other such reactions.[2513, 2514, 2541]

The reaction of aldehydo esters with urea, thiourea and their *N*-, *S*-, and *O*-alkyl derivatives has been used extensively in the past, e.g., in the synthesis of uracils and thiouracils. Among more interesting recent examples (not included in Table I) are the direct formation of 3-methyl-2-thiouracil (**13**) from ethyl ββ-diethoxypropionate and *N*-methylthiourea[2195] (the structure was checked by conversion into 3-methyluracil); a procedure for the synthesis of uracil from urea and diethyl malate (instead of malic acid);[2196] and the advantageous use of sodium hydride[2197] in place of sodium[1328] as a condensing agent in making (crude) ethyl formylbutyrate and its subsequent condensation with thiourea to yield 5-ethyl-4-hydroxy-2-mercaptopyrimidine. The danger inherent in the use of such unpurified formylated intermediates has been underlined in a related condensation: when methyl methoxyacetate is treated with methyl formate and the raw product condensed with thiourea,[2161] the resulting (unpurified) 4-hydroxy-2-mercapto-5-methoxypyrimidine contains some 7% of 4-hydroxy-2-mercapto-5-methoxy-6-methoxymethylpyrimidine arising from the methyl αγ-dimethoxyacetoacetate, which is formed from autocondensation of methyl methoxyacetate during preparation of the formylated intermediate.[2198]

Nitrouracil, its *N*-alkyl derivatives, and nitro-2-thiouracil have now been made[2263] by extending syntheses based on cyclization of ureidomethylene esters (*H* 42 *et seq.*). Thus ethyl nitroacetate, ethyl orthoformate, and urea give ethyl ureidomethylenenitroacetate which in ethanolic potassium ethoxide cyclizes to 2,4-dihydroxy-5-nitropyrimidine in 30% overall yield. By using *N*-methyl-, *N*-benzyl-, *N*-phenyl-, or *NN'*-dimethylurea, the corresponding 3-methyl-, 3-benzyl-, 3-phenyl-, or 1,3-dimethyl-5-nitrouracil is formed. Thiourea gives 4-hydroxy-2-mercapto-5-nitropyrimidine.[2263] The failure of such a synthesis from ethyl βββ-trifluoropropionate, ethyl orthoformate, and urea has been reported.[2264] Some interesting aspects of the mechanism of such reactions have been studied.[2263, 2265]

Pyrimidine-*N*-oxides have recently been made[2199, 2200] by semi-direct syntheses in this class: thus ethyl ββ-diethoxypropionate or ethyl β-ethoxyacrylate condenses with *N*-benzyloxyurea in ethanolic sodium ethoxide to yield 1-benzyloxyuracil (**14**) which on acid hydrolysis gives 1-hydroxyuracil (**15**), i.e., uracil-1-*N*-oxide.[2200] *N*-Methoxyurea can be used similarly to give 1-methoxyuracil (**16**) which can also be made by methylating 1-hydroxyuracil; an analogous synthesis from diethyl ethoxymethylenemalonate yields in sequence 1-benzyloxy-5-ethoxycarbonyluracil, the corresponding carboxylic acid, 5-carboxy-1-hydroxyuracil,[2199] and, finally by decarboxylation,[2200] 1-hydroxyuracil. Both 1- and 3-hydroxyuracil have been made also by unambiguous routes (see Ch. III, Sects. 2.E and 5.A).

A. Takamizawa and his colleagues have recently shown that ethyl β-ethoxy-α-methoxymethylenepropionate behaves, not as the expected aldehydo ester, but as an aldehydo ether in its reactions with urea, thiourea, and their *N*-alkyl derivatives. This stands in contrast to its normal reaction with amidines (*v.s*). Thus, with urea, 5-ethoxycarbonyl-1,6-dihydro-2-hydroxypyrimidine is formed;[2278] with thiourea, 5-ethoxycarbonyl-2-hydroxy-1,3-thiazine is formed *via* the corresponding 2-amino derivative; and with *NN'*-dimethylthiourea a mixture of appropriate thiazine and dihydropyrimidine results.[2279] The syntheses of uracil and thymine derivatives have been reviewed briefly.[2201]

Recent examples of the condensation of guanidine and its derivatives with aldehydo esters are in Table II. Of interest is the formation of 5-*p*-chlorophenylthio-2-dimethylamino-4-hydroxypyrimidine (50%) from *NN*-dimethylguanidine and crude ethyl α-isoamyloxymethylene-α-(*p*-chlorophenylthio)acetate;[2203] the formation of 5-carboxymethyl-2-cyanoamino-4-hydroxypyrimidine from diethyl formylsuccinate and dicyanodiamide;[2208] several substituted-guanidino derivatives such as 5-ethoxycarbonylmethyl-4-hydroxy-2-phenylguanidinopyrimidine made from the same three-carbon fragment and phenyl(or other) biguanide;[2208] and 2-amino-4-hydroxy-5-β-hydroxyethylpyrimidine from guanidine and the disguised aldehydo-ester, α-hydroxymethylene-γ-butyrolactone.[2514] A preparation of isocytosine (2-amino-4-hydroxypyrimidine) from formylacetic acid, guanidine, and fuming sulphuric acid is reported to give a 77% yield.[2588]

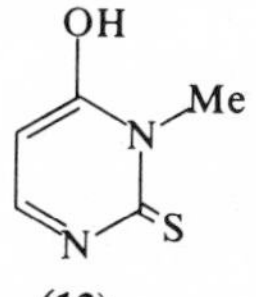

(13)

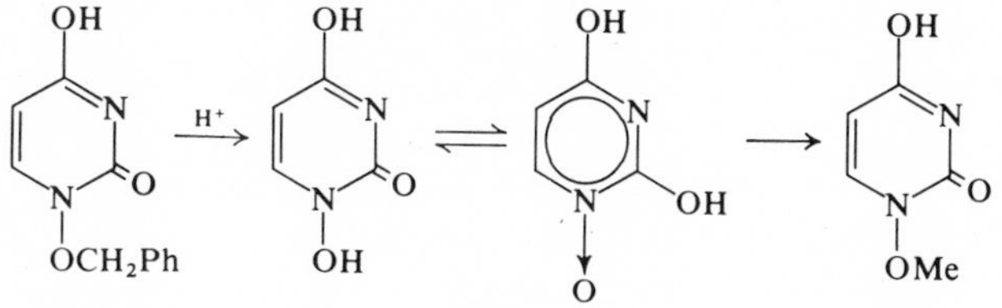

(14) (15) (16)

TABLE II. Additional Examples of the Use of Aldehydo Esters in the Principal Synthesis (*H* 40)

Three-carbon fragment	One-carbon fragment	Solvent and conditions	Pyrimidine and yield	Ref.
Ethyl fluoroformylacetate	thiourea	methanol; sodium methoxide; reflux; 2 hr.	5-fluoro-4-hydroxy-2-mercapto- (14%)	2191
Ethyl fluoroformylacetate	*S*-ethylthiourea	methanol; sodium methoxide	2-ethylthio-5-fluoro-4-hydroxy- (32%)	2515
α-Hydroxymethylene-γ-butyrolactone (Na)	thiourea	water; 12 hr.	4-hydroxy-5-β-hydroxyethyl-2-mercapto- (44%)	2202
α-Hydroxymethylene-γ-butyrolactone	urea	ethanol; sodium ethoxide; reflux; 5 days	2,4-dihydroxy-5-β-hydroxyethyl-[a] (2.5%)	2202
Ethyl *p*-chlorophenylthio-formylacetate[b]	thiourea	ethanol; sodium ethoxide; reflux; 6 hr.	5-*p*-chlorophenylthio-4-hydroxy-2-mercapto- (84%)	2203
Ethyl α-formylcaproate	thiourea	ethanol; sodium ethoxide; reflux; 7 hr.	5-butyl-4-hydroxy-2-mercapto-[c]	2204
Ethyl (or allyl) allylformylacetate	thiourea	ethanol; sodium ethoxide; reflux; 5 hr.	5-allyl-4-hydroxy-2-mercapto- (62%)	2077, 2194
Ethyl allylformylacetate	*S*-methylthiourea	ethanol; sodium ethoxide; 20°; 24 hr. *or* aq. ethanol; sodium hydroxide; 20°; 20 hr.	5-allyl-4-hydroxy-2-methylthio- [d] (24%; 64%)	2194
Diethyl ethoxymethylene-malonate	*O*-methylurea	methanol; sodium methoxide; reflux; 12 hr.	4-Hydroxy-2-methoxy-5-methoxycarbonyl- (*ca* 25%)	2205
Diethyl methoxymalonate	*S*-methylthiourea	aq. potassium hydroxide; 25° + 100°; 5 hr.	4-hydroxy-5-methoxy-2-methylthio- (39%)	2205
β-Ethoxyacryloyl chloride	*N*-methylthiourea	chloroform + pyridine; reflux; $2\frac{1}{2}$ hr. (then sodium hydroxide)	1-methyl-2-thiouracil (66%)	2311

Ethyl α-formyl-β-methyl-δ-pentenoate	*S*-methylthiourea	aq. ethanol; sodium hydroxide; 20°; 15 hr.	4-hydroxy-2-methylthio-5-α-vinylethyl-	2206
Ethyl formylacetate (Na)	*N*-β-amino-β-carboxyethylurea hydrobromide	water; 20°; 7 days	1-β-amino-α-carboxyethyl-uracil (willardiine)	2310
Ethyl *o*-hydroxyphenylazo-formylacetate	urea	ethanol; sodium ethoxide; reflux; 4 hr.	2,4-dihydroxy-5-*o*-hydroxyphenyl-[e] (16%)	2207
Ethyl fluoroformylacetate	guanidine	methanol; sodium methoxide; reflux; 2 hr.	2-amino-5-fluoro-4-hydroxy-	2209
Benzyl benzyloxyformylacetate	guanidine	ethanol; sodium ethoxide; reflux; 4.5 hr.	2-amino-5-benzyloxy-4-hydroxy- (48%)	2210
Ethyl formylacetate (or Na)	benzamidine	water; 2 days	4-hydroxy-2-phenyl- (31; 39%)	2799, 2554
Sodium formylacetate	propionamidine	water; 25°; 3 days	2-ethyl-4-hydroxy- (28%)	2554
Benzyl benzyloxyformylacetate	acetamidine	ethanol; sodium ethoxide; reflux; 8 hr.	5-benzyloxy-4-hydroxy-2-methylpyrimidine- (40%)	2210, *cf* 79
Ethyl allylformylacetate	guanidine	ethanol; sodium ethoxide; reflux; 4 hr.	5-allyl-2-amino-4-hydroxy- (37%)	2194
α-Hydroxymethylene-γ-butyrolactone (Na)	guanidine	ethanol; reflux; 5 hr.	2-amino-4-hydroxy-5-hydroxyethyl- (19%)	2202
Ethyl phenylazoformylacetate	guanidine	ethanol; sodium ethoxide; reflux; 3 hr.	2-amino-4-hydroxy-5-phenylazo-[f] (42%)	2211
Ethyl phenylazoformylacetate	*NN*-dimethyl-guanidine	ethanol; sodium ethoxide; reflux; 3 hr.	2-dimethylamino-4-hydroxy-5-phenylazo- (23%)	2211
Ethyl *p*-chlorophenylthio-formylacetate[b]	guanidine	ethanol; reflux; 6 hr.	2-amino-5-*p*-chlorophenyl-thio-4-hydroxy- (*ca.* 25%)	2203
Methyl methoxyformylacetate	guanidine	ethanol; sodium ethoxide; reflux; ½ hr.	2-amino-4-hydroxy-5-methoxy-[g] (66%)	2212
Ethyl 2′,3′-dimethoxybenzyl-formylacetate (crude)	guanidine	ethanol; reflux; 6–12 hr.	2-amino-5-2′,3′-dimethoxy-benzyl-4-hydroxy-[h]	2213

(*continued*)

TABLE II (*continued*).

Three-carbon fragment	One-carbon fragment	Solvent and conditions	Pyrimidine and yield	Ref.
Diethyl ethoxymethylene-malonate	*S*-methylthiourea	aq. ethanol; 25°; *ca.* 24 hr.	5-ethoxycarbonyl-4-hydroxy-2-methylthio- (70%)	2598
α-Formyl-γ-thiobutyrolactone	guanidine	ethanol; reflux; 6 hr.	2-amino-4-hydroxy-5-β-mercaptoethyl- (62%)	2908
α-Formyl-βγ-trimethylene-γ-butyrolactone (Na enolate)	urea	aq. HCl; then ethanolic sodium ethoxide; reflux	2,4-dihydroxy-5-2′-hydroxy-cyclopentyl-[i] (*ca.* 50%)	2952

[a] Better made indirectly from 2-mercapto analogue.[2202]
[b] As enol ether.
[c] And the hexyl homologue.
[d] And the benzylthio analogue.
[e] And the 2-mercapto analogue (32%); the reported failure to make 4-hydroxy-6-methyl-5-phenylazopyrimidine from ethyl phenylazoformylacetate and formamidine is less surprising than the actual product.
[f] And the *o*-benzyloxy derivative (78%).
[g] And several 5-alkoxy homologues.
[h] And analogues, *cf.* ref. 82.
[i] Acyclic intermediate isolated after acidic stage; 2-mercapto analogue made similarly.[2952]

6. Use of β-Keto Esters (*H* 48)

Keto esters have been so extensively used in the Principal Synthesis that little novelty can be expected in recent examples which are mainly in Table III.

Of note is the condensation of sodium diethyl oxalacetate with formamidine in aq. sodium hydroxide to yield 4-carboxy-6-hydroxy-pyrimidine in 63% yield.[2214] With formamidine acetate now so readily available,[2170] this method is superior to indirect preparations. With the same three-carbon fragment, guanidine yields 2-amino-4-carboxy-6-hydroxypyrimidine (34%),[2215] *S*-methylthiourea gives 4-carboxy-6-hydroxy-2-methylthiopyrimidine* (63%),[2214] and (it will be recalled: *H* 51) urea gives an hydantoin which can be slowly isomerized into 4-carboxy-2,6-dihydroxypyrimidine. It has now been shown[2216] that if disodium monoethyl oxalacetate is condensed with urea, the intermediate is not an hydantoin but simply uncyclized, and it may be relatively quickly converted into orotic acid in 61% yield. 4-Carboxy-6-hydroxy-2-mercaptopyrimidine may be made similarly.[2216]

Related pyrimidine aldehydes and ketones have also been made. Thus methyl γγ-dimethoxyacetoacetate with trifluoroacetamidine yields 4-dimethoxymethyl-6-hydroxy-2-trifluoromethylpyrimidine;[2217] with thiourea it yields the 2-mercapto-analogue;[2219] and with *N*-alkylthioureas it yields 4-dimethoxymethyl-1,2-dihydro-6-hydroxy-1-methyl-2-thiopyrimidine (**17**) and its 1-alkyl homologues.[2219] The α-alkylated derivatives of the three-carbon fragment similarly yield appropriate 5-alkylated pyrimidines,[2219] and isopropyl α-(4-phenylbutyl)-γγ-diethoxyacetoacetate with guanidine gives 2-amino-4-diethoxymethyl-6-hydroxy-5-δ-phenylbutylpyrimidine.[2978] Ethyl γγ-diethoxy-β-oxovalerate with thiourea gives 4-αα-diethoxyethyl-6-hydroxy-2-mercaptopyrimidine (**18**) and thence 4-acetyl-6-hydroxy-2-mercaptopyrimidine (**19**);[2218] the same fragment with guanidine gives 4-acetyl-2-amino-6-hydroxypyrimidine.[2218] Ethyl α-acetyl(or benzoyl)-δδ-diethoxyvalerate (**20**) and guanidine afford 2-amino-5-γγ-diethoxypropyl-4-hydroxy-6-methyl(or phenyl)pyrimidine.[2176, 2220]

Unusual 'keto esters' are exemplified in the oxalylketen acetals (**21**, X = S) and (**21**, X = O) which react with benzamidine to yield 4-ββ-bismethylthioacryloyl-6-ethoxy-2-phenylpyrimidine (**22**) and 4-ethoxy-6-(4-ethoxy-2-phenylpyrimidin-6-yl)-2-phenylpyrimidine (**23**)

* Appropriate diethyl α-alkyloxalacetates have been used similarly to give 4-carboxy-5-ethyl-6-hydroxy-2-methylthiopyrimidine and its 5-alkyl homologues.[3183]

TABLE III. Additional Examples of the Use of Keto Esters in the Principal Syntheses (*H* 44)

Three-carbon fragment	One-carbon fragment	Solvent and conditions	Pyrimidine and yield	Ref.
Ethyl α-acetamidoacetoacetate	acetamidine	ethanol; sodium ethoxide; reflux; 4 hr.	5-acetamido-4-hydroxy-2,6-dimethyl- (60%)	2223
Ethyl acetoacetate	butyramidine	methanol; sodium methoxide; reflux; 16 hr.	4-hydroxy-6-methyl-2-propyl-[a]	2224
Ethyl acetoacetate	pivalamidine	methanol; sodium methoxide; 5–20°; 16 hr.	2-t-butyl-4-hydroxy-6-methyl- (*ca.* 80%)	2224
Ethyl trifluoroacetoacetate	acetamidine	ethanol; sodium ethoxide; reflux; 4 hr.	4-hydroxy-2-methyl-6-trifluoromethyl- (38%)	2516
Ethyl trifluoroacetoacetate	butyramidine	methanol; sodium ethoxide; reflux; 5 hr.	4-hydroxy-2-propyl-6-trifluoromethyl- (*ca.* 75%)	2224
Ethyl ethoxalylchloroacetate	acetamidine	ethanol; sodium ethoxide; reflux; 2 hr.	5-chloro-4-ethoxycarbonyl-6-hydroxy-2-methyl- *and* 4-ethoxycarbonyl-5,6-dihydroxy-2-methyl	2191
Ethyl ethoxalylfluoroacetate	formamidine	ethanol; sodium ethoxide; reflux; 2 hr.	4-ethoxycarbonyl-5-fluoro-6-hydroxy-[b] (43%)	2191
Ethyl acetoacetate	phthalimido-acetamidine	methanol; sodium methoxide; 25°	4-hydroxy-6-methyl-2-phthalimidomethyl- (67%)	2875
Diethyl acetosuccinate	acetamidine	methanol; sodium methoxide; 25°; then aq. KOH	5-carboxymethyl-4-hydroxy-2,6-dimethyl-	2876
Ethyl acetoacetate	trifluoroacetamidine	ethanol; sodium ethoxide; reflux; 3 hr.	4-hydroxy-6-methyl-2-trifluoromethyl- (60%)	2193

Ethyl trifluoroacetoacetate	trifluoroacetamidine	ethanol; sodium ethoxide; reflux; 3 hr.	4-hydroxy-2,6-bistrifluoromethyl- (35%)	2193
Ethyl acetoacetate	fluoroacetamidine	[c]	2-fluoromethyl-4-hydroxy-6-methyl- (*ca.* 12%)	2116
Ethyl α-prop-2-ynylacetoacetate	acetamidine	ethanol; sodium hydroxide; 20°; 3 days (then reflux; 1 hr.)	4-hydroxy-2,6-dimethyl-5-prop-2′-ynyl-[d] (89%)	2307
Ethyl hexylacetoacetate	thiourea	ethanol; sodium ethoxide; reflux; 6 hr.	5-hexyl-4-hydroxy-2-mercapto-6-methyl-[e] (48%)	2204, 2225
Ethyl stearoylacetate	thiourea	ethanol; sodium ethoxide; reflux; 6 hr.	6-heptadecyl-4-hydroxy-2-mercapto- (93%)	2226
Ethyl phenylazoacetoacetate	thiourea	methanol; sodium methoxide; reflux[f]	4-hydroxy-2-mercapto-6-methyl-5-phenylazo-	2227
Methyl αγ-dimethoxyacetoacetate	thiourea	methanol; sodium methoxide; reflux; 2 hr.	4-hydroxy-2-mercapto-5-methoxy-6-methoxymethyl- (66%)	2198
2-Ethoxycarbonylcyclohexanone	urea	ethanol; sodium ethoxide; reflux; 2 hr.	2,4-dihydroxy-5,6-tetramethylene-[g] (36%)	2183
Ethyl acetoacetate	thiourea	plus potassium carbonate; 60°; 2 hr.; no solvent	4-hydroxy-2-mercapto-6-methyl- (80%)	2874
Ethyl ethoxalylfluoroacetate	*S*-ethylthiourea	ethanol; sodium ethoxide; reflux; 4 hr.	4-ethoxycarbonyl-2-ethylthio-5-fluoro-6-hydroxy-[h] (10%)	2517
Ethyl α-acetamidoacetoacetate	*S*-methylthiourea	aq. sodium hydroxide; 20°; 12 hr; (then 100°; 1 hr.)	5-acetamido-4-hydroxy-6-methyl-2-methylthio- (35%)	2223

(continued)

TABLE III (*continued*).

Three-carbon fragment	One-carbon fragment	Solvent and conditions	Pyrimidine and yield	Ref.
Ethyl (dichloroaceto)acetate	*S*-methylthiourea	methanol; sodium methoxide; 20°; 24 hr.	4-dichloromethyl-6-hydroxy-2-methylthio-[j] (30%)	2518
Ethyl methoxyacetoacetate	guanidine	aq. sodium hydroxide	2-amino-4-hydroxy-5-methoxy-6-methyl-	2228
α-Acetyl-γ-butyrolactone	guanidine	ethanol; sodium ethoxide; reflux; 4 hr.	2-amino-4-hydroxy-5-β-hydroxyethyl-6-methyl-[h] (39%)	2229
Ethyl acetoacetate (sodium)	guanidine	methanol	2-amino-4-hydroxy-6-methyl-[i] (80%)	2230, 2231
Ethyl α-(β-cyanoethyl)acetoacetate	guanidine	ethanol; reflux; 19 hr.	2-amino-5-β-cyanoethyl-4-hydroxy-6-methyl- (67%)	2232
Ethyl β-oxo-β-(2-thienyl)propionate	guanidine carbonate	130°; 3 hr.	2-amino-4-hydroxy-6-(2-thienyl)- (*ca.* 20%)	2181
Methyl γ-benzylacetoacetate	guanidine carbonate	ethanol; reflux; 18 hr.	2-amino-4-hydroxy-6-phenethyl-[j] (73%)	2308
α-Acetyl-γ-butyrolactone	*NN*-dimethylguanidine	ethanol; 5 hr.; reflux	2-dimethylamino-4-hydroxy-5-β-hydroxyethyl-6-methyl- (17%)	2638
Ethyl 2,4-diethoxy-3-oxobutyrate	guanidine carbonate	t-butanol; 18 hr.; reflux	2-amino-5-ethoxy-4-ethoxymethyl-6-hydroxy- (48%)	2980
Ethyl α-(α-naphthylazo)-acetoacetate	*NN*-dimethylguanidine	ethanol; sodium ethoxide; reflux; 3 hr.	2-dimethylamino-4-hydroxy-6-methyl-5-α-naphthylazo-[j] (76%)	2211
Ethyl α-*o*-hydroxyphenyl-azoacetoacetate	*N*-methylguanidine	ethanol; sodium ethoxide; reflux; 3 hr.	2-amino-1,6-dihydro-5-*o*-hydroxyphenylazo-1,4-dimethyl-6-oxo-[k] (53%)	2211
Ethyl α-acetamidoacetoacetate	guanidine	ethanol (or t-butanol); reflux; 8 hr. (or 15 hr.)	5-acetamido-2-amino-4-hydroxy-6-methyl- (37%; 72%)	2223, 2303

Ethyl acetoacetate	*p*-tolylbiguanide	methanol; reflux; 5 hr.	4-hydroxy-6-methyl-2-*p*-tolylguanidino-[l] (95%)	2233
Ethyl acetoacetate	decyloxybiguanide	aq. ethanolic sodium hydroxide; reflux; 1.5 hr.	2-decyloxyguanidino-4-hydroxy-6-methyl-[m] (*ca.* 50%)	2234
Ethyl α-δ′-phenylbutyl trifluoroacetoacetate	guanidine carbonate	t-butanol; reflux; 18 hr.	2-amino-4-hydroxy-5-δ-phenylbutyl-6-trifluoromethyl- (35%)	2391
Ethyl acetoacetate	δ-hydroxybutylguanidine	ethanol; reflux; 5 hr.	4-hydroxy-2-δ-hydroxybutylamino-6-methyl- and 2-amino-1,4-dihydro-1-δ-hydroxybutyl-6-methyl-4-oxo- (ratio 6:1)	2871
Ethyl α-butylbenzoylacetate	guanidine carbonate	t-butanol	2-amino-5-butyl-4-hydroxy-6-phenyl-[j] (39%)	2941
Ethyl α-*p*-ethoxycarbonylphenylbutylacetoacetate	guanidine carbonate	t-butanol; reflux; 60 hr.	2-amino-5-*p*-ethoxycarbonylphenylbutyl-4-hydroxy-6-methyl- (21%)	2942

[a] And twenty analogues some with a 5-alkyl substituent.

[b] Not analysed but characterized by hydrolysis to the carboxy analogue in 88% yield; also the 2-methyl analogue.

[c] Acyclic intermediate isolated from initial condensation in ethanolic sodium hydroxide at 20°; cyclized in methanolic sodium methoxide at 20°.

[d] And three homologues; triple bond reduced catalytically in each case to give 5-alkyl analogue, e.g., 4-hydroxy-2,6-dimethyl-5-propylpyrimidine.

[e] The melting points recorded by the Japanese and Russian authors differ by 10°; several 5-alkyl homologues are described (ref. 2204).

[f] 'A much cleaner product resulted' when methanolic sodium methoxide was used in place of ethanolic sodium ethoxide (*cf.* ref. 580).

[g] Also the 2-mercapto analogue (88%) for which the monocyclic intermediate was isolated and subsequently cyclized.

[h] And ten analogues.

[i] *cf.* *H* 49.

[j] And analogues.

[k] The synthesis is ambiguous and incomplete structural evidence is given.

[l] And the 2-*p*-bromophenylguanidino- analogue.

[m] Also seven 2-alkoxyguanidino and five 2-alkylguanidino analogues; beware of triazines as by-products!

respectively; also by ethyl *N*-cyano-α-ethoxycarbonylacetimidate, $EtO_2CCHC(:N.C \vdots N)OEt$, which with acetamidine or benzamidine gives 4-cyanoamino-6-hydroxy-2-methyl(or phenyl)pyrimidine.[2222]

An unexpected condensation is that of thiourea with ethyl diacetylacetate which yields 4-hydroxy-2-mercapto-6-methylpyrimidine (75%), rather than the expected 5-ethoxycarbonyl-2-mercapto-4,6-dimethylpyrimidine or 5-acetyl-4-hydroxy-2-mercapto-6-methylpyrimidine;[2519, 2520] guanidine carbonate behaves similarly in yielding

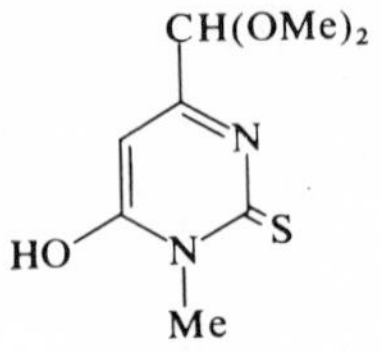

CH(OMe)₂ HO N N S Me

(17)

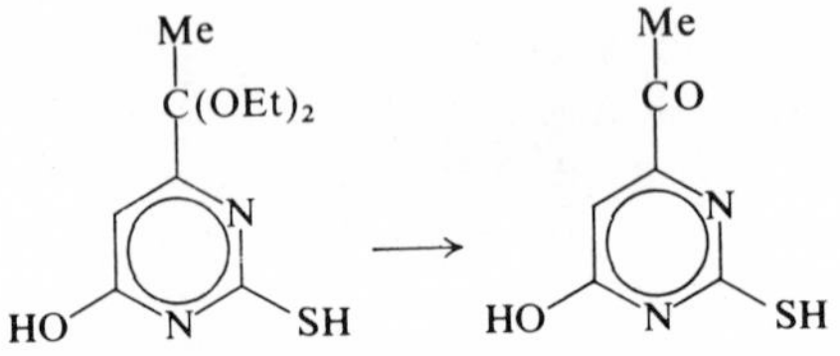

Me C(OEt)₂ N HO N SH → Me CO N HO N SH

(18) **(19)**

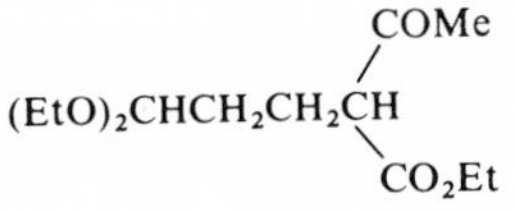

$(EtO)_2CHCH_2CH_2CH(COMe)(CO_2Et)$

(20)

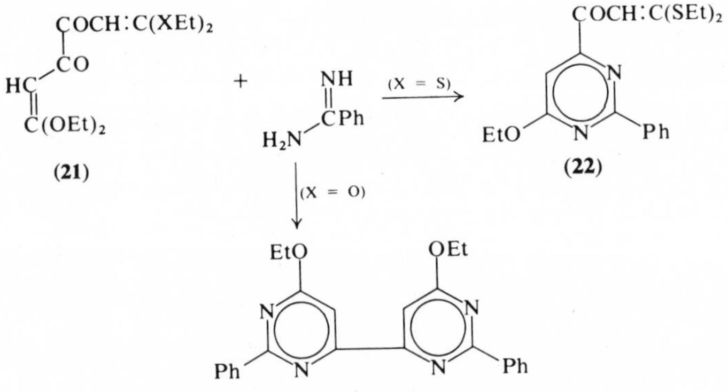

$COCH:C(XEt)_2$, HC(CO)=C(OEt)₂ + H_2NC(Ph)=NH —(X = S)→ COCH:C(SEt)₂ EtO N N Ph

(21) **(22)**

↓ (X = O)

EtO OEt N N Ph N N Ph

(23)

COR O O ≡ [COR COOEt OH]

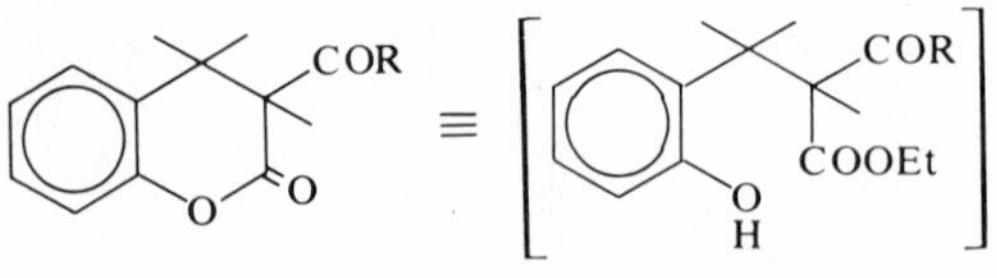

(24) **(24a)**

2-amino-4-hydroxy-6-methylpyrimidine.[1672, 2520] An analogous reaction is that of triethoxycarbonylmethane and thiourea which yields 4,6-dihydroxy-2-mercaptopyrimidine instead of its 5-ethoxycarbonyl derivative;[2519] urea behaves similarly to give 2,4,6-trihydroxypyrimidine.[2872]

The reaction of diketene with ureas to give 6-methyluracil and its derivatives (*H* 51) has been explored further. The yield of 6-methyluracil was first improved[1772] to 25% and then to 65% by introducing mercuric sulphate as a catalyst[3524] or a little pyridine into the chlorobenzene used as solvent;[2862] the mechanism was shown to involve acetoacetylurea ($AcCH_2CONHCONH_2$) by its isolation under gentle conditions and subsequent cyclization;[2863] and the reaction of *N*-monoalkylureas with diketene was shown to give a mixture of two isomeric uracils in which one predominated: thus *N*-butylurea gave mainly 1-butyl-6-methyluracil with a little of the 3-butyl isomer,[2863] identified by comparison with unambiguously synthesized material.[2864]

Ethyl γ-chloroacetoacetate has been condensed with *S*-methylthiourea in methanolic sodium methoxide to give 4-chloromethyl-6-hydroxy-2-methylthiopyrimidine;[2867] the unstable substance previously described as this pyrimidine,[76] corresponds neither in melting point nor behaviour.

The reduction of 3-acyl- or 3-ethoxycarbonyl-2-oxochromens by sodium borohydride gives the corresponding oxochromans, which lactones behave as keto esters or diesters, respectively, in the Principal pyrimidine Synthesis. Thus H. Wamhoff and F. Korte[2879] have shown that 3-acetyl-2-oxochroman (**24**; R = Me) reacts as the keto ester (**24a**; R = Me) with acetamidine in ethanolic sodium ethoxide to yield 4-hydroxy-5-*o*-hydroxybenzyl-2,6-dimethylpyrimidine in 88% yield, or with guanidine to give 2-amino-4-hydroxy-5-*o*-hydroxybenzyl-6-methylpyrimidine; analogues behave similarly, and the ester (**24**; R = OEt) with guanidine yields 2-amino-4,6-dihydroxy-5-*o*-hydroxybenzylpyrimidine.[2879] Examples of the use of other lactones are known (see Table III and ref. 2956).

Simple thiolactones have also been used as keto-esters in the Principal Synthesis. Thus α-acetyl-γ-thiobutyrolactone condenses with appropriate N—C—N fragments (guanidine, etc.) to yield 2-amino-4-hydroxy-5-β-mercaptoethyl-6-methylpyrimidine (62%), its 2,4-dihydroxy analogue (15%), its 4-hydroxy-2-mercapto analogue (50%), and 4-hydroxy-5-β-mercaptoethyl-2,6-dimethylpyrimidine (56%); α-benzoyl (or acetyl)-δ-thiovalerolactone may be used to give 2-amino-4-hydroxy-5-γ-mercaptopropyl-6-phenyl(or methyl)pyrimidine and related compounds.[2908]

Miscellaneous examples include 2-amino-5-γ-anilino-β-hydroxypropyl-4-hydroxy-6-methylpyrimidine and its analogues;[3184] and 6-methyl-3-phenyl(or tolyl)-2-thiouracil from ethyl acetoacetate and the *N*-arylurea or bis(arylformamidine) disulphide.[3211]

7. Use of β-Diesters: Malonic Esters (*H* 51)

There are a few interesting new types of condensation in this category. Benzoyloxyacetamidine condenses with diethyl malonate, its α-chloro derivative, or its α-methyl derivative to give (after acid hydrolysis of the protecting group) 4,6-dihydroxy-2-hydroxymethylpyrimidine (50%), its 5-chloro derivative, or its 5-methyl derivative.[2236, 2238] Free hydroxyacetamidine appears to be less troublesome and gives an 89% yield of the first of the above pyrimidines.[2237] However, it should be noted that the melting points given by the Russian and American authors differ by 100°. 5-Benzyl-4,6-dihydroxy-2-methoxymethylpyrimidine is made similarly.[2877]

At first sight, ethyl α-chloro-$\beta\beta$-diethoxyacrylate (**25**) might be mistaken for the equivalent of an aldehydo-ester. Because of its unsaturation, it is in fact the equivalent of a diester and could be thought of as the ortho ester, ethyl α-chloro-$\beta\beta\beta$-triethoxypropionate (**25a**). As such, it condenses in ethanol with free acetamidine or benzamidine to yield 5-chloro-4-ethoxy-6-hydroxy-2-methylpyrimidine (**26**), or its 2-phenyl-analogue; with guanidine it gives ethyl α-chloro-$\beta\beta$-diethoxy-β-guanidinopropionate which may be cyclized quantitatively to 2-amino-5-chloro-4-ethoxy-6-hydroxypyrimidine.[2239]

Diethyl ethylmalonate condenses conveniently with urea in dimethylformamide containing sodium methoxide, to give 5-ethyl-2,4,6-trihydroxypyrimidine in 80% yield.[2197] A simplified procedure for the formation of barbituric acid in 80% yield from diethyl malonate and urea, has been described.[2243] Two processes for making 1,3-diaryl-5-alkyl(or aryl)barbituric (and thiobarbituric) acids have been exemplified in twenty cases: thus, α-amylmalonic acid reacts with *NN'*-diphenylurea during 17 hr. in boiling chloroform containing acetyl chloride, to give 5-amyl-1,3-diphenylbarbituric acid in 20% yield; the alternative use of sodium ethoxide in dry xylene as condensing medium gives an 11% yield.[2244] The formation of 5-acetyl-4,6-dihydroxy-2-mercaptopyrimidine (58%) from malonic acid, thiourea, and acetic anhydride is interesting because, when the malonic acid is α-substituted,

O- instead of *C*-acetylation occurs, leading, e.g., to 4-acetoxy-5-benzyl-6-hydroxy-2-mercaptopyrimidine (**27**). *N*-Alkyl thioureas behave similarly in the reaction.[2245] *N*-Substituted-amino barbituric acids can be made by condensing an appropriately substituted semicarbazide (e.g., *N*-piperidinourea) with diethyl malonate. *N*-Piperidino-, *N*-morpholino-, and *N*-dimethylamino-barbituric acid have been made in excellent yield in this way.[2246]

Although phosphoryl chloride has been used to achieve condensation of ureas and malonic acids to barbituric acids (*H* 56), its use with dialkylureas and α-substituted malonic acids leads directly to 1,3,5-trisubstituted-4-chlorouracils. This 'chlorierende Kondensation' has been used to make, e.g., 5-butyl-4-chloro-1,2,3,6-tetrahydro-2,6-dioxo-1,3-diphenylpyrimidine (79% yield) from *NN'*-diphenylurea and α-butylmalonic acid.[2305]

The condensation of α-alkoxy-α-alkylmalonic esters with urea, thiourea, and guanidine has been studied in some detail by a Chinese group.[2878]

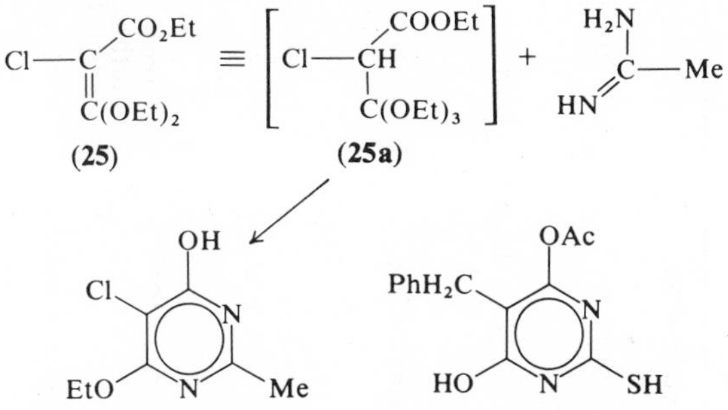

(**25**) (**25a**)

(**26**) (**27**)

The evident confusion (*H* 54) surrounding the condensation of *N*-alkylthioureas and malonic esters has been explained in part.[2247] While *N*-allylthiourea condenses normally with diethyl diethylmalonate in ethanolic sodium ethoxide to give 1-allyl-5,5-diethyl-2-thiobarbituric acid,[2248] *N*-(saturated-alkyl)thioureas do not. Thus *N*-methylthiourea, the ester, and the ethoxide in equimolecular amounts were reported[2248] to yield (among other unidentified products) a 'significant quantity' of 5,5-diethyl-1-methyl-2-thiobarbituric acid (**28**) with m.p. 123°. However, on more careful examination,[2249] the real product was shown to be formed (under the above conditions) in only minute yield and to melt about 80°. Its identity was checked by resynthesis in good yield from diethylmalonyl dichloride with *N*-methylthiourea and from the hydrolysis of the corresponding 4-aminobarbituric acid.[2249] When the molar

TABLE IV. Additional Examples of the Use of Malonic Esters in the Principal Synthesis (*H* 56)

Three-carbon fragment	One-carbon fragment	Solvent and conditions	Pyrimidine and yield	Ref.
Diethyl chloromalonate	acetamidine	ethanol; 20°; 24 hr.	5-chloro-4,6-dihydroxy-2-methyl- (35%)	2236
Diethyl chloromalonate	benzamidine[a]	ethanol; sodium ethoxide; 20°; 12 hr.	5-chloro-4,6-dihydroxy-2-phenyl-[b] (59%)	2238
Diethyl malonate	isobutyramidine	ethanol; sodium ethoxide; 20°; 2 hr.	4,6-dihydroxy-2-isopropyl- (25%)	2238
Diethyl phenoxymalonate	formamidine	ethanol; sodium ethoxide; 20°; 15 hr. (then reflux; 1 hr.)	4,6-dihydroxy-5-phenoxy-[c] (71%)	2240
Diethyl methylmalonate	benzamidine	ethanol; sodium ethoxide; reflux; 3 hr.	4,6-dihydroxy-5-methyl-2-phenyl- (49%)	2241
Diethyl diethylmalonate	acetamidine[d]	ethanol; sodium ethoxide; 120°; 20 hr.	5,5-diethyl-4,5-dihydro-6-hydroxy-2-methyl-4-oxo- (as ethanolate; *ca.* 60%)	2242
Diethyl malonate	fluoroacetamidine	methanol; sodium methoxide; 20°; 48 hr.	2-fluoromethyl-4,6-dihydroxy-[e] (21%)	2116
Diethyl benzylmalonate	benzamidine hydrochloride	heat at 220° for 25 min.	5-benzyl-4,6-dihydroxy-2-phenyl-[c] (43%)	2750
Triethoxycarbonylmethane	urea	ethanol; sodium ethoxide; 80°; 6 hr.	2,4,6-trihydroxy-[k] (poor)	2872

Diethyl malonate	trifluoroacetamidine	ethanol; reflux; 3 hr.	4,6-dihydroxy-2-trifluoromethyl-	2193
Diethyl butylmalonate	formamidine	ethanol; sodium ethoxide; 20°; 15 hr. (then reflux; 1 hr.)	5-butyl-4,6-dihydroxy-[f] (98%)	2242
Diethyl malonate	*O*-methylurea[g]	methanol; sodium methoxide; reflux; 4 hr.	4,6-dihydroxy-2-methoxy-[h] (*ca.* 64%)	2253
Dimethyl malonate	*O*-methylurea	methanol; sodium methoxide; 20°; 3 days	4,6-dihydroxy-2-methoxy-[h] (84%)	2254
Diethyl phenylmalonate	thiourea	methanol; sodium methoxide; reflux; 4 hr.	4,6-dihydroxy-2-mercapto-5-phenyl- (73%)	2255
Diethyl malonate	*N*-benzyloxyurea	ethanol; sodium ethoxide; reflux; 9 hr.	*N*-benzyloxybarbituric acid (64%)	2199
Diethyl malonate	*N*-methoxyurea	ethanol; sodium ethoxide; reflux; 4 hr.	*N*-methoxybarbituric acid (81%)	2199
Diethyl α-ethyl-α-*p*-fluorophenylmalonate	urea	methanol; sodium methoxide; reflux; 6 hr.	5-ethyl-5-*p*-fluorophenyl-barbituric acid-[i] (55%)	2256
Diethyl α-allyl-α-*o*-chlorophenylmalonate	urea	methanol; sodium methoxide; reflux; 6 hr.	5-allyl-5-*o*-chlorophenyl-barbituric acid-[i] (37%)	2256
Diethyl *p*-trimethylsilylbenzylmalonate	urea	ethanol; sodium ethoxide; reflux; 10 hr.	2,4,6-trihydroxy-5-*p*-trimethylsilylbenzyl-[k] (54%)	2866

(*continued*)

TABLE IV (*continued*).

Three-carbon fragment	One-carbon fragment	Solvent and conditions	Pyrimidine and yield	Ref.
α-Ethoxycarbonyl-γ-butyrolactone	urea	ethanol; sodium reflux; 4 hr.	2,4,6-trihydroxy-5-β-hydroxyethyl-[k] (34%)	2638
Diethyl heptadecylmalonate	urea	ethanol; sodium ethoxide; reflux (?)	5-heptadecylbarbituric acid (95%)	2239
Diethyl α-acetamido-α-propylmalonate	urea	ethanol; sodium ethoxide; reflux; 18 hr.	5-acetamido-5-propyl-barbituric acid[j] (67%)	2257
Diethyl α-acetamido-α-allyl-malonate	thiourea	ethanol; sodium ethoxide; reflux; 3 hr.	5-acetamido-5-allyl-barbituric acid[j] (66%)	2257
Diethyl α-ethyl-α-hydroxymalonate	urea	ethanol; sodium ethoxide; reflux; 5 hr.	5-ethyl-5-hydroxybarbituric acid (6%)	2258
Diethyl α-prop-2-ynylmalonate	urea	ethanol; sodium ethoxide; reflux; 3 hr.	2,4,6-trihydroxy-5-prop-2′-ynyl-[k] (66%)	2307
Diethyl α-methyl-α-β'-vinyloxy-ethylmalonate	urea	ethanol; sodium ethoxide; reflux; 10 hr.	5-methyl-5-β-vinyloxyethyl-barbituric acid[k] (43%)	2259

Diethyl α-*p*-chloroanilinomalonate	urea	ethanol; sodium ethoxide; reflux; 15 hr.	5-*p*-chloroanilino-2,4,6-trihydroxy-[l] (40%)	2260
Diethyl *m*-toluidinomalonate	guanidine	ethanol; sodium ethoxide; reflux; 6 hr.	2-amino-4,6-dihydroxy-5-*m*-toluidino-[k] (40%)	2260
Diethyl methoxymalonate	guanidine	ethanol; sodium ethoxide; reflux; 2 hr.	2-amino-4,6-dihydroxy-5-methoxy- (70%)	2212
Dimethyl malonate	guanidine	methanol; sodium methoxide; reflux; 1 hr.	2-amino-4,6-dihydoxy- (98%)	2286, *cf.* 170

[a] An improved method to produce anhydrous benzamidine hydrochloride is described.
[b] And four 2-alkyl analogues.
[c] Also the 2-methyl derivative.
[d] The analogous reaction with formamidine fails.
[e] Also the 5-methyl derivative (42%).
[f] And the 5-isopropyl homologue.
[g] Made by methylating urea and used without isolation.
[h] *Cf.* ref. 1741.
[i] And five other 5-halogenophenyl analogues.
[j] And other 5-alkyl homologues.
[k] And other analogues.
[l] Also the dichloroanilino, toluidino, and several 2-thio analogues.

quantity of *N*-methylthiourea in the original condensation was increased threefold, the predominant product (88%) was shown[2247] to be 5,5-diethyl-1,4,5,6-tetrahydro-1-methyl-2-(*N*-3-methylthioureido)-4,6-dioxopyrimidine (**29**) which could also be made by allowing '*N*-methylveronal' (**28**) to react with *N*-methylthiourea under similar conditions.[2247]

Malonic acid reacts immediately with *NN'*-dicyclohexylcarbodiimide (**30**; R = C_6H_{11}) in tetrahydrofuran to yield 1,3-dicyclohexylbarbituric (**31**, R = C_6H_{11}) and *NN'*-dicyclohexylurea.[2250, 2251] The 65% yield from this synthesis should be compared with the 5.5% obtained by an orthodox procedure from *NN'*-dicyclohexylurea and malonyl dichloride. Moreover, it is fairly general: mono- and di-alkylated malonic acids in combination with other dialkyl- and diaryl-carbodiimides are successful as a rule, although a few failures such as malonic acid or its monoethyl derivative with *NN'*-di-*p*-tolylcarbodiimide (**30**, R = *p*-C_6H_4Me) are recorded.[2251] Carbodiimides also react with malonyl dichloride or its monoalkyl derivatives to give similar barbituric acids *via* oxazine intermediates; the overall yields are 67–100%.[2252, 2535] (See Ch. II, Sect. 5.E(2).)

The preparations, properties, and uses of modern barbiturates have been neatly reviewed in 1961 by Z. Buděšínský and M. Protiva.[2261]

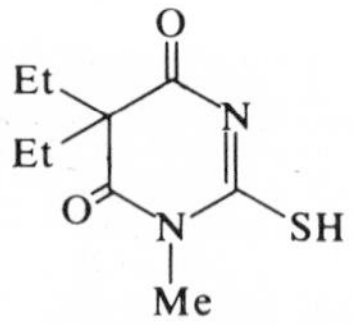

Et, Et, O, N, SH, O, N, Me

(**28**)

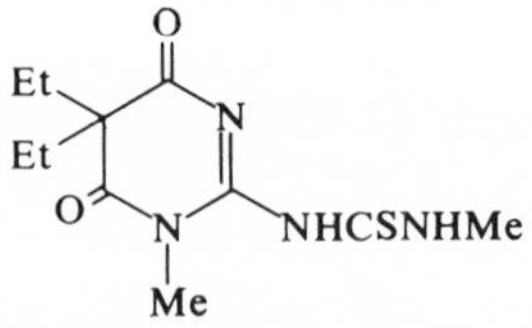

Et, Et, O, N, NHCSNHMe, O, N, Me

(**29**)

CO_2H–CH_2–CO_2H + RN=C=NR (**30**) ⟶ (**31**) [OH, N—R, O, N, O, R] + $OC(NHR)_2$

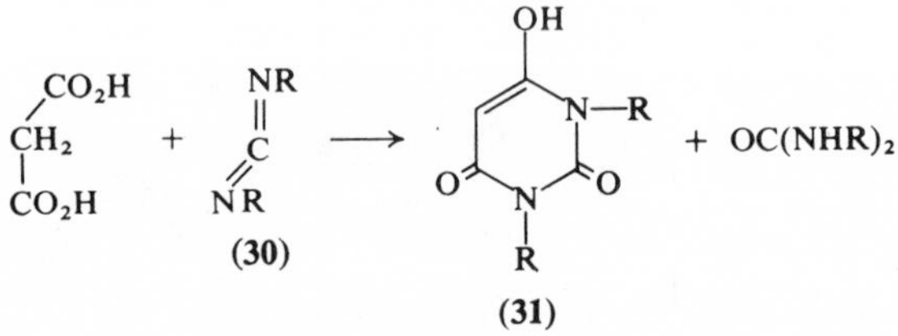

8. Use of β-Aldehydo Nitriles (*H* 59)

One of the cytosine-*N*-oxides has been made by the Principal Synthesis: either ββ-diethoxypropionitrile (**32**) or αβ-dibromopropionitrile condenses with *N*-benzyloxyurea (**33**) to give 4-amino-1-benzyloxy-1,2-dihydro-2-oxopyrimidine (**34**) which may be catalytically hydrogenated to 4-amino-1,2-dihydro-1-hydroxy-2-oxopyrimidine (**35**, cytosine-*N*-oxide).[2262] The same urea fails to yield a pyrimidine with ethyl ethoxymethylenecyanoacetate; instead, ethyl α-benzyloxyamino-methylene-α-cyanoacetate is formed.[2199]

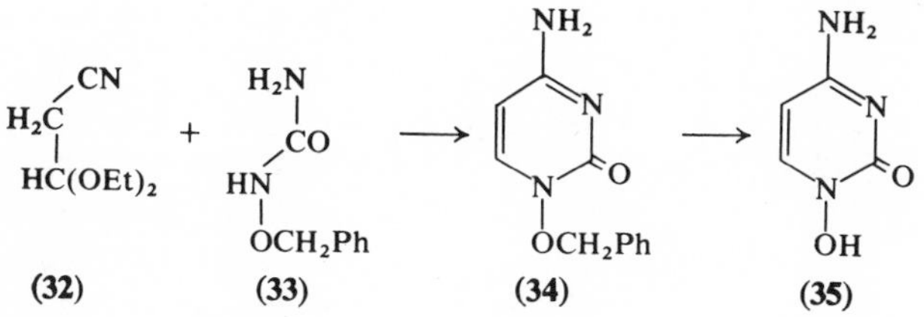

CN, H_2C, $HC(OEt)_2$ + H_2N, CO, HN, OCH_2Ph → NH_2, N, O, N, OCH_2Ph → NH_2, N, O, N, OH

(**32**) (**33**) (**34**) (**35**)

2,4-Diaminopyrimidine is formed directly by condensing β-ethoxy-acrylonitrile[2203] or *trans*-β-chloroacrylnitrile[2266] with guanidine; similar reactions yield 2,4-diamino-5-(3,4-dimethoxybenzyl)pyrimidine (from α-3′,4′-dimethoxybenzylidene-β-methoxypropionitrile, a tautomer of α-3′,4′-dimethoxybenzyl-β-methoxyacrylonitrile), and other 5-substituted-benzyl derivatives from appropriate 'benzal nitriles'.[2267] Similarly, βγ-diethoxyacrylonitrile gives 2,4-diamino-5-ethoxypyrimidine.[2521]

Although β-ethoxy-α-methoxymethylenepropionitrile (**36**) and its homologues react as aldehydo ketones with amidines to yield the expected pyrimidines[2189, 2190, 2270] (see Table IVa), with urea and *N*-alkylureas under acidic conditions they react as β-aldehydo ethers to give, for example, 5-cyano-1,6-dihydro-2-hydroxypyrimidine (**37**) which may then be dehydrogenated to 5-cyano-2-hydroxypyrimidine (**38**).[2276, 2277] With thiourea, the nitrile (**36**) again reacts as an aldehydo ether, but this time to form 5-cyano-2-hydroxy-1,3-thiazine (**39**).[2276]

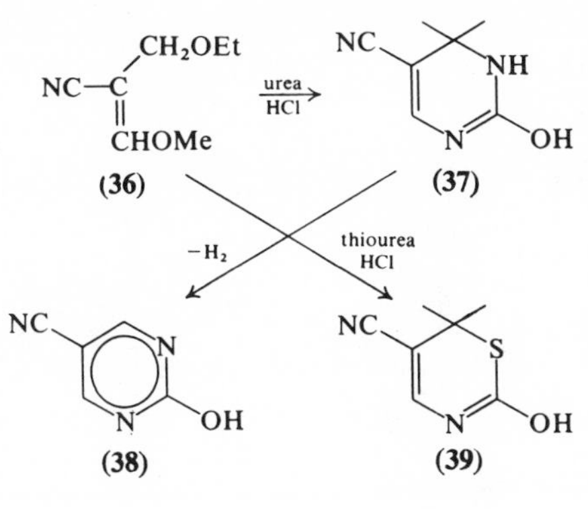

CH_2OEt, NC—C, CHOMe (**36**) —urea, HCl→ NC, NH, N, OH (**37**)

$-H_2$ → NC, N, N, OH (**38**)

thiourea, HCl → NC, S, N, OH (**39**)

TABLE IVa. Additional Examples of the Use of Aldehydo Nitriles in the Principal Synthesis (*New*)

Three-carbon fragment	One-carbon fragment	Solvent and conditions	Pyrimidine and yield	Ref.
Ethoxymethylene-malononitrile	pentafluoro-propionamidine	ethanol; 5°(?); 2 hr.	4-amino-5-cyano-2-pentafluoroethyl-[a] (69%)	2268
β-Ethoxy-α-methoxy-methylenepropionitrile	formamidine	ethanol; reflux; 5 hr.	4-amino-5-ethoxy-methyl[b] (61%)	2189, 2190
β-Methoxy-α-methoxy-methylenepropionitrile	acetamidine	ethanol; reflux; 5 hr.	4-amino-5-methoxy-methyl-2-methyl-[c] (56%)	2189
β-Ethoxy-α-methoxy-methylenepropionitrile	propionamidine	ethanol; 20°; 12 hr. (then) reflux; 3 hr.	4-amino-5-ethoxymethyl-2-ethyl- (69%)	2270
Ethoxymethylene-malononitrile	thiourea	aq. acetone; sodium hydroxide; 20°; ½ hr.	4-amino-5-cyano-2-mercapto- (12%)	2271, *cf.* 1698
Ethoxymethylene-malononitrile	*S*-benzylthiourea	aq. acetone; 20°; 12 hr.	4-amino-2-benzylthio-5-cyano-[d] (72%)	2271
ββ-Diethoxypropionitrile	urea	butanol; sodium butoxide; reflux; 2.2 hr.	4-amino-2-hydroxy-[e] (32%)	2272
Ethoxymethylene-malononitrile	*O*-methylurea	ethanol; reflux; 1 hr.	4-amino-5-cyano-2-methoxy- (*ca.* 50%)	2273
β-Ethoxy-α-phenyl-acrylonitrile	thiourea	ethanol; sodium ethoxide; reflux; 2 hr.	4-amino-2-mercapto-5-phenyl-[f] (79%)	2274
β-Anilino-α-phenyl-acrylonitrile	thiourea	sodium ethoxide; 135° *in vacuo*; 5 hr.	4-amino-2-mercapto-5-phenyl-[g] (68%)	2274
Ethyl α-ethoxymethylene-α-cyanoacetate	*O*-methylurea	methanol; sodium methoxide; 20°; 12 hr.	4-amino-2-methoxy-5-methoxycarbonyl-[h] (17%)	2205
Ethyl α-ethoxymethylene-α-cyanoacetate	thiourea	ethanol; sodium ethoxide; 20°; 2 hr.	4-amino-5-ethoxycarbonyl-2-mercapto- (91%)	2275

[a] Also the 2-heptafluoropropyl homologue.
[b] Also the 2-propyl (see also ref. 2269) and 2-phenyl derivatives; an excess of amidine must be avoided.
[c] And other 5-alkoxymethyl homologues.
[d] Also the methylthio and ethylthio analogues.
[e] An improved method of making the intermediate, 'cyanacetal', is described; *cf.* ref. 184.
[f] Also twelve analogues with 5-halogeno- or 5-methoxy- phenyl groups.
[g] And three 5-halogenophenyl analogues.
[h] The corresponding ethyl ester (14%) and 5-cyano-4-hydroxy-2-methoxypyrimidine (50%) were also isolated.

The Whitehead synthesis (*H* 62) has been used to prepare 4-amino-5-cyano-2-hydroxypyrimidine (5-cyanocytosine) from ureidomethylenemalononitrile.[2897]

αα-Dicyano-β-dimethylaminoethylene[$(CN)_2C{:}CHNMe_2$] or its 'hydrated' precursor, αα-dicyano-β-dimethylaminoethanol, have been condensed with formamidine, guanidine, etc., to give 4-amino-5-cyanopyrimidine and its 2-amino, 2-mercapto, 2-hydroxy, and other such derivatives.[3752, 3753]

9. Use of β-Keto Nitriles (*H* 65)

The iminoether (**40**) reacts as a keto nitrile with *O*-methylurea in methanol to give a good yield of 4-amino-2-methoxy-5,6-dimethylpyrimidine (**41**);[2280] the reaction appears to be unique. Other simple condensations include α-(α-ethoxyethylidene)malononitrile with formamidine to yield 4-amino-5-cyano-6-methylpyrimidine, with thiourea to yield 4-amino-5-cyano-2-mercapto-6-methylpyrimidine, and with *S*-methylthiourea (?; no details) to yield 4-amino-5-cyano-6-methyl-2-methylthiopyrimidine.[2273]

Most other recent examples involve the formation of 5-arylpyrimidines. Thus β-ethoxy(or anilino)-α-phenylcrotononitrile (**42**; **42a**) with thiourea gives 4-amino-2-mercapto-6-methyl-5-phenylpyrimidine;[2274] β-anilino-β-ethyl(or alkyl)-α-phenyl(or substituted-phenyl)acrylonitrile with thiourea gives 4-amino-6-ethyl(or alkyl)-2-mercapto-5-phenyl(or substituted-phenyl)pyrimidine;[2274] α-3,4-difluorophenyl-β-methoxycrotononitrile with guanidine gives 2,4-diamino-5-(3,4-difluorophenyl)-6-methylpyrimidine;[2281] α-*p*-cyanophenyl-β-ethyl-β-methoxyacrylonitrile (and homologues) with guanidine give 2,4-diamino-5-*p*-cyanophenyl-6-ethylpyrimidine (and appropriate homologues);[2282] β-benzyl-β-methoxy-α-phenylacrylonitrile (and analogues) with guanidine give 2,4-diamino-6-benzyl-5-phenylpyrimidine (and appropriate analogues);[2283] α-*p*-chlorophenyl-βγ-diethoxycrotononitrile with guanidine gives 2,4-diamino-5-*p*-chlorophenyl-6-ethoxymethylpyrimidine;[2302] and α-*p*-chlorophenyl-β-ethoxy-γγγ-trifluorocrotononitrile with guanidine gives 2,4-diamino-5-*p*-chlorophenyl-6-trifluoromethylpyrimidine.[2391] When α-*p*-chlorophenylthio-α-cyanoacetone is *O*-methylated to its enol ether, condensation becomes possible with guanidine to give 2,4-diamino-5-*p*-chlorophenylthio-6-methylpyrimidine;[2773] analogues were made similarly.[2773] Other examples are known, in which the phenyl group is separated from the pyrimidine.[2943, 2944, 2981]

An interesting, if complicated, example is provided by the condensation of *NN*-bis-(β-chloroethyl)-*p*-tricyanovinylaniline (**43**), which is equivalent to the keto nitrile (**43a**), with amidines to yield, for example, 4-amino-6-*p*-bis-(β-chloroethyl)aminophenyl-5-cyano-2-methylpyrimidine. The same intermediate with guanidine or *S*-benzylthiourea gives, respectively, the 2-amino- or 2-benzylthio-analogue of the above pyrimidine.[2906]

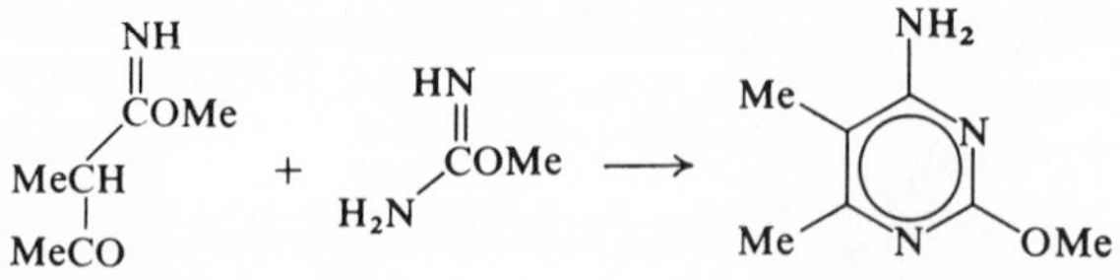

NH=C(OMe)–CH(Me)–COMe (**40**) + HN=C(OMe)–H_2N → 4-amino-5,6-dimethyl-2-methoxypyrimidine (**41**)

EtOC(Me):C(Ph)CN (**42**)

PhNHC(Me):C(Ph)CN (**42a**)

(*p*) $(ClCH_2CH_2)_2N—C_6H_4—C(CN):C(CN)_2$ (**43**)

[(*p*) $(ClCH_2CH_2)_2N—C_6H_4—C(:O)CH(CN)CN$ (**43a**)

10. Use of β-Ester Nitriles (*H* 67)

When $\alpha\beta\beta$-trichloroacrylonitrile (**44**) is allowed to react at room temperature in ethanol with acetamidine, 4-amino-5-chloro-6-ethoxy-2-methylpyrimidine (**46**) results. It appears that the nitrile (**44**) is first converted into the imino-ether (**45**) which subsequently condenses as an ester nitrile with the amidine; preconversion into the imino-ether improves the yield of pyrimidine.[2239] Benzamidine and the imino-ether (**45**) furnish 4-amino-5-chloro-6-ethoxy-2-phenylpyrimidine in 94% yield. Likewise, guanidine with either the imino-ether or trichloronitrile gives 2,4-diamino-5-chloro-6-ethoxypyrimidine.[2239]

Two other unusual intermediates are α-cyano-γ-butyrolactone (**47**; R = H) and its methyl derivative, α-cyano-γ-valerolactone (**47**; R = Me). The former reacts with thiourea or urea to give 4-amino-6-hydroxy-5-β-hydroxyethyl-2-mercaptopyrimidine or the corresponding 2-hydroxypyrimidine, respectively. The second lactone furnishes the 5-β-hydroxypropyl homologues.[2202] The corresponding thiolactones give 5-mercaptoalkyl derivatives.[2908]

Some recent examples in this bracket are noted in Table Va, and an unusual condensation is that of citrulline [5-ureidonorvaline: $NH_2CONH—(CH_2)_3CH(NH_2)CO_2H$] with ethyl cyanoacetate to give

6-amino-1-δ-amino-δ-carboxybutyluracil.[2860] 4-Amino-5-formamido-2,6-dihydroxypyrimidine and its 5-acetamido homologue have been made by acylation of the intermediate urea prior to cyclization.[2865]

Esters such as ethyl cyanoacetate may be replaced by a cyanoacyl chloride. Thus *N*-phenylurea and cyanoacetyl chloride in the presence of pyridine gave the urea, $PhNHCONHCOCH_2CN$, which underwent cyclization in alkali to give 6-amino-1-phenyluracil. The corresponding 2-thiouracil was made similarly; both pyrimidines were made also from ethyl cyanoacetate, but in less good yield.[2982]

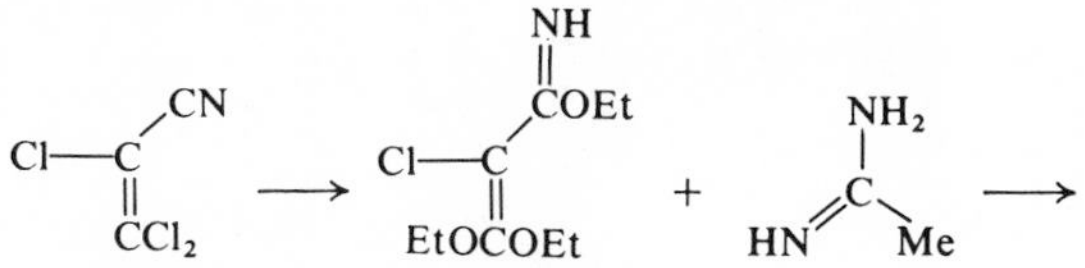

Cl—C(CN)=CCl_2 (**44**) → Cl—C(C(=NH)OEt)=C(OEt)OEt (**45**) + H_2N—C(=NH)Me →

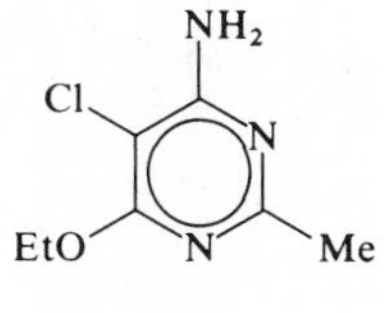

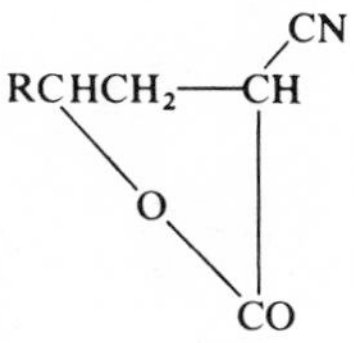

(**46**) [pyrimidine: NH_2, Cl, EtO, Me] (**47**) [RCHCH$_2$—CH(CN), O, CO]

11. Use of β-Dinitriles; Malononitriles (*H* 72)

The well recognized abnormal condensation of amidines with (unsubstituted) malononitrile has been recently illustrated by the formation of 4-amino-5-cyano-2,6-bistrifluoromethylpyrimidine (**48**) using trifluoroacetamidine.[2217] In contrast, phenylazomalononitrile reacts normally with the same amidine to give 4,6-diamino-5-phenylazo-2-trifluoromethylpyrimidine.[2217] Similarly, nitrosomalononitrile with butyramidine or trifluoroacetamidine yields 4,6-diamino-5-nitroso-2-propyl(or trifluoromethyl)pyrimidine, although the reactions are done in an unusual way by simply refluxing the solid nitrosomalononitrile-amidine salt in a dialkylpyridine for ten minutes.[2294, 2868] Using *NN*-dimethylguanidine in the same way, 4,6-diamino-2-dimethylamino-5-nitrosopyrimidine is obtained.[2297]

A 99% yield has been recorded[2296] in the well-known preparation of 4,6-diamino-2-mercaptopyrimidine from malononitrile and thiourea; a similar condensation using dimethylaminomalononitrile (**49**) yields

TABLE Va. Additional Examples of the Use of Ester Nitriles in the Principal Synthesis (*New*)

Three-carbon fragment	One-carbon fragment	Solvent and conditions	Pyrimidine and yield	Ref.
Ethyl cyanoacetate	benzamidine	ethanol; sodium ethoxide; reflux; 3 hr.	4-amino-6-hydroxy-2-phenyl-[a] (70%)	2284
Ethyl (methyl) cyanoacetate	guanidine	methanol; sodium methoxide; reflux; 3–5 hr.	2,4-diamino-6-hydroxy-[b] (*ca.* 90%)	2285, 2286
Ethyl cyanoacetate	thiourea	ethanol; sodium ethoxide; reflux; 2 hr.	4-amino-6-hydroxy-2-mercapto- (99%)	2287
Ethyl α-cyanopropionate	guanidine	ethanol; sodium ethoxide; reflux; 2 hr.	2,4-diamino-6-hydroxy-5-methyl- (56%)	2288
Ethyl α-cyano-α-decylacetate	thiourea	ethanol; sodium ethoxide; reflux; 5 hr.	4-amino-5-decyl-6-hydroxy-2-mercapto-[c] (*ca.* 80%)	2289
Ethyl α-cyanobutyrate	propionamidine	ethanol; reflux; 2 hr.	4-amino-2,5-diethyl-6-hydroxy- (91%)	2270
Ethyl cyanoacetate	trifluoroacetamidine	ethanol; sodium ethoxide; reflux; 3 hr.	4-amino-6-hydroxy-2-trifluoromethyl- (74%)	2193
Ethyl cyanoacetate	hydroxyacetamidine	methanol; sodium methoxide; reflux; 2 hr. (then 20°; 12 hr.)	4-amino-6-hydroxy-2-hydroxymethyl- (69%)	2290

Ethyl α-cyanobutyrate	thiourea	ethanol; sodium ethoxide; reflux.	4-amino-5-ethyl-6-hydroxy-2-mercapto-[d]	2204
Ethyl α-cyanopropionate	*N*-dimethylaminourea	ethanol; sodium ethoxide; 80°; 4 hr.	6-amino-1-dimethylamino-5-methyluracil[d] (*ca.* 50%)	2870
Ethyl cyanoacetate	*N*-benzyloxyurea	ethanol; sodium ethoxide; reflux; 6½ hr.	6-amino-1-benzyloxyuracil (64%)	2199
Ethyl cyanoacetate	*N*-dimethylaminourea	ethanol; sodium ethoxide; 75°; 5 hr.	6-amino-1-dimethylamino-uracil[e] (60–80%)	2291
Ethyl cyanoacetate	*N*-dimethylamino-thiourea	ethanol; sodium ethoxide; 75°; 7–8 hr.	6-amino-1-dimethylamino-2-thiouracil[e]	2291
Ethyl nitrosocyanoacetate	trifluoroacetamidine	ethanol; sodium ethoxide; 20°; 15 hr.	4-amino-6-hydroxy-5-nitroso-2-trifluoromethyl- (32%)	2217
α-Cyanopropionic acid	urea	acetic anhydride; 100°[f]	4-amino-2,6-dihydroxy-methyl- (58%)	2306
Cyanoacetic acid	*N*-methyl-*N*′-β-methylallylthiourea	glacial acetic acid; 55°; 2 hr.[f]	4-amino-1,2,3,6-tetrahydro-1-methyl-3-β-methylallyl-6-oxo-2-thio- (18%)	2309

[a] *Cf.* ref. 195 and ref. 1911 (58%).
[b] *Cf.* ref. 246.
[c] And eight similar pyrimidines.
[d] And four higher homologues.
[e] And similar compounds. No analyses given. Assignment of dimethylamino group to position-1 is based (ref. 2291) on spectra and on conversion into xanthines (ref. 2292), the structures of which are also spectrally based. *Cf.* an earlier assignment to position-3 (ref. 2293).
[f] Resulting intermediate cyclized by dissolution in aqueous base. (*Cf.* similar formation of 4-amino-1,2,3,6-tetrahydro-1,3-dimethyl 6-oxo-2-thiopyrimidine using *NN*′-dimethylthiourea.)[2935]

4,6-diamino-5-dimethylamino-2-mercaptopyrimidine (**50**).[2295] Malononitrile also condenses with thiosemicarbazide (*N*-aminothiourea!) to give 1,4,6-triamino-1,2-dihydro-2-thiopyrimidine (**51**) which undergoes a remarkable reaction with phosphoryl chloride in dimethylformamide to give 4,6-bis-(dimethylaminomethyleneamino)-2-thiocyanatopyrimidine.[2859]

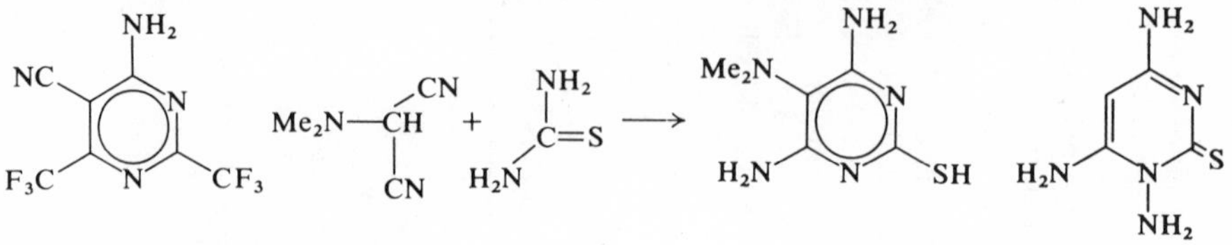

(**48**) (**49**) (**50**) (**51**)

A more complicated example involved the initial alkylation of malononitrile with 3-*p*-nitrophenoxypropyl bromide to give 1,1-dicyano-4-*p*-nitrophenoxybutane, $(CN)_2CH(CH_2)_3OC_6H_4NO_2(p)$, followed by condensation with guanidine to yield 2,4,6-triamino-5-*p*-nitrophenoxypropylpyrimidine in 51% yield.[2940]

CHAPTER III

Other Methods of Primary Synthesis (*H* 82)

1. General Remarks (*H* 82)

This chapter now covers a much wider range of synthetic procedures than did the original. Despite obvious difficulties, the former classification of methods has been retained where possible, in order to emphasize the supplementary nature of the present chapter. However, many new sections have become necessary, and others have been enlarged in scope to accommodate what are virtually new syntheses.

2. Synthesis Involving Preformed Aminomethylene Groups (*H* 82)

A. Aminomethylene Derivatives with Isocyanates (*H* 82)

There have been no recent classical examples of this synthesis but two new procedures are, for differing reasons, best classified in this group.

Methyl phenylacetate is formylated to methyl α-formylphenylacetate which may be condensed with *N*-methylurea by azeotropic removal in toluene of the water formed. The resulting methyl α-methylureidomethylenephenylacetate (**1**) is the same type of intermediate as that produced in the classical synthesis; it may be cyclized to 3-methyl-5-phenyluracil (**2**) by refluxing either in diphenyl ether (82%) or ethanolic sodium ethoxide (63%).[2312] The synthesis may be considered alternatively as a Principal Synthesis performed in two stages.

Behrend's original condensation[284, 285] of ethyl β-aminocrotonate (**3**) with phenyl isothiocyanate gave not only 6-methyl-3-phenyl-2-thiouracil but also the unwanted ethyl β-amino-α-phenylthiocarbamoylcrotonate (**4**). Such by-products also occurred in related syntheses (cf. *H* 83) and indicated[2313, 2542] the enamine nature of the original ethyl aminocrotonate. In an effort to use this property, benzoyl isothiocyanate was allowed to react with ethyl β-anilinocrotonate in ether. A good yield of 5-ethoxycarbonyl-1,4-dihydro-6-methyl-1,2-diphenyl-4-thiopyrimidine (**6**, R = CO_2Et) resulted presumably by dehydration of the intermediate ethyl β-anilino-α-benzoylthiocarbamoylcrotonate (**5**, R = CO_2Et).[2313, 2319] The reaction was applied successfully to a variety of αβ-unsaturated amino esters and acyl isothiocyanates,[2313, 2314, 2319] but when the oxygen analogue, benzoyl isocyanate, was used the intermediate could not be cyclized.[2313] In addition, αβ-unsaturated amino ketones and nitriles can also be used to yield for example 5-acetyl-1,4-dihydro-2,6-dimethyl-1-phenyl-2-thiopyrimidine (from 2-anilinopent-2-en-4-one and acetyl isothiocyanate) and 5-cyano-1,4-dihydro-6-methyl-1,2-diphenyl-4-thiopyrimidine (**6**, R = CN; from β-anilinocrotononitrile and benzoyl isothiocyanate *via* unisolated, **5**, R = CN), respectively.[2319, 2905] Other variations[2320] lead to replacement of the 6-methyl group by an alkoxy or amino group. (The exigencies of nomenclature cause these groups to appear as 4-substituents in the examples that follow). Thus the imino ether, best formulated as ethyl β-amino-β-ethoxyacrylate, and benzoyl isothiocyanate yield β-amino-β-ethoxy-α-ethoxycarbonyl-*N*-benzoylacrylothioamide, which on standing in ammonia cyclizes to 4-ethoxy-5-ethoxycarbonyl-6-mercapto-2-phenylpyrimidine; similarly, benzoylacetamidine eventually yields 4-amino-5-benzoyl-6-mercapto-2-phenylpyrimidine.[2320]

The above reaction is therefore general for making 5-alkoxycarbonyl (or acyl or cyano)-4-mercapto-6-alkyl (or alkoxy or amino)pyrimidines having an alkyl or aryl group in position 1 and/or 2.

An interesting extension of the enamine reaction has been briefly described.[2322] Dimethylaminomethylenenitromethane and benzoyl isocyanate give β-dimethylamino-α-nitro-*N*-benzoylacrylamide which with ammonia in dimethylformamide gives 4-hydroxy-5-nitro-2-phenylpyrimidine. In addition, β-amino-α-nitroacrylamide and *N*-diethoxymethyl-*NN*-dimethylamine give 4-hydroxy-5-nitropyrimidine.[2322]

A contribution has been made to the synthesis of uracils *via* dihydrouracils (*H* 83). 5-Bromodihydrouracil may be converted into uracil not only by thermal dehydrobromination but also by dissolution in cold concentrated aqueous sodium hydroxide or by boiling a suspension in dimethylaniline.[2196] Several 1-alkyl-5,6-dihydrouracils (or thiouracils) and 1-alkyl-5,6-dihydrocytosines have been made by cyclization of

N-alkyl-*N*-β-cyanoethylureas in acidic and basic media, respectively.[2963] Dehydrobromination of their 5-bromo derivatives led to such pyrimidines as 1-benzyl- and 1-butyl-uracil, but the formation of 1-alkylcytosines has not been successful.[2963]

$MeHNCONHCH{:}C(Ph)CO_2Me \longrightarrow$

(1) (2)

$MeC(NH_2){:}CHCO_2Et \longrightarrow MeC(NH_2){:}C(CO_2Et)CSNHPh$

(3) (4)

$[MeC(NHPh){:}CRCSNHCOPh] \longrightarrow$

(5) (6)

The 'unusual reaction' (*H* 84) of *N*-alkylcyanoacetamide and alkyl isocyanate to yield, e.g., 1-allyl-4-amino-3-ethyl-1,2,3,6-tetrahydro-2,6-dioxopyrimidine, has been extended in two directions. Useful thiobarbiturates have been made with its help.[2315] Thus ethyl dialkylcyanoacetate is hydrolysed to the corresponding acid and converted into the acid chloride. This with powdered potassium thiocyanate in boiling toluene gives a dialkylcyanoacetyl isothiocyanate (7) which reacts with an amine (or ammonia) to give a thioureido derivative (**8**). Cyclization yields a 5,5-dialkylpyrimidine (**9**, X = NH) which on hydrolysis gives a 2-thiobarbiturate (**9**, X = O).[2315]

Similarly, cyanoacetyl chloride and silver isocyanate give cyanoacetyl isocyanate which reacts with *O*-benzylhydroxylamine to give *N*-cyanoacetyl-*N*′-benzyloxyurea (**10**). This cyclizes to 6-amino-1-benzyloxyuracil (**11**, R = CH_2Ph) which by hydrolysis furnishes an unambiguous synthesis of 6-amino-1-hydroxyuracil (**11**, R = H; 6-aminouracil-*N*-1-oxide).[2199] The same *N*-oxide has been made by a distinct but rather analogous route: cyanoacetylurea reacts with hydroxylamine to give β-amino-β-hydroxyiminopropionylurea (**12**). On refluxing in dimethylformamide this eliminates ammonia to give the cyclized *N*-oxide (**11**, R = H).[2316]

B. Aminomethylene Derivatives with Imino-ethers or Imidoyl Chlorides (*H* 84)

The imino-ether reaction has been extended to the preparation of several 4-alkyl-2-aryl-6-hydroxypyrimidines. Thus ethyl benzimidate and ethyl β-aminocrotonate, $MeC(NH_2){:}CHCO_2Et$, yield 4-hydroxy-6-methyl-2-phenylpyrimidine (48%); and appropriate analogues of the reagents yield 4-hydroxy-6-methyl-2-α-thienylpyrimidine, 4-hydroxy-2-β-naphthyl-6-propylpyrimidine, and such like.[2892] Certain limitations are evident in the imino-ethers that will furnish pyrimidines: if too weakly basic, amidines result; if too strongly basic, no reaction occurs.[2892]

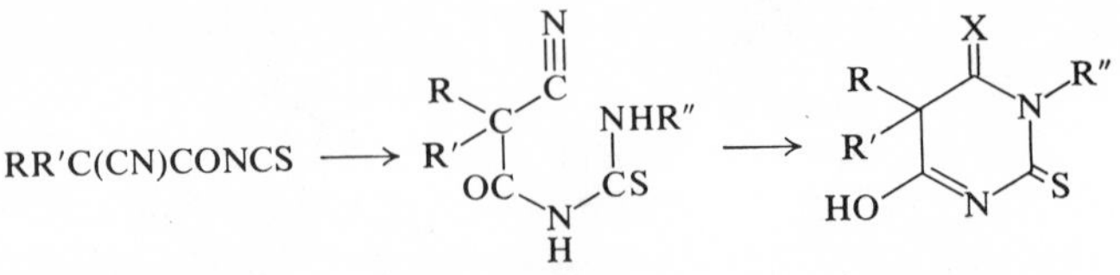

RR′C(CN)CONCS → (8) → (9)

(7) (8) (9)

(10) → (11) ← (12)

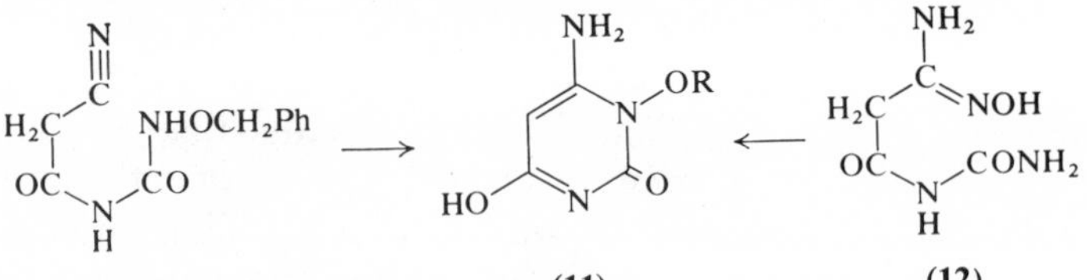

(10) (11) (12)

E. The Shaw Synthesis from Aminomethyleneacylurethanes (*H* 87)

G. Shaw has extended his synthesis so that 5-acetyluracils can now be readily prepared. Thus α-acetyl-β-ethoxy-*N*-ethoxycarbonylacrylamide (**13**) reacts with ammonia to give 5-acetyl-2,4-dihydroxypyrimidine (**15**, R = H, X = O) without isolation of the intermediate (**14**, R = H).[2317] The ammonia may be replaced by amines or α-amino acids which lead to appropriately 1-substituted 5-acetyluracils, such as the 1-methyl (from methylamine), 1-phenyl (from aniline), 1-α-carboxyethyl (from alanine) and 1-α-carboxy-β-*p*-hydroxyphenylethyl (from tyrosine) derivatives.[2317] The configuration of the amino acid is retained: 5-acetyl-1-α-carboxy-β-hydroxyethyluracil [**15**, R =

—$CH(CO_2H)CH_2OH$, X = O] from L-serine[2317] has m.p. 216° but that from DL-serine[2318] has m.p. 204°. In some cases an intermediate may be isolated and subsequently cyclized: e.g., cysteine gives α-acetyl-β-(α-carboxy-β-mercaptoethylamino)-*N*-ethoxycarbonylacrylamide [**14**, R = —$CH(CO_2H)CH_2SH$] and thence 5-acetyl-1-α-carboxy-β-mercaptoethyluracil [**15**, R = —$CH(CO_2H)CH_2SH$, X = O].[2317] Since these 5-acetyluracils are immune to hydrolysis by strong acids, they may be used with advantage in the study of *N*-terminal residues in proteins.[2317]

R. N. Warrener[2546] has further extended the synthesis to make 6-alkyl-5-cyanouracils substituted at *N*-1 by an amino, hydroxy, alkyl, or aryl group. The intermediate urethanes, $R'NHC(R){:}C(CN)CONHCO_2Et$, cyclize in base to pyrimidines such as 5-cyano-1,6-dimethyluracil or 1-amino-5-cyano-6-ethyluracil.[2546]

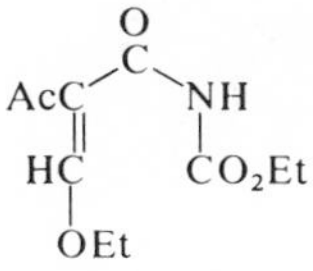

(13) 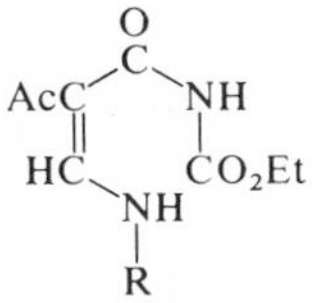 **(14)** 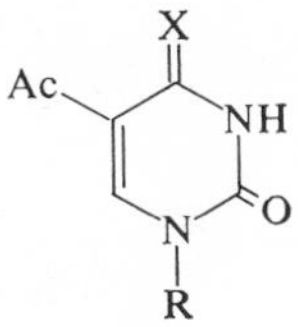 **(15)**

The synthesis has also been used to make several 1-substituted 4-thiouracils; the required intermediates are neatly made by an enamine-isothiocyanate reaction (*cf.* Sect. 2.A).[2319] Thus 1-methylaminobut-1-en-3-one reacts at C-2 with phenoxycarbonyl isothiocyanate (made *in situ* from phenyl chloroformate and sodium thiocyanate) to yield (after cyclization) 5-acetyl-1-methyl-4-thiouracil (**15**, R = Me, X = S); ethyl β-methylaminocrotonate similarly gives 5-ethoxycarbonyl-1,6-dimethyl-4-thiouracil; and β-phenyliminobutyronitrile gives 5-cyano 6-methyl-1-phenyl-4-thiouracil.[2319]

3-Hydroxyuracil (**17**, R = H; uracil-3-*N*-oxide) has been made unambiguously by a route which perhaps could be best described as a 'vinylogous Shaw Synthesis'. Methoxycarbonylaminoacrylic acid is converted into the acid chloride and allowed to react with *O*-benzylhydroxylamine to give β-methoxycarbonylamino-*N*-benzyloxyacrylamide (**16**) which cyclizes under alkaline conditions to 3-benzyloxyuracil (**17**, R = CH_2Ph). Acid hydrolysis yields the *N*-oxide (**17**, R = H).[2200]

When urethane is treated with triethyl orthoformate, ethoxymethyleneurethane (**19**) results. This condenses with ethyl β-aminocrotonate (**18**) to give 5-ethoxycarbonyl-4-hydroxy-6-methylpyrimidine (**20**).[2321]

Depending on the order of attachment, the unisolated linear intermediate may or may not be akin to that in a Shaw Synthesis.

The synthesis (*H* 89) involving treatment of an α-ethoxycarbonylaminomaleimide with alkali to give an orotic acid has been extended: for example the *N*-*p*-fluorophenyl derivative gave 6-carboxy-3-*p*-fluorophenyluracil.[3751]

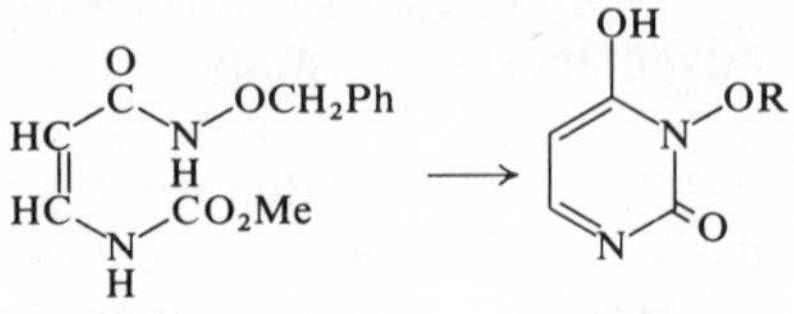

(16) (17)

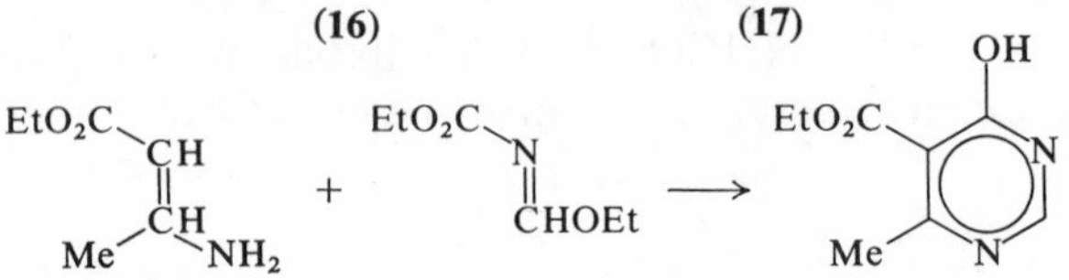

(18) (19) (20)

F. Use of β-Acylaminovinyl Ketones (*New*)

The formation of a pyrimidine from a β-acylaminovinyl alkyl (or aryl) ketone requires the supply of only one nitrogen atom at the ammonia level of oxidation. This possibility has been exploited to but a limited extent.[2329] Thus when β-acetamidovinyl phenyl ketone (**21**, R = Me, R′ = H) is heated at 200° with ammonia, 11% of 2-methyl-4-phenylpyrimidine (**22**, R = Me, R′ = H) results; better yields are obtained if the ammonia is replaced by formamide (39%), formamidine acetate (64%), acetamide (23%), or acetamidine hydrochloride (11%). By using one or other of the above reagents, β-acetamido-α-methylvinyl phenyl ketone (**21**, R = R′ = Me) gives 2,5-dimethyl-4-phenylpyrimidine (**22**, R = R′ = Me), β-acetamido-α-ethylvinyl phenyl ketone gives 5-ethyl-2-methyl-4-phenylpyrimidine, phenyl β-propionamidovinyl ketone gives 2-ethyl-4-phenylpyrimidine, phenyl α-methyl-β-propionamidovinyl ketone (**21**, R = Et, R′ = Me) gives 2-ethyl-5-methyl-4-phenylpyrimidine (**22**, R = Et, R′ = Me), α-ethyl-β-propionamidovinyl phenyl ketone gives 2,5-diethyl-4-phenylpyrimidine, β-butyramidovinyl phenyl ketone gives 4-phenyl-2-propylpyrimidine, β-butyramido-α-methylvinyl phenyl ketone gives 5-methyl-4-phenyl-2-propylpyrimidine, and β-butyramido-α-ethylvinyl phenyl ketone gives 5-ethyl-4-phenyl-2-propylpyrimidine.[2329] Although *p*-methoxybenzamidovinyl phenyl ketone yields 2-*p*-methoxyphenyl-4-phenylpyrimidine, benzamidovinyl phenyl ketone (**21**, R = Ph, R′ = H) is recorded[2329] as giving not 2,4-diphenylpyrimidine (**22**, R = Ph, R′ = H) but 4-phenylpyrimidine (**22**, R = R′ = H). When β-acetamidovinyl phenyl ketone reacts with guanidine carbonate, 2-amino-4-phenylpyrimidine results; β-acetamido-α-methylvinyl phenyl ketone similarly gives 2-amino-5-methyl-4-phenylpyrimidine.[2329]

Ph, CO, R′CH, HC, N, H, COR + NH_3 ⟶ Ph, R′, N, N, R

(**21**) (**22**)

3. Syntheses Involving an Aminomethylene Group Formed *in Situ* (*H* 90)

A. β-Dinitriles with Amidines (*H* 90)

Trifluoroacetamidine reacts abnormally with malononitrile as does acetamidine, but the yield of 4-amino-5-cyano-2,6-bistrifluoromethyl-pyrimidine is only 17%.[2217] With phenylazomalonitrile, a normal Principal Synthesis occurs.[2217]

B. Formamide with Compounds Containing an Active Methylene Group (*H* 91)

H. Bredereck, R. Gompper, and their colleagues have greatly extended the use of this reaction by employing trisformamidomethane[2324, 2325] in place of, or along with, formamide.* The mechanism seems to be still an open question, but for practical purposes it is best to think of trisformamidomethane as formylformamidine (**23**); with it, acetone would then be expected to yield 4-methylpyrimidine (**24**), and

* The use of formamide and related reagents in this and other syntheses of heterocycles, has been reviewed by Professor Bredereck and his colleagues.[2931]

TABLE VIa. Examples of Pyrimidine Syntheses from Compounds Containing an Active Methylene Group and Trisformamidomethane (*New*)

Starting material	Pyrimidine	Yield (%)	Ref.
Paraldehyde	unsubstituted	8	2163
Acetone	4-methyl-	39[a]	2163
Propanal	5-methyl-	8	2163
2-Methylpenan-4-one	4-isobutyl-	35	2163
Pinacolin	4-t-butyl-	17	2163
Acetophenone	4-phenyl-	72[b]	2163
α-Acetylnaphthalene	4-α-naphthyl-	18	2326
β-Acetylnaphthalene	4-β-naphthyl-	39	2326
4-Acetylbiphenyl	4-4′-biphenylyl-	55	2326
2-Acetylanthracene	4-2′-anthryl-	6	2326
2-Acetylfluorene	4-2′-fluorenyl-	10	2326
Butan-2-one	4,5-dimethyl-	47	2163
Pentan-3-one	4-ethyl-5-methyl-	37	2163
Propiophenone	5-methyl-4-phenyl-	53	2163
Butyrophenone	5-ethyl-4-phenyl-	26	2163
Cyclopentanone	4,5-trimethylene- (**25**)	52	2163
Cyclohexanone	4,5-tetramethylene-	36	2163
5-Acetyl-α-picoline	4-2′-methylpyridin-5′-yl-	43	2163
Diethyl malonate	5-ethoxycarbonyl-4-hydroxy-	41	2323
Dimethyl malonate	4-hydroxy-5-methoxycarbonyl-	47	2323, 2544
Dibutyl malonate	5-butoxycarbonyl-4-hydroxy-	36	2323
Benzyl cyanide	4-amino-5-phenyl-	61[c]	2327
p-Nitrobenzyl cyanide	4-amino-5-*p*-nitrophenyl-	60[d]	2327
m-Nitrobenzyl cyanide	4-amino-5-*m*-nitrophenyl-	47	2327
p-Aminobenzyl cyanide	4-amino-5-*p*-formamidophenyl-	49	2327
p-Diacetylbenzene	4-*p*-pyrimidin-4′-ylphenyl-	21	2326
Acetophenone[e]	2-methyl-4-phenyl-	30	2329
Propiophenone[e]	2,5-dimethyl-4-phenyl-	30	2329
Butyrophenone[e]	5-ethyl-2-methyl-4-phenyl-	18	2329
Acetophenone[f]	2-ethyl-4-phenyl-	58	2329
Acetophenone[g]	4-phenyl-2-propyl-	54	2329
Cyanoacetamide	5-cyano-4-hydroxy-[h]	7	2544
Ethyl carbamoylacetate	5-ethoxycarbonyl-4-hydroxy-[i]	11	2544
Malondiamide	5-carbamoyl-4-hydroxy-[j]	39	2544
Ethyl cyanoacetate	5-cyano-4-hydroxy-[k]	30	2544

[a] *Cf.* 2% with formamide and $ZnCl_2$.[296]
[b] *Cf.* 56% from formamide/dimethyl sulphate.
[c] *Cf.* 54% with formamide and ammonia.[297]
[d] *Cf.* 5% with formamide and ammonia.[298]
[e] With trisacetamidomethane.
[f] With trispropionamidomethane.
[g] With trisbutyramidomethane.
[h] Formamidine acetate gave 4-amino-5-carbamoylpyrimidine (53%).[2544]
[i] Formamidine acetate gave 5-carbamoyl-4-hydroxypyrimidine (23%).[2544]
[j] Same product and yield with formamidine acetate.
[k] With formamidine acetate only.

indeed does so. The reaction is catalysed by a little *p*-toluenesulphonic acid.[2163] Table VIa summarizes examples of this reaction.

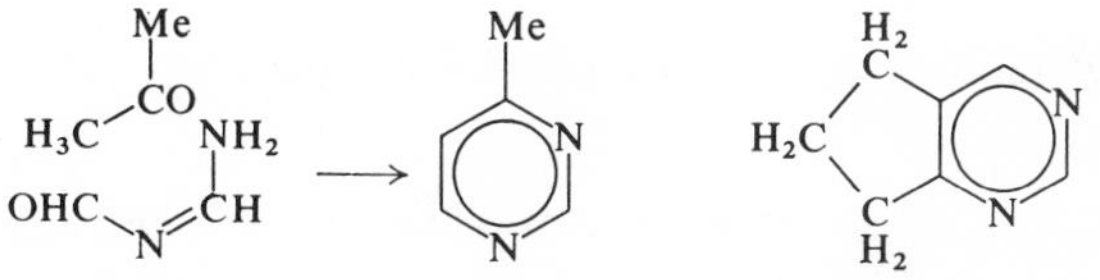

(23) (24) (25)

C. Formamide with β-Dicarbonyl and Related Compounds (*H* 92)

This synthesis has been extended by its discoverer and his colleagues to the preparation of pyrimidines bearing functional groups.[2323] Thus with formamide, 2-bromo-1,1,3,3-tetraethoxypropane gives 5-bromopyrimidine, and 2-ethoxycarbonyl-1,1,3,3-tetramethoxypropane gives 5-ethoxycarbonylpyrimidine.[2323] The use of a less obvious equivalent to a β-carbonyl compound is exemplified in the condensation of α-(β-acetoxyethyl)-β-chlorocrotonaldehyde with formamide to give 5-β-acetoxyethyl-4-methylpyrimidine.[2328]

D. The Frankland and Kolbe Synthesis from Nitriles (*H* 93)

The conditions for trimerization of acetonitrile and its homologues have been extended recently.[2895] Thus acetonitrile, iron carbonyl, and 3-hexyne* at 250° give 4-amino-2,6-dimethylpyrimidine (33%); and propionitrile, $Fe(CO)_5$, and 1-pentyne* give 4-amino-2,6-diethyl-5-methylpyrimidine.[2895]

H.-J. Kabbe has developed considerably what he calls 'Mischtrimerisierung' (*cf.* *H* 96).[2896] Thus *p*-chlorobenzyl cyanide (1 mole) with 4-cyanopyridine (2 moles) in butanolic sodium alkoxide at 115° gives 4-amino-5-*p*-chlorophenyl-2,6-di-4′-pyridylpyrimidine (97%), and a variety of analogues were made similarly.[2896] In addition, one molecule of nitrile can be replaced by an ester having an α-methylene group: this is illustrated in the condensation of two molecules of 4-cyanopyridine

* Presumably the alkyne takes no part in the reaction under these conditions: at least none of the expected products (*cf.* *H* 96) is mentioned.[2895]

with one of ethyl phenylacetate to yield 4-hydroxy-5-phenyl-2,6-di-4′-pyridylpyrimidine (80%), and in nineteen similar cases; however, if an excess of the ester is present, a pyridine is formed.[2896]

The synthesis in its classical form still proves useful today,[2791] for example, in making 4-amino-5-ethyl-2,6-dipropylpyrimidine.

E. Use of an Amino-oxadiazoline as a Reagent (*New*)

A few pyrimidines have been made by constructing a bicyclic oxadiazolopyrimidine from an amino-oxadiazoline, and then opening the five-membered ring to leave the required pyrimidine.[2330] Thus ethyl ethoxymethylenecyanoacetate (**26**) and 3-amino-5,5-dimethyl-2-phenyl-1,2,4-oxadiazoline (**27**) yield 3-β-cyano-β-ethoxycarbonylvinylamino-5,5-dimethyl-2-phenyl-1,2,4-oxadiazoline (**28**) which cyclizes during recrystallization to yield 6-cyano-1,5-dihydro-3,3-dimethyl-5-oxo-1-phenyl[1,2,4]oxadiazolo[4,3-*a*]pyrimidine (**29**). When cyclization is done in alcoholic sodium ethoxide, 5-cyano-2,4-dihydroxypyrimidine (**30**) is formed, presumably by degradation of (**29**); when alcoholic acid is used instead, 4-amino-5-ethoxycarbonyl-2-*N*-hydroxyanilinopyrimidine (**31**) results, presumably through an alternative cyclization involving the cyano group of (**28**) and subsequent scission of the oxygen ring. Similar reactions with diethyl ethoxymethylenemalonate or ethoxymethylenemalononitrile as starting materials eventually yield 5-carboxy-2,4-dihydroxypyrimidine and 4-amino-5-cyano-2-*N*-hydroxyanilinopyrimidine, respectively.[2330] The scope of the reaction is virtually unexplored.

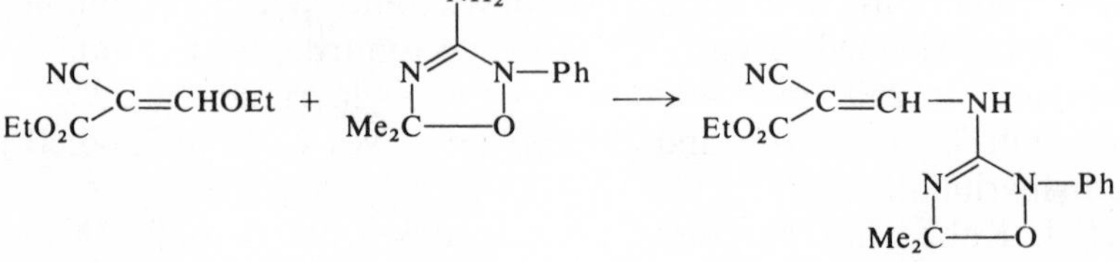

NC, EtO_2C >C=CHOEt + [NH_2; N, N—Ph; Me_2C—O] ⟶ NC, EtO_2C >C=CH—NH [N, N—Ph; Me_2C—O]

(**26**) (**27**) (**28**)

[H; NC; C; N; O; N; N—Ph; Me_2C—O] [NC; N; HO; N; OH] [EtO_2C; N; H_2N; N; N—Ph; OH]

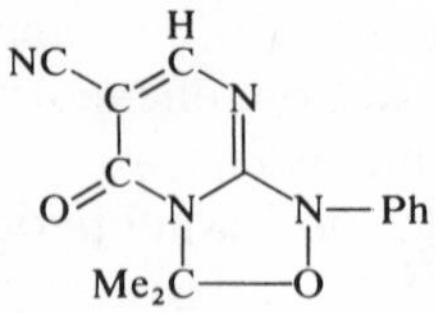

(**29**) 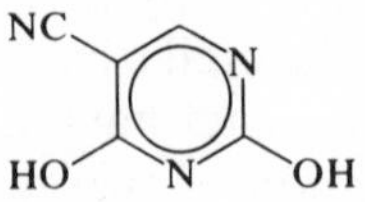 (**30**) 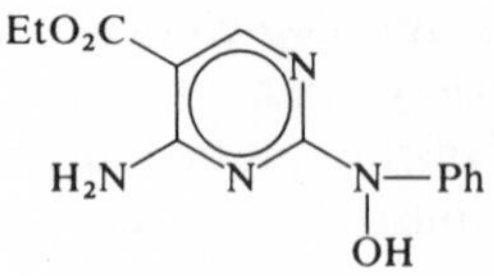 (**31**)

4. Syntheses from Malondiamides and Malondiamidines (*H* 97)

A. Malondiamides with Esters: The Remfry–Hull Synthesis (*H* 97)

Despite failure[2331] to condense malondithioamide with ethyl formate, the Remfry–Hull synthesis may be modified to use such thioamides with acyl chlorides.[2242] Thus γ-carbamoyl-γ-thiocarbamoylpentane (**32**) and acetyl chloride yields 5,5-diethyl-4,5-dihydro-6-hydroxy-2-methyl-4-thiopyrimidine (**33**), or tautomer. Similarly, γ-carbamoyl-γ-*N*-methylthiocarbamoyl- and γ-*N*-methylcarbamoyl-γ-thiocarbamoyl-pentane yield, respectively, 5,5-diethyl-1,4,5,6-tetrahydro-1,2-dimethyl-4-oxo-6-thiopyrimidine and its 6-oxo-4-thio isomer.[2242]

Appropriately α-substituted malondiamides have been used with ethyl formate to produce 4,6-dihydroxypyrimidines bearing at the 5-position a butyl, *o*-chlorophenyl, *p*-chlorophenyl, *p*-acetamidophenyl, or *p*-nitrophenyl group;[2522] other simple esters similarly gave the 2-methyl, 2-ethyl, 2-propyl, 2-methyl-5-phenyl, and 2-methyl-5-methylamino derivatives.[2523] Diethyl oxalate is also a satisfactory ester component in the synthesis; with malondiamide it yields 2-carboxy-4,6-dihydroxypyrimidine, and with carbamoylacetamidine, 4-amino-2-carboxy-6-hydroxypyrimidine.[2334]

B. Malondiamides with Amides (*H* 98)

The condensation[352] of malondiamide with formamide has been investigated in some detail as a step in adenine syntheses, and up to 80% yield of 4,6-dihydroxypyrimidine has now been reported.[2332, 2333, 2523, 3244] Appropriately α-substituted malondiamides similarly gave 5-methyl,[2240, 2523] 5-ethyl,[2523] 5-phenyl,[2523] 5-methoxy,[2524] 5-ethoxy,[2524] 5-isopropoxy,[2524] 5-butoxy,[2524] 5-methylamino,[2523] 5-piperidino,[2523] 5-β-piperidinoethyl,[2523] and other[3525, 3528, 3529] derivatives of 4,6-dihydroxypyrimidine.

D. The Use of Malondiamidines (*H* 100)

Phenylazomalondiamidine condenses with diethyl oxalate to yield 4,6-diamino-2-carboxy-5-phenylazopyrimidine;[2334] another example appears in Sect. 4.A.

The formation of *N*-substituted malondiamidines, which are of potential use in this type of synthesis, has been studied,[2335] and attention has been redrawn[2335] incidentally to the interesting (if unrelated) dimerizations[133] of α-cyano- and α-ethoxycarbonyl-acetamidine to yield 4,6-diamino-2-cyanomethylpyrimidine (**34**) and 4-amino-2-ethoxycarbonylmethyl-6-hydroxypyrimidine, respectively.

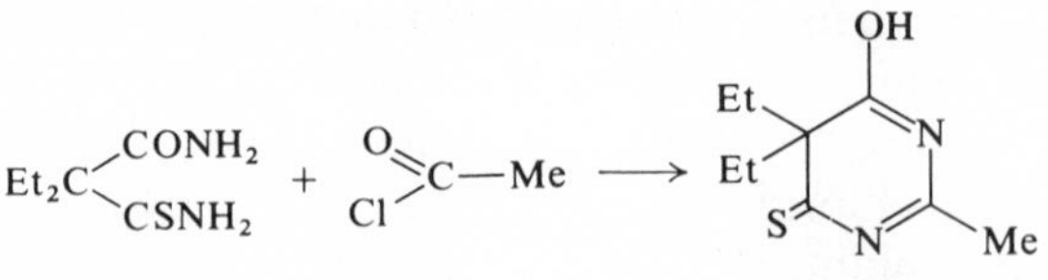

$$Et_2C(CONH_2)(CSNH_2) + ClC(O)Me \longrightarrow$$ (33)

(32) (33) 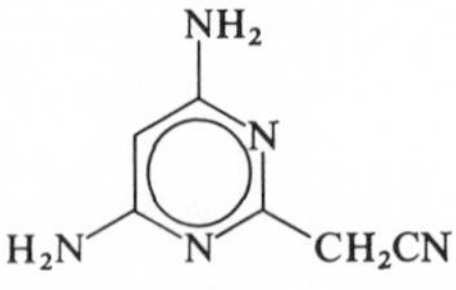 (34)

5. Other Syntheses of Pyrimidines (*H* 101)

A. Ethoxymethyleneacetic Acid to Uracil Derivatives (*H* 101)

This useful synthesis has been extended in scope. For example, sodium ethoxymethyleneacetate (i.e., sodium β-ethoxyacrylate) is readily converted into the acid chloride (**35**) and this reacts with silver cyanate to give β-ethoxyacryloyl isocyanate (**36**). Treatment with β-alanine ethyl ester yields *N*-β-ethoxyacryloyl-*N*′-ethoxycarbonylethylurea (**37**, R = $CH_2CH_2CO_2H$) which cyclizes quantitatively to 1-β-carboxyethyluracil (**38**, R = $CH_2CH_2CO_2H$);[2336] with ethyl γ-aminobutyrate the isocyanate gives the homologous 1-γ-carboxypropyluracil [**38**, R = $(CH_2)_3CO_2H$];[2336] with benzylamine, it gives 1-benzyluracil (**38**, R = CH_2Ph);[2337] and with ββ-diethoxyethylamine it gives first *N*-ββ-diethoxyethyl-*N*′-β-ethoxyacryloylurea [**37**, R = $CH_2CH(OEt)_2$] and then 1-ββ-diethoxyethyluracil [**38**, R = $CH_2CH(OEt)_2$] an intermediate in a synthesis of willardiine.[2338] The same isocyanate reacts with *O*-benzylhydroxylamine to give *N*-benzyloxy-*N*′-β-ethoxyacryloylurea (**37**, R = OCH_2Ph) and ultimately 1-benzyloxyuracil (**38**, R = OCH_2Ph), an unambiguous route to uracil-1-*N*-oxide.[2200]

When β-methoxy-α-methylacryloyl isocyanate is converted into β-methoxy-α-methylacryloylurethane and thence by heating with 4-aminocyclopentene into 1-cyclopenten-4′-yl-5-methyluracil, the yield is poor; when the intermediate, *N*-cyclopenten-4′-yl-*N*′-β-methoxy-α-methylacryloylurea, an analogue of (**37**), is isolated and cyclized by pyrolysis of its potassium derivative, the yield is 75%.[2344] 1-Butyl-5-methyluracil may be made similarly.[2344]

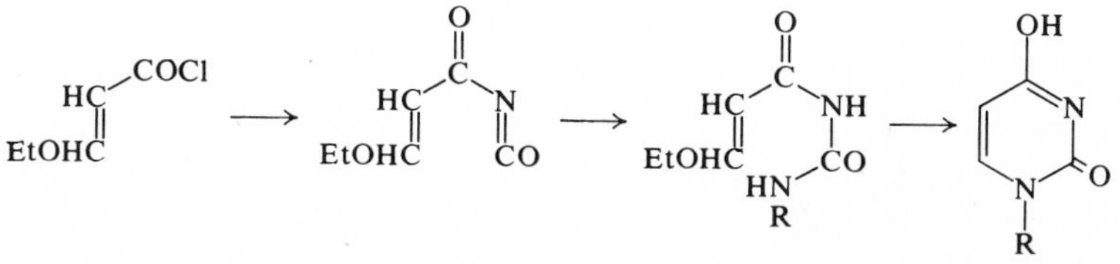

(35) (36) (37) (38)

B. Maleic Diamide and Other Unsaturated Amides to Uracil Derivatives (*H* 102)

This synthesis has not been used recently, but a more general procedure, that might well be considered unrelated, is analogous in that final cyclization occurs between an isocyanate and amide residue.[2339]

The starting material is an $\alpha\beta$-dialkyl-α-morpholinoethylene, e.g., γ-morpholino-β-pentene (**39**),[2340] which is allowed to react as an enamine with phenyl isothiocyanate to give a substituted acrylothioanilide, e.g., β-ethyl-β-morpholino-α-methylacrylothioanilide (**40**). This easily reacts with two molecules of phenylisocyanate; one removes the morpholino group by forming the urea (**42**); the other completes the pyrimidine ring to give a substituted 2-thiouracil, e.g., 4-ethyl-1,2,3,6-tetrahydro-5-methyl-6-oxo-1,3-diphenyl-2-thiopyrimidine (**41**).[2339] The scope of the synthesis is almost unexplored.

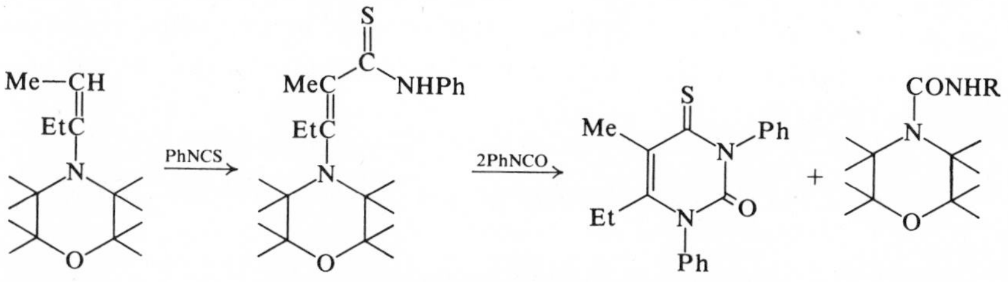

(39) PhNCS → (40) 2PhNCO → (41) + (42)

D. Amidines with Ketones (*H* 102)

The original synthesis based on the reaction of $\alpha\beta$-unsaturated ketones with amidines has not been further explored, but a fundamentally new reaction of saturated ketones with dicyandiamide has proved useful in the hands of E. J. Modest and his colleagues.

The first example, described in a patent,[2341] was the condensation at 150° of dicyandiamide (**44**) with cyclohexanone to yield 2,4-diamino-5,6-tetramethylenepyrimidine (2,4-diamino-5,6,7,8-tetrahydroquinazoline). This can be extended to other cyclic ketones giving, for example, the homologous tri-, penta-, hexa-, and even tridecamethylenepyrimidines.[2342, 2543] In addition, *N*-methyl- and *NN*-dimethyl-dicyandiamide can be used to give with cyclohexanone, respectively, 4-amino-2-methylamino- and 4-amino-2-dimethylamino-5,6-tetramethylenepyrimidine.[2342] The reaction may also be applied to acyclic ketones to yield simple pyrimidines. Phenylacetone (**43**) yields 2,4-diamino-6-methyl-5-phenylpyrimidine (**45**), benzyl ethyl ketone yields the 6-ethyl-homologue, benzylacetone yields 2,4-diamino-5-benzyl-6-methylpyrimidine, and acetophenone yields 2,4-diamino-6-phenylpyrimidine.[2343]

PhCH$_2$–COMe (**43**) + N≡C–NH–C(=NH)–NH$_2$ (**44**) ⟶ 2,4-diamino-6-methyl-5-phenylpyrimidine (Ph, Me, NH$_2$, NH$_2$) (**45**) + H_2O

E. Synthesis of Pyrimidines from Other Ring Systems (*H* 103)

The number of known pyrimidine syntheses from other ring systems has increased following recent interest in unusual heterocycles. However, with two exceptions, the methods remain of little practical importance.

(1) *Pyrimidines from Hydantoins and Other Imidazoles* (*H* 103)

Unlike the reaction of diethyl oxalacetate (**46**, R = Et, R′ = H) with urea (**47**, *R*″ = H) which proceeds (*cf. H* 104) to orotic acid (**51**, R′ = R″ = H, X = OH) *via* hydantoin intermediates (**48**; **49**), mono-ethyl oxalacetate (**46**, R = R′ = H) and urea appear to give directly the acyclic intermediate, ethyl β-carboxy-β-ureidoacrylate (**50**, R′ = R″ = H), which then cyclizes under gentle alkaline conditions to orotic acid (**51**, R′ = R″ = H, X = OH).[2216] In a similar way, *N*-methylurea (**47**, R″ = Me) yields 6-carboxy-3-methyluracil (**51**, R′ = H, R″ = Me,

X = OH), the same compound as that obtained (*H* 370) by methylating orotic acid with dimethyl sulphate.[2216]

The synthesis *via* hydantoins has been extended in scope. Appropriate dialkyl *C*-substituted-oxalacetates, e.g., (**46**, R = Et, R′ = alkyl), and *N*-alkylureas (**47**, R″ = alkyl) yield 3,5-dialkyl-6-carboxyuracils (**51**, R′ = alkyl, R″ = alkyl, X = OH) and the two hydantoin intermediates (**48**; **49**) may be isolated in each case.[2345, 2346] Orotic acids produced in this way are typified in 5-butyl-6-carboxy-3-phenyluracil (**51**, R′ = Bu, R″ = Ph, X = OH), 6-carboxy-5-methyl-3-phenyluracil, and 6-carboxy-3-ethyluracil;[2345] in 6-carboxy-3-methyl-5-phenyluracil and 6-carboxy-3-phenyluracil;[2346] and in 6-carboxy-3,5-dimethyluracil, 6-carboxy-5-ethyl-3-methyluracil, and 5-butyl-6-carboxy-3-methyluracil.[2347] 5-Butyl-4-carboxy-2,6-dihydroxypyrimidine has been made twice in this way but the reported melting points are 33° apart.[2345, 2346]

A rather different approach to the synthesis of orotic acid is exemplified in the condensation of acetylenedicarboxylic acid and urea to give 5-carboxymethylenehydantoin (**49**, R′ = R″ = H) which is then converted into orotic acid by the usual alkaline treatment.[2348] Thiourea similarly gives 4-carboxy-6-hydroxy-2-mercaptopyrimidine *via* a thiohydantoin.[2348]

Imidazoles other than hydantoins have also been implicated as intermediates in pyrimidine syntheses. Thus if diethyl methyloxalacetate (**46**, R = Et, R′ = Me) is condensed with guanidine instead of urea, 2-amino-4-α-ethoxycarbonylethylidene-5-hydroxyimidazole, a tautomer of (**52**, R′ = Me), is formed; it may be converted with alkali into 2-amino-4-carboxy-6-hydroxy-5-methylpyrimidine, a tautomer of (**51**, R′ = Me, R″ = H, X = NH_2).[2391] 2,4-Diamino-6-carboxy-5-*p*-chlorophenylpyrimidine may be prepared by an analogous route.[2391]

(**46**) + (**47**) —when R = Et→ (**48**) → (**49**) → (**50**) → (**51**)

(**47**) —when R = H→ (**50**)

(**46**) —+guanidine→ (**52**) → (**51**)

(2) *Pyrimidines from Isoxazoles, Oxazoles, Oxazines, and Thiazines* (*H* 104)

5-Amino-3,4-dimethylisoxazole may be acylated to give the 5-formamido (**53**) or valeramido analogues. On catalytic hydrogenation these yield, respectively, 4-hydroxy-5,6-dimethylpyrimidine (**54**) and its 2-butyl derivative.[2349]

When 5-acetyl-4-methyloxazole is heated with ammonia under pressure, a good yield of 5-hydroxy-4,6-dimethylpyrimidine results; homologues may be made similarly.[2533]

1,3-Oxazine intermediates have already been mentioned (Ch. II, Sect. 7) in the formation of barbiturates from malonyl dichloride and carbodiimides. Thus diphenylcarbodiimide and α-benzylmalonyl dichloride yield 67% of 5-benzyl-6-chloro-3,4-dihydro-4-oxo-3-phenyl-2-phenylimino-1,3-oxazine (**55**) which may be converted easily into 5-benzyl-1,2,3,4-tetrahydro-6-hydroxy-2,4-dioxo-1,3-diphenylpyrimidine (**56**) in almost quantitative yield; other analogues may be made similarly.[2252, 2535]

A related rearrangement of 3-benzyl-3,4-dihydro-6-methyl-4-oxo-2-phenylimino-1,3-oxazine in benzene containing *p*-toluenesulphonic acid gives 1-benzyl-1,2,3,6-tetrahydro-4-methyl-2,6-dioxo-3-phenylpyrimidine.[2966]

Suitable oxazines have been used in another way: by conversion into pyrimidines with ammonia or primary amines. For example, the easily made 6-oxo-2,5-diphenyl-1,3-oxazine (**57**) with ammonia yields 4-hydroxy-2,5-diphenylpyrimidine (**58**), and with aniline yields 1,6-dihydro-6-oxo-1,2,5-triphenylpyrimidine; the structure of the second pyrimidine was checked by a Principal Synthesis.[2350] Similarly, 4-methyl-6-oxo-2-phenyl-1,3-oxazine and alcoholic ammonia at room temperature afford 4-hydroxy-6-methyl-2-phenylpyrimidine.[2351] Other examples are known.[2536, 2904, 2932, 3522, 3755]

Suitable 1,3-thiazines may also be converted into pyrimidines by treatment with primary amines. For example, 2,3-dihydro-3-methyl-4-oxo-2-thio-1,3-thiazine (**59**; made from propiolic acid and *N*-methyldithiocarbamic acid) reacts with ethylamine to yield 70% of 1-ethyl-1,2,3,4-tetrahydro-3-methyl-4-oxo-2-thiopyrimidine (**60**). Since the 1- and 3-substituents may be easily varied,[2965] the synthesis is general for 1- and/or 3-alkyl-2-thiouracils,[2352] and has been extended to 1-hydroxy-2-thiouracil.[2964]

It will be realized that the oxazines and thiazines above must carry an oxo, thio, or imino grouping in addition to two double bonds normally associated with such ring systems. If this is not so, they will

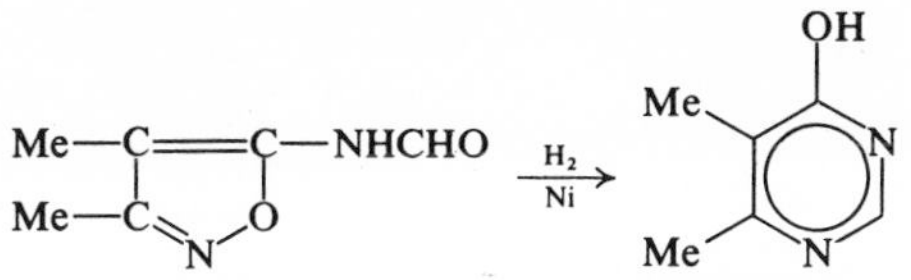

Me—C═C—NHCHO
Me—C═N—O

$\xrightarrow[Ni]{H_2}$

OH, Me, N, Me, N

(53) (54)

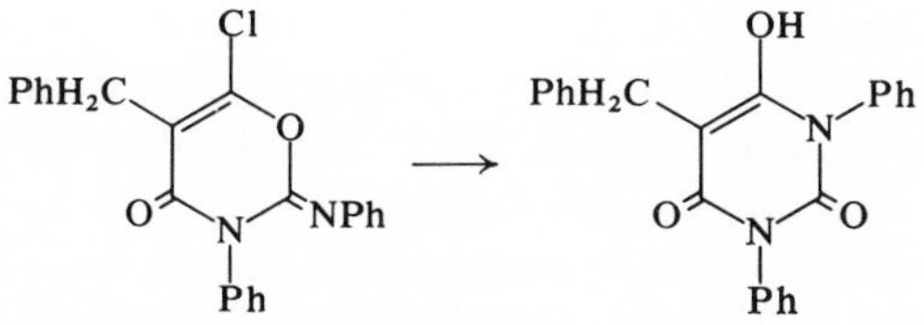

Cl, PhH_2C, O, O, N, NPh, Ph

$\longrightarrow$

OH, PhH_2C, N, Ph, O, N, O, Ph

(55) (56)

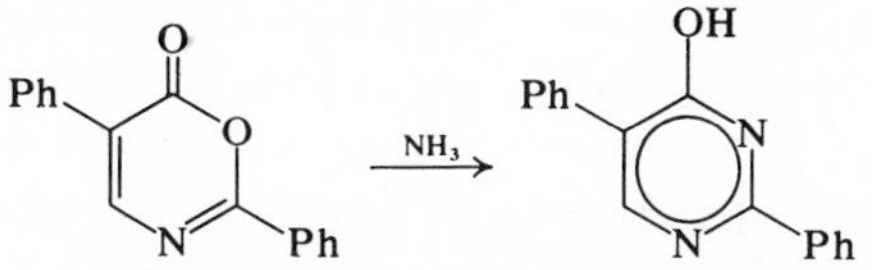

O, Ph, O, N, Ph

$\xrightarrow{NH_3}$

OH, Ph, N, N, Ph

(57) (58)

O, N, Me, S, S

$\xrightarrow{EtNH_2}$

O, N, Me, N, S, Et

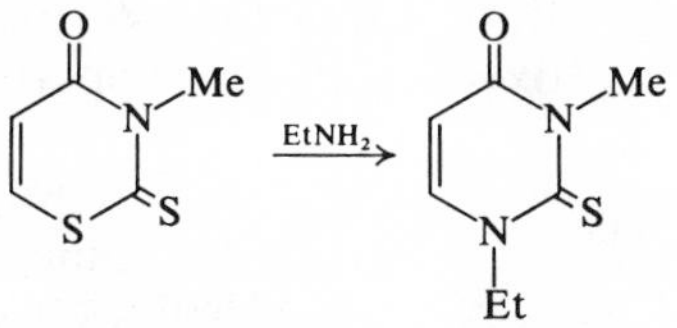

(59) (60)

yield dihydropyrimidines,[2279] which have indeed been so synthesized (see Ch. XII, Sect. 1.G).

(3) *Pyrimidines from Pyrroles* (*H* 105)

A useless but interesting synthesis of 2,4,6-triphenylpyrimidine is provided by prolonged radiation of 2,3,5-triphenylpyrrole (or its *N*-methyl, phenyl, or *p*-tolyl derivative) in alcoholic ammonia open to the air;[1648] other products are benzamide and an amine corresponding to the *N*-substituent. The mechanism is discussed[1648] and other papers[2353, 2354] are relevant.

(4) *Pyrimidines from Purines, Pyrazolopyrimidines, and 8-Azapurines* (*H* 105)

The mechanism of uric acid degradation by acetic anhydride in pyridine has again been discussed,[2355] this time in the light of changes in electrical resistance of the reaction mixture. The initial step is said to be the formation of 7-acetyluric acid. The only hydrolytic product recorded[2355] was a mixture which contained 21.15% of mono- and 78.85% of di-acetylated 4,5-diamino-2,6-dihydroxypyrimidine as 'determined by conductometric titration'.

A practical synthesis of 4-amino-5-(substituted-amino)-pyrimidines from purines has recently emerged. For example, 7-methylpurine (**61**) is now readily made by heating methylaminoacetonitrile with formamidine acetate and formamide, and it then easily undergoes alkaline hydrolysis to 4-amino-5-methylaminopyrimidine (**62**) in 50% overall yield.[2356] Similarly, ethylaminoacetonitrile yields 4-amino-5-ethylaminopyrimidine, β-hydroxyethylaminoacetonitrile yields 4-amino-5-β-hydroxyethylaminopyrimidine, and anilinoacetonitrile yields 4-amino-5-anilinopyrimidine.[2356] An elegant variation[2357] is exemplified in the quaternization of 9-benzylxanthine to 9-benzyl-2,6-dihydroxy-7-methylpurinium *p*-toluenesulphonate which may be isolated also as the betaine (**63**). Alkaline treatment gives 4-benzylamino-2,6-dihydroxy-5-*N*-methylformamidopyrimidine (**64**) which may be debenzylated by catalytic hydrogenation to 4-amino-2,6-dihydroxy-5-*N*-methylformamidopyrimidine. Deformylation with methanolic hydrogen chloride thence gives 4-amino-2,6-dihydroxy-5-methylaminopyrimidine;[2357] 2,4-diami-

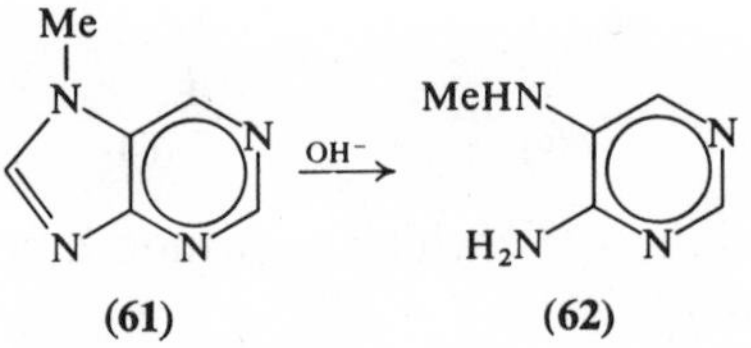

Me N N N N $\xrightarrow{OH^-}$ MeHN N H$_2$N N

(61) **(62)**

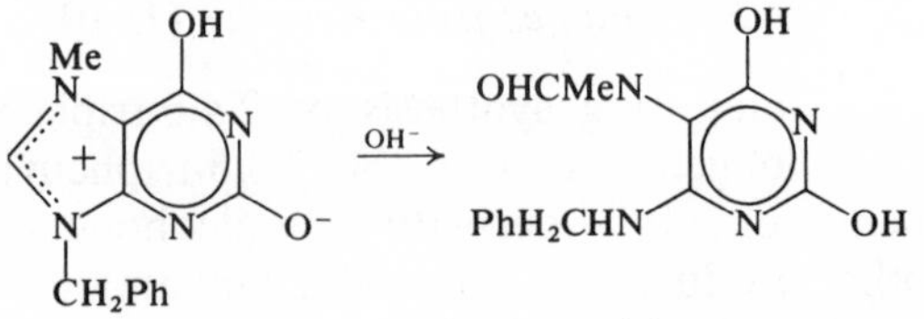

OH Me N + N N O$^-$ CH$_2$Ph $\xrightarrow{OH^-}$ OH OHCMeN N PhH$_2$CHN N OH

(63) **(64)**

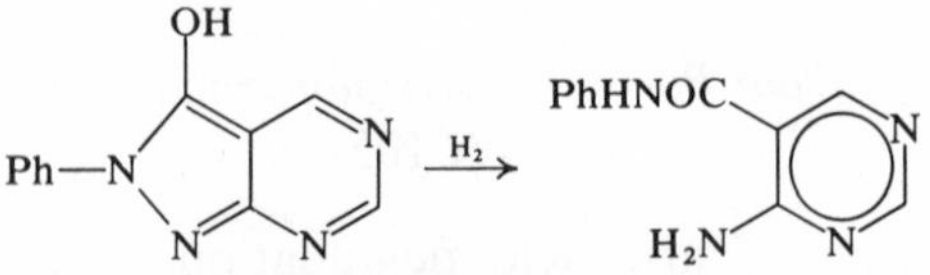

OH Ph—N N N N $\xrightarrow{H_2}$ PhHNOC N H$_2$N N

(65) **(66)**

no-6-hydroxy-5-methylaminopyrimidine may be prepared similarly[2357] or from 7-methylguanosine.[2358]

2-Hydroxy-8-trifluoromethylpurine has been shown spectrometrically to undergo rapid hydrolytic fission at pH 0 to yield 4-amino-2-hydroxy-5-trifluoroacetamidopyrimidine,[2360] and other degradations to pyrimidines are known.[2900, 2901]

A convenient preparation of 4-amino-5-carboxypyrimidine starts from 3-amino-5-hydroxy-1-phenylpyrazole (3-amino-1-phenyl-5-pyrazolone). Treatment at 190° with formamide yields 3-hydroxy-2-phenylpyrazolo[3,4-*d*]pyrimidine (**65**) which, on catalytic hydrogenation, undergoes fission to 4-amino-5-phenylcarbamoylpyrimidine; alkaline hydrolysis of the anilide completes the synthesis.[2359] 3-Hydroxy-6-methylthiopyrazolo[3,4-*d*]pyrimidine is oxidized by chlorine in water, ethanol, or butanol to give, respectively, 5-carboxy-, 5-ethoxycarbonyl-, and 5-butoxycarbonyl-2,4-dihydroxypyrimidine in good yields.[2537]

8-Azaguanine is recorded as yielding 2,4,5-triamino-6-hydroxypyrimidine by treatment with acid; 8-azahypoxanthine, on the other hand, gives 4-amino-5-carbamoyl-1,2,3-triazole.[2538]

(5) *Pyrimidines from Pteridines and Other Polyazanaphthalenes* (*H* 106)

Although pteridines provide a highway to pyrazines, the rarity of the alternative fission to pyrimidines is indicated by the lack of recent examples. However 4-hydroxypteridine and hydrazine hydrate give 4,5-diaminopyrimidine and some 2-amino-3-hydrazinocarbonylpyrazine; 4-hydroxy-6-methylpteridine gives the same pyrimidine; and 4-hydroxy-2-methyl- and 4-hydroxy-2,6-dimethyl-pteridine give 4,5-diamino-6-hydroxy-2-methylpyrimidine. These syntheses are more interesting than important.[3196]

A useful route to 4-amino-5-aminomethylpyrimidine and its 2-alkyl derivatives involves the initial formation of appropriate dihydropyrimido[4,5-*d*]pyrimidines. Thus 2-cyano-1,3-diethoxy-1-methoxypropane and butyramidine furnish 5,6-dihydro-2,7-dipropylpyrimido-[4,5-*d*]pyrimidine (**67**) which undergoes alkaline hydrolysis to 4-amino-5-butyramidomethyl-2-propylpyrimidine (**68**, R = PrCO). Deacylation in ethanolic hydrogen chloride yields 4-amino-5-aminomethyl-2-propylpyrimidine (**68**, R = H).[2190]

In any attempt to 4-decarboxylate 2-amino-4-carboxy-1-β-carboxyethyl-1,6-dihydro-6-oxopyrimidine (**69**, R = CO_2H) cyclization occurs to 6,7-dihydro-8-hydroxy-4-oxopyrimido[1,2-*a*]pyrimidine (**70**); subsequent alkaline treatment yields the required monocyclic product

(**69**, R = H). The isomeric 2-amino-1-β-carboxyethyl-1,4-dihydro-4-oxopyrimidine is formed similarly from 6,7-dihydro-8-hydroxy-2-oxopyrimido[1,2-*a*]pyrimidine.[2215]

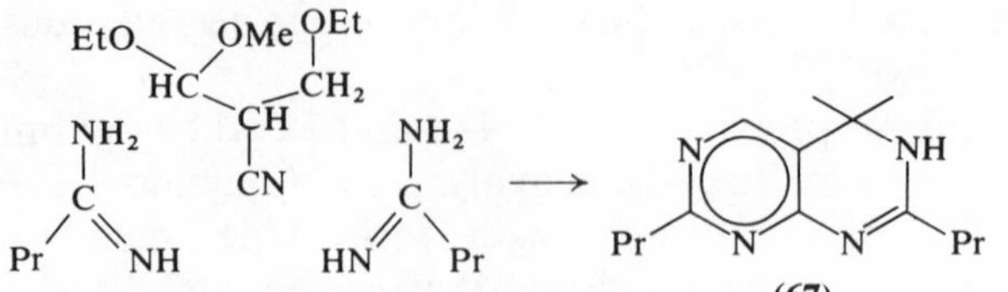

(**67**)

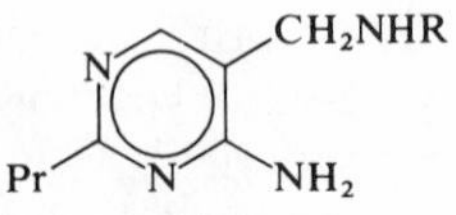

(**68**) 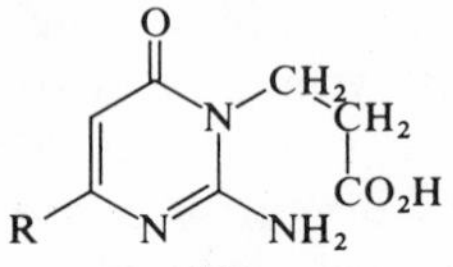 (**69**) 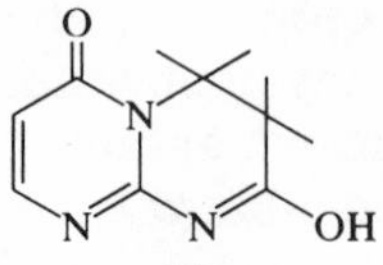 (**70**)

(6) *Pyrimidines from Oxa or Thia Bicyclic Systems* (*New*)

When 1,2,3,4-tetrahydro-6-hydroxy-1,3-dimethyl-2,4-dioxopyrimidine (1,3-dimethylbarbituric acid) is treated with malonic acid or methylmalonic acid in acetic anhydride, 1,2,3,4,5,6-hexahydro-1,3-dimethyl-2,4,5,7-tetraoxopyrano[2,3-*d*]pyrimidine (**71**) and its 1,3,6-trimethyl homologue are formed, respectively.[2361] An analogous 2-thio analogue is similarly made.[2362] Such bicyclic lactones easily revert to pyrimidines not otherwise available so easily. Thus (**71**) with boiling ethanol yields 5-ethoxycarbonylacetyl-1,2,3,4-tetrahydro-6-hydroxy-1,3-dimethyl-2,4-dioxopyrimidine (**72**, R = OEt) and thence with alkali the simple 5-acetyl-analogue; with isopropanol it yields the 5-isopropoxycarbonylacetyl homologue (**72**, R = OPr) and with aqueous ammonia, the 5-carbamoylacetyl analogue (**72**, R = NH_2).[2361]

6-Acetyl-3-ethoxycarbonylmethylcarbamoyl-2,3-dihydro-7-oxo-oxazolo[3,2-*a*]pyrimidine (**73**) is fairly easily made.[2318] In hot concentrated ethanolic ammonia it yields 5-acetyl-1-β-amino-α-(ethoxycarbonylmethylcarbamoyl)ethyluracil (**74**, R = OH, R′ = NH_2), but in a cold dilute solution it gives 5-acetyl-2-amino-1-α-ethoxycarbonylmethylcarbamoyl-β-hydroxyethyl-1,4-dihydro-4-oxopyrimidine (**74**, R = NH_2, R′ = OH); benzylamine gives the 2-benzylamino homologue (**74**, R = $PhCH_2NH$, R′ = OH).[2318] Unhydrogenated oxazolopyrimidines can also yield pyrimidines. For example, hydrogenation of 4-amino-

3-methylisoxazolo[5,4-*d*]pyrimidine followed by boiling with water yields 5-acetyl-4-amino-6-hydroxypyrimidine in 91% yield; the 5-benzoyl analogue is made similarly.[2389]

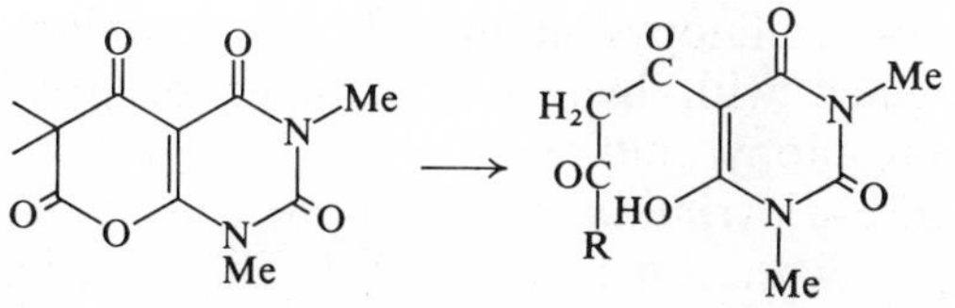

(71) (72)

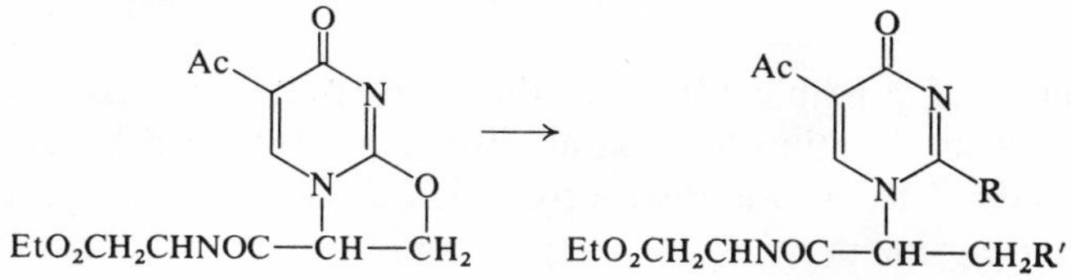

(73) (74)

6-Ethoxy-3-ethyl-1,2,3,4-tetrahydro-7-hydroxy-2,4-dioxo-1-propyl-pyrimido[5,4-*b*][1,4]thiazine (**75**, R = OEt, R′ = H) and its 6,6-diethoxy analogue (**75**, R = R′ = OEt) may be prepared in several steps from 4-amino-5-chloro-1-ethyl-1,2,3,6-tetrahydro-2,6-dioxo-3-propyl-pyrimidine.[2363] Treatment of each bicyclic compound with ammonia yields 4-amino-5-α-carbamoyl-α-ethoxymethylthio-1-ethyl-1,2,3,6-tetra-hydro-2,4-dioxo-3-propylpyrimidine (**76**, R = OEt, R′ = R″ = H) and its αα-diethoxy analogue (**76**, R = R′ = OEt; R″ = H), respectively; treatment of (**75**, R = OEt, R′ = H) with methylamine yields the *N*-methylcarbamoyl (**76**, R = OEt, R′ = H, R″ = Me). The structures (**76**) were confirmed by removal of the 5-substituent from each with Raney nickel to give the known 4-amino-1-ethyl-1,2,3,6-tetrahydro-2,6-dioxo-3-propylpyrimidine.[2363]

A synthesis of rather related pyrimidines is exemplified in the formation of 4-formamido-1,2,3,6-tetrahydro-1,3-dimethyl-5-methyl-sulphamoyl-2,4-dioxopyrimidine (**78**, R = CHO) by hydrolytic cleavage of 5,6,7,8-tetrahydro-2,5,7-trimethyl-6,8-dioxopyrimido[4,5-*e*][1,2,4]-thiadiazine 1,1-dioxide (**77**) which is, however, best made by cyclizing 4-amino-1,2,3,6-tetrahydro-1,3-dimethyl-5-methylsulphamoyl-2,6-di-oxopyrimidine (**78**, R = H) with triethyl orthoformate.[2364] The synthesis is therefore more of interest than importance.

In the same category is the reductive cleavage of 7-butylamino-[1,2,5]thiadiazolo[3,4-*d*]pyrimidine (**79**, R = Bu) to 4,5-diamino-6-butylaminopyrimidine (**80**), from which it is made most easily in the first place.[2365] However, the bicyclic compound may also be conveniently made by transamination of the corresponding 7-amino-analogue (**79**, R = H), made in turn from commercially available 4,5,6-triaminopyrimidine.[2365] The homologue (**79**, R = Ph) similarly gave 4,5-diamino-6-anilinopyrimidine.[2933] The amino homologue (**79**, R = H), treated with hydrogen sulphide in pyridine, yielded 4,5-diamino-6-mercaptopyrimidine.[2933]

Several thieno[2,3-*d*]pyrimidines have been converted into pyrimidines by treatment with Raney nickel.[2531] This reaction gave such simple derivatives as 4-amino-5-isopropyl-, 4-amino-5-2′-naphthyl-, 4-amino-5-4′-aminopyrimidin-5′-yl-, and 4-amino-5-cyclohex-1′-enylpyrimidine.[2898]

A sequence of possible utility is the formation of thiazolo[4,5-*d*]-pyrimidines from 4-amino-5-carbamoylthiazoles, followed by degradation with alkali to a 5-mercaptopyrimidine: 4-anilino-6-hydroxy-5-mercaptopyrimidine is so made in 85% yield.[2899]

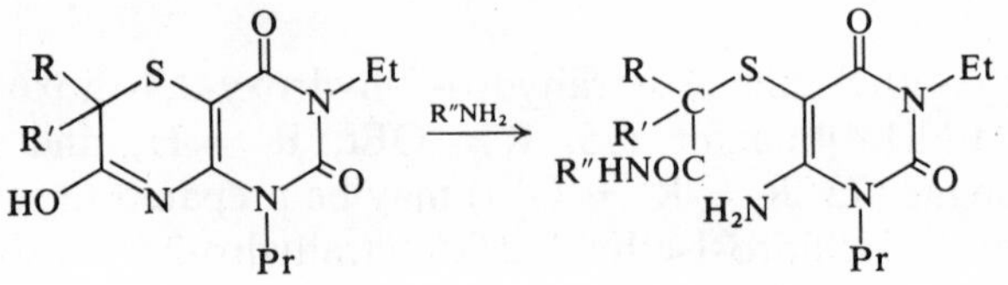

R″NH₂ →

(**75**) (**76**)

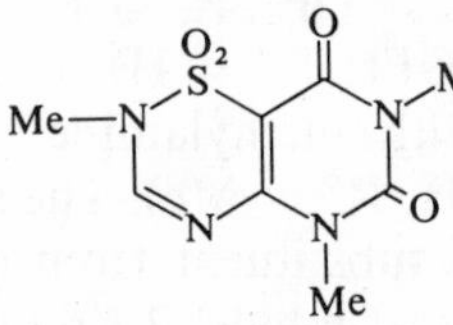

(**77**) 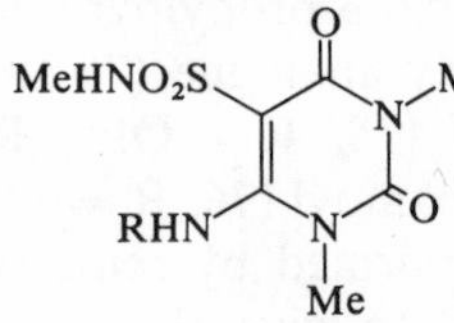 (**78**)

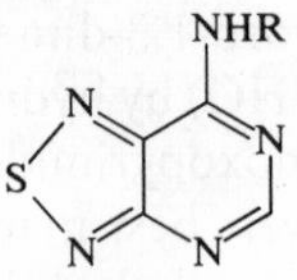

(**79**) 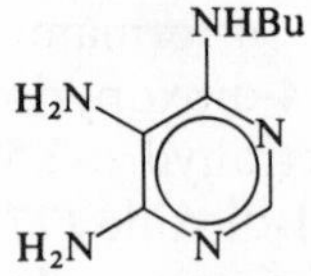 (**80**)

(7) *Pyrimidines from Triazines* (*New*)

The reaction of 1,3,5-triazine (**81**) with imidates, amidines, and amidine salts (which have an acidic α-methylene group) is a useful route to 4,5-disubstituted pyrimidines. Thus triazine (**81**) reacts with α-ethoxycarbonylacetamidine hydrochloride (**82**) in acetonitrile to give 4-amino-5-ethoxycarbonylpyrimidine (**83**) in good yield;[2366] with ethyl α-ethoxycarbonylacetimidate (base; **84**) it gives the same pyrimidine (**83**) but with the acetimidate hydrochloride (**85**) it gives 4-ethoxy-5-ethoxycarbonylpyrimidine (**86**); with ethyl benzoylacetimidate hydrochloride it gives 5-benzoyl-4-ethoxypyrimidine; with α-carbamoylacetamidine (base or hydrochloride), 4-amino-5-carbamoylpyrimidine; with phenylacetamidine, 4-amino-5-phenylpyrimidine; with ethyl α-cyanoacetimidate, 4-amino-5-cyanopyrimidine; with ethyl α-carbamoylacetimidate, 5-carbamoyl-4-ethoxypyrimidine; and with methyl α-cyanothioacetimidate, 5-cyano-4-methylthiopyrimidine.[2366] On the other hand, triazine and ethyl α-cyanoacetimidate hydrochloride (*cf.* base above) give 3,5-dicyano-2,6-diethoxypyridine,[2367] while triazine and phenylacetamidine hydrochloride (*cf.* base above) give a mixture of mono- and di-benzyl-1,3,5-triazine.[2368]

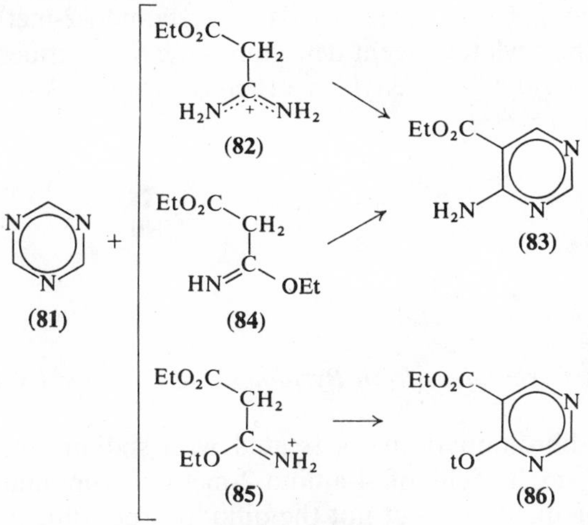

The reaction of triazine with other compounds having an active methylene group is less predictable, although it often does lead to pyrimidines. Thus triazine and malononitrile in ethanolic sodium ethoxide yield 4-amino-5-cyanopyrimidine (**87**; *cf.* formamidine and malononitrile; *H* 90), but in ethanol, *N*-ββ-dicyanovinylformamidine (**88**) and aminomethylenemalononitrile (**89**) are the chief products, and in dimethylformamide the second of these alone is formed.[2367] The dicyanovinylformamidine (**88**) is isomerized so easily into 4-amino-5-cyanopyrimidine (**87**)[2367] that the latter was mistaken[2369] earlier as the primary product of the reaction. Warmed in ethanolic sodium ethoxide, triazine reacts with diethyl malonate to give 5-ethoxy-carbonyl-4-hydroxypyrimidine;[2367, 2369] with malondiamide to give 5-carbamoyl-4-hydroxypyrimidine;[2367] with phenylacetamide to give 4-hydroxy-5-phenylpyrimidine;[2367] with ethyl benzoylacetate to give 5-ethoxycarbonyl-4-phenylpyrimidine;[2369] with 2-cyanoacetamide to give 4-amino-5-carbamoylpyrimidine (**90**; R = NH_2);[2367] with 2-cyanothioacetamide to give (!) 5-cyano-4-mercaptopyrimidine (**90**, R = SH);[2367] and with benzoylacetonitrile to give 5-cyano-4-phenyl-pyrimidine.[2367] A variety of other products, both cyclic and acyclic, are formed from such reactions under differing conditions, and the mechanisms are discussed; some of the extensive footnotes are particularly rewarding.[2366, 2367, 2369]

A useful isomerization of several triazinyl ketones to 4-acetamido-pyrimidine derivatives has been described.[2539, 2540] Thus 2,4,6-tri-methyl-1,3,5-triazine may be acylated easily in the presence of sodium amide to give, for example, 2,4-dimethyl-6-phenacyl-1,3,5-triazine, which on boiling in water yields 4-acetamido-2-methyl-6-phenyl-pyrimidine; the acylating agent determines the 6-substituent which may be propyl, β-pyridyl, 2′-thienyl, or other groups.[2539] A mechanism has been postulated.[2539]

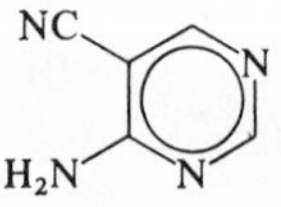

NC, H_2N (**87**) NC, H, N, NC, CH, H_2N (**88**) NC, H, NH_2, NC (**89**) H_2NOC, N, R, N (**90**)

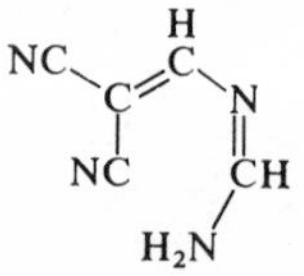

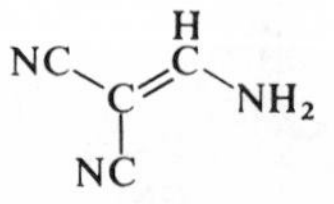

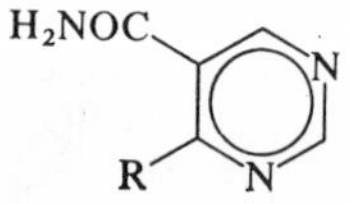

(8) *Pyrimidines from Pyridines and Pyrazine* (*New*)

When 2,6-dibromopyridine is treated with sodium amide in liquid ammonia, a small yield of 4-amino-2-methylpyrimidine results; the corresponding dichloro-, but not the difluoro-, pyridine gives the same product.[2541, 2890]

The irreversible transformation of pyrazine into pyrimidine has been observed when an iso-octane solution of the former is irradiated with ultra-violet light of 254 mμ wavelength. The yield and quantum yield are minute.[2891]

F. Miscellaneous Sequences (*New*)

The syntheses that follow do not fit exactly into any previous class because of a mechanistic doubt or other reason.

(1) *The Cyclization of Acylaminomethylenemalononitriles* (*New*)

The unstable chloro-compound (BzN=CClPh) from *N*-benzoylbenzamide (BzNHBz) and phosphorus pentachloride[2371] condenses with malononitrile to give the acylaminomethylenemalononitrile, α-benzamido-ββ-dicyanostyrene, formulated here as its tautomer (**91**). On boiling this in acidic ethanol, 5-cyano-4-hydroxy-2,6-diphenylpyrimidine (**93**) is formed,[2370] presumably through rearrangement of an oxazine (?**92**) or other intermediate. The reaction has been extended only to the *p*-chlorophenyl- and *p*-tolyl-analogues.[2370]

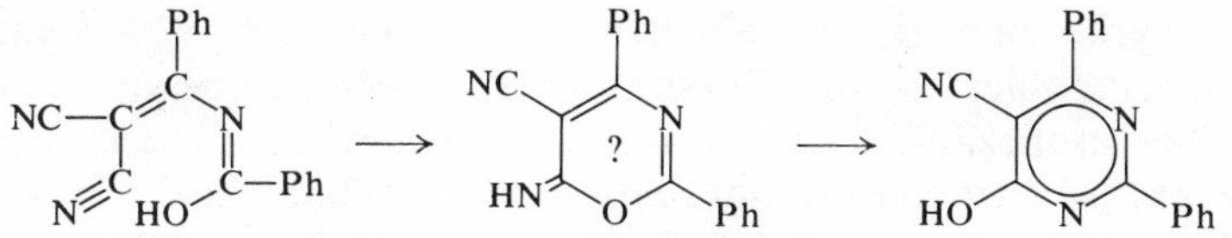

(**91**) → (**92**) → (**93**)

(2) *Syntheses from Fragments with Inbuilt Oxidation Capacity* (*New*)

The following syntheses lead directly to dihydropyrimidines with substituents that assure final automatic dehydrogenation of the ring.

In the first type, an attached benzylidene group performs the dehydrogenation and appears finally as a benzyl group.*[2372] Thus 3-benzylideneacetylacetone (**94**) condenses with two molecules of ammonia and one of benzaldehyde to yield 5-benzyl-4,6-dimethyl-2-phenylpyrimidine (**96**) in 54% yield, presumably *via* 5-benzylidene-2,5-dihydro-4,6-dimethyl-2-phenylpyrimidine (**95**). Also made in this way were 5-*m*-nitrobenzyl-2-*m*-nitrophenyl-, 5-benzyl-2-*m* (and *p*-)nitrophenyl-, 5-*m*-nitrobenzyl-2-phenyl-, 2-*p*-chlorophenyl-5-α-pyridyl-

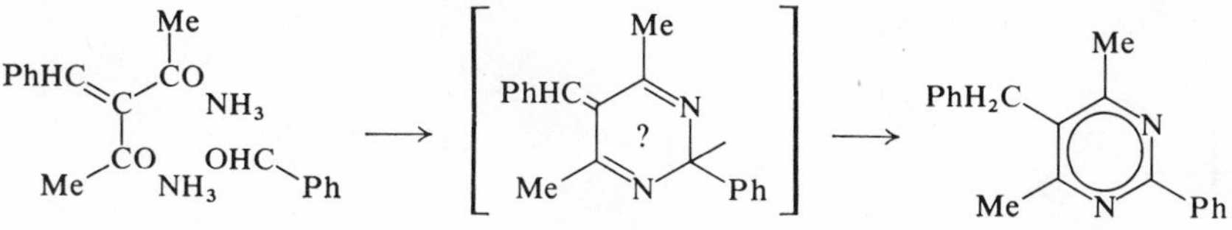

(**94**) → [(**95**)] → (**96**)

* *Cf.* the reaction[384] of αβ-unsaturated ketones with amidines in which the dihydropyrimidine is oxidized by the excess of ketone (*H* 102).

methyl-, and 2-2′-quinolyl-5-2′-quinolylmethyl-, 4,6-dimethylpyrimidine; as well as 5-benzyl-4-ethyl-6-methyl-2-β-pyridylpyrimidine.[2372]

In the second type, oxidation is provided by spontaneous elimination of hydrogen chloride from an intermediate 5-chlorodihydropyrimidine.[2372] Thus dibenzoylbromomethane, benzaldehyde, and ammonium acetate in acetic acid give 2,4,6-triphenylpyrimidine (**98**), presumably *via* 5-chloro-2,5-dihydro-2,4,6-triphenylpyrimidine (**97**). When 3-chloroacetylacetone is used, the product reacts further with benzaldehyde to yield finally 2-phenyl-4,6-distyrylpyrimidine, but with *p*-nitrobenzaldehyde in a limited excess, 4,6-dimethyl-2-*p*-nitrophenylpyrimidine may be isolated in good yield. Other analogues made directly with appropriate aldehydes include 4-methyl-2-*p*-nitrophenyl-6-phenylpyrimidine (from α-benzoyl-α-chloroacetone), 2,4-diphenyl-6-styrylpyrimidine, 4-ethyl-2-*m*-nitrophenyl-6-*m*-nitrostyrylpyrimidine (from 3-chlorohexan-2,4-dione), 4-*p*-bromophenyl-2,6-diphenylpyrimidine (from α-benzoyl-α-bromo-α-*p*-bromobenzoylmethane), 2-*m*-nitrophenyl-4,6-diphenylpyrimidine (from dibenzoylbromomethane), and 4-hydroxy-2-*m*-nitrophenyl-6-phenylpyrimidine (from ethyl benzoylchloroacetate).[2372] *p*-Methoxybenzaldehyde, 2,4-dichlorobenzaldehyde, and *p*-chlorobenzaldehyde may be used similarly.[2373] The chloro substituent may be replaced by an acetoxy group: e.g., dibenzoylmethyl acetate with benzaldehyde gives a small yield of 2,4,6-triphenylpyrimidine.[2372] When chloromalondiamide is similarly treated with *p*-methoxybenzaldehyde the major product is an oxazoline with the expected 4,6-diamino-2-*p*-methoxyphenylpyrimidine as a minor product.[2372]

In the third type, a tetrahydropyrimidine is oxidized to the dihydro stage by spontaneous removal of a quaternary pyridinium group as pyridine salt and the oxidation is probably completed by the excess of

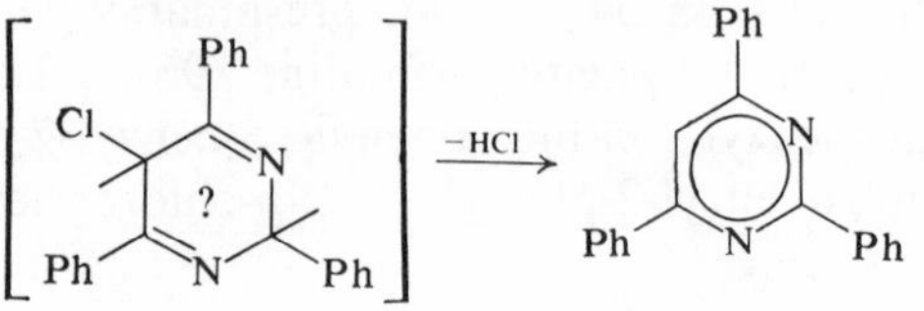

Cl, Ph, N, ?, Ph, N, Ph — (**97**) $\xrightarrow{-HCl}$ Ph, N, Ph, N, Ph — (**98**)

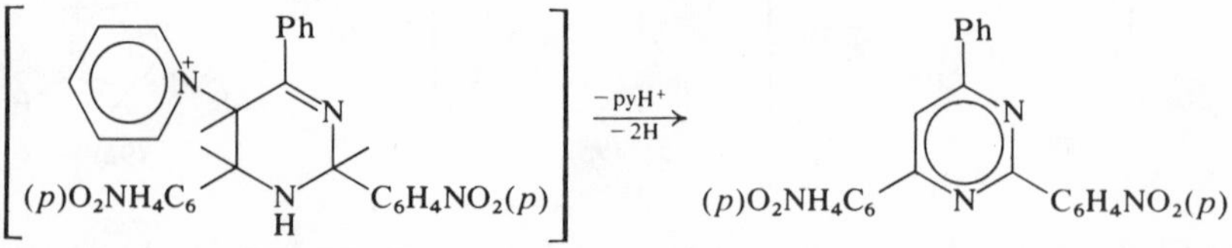

Ph, N^+, N, $(p)O_2NH_4C_6$, N H, $C_6H_4NO_2(p)$ — (**99**) $\xrightarrow[-2H]{-pyH^+}$ Ph, N, $(p)O_2NH_4C_6$, N, $C_6H_4NO_2(p)$ — (**100**)

aromatic aldehyde present. It is exemplified in the condensation of *N*-phenacylpyridinium bromide with two molecules of ammonia and two of *p*-nitrobenzaldehyde to give 2,4-di-*p*-nitrophenyl-6-phenylpyrimidine (**100**) in 76% yield, probably *via* the tetrahydropyrimidine (**99**). *N*-Phenacylisoquinolinium bromide, *N*-phenacylquinolinium bromide, or even dimethyl phenacyl sulphonium bromide may be substituted for *N*-phenacylpyridinium bromide without loss of yield.[2372]

(3) *Syntheses Involving Pyrimidine Rearrangements* (*New*)

In a pyrimidine rearrangement that involves ring fission to an aliphatic intermediate and subsequent recyclization to a different pyrimidine, the second stage may be considered as a 'synthesis'.

This is exemplified in the Dimroth rearrangement (see Ch. X, Sect. 2.B for a full treatment) of 1,2-dihydro-2-imino-1-methylpyrimidine (**101**, R = H).[2374] In alkali it first undergoes ring fission to *N*-β-formylvinyl-*N*′-methylguanidine (**102**, R = H) which may be isolated as its oxime or recyclized either in acid to the imine (**101**, R = H), or in alkali to 2-methylaminopyrimidine (**103**, R = H).[2375] Similarly, from 5-cyano-1,2-dihydro-2-imino-1-methylpyrimidine (**101**, R = CN) at pH 10, the acyclic *N*-β-cyano-β-formylvinyl-*N*′-methylguanidine (**102**, R = CN) precipitates. Treatment of this with acid causes recyclization to the initial imine (**101**, R = CN), but treatment with ammonia or sodium hydroxide causes cyclization, respectively, to 5-cyano-2-methylaminopyrimidine (**103**, R = CN) or to 4-amino-5-formyl-2,3-dihydro-2-imino-3-methylpyrimidine (**104**).[2376]

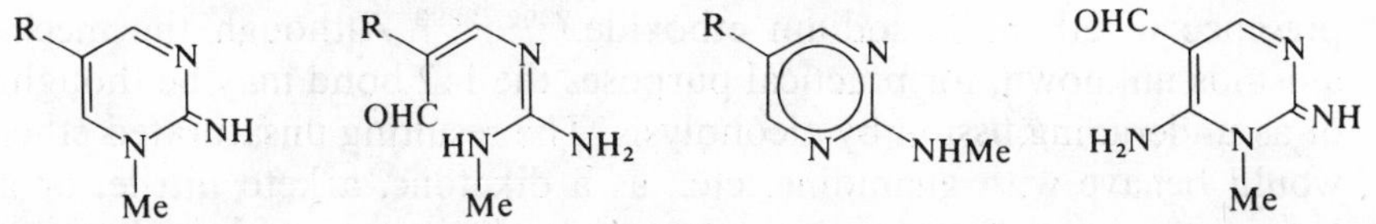

(**101**) (**102**) (**103**) (**104**)

Another rearrangement-synthesis is that described by J. A. Carbon.[2164] 4-Amino-6-guanidino-5-nitropyrimidine (**105**) undergoes fission in acid and the resulting (unisolated) aliphatic guanidine (**106**) partly recyclizes *in situ* to 2,4,6-triamino-5-nitropyrimidine (**109**, R = R′ = NH_2) and partly hydrolyses to β-guanidino-β-imino-α-nitropropionamide (**107**) which is precipitated and characterized as its hydrochloride. The latter may be cyclized to 2,4-diamino-6-hydroxy-5-nitropyrimidine (**109**, R = OH, R′ = NH_2). In alkali the reaction takes an additional

(105) → (106) → (107) → (109); (105) → (108) → (109)

course to form 4-amino-5-nitro-6-ureidopyrimidine (**108**) which partly hydrolyses further to 4,6-diamino-5-nitropyrimidine (**109**, R = NH_2, R′ = H) and partly rearranges, presumably *via* the ureido analogue of (**106**), to give 4,6-diamino-2-hydroxy-5-nitropyrimidine (**109**, R = NH_2, R′ = OH).

The Dimroth rearrangement is often of preparative value, but Carbon rearrangement(!) is more interesting than valuable.

(4) *Syntheses Involving Benzofurans* (*New*)

Some 5-*o*-hydroxyphenylpyrimidines, e.g., (**110a**), can be made by treating a benzofuran bearing a 3-acyl, cyano, carboxy, alkoxycarbonyl, or carbamoyl substituent with guanidine, thiourea, or urea in the presence of ethanolic sodium ethoxide.[2392, 2893] Although the mechanism is unknown, for practical purposes the 1:2 bond may be thought of as undergoing fission by alcoholysis. The resulting unsaturated ether would behave with guanidine, etc., as a diketone, a keto nitrile, or a keto ester according to the original 3-substituent, and undergo a Principal Synthesis. Thus 2-ethyl-3-formylbenzofuran (**110**, R = H) and guanidine yield 2-amino-4-ethyl-5-*o*-hydroxyphenylpyrimidine (**110a**, R = H) and appropriate variations of R in the benzofuran (**110**) give 2-amino-4,6-diethyl-, 2-amino-4-ethyl-6-methyl-, 2-amino-4-ethyl-6-*p*(?)-methoxyphenyl-, 2,4-diamino-6-ethyl-, and 2-amino-4-ethyl-6-hydroxy-5-*o*-hydroxyphenylpyrimidines. The 2-mercapto and 2-hydroxy analogues may be similarly made with thiourea and urea, respectively, but the latter reagent gives poor yields.[2392, 2893] The corresponding 2-methylpyrimidines may be made using acetamidine.[2894]

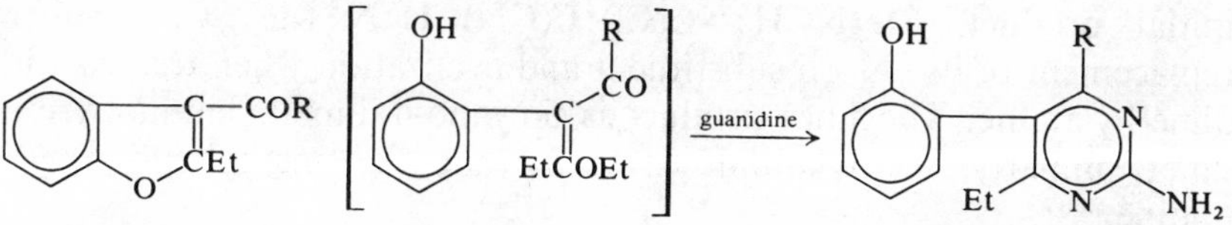

(110) → guanidine → (110a)

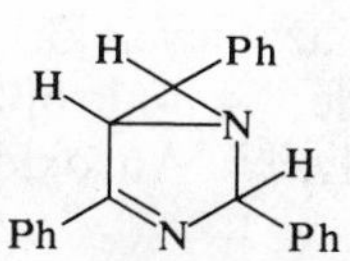

(111) 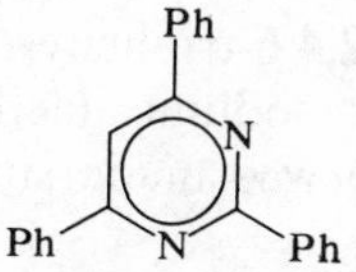 (111a)

(5) *Syntheses from Aryl Cyanates and Ethyl Aroyl Acetate or Ethyl Cyanoacetate or Malononitrile* (*New*)

When ethyl cyanoacetate is allowed to react with *p*-chlorophenyl cyanate (presumably as trimer?), an intermediate (p)$ClC_6H_4OC(:NH)\cdot CH(CN)CO_2Et$, is formed. In the presence of triethylamine, this reacts with a second molecule of the cyanate to give 2,4-di-*p*-chlorophenoxy-5-cyano-6-hydroxypyrimidine. By using other aryl cyanates, which may differ in each stage, analogues such as 5-cyano-4-hydroxy-2-phenoxy-6-*p*-tolyloxypyrimidine can be prepared in good yield.[2545, 2902] When malononitrile is used initially, similar two-stage syntheses yield amino analogues such as 4-amino-6-*p*-chlorophenoxy-5-cyano-6-phenoxy-pyrimidine.[2545, 2902] Likewise, ethyl *p*-nitrobenzoylacetate with two molecules of *o*-tolyl cyanate affords 5-ethoxycarbonyl-4-*p*-nitrophenyl-2,6-di-*p*-tolylpyrimidine; and other examples are given.[2903]

(6) *Some Other Syntheses* (*New*)

An interesting synthesis, as yet undeveloped, involves the cyclization of β-(β-aminocrotonamido)crotonamide [$MeC(NH_2):CHCONHC\cdot(Me):CHCONH_2$] in alkali to give 4-hydroxy-2,6-dimethylpyrimidine (88%) with loss of water and acetamide.[2888, 2889] The intermediate amide can be made from diketene by two distinct and reasonably simple routes.[2889] β-Acetamidocrotonamide cyclizes to give 4-hydroxy-2,6-dimethylpyrimidine at 220° in 52% yield.[3763]

Another novel synthesis is that starting from malononitrile and two molecules of chloromethylene dimethyl ammonium chloride. The initial product, $Me_2NCH{:}NC(Cl){:}C(CN)CH{:}N^+Me_2\ Cl^-$, suffers replacement of its chloro substituent and cyclization when treated with dimethylamine. The final product is 5-cyano-4-dimethylaminopyrimidine; similarly, methylaniline gives 5-cyano-4-(*N*-methylanilino)pyrimidine.[2907]

H. W. Heine and his colleagues have reported[2977] an interesting and unique synthesis of 2,4,6-triphenylpyrimidine (**111a**) by treatment of the fused aziridine, 2,4,6-triphenyl-1,3-diazabicyclo[3,1,0]hex-3-ene (**111**), with methanolic sodium methoxide; 4-*p*-nitrophenyl-2,6-diphenylpyrimidine (69%) was made similarly.[2977] An oxidation must follow rearrangement.

An interesting synthesis of 1,2,3,4-tetrahydro-2,4-dioxo-1,3,6-triphenylpyrimidine has yet to be described in detail: it involves the reaction of two molecules of phenyl isocyanate with one of triethylphenylethynyl-lead ($Et_3PbC{⋮}CPh$) followed by an hydrolysis.[3761]

6. Formation of Pyrimidine Ring in Fused Heterocycles (*H* 107)

With the advent of a bracket of volumes, 'The Fused Pyrimidines', in this series,* a full supplementary treatment of this section is unnecessary. However, a few of the more important recent papers on the completion of the pyrimidine ring in fused systems are briefly mentioned: several come from E. C. Taylor and his colleagues.

A good route to *quinazolines* is exemplified in the cyclization of *N*-*o*-cyanophenyl-*N*′-phenylthiourea to 4-amino-2,3-dihydro-3-phenyl-2-thioquinazoline (**112**) from which numerous other quinazolines may be prepared, e.g., by Dimroth rearrangement (4-anilino-2-mercaptoquinazoline), hydrolysis, methylation, etc.[2377] Benzonitrile and *o*-aminobenzonitrile yield 4-amino-2-phenylquinazoline, and analogues are made similarly.[2386] Other examples are described.[2958–2961]

The formation of *purines* from imidazole derivatives has been well reviewed.[2378, 3198] Of particular interest is the formation of 9-aminopurines from 1-aminoimidazoles,[2525, 2526] e.g., 9-amino-6-hydroxy-8-methylpurine (**113**) from 1,5-diamino-4-carbamoyl-2-methylimidazole;[2379] the formation of 6-amino-2-mercaptopurine-1-*N*-oxide (**114**) from 4-amino-5-*C*-aminohydroxyiminomethylimidazole in carbon bi-

* A monograph on quinazolines (W. L. F. Armarego) has appeared already;[3197] others on purines (J. H. Lister), pteridines, and miscellaneous systems are in preparation.

sulphide and pyridine;[2380] the simultaneous completion of both rings in 6-aminopurine by treating α-aminomalondiamidine with triethyl orthoformate;[2381] the formation of imines such as 1-butyl-1,6-dihydro-6-imino-7-methylpurine from an imidazole;[2385] the reaction of 4-amino-5-cyano-1-methylimidazole with benzonitrile to give 6-amino-7-methyl-2-phenylpurine;[2386] and like reactions.[2527]

A few *pteridines* have been made from pyrazine intermediates. The most interesting are 3-amino-3,4-dihydro-2-methyl-4-oxopteridine (**115**) and its homologues, made by heating 2-acetamido-3-hydrazinocarbonylpyrazine, or appropriate homologues, in isopropanol.[2382] 3,4-Dihydro-3,7-dimethyl-4-oxopteridine may be made from 2-amino-6-methyl-3-methylcarbamoylpyrazine with triethyl orthoformate and acetic anhydride.[2383] 3-Hydroxylumazine (2,4-dihydroxypteridine-3-oxide) can be made by treating 2,3-dimethoxycarbonylpyrazine with hydroxylamine to give 2,3-bis-hydroxycarbamoylpyrazine, which is cyclized with benzenesulphonyl chloride to 3-benzenesulphonyloxylumazine yielding the *N*-oxide by alkaline hydrolysis.[2962] 2-Amino-3-carbamoyl-5-methylpyrazine has been converted into 4-hydroxy-2,6-dimethylpteridine by triethyl orthoacetate and acetic anhydride.[3196]

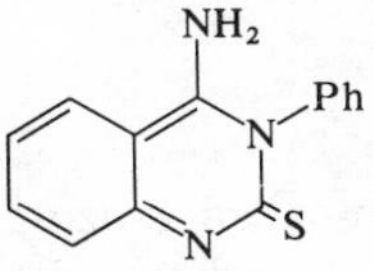

(112) 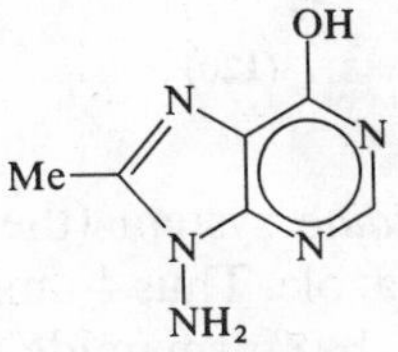 (113) 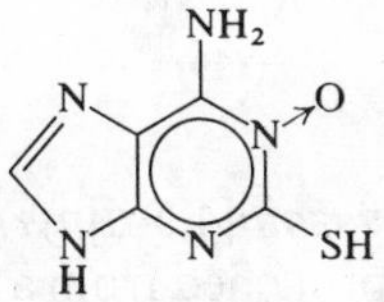 (114) 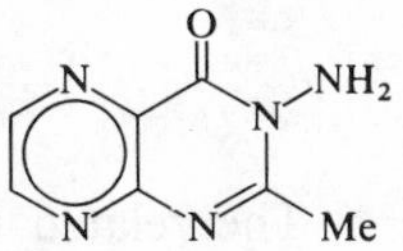 (115)

A similar type of reaction converts 2-amino-3-carbamoyl-5-cyanopyridine by diethyl carbonate into 6-cyano-2,4-dihydroxy*pyrido*[2,3-d]-*pyrimidine* (**116**) and guanidine is used to convert 2-amino-5-carboxy-3-ethoxycarbonylpyridine into 2-amino-6-carboxy-4-hydroxypyrido-[2,3-*d*]pyrimidine.[2384] Several 4-hydroxy*pyrido*[4,3-d]*pyrimidines* have been made from pyridines.[3200]

Some *pyrazolo*[3,4-d]*pyrimidines* have been made from pyrazoles. Thus 4-cyano-1-methyl-5-methylaminomethyleneaminopyrazole in boiling benzene slowly gives 4,5-dihydro-4-imino-1,5-dimethylpyrazolo-[3,4-*d*]pyrimidine (**117**) which can undergo Dimroth rearrangement to 1-methyl-4-methylaminopyrazolo[3,4-*d*]pyrimidine.[2385] A different type of synthesis is exemplified in the condensations of 3-amino-4-cyano-2-methylpyrazole with benzonitrile to give 4-amino-1-methyl-6-phenylpyrazolo[3,4-*d*]pyrimidine,[2386] and of 3-amino-4-cyano-2-cyclohexylpyrazole with formamide to give 4-amino-1-cyclohexylpyrazolo-[3,4-*d*]pyrimidine.[2387] Other examples are known.[2528, 2958–2961]

Pyrimido[4,5-e][1,2,4]*triazines* have been made from triazines. For example, 3,5-diamino-6-carbamoyl-1,2,4-triazine and diethyl carbonate yield the 3-amino-6,8-dihydroxy derivative (**118**), which has also been made from a pyrimidine; formamide and the same triazine gave the 3-amino-8-hydroxy derivative.[2388, 2529]

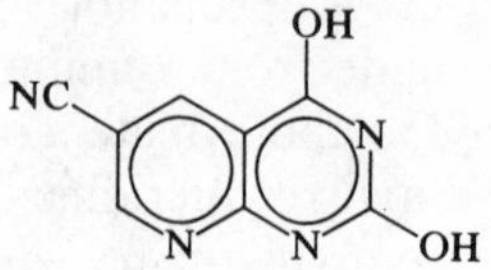

(**116**) 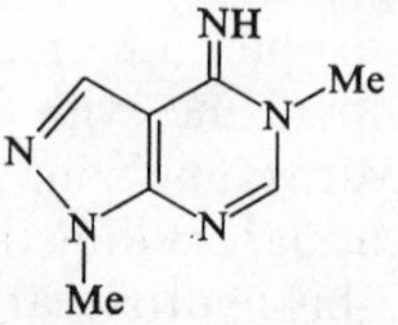 (**117**) 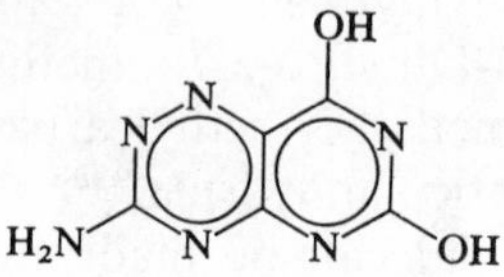 (**118**)

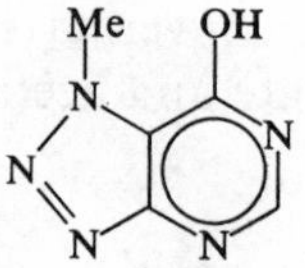

(**119**) 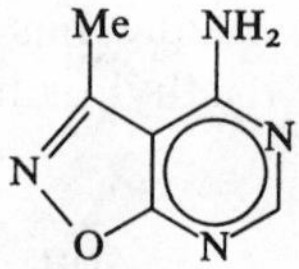 (**120**)

The related v-*triazolo*[4,5-d]*pyrimidine* system (the so-called 8-*aza-purines*) may be approached from a triazole. Thus 4-amino-5-carbamoyl-1-methyl-1,2,3-triazole is converted by formamide into 7-hydroxy-1-methyl-*v*-triazolo[4,5-*d*]pyrimidine (**119**),[2530, 3249] and 5-amino-4-carbamoyl-1-phenyl-1,2,3-triazole (and analogues) by triethyl orthoformate/acetic anhydride into 7-hydroxy-3-phenyl-*v*-triazolo[4,5-*d*]-pyrimidine and its analogues.*[2534, 3252] The 7-hydroxy-2-methyl-*v*-triazolo[4,5-*d*]pyrimidine and its analogues have been made also.[3250,3251]

The formation of *isoxazolo*[5,4-d]*pyrimidines* by completing the pyrimidine ring is exemplified in the reaction of 5-amino-4-cyano-3-methylisoxazole (readily prepared from α-ethoxyethylidenemalononitrile and hydroxylamine) with triethyl orthoformate and acetic anhydride to give 4-cyano-5-ethoxymethyleneamino-3-methylisoxazole which on treatment with ethanolic ammonia (or an amine) gives 4-amino-

* 3-Benzyl-7-mercapto-*v*-triazolo[4,5-*d*]pyrimidine underwent the fascinating 'Christmas rearrangement' to 7-benzylamino[1,2,3]thiadiazolo[5,4-*d*]pyrimidine involving ring fission.[3252] The kinetics of this equilibrium reaction have been studied in several analogues.[2887]

3-methylisoxazolo[5,4-*d*]pyrimidine (**120**) or its 4-alkylamino- homologues.[2389]

Thieno[2,3-d]*pyrimidines* have been made from appropriate thiophens,[2531] some *thiazolo*[4,5-d]*pyrimidines* from 4-aminothiazole derivatives,[2527] and *thiazolo*[5,4-d]*pyrimidines* from 5-aminothiazoles.[2532]

CHAPTER IV

Pyrimidine and Its *C*-Alkyl and *C*-Aryl Derivatives (*H* 116)

Remarkably little has been published recently on the chemistry of pyrimidine and its homologues.

1. Pyrimidine (Unsubstituted) (*H* 116)

Recent work on unsubstituted pyrimidine has been concerned almost entirely with its physical properties, mainly in relation to those of the other azalogues of benzene. A little work on its preparation has been recorded.[2163]

B. Properties of Pyrimidine (*H* 117)

The crystal structure of pyrimidine (**1**) has been determined by a three-dimensional least-squares analysis. Its orthorhombic $P_{n\ a}2_{l}$ crystals have a unit cell of four molecules, and the corrected bond lengths are 1–2, 1.34 Å; 2–3, 1.33 Å; 3–4, 1.36 Å; 4–5, 1.38 Å; 5–6, 1.41 Å; and 1–6, 1.35 Å.[2393] The bond angles derived from these data are a welcome confirmation of those calculated less directly.[2394–2396] Localization energy and π-electron density have been recalculated.[2397, 2913, 2914, 3462]

A determination of the heat of combustion of pyrimidine has been used to recalculate its resonance energy and heat of formation.[2398] A rapid method of calculating the dipole moments of simple heterocycles has been applied to pyrimidine;[2399] the resulting value (2.13 D) is in good agreement with earlier observed[2400] (2.10) and calculated[2401] (2.19) figures.

The ultra-violet absorption of pyrimidine vapour in the 150–200 mμ region has been studied,[2402] and detailed assignments have been made in its infra-red, Raman, and ultra-violet spectra.[2403, 3175] Spectral studies of pyrimidine have also included its $n \rightarrow \pi$ transition,[2404] the absorption of its unstable anion,[2406] its fluorescent emission in aqueous solutions over a range of pH values,[2407] and other aspects.[2405, 2408, 2409, 2975] The infra-red spectrum of solid pyrimidine has been compared with those of solutions.[3526]

The nuclear magnetic resonance spectrum of pyrimidine has been measured on a 40 Mc./s. instrument and analysed by first-order methods;[2410, 2411] coupling constants have been calculated.[3177] Attempts to study the electron spin resonance of the anion of pyrimidine (potassium at a low temperature) suggested the formation of a dipyrimidinyl.[2412]

In the mass spectrograph, pyrimidine undergoes logical fragmentation after initial molecular ion formation (*m*/*e* 80). Several derivatives were also studied.[2428, 3176]

C. Reactions of Pyrimidine (*H* 118)

When pyrimidine is heated with aqueous hydrazine hydrate at 130° for 5 hr., pyrazole (80%) is formed; 4,6-dimethylpyrimidine behaves similarly at 190° to give 3,5-dimethylpyrazole.[3246] The quaternized pyrimidines give the same pyrazoles even at 45°, and a mechanism has been proposed.[3246]

2. *C*-Alkyl and *C*-Ayrl Pyrimidines (*H* 119)

Molecular orbital calculations have been made for 4-phenylpyrimidine in respect of π-electron densities, bond orders, etc.[2854]

A. Preparation of Alkyl Pyrimidines (*H* 119)

Some new direct syntheses of simple alkylpyrimidines[2324, 2325, 2329] have already been discussed (Ch. III, Sects. 2.F and 3.B) with examples in Table VI. Three other new syntheses[2372] leading to less simple alkylpyrimidines are described in Ch. III, Sect. 5.F(2).

The interconversion of alkyl groups has been exemplified[2413] (albeit with the addition of irrelevant groups) in the condensation of 2,4-diamino-6-methyl-5-nitropyrimidine (**2**) with *p*-dimethylaminobenzaldehyde to give the 6-*p*-dimethylaminostryrylpyrimidine (**3**) which on hydrogenation (over Pd) gave 2,4,5-triamino-6-*p*-dimethylaminophenylethylpyrimidine (**4**). Likewise, 2,4-dihydroxy-5-nitro-6-syrylpyrimidine gave the corresponding 5-amino-6-phenethylpyrimidine on hydrogenation over platinum,[2655] and 4,5-diamino-6-chloro-2-styrylpyrimidine with hydrazine hydrate and palladized strontium carbonate gave 4,5-diamino-2-phenethylpyrimidine.[2562] Other such reductions are recorded.[2978–2981]

Direct alkylation can also be used to increase the size of a suitably activated alkyl grouping: 2-*p*-chlorophenyl-4-methoxy-6-methoxycarbonylmethylpyrimidine with methyl iodide in the presence of sodium amide in ammonia at $-70°$ gave the corresponding 6-α-methoxycarbonylethyl or the 6-α-methoxycarbonyl-α-methylethyl analogue, according to the excess of reagents employed; other such alkylations were successful.[3481]

A photo-induced *C*-methylation of pyrimidines has been described.[2974] Thus irradiation of 4-amino-5-cyanopyrimidine (or its 2-methyl derivative) in 2% methanolic hydrogen chloride for 6 hr. produced 4-amino-5-cyano-2,6-dimethylpyrimidine in 60% and 80% yield, respectively; 2,4-diamino-5-cyano-6-methyl- and 4-amino-5-aminomethyl-2,6-dimethyl-pyrimidine were made similarly.[2974] Using higher alcohols, the yields of appropriate alkylpyrimidines declined.[2974]

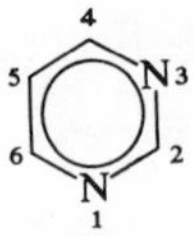

(1) 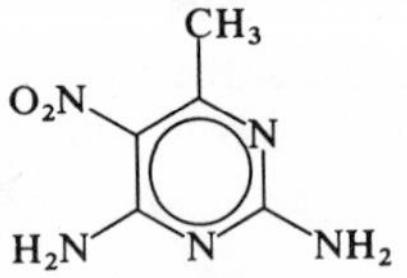 **(2)**

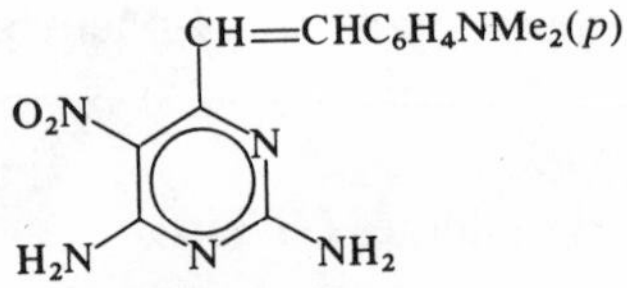

(3) 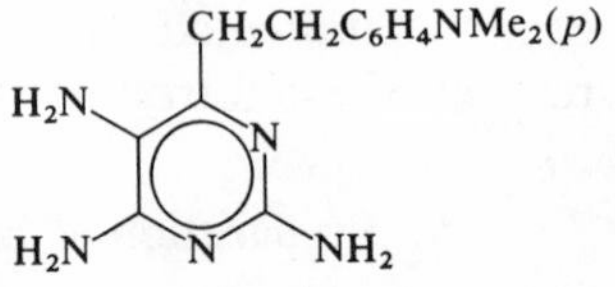 **(4)**

The formation of alkylpyrimidines by dehalogenation is represented in the hydrogenation of 5-benzyl-2,4,6-trichloropyrimidine over Raney nickel in the presence of sodium carbonate;[2241] 5-methyl-2(and 4)*-phenylpyrimidine were made similarly from the 4,6-dichloro and 6-chloro derivative, respectively.[2241]

* The experimental section sub-heading[2241] is incorrectly numbered.

C. Reactions of Alkyl and Aryl Pyrimidines (*H* 124)

Good yields of nine styrylpyrimidines have been obtained[2413] by condensing the corresponding methylpyrimidines with *p*-dialkyl-aminobenzaldehydes in the presence of hydrochloric acid, a condensing agent seldom before used the series (*cf.* *H* 125). The products are exemplified[2413] in 4-amino-2-cyclohexylamino-6-*p*-dimethylaminostyryl-5-nitropyrimidine and 2,4-diamino-6-*p*-aminostyryl-5-nitropyrimidine (**5**). In the presence of piperidine, 1,2,3,4-tetrahydro-1,3,6-trimethyl-5-nitro-2,4-dioxopyrimidine gives with anisaldehyde the corresponding 6-*p*-methoxystyryl derivative;[2414] likewise, 4,6-dihydroxy(or 4-amino)-2-methyl-5-nitropyrimidine with benzaldehyde gives the 2-styryl analogue.[2334, 2562] 4-Methyl-2-methylthiopyrimidine has been condensed with *p*-bromobenzaldehyde and analogues in concentrated sulphuric acid to give 4-*p*-bromostyryl-2-methylthiopyrimidine and related compounds.[2174, 2449, 2450, 2912] 4-Methylpyrimidine has been condensed with benzaldehyde in acetic anhydride to give 4-styrylpyrimidine;[3201] with cinnamaldehyde to give 4-4′-phenylbuta-1′,3′-dienylpyrimidine;[3202] and with appropriate aldehydes to give higher homologous polyenes.[3202]

Direct oxidation of methyl- to carboxy-pyrimidines has long been used (*H* 126) but only recently has a practical oxidation of methyl- to formyl-pyrimidines been achieved.[2947] Thus 2,4-dihydroxy-6-methylpyrimidine is oxidized by selenium dioxide in acetic acid to give 4-formyl-2,6-dihydroxypyrimidine (58%); 2,4-dihydroxy-5,6-dimethylpyrimidine undergoes ‘similar oxidation’ only at the ‘active’ 6-methyl group to give 4-formyl-2,6-dihydroxy-5-methylpyrimidine (94%).[2947] Permanganate oxidation of 4-methyl- gave 4-carboxy-2-phenylpyrimidine (42%).[2603]

Sometimes a methylpyrimidine may be converted into an oxime of the corresponding aldehyde by nitrosation. Thus 2-hydroxy-4,6-dimethylpyrimidine gives, not the 5-nitroso derivative as previously thought (*H* 148), but 2-hydroxy-4-hydroxyiminomethyl-6-methylpyrimidine (**7**, R = Me);[2592, 2954] 2-hydroxy-4-hydroxyiminomethylpyrimidine and its 6-phenyl derivative (**7**, R = Ph) were made similarly.[2592, 2954] All three oximes were proven in structure by treatment with phosphoryl chloride to give 2-chloro-4-cyanopyrimidine,[2954] its 6-methyl,[2592, 2954] and its 6-phenyl derivative.[2592] Such extranuclear nitrosations may be accompanied by 5-nitrosation if there are sufficient electron-releasing groups present (*cf.* *H* 148), and should not be confused with Claisen reactions leading to similar products (see below and *H* 129).

It is now known that peroxide oxidation of 4-methylpyrimidine gives not a single *N*-oxide[438, 469, 470, 2415–2417] but a mixture of the *N*-1-oxide (m.p. 45–47°) and the *N*-3-oxide (m.p. 82–83°), which may be separated chromatographically.[2418] Their structures have been allotted beyond doubt by n.m.r. spectra; their respective dipole moments (*cf. H* 128) are 4.05 and 3.72 D.[2418]

Another oxidation of 4-methylpyrimidine is that achieved by boiling with sulphur in aniline or other aromatic amine. Good yields of 4-anilinothiocarbonylpyrimidine (**6**) and analogous thiotoluides, etc., are obtained. 4,6-Dimethylpyrimidine, sulphur, and aniline yield 4-anilinothiocarbonyl-6-methylpyrimidine. Such thioanilides cyclize in part to 4-benzothiazol-2′-ylpyrimidine (**6a**) and its derivatives.[2427]

Direct chlorination of the methyl group in 4-chloro-2-methyl-, 2-chloro-4-methyl-, and 4,6-dichloro-2-methyl-pyrimidine can be done in ultra-violet light at 150°. No substitution takes place at position 5, and the yields of the respective trichloromethyl derivatives reach about 90% in 10 hr.[2419–2421]

The Claisen condensation of methylpyrimidine (*cf. H* 131) has been further exemplified in the formation of 4-ethoxalylmethyl-2,6-diethoxy-5-nitropyrimidine, 2,4-diallyloxy-6-ethoxalylmethyl-5-nitropyrimidine, and 4-amino-6-ethoxalylmethyl-2-ethoxy-5-nitropyrimidine (**8**) from the corresponding methylpyrimidines with diethyl oxalate and sodium;[2223, 2422] 2,4-diethoxy(and dibenzyloxy)-6-ethoxalylmethylpyrimidine are formed rather similarly by using a mixture of ether and pyridine as solvent and potassium ethoxide as condensing agent.[2423]

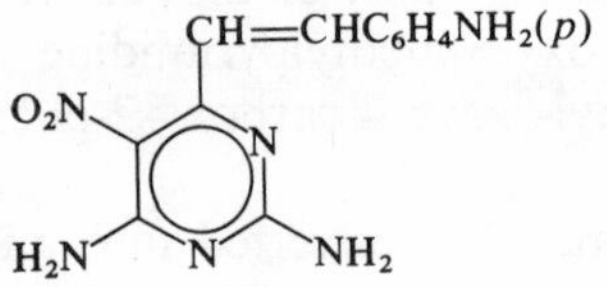

$CH{=}CHC_6H_4NH_2(p)$; O_2N; H_2N; NH_2

(5)

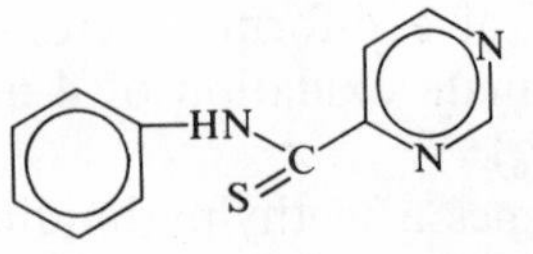

(6)

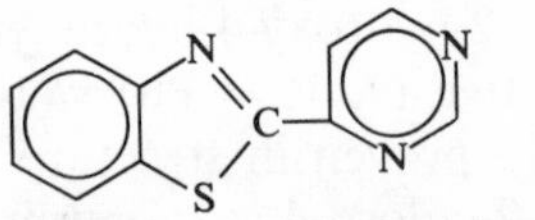

(6a)

CH:NOH; R; OH

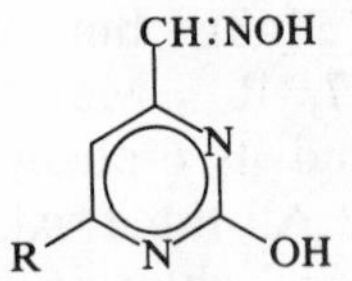

(7)

The type of Claisen condensation employing a nitrite ester (*cf. H* 129) has been performed on 4-methylpyrimidine in the presence of potassium t-butoxide to give an 80% yield, or with sodium amide or sodium hydride to give a lower yield of 4-nitrosomethylpyrimidine, tautomeric with 4-hydroxyiminomethylpyrimidine (**9**).[2426, 3242] Other classical examples are recorded.[2832, 2833, 2955]

The reaction[98] of 4-methylpyrimidine with paraformaldehyde to give 4-β-hydroxyethylpyrimidine (*cf.* *H* 122) has been repeated and extended[2424] to the formation from appropriate methylpyrimidines of 4-β-hydroxyethyl-6-methyl-2-propoxypyrimidine (poor yield) and 4-β-hydroxyethyl-2,6-dipropoxypyrimidine (24%); 2-chloro-4,6-dimethyl- and 2,4-dichloro-6-methylpyrimidine failed to yield hydroxyethyl analogues.[2425]

New and interesting reactions of 4-methylpyrimidine are its conversion into 4-β-dimethylaminovinylpyrimidine (**10**) by the reagent dimethylamino-diethoxymethane, 'dimethylformamide diethyl acetal', $Me_2NCH(OEt)_2$, or by the dimethylformamide/phosgene complex in the presence of acid; with the same complex under neutral conditions (followed by hydrolysis) it gives 4-diformylmethylpyrimidine (**11**).[2426]

One type of Mannich reaction is represented in the condensation of uracil or 2,4-dihydroxy-6-methylpyrimidine with paraformaldehyde and β-chloroethylamine to give 5-β-chloroethylaminomethyl-2,4-dihydroxypyrimidine and its 6-methyl derivative, respectively.[2830] In a similar way,

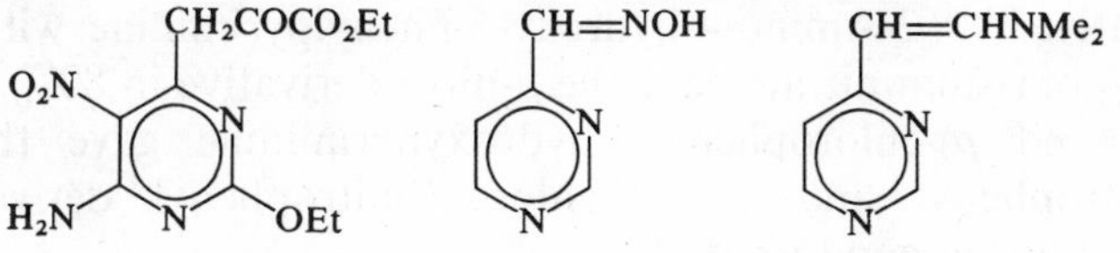

CH_2COCO_2Et ; O_2N ; H_2N ; N ; N ; OEt
(**8**)

CH=NOH ; N ; N
(**9**)

$CH{=}CHNMe_2$; N ; N
(**10**)

HC(CHO)CHO ; N ; N
(**11**)

$NHCOCH_2CH_2N(CH_2CH_2Cl)_2$; N—Me ; O ; N ; H ; O
(**12**)

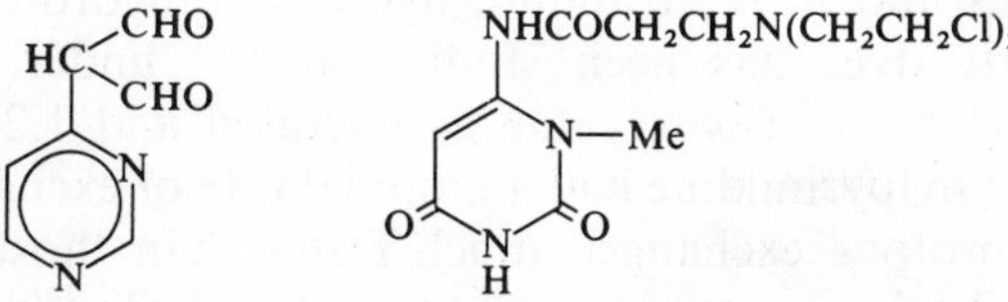

bis-(β-chloroethyl)amine gives 5-bis-(β-chloroethyl)aminomethyl-2,4-dihydroxypyrimidine, its 6-methyl derivative, and 5-bis-(β-chloroethyl)-aminomethyl-4-hydroxy-2-mercaptopyrimidine.[2658, 2831] The same reagents attack 6-acetamido-1-methyluracil to give a product formulated[2831] as 6-β-[bis-(β-chloroethyl)amino]propionamido-1-methyluracil (**12**), although the alternative possibility of attack at the 5-position does not seem to be excluded.

D. E. O'Brien, R. H. Springer, and C. C. Cheng have reported[2953,3476] a new type of Mannich reaction involving the 6-position. Thus

2,4,5-trihydroxypyrimidine, formaldehyde, and piperidine gave the 6-piperidinomethyl derivative (**13**) in 57% yield. When piperidine was replaced by the primary amine, methylamine, the Mannich reaction was followed by cyclization involving the 5-hydroxy group to give 1,2,3,4-tetrahydro-6,8-dihydroxy-3-methyl-1-oxa-3,5,7-triazanaphthalene (**14**, R = OH); similarly, 4,5-dihydroxy-2-methylpyrimidine gave the 8-hydroxy-3,6-dimethyl analogue (**14**, R = Me).[2953]

4-Phenypyrimidine has been reported to yield 4-*m*-nitrophenylpyrimidine (*H* 134).[34] This work has been reinvestigated with the aid of modern tools, and the picture is less simple:[2854] the nature of the products is dependent upon the nitrating agent. Thus nitric acid–sulphuric acid mixture yields 4-*o*- and 4-*m*-nitrophenylpyrimidine in the ratio 2:3; nitric acid–trifluoroacetic anhydride yields 4-*o*-, 4-*m*-, and 4-*p*-nitrophenylpyrimidine in the ratio 45:30:25; and nitric acid–acetic anhydride yields what appears to be 2,4-diacetoxy-1,2,3,4-tetrahydro-1,3,5-trinitro-6-phenylpyrimidine. The *o*-nitrophenyl isomer cyclized to give 2,9,9*a*-triazafluorene.[2854] When 4-2′-furylpyrimidine was similarly treated, a different reaction ensued involving nitration of the pyrimidine ring and subsequent degradation to 2-α-formamido-β-formylvinylfuran.[3759] Nitration of 5-acetamido-2-phenylpyrimidine gave the *m*-nitro derivative.[3463] 2-Amino-4-hydroxy-5-phenylpyrimidine with sulphuric acid/potassium nitrate gave the *p*-nitro derivative in 80% yield; 2-amino-5-*o*(and *p*)-chlorophenyl-4-hydroxypyrimidine gave the 2′-chloro-5′-nitrophenyl and the 4′-chloro-3′-nitrophenyl derivatives, respectively, both in good yield.[3532]

The deuterium exchange of *C*-methyl protons in simple methylpyrimidines and in their amino, hydroxy, dihydro-oxo, and dihydro-imino derivatives has been studied at 33° under acidic and basic conditions.[2861] Acid-base catalysis operated and 1,2-dihydro-2-imino-1,4,6-trimethylpyrimidine had a minimal rate of exchange *ca.* pH 4; the 6-methyl protons exchanged much faster than those of the 4-methyl group, while ring protons were apparently stable.[2861] However, *H*-2 in 4-hydroxy- and 4-mercapto-pyrimidine exchanged at a reasonable rate in D_2O at 90°;[2976] several 1- and 3-alkyl derivatives were included in

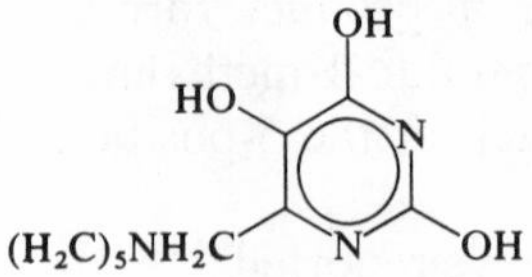

OH, HO, N, $(H_2C)_5NH_2C$, N, OH

(**13**)

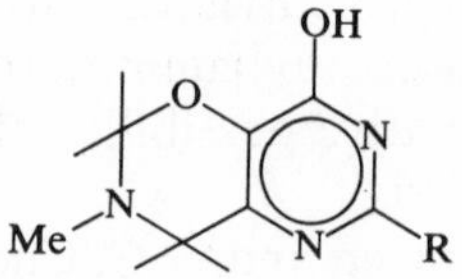

OH, O, N, Me, N, N, R

(**14**)

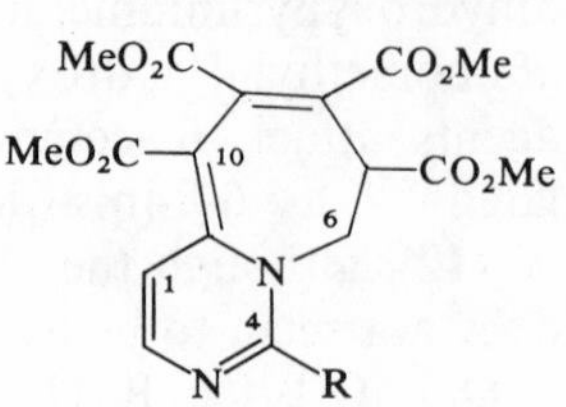

MeO_2C, CO_2Me, MeO_2C, 10, CO_2Me, 6, N, 1, 4, N, R

(**15**)

the study, but such substitution had little effect. Acid-base catalysis did not operate and only in the quaternary compound, 1,6-dihydro-1,3-dimethyl-6-oxopyrimidinium iodide, was exchange really rapid.[2976]

4,6-Dimethylpyrimidine condenses with two molecules of dimethyl acetylenedicarboxylate to give 6,7-dihydro-7,8,9,10-tetramethoxycarbonyl-2-methylazepino[1,2-*c*]pyrimidine (**15**, R = H); 2,4,6-trimethylpyrimidine gives the 2,4-dimethyl homologue (**15**, R = Me).[3213] It is evident that rearrangement is involved in the formation of the products as recorded in related series.[3248]

CHAPTER V

Nitro-, Nitroso-, and Arylazo-pyrimidines (*H* 138)

The introduction of a 5-amino group remains the chief function of the above pyrimidines. Three interesting developments have been the preparation of 5-nitropyrimidine, the nitration of pyrimidines bearing only one electron releasing group, and the direct use of nitrosopyrimidines in purine syntheses and in making nitropyrimidines.

The electrochemical characteristics of nitropyrimidines as cathodes (as well as amino- and hydroxy-pyrimidines as anodes) have been studied.[2973]

1. The Nitropyrimidines (*H* 138)

The powerful electron-withdrawal occasioned by a 5-nitro substituent is not only evident in physical properties such as the lowering of basic pK_a values but also in two other ways: the first is the activation of chloro, methoxy, or methylthio groups in the same molecule towards nucleophilic replacement; the second is the facilitation of ring-fission reactions. The activating effect is placed on a semi-quantitative basis by recent comparative figures[2668] showing that the reactivity of 2-chloro-, 2-methoxy-, or 2-methylthiopyrimidine towards aminolysis is increased 1–3 millionfold by an added 5-nitro substituent. Facilitation of ring-fission is seen in the Dimroth rearrangement, where a nitro substituent reduces the $t_{1/2}$ for rearrangement of 4-amino-1,6-dihydro-6-imino- or 4-dimethylamino-1,2-dihydro-2-imino-1-methylpyrimidine from 15 and 2000 min., respectively, to a few seconds in each case;[2626, 2855] in the instability of 5-nitropyrimidine and its 2-alkyl derivatives to alkali* in

* 5-Nitropyrimidine is relatively stable in water undergoing a first-order change with $t_{1/2} \sim 7.3$ days at 25°; also in acidic solution of H_0-2 in which it exists as a covalently hydrated cation.[2688 (cf. 2945, 2946)]

which they decompose rapidly ($t_{1/2} < 1$ min.);[2688] and in the ring cleavage of 4-chloro-6-dimethylamino-5-nitropyrimidine in acid to give 1-amino-2-cyano-1-dimethylamino-2-nitroethylene [$H_2N(Me_2N)C{:}C \cdot (NO_2)CN$], and related reactions.[2856]

A. Preparation of Nitropyrimidines (*H* 138)

(1) *Nitropyrimidines by Direct Synthesis* (*H* 139)

When nitromalondialdehyde was condensed with ethoxycarbonylacetamidine, the expected 2-ethoxycarbonylmethyl-5-nitropyrimidine (**1**) did not result; instead, 2-amino-3-ethoxycarbonyl-5-nitropyridine (**2**) was formed in good yield.[2429] This inevitably cast doubt on the so-called 2-methyl-5-nitropyrimidine, 2-benzyl-5-nitropyrimidine, and other nitropyrimidines prepared similarly from acetamidine or its *C*-substituted derivatives;[26, 499] these doubts were fully justified by subsequent indirect synthesis [see Sect. A(3) below] of such pyrimidines and the confirmation of their structures by p.m.r. spectra, and, in some cases, by reduction to 5-aminopyrimidines.[2562, 2688] There is no reason to doubt the structures of 5-nitro-2-phenylpyrimidine and its derivatives;[26] indeed the structure of 5-nitro-2-*m*-nitrophenylpyrimidine has been confirmed recently by independent synthesis.[3463]

O_2N, N, N, CH_2CO_2Et (**1**) ⇸ O_2N–CH(CHO)–CHO + H_2N–C(=NH)–CH_2CO_2Et ⟶ O_2N, CO_2Et, N, NH_2 (**2**)

The successful primary synthesis[2263] of 5-nitrouracil, 5-nitro-2-thiouracil, and a variety of derivatives has already been discussed (Ch. II, Sect. 5). In addition, a number of pyrimidines bearing nitro groups on substituents have been made directly: for example, ethyl cyanoacetate and *p*-nitrophenylbiguanide yield 4-amino-6-hydroxy-2-*p*-nitrophenylguanidinopyrimidine;[2430] acetylacetone and 5-nitro-2-furamidine yield 4,6-dimethyl-2,5′-nitro-2′-furylpyrimidine;[2181] and 4-hydroxy-5-*p*-nitrophenylpyrimidine is formed from nitrophenylacetamide and trisformamidomethane in formamide.[2327]

(2) *Nitropyrimidines by Nitration* (*H* 139)

The presence of at least two electron-releasing groups in pyrimidine is now known to be unnecessary for a successful nitration (*cf.* *H* 139).

Thus 2-hydroxypyrimidine (**3**) may be nitrated under very vigorous conditions to give 2-hydroxy-5-nitropyrimidine (**4**), which is also formed by similar treatment of 2-aminopyrimidine (**5**).[2431, 3483] Likewise 1,2-dihydro-1-methyl-2-oxopyrimidine yields its 5-nitro derivative (**6**).[2431, 3483] Another surprise (*cf. H* 140, footnote) has been the nitration of 2,4-diaminopyrimidine to give, albeit in poor yield, 2,4-diamino-5-nitropyrimidine.[2432] Quite normal nitrations are exemplified in the formation of 4-amino-6-methylamino-,[2433] 4-amino-6-dimethylamino-,[2433] 4-dimethylamino-6-methylamino-,[2433] 4,6-dihydroxy-,* [2332, 2333, 2434] 2,4-dihydroxy-6-isopropyl-,[2435] 2-amino-4-hydroxy-6-methyl-,[2436] 2,4-diamino-6-hydroxy-,[2164] 4,6-diamino-2-hydroxy-,[2164] and 2,4,6-triamino-5-nitropyrimidine;† [2164] also in 1-β-carboxyethyl-5-nitrouracil,[2336] 1,2,3,4-tetrahydro-1,3,6-trimethyl-5-nitro-2,4-dioxopyrimidine,[2437, 2850] and other examples.[2851, 2909]

New examples (*cf. H* 141) of successful nitrations in the presence of easily-oxidized alkylthio groups or easily-hydrolysed alkoxy groups are furnished in the preparation of 4,6-dihydroxy-2-methylthio-,[2454] 4,6-dihydroxy-2-methoxy-,[2253] and 2,4-dimethoxy-6-methyl-5-nitropyrimidine.[2437] Nitration in the presence of an active chloro substituent has also been successful. Thus 6-chlorouracil,[2438] 6-chloro-3-methyluracil,[2439, 2440] 6-chloro-1,3-dimethyluracil (**7**),[2441, 2718, 2882] 2-amino-4-chloro-6-hydroxypyrimidine,‡ [2442, 2444, 2457] and 2-amino-4-chloro-1,6-dihydro-1-methyl-6-oxopyrimidine[2444] may all be nitrated under sufficiently gentle conditions to furnish their 5-nitro derivatives; likewise, 4-chloro-2-dimethylamino-6-hydroxy- and 2-amino-4-chloro-6-methoxy-pyrimidine.[2717, 2718]

However, although similar treatment of 2,4-diamino-6-chloropyrimidine (**8**) was first reported[2443] to yield 2,4-diamino-6-chloro-5-nitropyrimidine (**10**) the product has since been shown[2432] to be 2-amino-4-chloro-6-nitroaminopyrimidine (**9**). Only in the presence of an excess of concentrated sulphuric acid, in which the nitramine (**9**) rearranges, does nitration afford (in good yield) the 5-nitropyrimidine (**10**).[2432] 4-Nitramines have also been isolated from 4-amino-6-methylamino- and 4-amino-6-dimethylamino-pyrimidine when the excess of sulphuric acid and/or the temperature was insufficiently high;[2433] and 4-hydroxy-6-nitroamino-2-trifluoromethylpyrimidine was isolated by gentle nitration of the corresponding aminopyrimidine,[2217]

* Reported as improved procedures (*cf.* 506, 511).

† The method of Gabriel[512] is said[2164] to yield 4,6-diamino-2-hydroxy-5-nitropyrimidine. However, Gabriel does give good N analyses for his nitrotriamine and its derived tetramine.

‡ Described as a hydrate[2457] (m.p. ≮ 360°) or as anhydrous material (m.p. 275–276°).

although the related 4,6-dihydroxy-2-trifluoromethylpyrimidine undergoes normal nitration at the 5-position.[2193]

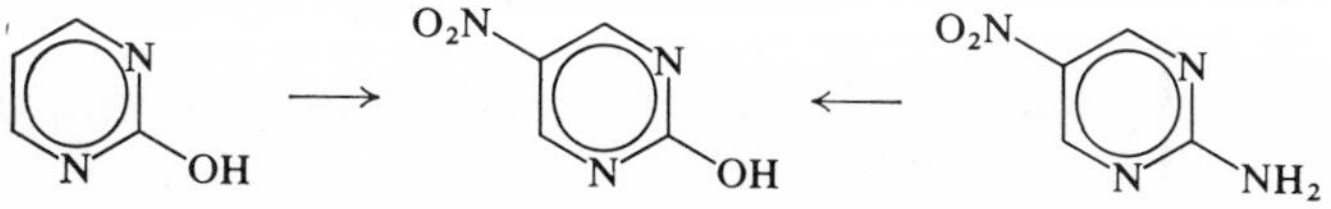

(3) *Nitropyrimidines by Indirect Syntheses* (*H* 142)

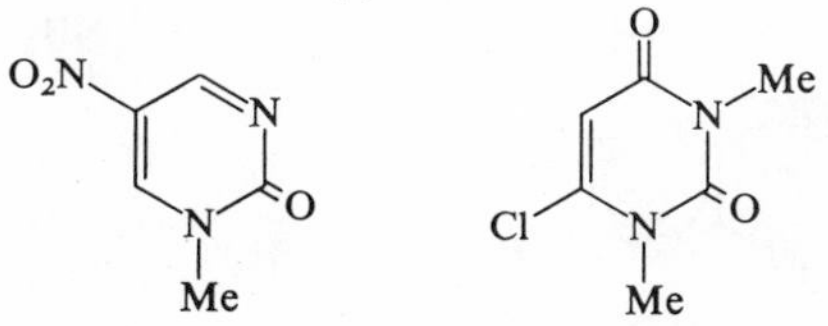

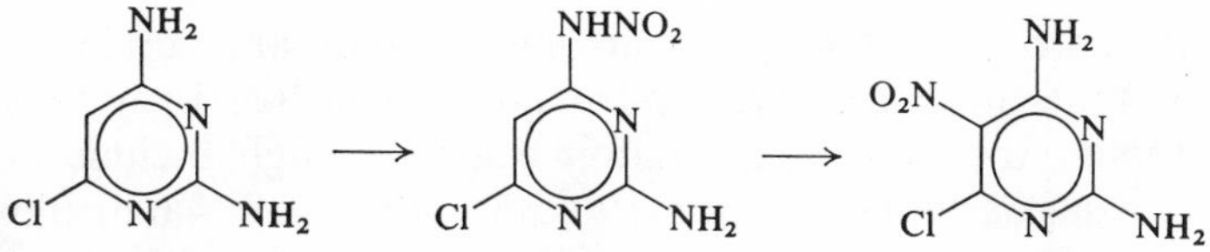

E. C. Taylor and A. McKillop have recently described an important new synthesis of nitropyrimidines by the oxidation of readily available nitrosopyrimidines with hydrogen peroxide in trifluoroacetic acid.[2445] The reaction is general and yields are excellent. Thus 4,6-diamino-5-nitroso-2-phenylpyrimidine (**11**) yields the 5-nitro-analogue (**12**) and from the corresponding 5-nitroso-compounds were also obtained 4,6-diamino-2-methyl-, 4-amino-2,6-dihydroxy-, 4,6-diamino-2-hydroxy-, 2-amino-4-ethylamino-6-hydroxy-, 4-amino-2-dimethylamino-6-hydroxy-, and other 5-nitropyrimidines. In addition, nitrosopyrimidines bearing a mercapto- or methylthio- group were converted into the corresponding hydroxynitropyrimidines, e.g., 4,6-diamino-2-methylthio-5-nitroso- became 4,6-diamino-2-hydroxy-5-nitro-pyrimidine.[2445] The process has also been used to oxidize 2,4-diamino-6-*p*-bromoanilino-5-nitrosopyrimidine to the corresponding 5-nitropyrimidine-*N*-oxide in 73% yield,[2432] and to make 2,4-diamino-3,6-dihydro-3-methyl-5-nitro-6-oxopyrimidine in 50% yield.[2909]

Several simple nitropyrimidines have been made from their hydrazino derivatives by oxidation with silver oxide or silver acetate. Thus

5-nitropyrimidine* was made from its 4,6-dihydrazino derivative,[2688] or from its 4-hydrazino derivative;[2857, 2858] 2-benzyl(or methyl)-5-nitropyrimidine from the 4,6-dihydrazino derivative;[2688] 4-amino-5-nitropyrimidine, its 2-methyl, and its 2-styryl derivative by removal of a 6-hydrazino group in each case;[2562] and 4-methoxy-5-nitropyrimidine from its 6-hydrazino derivative.[2562]

NH_2, ON, N, H_2N, N, Ph — $\xrightarrow[F_3CCO_2H]{H_2O_2}$ — NH_2, O_2N, N, H_2N, N, Ph

(11) (12)

Other indirect syntheses of nitropyrimidines are confined to the formation of those bearing a nitro group attached to a substituent rather than to the 5-position. Random examples might include 4-chloro-6-*p*-nitrobenzenesulphonamidopyrimidine (**13**; from 4,6-dichloropyrimidine and sodio *p*-nitrobenzenesulphonamide),[2446, 2447] 4-methoxy-6-*p*-nitrobenzenesulphonamidopyrimidine (from 4-amino-6-methoxypyrimidine and *p*-nitrobenzenesulphonyl chloride),[2446] 5-fluoro-4-methoxy-2-*p*-nitrobenzenesulphonamidopyrimidine (from 2-amino-5-fluoro-4-methoxypyrimidine and *p*-nitrobenzenesulphonyl chloride),[2209] 5-2′,4′-dinitroanilino-2,4-dihydroxypyrimidine (**14**; from 5-amino-2,4-dihydroxypyrimidine and 1-bromo-2,4-dinitrobenzene; structure probable but unconfirmed),[2260] hexahydro-1,3-dimethyl-5-*o*-nitroanilinomethylene-2,4,6-trioxopyrimidine (**15**; from 1,3-dimethylbarbituric acid, ethyl orthoformate, and *o*-nitroaniline or by three other routes),[2120] 5,5-diethyl-1-*p*-nitrobenzoylbarbituric acid (**16**; and other such compounds from *p*-nitrobenzoyl chloride and silver salts of 5,5-diethyl- and other barbituric acids; complicated and interesting reactions),[2448] 2-amino-4-*p*-nitrostyrylpyrimidine (from 2-amino-4-methylpyrimidine and *p*-nitrobenzaldehyde in sulphuric acid),[2449] 2-amino-4-5′-nitro-2′-furylvinylpyrimidine (from 2-amino-4-methylpyrimidine and 5-nitrofurfural in acid; also many analogues),[2174, 2449–2451] 4-chloro-6-methyl-2-5′-nitro-2′-furfurylidenehydrazinopyrimidine (**17**, from the corresponding benzylidene derivative and 5-nitrofurfural; also analogues),[2452] 2,4-dihydroxy-6-methyl-5-*p*-nitrobenzenediazoaminopyrimidine (**18**; from 5-amino-2,4-dihydroxy-

* The cations of 5-nitropyrimidine and its 2-methyl and 2-benzyl derivatives were the first pyrimidines shown to be covalently hydrated in aqueous solution.[2688] This phenomenon was further discussed in the context of other series which commonly hydrate.[3472]

6-methylpyrimidine and *p*-nitrobenzenediazonium chloride),[2453] and 1-δ-*p*-nitrobenzoyloxybutyluracil and homologues (from uracil and δ-chlorobutyl *p*-nitrobenzoate).[2547]

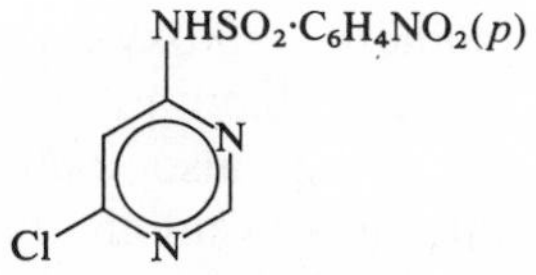

$NHSO_2{\cdot}C_6H_4NO_2(p)$ Cl N N

(13)

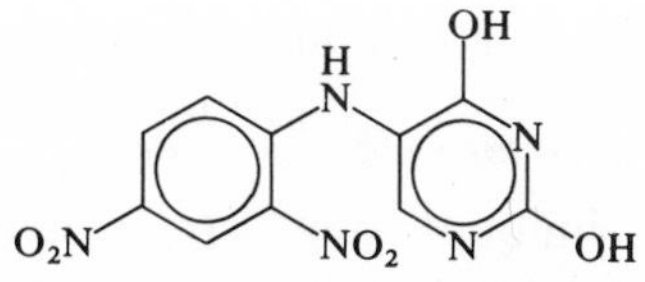

OH H N O_2N NO_2 N N OH

(14)

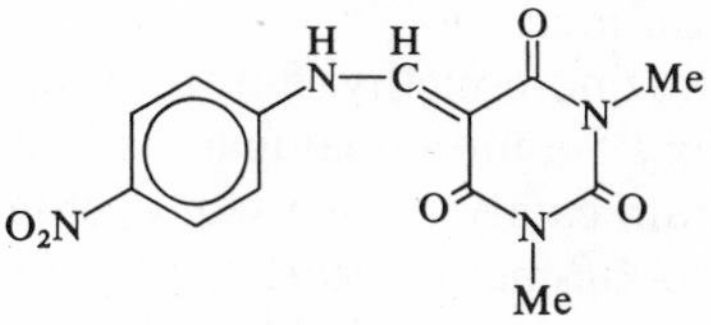

H H N—C O Me N O_2N O N O Me

(15)

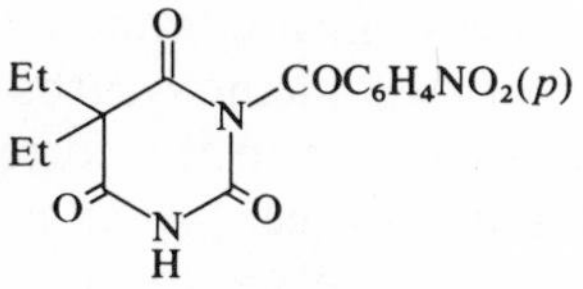

O Et Et $COC_6H_4NO_2(p)$ N O N O H

(16)

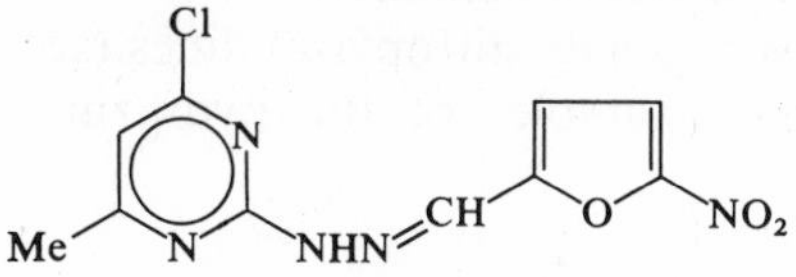

Cl N Me N NHN=CH O NO_2

(17)

OH N=N·HN N O_2N Me N OH

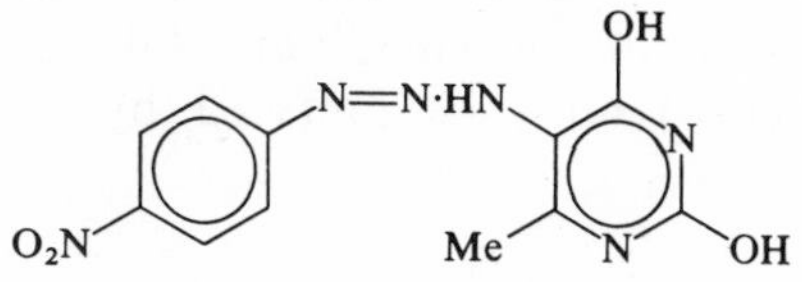

(18)

B. Reactions of Nitropyrimidines: Mainly Reduction (*H* 143)

Attempts to prepare fused pyrimidines by involving a nitro group in direct intramolecular reaction with an adjacent guanidino group have failed in this series.[2164] However, cyclization of 1,2,3,4-tetrahydro-6-*p*-methoxystyryl-1,3-dimethyl-5-nitro-2,4-dioxopyrimidine (**19**) occurs in triethyl phosphite to yield the pyrrolopyrimidine (**20**), possibly *via* a nitrene intermediate.[2414]

No attempt will be made to list all recent examples of the catalytic hydrogenation of nitropyrimidines. The most popular catalyst was again Raney nickel, and its surprisingly successful use in the presence of alkylthio and activated halogeno substituents was further exemplified in the reduction to their 5-amino-analogues of 5-nitro-,[2688] 4,6-bis-methylamino-2-methylthio-5-nitro-,[2454] 4-methoxy-5-nitro-,[2562] 4-chloro-2-dimethylamino-6-hydroxy-5-nitro-,[2444] 4-amino-2-chloro-6-ethoxycarbonyl-5-nitro-,[2467] 2-amino-4-chloro-1,6-dihydro-1-methyl-5-nitro-6-oxo-,[2444] 4,6-dichloro-5-nitro-,[2227, *cf.* 616] and 4-amino-6-fluoro-5-nitro-pyrimidine.[2455] The result of using a massive amount of this catalyst in such a case is shown by the direct transformation of

4-amino-6-dimethylamino-2-methylthio-5-nitropyrimidine (**21**) into 4,5-diamino-6-dimethylaminopyrimidine.[2433] Several hydrogenations in the presence of sugar groups have been described,[2456, 2457] but the products cyclized to dihydropteridines before isolation. Other Raney nickel hydrogenations include some alkoxy-nitropyrimidine[2444, 2458, 2459, 2725] and amino- or diamino-nitropyrimidines;[827, 2110, 2460–2463, 2851, 2852] the latter group also includes two unusual techniques, one using ethyl acetate as solvent in the hydrogenation of 4-amino-6-butylamino-5-nitropyrimidine,[2365] and the other bubbling hydrogen through a stirred suspension of Raney nickel and 4,6-diamino-5-nitropyrimidine in warm methanol.[2433]

2-Chloro-4-β-methoxycarbonylhydrazino-6-methyl-5-nitropyrimidine has been reduced to the corresponding 5-amino analogue with Raney nickel and hydrogen; when a palladium catalyst was used, dechlorination also took place to give 5-amino-4-β-methoxycarbonylhydrazino-6-methylpyrimidine.[3214] 4-β-Benzoylhydrazino-2-chloro-6-methyl-5-nitropyrimidine and related compounds behaved similarly.[3214]

The use of palladium catalysts to hydrogenate nitropyrimidines (see below) must be prefaced by two rare examples of the hydrazine/

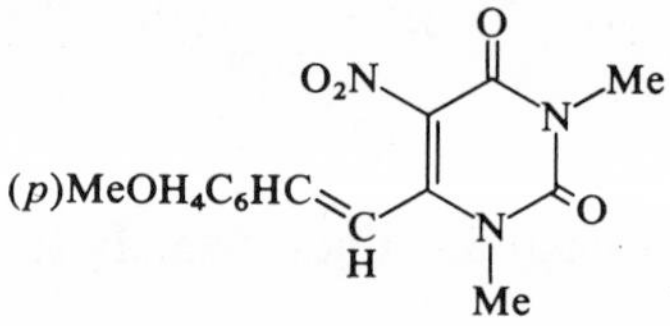

(**19**)

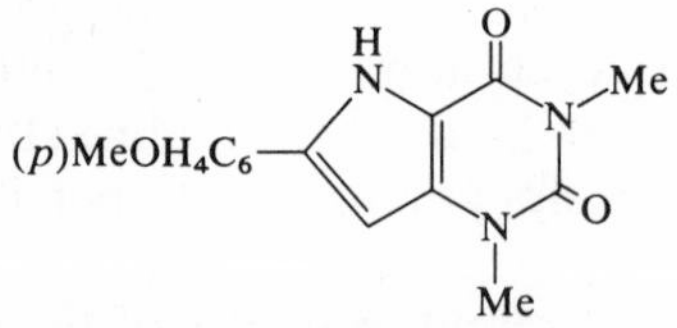

(**20**)

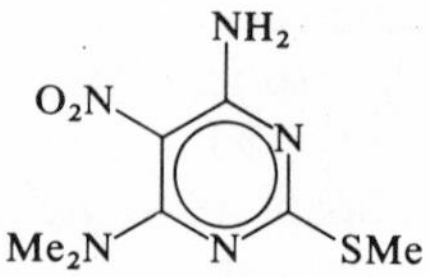

(**21**)

palladium reduction of nitropyrimidines. Thus 5-nitro-2-phenylpyrimidine was boiled in benzene with hydrazine hydrate and palladium-on-charcoal to give 5-amino-2-phenylpyrimidine (67%);[3463] similarly, 5-acetamido-2-*m*-nitrophenylpyrimidine gave the 2-*m*-aminophenyl analogue, characterized as 5-acetamido-2-*m*-acetamidophenylpyrimidine.[3463]

It is clear that hydrogenation of nitropyrimidines with palladium catalysts may be done in a variety of solvents. Thus 2-acetamido-5-nitro-4-hydroxypyrimidine may be so hydrogenated in dimethylformamide;[2464] 3-β-ethoxycarbonylethyl-5-nitrouracil in 2-methoxyethanol;[2336] 4-amino-5-nitro-6-D-ribitylaminopyrimidine, 2-amino-4-hydroxy-5-nitro-6-D-sorbitylaminopyrimidine, and analogues in 50% acetic or 90% formic acid;[2442] 2-amino-4-hydroxy-6-methyl-5-nitropyrimidine in *N*-hydrochloric acid;[2436] and 2-amino-4-β-ethoxycarbonylpropylamino-6-hydroxy-5-nitropyrimidine,[2465] 4-hydrazino-6-hydroxy-5-nitropyrimidine,[2466] and 4-amino-6-hydrazino-5-nitropyrimidine[2466] in water. The hydrazino groups were almost unaffected by hydrogenation under these conditions.[2466] Alcohol may also be used.[2850]

The use of platinum catalysts (which may be used with ammoniacal solutions) is exemplified in the hydrogenation of 2,4-dihydroxy-4-β-hydroxyethyl-5-nitropyrimidine,[2438] 4-chloro-1,2,3,6-tetrahydro-1,3-dimethyl-5-nitro-2,6-dioxopyrimidine,[2441] 6-chloro-3-methyl-5-nitrouracil,[2439, 2440] and 2-dimethylamino-5-nitropyrimidine.[3519]

The unpopularity of sodium dithionite (*H* 145) for reducing nitropyrimidines probably stems from its apparently erratic results with nitrouracil; however, in the presence of ammonia, a 90% yield of 5-aminouracil has been reported,[994] thus making it superior to other methods. More recently dithionite has been successfully used to reduce 4-mercapto-6-methoxy-,[2227] 4-amino-6-ethoxycarbonyl-2-hydroxy (or mercapto)-,[2467] 4-amino-2-ethoxy-6-ethoxycarbonyl-,[2467] 4-amino-6-ethoxycarbonyl-2-methylthio- (in aqueous acetone),[3458] 2,4-diamino-6-ethoxycarbonyl-,[2467] 2-amino-4-cyanomethylamino-6-hydroxy-,[2465] 2,4-dihydroxy-6-β-hydroxyethylamino-,[2438] 2,4-dihydroxy-6-D-ribitylamino-,[2468] and 2-chloro-4-ethoxycarbonyl-6-ethoxycarbonylmethylamino-5-nitropyrimidine.[2469]

Among other classical reducing agents used recently to reduce nitropyrimidines are iron and acetic acid for converting 4,6-diamino-5-nitro-2-trifluoromethylpyrimidine to the corresponding triamine in 93% yield;[2193] zinc and concentrated formic acid for 2,4-dihydroxy-6-methylamino-5-nitro- into 5-formamido-2,4-dihydroxy-6-methylamino-pyrimidine;[2470] hydriodic acid and red phosphorus for 2-chloro-4-ethoxycarbonylmethylamino-5-nitropyrimidine (**22**, R = O) into its 5-amino analogue (**22**, R = H) in >77% yield;[2471] stannous chloride in ethanolic hydrochloric acid for 4,6-dichloro- and 2,4-dichloro-6-methyl-5-nitropyrimidine into their 5-amino analogues in >80% yield;[2435, 2886] zinc dust in boiling water for 4-amino-2-hydroxy-6-methyl-5-nitropyrimidine into its 5-amino analogue;[2472] and stannous chloride in aqueous hydrochloric acid* for reducing 4-amino-2-

* The same reagent satisfactorily reduced the nitro group in 4-amino-5-nitro- and 4-amino-6-chloro-5-nitro-2-styrylpyrimidine without affecting other groups.[2562]

diethylamino-6-*p*-dimethylaminostyryl-5-nitropyrimidine (**23**, R = O, R′ = NEt_2) to its 5-amino analogue (**23**, R = H, R′ = NEt_2) without reducing the ethylenic double bond.[2413] Analogous selectivity may be achieved by hydrogenation over Raney nickel but not over palladium. With the latter catalyst, for example, 2,4-diamino-6-*p*-dimethylaminostyryl-5-nitropyrimidine (**23**, R = O, R′ = NH_2) yields 2,4,5-triamino-6-*p*-dimethylaminophenethylpyrimidine (**24**).[2413]

2-Hydroxy-5-nitropyrimidine underwent an acid-catalysed covalent addition of acetone to the 3,4-bond to give 4-acetonyl-3,4-dihydro-2-hydroxy-5-nitropyrimidine, which in aqueous sodium hydroxide at room temperature gave *p*-nitrophenol in 93% yield![3483] 1,2-Dihydro-1-methyl-5-nitro-2-oxopyrimidine similarly gave a separable mixture of 4-acetonyl-1,2,3,4-tetrahydro-1(and 3)-methyl-5-nitro-2-oxopyrimidine, and each isomer gave *p*-nitrophenol in alkali.[3483]

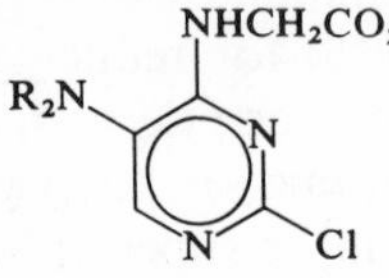

(22) 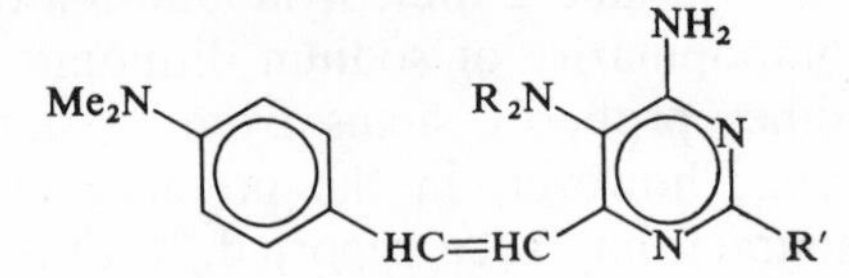 **(23)**

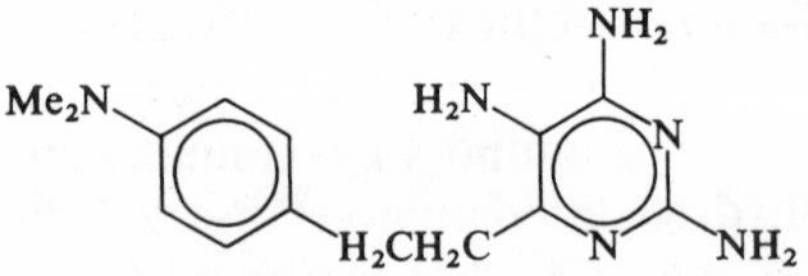

(24)

2. Nitrosopyrimidines (*H* 146)

There is still no real knowledge on the predominating tautomeric state of nitrosopyrimidines, nor has there been any systematic work on their electronic effect. However, R. M. Cresswell and T. Strauss[2473] have studied preparatively the surprisingly easy aminolysis of the methylthio groups in 4-amino-6-hydroxy- and 4,6-diamino-2-methylthio-5-nitrosopyrimidines; although controls with un-nitrosated analogues (**25**, R = NH_2 or OH) are not recorded, it is beyond doubt (*cf. H* 289) that their aminolysis would be very sluggish. Some aspects of the ultra-violet spectra of nitrosopyrimidines have been discussed.[2881]

A. Preparation of 5-Nitrosopyrimidines (*H* 147)

A great many 5-nitrosopyrimidines have recently been made, mostly by addition of sodium nitrite to an acidic solution of the pyrimidine to be nitrosated. This basic technique is exemplified in the nitrosation of 2,4-dihydroxy-6-β-hydroxyethylamino-,[2438, 2474] 4-amino-2-hydroxy-6-methylamino-,[2475] 2-dimethylamino-4-hydroxy-6-methylamino-,[2476] and many other pyrimidines,[2243, 2246, 2285, 2291, 2430, 2433, 2444, 2475, 2477–2482, 2485, 2487, 2949, 3182] including the *N*-oxide, 6-amino-1-hydroxyuracil.[2316]

Some interesting and useful variants on the above procedure are provided in the nitrosation of 2,4-dihydroxy-6-D-ribitylaminopyrimidine using barium nitrite and subsequent isolation of its 5-nitroso derivative as a barium salt;[2442] in the isolation of 4,6-dihydroxy-5-nitroso-2-trifluoromethylpyrimidine (**26**) by its extraction into ether;[2217] in the nitrosation of 2-amino-4-hydroxy-6-methylaminopyrimidine by treatment of a suspension in formamide with sodium nitrite and formic acid;[2483] in the use of organic solvents (ethanol, dioxane, etc.) with isoamyl nitrite for nitrosating water-insoluble pyrimidines such as 6-amino-1-benzyl-2-thiouracil (**27**),[2484] 6-amino-1-dibenzylaminouracil,[2291] 6-amino-1-dimethylamino-2-thiouracil,[2291] 4-cyclohexylamino-1,2,3,6-tetrahydro-1-methyl-2,6-dioxo-3-piperidinopyrimidine (also eight related compounds),[2246] and 1,2,3,4-tetrahydro-6-hydroxymethylamino-1,3-dimethyl-2,4-dioxopyrimidine (also related derivatives).[2485, 2486] In at least some cases, e.g., 1,2,3,4-tetrahydro-1,3-dimethyl-2,4-dioxo-6-phenethylaminopyrimidine (**28**), nitrosations in aqueous and in non-aqueous media produce comparable yields.[2485]

The primary synthesis of nitrosopyrimidines (*H* 149) has been represented recently by the condensation of *NN*-dimethylguanidine with nitrosomalononitrile to give 4,6-diamino-2-dimethylamino-5-nitrosopyrimidine,[2297] and the similar formation of other analogues.[1169]

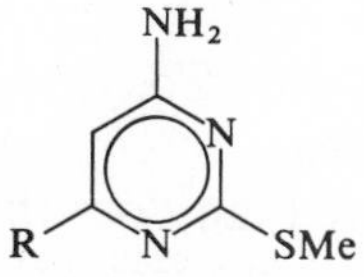

(**25**) 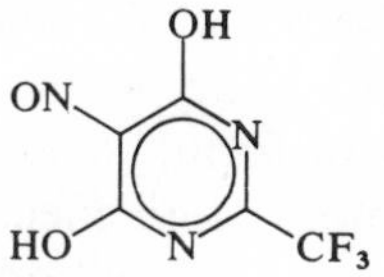 (**26**) 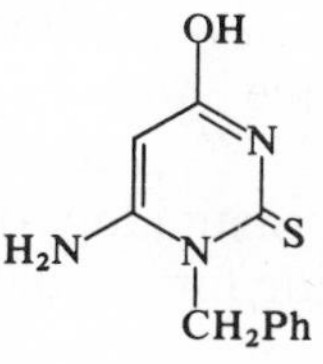 (**27**) 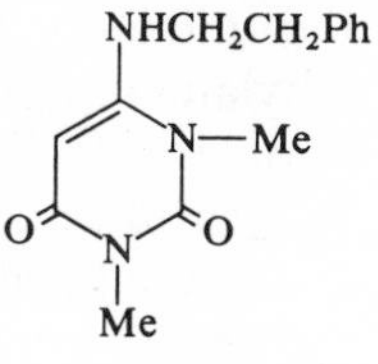 (**28**)

B. Reactions of 5-Nitrosopyrimidines (*H* 149)

Reduction of 5-nitrosopyrimidines to their 5-amino analogues has been used more extensively than ever; a few other reactions have been described.

(1) *Reduction of* 5-*Nitrosopyrimidines* (*H* 149)

Sodium (or ammonium) hydrogen sulphide is still occasionally used to reduce nitroso derivatives such as 2,4-diamino-6-mercapto-5-nitrosopyrimidine,[2487] its *S*-methyl derivative,[2487] 6-amino-1-dimethylamino-5-nitrosouracil,[2291] and ten other such compounds.[2291]

In contrast, sodium dithionite continues to be extensively used, generally in alkaline solution. 5-Amino derivatives formed in this way include 2,5-diamino-4-ethylamino-6-hydroxy,[2434] 2,4,5-triamino-6-benzyloxy-,[2285] 4,5,6-triamino-2-methylthio-,[2433] 4,5-diamino-6-hydroxy-2-mercapto- (improved procedure),[2478] 4,5,6-triamino-2-morpholino-,[2473] 4,5,6-triamino-1,2-dihydro-1-methyl-2-oxo-,[2481] and many other pyrimidines.[2473, 2477, 2481] Also included are 5,6-diamino-1-methyl-2-thiouracil,[2492] its 1-benzyl homologue (by an unusual procedure),[2484] the *N*-oxide formulated as 5,6-diamino-1-hydroxyuracil,[2316] 5-formamido-2,4-dihydroxy-6-methyl(or ethyl)aminopyrimidine (from the respective 5-nitroso analogue in boiling formamide and formic acid with dithionite),[2243] and 2-amino-5-formamido-4-hydroxy-6-methylaminopyrimidine (similarly).[2483]

Although catalytic hydrogenation offers the advantage of a neutral non-aqueous medium for reduction of nitrosopyrimidines bearing labile groups, it is seldom used. Examples are the preparation of 4,5-diamino-2-dimethylamino-6-hydroxypyrimidine (Ni),[2476] 5-amino-2-dimethylamino-4-hydroxy-6-methylaminopyrimidine (Ni),[2476] 5-amino-2,4-dihydroxy-6-D-ribityl(and sorbityl)aminopyrimidine (Pd),[2442] 2,4,5-triamino-6-methoxypyrimidine (Ni),[2493] and 2,5-diamino-4-methoxy(and isopropoxy)-6-methylaminopyrimidine (Ni).[2482]

Sodium hydrogen sulphite is an unsuspected reagent for reducing the nitrosopyrimidines yielding 5-amino-2,4-dihydroxy-6-β-hydroxyethylaminopyrimidine and 5-amino-4-butylamino-2,6-dihydroxypyrimidine in 95% yield.[2474] Zinc dust in formic acid appears to convert 5-nitroso- into 5-formamidopyrimidines more effectively than does the dithionite/formamide/formic acid mixture mentioned above.[2243, 2480, 2489, 2928]

(2) *Other Reactions of* 5-*Nitrosopyrimidines* (*H* 151)

The metal salts and complexes of 4-amino-1,2,3,6-tetrahydro-1,3-dimethyl-5-nitroso-2,6-dioxopyrimidine have been described, and a gravimetric micro-method for so determining nickel and copper has been developed.[2494]

A novel rearrangement of 4-amino-5-nitrosopyrimidines into 1,3,5-triazines occurs in the presence of benzenesulphonyl chloride and pyridine; thionyl chloride; phosphoryl chloride and pyridine; trifluoroacetic anhydride; and even acetic anhydride.[2297, 2495] Thus 4,6-diamino-5-nitroso-2-phenylpyrimidine (**29**) affords 2-amino-4-cyano-6-phenyl-1,3,5-triazine (**30**) in 30% yield, probably by the route shown, although another intermediate has been suggested[2951] since. It is interesting that when this rearrangement was attempted in phosphoryl chloride alone, 4,6-diamino-5-chloro-2-phenylpyrimidine (**31**) resulted,[2297] and, by a strange coincidence, it had the same m.p. as the triazine (**30**) obtained[2495] by using phosphoryl chloride in an excess of pyridine. Similar rearrangement converted 4,6-diamino-2-methylthio-5-nitrosopyrimidine into 2-acetamido-4-cyano-6-methylthio-1,3,5-triazine (acetic anhydride), 4-amino-2-dimethylamino-6-hydroxy-5-nitrosopyrimidine into 2-cyano-4-dimethylamino-6-hydroxy-1,3,5-triazine (acetic anhydride), 2,4-diamino-6-hydroxy-5-nitrosopyrimidine into 2-amino-4-cyano-6-hydroxy-1,3,5-triazine (trifluoroacetic anhydride), and there are other examples.[2297, 2495]

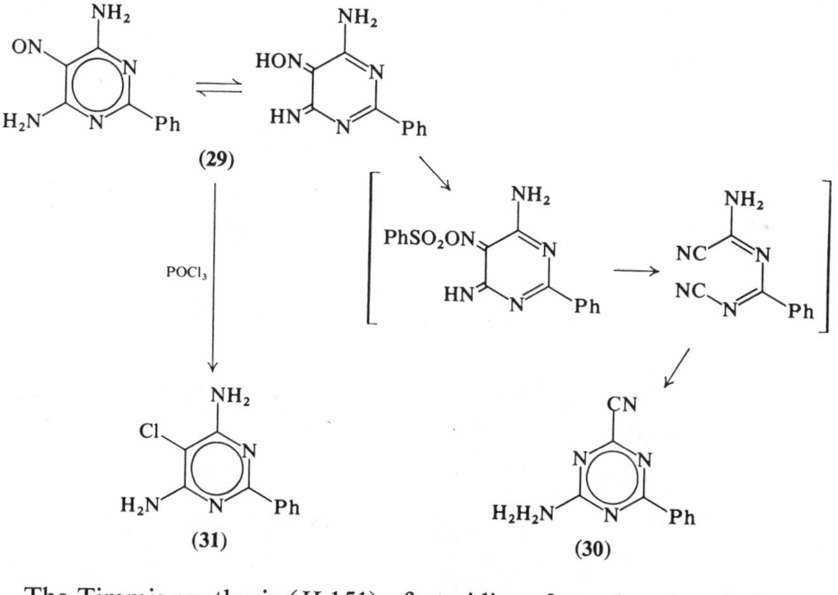

The Timmis synthesis (*H* 151) of pteridines from 4-amino-5-nitrosopyrimidines has been reviewed,[2496] and continues to be used quite widely.[2868, 2949, 3331, 3333] It has also been drastically modified (initially by the original discoverer and his colleagues[2501]) to yield purines. Thus 4-amino-1,2,3,6-tetrahydro-1,3-dimethyl-5-nitroso-2,6-dioxopyrimidine (**32**) with benzyltrimethylammonium iodide yields 1,2,3,6-tetrahydro-1,3-dimethyl-2,6-dioxo-8-phenylpurine (**34**) by elimination of trimethylamine from the assumed intermediate (**33**);[2497] the nitroso-pyrimidine (**32**) with benzaldehyde yields the Schiff's base (**35**) and thence the purine-7-*N*-oxide (**36**)[2498, 2499] which is probably best formulated[2485] as the 7-hydroxypurine (**36a**), and which may be subsequently reduced to the purine (**34**).[2485, 2499] Other important variants of this valuable general reaction have been described.[2485, 2486, 2497–2500, 2950]

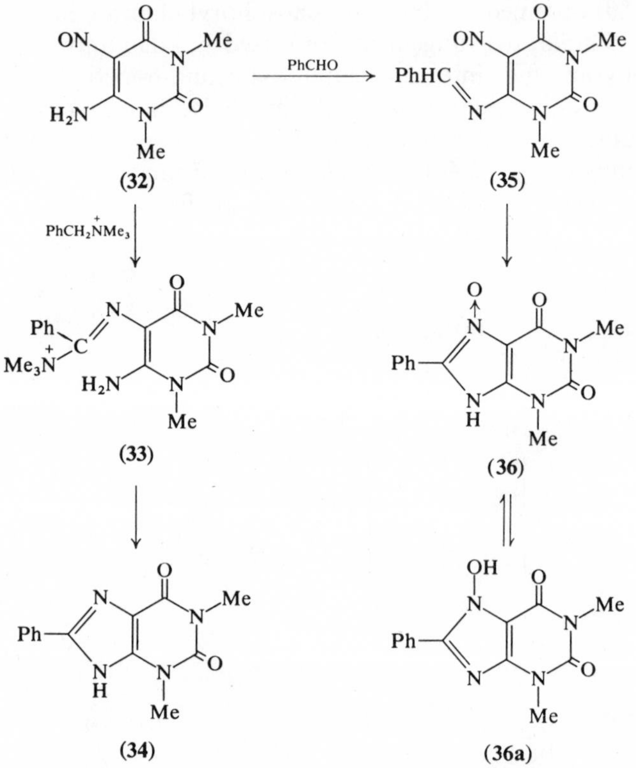

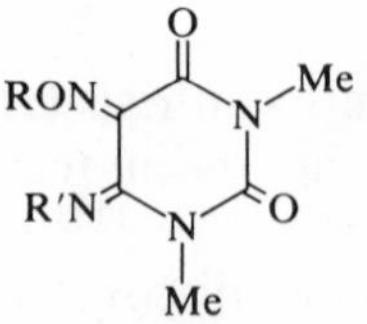

PhCHO

(**32**) (**35**)

$PhCH_2\overset{+}{N}Me_3$

(**33**) (**36**)

(**34**) (**36a**)

(**37**)

Treatment of a 4-alkylamino-5-nitrosopyrimidine, lacking other tautomerizable groups, with an acylating agent causes acylation of the (iso)nitroso group rather than the alkylamino group. Thus 1,2,3,4-tetrahydro-1,3-dimethyl-6-methylamino-5-nitroso-2,4-dioxopyrimidine (isonitroso form: **37**; R = H, R′ = Me) and benzoyl chloride give 5-benzoyloxyiminohexahydro-1,3-dimethyl-4-methylimino-2,6-dioxopyrimidine (**37**; R = Bz, R′ = Me).[2950] When a 4-(primary)amino-5-nitrosopyrimidine is treated similarly, a diacylated derivative is often formed: 4-amino-1,2,3,6-tetrahydro-1,3-dimethyl-5-nitroso-2,6-dioxopyrimidine (**37**; R = R′ = H) and benzoyl chloride give 4-benzoylimino-5-benzoyloxyiminohexahydro-1,3-dimethyl-2,6-dioxopyrimidine (**37**; R = R′ = Bz).[2951] Both the mono- and di-acylated derivatives are themselves powerful acylating agents.[2950, 2951]

C. 4-Nitrosopyrimidines (*H* 151)

The preliminary record (*H* 151) of the formation of 2-amino-4,5-dihydroxy-6-nitrosopyrimidine and its reduction to divicine (**38**) has now been expanded.[2210] The reduction is more effective with dithionite than with hydrogen over palladium.[2210]

D. (*N*-Nitrosoamino)pyrimidines (*New*)

N-Nitroso derivatives of secondary-aminopyrimidines may be made in the usual way. Thus if 2- or 4-methylaminopyrimidine is treated in an excess of 2*N*-hydrochloric acid with sodium nitrite, the respective methylnitrosoaminopyrimidine (**39**) is formed; each is fairly stable as cation but reverts in mildly acidic or neutral aqueous solution to the

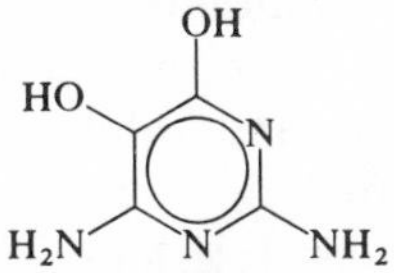

OH HO H_2N N N NH_2

(**38**)

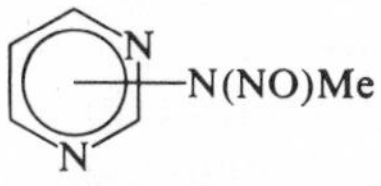

N N(NO)Me N

(**39**)

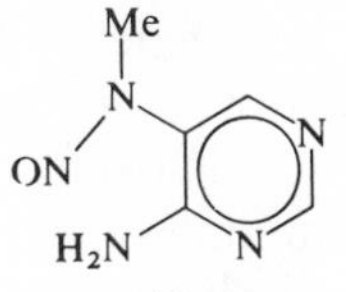

Me N ON N H_2N N

(**40**)

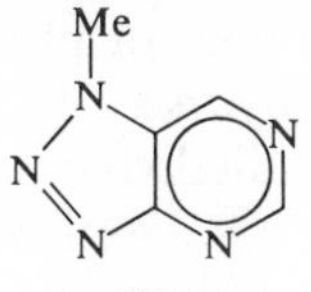

Me N N N N N

(**41**)

parent amine.[3491] Similarly formed are 4-amino-5-benzylnitrosoamino-2,6-dihydroxypyrimidine,[2489] 6-amino-5-benzylnitrosoamino-3-methyluracil,[2489] 2,4,-dihydroxy-5-methylnitrosoaminomethylpyrimidine,[2490] and 4-hydroxy-6-methyl-5-methylnitrosoaminomethyl-2-piperidinopyrimidine.[2490] Treatment of 4-amino-5-methylaminopyrimidine in boiling ethanol with isoamyl nitrite yielded 4-amino-5-methylnitrosoaminopyrimidine (**40**) which gave a typical Liebermann nitrosamine test, but was unaffected by sodium ethoxide, gave only the corresponding 4-acetamido compound with acetic anhydride, and resisted other attempts to cyclize it to the penta-azaindene (**41**).[2491]

3. The Arylazopyrimidines (*H* 152)

There are no recent fundamental advances in the chemistry of arylazopyrimidines.

A. Preparation of Arylazopyrimidines (*H* 152)

Primary synthesis is exemplified in the condensation of phenylazomalondiamidine with diethyl oxalate to give 4,6-diamino-2-carboxy-5-phenylazopyrimidine;[2334] other successful syntheses have been recorded in Ch. II, and the failure of phenylazoacetylacetone and acetamidine to yield a pyrimidine has been discussed in Ch. II, Sect. 4 (*cf.* 2518).

The coupling of 5-unsubstituted pyrimidines with diazonium compounds to give 5-phenylazopyrimidines is illustrated in the preparation of 2,4,6-triamino-5-*p*-ethoxycarbonylphenylazopyrimidine (and analogues),[2502] 2,4-diamino-6-chloro(and mercapto)-5-phenylazopyrimidine,[2503] 2-amino-4-chloro-5-*p*-chlorophenylazo-6-ethylaminopyrimidine,[2505] its 6-β-hydroxyethylamino analogue (**42**),[2504] 4-chloro-2,6-dihydroxy-5-phenylazopyrimidine,[2506] and other such compounds.[2296, 2487, 2505, 2507, 3785] The formation of 4-*p*-chlorophenylazo-5,6-dihydroxy- and 2-amino-4,5-dihydroxy-6-*p*-sulphophenylazo-pyrimidine using diazotised *p*-chloroaniline and sulphanilic acid, respectively, has now been described in detail.[2210] A valuable paper by E. J. Modest, M. Israel, and their colleagues describes 25 couplings by 7 procedures.[2853]

Reverse coupling (*H* 150) between diazotised 5-amino-2,4-dihydroxypyrimidine and amines or phenols has been extended in the formation of 5-1′-amino-7′-hydroxy-4′-naphthylazo-,[2507] 5-8′-amino-5′-quinolylazo-,[2507] 5-2′-β-diethylaminoethylamino-1′-naphthylazo-,[2508] and many other[2507–2509] 2,4-dihydroxypyrimidines. Diazotised 5-amino-2,4-dimethoxypyrimidine has been similarly used in preparing 5-2′-

β-diethylaminoethylamino-1'-naphthylazo-2,4-dimethoxypyrimidine (**43**).[2508]

B. The Reduction of Arylazopyrimidines (*H* 156)

The use of various reducing agents is exemplified in the reduction to aminopyrimidines of 2,4-diamino-6-methylthio-5-phenylazopyrimidine (tin and hydrochloric acid),[2487] 2,4-diamino-6-mercapto-5-phenylazopyrimidine (sodium hydrogen sulphide, 53% yield;[2487] tin and hydrochloric acid, 37% yield[2503]), 4,6-diamino-5-phenylazo(or *p*-nitrophenylazo)pyrimidine (Raney nickel),[2296] 2,4-dihydroxy-5-phenylazo-6-D-ribitylaminopyrimidine (zinc dust and sulphuric acid),[2506] 2-amino-5-*p*-chlorophenylazo-4-β-hydroxyethylamino-6-mercaptopyrimidine (sodium dithionite),[2504] 2,4-diamino-6-chloro-5-phenylazopyrimidine (zinc and acetic acid),[2503] 2-amino-4-chloro-5-*p* chlorophenylazo-6-ethylaminopyrimidine (zinc dust and acetic acid, 74% yield; stannous chloride, 76% yield; iron and hydrochloric acid, 71% yield),[2505] 4-*p*-chlorophenylazo-5,6-dihydroxypyrimidine (sodium dithionite),[2210] and others.[2210, 2227, 2505] 4-Amino-6-hydroxy-5-phenylazopyrimidine has been reduced electrolytically in formic acid to yield 4-amino-5-formamido-6-hydroxypyrimidine (**44**) and thence hypoxanthine; other purines have been made similarly.[2510] Raney nickel has been used without hydrogen to convert 4-hydroxy-2-mercapto-6-methyl-5-phenylazo- into 5-amino-4-hydroxy-6-methyl-pyrimidine (40%).[2563]

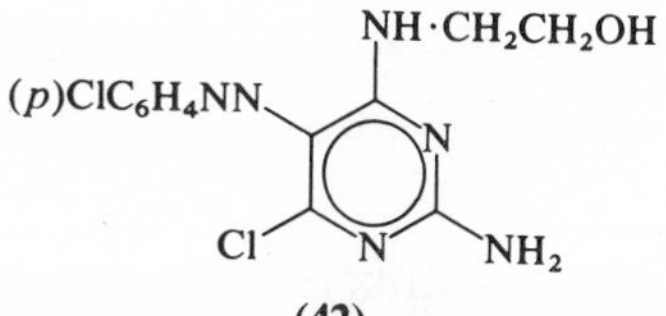

NH·CH_2CH_2OH (*p*)ClC_6H_4NN N Cl N NH_2

(**42**)

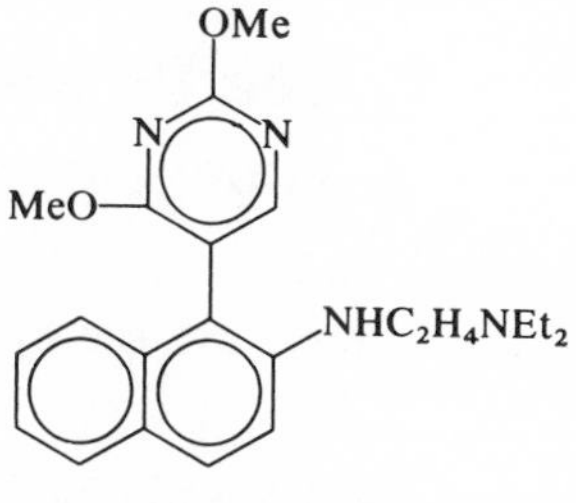

OMe N N MeO $NHC_2H_4NEt_2$

(**43**)

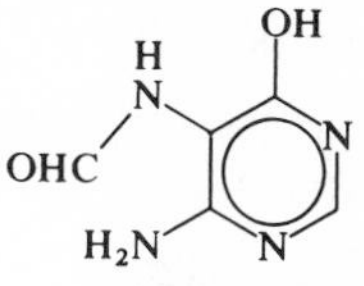

OH H N OHC N H_2N N

(**44**)

CHAPTER VI

Halogenopyrimidines (*H* 162)

Halogenopyrimidines have continued to be the most useful type of intermediate in the series. In this chapter, examples of their formation and subsequent use in the presence of all sorts of other groups are given. In addition, many halogenopyrimidines have marked biological activity: 2-chloro-4-dimethylamino-6-methylpyrimidine, Castrix, is an effective rat poison;[2548, 2549] 5-bromo-3-s-butyl-6-methyluracil, Bromacil,* is a herbicide especially good for grasses;[2550] several multichloro derivatives, e.g., 2,4,5-trichloropyrimidine, show fungicidal activity;[2551] and many 5-halogenated derivatives of uracil, uridine, and related pyrimidines are incorporated during nucleic acid biosyntheses or undergo some other form of 'lethal synthesis' *in vivo*, thus affording the possibility of selective toxicity towards invading bacteria or tumor cells.[2552, 2883, 3180] The whole subject has been reviewed recently and expertly by C. C. Cheng.[2553] The tendency of 2,4- and 4,6-dichloro-5-nitropyrimidine and related compounds to produce dermatitis in susceptible people is well known and has been studied recently.[3194]

1. The Preparation of 2-, 4-, and 6-Halogenopyrimidines (*H* 162)

The range of known halogenopyrimidines has been greatly extended. Of particular interest are the first examples of fluoro derivatives in this category.

A. Phosphoryl Chloride on 2-, 4-, and 6-Hydroxypyrimidines (*H* 162)

(1) *Simple Cases* (*H* 162)

Boiling phosphoryl chloride has been used to convert appropriate monohydroxy derivatives into 4-chloro-2-methylpyrimidine (>90%;

* This pyrimidine and its homologue, Isocil (5-bromo-3-isopropyl-6-methyluracil)[2967] appear to act by inhibiting photosynthesis in higher plants.[2968,2969]

cf. old method[126]),[2541, 2554] 4-chloro-2-ethylpyrimidine,[2554] 4-chloro 2-phenylpyrimidine (67%; *cf.* old method[630]),[2554, 2555] 2-chloro-4,6-diphenylpyrimidine,[2556] 4-chloro-2-methyl-6-phenylpyrimidine,[2556] 4-chloro-6-methyl-2-propylpyrimidine,[2224] and other such pyrimidines.[2224, 3754] Alkenyl or alkynyl substituents are no hindrance: 5-allyl-2-chloro-4,6-dimethylpyrimidine (60%), 2-chloro-4,6-dimethyl-5-prop-1′-ynylpyrimidine (50%), its prop-2′-ynyl isomer (46%), and its 5-propyl analogue (73%; for comparison) were all made from the corresponding hydroxypyrimidines by boiling in phosphoryl chloride for 10 hr.[2700]

The same procedure has been used sometimes to convert di- and tri-hydroxypyrimidines into corresponding chloro derivatives such as 2,4-dichloro-5-methylpyrimidine,[2703] 4,6-dichloro-5-isopropylpyrimidine,[2242] 2-butyl-4,6-dichloropyrimidine,[2242] or 2,4-dichloro-5,6-tetramethylenepyrimidine.[2183] However it is generally necessary to add a tertiary organic base (e.g., diethylaniline) as 'catalyst' in order to obtain good yields, and in this way 2,4-dichloro-5-methylpyrimidine (91%),[2557] 5-butyl-4,6-dichloropyrimidine (82%;[2522] *cf.* 77% without base[2242]), 4,6-dichloro-5-phenylpyrimidine (84%),[2522] 2,4,6-trichloropyrimidine (>80%; *cf.* earlier methods[553, 615]),[2243, 2558, 2610] 2,4,6-trichloro-5-ethylpyrimidine,[2197] and 5-higher-alkyl derivatives[2559] have been made. When *N*-alkylbarbituric acids are allowed to react with phosphoryl chloride to which a little water has been added, one 4(6)-monochloro derivative is formed, apparently without any other isomer or (where possible) a dichloro derivative. Thus only 6-chloro-3-methyluracil (**2**) results from *N*-methylbarbituric acid (**1**),[2560] and 4-chloro-1,2,3,6-tetrahydro-1,3,5-trimethyl-2,6-dioxopyrimidine (**4**) from the corresponding 4-hydroxy compound (1,3,5-trimethylbarbituric acid; **3**);[2305] other examples are known.[2246, 2305, 2569]

Phosphoryl chloride may sometimes be replaced conveniently by the higher boiling phenylphosphonic dichloride ($PhPOCl_2$). This reagent, introduced by M. M. Robison, has been used in the pyrimidine series.[2561]

(2) *In the presence of a Nitro Group* (*H* 163)

Treatment of 4,6-dihydroxy-5-nitro-2-styrylpyrimidine with phosphoryl chloride and diethylaniline to give the dichloro compound (**5**) provides a new example in this group.[2562] Variations in procedure to obtain 4,6-dichloro-5-nitropyrimidine[2332] and 2,4-dichloro-6-methyl-5-nitropyrimidine[2563] have been described.

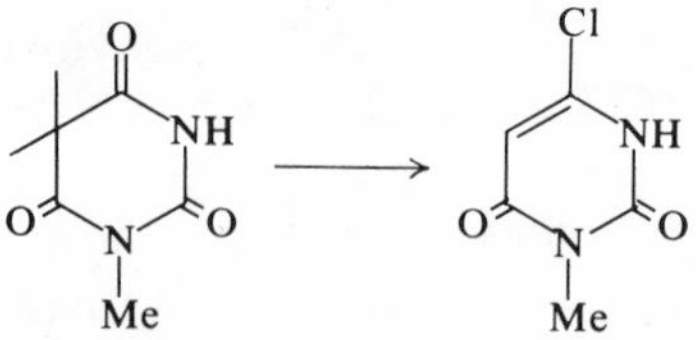

(1) (2)

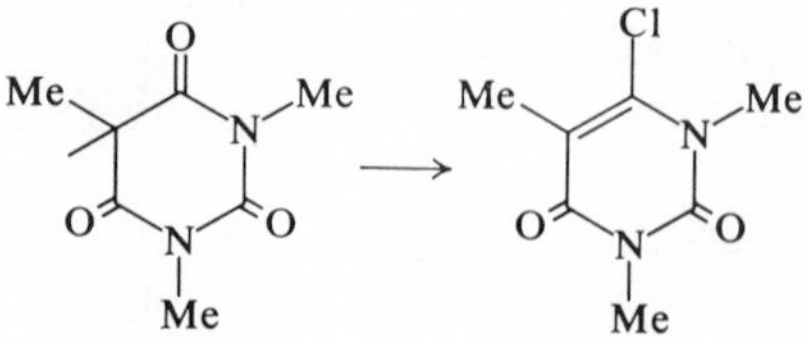

(3) (4)

(3) *In the Presence of Amino Groups* (*H* 163)

Successful examples of the use of phosphoryl chloride (generally with dialkylaniline) in the presence of amino groups have continued to multiply. Among these are the preparation of 2-amino-4-chloro-6-methylpyrimidine (84%),[2230] 4-chloro-6-methylaminopyrimidine (*ca.* 50%),[2433] 5-amino-4-chloro-6-methylpyrimidine (13%; good yields are seldom obtained in the presence of a 5-amino group),[2227] 4-amino-2-benzyl-6-chloropyrimidine,[3504] 4-chloro-2-ethylamino(or dimethylamino)-6-methylpyrimidine,[2564] 2-amino-4-chloro-6-ethylpyrimidine,[2564] 4-chloro-6-methyl-2-methylaminopyrimidine,[2564] 2-amino-4-chloro-6-phenethylpyrimidine (95%; also related compounds),[2308] 2-amino-4-chloro-6-δ-phenylbutyl-6-methylpyrimidine (**6**; 110%!),[2565] 4-amino-2-chloro-5,6-tetramethylenepyrimidine (96%),[2183] and 5-allyl-2-amino-4-chloro-6-methylpyrimidine (13%).[2566] Dichlorinated examples are provided by 4-amino-2,6-dichloropyrimidine (64–80%),[2280, 2620, 2920] 5-anilino-2,4-dichloropyrimidine (34%),[2567] 2-amino-4,6-dichloropyrimidine (50%),[2503] and such like.[2567] Monochlorination of 2-dimethylamino-4,6-dihydroxypyrimidine is achieved by brief heating in phosphoryl chloride to give 4-chloro-2-dimethylamino-6-hydroxypyrimidine (**7**; *ca.* 70%).[2476] Chlorination is also successful in the presence of two amino groups, as in the formation of 4-amino-6-chloro-2-methylaminopyrimidine (35%),[2288] 2,4-diamino-6-chloropyrimidine (*ca.* 65%),[2285] and 4,5-diamino-6-chloropyrimidine (15%; effect of the 5-amino group?).[2568]

Chlorination of 2-amino-4-hydroxy-5-nitropyrimidine with phosphoryl chloride has been shown[2462] to give the stable 4-chloro-2-

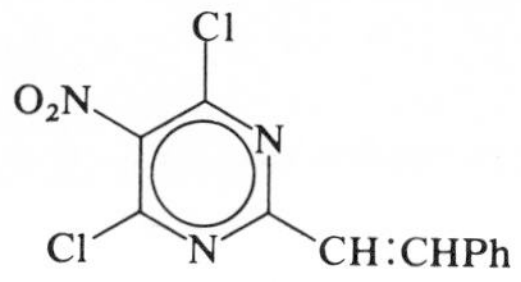

(5) 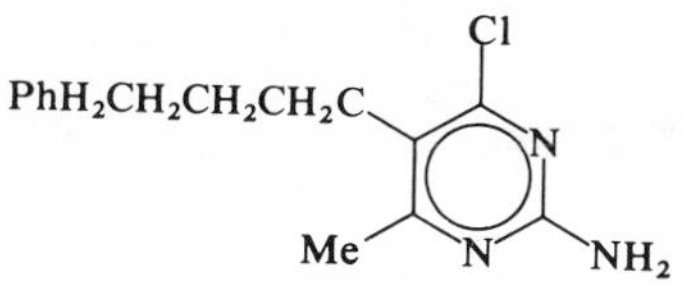 (6)

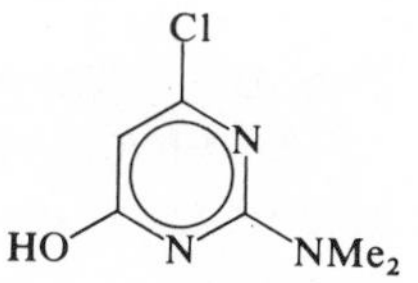

(7) 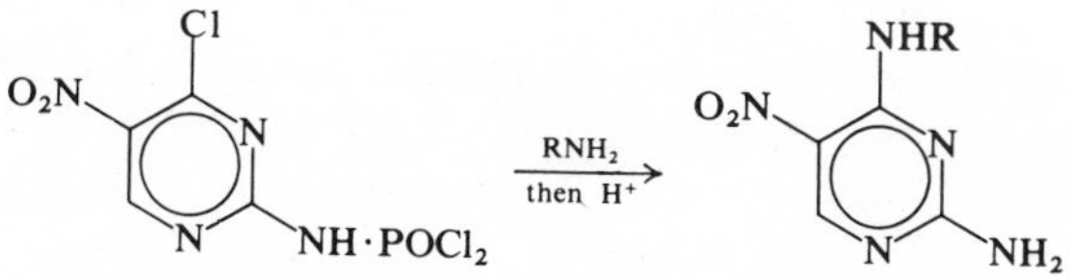 (8) (9)

dichlorophosphinylamino-5-nitropyrimidine (**8**) which on amination and hydrolysis gave several 2-amino-4-(substituted-amino)-5-nitropyrimidines (**9**). 4-Chloro-2-ethylthio-5-dichlorophosphinylaminopyrimidine was described many years ago.[1261]

(4) *In the Presence of Any Other Group* (*New*)*

The action of phosphoryl chloride on hydroxypyrimidines already having a halogeno substituent is exemplified in the formation of 2,4-dichloro-5-fluoropyrimidine (phosphoryl chloride and pyridine),[2570, 2572] and 2,4,5-trichloropyrimidine,[2571] from 5-fluoro- and 5-chloro-2,4-dihydroxypyrimidine, respectively; in the preparation of 5-bromo-2-chloro-,[2573] 5-bromo-4-t-butyl-6-chloro-,[2574] 5-bromo-4-chloro-6-methyl(or phenyl)-,[2574] 4-chloro-2-methyl-6-trifluoromethyl-,[2516] 4,6-dichloro-2-trifluoromethyl-,[2193] and 2-benzyl-5-bromo-4,6-dichloro-pyrimidine;[2575] and in other examples.[2193, 2226, 2236, 2575]

A cyano group was present in making 4-chloro-5-cyano-2,6-diphenyl(or substituted-phenyl)pyrimidines,[2370] 2,4-dichloro-5-cyanopyrimidine,[2680] and 2-chloro-5-cyanopyrimidine;[2276] an ester grouping in making 4-chloro-5-ethoxycarbonylpyrimidine (phosphoryl chloride and triethylamine),[2323] 5-butoxycarbonyl-4-chloropyrimidine,[2323] the lactone of 4-carboxy-2,6-dichloro-5-hydroxymethylpyrimidine,[2884] 2-chloro-4-methoxycarbonyl-6-methylpyrimidine,[2184] 2,4-dichloro-5-ethoxycarbonyl-6-methylpyrimidine,[2576] and 4-chloro-5-ethoxycarbonylmethylpyrimidine;[2577] a *C*-acyl group in 5-acetyl-2,4-dichloropyrimidine;[2577] and an acetamido group in 2-acetamido-4-chloro-6-methylpyrimidine.[2710]

* The original fourth section (*H* 164) had a wider scope now covered by Sections (4) and (5).

An ether or thioether grouping was present while preparing the following: 4,6-dichloro-2-methoxy-,[2253, 2578] 2,4-dichloro-5-methoxymethyl-,[2579] 5-benzyloxy-2,4-dichloro- (only 5%),[2210] 2,4-dichloro-5-phenoxy(or methoxy)-,[2240, 2586] 1-benzyloxy-4-chloro-1,2-dihydro-2-oxo-,[2262] 4-chloro-2-ethylthio-6-methyl-(and 2-alkyl homologues),[2580] 2-chloro-4-methylthio-,[2581] 4-chloro-2-ethylthio-5-methyl-,[2582] 5-allyl-2-benzylthio-4-chloro-,[2555] 2,4,6-trichloro-5-methoxy- (at 130°),[2586] and 4-chloro-1,6-dihydro-1-methyl-2-methylthio-6-oxo-pyrimidine;[2439, 2440] also other examples.[2583, 2586, 2589, 3527] A 5-hydroxy group survived the preparation of 2,4-dichloro-5-hydroxypyrimidine.[2585]

(5) *In the Presence of Two Different Groups* (*New*)

Phosphoryl chloride (generally with a dialkylaniline) affords appropriate chloro- or dichloro-pyrimidines from the following hydroxy derivatives: 4,6-dihydroxy-2-methoxy(or methylthio)-5-nitro-,[2253, 2454] 4-hydroxy-5-methoxy-2-methylthio- (67–86%),[2205, 2584, 2586] 5-benzyloxy-4-hydroxy-2-methylthio- (81%),[2585] 4-amino-6-hydroxy-2-methylthio-,[2475, 2587] 4-amino-6-hydroxy-5-methyl(or other alkyl)-2-methylthio-,[2204] 5-anilino-4,6-dihydroxy-2-methylthio-,[2260] 2-amino-4,6-dihydroxy-5-methoxy-,[2212] 2-amino-4-hydroxy-5-iodo- (90%),[2588] 4-hydroxy-6-methoxycarbonyl-2-methylthio- (84%),[2214] 5-chloro-2,4-dihydroxy-6-methoxycarbonyl- (47%),[2591] and 4-carboxy-2,6-dihydroxy-5-nitro(or chloro)-pyrimidine;[2590, 2591] as well as from other related compounds.[2116, 2204, 2255, 2260, 2587]

When 4,6-dihydroxy-2-hydroxymethyl-5-methylpyrimidine is treated with phosphoryl chloride, the trichloropyrimidine (**10**) is formed;[2238] 4,5,6-trichloro- (91%),[2236] 4-amino-6-chloro- (51%),[2290] 4-chloro- (75%),[2192] and 4,6-dichloro-2-chloromethylpyrimidine[2237] are all formed similarly. It is interesting that 2-amino-5-diethoxypropyl-4-hydroxy-6-methylpyrimidine (**11**) failed to react cleanly with phosphoryl chloride under a variety of conditions.[2566]

Phosphoryl chloride can bring about additional reactions during chlorination. 2-Hydroxy-4-hydroxyiminomethyl-6-methylpyrimidine (**12**) undergoes both chlorination and dehydration to give 2-chloro-4-cyano-6-methylpyrimidine (**13**).[2592] 4,6-Dihydroxypyrimidine undergoes a Vilsmeier reaction with phosphoryl chloride in dimethylformamide to give 4,6-dichloro-5-formylpyrimidine (**14**); 4-amino-6-chloro-5-formyl-2-methylthiopyrimidine and several analogues are formed similarly;[2593] but the same reagents react differently with 4-amino-6-hydroxypyrimidine to give 4-chloro-6-dimethylaminomethyleneaminopyrimidine

(**15**),[2593] and with uracil to give an open-chain product.[2594] Similar chloro aldehydes may be made in two stages [see Ch. XI, Sect. 5.A(2)] and the mechanism has been explored.[2594]

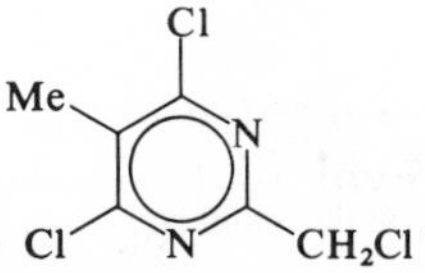

(**10**) 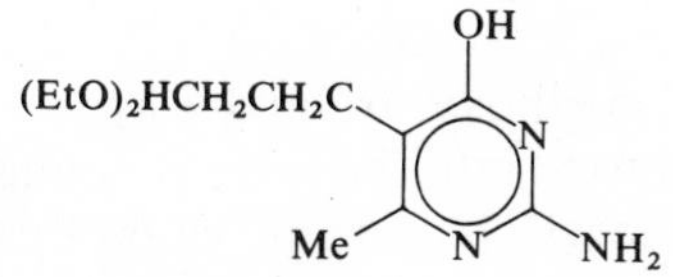 (**11**) 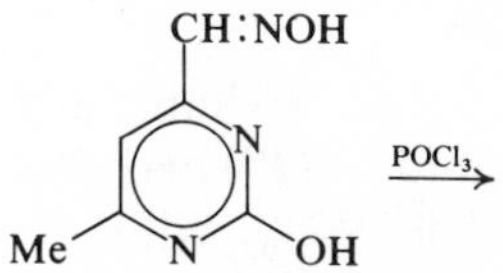 (**12**) 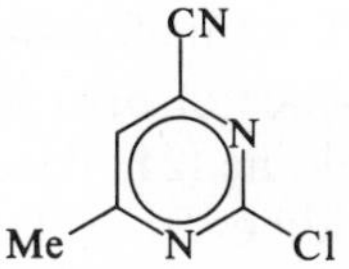 (**13**) 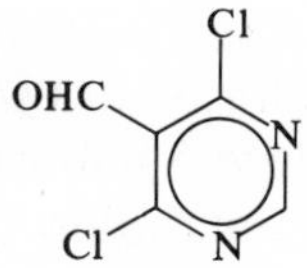 (**14**) 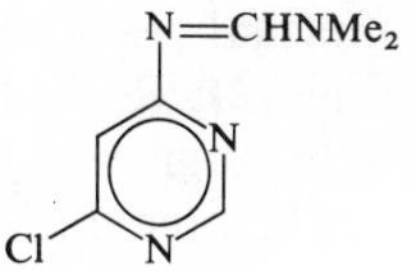 (**15**)

B. The Use of Phosphorus Pentachloride to Produce Chloropyrimidines (*H* 165)

Phosphorus pentachloride is now seldom used to make 2-, 4-, or 6-chloropyrimidines save when additional 5- or ω-chlorination is desirable. As examples, 4-chloro-6-phenylpyrimidine has been made in 90% yield from the corresponding hydroxypyrimidine with phosphorus pentachloride;[2595] the same reagent has been used to prepare 2,4-dichloro-5-chloromethylpyrimidine from either 2,4-dihydroxy-5-hydroxymethyl- or 5-chloromethyl-2,4-dihydroxy-pyrimidine;[2596] 2-chloro-4-trichloromethylpyrimidine* (**16**; 80%) is formed from 2-hydroxy-4-methylpyrimidine ($POCl_3 + PCl_5$ at 135°);[2597] a small yield of 2,4,5,6-tetrachloropyrimidine is obtained from 4-carboxy-5-chloro-2,6-dihydroxypyrimidine with the same reagents;[2591] and 2,4-dichloro-

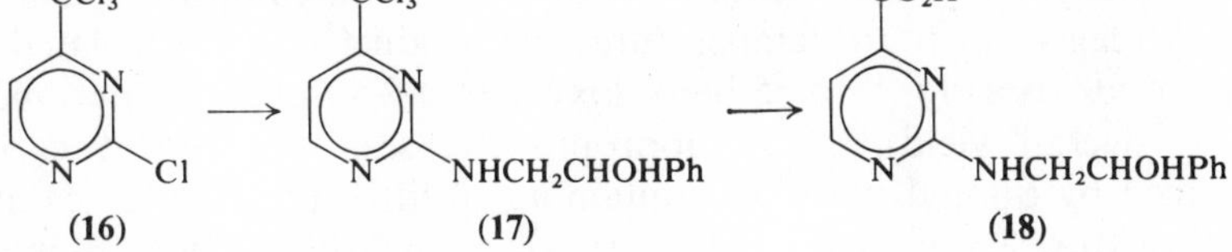

(**16**) → (**17**) → (**18**)

* J. R. Marshall and J. Walker[64] first made this compound and rightly called it 2,x,x,x-tetrachloro-4-methylpyrimidine because of lack of structural evidence. D. E. Heitmeier *et al.*[2597] now record that one chlorine atom was replaced by β-hydroxyphenethylamine and that the product (**17**) lost the remaining three on reaction with silver nitrate to give the corresponding carboxylic acid (**18**), thus proving the structure.

6-chlorocarbonylpyrimidine (64%) is formed similarly from 4-carboxy-2,6-dihydroxypyrimidine.[2591]

C. Other Ways of Preparing 2-, 4-, and 6-Chloropyrimidines (*H* 166)

The sodium salt of 5-ethoxycarbonyl-4-hydroxy-2-methylthiopyrimidine (**19**) may be converted into the corresponding 4-chloro derivative (**20**; 90%) with phosphoryl chloride in chloroform,[2598, 2599] but the free hydroxy ester (**19**) is exceptional in yielding the same chloro derivative (84%) by treatment with thionyl chloride.[2100, 2599] Another unusual reaction is direct chlorination of 2,4-dichloro-5-fluoropyrimidine (**21**, R = H) by chlorine in the presence of benzoyl peroxide and light to give 2,4,6-trichloro-5-fluoropyrimidine (**21**, R = Cl).[2600]

The preparation of simple chloropyrimidines from the corresponding aminopyrimidines by treatment with nitrous acid in concentrated hydrochloric acid continues to prove useful. The corresponding hydroxypyrimidine is invariably formed as a major by-product. In such a way have been made 2-chloro-4-methoxy(or ethoxy, etc.)-6-methyl-,[2580, 2601] 2-chloro-4,6-dimethoxy-,[2601] 2-chloro-4,6-bismethylthio-,[2601] 2-chloro-5-nitro-,[2431] 4-t-butyl-2-chloro-,[2602] 2-chloro-4-ethylthio(or other alkylthio)-6-methyl-,[2580] and 4-chloro-2,6-dimethyl-pyrimidine.[2603]

A thiocyanatopyrimidine has been converted into the corresponding chloro derivative: although bromine simply gives a 5-bromo derivative of 4,6-bis-(dimethylaminomethylenamino)-2-thiocyanatopyrimidine, chlorine in acetic acid yields 4,6-diamino-2,5-dichloropyrimidine from the same substrate.[2859]

Of recent years, F. Šorm and his colleagues have introduced chloromethylene dimethyl ammonium chloride* ($ClCH{:}N^+Me_2\,Cl^-$) as a reagent for converting hydroxy- into chloro-pyrimidines when the strongly acidic conditions resulting from the use of phosphorus halides are contraindicated.[2845–2847] Its use with a simple pyrimidine is illustrated by the conversion of 5-benzyloxymethyl-2,4-dihydroxy- into 5-benzyloxymethyl-2,4-dichloro-pyrimidine with a chloroform solution of the reagent at room temperature; the product was not isolated but converted directly into 5-benzyloxymethyl-2,4-dimethoxypyrimidine (72% overall yield).[2847] In appropriate cases, the reagent can be replaced by thionyl chloride containing a little dimethylformamide: this mixture converted 2′,3′,5′-tri-*O*-benzoyluridine into 4-chloro-1,2-dihydro-2-oxo-1-(2,3,5-tri-*O*-benzoyl-β-D-ribofuranosyl)pyrimidine in almost quantitative yield.[2846]

* This crystalline reagent may be made from dimethylformamide with phosgene, thionyl chloride, phosphorus pentachloride, or even oxalyl chloride.[2848, 2849]

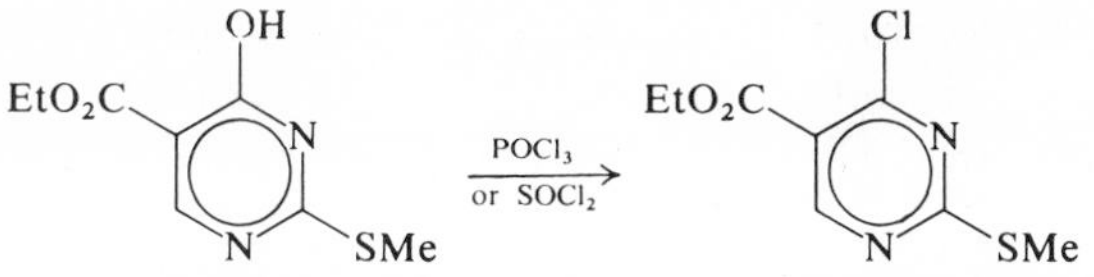

$\xrightarrow[\text{or } SOCl_2]{POCl_3}$

(19) **(20)** 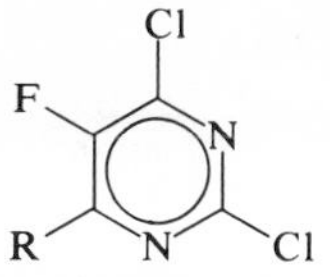 **(21)**

D. The Preparation of 2-, 4-, and 6-Bromopyrimidines (*H* 167)

For most purposes, bromopyrimidines have little advantage to chloropyrimidines as intermediates. However it has been pointed out[2604] that 2-bromopyrimidine undergoes an Ullmann reaction whereas 2-chloropyrimidine does not, and that the bromopyrimidine reacts more rapidly with trimethylamine than does the chloropyrimidine. Hence it is sometimes thought worthwhile to make such bromo(or iodo)pyrimidines.

The use of phosphoryl bromide is exemplified in the formation of tetrabromopyrimidine from 5-bromo-2,4,6-trihydroxypyrimidine,[2600] 2-bromopyrimidine (**24**; only 5% yield) from 2-hydroxypyrimidine (**22**),[2604] and of 4-bromo-5-ethoxycarbonyl-2-methylthiopyrimidine (61%).[2599] In the last example dilution of the phosphoryl bromide with toluene was advantageous,[2599] and this was also true in making 2,4-dibromopyrimidine.[2784]

A new alternative method for preparing such a bromo derivative (in rather poor yield) is by treatment of an aminopyrimidine, e.g., (**23**), with nitrous acid in the presence of an excess of bromide ion; the corresponding hydroxypyrimidine is also formed in appreciable amount. The technique is outlined for the formation of 2-bromopyrimidine (**24**; 27%),[2604] its 4,6-dimethyl derivative (24%),[2605] and 2-bromo-4-chloro-6-methylpyrimidine (16% from 2-amino-4-chloro-6-methylpyrimidine).[2605] Yields are sometimes improved by first forming a perbromide of the amine followed by addition of nitrous acid but the method is applicable only to 5-substituted aminopyrimidines because of rapid 5-bromination when the position is free. Thus 2-bromo-4,5-diethoxy- (70%), 2-bromo-4-chloro-5-ethoxy- (69%), and 2-bromo-4-chloro-5-propyl-pyrimidine (9%) are formed satisfactorily,[2606] but application to 2-amino-4-methoxypyrimidine gave a mixture of its 5-bromo derivative and 2,5-dibromo-4-methoxypyrimidine, and 2-amino-5-bromopyrimidine was the only product from 2-aminopyrimidine.[2606]

Perhaps the most satisfactory way to make bromopyrimidines may yet prove to be replacement of a chloro substituent. Thus J. F. W.

McOmie *et al.*[2603] treated 2-chloropyrimidine (**25**) with phosphorus tribromide to give 2-bromopyrimidine (**24**; 52%) and its 4,6-dimethyl derivative (58%) was made similarly.[2603]

E. The Preparation of 2-, 4-, and 6-Iodopyrimidines (*H* 168)

Iodopyrimidines are still made only from their chloro analogues by treatment with iodide ion. The reagents and techniques are illustrated in the following examples.

N OH
(**22**)
$POBr_3$ (5%)
N NH$_2$ — HBr/HNO_2 (27%) → N Br ← PBr_3 (52%) — N Cl
(**23**) (**24**) (**25**)

Hydriodic acid was used to make 4-iodo-2-methylthio- (60%),[2608] 5-ethoxycarbonyl-4-iodo-2-methylthio- (79%),[2608] 2-iodo-4,6-dimethyl- (60%),[2603] 4-iodo-6-methyl-2-phenyl- (52%),[2603] 4-iodo-6-methyl-2-methylthio- (59%),[2603] 4-iodo-2,6-dimethyl- (32%),[2603] 4-iodo-5-phenyl- (85%),[2603] and 2-iodo-4-methyl-6-phenyl-pyrimidine (5%; dehalogenation occurred to some extent in this example);[2603] potassium iodide in boiling dimethylformamide to make 2-iodo- (from 2-*bromo*-pyrimidine in 50% yield),[2607] 4-iodo-2,6-dimethoxy- (42%),[2611] and 2,4-dihydroxy-5-iodo-pyrimidine (73%);[2611] sodium iodide in glacial acetic acid to make 5-ethoxycarbonyl-4-iodo-2-methylthiopyrimidine (25%; *cf.* above);[2599] and sodium iodide with hydriodic acid in acetone to make 4,6-di-iodopyrimidine (39%).[2609]

F. The Preparation of 2-, 4-, and 6-Fluoropyrimidines (*New*)

A respectable number of such fluoropyrimidines have now been made from the corresponding chloropyrimidines with silver fluoride, potassium fluoride, or sulphur tetrafluoride. The first of these reagents appears to be the most convenient for laboratory preparations, but repeated treatments are often required. The following fluoropyrimidines were made as indicated: 2,4,6-trifluoro- (AgF,

> 70%;[2610, 2612, 2618] KF at 260° in presence of Sb_2O_3, 48%;[2613, 2614] KF at 300°, 90%[2615]), 2,4- and 4,6-difluoro- (SF_4 at 150°, *ca.* 70%),[2157] 2,4-difluoro-6-methyl- (KF + Sb_2O_3 at 240°, *ca.* 30%),[2613] 5-bromo-2,4-difluoro-6-methyl- (KF + Sb_2O_3),[2613] 4,6-difluoro-5-nitro- (AgF, *ca.* 80%),[2455] and 5-chloro-2,4,6-trifluoro-pyrimidine (KF* at 340°, 35%;[2616] AgF, *ca.* 70% yield[2618]). Other examples using potassium fluoride in dimethylformamide are described.[3520]

When 2,4-dimethoxypyrimidin-6-yl trimethyl ammonium chloride (**26**) is treated with potassium fluoride in diethylene glycol, trimethylamine is lost, and 4-fluoro-2,6-dimethoxypyrimidine (**27**) results in good yield.[2611] Sodium fluoride failed to react with 2-methylsulphonylpyrimidine.[2619] 4-Amino-6-methylpyrimidin-2-yl trimethyl ammonium chloride reacted with butanolic potassium fluoride to give 4-amino-2-fluoro-6-methylpyrimidine in rather poor yield.[3520]

2. The Preparation of 5-Halogenopyrimidines (*H* 168)

Previously unused routes to 5-halogenopyrimidines include direct fluorination and use of *N*-iodosuccinimide for direct iodination.

A. 5-Halogenopyrimidines by Direct Halogenation (*H* 169)

Direct halogenation remains the best way to make 5-halogenopyrimidines, and even direct fluorination has now been achieved: 2,4,6-trifluoropyrimidine (**28**) and silver difluoride in triperfluorobutylamine at 90° gives tetrafluoropyrimidine (**29**) smoothly.[2612, 2618]

Chlorination using elemental chlorine is seldom easy in practice and yields are rather indifferent. However, 2,4-diamino-6-chloropyrimidine gives its 5-chloro derivative where pH is carefully controlled;[2621] 1-benzyloxyuracil,[3464] and 2-amino-4-methoxy-[2571] and 4,6-dihydroxy-2-phenyl-pyrimidine[2575] may be 5-chlorinated in acidic solution; methanolic chlorine may be used.[3531] Sulphuryl chloride is often more satisfactory. Thus 5-chloro-2,4-dihydroxy-6-methylpyrimidine resulted in 97% yield using sulphuryl chloride in acetic acid with a little ferric chloride;[2236] and 5-chloro-2,4,6-trihydroxypyrimidine (80%; in water),[2622] 5-chloro-4-heptadecyl-2,6-dihydroxypyrimidine (89%; in acetic acid and acetic anhydride with some ferric chloride),[2226] and some of its 6-alkyl homologues[2226] were all made with sulphuryl

* Also 15% of tetrafluoropyrimidine,[2616] which is the only product when CsF at 340°,[2616] or KF at 500°(!) is used.[2617] However, tetrafluoropyrimidine (57%) is best made from tetrachloropyrimidine and KF at 410°.[3521]

(26) $\xrightarrow{KF}$ (27)

(28) $\xrightarrow{AgF_2}$ (29)

chloride. *N*-Chlorosuccinimide in chloroform or acetic acid may also be used to make, for example, 5-chloro-4-hydroxy-6-methyl-2-methylthio- (51%; benzoyl peroxide),[2160] 2(or 4)-amino-5-chloro-4(or 2)-hydroxy- (43%; 69%),[2610] 2-amino-5-chloro-4-methoxy(or *p*-chloroanilino)-6-methyl- (75%; 100%),[2623] 2-amino-4,5-dichloro-6-methyl- (100%),[2623] 2(or 4)-amino-5-chloro-4,6(or 2,6)-dimethyl- (81%; 82%),[2623] 4-amino-5-chloro-6-methyl- (50%),[2623] and 5-chloro-4-hydroxy-6-methyl-pyrimidine (70%);[2623] and their analogues.[2623]

Bromination, using bromine in acetic acid or other solvent, usually gives >80% yield and is exemplified in the formation of 2(or 4)-amino-5-bromo-4(or 2)-methylamino-,[2624] 5-bromo-2-dimethylamino-4-methylamino-,[2624] 5-bromo-2,4-dihydroxy- (71%),[2182] 5-bromo-2,4-dihydroxy-6-methyl- (86%),[2625] 2,4-diamino-5-bromo-6-chloro- (88%; in acetic acid/sodium acetate at 90°;[2443] 62% in aq. methanolic sodium bicarbonate[2621]), 5-bromo-4-t-butyl-6-hydroxy- (75%),[2574] 5-bromo-4-hydroxy-6-methyl(or phenyl)- (91%; 69%),[2574] 5-bromo-2-*o*(or *p*)-chlorobenzylthio-4-hydroxy- (in pyridine),[2255] 2-amino-5-bromo-4-methyl(or t-butyl)- (72%, 56%; in an aqueous suspension of calcium carbonate),[2602] 2-amino-5-bromo-4,6-dimethyl- (86%; aq. $CaCO_3$),[2626] 5-bromo-2-methylamino- (90%; aq. $CaCO_3$)[2627] 4-amino-5-bromo-2,6-dihydroxy- (90% formamide),[2628] and 5-bromo-2,4,6-trihydroxy-pyrimidine (78%; formamide);[2628] as well as other examples.[2255, 2575, 2624, 2628, 2655, 2880, 3464] Bromination with *N*-bromosuccinimide (with or without added peroxide) also has a certain following; it is most useful in the presence of acid-sensitive groups.* In this

* Normal 5-bromination of 4-methoxy-6-methyl-2-methylthiopyrimidine occurs with *N*-bromosuccinimide or with bromine in acetic acid at 20°. However, the latter reagent at 100° yields an isomer, formulated as 5-bromo-1,2-dihydro-6-methoxy-1,4-dimethyl-2-thiopyrimidine 'or an isomer'.[2575] A more likely structure might be 5-bromo-1,6-dihydro-1,4-dimethyl-2-methylthio-6-oxopyrimidine.

way there have been made 5-bromo derivatives of 4,6-dimethoxy-,[2575] 2,4,6-trimethoxy-,[2575] 2-amino-4-methoxy-6-methyl-,[2629] 1,6-dihydro-1-methyl-6-oxo-,[2630] 2-amino-4-anilino-6-methyl-,[2631] and 4,6-dimethyl-2-morpholino-pyrimidine.[2623] Other examples abound.[2575, 2623, 2631, 2842]

Direct iodination is best done with *N*-iodosuccinimide as exemplified in the papers of T. Nishiwaki[2623, 2629] where procedures for so making 4-hydroxy-5-iodo-6-methylpyrimidine, 2,4-dihydroxy-5-iodopyrimidine (**31**), its 6-methyl derivative, and 2-amino-5-iodo-4,6-dimethylpyrimidine are described. Iodine monochloride has been used to make 2,4-dihydroxy-5-iodopyrimidine (**31**) from uracil (**30**),[2632] for obtaining 2-amino-5-iodo-4-methoxypyrimidine,[2571] and for converting 1(or 3)-benzyloxyuracil into the 5-iodo derivatives.[3464] More direct use of iodine gave rise to 5-iodo-2-methylaminopyrimidine (iodine in aq. dioxan with some mercuric acetate),[2633] 4-hydroxy-5-iodopyrimidine (I/NaOH),[2191] 2-amino-4-hydroxy-5-iodopyrimidine,[2588] 1- and 3-hydroxy-5-iodouracil (I/NaOH; phosphorus trichloride removed the *N*-oxide groupings),[3464] and related pyrimidines.[2634, 2655]

The formation of 5,5-dichlorobarbituric acid (**32**; R = Cl) has been described (H_2O_2 + HCl on 2,4,6-trihydroxypyrimidine),[2622] and a paper on 5-chloro-5-nitrobarbituric acid (**32**; R = NO_2) has appeared.[2635] When 1-benzyloxy-5-bromouracil was treated with bromine

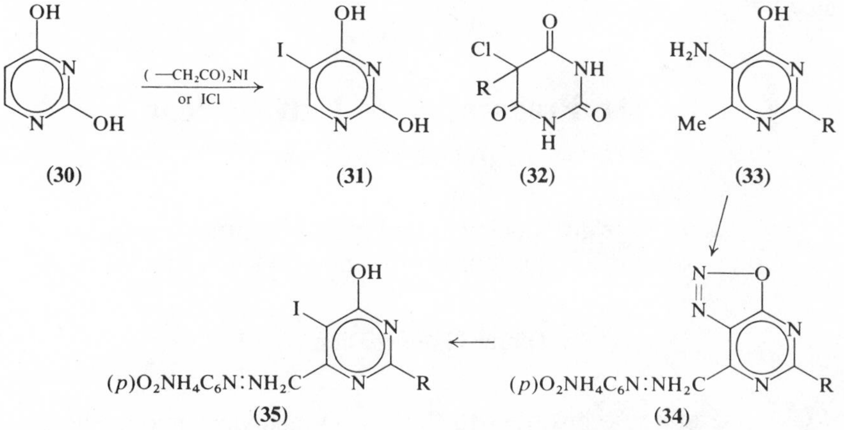

$(—CH_2CO)_2NI$ or ICl

(30) **(31)** **(32)** **(33)**

(35) **(34)**

in methanol, 1-benzyloxy-5,5-dibromo-5,6-dihydro-6-methoxyuracil was formed; the 3-benzyloxy isomer was prepared similarly, but analogous treatment with chlorine gave aliphatic products.[3464] Little else has been added to chemical knowledge (*cf. H* 172) of such compounds; most other relevant work is rather biochemical.[2636]

C. Other Methods of Preparing 5-Halogenopyrimidines (*H* 175)

The Sandmeyer reaction has been used to convert 5-amino-2,4-dihydroxy-6-methylpyrimidine into its 5-iodo (potassium iodide), chloro (cuprous chloride), and bromo (cuprous bromide) analogues in reasonable yield.[3507]

The normal primary syntheses of 5-halogenopyrimidines have been covered (Ch. II; Ch. III), but one interesting and rather specialized route must be mentioned here. When 5-amino-2,4-dihydroxy-6-methylpyrimidine (**33**, R = OH) is coupled with diazotised *p*-nitroaniline, and then treated with nitrous acid, 6-hydroxy-4-*p*-nitrobenzeneazomethyl-1-oxa-2,3,5,7-tetra-azaindene (**34**, R = OH) is formed (by a route which is discussed). Treatment of this with aq. hydriodic acid gives 2,4-dihydroxy-5-iodo-6-*p*-nitrobenzeneazomethylpyrimidine (**35**, R = OH). The amino analogues (**34**, R = NH_2; **35**, R = NH_2) were made similarly,[2453] but attempts to prepare 1-thia analogues of compound (**34**) failed.[2453] 1-Benzyloxy-5-fluorouracil has been made by two primary syntheses; hydrogenolysis gave the 1-hydroxy analogue.[3464]

A particularly easy change of 5-halogeno substituent is the conversion of 5-bromo- into 5-chloro-4,6-dihydroxypyrimidine which takes place in 82% yield by warming with hydrogen chloride in dimethylformamide for 1 min. at 100°. The chloro compound is best made in this way.[3181]

3. The Preparation of Extranuclear Halogenopyrimidines (*H* 176)

New examples are given below for all types of syntheses used to make pyrimidines bearing a halogeno substituent on a side chain.

A. By Direct Halogenation (*H* 176)

Chlorine at 150° converted to 2-chloro-4-methylpyrimidine[2419] and related compounds[2420, 2421] into their corresponding trichloromethyl analogues. Direct ω-halogenation has also been done with *N*-bromosuccinimide to yield 5-bromomethyl-4,6-dichloro-2-chloromethylpyrimidine (from its 5-methyl analogue);[2238] 2-diacetylamino-4-dibromomethylpyrimidine;[2955] and 2(or 4; or 5)bromomethyl-4,5,6-(or 2,5,6;

or 2,4,6)trichloropyrimidine;[2236] elemental bromine has been used to convert 5-acetyl-2,4-dihydroxypyrimidine (**36**, R = R′ = H) into the corresponding bromoacetyl derivative (**36**; R = H, R′ = Br),[2218] and thence into the dibromoacetyl derivative (**36**; R = R′ = Br).[2637] Bromination of 4,5-diamino-2-methylpyrimidine failed under vigorous conditions, probably owing to powerful deactivation of the methyl group by the amino substituents.[2562] Halogenation of activated methylene groups attached to pyrimidines is known.[2786–2788] When 4-methylpyrimidine is boiled with iodine in pyridine, pyrimidin-4-ylmethylpyridinium iodide (55%) results, presumably *via* 4-iodomethylpyrimidine.[2834]

B. From the Corresponding Hydroxyalkylpyrimidine (*H* 178)

Thionyl chloride has been increasingly used to convert alcoholic hydroxy groups into chloro substituents. Simple examples include the formation of 5-β-chloroethyl-4-hydroxy-2-mercapto-6-methylpyrimidine (91%),[2638] 4-amino-5-chloromethyl-2-hydroxypyrimidine (95%),[2639] 5-chloromethyl-2,4-dihydroxypyrimidine (100%),[2160, 2640] and 2-amino-5-β-chloroethyl-4-hydroxypyrimidine.[2514] Less simple compounds also prepared with thionyl chloride include 2-β-chloroethylamino-,* [2642] 4-β-chloroethylamino-2-hydroxy-,[2910, cf. 2641.] 4-β-

OH
R′RHCOC
N
N OH
(36)

N
N NHCH$_2$CH$_2$Cl
(37)

→

N
N N
H$_2$C——CH$_2$
(38)

N(CH$_2$CH$_2$Cl)$_2$
N
N SH
(39)

→

CH$_2$CH$_2$Cl
N——CH$_2$
CH$_2$
N
N S
(40)

* This pyrimidine (**37**) has m.p. 65–66°.[2642] It easily cyclizes thermally to the isomeric hydrochloride of 1,3a,7-triazindane (**38**) with m.p. *ca.* 280°,[2640] which appears to have been formulated elsewhere[2644] as its isomer (**37**). Similarly, a compound formulated[2648] as 4-bis-(β-chloroethyl)amino-2-mercaptopyrimidine (**39**) was later shown[2649] to be the hydrochloride of 1-β-chloroethyl-1,4-dihydro-4-thio-1,3a,5-triazaindane (**40**); its 4-oxo isologue is made by the same route.[2910]

chloroethylamino-1,2-dihydro-1-methyl-2-oxo-,[2641, cf. 2910] 5-β-chloroethylamino-2,4-dihydroxy-,[2643] 4-bis-(β-chloroethyl)amino-2,6-dihydroxy(or dimethoxy)-,[2645, 2885] 5-bis-(β-chloroethyl)aminomethyl-2,4-dihydroxy-,[2646, 2647] and 5-(*N*-β-chloroethyl-*N*-β-fluoroethylamino)methyl-2,4-dihydroxy-pyrimidine;[2625] as well as 5-bis-(2-chloroethyl)amino-1(or 3)methyluracil,[2643] and other compounds.[2828] (Are any bicyclic?)*

Hydrobromic acid, and even hydrochloric and hydriodic acids, are also convenient in making extranuclear halogeno derivatives such as 5-β-bromoethyl-2,4-dihydroxy-,[2514] 5-bromo(or chloro)methyl-2,4-dihydroxy-,[2160, 2650] 2,4-dihydroxy-5-iodomethyl-,[2650] 5-bromomethyl-2-methyl-,[2651, 2654] 5-bromomethyl-2,4-dihydroxy-6-methyl-,[2652] 4-amino-5-bromomethyl-2-trifluoromethyl-,[2516] and 5-bromomethyl-2-methyl-4-methylamino-pyrimidine;[2653] also related compounds.[2651, 2980]

The capacity of phosphorus halides to replace ω-hydroxy groups at the same time as 2-, 4-, or 6-hydroxy groups has been mentioned. Thus were made 4-amino-6-chloro-2-chloromethyl- ($POCl_3$)[2290] 4-chloro-2-chloromethyl- ($POCl_3$),[2192] 4,6-dichloro-2-chloromethyl- ($POCl_3$),[2237] 4,6-dichloro-2-chloromethyl-5-methyl- ($POCl_3/PCl_5$),[2238] and 2,4-dichloro-5-chloromethyl-pyrimidine ($POCl_3/PCl_5$).[2596]

C. From the Corresponding Alkoxyalkylpyrimidine (*H* 179)

The few recent examples of the preparative route are typified by the conversion of 5-methoxymethyl- into 5-bromomethyl-pyrimidine by hydrobromic acid,[2579] 4-amino-5-ethoxymethyl- into 4-amino-5-chloromethyl-2-methylpyrimidine by alcoholic hydrogen chloride,[2231] and 4-amino-5-ethoxymethyl- into 4-amino-5-bromomethyl-2-ethylpyrimidine with hydrobromic acid;[2270] also by the rupture of 4-amino-5-ethoxymethyl- to 4-amino-5-iodomethyl-2-methylpyrimidine by hydriodic acid.[2836]

D. By Direct Synthesis (*H* 179)

A number of primary syntheses leading to ω-halogenopyrimidines have been discussed in Chapters II and III. Typical examples are the condensation of 3-benzylideneacetylacetone with *p*-chlorobenzaldehyde

* Such cyclization occurs readily: even β-hydroxyethylamino[2910] and related derivatives[2909] undergo ring closure to analogues of (**40**) under acidic conditions (*cf.* 2700, 2911).

and ammonia to give finally 5-benzyl-2-*p*-chlorophenyl-4,6-dimethylpyrimidine (*cf.* Ch. III, Sect. 5.F(2));[2372] the condensation of ethyl γ-fluoroacetoacetate with formamidine to give 4-fluoromethyl-6-hydroxypyrimidine;[2518] the condensation of ethyl $\gamma\gamma$-dichloroacetoacetate with acetamidine or thiourea to give, respectively, 4-dichloromethyl-6-hydroxy-2-methyl(or mercapto)pyrimidine;[2518] and the reaction of *NN'*-di-*p*-chlorophenyl carbodiimide with ethylmalonyl dichloride to give 1,3-di-*p*-chlorophenyl-5-ethyl-4-hydroxy-1,2,3,6-tetrahydro-2,6-dioxopyrimidine.[2535]

E. By Other Means (*H* 180)

Extranuclear halogeno substituents may be introduced by making a halogen-containing derivative of an existing pyrimidine. For example, 5-chloromethyl-2,4-dihydroxypyrimidine undergoes aminolysis by bis-(β-chloroethyl)amine to give 5-bis-(β-chloroethyl)aminomethyl-2,4-dihydroxypyrimidine (**41**; R = R' = Cl),[2640] by bis-(β-fluoroethyl)-amine to give the analogue (**41**; R = R' = F),[2656] and by *N*-β-chloroethyl-*N*-β-fluoroethylamine to give another analogue (**41**; R = Cl, R' = F).[2657] The compound (**41**; R = R' = Cl) has also been made by allowing uracil to react under Mannich conditions with formaldehyde and bis-(β-chloroethyl)amine,[2658] but the melting points differ by *ca.* 20°. Another such aminolysis leads to 2,4-bis-*p*-chloroanilinopyrimidine and related compounds from 2,4-dichloropyrimidine.[2659] The potential extranuclear halogen may be supplied to the pyrimidine in an acylating agent, to yield such compounds as 4-amino-5-$\alpha\alpha$-dichloropropionamidopyrimidine (**42**; from $\alpha\alpha$-dichloropropionyl chloride and 4,5-diaminopyrimidine),[2660] 4-amino-2-hydroxy-5-trifluoroacetamidopyrimidine,[2661] 4-*p*-bromobenzamido-2,6-dimethoxypyrimidine,[2662] and related derivatives.[2662, 2663, 2751]

Di- and trifluoromethylpyrimidines may be made from carboxy- and formylpyrimidines, respectively, by treatment with sulphur tetrafluoride. In this way, 5-carboxy- gave 5-trifluoromethyl-2,4-dihydroxypyrimidine (**43**; R = F) in 77% yield,[2264, 2664] and 5-formyl- gave 5-difluoromethyl-2,4-dihydroxypyrimidine (**43**; R = H) in 60% yield.[2664, 2665]

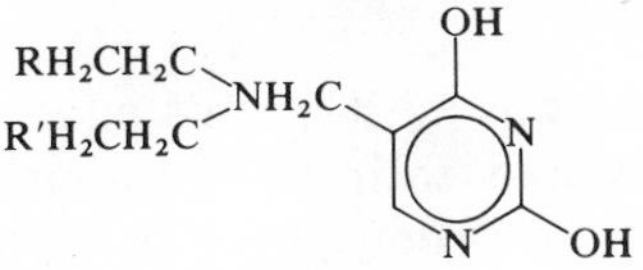

RH_2CH_2C, $R'H_2CH_2C$ NH_2C; OH; N; N; OH

(41)

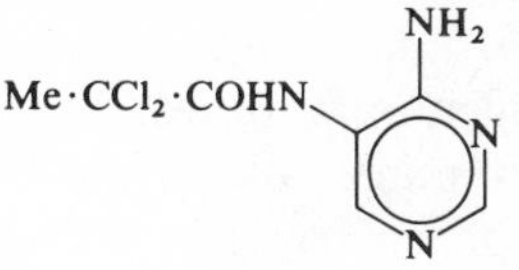

$Me \cdot CCl_2 \cdot COHN$; NH_2; N; N

(42)

RF_2C; OH; N; N; OH

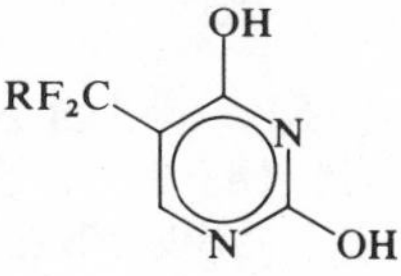

(43)

Both reactions are reversed completely even in aq. sodium bicarbonate, the former in 24 hr.,[2667] the latter within 2 min.[2666]

The replacement of an ω-amino group by a chloro substituent can be achieved by treatment with nitrous acid in the presence of an excess of chloride ion. In this way 4-amino-5-chloromethyl-2-heptafluoropropyl (or pentafluoroethyl or trifluoromethyl)pyrimidine may be derived from the corresponding 5-aminomethyl derivatives.[2268]

Halogen interchange is seen in the conversion of 5-chloromethyl-2,4-dihydroxypyrimidine into its iodo analogue by hydriodic acid in 89% yield,[2650] and in the change from 2,4-dichloro-5-chloromethyl- to 2,4-dichloro-5-iodomethyl-pyrimidine in 99% yield using sodium iodide in acetone.[2596]

An interesting route to pyrimidines bearing β-chloroethylamino groups is provided by nucleophilic displacement of a 2- or 4-chloro group by aziridine followed by fission of the three-membered ring with dry ethereal hydrogen chloride. Thus 2,4-dichloro-5-fluoropyrimidine gave 2,4-diaziridino-5-fluoropyrimidine which underwent fission to 2,4-bis-(β-chloroethylamino)-5-fluoropyrimidine in 97% yield.[2570] This synthesis will undoubtedly be extended.

Uracil undergoes 5-chloromethylation in reasonable yield by treatment with aqueous formaldehyde and hydrochloric acid.[2071, 2115, 3193]

5. Reactions of 2-, 4-, and 6-Halgenopyrimidines (*H* 183)

Pyrimidines bearing active halogeno substituents have held their place as the most useful intermediates in the series. On the other hand, it is now recognized that the alkylsulphonyl and, to a lesser extent, the alkylsulphinyl groups are potential rivals; they are sometimes more easily made, more stable, and undergo nucleophilic displacement even more easily than do halogeno substituents in the pyrimidine[2619, 2668] and related series.[2669–2672]

A. Removal of 2-, 4-, and 6-Halogens (*H* 183)

Removal of halogen in favour of hydrogen was initially used mainly (*H* 183 *et seq.*) to make alkylpyrimidines and their simple amino derivatives, but emphasis has now shifted to the production of a wider variety of derivatives.

The older type is exemplified in the preparation of 2-chloro- from 2,4-dichloro-pyrimidine by zinc and aqueous ammonium chloride,[2231] 2-chloro- from 2,4,6-trichloro-5-methylpyrimidine in 85% yield using zinc,[2630] 2-amino-4-dimethylaminopyrimidine from its 6-chloro derivative (Pd*/H_2; 80%),[2627] several diamines analogous to the last,[2288, 2624] 5-amino-4-hydrazinopyrimidine from its 6-chloro derivative (Pd/H_2; 86%),[2673] 4-benzylaminopyrimidine and its 5-methyl derivative from their respective 2-chloro derivatives (Pd/H_2; 73%),[2674] and 5-amino-4-benzylaminopyrimidine from its 6-chloro derivative (Pd/H_2 + MgO; 70%).[2675]

A wider spectrum of application for dehalogenation is evident in the formation of the following pyrimidines, generally using catalytic hydrogenation over palladium in the presence of a base, but occasionally (where stated) zinc dust: 2-chloro-5-hydroxypyrimidine from its 4-chloro derivative (zinc),[2585] 2-ethyl-4-hydroxy- from 4-chloro-2-ethyl-6-hydroxypyrimidine,[2554] 4-amino-2-hydroxypyrimidine-3-oxide (**45**) from 4-amino-3-benzyloxy-6-chloro-2,3-dihydro-2-oxopyrimidine (**44**),[2262] 5-benzyloxy-2-methylthiopyrimidine (zinc),[2585] 4-dimethylamino-2-methoxypyrimidine,[2676] 1-benzyloxyuracil from its 5-iodo derivative (no base),[3464] 4-amino-2-ethoxy(or methoxy)pyrimidine (NaOH or $CaCO_3$; 74–84%),[2677, 2678] 5-fluoro-2-methylthiopyrimidine (zinc; 61%),[2679] 5-ethyl-2,4-dimethoxypyrimidine (MgO),[2197] 2-amino-5-phenoxypyrimidine (zinc; 35%),[2212] 5-methoxymethylpyrimidine (**46**) from its 2,4-dichloro derivative (**47**) ($MeCO_2Na$),[2579] 5-cyanopyrimidine from its 2,4-dichloro derivative (CaO; 9%),[2680] 5-cyano-4-dimethylamino-2-methylpyrimidine ($CaCO_3$; 86%),[2681] 5-ethoxycarbonylpyrimidine from its 2,4-dichloro derivative (CaO; 63%),[2680] the same compound from its 4-chloro derivative (MgO; 75%),[2323] its 2-methyl derivative similarly (67%),[2681] 5-acetylpyrimidine from its 2,4-dichloro derivative (MgO; 47%),[2577] 5-formylpyrimidine (79%),[2732] 5-ethoxycarbonylmethylpyrimidine from its 4-chloro derivative similarly (53%),[2577] its 4-methyl derivative,[2576] 4-sulphanilamidopyrimidine from its 6-chloro derivative (NaOH; 18%),[2682] and other examples,[2231, 2683] including some where subsequent reduction of the ring has taken place.[2162, 2684–2686]

With the advent of n.m.r. and mass spectral studies, it is often convenient to insert deuterium atoms at specific points. This can be done by shaking a chloropyrimidine with deuterium in the presence of a palladium catalyst. The technique is illustrated in the formation of 4,5-diamino-2-deuteropyrimidine, its 6-deutero isomer, and 4,5-di-

* The convenient preparation of palladium-on-charcoal catalyst, which can be stored active under water, has been described.[2357, 2930]

amino-6-deutero-2-methylpyrimidine from their respective chloro analogues; they were converted into deuteropurines and pteridines.[2835]

A unique contraction of the pyrimidine to the pyrrole ring during zinc dust dehalogenation has been described recently.[3473] Thus brief treatment of 4-chloro-6-methyl-2-phenylpyrimidine with zinc dust in hot aqueous acetic acid gave some of the expected 4-methyl-2-phenylpyrimidine, but on continued treatment this was transformed into a mixture of 2- and 4-methyl-5-phenylpyrrole. The same reagents converted 4,6-dichloro-2-*p*-chlorophenylpyrimidine into 2-*p*-chlorophenylpyrrole, and 5-methyl-2-phenylpyrimidine into 3-methyl-5-phenylpyrrole.[3473]

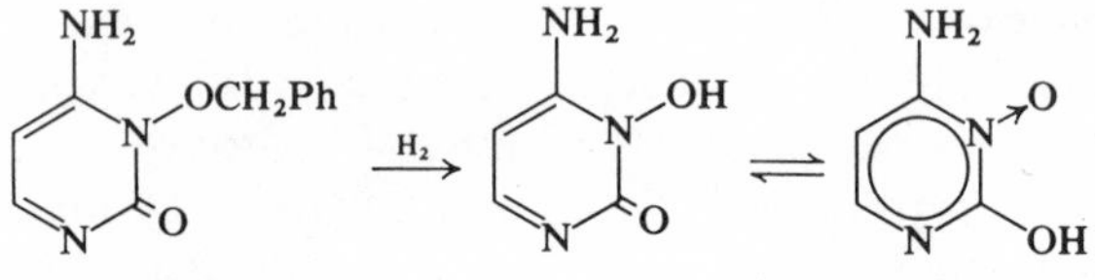

NH_2 OCH_2Ph N N O $\xrightarrow{H_2}$ NH_2 OH N N O $\rightleftharpoons$ NH_2 O N N OH

(44) **(45)**

$MeOH_2C$ N N $\xleftarrow{H_2}$ $MeOH_2C$ Cl N N Cl

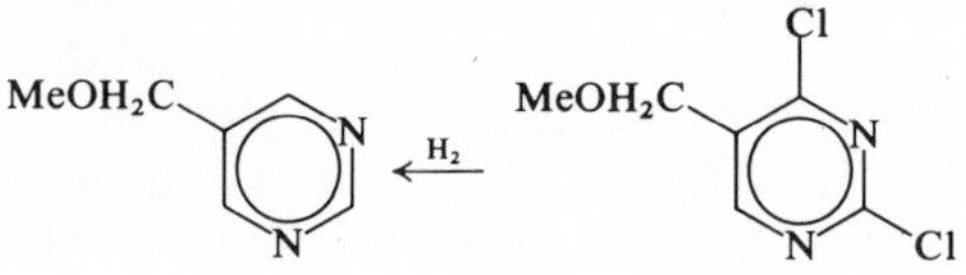

(46) **(47)**

The indirect removal of chloro substituents (*H* 187) is now possible under oxidative conditions permitting retention of groups, such as nitro or azo, which are altered by the usual reductive procedures for dehalogenation. The chloro substituent is replaced by a hydrazino group which is then removed oxidatively as nitrogen. The various techniques and reagents for the latter step have been investigated by A. Albert and G. Catterall in a variety of heterocycles,[2687] and silver oxide in methanol or silver acetate have emerged as the most generally effective reagents for sensitive compounds in the pyrimidine series. This route has been used to convert 4,6-dichloro-5-nitropyrimidine (**48**) into 5-nitropyrimidine (**50**) *via* the intermediate (**49**);[2688] in the analogous preparations of 2-methyl(or benzyl)-5-nitropyrimidine;[2688] in making 4-amino(or methoxy)-5-nitropyrimidine and 4-amino-5-nitro-2-styrylpyrimidine from their 6-chloro derivatives;[2562] in making 5-bromo-4-t-butyl(or methyl or phenyl)pyrimidine from 6-chloro derivatives;[2574] and in making 2,4-dimethylpyrimidine,[2554] 5-nitropyrimidine from its 4-hydrazino derivative,[2858] and 4-methoxy-2-methyl-5-nitropyrimidine from its hydrazino derivative.[3312] Yields are generally improved by using silver oxide in a secondary or tertiary alcohol instead of in a primary alcohol.[1391]

(48) $\xrightarrow{NH_2NH_2}$ (49) $\xrightarrow{AgO}$ (50)

B. Replacement of 2-, 4-, and 6-Halogens by Amino Groups (*H* 187)

The aminolysis of halogenopyrimidines has been so widely used in syntheses that only the more significant recent examples can be detailed below. Despite this apparent wealth of information only a few quantitative data[2690–2693] on the effect of position, substitution,* and type of amine on the rate of aminolysis followed the valuable but limited contributions of N. B. Chapman's group (*H* 181). In a wider field, the ease of nucleophilic displacements in nitrogen heterocycles has been approached pragmatically by R. G. Shepherd and J. L. Fedrick,[2689] who reviewed existing qualitative data with rare insight. In addition, a start has now been made in Canberra to systematically measure displacement rates for halogeno, alkoxy, alkylthio, alkylsulphonyl, alkylsulphinyl, and other groups from a variety of heterocyclic systems by a number of amines and other nucleophiles; the results to 1967 have been summarized,[2668] and as applied to pyrimidines, will be mentioned at appropriate points in this *Supplement*.

(1) *Aminolysis*† *of Simple Halogenopyrimidines* (*H* 188)

(a) *With One Active Halogen* (*H* 188). From what has been written above, it is not surprising that the very ubiquity of aminolysis of

* V. P. Mamaev and his colleagues record[2693] that piperidine reacts with the following substituted 2-chloropyrimidines at the given rates relative to unsubstituted 2-chloropyrimidine: 4-phenyl, 46%; 4-methyl, 43%; 4,6-dimethyl, 22%; 4,6-diphenyl, 16%; and 5-methyl-4,6-diphenyl, 5%. The same order is maintained when methoxide is the nucleophile.[2692]

† The term 'aminolysis' seems more appropriate than 'amination' used in the original work. The latter term is now reserved for the direct replacement of hydrogen by an amine residue, e.g., by using sodium amide.

chloropyrimidines has led to widespread guesswork in seeking optimum conditions and hence to poor yields, either from incomplete reaction or from the formation of by-products in a too vigorous reaction. This led J. M. Lyall to develop a simple experimental method for predicting optimum conditions of temperature and time for such aminolyses.[2573] For a meaningful prediction it was necessary to know in advance the order of the reaction, its rate constant, and the temperature coefficient of the latter. A first-order reaction was assured by a reasonable excess of amine; the rate constant could be read from an existing nomograph[2696] after determining the percentage completed at a given temperature in a given time by titration of liberated chloride ion; and for practical purposes the temperature coefficient proved to be constant within a defined framework.[2573] Prediction of optimum conditions in any particular case now requires only a single small-scale experiment culminating in a titration of chloride ion; the truth and utility of this procedure has been proven in a variety of examples.[2334, 2573, 2694]

Some interesting and important facts emerge[2668] from the published rates for aminolysis of simple chloropyrimidines under preparative conditions.[2573, 2694, 2695] The rate constant for aminolysis of a given chloropyrimidine by an n-alkylamine is almost unaffected by increasing the chain length or by γ-branching of the chain. A β-branch has a small, and an α-branch a profound slowing effect: one α-branch reduces the rate to *ca.* 5%, and two such branches to *ca.* 0.1% of that for the corresponding n-alkylamine. Di-n-alkylamines approximate in rate to primary amines with an α-branch. A 4-chloropyrimidine is usually a little more reactive than its 2-chloro isomer, but inversion of this relative reactivity takes place at the isokinetic temperature, which may or may not fall within the temperature range over which measurements are possible.[2695] The general effect of substitution in the pyrimidine ring is exemplified in the following approximate relative reactivities:[2668] 2-chloro-4,6-dimethylpyrimidine, 1; 2-chloropyrimidine, 10; 5-bromo-2-chloropyrimidine, 200; and 2-chloro-5-nitropyrimidine, 3,000,000. The addition of copper powder or copper salts appears to have no effect on the rate of aminolysis.[2573]

The kinetics of aminolysis of 4-chloro-2(and 6)-methyl-, 2-t-butyl-4-chloro, and 4-t-butyl-6-chloro-pyrimidine by piperidine have been measured in toluene and in ethanol. A marked steric hindrance to solvation of the aza groups seems to be evident in the t-butyl derivatives.[3754]

Preparative aminolysis of simple chloropyrimidines is exemplified in the formation of the following amino derivatives from the corresponding chloropyrimidines with 3 moles of amine (without solvent), under optimum conditions determined as above:[2573, 2694] 2-butylamino- (40°/60 min.; 82%), 2-amylamino- (40°/90 min.; 80%), 2-hexylamino- (40°/75 min.; 80%), 2-ethyl(to heptyl)amino-4,6-dimethyl-, 4-ethyl(to hexyl; also decyl)amino-2,6-dimethyl, 2-isopropyl(to isoamyl)amino-4,6-dimethyl-, 4-isopropyl(to isoamyl)amino-2,6-dimethyl-, 2(or 4)-s-butylamino-4,6(or 2,6)dimethyl- (100°/3.5 or 3 hr.; 79, 75%), 2-(or 4)-t-butylamino-4,6(or 2,6)-dimethyl- (150°/4.5 or 6 hr.; 61, 55%), 2-diethyl(or dipropyl)amino-4,6-dimethyl- (130°/1 hr.; 75, 83%), and 2-t-butylamino-pyrimidine, as well as by other instances.[2573, 2694]

Aminopyrimidines prepared by aminolysis in a solvent, but not necessarily under optimum conditions, are typified by 4-butylamino-,[2697] 2-ethyl(or propyl, or amyl, or heptyl)amino-,[2627] 2-benzyl(or isopropyl)amino-,[2626] 4-dimethylamino-2-methyl-,[2698] 4-t-butyl(or methyl)-2-methylamino-,[2602] 4,6-dimethyl-2-piperidino(or 4′-ethylpiperidino),[2623, 2699] 4,6-dimethyl-2-morpholino-,[2623] 2,4-dimethyl-6-morpholino-,[2695] 4-amino-2-ethyl-,[2554] 4-amino-6-butyl-,[2574] 4,6-dimethyl-2-methylamino-5-propyl(or prop-1′-ynyl or prop-2′-ynyl*)-,[2700] 5-allyl(or allenyl*)-4,6-dimethyl-2-methylamino-,[2700] 4-dimethylamino-5,6-tetramethylene-,[2183] 4-methylamino-5-phenyl-,[2327] 4-phenethylamino-,[2231] 2(or 4)-β-hydroxyethylamino-,[2642, 2644] 2-ββ-diethoxyethylamino-,[2642] 2-ββ-dibenzyl-β-hydroxyethylamino-,[2597] and other substituted-pyrimidines.[2183, 2597, 2623, 2642, 2644, 2701, 2702, 3247]

(b) *With Two or More Active Halogens* (*H* 188). There is little new on the reactions of dichloropyrimidines with ammonia, but welcome information on the selective alkylaminolysis of 2,4-dichloropyrimidines has emerged (*cf. H* 190). Treatment of 2,4-dichloropyrimidine with 2.5 moles of methylamine in ethanol at 20° gave a mixture (separable by steam distillation) of 2-chloro-4- and 4-chloro-2-methylamino-pyrimidine with the former predominating;[2288] in contrast 2,4-dichloropyrimidine and aq. butanolic benzylamine (2 moles) at 100° apparently gave only 4-benzylamino-2-chloropyrimidine,[2674] and 2,4-dichloro-5-methylpyrimidine with 4 moles of aq. methylamine or dimethylamine gave 2-chloro-5-methyl-4-methylamino- and 2-chloro-4-dimethylamino-5-methyl-pyrimidine, respectively, without contamination by isomers.[2703] However, aq. dimethylamine in excess at 20° gave 2,4-bisdimethylaminopyrimidine,[2676] and 2,4-dichloro-6-methylpyrimidine

* During methylaminolysis of 2-chloro-4,6-dimethyl-5-prop-2′-ynylpyrimidine (**51**), a partial isomerization led to a mixture of the prop-2′-ynylpyrimidine (**52**) and its allenyl isomer (**53**). On buffering the methylamine with acetic acid much less prototropic change took place at the 5-position. The prop-1′-ynyl isomer (**56**) was obtained by completing the second stage of prototropy during hydrolysis of the chloropyrimidine with strong alkali. The hydroxypyrimidine (**54**) was then treated with phosphoryl chloride and the chloroprop-1′-ynylpyrimidine (**55**) submitted to methylaminolysis.[2700]

with an excess of hot piperidine gave 4-methyl-2,6-dipiperidinopyrimidine.[2699] According to H. Ballweg[2674] the secret for preferential 4-aminolysis in such compounds is to use an aq. alcoholic solution of the amine: the technique is further illustrated[2674] in the preparation of 4-amino-2-chloropyrimidine (63%), 4-benzylamino-2-chloro-5-methylpyrimidine (73%), 2-chloro-4-β-imidazol-4′-ylethylaminopyrimidine (**57**; 60%; from histamine), 2-chloro-4-β-2′-chloropyrimidin-4′-ylaminoethylaminopyrimidine (**58**; 35%), and such like.[2910] The

Me; HC⋮CH_2C; N; Me; N; Cl — **(51)** → Me; HC⋮CH_2C; N; Me; N; NHMe — **(52)** + Me; H_2C:C:HC; N; Me; N; NHMe — **(53)**

(51) ↓

Me; MeC⋮C; N; Me; N; OH — **(54)** → Me; MeC⋮C; N; Me; N; Cl — **(55)** → Me; MeC⋮C; N; Me; N; NHMe — **(56)**

second stage of the aminolysis of 2,4-dichloropyrimidines is seen in isolation in the conversion of 2-chloro-4-dimethylamino- into 4-dimethylamino-2-methylamino-pyrimidine (ethanolic methylamine at 100°),[2627] 2-amino-4-chloro- into 2,4-diamino-6-phenethylpyrimidine (ethanolic ammonia at 140°),[2308] 2-amino-4-chloro- into 2,4-diamino-6-methyl-5-δ-phenylbutylpyrimidine (similarly),[2565] 2-amino-4-chloro- into 2-amino-4-4′-amino-2′-methylquinazolin-6′-ylamino-6-methylpyrimidine,[2709] the same chloro compound into 2-amino-4-β-2′-amino-4′-methylpyrimidin-6′-ylaminoethylamino-6-methylpyrimidine,[2710] 4-amino-2-chloro- into 4-amino-2-dimethylamino-pyrimidine,[2837] and by other examples,[2837] including some in which natural amino acids were used as the aminolytic agents.[2970–2972]

Because of symmetry, preferential mono-aminolysis of 4,6-dichloropyrimidines presents no difficulty.[2782] The process is exemplified in the formation of 4-chloro-6-piperidinopyrimidine[2704] using piperidine in acetone at 20° (98%; a marginal improvement over 93% using aq. piperidine[623]) and 4-amino-6-chloro-5-phenyl(or *o*-chlorophenyl)pyrimidine[2522] (96%, 77%; using ethanolic ammonia at 100°). A second-stage aminolysis is exemplified by the conversion of 4-chloro- into

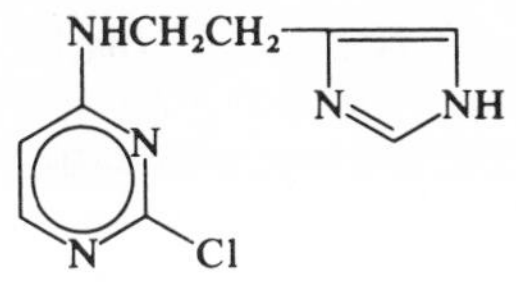

(57)  (58)

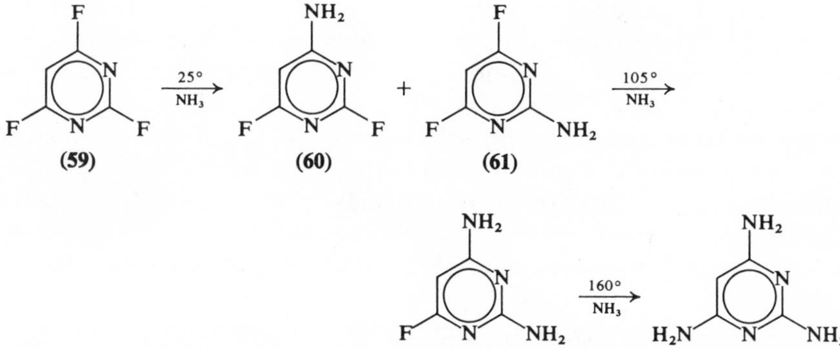

(59) (60) (61)

(62) (63)

4-dimethylamino-6-methylaminopyrimidine by aq. dimethylamine at 125°.[2433] 4,6-Dichloro- is converted into 4,6-bisbenzylamino-2-methylpyrimidine by amine at 200°.[2747]

The aminolysis of 2,4,6-trichloropyrimidine has been quite carefully studied (*H* 191): under gentle conditions a mixture of the 2- and 4-amino-dichloropyrimidines is formed which may be transformed under more vigorous conditions into the 2,4-diamino-chloropyrimidine (e.g., 4-chloro-2,6-dipiperidinopyrimidine,[2922]) and finally into the triamine.* 2,4,6-Trifluoropyrimidine (**59**) behaves quite similarly: the first step to the monoamines, (**60**) and (**61**), occurs in alcoholic ammonia at 25°, the second at 105° to give 2,4-diamino-6-fluoropyrimidine (**62**), and the third at 160° to afford triaminopyrimidine (**63**).[2610] When a dichloro dialkylaminopyrimidine is desired, either a dialkylamine or a trialkylamine may be used for aminolysis; indeed there is less chance of a second aminolysis occurring to give a 4-chloro-2,6-bisdialkylaminopyrimidine if a tertiary base is used. Thus 2,4,6-trichloropyrimidine with dibutylamine at 100° gives 4-chloro-2,6-bisdibutylaminopyrimidine,[2705] but with tributylamine at 110° or even 185° only

* The action of aziridine on trichloropyrimidine in the presence of triethylamine has been reinvestigated:[2255] the structures tentatively assigned[617] to 4-aziridino-2,6-dichloro- and 2,4-diaziridino-6-chloropyrimidine have been confirmed, but the existence of 2-aziridino-4,6-dichloropyrimidine in the products is quite doubtful.[2255] Despite reported[617, 2255] failures, the direct(?) preparation of 2,4,6-triaziridinopyrimdine has been claimed.[2727]

4,6-dichloro-2-dibutylaminopyrimidine* plus butyl chloride;[2706] 2,4,5,6-tetrachloropyrimidine is similarly mono-aminolysed by tertiary amines to give, for example, 4,5,6-trichloro-2-diethyl(or dibutyl)-aminopyrimidine.* [2706, 2707] *N*-Methylmorpholine is exceptionally active towards the trichloropyrimidine giving 2,4,6-trimorpholinopyrimidine, but not apparently towards tetrachloropyrimidine with which it gives (for a steric reason?) only 4,5,6-trichloro-2-morpholinopyrimidine.* [2706] Tetrafluoropyrimidine is aminolysed by ammonia or amines to give 4-amino- or 4,6-diamino-analogues according to conditions, all in reasonable yield.[3521]

Second-stage aminolyses in isolation are typified in the conversions of 2-amino-4,6-dichloropyrimidine into 2-amino-4-chloro-6-methylaminopyrimidine (propanolic or aqueous methylamine 80–100°),[2482] 2-amino-4-chloro-6-β-hydroxyethylaminopyrimidine,[2482, 2504] 2-amino-4-butylamino-6-chloropyrimidine,[2505] and its ethylamino homologue.[2505]

Third-stage aminolyses are seen in the conversion of 2-amino-4,6-dichloropyrimidine into 2-amino-4,6-di-*p*-anisidino(or bis-*p*-chloroanilino)pyrimidine,[2505] of 4-chloro-2,6-bisdibutylaminopyrimidine into 2,4-bisdibutylamino-6-dihexylaminopyrimidine by dihexylamine at 210°,[2705] and of 2,4-diamino-6-chloropyrimidine into 2,4-diamino-6-*p*-bromoanilinopyrimidine and related compounds.[2739]

(2) *Aminolysis of Halogenonitropyrimidines* (*H* 193)

Although recent examples of aminolysis of monochloronitropyrimidines are few, the di- and trichloro derivatives continue to be widely used in syntheses of aminopyrimidines.

(a) *Aminolyses of Monochloronitro Derivatives* (*H* 193). 2-Chloro-5-nitropyrimidine reacts so easily with ethanolic butylamine, or even t-butylamine, that mere warming to 60° is sufficient to complete the formation of 2-*n*(or t)-butylamino-5-nitropyrimidine.[2746]

(b) *Aminolyses of 2,4-Dichloronitro Derivatives* (*H* 193). The well-known preferential aminolysis of the 4-chlorine in 2,4-dichloro-5-nitropyrimidines (*H* 194) has been extended to the preparation in good yields of 2-chloro-4-dimethylamino-5-nitropyrimidine,[2626] 2-chloro-4-

* The authors' confident formulation[2706, 2707] of this and other products as exclusively the 2-aminolysed isomers must be viewed in the light of their statement,[2706] 'It is tentatively postulated that the substitution occurred in the 2-position, since the electron-withdrawing effect of two adjacent nitrogen atoms should make the 2-chlorine atom more reactive than those in 4- or 6-position,' the certainty that trichloropyrimidine gives a mixture of 2,4-dichloro-6- and 4,6-dichloro-2-diethylaminopyrimidine on mono-aminolysis with diethylamine,[2708] and the fact that tetrafluoropyrimidine gives 4-dimethylamino-2,5,6-trifluoropyrimidine beyond doubt.[3521]

methyl-6-methylamino-5-nitropyrimidine,[2334, 2563] 2-chloro-4-ethoxycarbonylmethylamino-5-nitropyrimidine,[2471] and 4-amino-2-chloro-6-ethoxycarbonyl-5-nitropyrimidine.[2467, 3199] Further observations on the similar formation of 2-chloro-4-methylamino-5-nitropyrimidine (**65**) from 2,4-dichloro-5-nitropyrimidine (**64**) and methylamine acetate[292] have revealed some of the 4-chloro-2-methylamino isomer (**66**) in the mother liquors;[2711] the described[292] by-product, 2,4-bismethylamino-5-nitropyrimidine (**67**), was also confirmed.[2711] A more precise

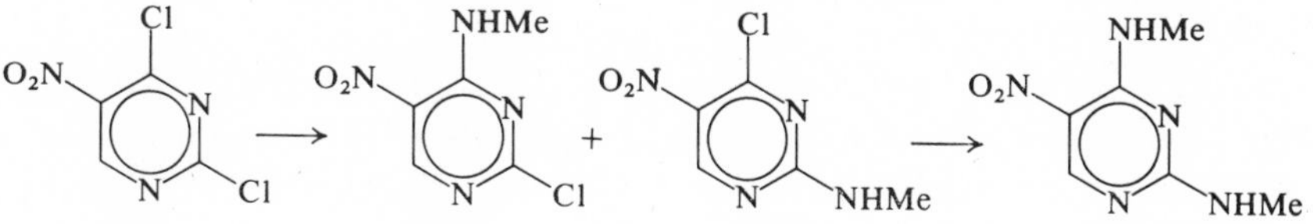

(**64**) (**65**) (**66**) (**67**)

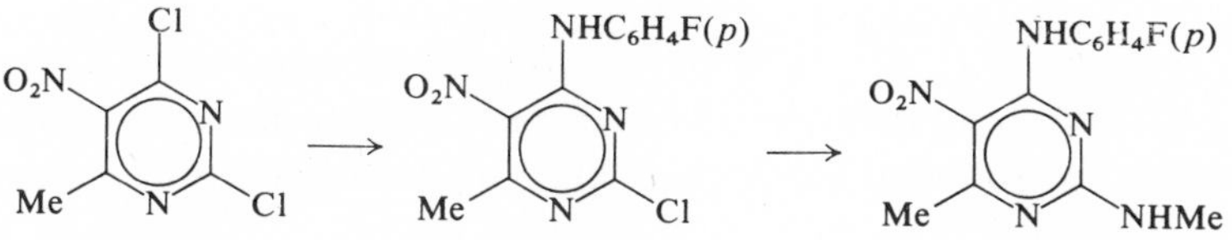

(**68**) (**69**) (**70**)

study[3783] later showed that the products consisted of (**65**) 41%, (**66**) 5%, and (**67**).

Complete aminolysis of 2,4-dichloro-5-nitropyrimidines is exemplified by the preparation in good yield of 2,4-diamino- (improved procedure using boiling alcoholic ammonia),[2711] 2,4-bisdimethylamino-,[2712, *cf.* 1656] 2,4-bisdiethylamino-,[2712] 2,4-bismethylamino-,[2463] 2,4-bisethylamino-,[2461, 2463] 2,4-dianilino-,[2463] 2,4-bismethylamino-6-methyl-,[2463, 2713] 2,4-bisethylamino-6-methyl-,[2461, 2463] 2,4-diaziridino-6-methyl-,[2714] and 2,4-dipyrrolidino-6-methyl-5-nitropyrimidine;[2714] as well as many other cases.[2463, 2712, 2714] Diaminolyses may be done in two stages to afford a nitropyrimidine with different amine residues in the 2- and 4-positions. The second stage of such a process* is seen in the conversion of 4-chloro-2-dimethylamino- into 2-dimethylamino-4-ethylamino-5-nitropyrimidine (aq. ethylamine for 5 min. at *ca.* 100°),[2461] 4-amino 2-chloro- into 4-amino-2-ethylamino-5-nitropyrimidine (aq. ethylamine for 4 hr. at 100°),[2461] 2-chloro- into 2-amino-4-methylamino-5-nitropyrimidine (ethanolic ammonia for 4 hr. at 25°),[2711] 2-chloro- into 2-amino-4-dimethylamino-5-nitropyrimidine,[2626] 2-amino-4-chloro-

* Some of the mono-amino-chloropyrimidines used as substrates for the second-stage were not themselves made by aminolysis.

into 2-amino-4-cyanomethylamino(or ββ-diethoxyethylamino)-6-hydroxy-5-nitropyrimidine,[2465, 2715] 2-amino-4-chloro- into 2-amino 4-t-butylamino-1,6-dihydro-1-methyl-5-nitro-6-oxopyrimidine (refluxing t-butylamine for 10 min.; a surprisingly facile reaction),[2716] 4-chloro-2-dimethylamino-6-hydroxy- into 2-dimethylamino-4-hydroxy-6-hydroxyethylamino-5-nitropyrimidine,[2717] dichloro-6-methyl-5-nitropyrimidine (**68**) into 2-chloro-4-*p*-fluoroanilino-6-methyl-5-nitropyrimidine (**69**) and thence into 4-*p*-fluoroanilino-6-methyl-2-methylamino-5-nitropyrimidine (**70**),[2714] 4-chloro-2-dimethylamino- into 2-dimethylamino-4-guanidino-5-nitropyrimidine,[2718] and such like.[2472, 2716–2720, 2796–2798, 2927]

(c) *Aminolyses of 4,6-Dihalogenonitro Derivatives* (*H* 195). Although they are extremely reactive towards amines, 4,6-dichloro-5-nitropyrimidines can usually be aminolysed in two stages without difficulty; on account of symmetry, only one mono-amino isomer is possible. First stage aminolyses are typified by an improved procedure for making the invaluable intermediate, 4-amino-6-chloro-5-nitropyrimidine, in 82% yield by slowly adding alcoholic ammonia to a solution of 4,6-dichloro-5-nitropyrimidine in tetrahydrofuran at 50° in the presence of sodium bicarbonate;[2721] also in the preparation of 4-chloro-6-diethylamino-5-nitropyrimidine (amine in ether at −40°),[2460] 4-chloro-6-ββ-diethoxyethylamino-5-nitropyrimidine (isolated after alkaline hydrolysis to the corresponding hydroxy compound),[2722] 4-chloro-2-methyl-6-methylamino-5-nitropyrimidine,[2458] and 4-amino-6-fluoro-5-nitropyrimidine (from 4,6-difluoro-5-nitropyrimidine and ammonia gas in ether at −60°).[2455]

Complete diaminolysis has been used to give 2-methyl-4,6-bismethylamino-,[2454] 4,6-diamino- (ethanolic ammonia at 60°),[2332] 4,6-dimorpholino-,[2712] 4,6-di-*p*-toluidino-,[2887] 4,6-bisisopropylamino-,[2463] and 4,6-bis-β-diethylaminoethylamino-5-nitropyrimidine;[2723] also 4,6-diamino-5-nitro-2-trifluoromethylpyrimidine (ammonia in benzene at 50°),[2193] and other compounds.[2463] Isolated second-stage aminolyses are exemplified in the conversion of 4-chloro- into 4-diethylamino-6-methylamino-5-nitropyrimidine,[2460] 4-amino-6-chloro- into 4-amino-6-γ-dimethylaminopropylamino-5-nitropyrimidine (and homologues),[2721] 4-amino-6-chloro- into 4-amino-6-β-hydroxyethylamino-5-nitropyrimidine,[2724] 4-chloro-6-dimethylamino-5-nitro- into 4-dimethylamino-5-nitro-6-*p*-toluidino-pyrimidine (alcoholic *p*-toluidine in presence of triethylamine as proton scavenger),[2725] 4-amino-6-chloro- into 4-amino-6-guanidino-5-nitropyrimidine,[2164] and 4-amino-6-chloro- into 4-amino-6-ζ-4′-amino-5′-nitropyrimidin-6′-ylaminohexylamino-5-nitropyrimidine (**71**).[2726]

(d) *Aminolyses of Trichloro-5-nitropyrimidine* (*H* 196). The action of aziridine (ethyleneimine) and other amines on trichloronitropyrimidine and related compounds is discussed in a series of publications.[2727–2731] Products such as 2,4,6-triaziridino-5-nitropyrimidine and 2,4-dichloro-4-diethylamino-5-nitropyrimidine, and a variety of 4-aziridino-2,5,6-trihalogenopyrimidines are described. When 2,4,6-trichloro-5-nitropyrimidine was allowed to react at −22° with *p*-bromoaniline in ether, the reaction could not be stopped before the second stage; subsequent treatment with ammonia gave a single amino-bis-*p*-

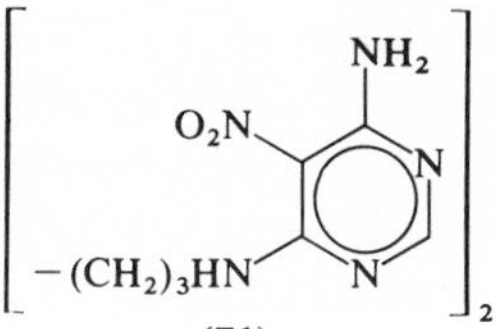

(71)

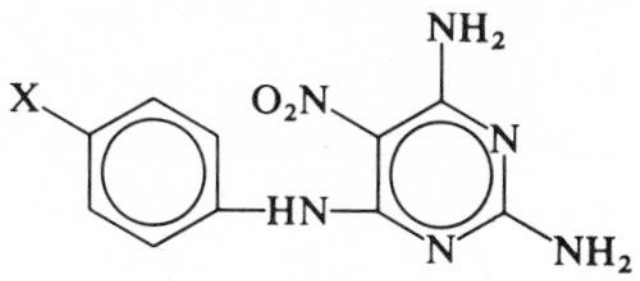

(72)

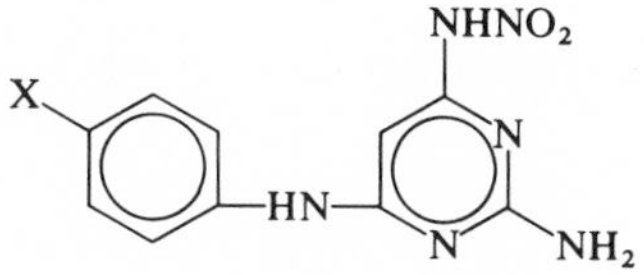

(73)

bromoanilino-5-nitropyrimidine.[2432] The third stage in such aminolyses was better studied in the conversion of 2,4-diamino-6-chloro-5-nitropyrimidine into its 6-*p*-bromo(or iodo)anilino derivative (**72**, X = Br or I) by refluxing with alcoholic halogenoaniline for 2 hr.[2432] Compounds first described[2443] as these derivatives (**72**) turned out to be the interesting isomeric 2-amino-4-*p*-bromo(or iodo)anilino-6-nitroaminopyrimidines (**73**; X = Br or I).[2432]

(3) *Aminolysis of 5-Aminochloropyrimidines* (*H* 196)

In contrast to the chloronitropyrimidines above, 5-aminochloropyrimidines react sluggishly with amines. This is advantageous in achieving mono-aminolysis of 4,6-dichloropyrimidines. However, any such advantage with the unsymmetrical 2,4-dichloropyrimidines is offset by the necessity to separate the resulting mixture of isomers.*

* Aminolyses of 5-amino-2,4-dichloro-6-trifluoromethylpyrimidine appear to give only the 2-chloro isomer, e.g., 5-amino-2-chloro-4-methylamino-6-trifluoromethylpyrimidine.[2735]

Thus heating 5-amino-2,4-dichloropyrimidine and aq. ammonia under pressure gives 2,5-diamino-4-chloropyrimidine and 4,5-diamino-2-chloropyrimidine which can be separated by the greater solubility of the 4-chloro isomer in boiling ethanol;[2590] under sufficiently vigorous conditions, such a mixture of isomers is converted into a single 2,4-diamino derivative, as exemplified in the formation of 2,4-diamino-5-anilinopyrimidine and several analogous compounds.[2567]

Useful mono-aminolyses of 5-amino-4,6-dichloropyrimidines led to 4,5-diamino-6-chloro- (ethanolic ammonia at 150° for 3 hr.),[2733] 5-amino-4-benzylamino-6-chloro- (83%; benzylamine at 100° for 30 min.),[2675] 5-amino-4-chloro-6-propyl(or higher)amino-,[2734] 5-amino-4-chloro-6-methylamino(or higher alkylamino)-2-trifluoromethyl- (92%; refluxing methanolic methylamine for 2 hr.),[2735] 5-amino-4-chloro-6-β-hydroxyethylamino-,[2724] 5-amino-4-chloro-6-*cis*(or *trans*)-2′-hydroxycyclopentylamino- (91%; 74%),[2736] 5-amino-4-chloro-6-β-diethylaminoethylamino-,[2723] and such like pyrimidines.[2726, 2887]

(4) *Aminolysis of Halogeno-(hydroxy-, alkoxy-, or oxo-)pyrimidines* (*H* 197)

Aminolysis of chlorohydroxypyrimidines is represented by the transformation of 4-chloro-2,6-dihydroxypyrimidine into 4-amino-2,6-dihydroxy- (ethanolic ammonia at 145° for 7 hr.),[2737] 2,4-dihydroxy-6-methylamino- (aq. methylamine at reflux for 3.5 hr.,[2737] or with pure amine at 100–120° for 1 hr.[2243]), 4-dimethylamino-2,6-dihydroxy- (ethanolic dimethylamine at 120° for 4 hr.),[2438, 2926] 4-benzylamino-2,6-dihydroxy-,[2243] 2,4-dihydroxy-6-*p*-toluidino-,[2260] and 2,4-dihydroxy-6-β-hydroxyethylamino-pyrimidine (boiling ethanolamine* for 5 min.,[2243] or aq. ethanolamine at 140° for 2 hr.[2438]); of 4-fluoro-2,6-dihydroxypyrimidine into 4-amino-2,6-dihydroxy- and 2,4-dihydroxy-6-methylamino-pyrimidine (same conditions as for chloro isologue above);[2737] and of 4-chloro-2,6-dihydroxy-5-nitropyrimidine into 2,4-dihydroxy-6-methylamino-5-nitropyrimidine and its analogues (momentary boiling in ethanolic methylamino, etc.; *cf.* conditions needed above for analogue without nitro group).[2438, 2909] Second-stage aminolyses of dichloro-hydroxypyrimidines are seen in the conversion of 2-amino-4-chloro-6-hydroxypyrimidine into 2-amino-4-hydroxy-6-β-hydroxyethylamino- (boiling ethanolamine for 30 min.),[2444]

* When dimethylformamide was used as solvent, 4-dimethylamino-2,6-dihydroxypyrimidine was formed instead. A control experiment without any amine present gave the same result.[2438] Pure diethanolamine at 170° gave 4-(bis-β-hydroxyethyl)amino-2,6-dihydroxypyrimidine.[2885]

2-amino-4-hydroxy-6-methylamino-,[2483] and 2-amino-4-hydroxy-6-propylamino-pyrimidine (propylamine in boiling ethoxyethanol),[2738] as well as related compounds;[2444, 2738] of 4-chloro-2-dimethylamino-6-hydroxy- into 2-dimethylamino-4-hydroxy-6-methylaminopyrimidine ('methylamine solution' at 140° for 3 hr.);[2476] of 4-amino-6-chloro-2-hydroxy- into 4-amino-2-hydroxy-6-methylamino- (aq. methylamine under reflux),[2475] 4-amino-6-ethylamino(or diethylamino or benzylamino, etc.)-2-hydroxy-,[2475] and 4-amino-2-hydroxy-6-*p*-toluidino-pyrimidine;[2260] and of 2-amino-4-chloro- into 2-amino-4-benzylamino-6-hydroxy-5-nitropyrimidine (benzylamine in refluxing ethanol for 15 min.) and related compounds.[2444]

Aminolysis of alkoxychloropyrimidines almost invariably leads to attack of the halogen rather than the alkoxy group. Thus 4,6-dichloro-gives 4-amino-6-chloro-5-methoxypyrimidine (82%) in liquid ammonia at room temperature,[2740] 4-chloro-5-methoxy- gives 4-amino-5-methoxy-6-methylthiopyrimidine in ethanolic ammonia at 135°,[2205] 4-chloro-2,6-dimethoxy- gives 2,4-dimethoxy-4-γ-morpholinopropyl-amino-pyrimidine in refluxing ethanolic amine for 4–12 hr.,[2741] and other examples are described.[2741, 3503] But chemistry is full of surprises, and 2,5-dichloro-4-methoxypyrimidine (**75**) in liquid ammonia at 20° or in methanolic ammonia at 80° gives 4-amino-2,5-dichloropyrimidine (**76**) rather than the expected amine (**74**).[2571]

An oxo substituent was present during the aminolytic formation (from the corresponding chloro compound) of 6-benzylamino-1-methyl(or ethyl)uracil (refluxing benzylamine for 1 hr.),[2560, 2928] 1-methyl-6-methylaminouracil (**77**),[2560] 1,3-diethyl-1,2,3,4-tetrahydro-6-methylamino-2,4-dioxopyrimidine (and related compounds),[2305, 2569, 2929] 6-anilino(or benzylamino)-1-dimethylamino-uracil (**78**, R = Ph or CH_2Ph; and related compounds),[2246] 4-amino-

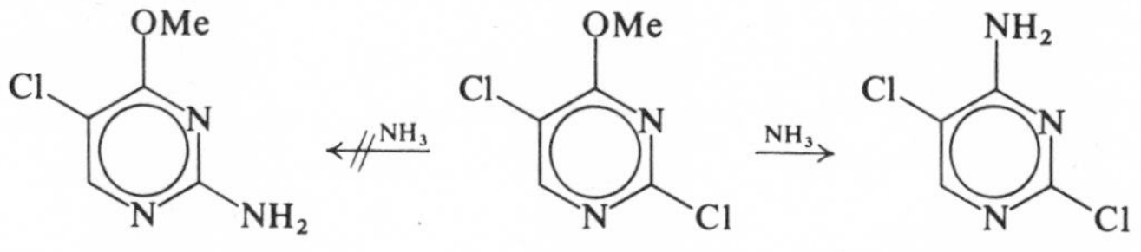

OMe Cl N N NH_2 ⇷ NH_3 OMe Cl N N Cl $\xrightarrow{NH_3}$ NH_2 Cl N N Cl

(**74**) (**75**) (**76**)

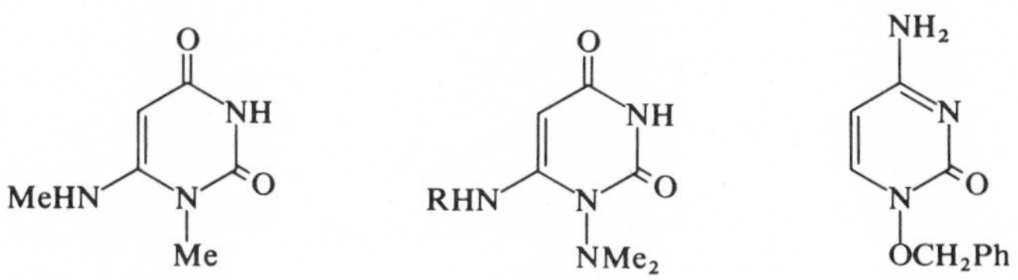

O NH MeHN N O Me O NH RHN N O NMe_2 NH_2 N N O OCH_2Ph

(**77**) (**78**) (**79**)

1-benzyloxy-1,2-dihydro-2-oxopyrimidine (**79**),[2262] and 6-cyclohexylamino-1-methyluracil.[2838]

(5) *Aminolysis of Chloro-alkylthiopyrimidines* (*H* 198)

Chloro-alkylthiopyrimidines are convenient intermediates in that the chlorine can be aminolysed without affecting the alkylthio group which subsequently can be removed by Raney nickel or hydrolysed to an hydroxy group. The aminolysis step is exemplified by the conversion of 4-chloro- into 4-hexylamino(or dimethylamino)-2-methylthiopyrimidine (ethanolic amine at 25°),[2573, 2745] 5-allyl-2-benzylthio-4-chloro- into 5-allyl-4-amino-2-benzylthio-pyrimidine (ethanolic ammonia at 150° for 20 hr.),[2555] 4-chloro-2-*o*-chlorobenzylthio- into 2-*o*-chlorobenzylthio-4-dimethylamino-5-methylpyrimidine (aq. alco-

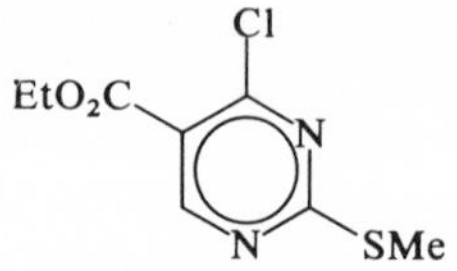

Cl, EtO_2C, N, N, SMe
(**80**)

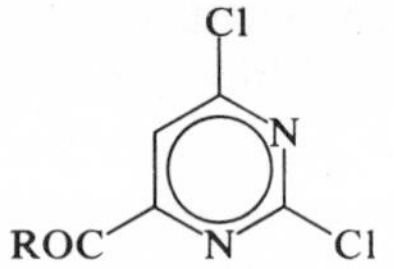

NMe_2, EtO_2C, N, N, SMe
(**81**)

Cl, N, ROC, N, Cl
(**84**)

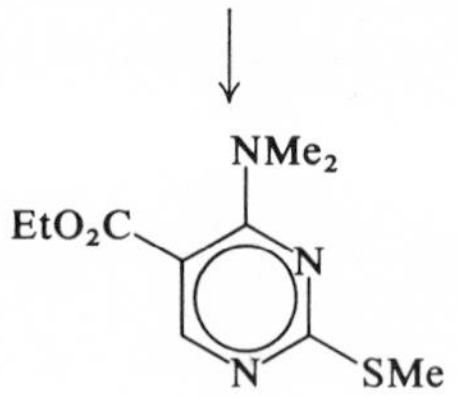

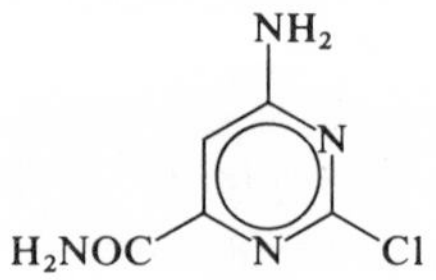

NH_2, N, H_2NOC, N, Cl
(**85**)

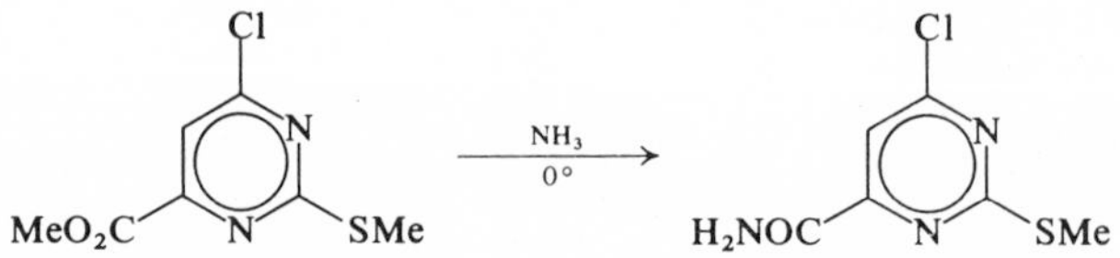

Cl, N, MeO_2C, N, SMe
(**82**)

NH_3 0°

Cl, N, H_2NOC, N, SMe
(**86**)

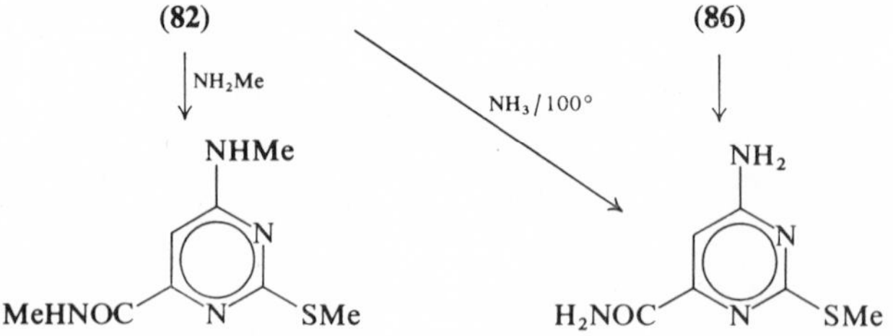

NH_2Me

NH_3/100°

NHMe, N, MeHNOC, N, SMe
(**83**)

NH_2, N, H_2NOC, N, SMe
(**87**)

holic amine at 130°),[2583] 2-chloro-4-methyl- into 4-methyl-2-piperidino-6-*o*-tolylthiopyrimidine and related compounds,[2742] 4-amino-6-chloro- into 4-amino-6-dibenzylamino-2-methylthiopyrimidine (180° for 4 hr.),[2743] 4,6-dichloro- into 4,6-dianilino-2-methylthiopyrimidine,[2589] 4,6-dichloro- into 4,6-bismethylamino-2-methylthio-5-nitropyrimidine (ethanolic amine at 20° for 1 hr.),[2454] and 4-chloro- into 4-anilino-5-ethoxycarbonylmethyl-2-methylthiopyrimidine;[2744] also in the preparation from 4-chloro-5-ethoxycarbonyl-2-methylthiopyrimidine (**80**) of the corresponding 4-dimethylamino (**81**),[2599] aziridino,[2255] anilino,[2598] allylamino,[2744] and many other analogues.[2598, 2744] In contrast, when 4-chloro-6-methoxycarbonyl-2-methylthiopyrimidine (**82**) is treated under comparable conditions with methylamine, 4-methylamino-6-methylcarbamoyl-2-methylthiopyrimidine (**83**) results.[2214] In fact the ammonia or amine reacts first with the ester group and subsequently with the chloro substituent: thus at 0° ethanolic ammonia gives 4-carbamoyl-6-chloro-2-methylthiopyrimidine (**86**), but at 100° it gives 4-amino-6-carbamoyl-2-methylthiopyrimidine (**87**).[2214] Rather similarly, 2,4-dichloro-6-methoxycarbonylpyrimidine (**84**, R = OMe) and ice-cold ethanolic ammonia give a separable mixture of 4-carbamoyl-2,6-dichloropyrimidine (**84**, R = NH_2; in 43% yield) and 4-amino-6-carbamoyl-2-chloropyrimidine (**85**; in 38% yield), but the latter is the sole product (83%) when the reaction is done at room temperature.[2214]

The aminolysis of 4-chloro-2-methylthiopyrimidine by amino acids has been done in two ways: one simply involved condensation in aqueous alcoholic sodium carbonate solution; the other used the amino acids in the form of the trimethylsilyl esters of the *N*-trimethylsilylamino acids and the condensation was done in triethylamine/tetrahydrofuran. The first method gave better yields: gylcine gave 75% of 4-carboxymethylamino-2-methylthiopyrimidine by the simple procedure, but only 55% by the esoteric route.[3769]

(6) *Aminolysis of Chloropyrimidines Substituted by Other Groups* (*H* 199)

The mildly activating effect of a 5-bromo or chloro substituent on other halogeno substituents is seen in the transformation of 5-bromo-2-chloro- into 5-bromo-2-hexylamino-pyrimidine (94%; hexylamine at 25° for 5 min.),[2573] of 5-bromo-2-chloro- into 5-bromo-2-diethylamino-4,6-dimethylpyrimidine (72%; refluxing ethanolic amine),[2519] of 2-amino-5-bromo-4-chloro- into 2-amino-5-bromo-4-*o*(or *m* or *p*)-chloroanilino-6-methylpyrimidine (*ca.* 90%),[2631] of 2,4-diamino-5-bromo-6-chloro- into 2,4-diamino-6-anilino-5-bromo-pyrimidine (aniline hydrochloride at 100°),[2621] of 2,4-dichloro-5-fluoropyrimidine into 4-amino-2-chloro-5-fluoropyrimidine (39% by alcoholic ammonia; 97% by liquid ammonia at 25°)[2209, 2748] and thence into 2,4-diamino-5-fluoropyrimidine (ethanolic ammonia at 140°),[2748] of the same dichlorofluoropyrimidine into 4-aziridino-2-chloro- or 2,4-diaziridino-5-fluoropyrimidine according to conditions,[2570] of 5-bromo-2,4-dichloro- and 2,4,5-trichloro-6-methylpyrimidine into amines like 4-aziridino-5-bromo-2-chloro- or 2,4-dianilino-5-chloro-6-methylpyrimidine,[2714] and of tetrachloro- or tetrafluoro-pyrimidine into a variety of amines ranging from 4-anilino-2,5,6-trichloro- to 2,4,5,6-tetrakisdibutylamino-pyrimidine under various conditions.[2612, 2618, 2690, 2691]

(88) $\xleftarrow[25°]{NH_3}$ (89) $\xrightarrow[0°]{MeNH_2}$ (90) $\xrightarrow[25°]{MeNH_2}$ (91)

(88): 4-NH_2, 5-OHC, 6-Cl pyrimidine; (89): 4-Cl, 5-OHC, 6-Cl pyrimidine; (90): 4-NHMe, 5-OHC, 6-Cl pyrimidine; (91): 4-NHMe, 5-MeN:HC, 6-MeHN pyrimidine

Other activating groups are seen at work during the conversion of 4,6-dichloro-5-formylpyrimidine (**89**) into 4-amino-6-chloro-5-formylpyrimidine (**88**) by ammonia in benzene at 25°, into 4-chloro-5-formyl-6-methylaminopyrimidine (**90**) by methylamine in benzene at 0°, and into 4,6-bismethylamino-5-methyliminomethylpyrimidine (**91**) by methylamine in benzene at 25°;[2594] of 4-carboxy-2,6-dichloro- into 2,4-diamino-6-carboxy- or 4-carboxy-2,6-bismethylamino-pyrimidine by ammonia at 130° or methylamine at 0°(!), respectively,[2214] of 2,4-dichloro-6-methoxycarbonylpyrimidine into products mentioned in the previous section;[2214] of 4-chloro- into 4-dimethylamino-5-ethoxycarbonyl-2-methylpyrimidine by ethanolic amine at 0°;[2698] of 4-chloro-5-cyano- into 5-cyano-4-dimethylamino-2-methylpyrimidine;[2698] of 2-chloro- into 2-amino-5-ethoxycarbonyl(or cyano)pyrimidine by ethanolic ammonia at 100°;[2276, 2278] 4,6-dichloro-5-cyano- into 4-chloro-5-cyano-6-dimethylamino- (2 moles of dimethylamine in dioxane at 10–25°) or 5-cyano-4,6-bisdimethylamino-2-methylpyrimidine (an excess of pure dimethylamine at 25° for 16 hr.), or related compounds;[2681] of 4-chloro-5-cyano- into 4-amino-5-cyano- (ethanolic ammonia at 100° for 36 hr.) or 5-cyano-4-morpholino-2,6-diphenylpyrimidine (morpholine at 80° for 10 min.);[2370] and of 2,4-diamino-6-chloro- into 2,4-diamino-6-diethylamino-5-*p*-ethoxycarbonylphenylazopyrimidine (**92**) by ethanolic diethylamine at 100° for 12 hr.[2502] Other examples are also known,[2884, 3214, 3461] including the reaction of 4-chloro-5-cyanopyrimidine with a variety of amines under very mild conditions to give, e.g., 4-amino-5-cyano-, 4-benzylamino-5-cyano-, and 5-cyano-4-hydrazino-pyrimidine.[3746]

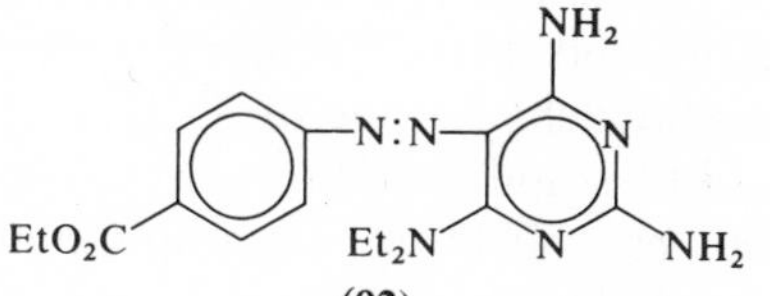

NH_2 N:N N EtO_2C Et_2N N NH_2

(**92**)

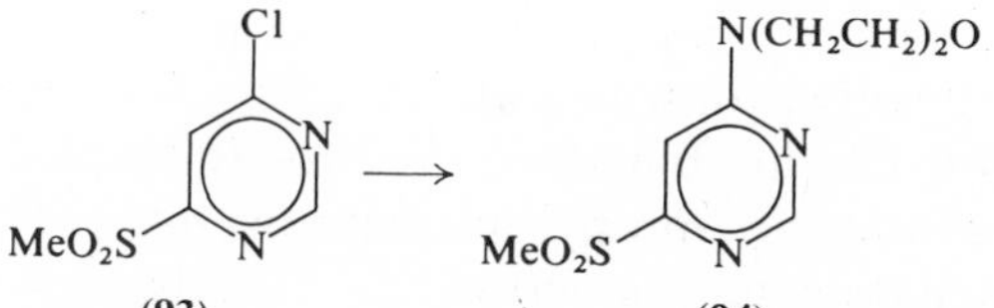

Cl N MeO_2S N → $N(CH_2CH_2)_2O$ N MeO_2S N

(**93**) (**94**)

It is now recognized[2668–2670] that methylsulphonyl is a slightly better leaving group than is chlorine in the same position. However, 4-chloro-6-methylsulphonylpyrimidine (**93**) undergoes preferential aminolysis of its chlorine to give 4-aziridino(or *p*-bromoanilino)-6-methylsulphonylpyrimidine or 4-methylsulphonyl-6-morpholinopyrimidine (**94**).[2749] Since the potential leaving groups in the substrate (**93**) are disposed symmetrically *vis à vis* the activating ring nitrogen atoms, it is clear that the chlorine is additionally activated by the powerfully electron-withdrawing sulphone group while the sulphone is virtually unaffected by the weakly electron-withdrawing chlorine.

C. Replacement of 2-, 4-, and 6-Chloro by Hydrazino, Hydroxyamino, Azido, and Related Groups (*H* 199)

Hydrazine reacts more readily with chloropyrimidines than might be expected from its mediocre basic strength (pK_a 8.1), and oxidation of the resulting hydrazinopyrimidine is a convenient indirect way to remove a chloro substituent in favour of hydrogen (see Sect. 5.A above).

Such aminolysis is seen in the formation of 4-hydrazino-2,6-dimethyl- (aq. hydrazine at 60°),[2224] 2-chloro-4-hydrazino- (ethanolic hydrazine, 25°),[2753] 4,6-dihydrazino- (ethanolic hydrazine, 80°),[2754] 4-dimethylamino-2-hydrazino-5-nitro- (ethereal hydrazine at 25°),[2712] 2,4(or 4,6)-dihydrazino-5-nitro-,[2712] 2-hydrazino-5-nitro- (or 5-nitro-2-5′-nitropyrimidin-2′-ylhydrazinopyrimidine with a limited amount of hydrazine),[3463] 4-amino-6-hydrazino-5-nitro-,[2466] 4-hydrazino-5-phenyl- (hydrazine hydrate warmed for a few minutes),[2327] 5-bromo-4-t-butyl-6-hydrazino- (ethanolic hydrazine *ca.* 80° for 1 hr.),[2574] 5-bromo-4-hydrazino-6-methyl(or phenyl)-,[2574] 5-amino-4-chloro-6-hydrazino- (25°),[2673] 5-amino-4-α-benzylhydrazino-6-chloro- (benzylhydrazine in refluxing benzene)-,[2673] 5-amino-4-benzylidenehydrazino-6-chloro- (**95**; benzylidenehydrazine at 25°),[2673] 5-amino-4-chloro-6-α-methylhydrazino- (and some 5-amino-4,6-bis-α-methylhydrazino-; aq. methylhydrazine at *ca.* 50°),[2673] 4-amino-6-hydrazino-5-nitro- (and 2-methyl and 2-styryl derivative),[2562] 4-hydrazino-6-methoxy-5-nitro- (ethanolic hydrazine below 0°),[2562] 2-benzyl-4,6-dihydrazino-5-nitro- (and 2-methyl homologue),[2688] 5-amino-4-chloro-6-αβ-dimethylhydrazino- (aq. dimethylhydrazine at 70°),[2752] 5-amino-4-hydrazino-6-methyl-,[2227] 4-hydrazino-6-hydroxy-5-nitro- (methanolic hydrazine at 25°),[2466] 4-hydrazino-2-propyl-6-trifluoromethyl-,[2224] 5-ethoxycarbonyl-4-hydrazino-2-trifluoromethyl-,[2217] 5-benzyloxy-4-hydrazino- (ethanolic hydrazine),[2584] 4-hydrazino-5-methoxy-2-methylthio-,[2584] 2,4-dibenzyloxy-6-hydrazino-,[2558] 5-amino-4-chloro-6-β-diphenylmethylhydrazino- (propanolic amine),[3772] 2-hydrazino-4-methyl-6-propoxy(or propylthio)- (and homologues),[2580] and 4-cyano-2-hydrazino-6-methyl-pyrimidine;[2592] also in the conversion of 6-chloro-3-methyluracil into 6-hydrazino-3-methyl-[2755] or 3-methyl-6-α-methylhydrazinouracil,[2439, 2440] of 6-chloro-3-methyl-5-nitro- into 6-hydrazino-3-methyl-5-nitro-uracil,[2755] of 4,6-dichloro-5-ααα-trifluoro-*N*-methylacetamido- (**96**) into 4-chloro-6-hydrazino-5-methylamino-pyrimidine (**97**; aq. hydrazine inducing deacylation as well as aminolysis),[2752] and such like.[2230, 2440, 2601, 2756, 2769, 2770]

Treatment of 4-chloro-2,6-dihydroxypyrimidine with warm ethanolic hydroxylamine gave 2,4-dihydroxy-6-hydroxyaminopyrimidine, but 4-chloro-2,6-dimethoxypyrimidine failed to react similarly.[2757]

Some azidopyrimidines have also been made by nucleophilic displacement: 4-amino-2-chloro-5-nitropyrimidine and sodium azide in refluxing ethanol gave 4-amino-2-azido-5-nitropyrimidine;[2712] 4-chloropyrimidine hydrochloride and sodium azide in dimethylformamide at 85° gave 4-azidopyrimidine (**98**), in equilibrium with the bicyclic form, 1,2,3,3a,5-penta-azaindene (**99**);[2466] 5-acetamido-4,6-dichloropyrimidine similarly gave 5-acetamido-4,6-bisazidopyrimidine, also in equilibrium with a bicyclic form;[2466] 4-amino-6-azido-2-hydroxypyrimidine was made in *ca.* 60% yield from the 6-chloro analogue and aqueous sodium azide;[2475] and 5-amino-4,6-dichloro- gave 5-amino-4-azido-6-chloro-pyrimidine on treatment with sodium azide (1 mole) in dimethylformamide.[2934]

Being a respectable base of pK_a 3.65 (*cf.* urea, pK_a 0.1),[2937] semicarbazide may be used as an aminolytic agent: 4-chloro-6-methyl-2-methylthio- gave 4-methyl-2-methylthio-6-semicarbazido-pyrimidine,[2938] and 2-amino-4-chloro-6-methyl- gave 2-amino-4-methyl-6-semicarbazido-pyrimidine.[2939] Similarly acetohydrazide reacted with 2,4-dichloro-6-methyl-5-nitropyrimidine to give 4-β-acetylhydrazino-2-chloro-6-methyl-5-nitropyrimidine (49%), and benzohydrazide gave the 4-β-benzoylhydrazino homologue; methoxycarbonylhydrazine likewise gave 2-chloro-4-β-methoxycarbonylhydrazino-6-methyl-5-nitropyrimidine (56%).[3214]

D. Replacement of 2-, 4-, and 6-Halogens by Alkoxy Groups (*H* 201)

Of the commonly used nucleophiles, alkoxide ion is the sledge hammer: no matter how deactivated a chloropyrimidine may be, it will almost certainly yield its alkoxy analogue by the action of sodium alkoxide in the appropriate alcohol, although the activity of the chloro substituent may be reflected to some extent in the conditions required. Indeed, a considerable selectivity in attacking a di- or trichloropyrimidine with alkoxide can be achieved by controlled reaction conditions and by the amount, concentration, and type of alkoxide ion. Such selectivity is evident in the conversion of 4,6-dichloro- into 4-chloro-6-methoxypyrimidine (80%),[2682, *cf.* 1379] of 2,4-dichloro- into 2-chloro-4-methoxy(or other alkoxy)-6-methylpyrimidine (80–90%),[2699] of 2,4,6-trichloro(or bromo or fluoro)- into 4-chloro(or bromo or fluoro)-2,6-dimethoxy-pyrimidine,* [2243, 2610, 2611, 2614] of 2,4,6-trichloro- into 2,4-dibenzyloxy-6-chloro-pyrimidine,[2558] of 4-amino-2,6-dichloro- into 4-amino-6-chloro-2-ethoxy(or methoxy)-pyrimidine,[2677, 2678, 2837, 2839] of 2,4-dichloro-6-dimethylamino- into 4-chloro-6-dimethylamino-2-methoxy-pyrimidine,[2676] of 2-amino-4,6-difluoro- into 2-amino-4-ethoxy(or benzyloxy)-6-fluoro-pyrimidine,[2610] of 2,4,5-trichloro- into 2,5-dichloro-4-methoxy-pyrimidine (84%),[2571] of 5-bromo-4,6-dichloro- into 5-bromo-4-chloro-6-methoxy-pyrimidine,[2519] of 2,4-dichloro- into 2-chloro-4-ethoxy-5-fluoropyrimidine (96%),[2748] of 2,4,5,6-tetrafluoro-

* Trifluoropyrimidine has been shown to give at least some 2-benzyloxy-4,6-difluoropyrimidine which was not purified but converted into 4-amino-2-benzyloxy-6-fluoropyrimidine (33% yield overall).[2610]

into 4-ethoxy-2,5,6-trifluoro- or 2,4-diethoxy-5,6-difluoro-pyrimidine according to conditions,[2618] of the same tetrafluoropyrimidine into the derived 4-methoxy, 4,6-dimethoxy, or 2,4,6-trimethoxy compounds according to the conditions used (*cf.* ethoxy homologues above),[3521] of 2-bromo-4-chloro-5-ethoxy- into 2-bromo-4,5-diethoxy- or 2,4,5-tri-ethoxy-pyrimidine,[2606] of 4,6-dichloro-5-formyl- (**100**) into equal parts of 5-formyl-4,6-dimethoxy- and 5-formyl-4-hydroxy-6-methoxypyrimidine (an unusual mechanism is proposed to account for this),[2758] of 4-anilino-2,5,6-trichloro- into a mixture of 4-anilino-2,5-dichloro-6-ethoxy-, 4-anilino-5,6-dichloro-2-ethoxy- and 4-anilino-5-chloro-2,6-diethoxy-pyrimidine (rates measured),[2690] and in other examples.[2815, 2886]

The many nonselective mono-alkoxylations are typified in the formation, from the corresponding chloropyrimidine, of 2-methoxy-,*[2511] 2-ethoxy (and other n-, iso-, and s-alkoxy)-,[2511, 2630, 2697] 4-ethoxy-6-methyl(or 2,6-dimethyl)-,[2451] 4-methoxy(or isopropoxy)-2-methyl-,[2698] 2-*p*-tolyloxy- (*p*-cresol plus potassium carbonate at 175°),[2630] 2-methyl-4-phenoxy- (potassium phenoxide with copper powder in dioxane at 150°),[2698] 2-methoxy(or 2-ethoxy or 4-methoxy)-5-methyl-,[2630] 5-bromo-2-ethoxy-,[2630] 5-bromo-2(or 4)-methoxy-,[2746] 5-bromo-4-methoxy-6-phenyl-,[2519] 2-methoxy(or ethoxy)-5-nitro-,[2746, 2762] 4-methoxy-2-methylamino-,[2288] 1,6-dihydro-4-methoxy-

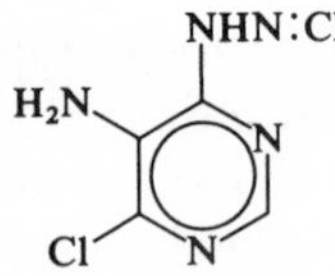

NHN:CHPh
H_2N
Cl

(**95**)

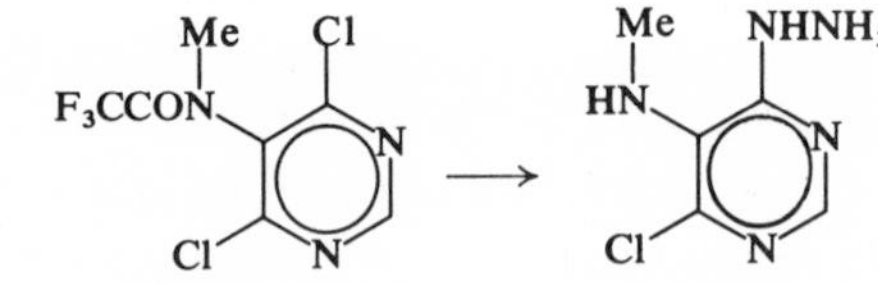

Me Cl
F_3CCON
Cl

(**96**)

Me $NHNH_2$
HN
Cl

(**97**)

N_3

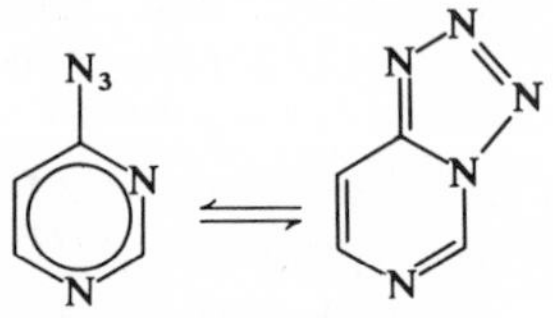

(**98**)

(**99**)

Cl
OHC
Cl

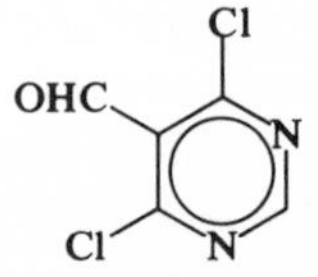

(**100**)

1-methyl-6-oxo-,[2760] 2-amino-4-butylamino-5-*p*-chlorophenylazo-6-methoxy-,[2761] 2-methoxy-4-methylthio-,[2762] and 4-allyloxy-2-amino(or methyl or methylthio or phenyl or trifluoromethyl, etc.)pyrimidine;[2194, 2555] as well as by other well described examples.[2212, 2214, 2262,]

* A method using only 1 molar quantity of methanol-^{18}O and giving 89% yield has been described.[2759]

2467, 2571, 2578, 2582, 2587, 2589, 2623, 2629, 2631, 2683, 2711, 2758, 2763–2765, 2837, 2840, 2841, 2921

Some notable examples of difficult alkoxylation requiring vigorous conditions are the preparation of 2,4-diallyloxy-6-aminopyrimidine (100° in sealed tube),[2678] 4-amino-2,6-dimethoxypyrimidine (125° for 12 hr.),[2280] 2,4-diamino-6-methoxy(or isopropoxy or benzyloxy)-pyrimidine (120°, 150°, 160°, respectively),[2285] and 2-amino-4-methoxy-(or isopropoxy or benzyloxy)-6-methylaminopyrimidine (160°).[2482, 2925] A most useful route to 2-amino-4-methoxy- (**102**) and 4-amino-2-methoxy-pyrimidine (**103**), useful for making isocytosine (**105**; R = H) and cytosine (**106**; R = H), respectively, involves mono-aminolysis of 2,4-dichloropyrimidine (**101**), methoxylation of the resulting mixture (**102** + **103**), and a simple separation of the isomers (**105**, R = Me; **106**; R = Me) by solubility of the former in dioxane;[2766] 4-dimethyl-amino-2-methoxypyrimidine (**104**) has been made rather similarly.[2745]

Complete alkoxylation of di(or tri)chloropyrimidines is seen in the preparation of 4,6-dimethoxy-,[2682] 2,4-diethoxy-5-fluoro-,[2767] 4,6-di-benzyloxy-2-methyl-,[2423] 5-fluoro-2,4-dimethoxy-,[2572] 4,6-diethoxy-5-nitro-,[2164] 5-bromo-2,4-dimethoxy-6-methyl-,[2714] 4-methyl-2,6-dipropoxy(or diphenoxy)-,[2699] 4-amino-2,6-dimethoxy-,[2920] 2,4-diallyl-oxy-,[3479] 2,4-dimethoxy-6-methoxycarbonyl- (from 2,4-dichloro-6-chlorocarbonylpyrimidine),[2591] 2,4,6-triallyloxy-,[2768] 2,4-di-t-butoxy-,[2829] and other substituted-pyrimidines.[2829, 2886]

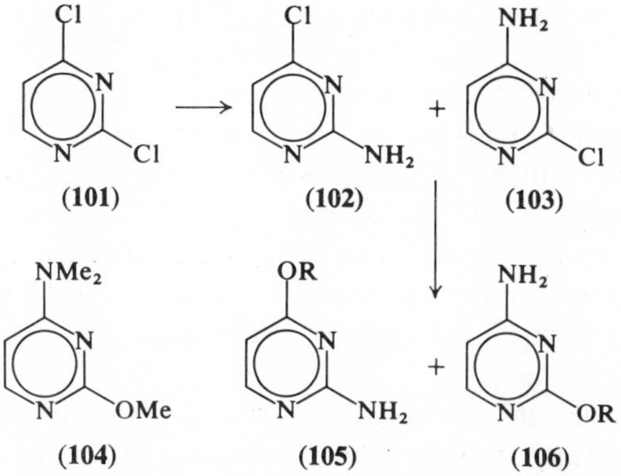

A kinetic investigation has been made of the reaction of *p*-nitro-phenoxide ion in methanol with six chlorodiazines, including 2-, 4-, and 5-chloropyrimidine. In this system, 2-chloropyrimidine is more reactive than its 4-isomer; 5-chloropyrimidine is relatively unreactive. The original should be consulted for some interesting comparisons.[3195]

E. Replacement of 2-, 4-, and 6-Halogens by Hydroxy Groups (*H* 203)

The direct hydrolysis of a chloro- to an hydroxy-pyrimidine was avoided for many years (*H* 203), but no such inhibition seems to be operating now. Acid or alkali may be used, although the latter is more usual, and preferential monohydrolysis of a di- or trihalogenopyrimidine is quite possible.

Acid hydrolysis is represented by the change of 4-chloro- into 4-hydroxy-2-methyl-6-methylamino-5-nitropyrimidine (refluxing 10*N*-HCl for 15 min.),[2458] of 4-fluoro-2,6-dihydroxy- into 2,4,6-trihydroxypyrimidine (*N*-HCl at 100° for 30 min.),[2737] of 5-amino-4,6-dichloro- into 5-amino-4-chloro-6-hydroxy-pyrimidine (boiling 98% formic acid for 1.5 hr.),[2466] and of 2,4,5,6-tetrafluoro- into 2,4,5-trifluoro-6-hydroxypyrimidine (water plus tetrahydrofuran for 2 hr. at 20°),[2618] etc.[2554]

Examples of alkaline hydrolysis are more numerous: simple hydrolyses include, for example, the transformation of 2-chloro-4,6-dimethyl-5-prop-2′-ynyl- into 2-hydroxy-4,6-dimethyl-5-prop-1′-ynyl-pyrimidine (*N*-sodium hydroxide at 100°; note prototropic change in 5-group),[2700] of 2-chloro-4-ββ-diethoxyethylamino- into 4-ββ-diethoxy-ethylamino-2-hydroxy-5-nitropyrimidine (*N*-sodium hydroxide at 100°),[2948] of 4-chloro-5-fluoro- into 5-fluoro-4-hydroxy-6-methyl-2-methylthiopyrimidine (*N*-sodium hydroxide at 100°),[2679] of 4-amino-6-chloro-1,2-dihydro-1-methyl-2-oxopyrimidine (**107**) into 6-amino-3-methyluracil (**108**; *N*-alkali at 100° for 15 min.),[2475] of 2-amino-4-ethoxy-6-fluoro- into 2-amino-4-ethoxy-6-hydroxy-pyrimidine (*N*-sodium hydroxide at 100° for 30 min.),[2610] of 2-chloro- into 2-hydroxy-4-methylamino-5-nitropyrimidine (*N*-alkali at 100° for 1 hr.),[2711] and of 2-amino-4-chloro- into 2-amino-4-hydroxy-6-*o*-toluidinopyrimidine (and eight analogues by sodium hydroxide in ethylene glycol at 175°).[2739]

Preferential alkaline hydrolysis is seen in the formation of 4-chloro-6-hydroxy-5-nitropyrimidine (aq. sodium carbonate at 40° for 5 hr.),[2240] 4-chloro-6-hydroxy-5-phenoxypyrimidine (aq. alkali plus dioxan at 100°),[2240] 4-chloro-2,6-dihydroxy- from 2,4,6-trichloropyrimidine (2.5*N*-alkali at 100° for 1 hr.),[2442] 4-fluoro-2,6-dihydroxypyrimidine (**109**) from 2,4,6-trifluoropyrimidine (**110**; 2.5*N*-alkali at 80° for 1 hr.),[2614] 4,6-difluoro-2-hydroxypyrimidine (**111**) from 2,4,6-trifluoropyrimidine (**110**; 0.5*N*-potassium hydroxide in 50% aq. acetone at <30°; the product was identified by aminolysis to the known 4,6-diamino-2-hydroxypyrimidine),[2614] 2-chloro-5-fluoro-4-hydroxypyrimidine (1 mole of 2*N*-sodium hydroxide at 45°),[2748] and 2(or 5)-butyl-4-chloro-6-hydroxypyrimidine (and homologues; 1.25*N*-alkali at 100° for 2–3 hr.).[2242]

(107) → NaOH → (108)

(109) ← hot NaOH ← (110) → cold NaOH → (111)

F. Replacement of 2-, 4-, and 6-Chloro by Alkylthio and Arylthio Groups (*H* 205)

Apart from primary syntheses, alkylthiopyrimidines are made most often by *S*-alkylation of the corresponding mercaptopyrimidines, but sometimes by an alternative route from chloropyrimidine with sodium alkyl mercaptide. For making arylthiopyrimidines, *S*-arylation is precluded and the second route therefore becomes of the utmost importance. It is exemplified in the formation of 2-phenylthiopyrimidine (ethanolic sodium thiophenate at *ca.* 80° for 2 hr.),[2619] its 4,6-dimethyl derivative,[2771] 2,4-dimethoxy-6-phenylthiopyrimidine,[2771] 2-*p*-chlorophenylthiopyrimidine (sodium *p*-chlorothiophenate in diethylene glycol monoethyl ether at 120° for 4 hr.),[2772] its 5-chloro derivative,[2772] 2,4-diamino-5-bromo-6-phenylthiopyrimidine (ethanolic thiophenol plus triethylamine at *ca.* 80° for 2 hr.),[2621] its *o*-amino derivative,[2621] 2,4-diamino-6-*p*-chlorophenylthiopyrimidine (*p*-chlorothiophenol and potassium carbonate first in refluxing glycol and then at 100° for 12 hr.),[2773] 2-chloro-4-methyl-6-phenylthiopyrimidine (1 mole of ethanolic potassium thiophenate at $<0°$),[2742] 4-methyl-2,6-bisphenylthiopyrimidine (2 moles ethanolic potassium thiophenate at 100°),[2742] and related compounds.[2742]

The more frequent, but less important, use of sodium alkyl mercaptides is illustrated in the preparation of 5-cyano-4-ethylthiopyrimidine (isopropanolic sodium methyl mercaptide under reflux for 4 hr.;

61% yield), 5-bromo-2(or 4)-methylthiopyrimidine (aq. alcoholic sodium methyl mercaptide at 25°),[2746] 4-methoxy-6-methylthiopyrimidine (methanolic mercaptide at 60°),[2682] 4-ethyl(or benzyl)thio-2,6-dimethoxypyrimidine,[2771, 2774] 4-butylthio-5,6-tetramethylenepyrimidine,[2183] 4-amino-6-chloro-2-ethylthio- from 4-amino-2,6-dichloropyrimidine (1 mole mercaptide at 50°),[2678] 4-amino-2,6-bisethylthiopyrimidine from the same substrate (excess of mercaptide at 110°),[2678] 4-allylthio-2-benzylthio(or amino)pyrimidine from the 4-chloro analogue (allyl mercaptan in aq. alcoholic sodium carbonate at 100°),[2555] 4-n(or t)-butylthio-5-ethoxycarbonyl-2-methylthiopyrimidine from its 4-chloro analogue (similarly),[2744] 4-iodo-6-methylthio- from 4,6-di-iodo-pyrimidine (40°),[2609] 4,6-bismethylthio- from 4,6-dichloro-pyrimidine (refluxing methanol),[2609] and such like.[2240, 2288] The mercaptide may be generated within the actual reaction mixture: 2-amino-4,6-dichloropyrimidine, *S*-methylthiourea sulphate, and aq. methanolic alkali at 100° give 2-amino-4,6-bismethylthiopyrimidine (>80%).[2601] An odd variant is exemplified in the treatment of 4-chloro-5-ethoxy-carbonyl-2-methylthiopyrimidine with sodium cyanide, lithium cyanide, or sodium sulphite in dimethyl sulphoxide or dimethylformamide to give mainly 5-ethoxycarbonyl-2,4-bismethylthiopyrimidine; similarly, 4-chloro-2-methylthio- gives 2,4-bismethylthio-pyrimidine.[2599] More complicated examples include the formation of 2,4-bis-β-diethylamino-ethylthio-5-nitropyrimidine (and analogues), [3234] and 2-amino-6-5′-methylpyrimidin-4′-ylthiopurine (and six analogues).[3235]

Carboxymethylthiopyrimidines, so familiar as intermediates during the conversion of mercapto- into hydroxypyrimidines, may be made also by treating a chloropyrimidine with sodium thioglycollate. Thus 4,6-dichloro- gives 4-carboxymethylthio-6-chloro-5-phenylpyrimidine,[2522] and 4-chloro- gives 4-carboxymethylthio-6-methoxy-5-nitro-pyrimidine under gentle conditions.[2775]

G. Replacement of 2-, 4-, and 6-Halogens by Mercapto Groups (*H* 205)

Some recent examples of the use of alcoholic or aq. sodium hydrogen sulphide in converting a chloro- into a mercapto-pyrimidine (in the presence of a variety of groups) include the preparation of 2-amino(or dimethylamino)-4-mercapto-,[2776] 4-amino-6-mercapto-,[2777] 4-dimethylamino-2-mercapto-5-methyl- (in ethoxyethanol at 140°),[2583] 4-mercapto-6-methoxy-,[2760] 1,4-dihydro-6-hydroxy-1-methyl-4-thio-(or tautomer),[2760] 4-amino-2-hydroxy-6-mercapto-,[2475, 2777, *cf.* 2165] 2(or

5)-butyl-4-hydroxy-6-mercapto-,[2242] and 4,6-dimercapto-2-trifluoromethyl-pyrimidine;[2193] unusually vigorous conditions were needed to make 2,4-diamino-6-mercaptopyrimidine (ethanolic NaHS at 80° for 7 days,[2503] or NaHS in glycol at 150° for 1 hr.[2778]) and 2-amino-4-ethyl(or benzyl)amino-6-mercaptopyrimidine (in glycol as above).[2738] Some unusual metatheses in this group are 4-chloro-6-dimethylamino-3-methylpyrimidinium iodide (**112**) into 4-dimethylamino-1,6-dihydro-1-methyl-6-thiopyrimidine (**113**),[2776] 2,4-dichloro-5-methyl- into 5-methyl-

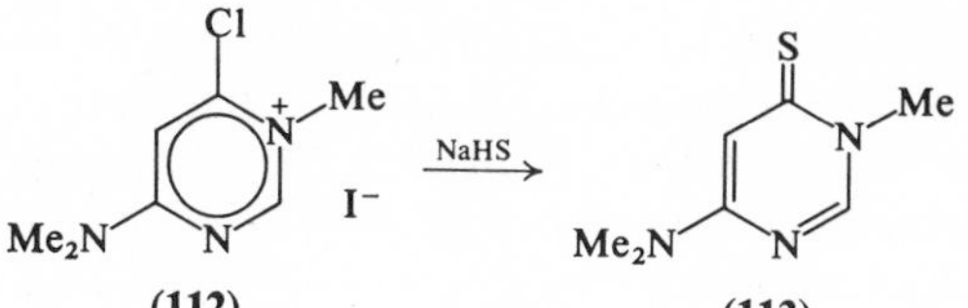

(**112**) (**113**) 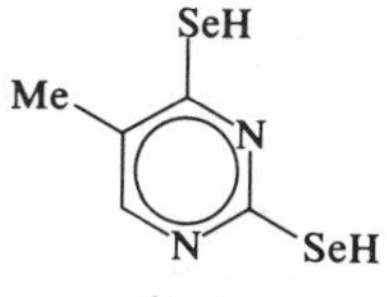 (**114**)

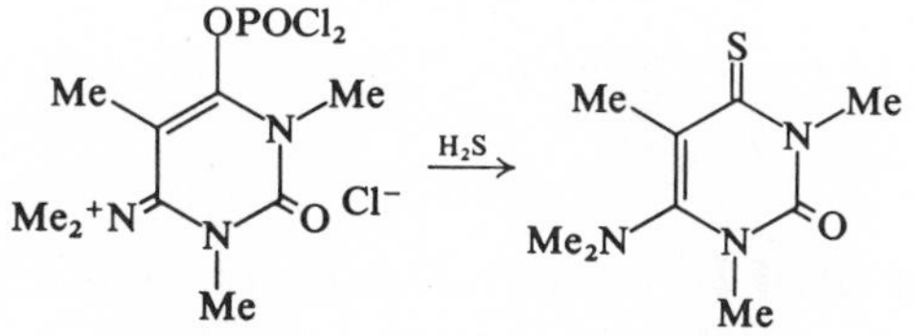

(**115**) (**116**)

2,4-diselenyl-pyrimidine (**114**),[2779] and a phosphoryl chloride 'complex' formulated as (**115**) into 4-dimethylamino-1,2,3,6-tetrahydro-1,3,5-trimethyl-2-oxo-6-thiopyrimidine (**116**) by introducing hydrogen sulphide;[2569] also other examples.[3214]

Other reactions may occur during such treatment. The well-known (*H* 206) and often useful reduction of a 5-nitro group by sodium hydrogen sulphide is seen again in the conversion of 2-chloro-4-methyl-6-methylamino-5-nitropyrimidine into 5-amino-2-mercapto-4-methyl-6-methylaminopyrimidine,* [2563, 2675] and in other examples.[3199] Another change, thioether to mercaptan, is illustrated by the conversion of 4-chloro-2-methylthiopyrimidine into 2,4-dimercaptopyrimidine by sodium hydrogen sulphide in glycol at 150°; similarly, 4-chloro-6-hydroxy-2-methylthio- gives 4-hydroxy-2,6-dimercapto-pyrimidine,[2165] and 4-chloro-6-methoxycarbonyl-2-methylthio- gives 4-carboxy-2,6-dimercapto-pyrimidine.[2214] On the other hand, the reduction which might be expected occurs neither in converting 2-amino-4-chloro-5-*p*-

* The melting point given for this compound is not the same in both papers; nor is that for the desulphurized product, 5-amino-4-methyl-6-methylaminopyrimidine. However, the m.p. and spectra of the derived 6,9-dimethylpurine are satisfactorily similar in both papers.[2563, 2675]

chlorophenylazo-6-β-hydroxyethylaminopyrimidine into its mercapto analogue with alcoholic sodium hydrogen sulphide,[2504] nor in the similar formation of 2-amino-4-butylamino-5-*p*-chlorophenylazo-6-mercaptopyrimidine.[2761] Even stranger is the survival of the nitro group during treatment of 2,4-dichloro-5-nitropyrimidine with alcoholic potassium hydrogen sulphide at 100° to give 2,4-dimercapto-5-nitro-pyrimidine, and in the similar preparation of 4-carboxy-2,6-dimer-capto-5-nitropyrimidine.[2590]

The use of a thiouronium intermediate in converting a chloro- into a mercaptopyrimidine has been less used than previously, but is illustrated in the formation of 2-mercapto- (55%),[2780] 4-mercapto-5,6-tetramethylene- (86%; no intermediate isolated),[2183] 5-bromo-4-mercapto- (*ca.* 80%; intermediate treated alkali),[2746] 2-mercapto-5-nitro- (*ca.* 80%; thiouronium chloride analysed and then hydrolysed),[2746] 2-mercapto-4-methylthio-* (70%; intermediate isolated),[2581] 4-mercapto-6-methylamino-,[2781] 5-bromo-2,4-dimercapto-,[2519] and 4-carboxy-2,6-dimercapto-pyrimidine,[2214] as well as other such compounds.[2816, 3480]

H. Replacement of 2-, 4-, and 6-Halogens by Other Sulphur-Containing Groups (*H* 207)

Sodium sulphite has been used to convert 4-amino-6-chloro-2-hydroxy- into 4-amino-2-hydroxy-6-sulphopyrimidine (**117**).[2475] A number of benzenesulphonamidopyrimidines has been made by treating chloropyrimidines with sodio sulphanilamide and related compounds, often in acetamide solution. This far from minor process is exemplified in the formation of 4-chloro(or methoxy)-6-sulphanil-amidopyrimidine (**118**; R = Cl or OMe),[2682] and similar reactions.[2349, 2446, 2447 2571, 2578, 2683, 2758, 3241,] Alkylsulphonylpyrimidine may be made from the corresponding chloro derivative by treatment with potassium benzenesulphinate or related compound. Although this method has not been used much, it provides an alternative to the oxidative route to sulphones. The process is illustrated in the formation of 2,4-dimethoxy-6-phenylsulphonylpyrimidine (35%; *cf.* 85% by oxidation of the corresponding phenylthiopyrimidine),[2771] and of 2-phenylsulphonylpyrimidine (**119**; *ca.* 60%).[2619]

Only a few thiocyanatopyrimidines have been made recently by treating chloro(or bromo)pyrimidines in ethanol with ammonium

* *Cf.* the formation of 2,4-dimercaptopyrimidine from 4-chloro-2-methylthiopyrimidine and sodium hydrogen sulphide.[2165]

thiocyanate: typical examples are 2-thiocyanato- (**120**; 25% from chloro- and 41% from bromopyrimidine),[2607, 2783] 4,6-dimethyl-2-thiocyanato- (18%),[2783] and 2-amino-4,5-dimethyl-6-thiocyanato-pyrimidine (50%);[2783] in each case an alternative route using cyanogen bromide with the corresponding mercaptopyrimidine gave better yields.[2783] Another example is the formation of 5-ethoxycarbonyl-2-methylthio-4-thiocyanatopyrimidine from its chloro analogue with potassium thiocyanate in dimethyl sulphoxide.[2936]

I. Other Reactions of 2-, 4-, and 6-Chloropyrimidines (*H* 208)

The condensation of a chloropyrimidine with an activated methylene compound occurs when 4,6-dichloro-5-nitropyrimidine reacts with ketene diethylacetal to give 4-chloro-6-ββ-diethoxyvinyl-5-nitropyrimidine (**121**), a usefully reactive intermediate;[2785] when 2-amino-4-chloro-6-methyl-5-nitropyrimidine reacts with diethyl acetamidomalonate to give 4-α-acetamido-α-diethoxycarbonylmethyl-2-amino-6-methyl-5-nitropyrimidine* (**122**);[2786, 2787] when the same substrate reacts

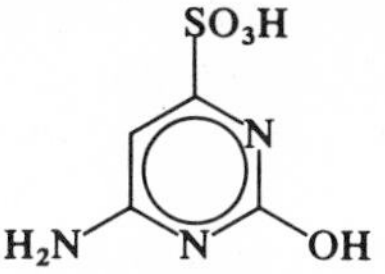

SO_3H / H_2N / N / N / OH

(117)

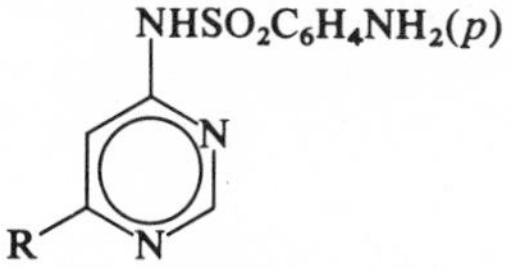

$NHSO_2C_6H_4NH_2(p)$ / R / N / N

(118)

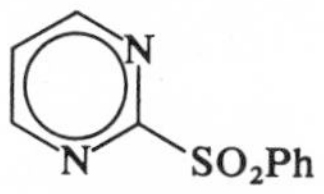

N / N / SO_2Ph

(119)

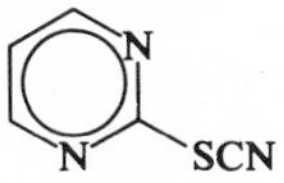

N / N / SCN

(120)

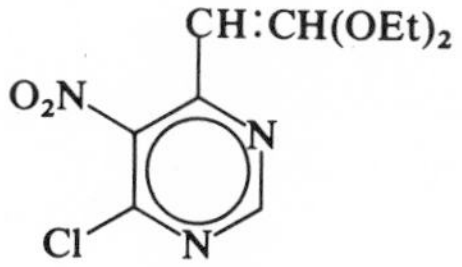

$CH{:}CH(OEt)_2$ / O_2N / Cl / N / N

(121)

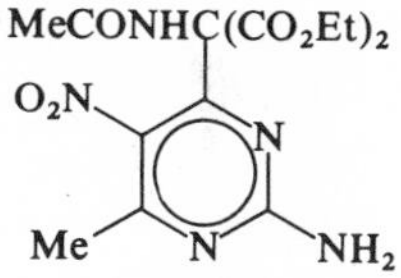

$MeCONHC(CO_2Et)_2$ / O_2N / Me / N / N / NH_2

(122)

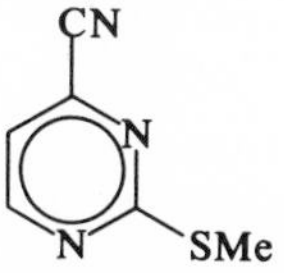

CN / N / N / SMe

(123)

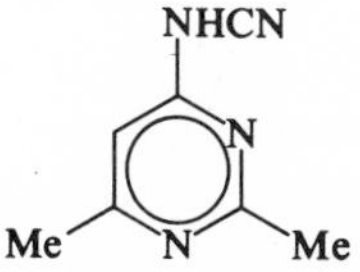

NHCN / Me / N / N / Me

(124)

N / N / $PO(OR)_2$

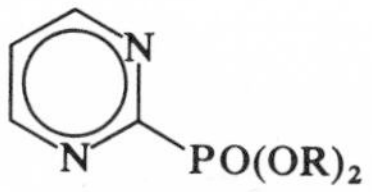

(125)

* Unlike 2-amino-4-diethoxycarbonylmethyl-6-methyl-5-nitropyrimidine, which undergoes hydrolysis to 2-amino-4,6-dimethyl-5-nitropyrimidine,[536, 2787] the α-acetamido derivative (**122**) gives 2-amino-4-hydroxy-6-methyl-5-nitropyrimidine, on acidic or alkaline hydrolysis.[2786, 2787] On the other hand, aminolysis of the 4-diethoxycarbonylmethyl derivative causes rupture of the C—C bond to yield 2,4-diamino-6-methyl-5-nitropyrimidine.[2790]

with dimethyl malonate to give 2-amino-4-dimethoxycarbonylmethyl-6-methyl-5-nitropyrimidine;[2788, 2789] when the same substrate reacts with ethyl cyanoacetate to give 2-amino-4-α-cyano-α-ethoxycarbonyl-methyl-6-methyl-5-nitropyrimidine;[2788, 2789] and when 2-chloro-4,6-diphenylpyrimidine reacts with malononitrile to give 2-dicyanomethyl-4,6-diphenylpyrimidine which appears to exist in the tautomeric form, 2-dicyanomethylene-1,2-dihydro-4,6-diphenylpyrimidine.[2795] 4-Chloro-2-*p*-chlorophenyl-6-methylpyrimidine and ethyl cyanoacetate in dimethylformamide with sodium hydride gave the 4-α-cyano-α-ethoxy-carbonylmethyl analogue.[3481]

The direct conversion of 4-iodo- into 4-cyano-2-methylthiopyrimidine (**123**; 54%) by cuprous cyanide in pyridine is probably unique in the series;[2608] such replacement is done usually *via* a trimethylammonio or sulpho grouping. The cyanoamino group may be introduced by the action of sodium cyanamide on a chloro pyrimidine. Thus 4-chloro-gives 4-cyanoamino-2,6-dimethylpyrimidine (**124**) by refluxing with the ethanolic reagent for 63 hr.[2791]

Despite early failures to produce simple pyridyl-[2792] or pyrimidinyl-phosphonic esters,[1396] 2-chloropyrimidine and tri-isopropyl phosphite evolved isopropyl chloride and gave 2-di-isopropoxyphosphinylpyrimidine (**125**; R = Pr^i).[2182] Rather similar procedures gave 2-diethoxy-phosphinyl-4,6-dimethylpyrimidine and 2-chloro-4-di-isopropoxyphos-phinylpyrimidine* (by selective attack of the 4-chloro substituent in 2,4-dichloropyrimidine),[2182, 2793] but attempts to replace the bromo atom in 5-bromouracil by similar means failed. The free acids, e.g., 2-dihydroxyphosphinylpyrimidine (**125**; R = H), were made by treating the esters with dry hydrogen bromide.[2182]

The reactions of 2-bromopyrimidine with pyridine-*N*-oxide to give 2-β-pyridyloxypyrimidine and α-ureidopyridine have been explored.[2794] 2-Bromopyrimidine, unlike its chloro isologue, undergoes the Ullmann reaction (copper in dimethylformamide) to give 2-pyrimidin-2′-ylpyrimidine (**126**; R = H) in 35% yield;[2604] 2-4′,6′-dimethylpyrimidin-2′-yl-4,6-dimethylpyrimidine (**126**; R = Me) may be made similarly from 2-bromo-4,6-dimethylpyrimidine,[2605] but not from 2-iodo-4,6-dimethylpyrimidine which is simply dehalogenated instead.[2603] On the other hand, 4-iodo-6-methyl-2-phenylpyrimidine yields 4-methyl-6-4′-methyl-2′-phenylpyrimidin-6′-yl-2-phenylpyrimidine (35%) under Ullmann conditions. The Busch biaryl synthesis (aryl halide in refluxing methanolic potassium hydroxide with hydrazine hydrate and palladium

* The phosphonic ester grouping survived the replacement of the chloro substituent by hydrogen (hydrogenation over Pd/MgO), by ethoxy (sodium ethoxide), and by dimethylamino (free amine).[2182]

on calcium carbonate) was successful in converting 4-iodo-2,6-dimethylpyrimidine into 4-2′,4′-dimethylpyrimidin-6′-yl-2,6-dimethylpyrimidine.[2603]

The fascinating ring transformations that occur during reactions of halogeno-heterocycles with nucleophiles are being explored by H. J. den Hertog, H. C. van der Plas, and their colleagues at Wageningen. When 4-chloro-2-phenylpyrimidine (**127**) is treated with potassium amide in liquid ammonia, 2-methyl-4-phenyl-1,3,5-triazine (**132**) and a little 4-amino-2-phenylpyrimidine (**131**) are formed.[2554] If a 4-^{14}C tag is incorporated in the substrate (**127**), the tag appears at the corresponding position in the triazine (**132**), strongly suggesting that ring fission occurs at the 5,6-bond of the pyrimidine (**127**) to give some such intermediate at (**128**) which recyclizes to the triazine (**132**).[2799] The tag also appears at the 4-position in the minor product (**131**), which must therefore be formed by a more normal addition-elimination (?) mechanism at the 4-position.[2800] 4-Chloro-2-ethyl(or methyl)pyrimidine gives analogous products; 4-chloro-2,5-dimethylpyrimidine (**129**) gives 2-ethyl-4-methyl-1,3,5-triazine (**130**) in which the ethyl group has arisen from the 5-carbon plus the 5-methyl group in the pyrimidine (**129**);[2554] in contrast, 4-chloro-2,6-dimethylpyrimidine yields only 4-amino-2,6-dimethylpyrimidine, showing that a free 6-position is necessary for the attack and fission needed to produce a triazine.[2554]

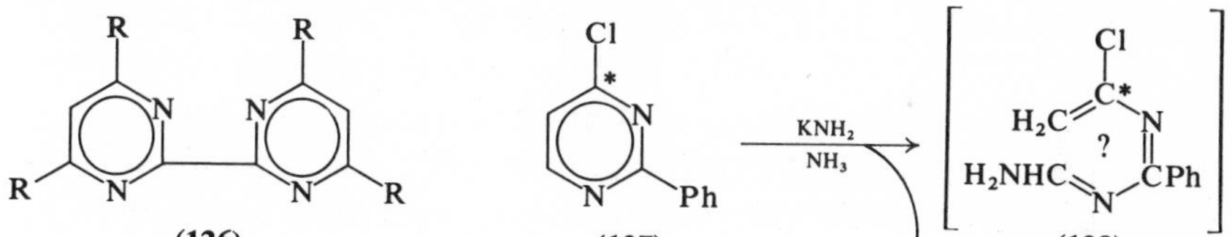

(**126**) (**127**) (**128**)

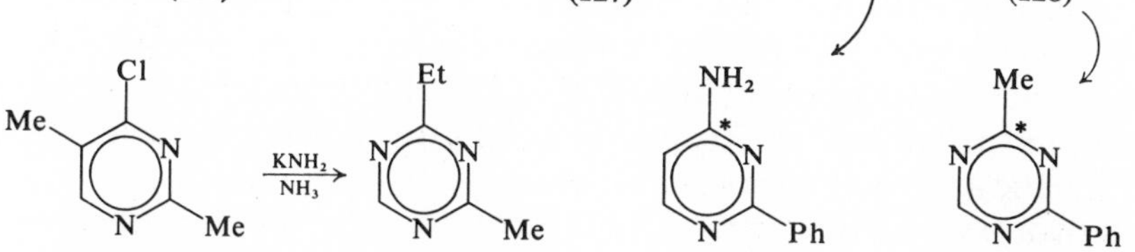

(**129**) (**130**) (**131**) (**132**)

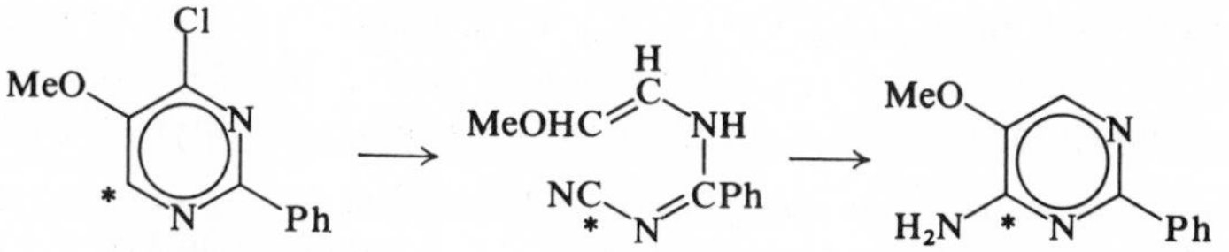

(**133**) (**134**) (**135**)

On treating 4-chloro-5-methoxy-2-phenylpyrimidine (**133**) similarly, no triazine is formed, but an acyclic intermediate (**134**) is isolated;[2801] this can be isomerized in boiling toluene to give 4-amino-5-methoxy-2-phenylpyrimidine (**135**), and tagging shows that the original C-6 of the substrate (**133**) is now C-4 in the product (**135**).* [2800, 2801] Such a fission mechanism is in sharp contrast to that operating in the formation of the simpler analogue (**131**).[2800, 2811] Some interesting imidazoles arise when 4-chloro-5-aminopyrimidines are treated similarly.[2801]

6. Reactions of 5-Halogenopyrimidines (*H* 210)

Although there have been few recent papers mentioning reactions of 5-halogenopyrimidines, at least eight distinct types of reaction have been described.

A. The Action of Amines on 5-Halogenopyrimidines (*H* 210)

Simple aminolysis by aq. ammonia at 135° for 50 hr. has been used to convert 5-bromo- into 5-amino-4-t-butylpyrimidine in 85% yield;[2574] 5-amino-4-methyl (or phenyl)pyrimidine were made similarly.[2574] The aminolysis of 5-bromo-2,4-dihydroxypyrimidine (**136**) by appropriate aromatic amines in ethylene glycol at 180° for 2–4 hr. gave 5-anilino-2,4-dihydroxypyrimidine (**137**) and seventeen analogues;[2462, 2567] *N*-alkylanilines similarly gave 2,4-dihydroxy-5-*N*-methylanilinopyrimidine and five other *N*-alkyl analogues;[2567] 2-amino-5-anilino-4-hydroxypyrimidine (**138**) and four analogues were made similarly from 2-amino-5-bromo-4-hydroxypyrimidine (**139**) but required the addition of sodium acetate to the reaction mixture to obviate conversion into the corresponding dihydroxypyrimidines, e.g., (**137**), under the acidic conditions pertaining.[2567] Analogous aminolyses by aqueous solutions of aliphatic amines at 100° gave 5-cyclohexylamino(or dimethylamino)-2,4-dihydroxypyrimidine and 2,4-dihydroxy-5-methylaminopyrimidine in good yield;[2462] and ethanolamine (or related amines) at 160° converted appropriate bromopyrimidines into 2,4-dihydroxy-5-β-hydroxyethylaminopyrimidine (80%), its 5-ethylamino analogue, 5-β-hydroxyethylamino-1(or 3)-methyluracil, and 5-bis-(β-hydroxyethyl)-amino-3-methyluracil.[2643] 5-Bromo-4-carboxymethylpyrimidine in

* The valuable and relevant observations of N. Okuda and I. Kuniyoshi[2765, 2843, 2844] should not be overlooked: for example, 4-chloro-2,6-dimethoxypyrimidine and sodium amide in liquid ammonia give mainly 2-amino-4,6-dimethoxypyrimidine.

refluxing morpholine (or piperidine) gives 4-methyl-5-morpholino-pyrimidine and its piperidine analogue in *ca.* 40% yield.[2880]

When 4-amino-5-bromo-2,6-dihydroxypyrimidine (**141**) was treated with piperidine or morpholine, the expected 4-amino-2,6-dihydroxy-5-piperidino(or morpholino)pyrimidine (**140**) resulted;[2802] when benzyl-amine, aniline, or *p*-toluidine was used, 4-benzylamino(or anilino)-2,6-di-hydroxy- or 2,4-dihydroxy-6-*p*-toluidino-pyrimidine (**142**) resulted,[2802] and these were later made unambiguously from 4-chloro-2,6-dihydroxy-pyrimidine (**143**).[2803] These facts remain unexplained.

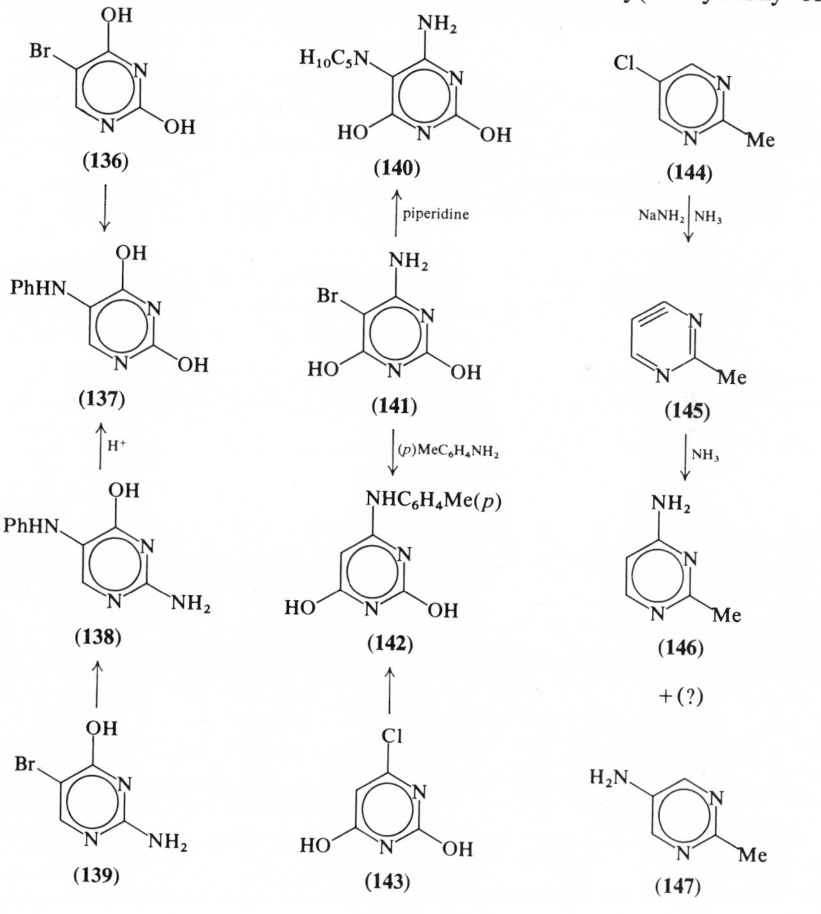

'Pyrimidyne' intermediates have been inferred in some rather related reactions. Thus 5-chloro-2-methylpyrimidine (**144**) and sodium amide in liquid ammonia give 4-amino-2-methylpyrimidine (**146**) and (prob-ably) 4-amino-5-methylpyrimidine (**147**), a result consistent with intermediate (**145**).[2804] Similarly, 5-bromo-4-methoxy(or hydroxy or phenyl)pyrimidine gives respectively 4-amino-6-methoxy(or hydroxy or phenyl)pyrimidine, but without any evidence of the 5-amino isomer.[2595]

B. Other Reactions of 5-Halogenopyrimidines (*H* 211)

Deliberate removal of a 5-bromo substituent in favour of hydrogen may be done by catalytic hydrogenation over palladium, as in the debromination of 2-amino-5-bromo-1,4-dihydro-4-imino-1-methylpyrimidine (hydrochloride) or 5-bromo-4-dimethylamino-1,2-dihydro-2-imino-1-methylpyrimidine (hydriodide);[2624] an alternative method is illustrated by the debromination (60% yield) of 2-amino-5-bromopyrimidine with hydrazine and a palladium catalyst.[2603] Unexpected removal of the 5-halogeno substituent occurred on boiling a glycol solution of 5-bromo-2,4-dihydroxypyrimidine or its 6-methyl derivative for several hours.[2805, *cf.* 2203]

Replacement of a 5-bromo by a cyano group is well illustrated in the formation (cuprous cyanide in boiling quinoline) of 5-cyanopyrimidine (previously made[2806] by a praiseworthy but tedious 10-stage route),[2607] its 2-methylamino derivative,[2376] and 2-amino-5-cyano-4-methyl(or 4,6-dimethyl)pyrimidine.[2602, 2633]

Examples of the replacement of a 5-halogeno by an alkoxy group are largely confined to sulphonamides. Thus 2-*p*-amino(or acetamido)-benzenesulphonamido-5-bromo(or iodo)pyrimidine reacts with sodium alkoxide in an alcohol at *ca.* 140° with a copper catalyst to give the corresponding 5-methoxy (e.g., **148**) ethoxy, or other alkoxy analogue.[2807–2810] In addition, 5-chloro- gives 5-ethoxy-4,6-dihydroxy-2-hydroxymethylpyrimidine by refluxing in ethanolic sodium ethoxide for 12 hr.; the yield was 80%.[2236]

In contrast, a 5-halogeno is frequently replaced by an alkylthio group as in the transformation of 5-bromo- into 5-methylthio-pyrimidine (refluxing ethanolic sodium methyl mercaptide for 12 hr.),[2619] of 2-amino-5-bromo-4-hydroxy- into 2-amino-4-hydroxy-5-phenyl(or substituted-phenyl)thiopyrimidine (thiophenol or an appropriate derivative with potassium carbonate in glycol at 150°),[2203, 2811] of 5-bromo- into 5-*p*-chlorophenylthio-2,4-dihydroxypyrimidine (**149**)[2203] and related compounds,[2811] of 5-bromo- into 5-2′-amino-4′-chlorophenylthio-4-hydroxy-2-piperidinopyrimidine (aq. ethanol at 80–90°),[2811] of 5-chloro-2,4-dihydroxy- into 2,4-dihydroxy-5-phenylthiopyrimidine (as from bromo isologue),[2812] of 4-amino-5-chloro- into 4-amino-5-carboxymethylthio-1-ethyl-1,2,3,6-tetrahydro-2,6-dioxo-3-propylpyrimidine (**150**; aq. thioglycollic acid at 100°),[2363] and in related cases.[2363, 2813]

5-Bromo-2-methylpyrimidine (**151**) could not be converted into 5-dibutoxyphosphinyl-2-methylpyrimidine (**153**) by the action of sodium dibutyl phosphite in refluxing toluene, but treatment of the bromopyrimidine with butyl-lithium (to form the intermediate, **152**) followed by dibutyl phosphorochloridate gave the desired product.[2182, 2793] Similarly, treatment of appropriate 5-bromopyrimidines with butyl-lithium furnishes solutions of 2,4-dibenzyloxypyrimidin-5-yl-lithium, 4,6-dimethoxypyrimidin-5-yl-lithium, the 2,4,6-trimethoxy analogue, 4,6-bismethylthiopyrimidin-5-yl-lithium, and other analogues having in general two electron-releasing groups.[2575, 2814, 2923, 3756] Without such groups, 5-bromopyrimidine yields its 5-lithium analogue only at $-110°$; at higher temperatures 4,5-addition* of the butyl-lithium occurs.[2924] These lithiated compounds may be converted by carbon

MeO; NHSO$_2$C$_6$H$_4$NH$_2$(p)

(148)

(p)ClH$_4$C$_6$S; OH; OH

(149)

HO$_2$CH$_2$CS; NH$_2$; Pr; O; O; Et

(150)

Br; Me — BuLi → [Li; Me]

(151) **(152)**

NaPO(OBu)$_2$ (crossed out); ClOP(OBu)$_2$

(BuO)$_2$OP; Me

(153)

dioxide into carboxylic acids, e.g., 5-carboxypyrimidine, 5-carboxy-4,6-dimethoxypyrimidine, or 5-carboxy-4,6-bismethylthiopyrimi-

* Such addition can be useful: 2-thienyl-lithium and 5(or 2)-bromopyrimidine yield adducts which can be aromatized with potassium permanganate to yield 5(or 2)-bromo-4-2′-thienylpyrimidine in good yield.[2924]

dine;[2575, 2924] by dimethylformamide into aldehydes, e.g., 5-formyl-2,4,6-trimethoxypyrimidines;[2575] by sulphur into polysulphides which on treatment with chloroacetic acid yield 5-carboxymethylthio-4,6-dimethoxypyrimidine and such like;[2575] and by benzaldehyde (or other aryl aldehyde) into 2,4-diethoxy-5-α-hydroxybenzylpyrimidine (and related compounds) which undergo oxidation to 5-benzoyl-2,4-diethoxypyrimidine, etc.[2814] In addition, 2,4-dibenzyloxypyrimidin-5-yl-lithium reacts with trimethyl borate to give 2,4-dibenzyloxy-pyrimidin-5-ylboronic acid, R—B(OH)$_2$, which on hydrogenolysis gives uracil-5-boronic acid, converted by subsequent oxidation into 2,4,5-trihydroxypyrimidine.[2923]

5-Bromopyrimidines undergo both the Ullmann and the Busch reaction to give pyrimidinylpyrimidines. Thus 5-bromo-4-carboxy-2-methylthiopyrimidine and 'copper bronze' in dimethylformamide gave 2-methylthio-5-2′-methylthiopyrimidin-5′-ylpyrimidine (40% with decarboxylation occurring during the reaction); and 5-bromo-2-phenyl-pyrimidine with hydrazine hydrate plus a palladium catalyst, or with copper bronze in dimethylformamide, gave 2-phenyl-5-2′-phenyl-pyrimidin-5′-ylpyrimidine (46%, 70%, respectively).[2603]

7. Reactions of Extranuclear Halogenopyrimidines (*H* 214)

For want of examples this section is no longer sub-divided, but reactions are treated in the same order as previously.

Preferential ω-alkoxylation is illustrated by the conversion of 2,4-dichloro-5-chloromethylpyrimidine (**154**) into 2,4-dichloro-5-ethoxy-methylpyrimidine (**155**; 87%, by 1 mole of sodium ethoxide in ethanol), into 2-chloro-4-ethoxy-5-ethoxymethylpyrimidine (**156**; 86%, by 2 moles of sodium ethoxide), or into 2,4-diethoxy-5-ethoxymethyl-pyrimidine (**157**; 88%, by an excess of ethoxide);[2596] the same type of preference can be achieved with other alkoxides and aryloxides.[2596] 5-Bromo(or chloro)methyl-2,4-dihydroxypyrimidine with sodium alk-oxide in an alcohol gives 5-ethoxymethyl-2,4-dihydroxypyrimidine or its methoxymethyl homologue,[2160, 2579] but a similar result can be achieved (more slowly?) by simply refluxing 5-bromomethyl-2,4-dihydroxy-6-methylpyrimidine in anhydrous methanol to give the corresponding methoxymethylpyrimidine in 90% yield;[2652] similarly, 4-chloro-2-chloromethylpyrimidine and ethanol give 4-chloro-2-ethoxy-methylpyrimidine.[2819]

The ease of alkaline hydrolysis of an ω-bromo group is shown in the formation of 4-amino-5-hydroxymethyl- from 4-amino-5-bromomethyl-2-methylpyrimidine on boiling in aq. sodium carbonate solution,[2817] and in the hydrolysis of 5-bromoacetyl- (**158**; R = Br) into 5-glycolloyl-2,4-dihydroxypyrimidine (**158**; R = OH).[2637]

The change from an ω-bromo to an acetoxy group is seen in the preparation of 5-acetoxymethyl- from 5-bromomethyl-2,4-dihydroxypyrimidine by treatment with anhydrous sodium acetate in acetic acid.[2160]

Aminolysis of ω-halogen substituents occurs very easily and it has been used more of recent years: 2,4-dichloro-5-chloromethylpyrimidine (**154**) and phenethylamine or morpholine in toluene at 25° give good yields of 2,4-dichloro-5-phenethylaminomethylpyrimidine* and its morpholinomethyl analogue* [**159**; R = $N(CH_2CH_2)_2O$], respectively;[2596] 5-chloromethyl- gives 5-dimethylaminomethyl-2,4-dihydroxypyrimidine by treatment with cold dimethylamine or at 150° with dimethylformamide (carbon monoxide evolved);[2640] its piperidinomethyl analogue is made similarly by piperidine in dioxane;[2160] 5-β-bromoethyl- gives 5-β-aminoethyl-2,4-dihydroxypyrimidine by initial treatment with potassium phthalimide followed by acidic hydrolysis

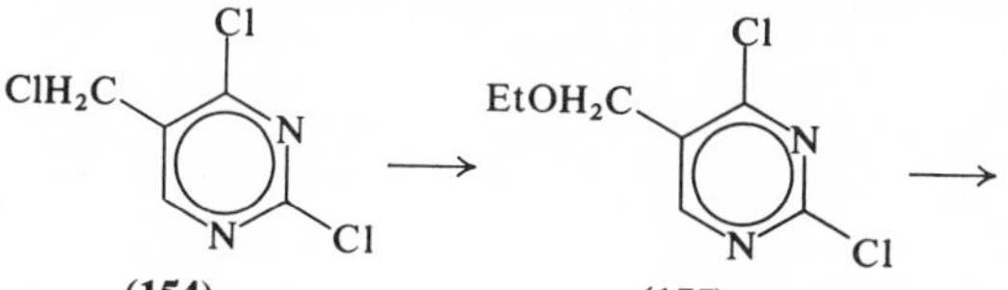

(**154**) (**155**)

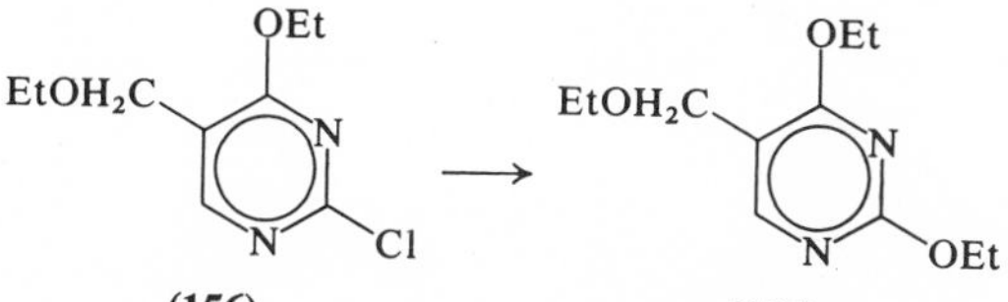

(**156**) (**157**)

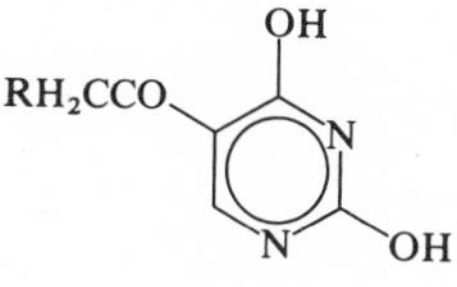

(**158**) 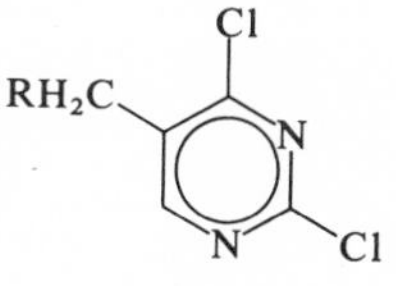 (**159**) 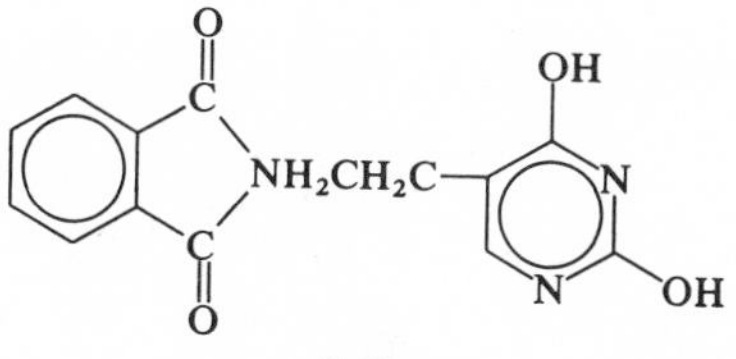 (**160**)

* These chloropyrimidines are very prone to hydrolysis: even standing in aq. acetone at room temperature gives the 2-chloro-4-hydroxy analogues.[2596]

of the intermediate (**160**);[2514] the corresponding 6-methyl derivative is made similarly;[2514] and other simple examples are described.[2231, 2656, 2674, 2751, 3193] 5-Chloromethyl-2,4-dihydroxypyrimidine (**161**; R = Cl) reacts in an interesting way with hydrazine or hydroxylamine: A. Giner-Sorolla and A. Bendich[2820] have shown that the products are, respectively, 5-hydrazonomethyl-2,4-dihydroxy- (**163**) and 2,4-dihydroxy-5-hydroxyiminomethyl-pyrimidine, presumably formed by dehydrogenation of the (expected) hydrazinomethyl- (**162**) and hydroxyaminomethyl-derivatives.[2820] These oxidations might be brought about by additional molecules of hydrazine or hydroxylamine which would thereby undergo reduction to two molecules of ammonia in one case and to one of ammonia plus one of water in the other,[2820] as in analogous reactions with phenacyl bromides.[2821] 2,4-Dihydroxy-5-mercaptomethylpyrimidine (**161**; R = SH) undergoes an analogous reaction to give the hydrazone (**163**).[2820] In appropriate compounds, an ω-chloro substituent attached to a pyrimidine can react with a ring nitrogen of the same molecule. Such a reaction is the cyclization of 4-β-chloroethylamino-1,2-dihydro-1-methyl-2-oxopyrimidine (**164**) by boiling in pyridine to give 4,5-dihydro-5-methyl-4-oxo-1,3a,5-triaza-indane (**165**), confirmed in structure by alkaline hydrolysis to 1-β-aminoethyl-1,2,3,6-tetrahydro-3-methyl-2,6-dioxopyrimidine (**166**);[2641] other such cyclizations are known.[2641, 2642, 3758]

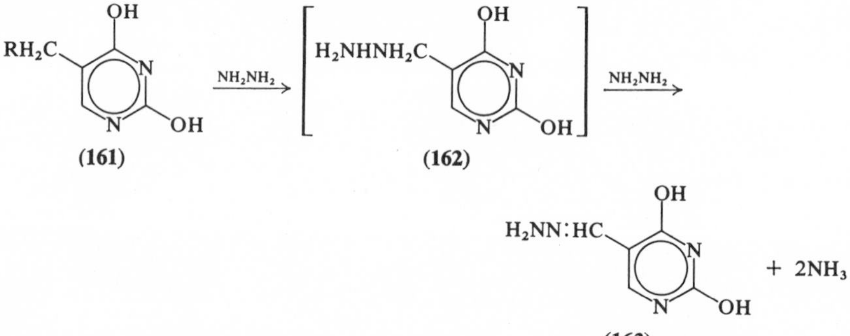

RH_2C, OH, N, N, OH $\xrightarrow{NH_2NH_2}$ [H_2NHNH_2C, OH, N, N, OH] $\xrightarrow{NH_2NH_2}$

(**161**) (**162**)

$H_2NN{:}HC$, OH, N, N, OH + $2NH_3$

(**163**)

HN—CH_2CH_2Cl, N, N, O, Me $\xrightarrow{\text{pyridine}}$ N, N, N, O, Me $\xrightarrow{NaOH}$ O, N, $CH_2CH_2NH_2$, N, O, Me

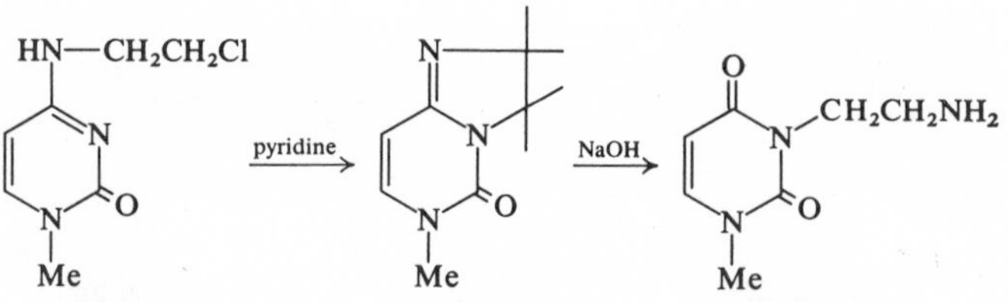

(**164**) (**165**) (**166**)

The aminolysis of a different type of halogenopyrimidine is seen in the conversion of 2-amino-5-bromoacetamido- into 2-amino-5-anilino-acetamido-4-hydroxy-6-methylpyrimidine by aniline in dimethyl sulphoxide at room temperature.[2303]

That an extranuclear bromo substituent may be replaced satisfactorily by a cyano group is indicated by the conversion (63%) of 2-amino-4-bromomethyl- into 2-amino-4-cyanomethyl-6-hydroxy-5-δ-phenylbutylpyrimidine at 100° with sodium cyanide in dimethylformamide.[2980]

Replacement of ω-chloro by sulphur-containing groups is exemplified in the conversion of 2,4-dichloro-5-chloromethylpyrimidine into

2,4-dichloro-5-methylthiomethyl- and 2,4-bismethylthio-5-methylthiomethyl-pyrimidine by appropriate proportions of alcoholic sodium methyl mercaptide;[2596] of 5-bromoacetyl-2,4-dihydroxy- into 2,4-dihydroxy-5-thiocyanatoacetyl-pyrimidine (alcoholic potassium thiocyanate at 50°);[2218] and of 5-chloromethyl-2,4-dihydroxy- (**168**) into 2,4-dihydroxy-5-thiocyanatomethyl- (**167**; potassium thiocyanate at 80°), 5-benzylthiomethyl-2,4-dihydroxy-* (**169**; benzyl mercaptan at 150°), 5-carboxymethylthiomethyl-2,4-dihydroxy- (thioglycollic acid at 150°), 5-acetimidoylthiomethyl-2,4-dihydroxy- [**173**; thioacetamide in dimethylformamide at 40°; converted by refluxing methanol into 2,4-dihydroxy-5-mercaptomethylpyrimidine (**170**) in 84% yield, and by hot water into 5-acetylthiomethyl-2,4-dihydroxypyrimidine (**174**; in 88% yield)], and 5-amidinothio-2,4-dihydroxy-pyrimidine (**172**; thiourea in dimethylformamide),[2640, 2650] as well as other cases.[2830]

The conversion of the trichloromethyl into the carboxy group has proved useful in making 4-carboxy-2-β-hydroxyphenethylamino- from 2-β-hydroxyphenethylamino-4-trichloromethyl-pyrimidine by the action of silver nitrate;[2597] the conversion of 5-bromo-2-tribromomethyl- into 5-bromo-2-carboxy-pyrimidine has now been described in detail,[2806]

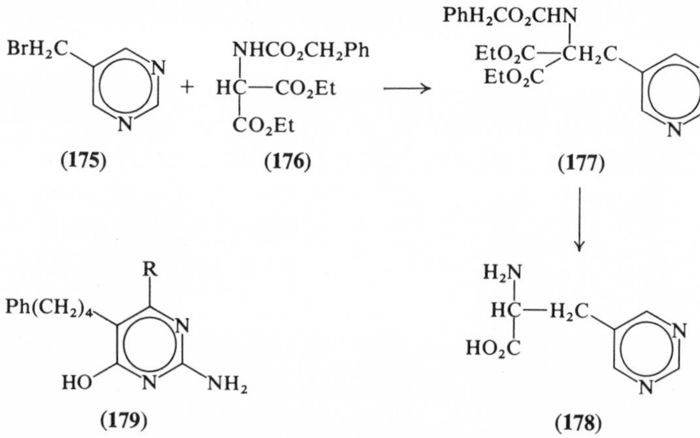

(**175**) (**176**) (**177**)

(**179**) (**178**)

but in addition, partial debromination with acetone gives 5-bromo-2-dibromomethylpyrimidine which gives 5-bromo-2-formylpyrimidine by treatment with silver nitrate in boiling aq. ethanol; and a related metathesis is 5-bromoacetyl- into 5-carboxy-2,4-dihydroxypyrimidine by warming in pyridine.[2218]

* Also prepared by benzyl chloride with the 5-mercaptomethyl derivative (**170**).[2650]

Reductive removal of ω-bromine is seen in the hydrogenolysis of 4-amino-5-bromomethyl-2-ethylpyrimidine to give the 5-methyl-analogue;[2270] and of chlorine from 5-chloro-methyluracil by zinc/DCl to give monodeuterated thymine.[2196]

Several extranuclear azido derivatives, each made from the corresponding halogeno analogue, are described: 5-azidomethyl-2,4-dihydroxypyrimidine (**171**; sodium azide in boiling acetonitrile),[2640] 5-azidoacetyl-2,4-dihydroxypyrimidine* (sodium azide in aq. alcohol at 25°),[2218] and the three azido derivatives from 5-chloromethyl-2,4-dichloropyrimidine, *viz.*, 5-azidomethyl-2,4-dichloro- (first), then 4-azido-5-azidomethyl-2-chloro, and finally 2,4-bisazido-5-azidomethylpyrimidine.[2596]

The condensation of ω-halogenopyrimidines with compounds containing an activated methylene group is most simply illustrated by the reaction of 5-bromomethylpyrimidine (**175**) with diethyl benzyloxycarbonylaminomalonate (**176**) in the presence of sodium alkoxide to give 5-β-benzyloxycarbonylamino-ββ-diethoxycarbonylethylpyrimidine (**177**) which can be degraded to 5-β-amino-β-carboxyethylpyrimidine (**178**; '5-pyrimidinylalanine').[2579] Such reactions have been thoroughly explored by Y. P. Shvachkin and his colleagues in Moscow, who have used them to produce a combination of several pyrimidines with a variety of amino acids.[2192, 2237, 2290, 2652, 2819, 2822–2827]

Another way to replace the C—Br bond by a C—C bond is by making use of the Wittig reaction. Thus 2-amino-4-bromomethyl-6-hydroxy-5-δ-phenylbutylpyrimidine (**179**, R = CH_2Br) was treated with triphenylphosphine to give the Wittig reagent, 2-amino-4-hydroxy-5-δ-phenylbutylpyrimidin-6-ylmethyl triphenyl phosphonium bromide (**179**, R = $CH_2P^+Ph_3\ Br^-$), which reacted with *p*-nitrobenzaldehyde to give 2-amino-4-hydroxy-6-*p*-nitrostyryl-5-δ-phenylbutylpyrimidine (**179**, R = $CN{:}CHC_6H_4NO_2$), or with *p*-nitrocinnamaldehyde to give the 6-*p*-nitrophenyl-1′,3′-butadien-1′-yl analogue (**179**, R = $CH{:}CHCH{:}CHC_6H_4NO_2$).[2980] The same type of product may be made in the reverse manner, by treating a pyrimidine aldehyde with a preformed Wittig reagent.

* Hydrogenation gives 5-glycyl-2,4-dihydroxypyrimidine.[2218]

CHAPTER VII

Hydroxy- and Alkoxy-pyrimidines (*H* 227)

In recent years much interest has been maintained in the fine (tautomeric) structure of hydroxypyrimidines in general, and of 4,6-dihydroxypyrimidine and barbituric acid derivatives in particular. This has involved X-ray and neutron diffraction studies, n.m.r. spectra, and more conventional spectral measurements.

Another area of activity has been the photo-dimers of uracil, thymine, and their derivatives. The process of dimerization appears to be of vital importance in the untoward effects produced in living tissue by irradiation.

1. Preparation of 2-, 4-, and 6-Hydroxypyrimidines (*H* 227)

No new methods of any importance for the synthesis of these hydroxypyrimidines have emerged recently. However, known methods have been extended considerably.

A and B. By the Principal and Other Primary Syntheses (*H* 227)

The formation of hydroxypyrimidines by a variety of known and new primary syntheses has been treated fully in Chs. II and III.

The direct introduction of a 2-hydroxy group occurred during the peroxide oxidation of 4,6-diamino-5-nitroso- to 4,6-diamino-2-hydroxy-5-nitro-pyrimidine (**1**) in 60% yield.[2445] The structure of the product was confirmed by unambiguous nitration of 4,6-diamino-2-hydroxypyrimidine.[2445]

C. By Hydrolysis of Halogenopyrimidines (*H* 228)

The hydrolytic conversion of halogeno- into hydroxy-pyrimidines has been discussed already (Ch. VI, Sect. 5.E). The odd, but quite practical, preparation of 4-hydroxypyrimidine (58%) by boiling 2,4-dichloropyrimidine in hydriodic acid containing red phosphorus (*H* 204) has been confirmed.[2231]

D. From Aminopyrimidines (*H* 229)

The three methods for converting amino- into hydroxy-pyrimidines (acid or alkaline hydrolysis; treatment with nitrous acid) have been compared using 2,4-diamino-5-*p*-chlorophenyl-6-ethylpyrimidine (**2**; R = *p*-Cl, R′ = Et) and several analogous fused-pyrimidines as substrates.[2983] The 2-amino group was removed by nitrous acid but was resistant to acid and to alkaline hydrolysis; the 4-amino group was unreactive towards nitrous acid but was removed by acid (and in two fused-pyrimidines, by alkaline) hydrolysis.[2983] These observations clearly apply only to such 2,4-diamino derivatives, and even then, only in general terms: thus 2,4-diamino-5-*o*-hydroxyphenylpyrimidine (**2**; R = *o*-OH, R′ = H) underwent acid hydrolysis to give apparently only 2-amino-4-hydroxy-5-*o*-hydroxyphenylpyrimidine (in 40% yield), but the 6-ethyl derivative (**2**; R = *o*-OH, R′ = Et) under similar conditions gave a 2:1-mixture of the isomers 2-amino-4-ethyl-6-hydroxy- and 4-amino-6-ethyl-2-hydroxy-5-*o*-hydroxyphenylpyrimidine.[2893]

Acid hydrolysis of aminopyrimidines is exemplified further in the formation of 4-hydroxy-2-phenyl- (>90%; by conc. hydrochloric acid at 150°),[2800] 4-hydroxy-2-methyl-6-phenyl- (in refluxing 20% v/v sulphuric acid),[2539] and 5-cyano-2,4-dihydroxy-pyrimidine (from 4-amino-5-cyano-2-hydroxypyrimidine by boiling dilute hydrochloric acid);[2897] also in the formation of 3-methyluracil from 4-acetamido-2,3-dihydro-3-methyl-2-oxopyrimidine[2984] or other amine,[2918] 2-amino-5-*p*-chlorophenyl-4-ethyl-6-hydroxypyrimidine (**2**; R = *p*-Cl, R′ = Et) in 95% yield from the corresponding diaminopyrimidine in boiling hydrochloric acid,[2983] and 3-benzyl-6-methyluracil (**3**; R = OH; >80%) from 1-benzyl-1,6-dihydro-2-hydrazino-4-methyl-6-oxopyrimidine (**3**; R = $NHNH_2$).[2985] The deamination of 4-amino-2-hydroxypyrimidine to 2,4-dihydroxypyrimidine has been reported[2916] to occur in aq. sodium thioglycollate at room temperature; acetic, trichloroacetic, or hydrochloric acid do not promote such deamination under comparable conditions.[2916]

The alkaline hydrolysis of aminopyrimidines has not been used much of late for preparative purposes: 2-methylamino- has been converted into 2-hydroxy-pyrimidine in good yield by alkali at 165°,[2986] and the original procedure[984] for preparing 2-hydroxy-* from 2-amino-pyrimidine by alkali at 120° (*H* 229) has been modified in detail to facilitate isolation.[2604, 2987, 2988] However two rather important studies of such alkaline hydrolytic deaminations have been reported. In the first,[2991, 2992] cytosine (**4**; R = H) and its derivatives were treated with *N*-alkali at 100° for an hour. The products were identified and estimated chromatographically and spectrometrically.[2991, 2992] Under these conditions, for example, cytosine (**4**; R = H) gave 2,4-dihydroxy-pyrimidine (4%), 4-amino-2-mercaptopyrimidine (thiocytosine) was unchanged, 4-amino-2-hydroxy- gave 2,4-dihydroxy-5-hydroxymethyl-

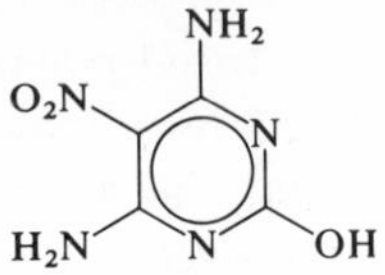

(**1**) 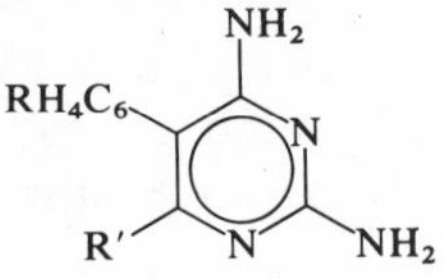 (**2**)

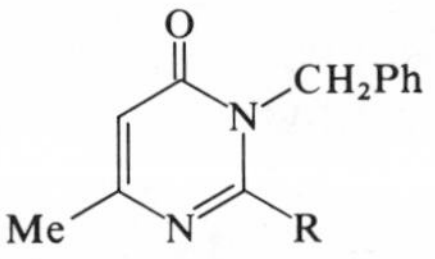

(**3**) 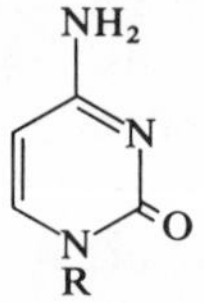 (**4**)

pyrimidine (7%), 4-amino-1,2-dihydro-1-methyl-2-oxopyrimidine (**4**, R = Me) gave 1-methyluracil (31%), and 4-amino-1,2-dihydro-1-*p*-nitrobenzyl-2-oxopyrimidine was completely deaminated to 1-*p*-nitrobenzyluracil; these and other figures indicated that electron-withdrawal by the substituent tended to increase the extent of deamination.[2992] In the second study, the precise kinetics for hydrolytic deamination by 1–6*M*-potassium hydroxide were measured over a range of temperatures for several aminopyrimidines.[2993] A convenient comparison may be made by abstracting the pseudo first-order rate constants (10^4k

* 2-Hydroxypyrimidine was first obtained[894] in a London laboratory as a solid, m.p. *ca.* 160°, but this changed spontaneously into a form with m.p. 178–180° during the third recrystallization prior to analysis. Thereafter, the lower-melting form was never obtained from subsequent batches prepared in the same laboratory during the next five years.[171] However, later batches prepared in Canberra,[171] Lafayette,[2604, 2989] Cambridge[2987, 2990] and elsewhere all initially gave the lower-melting form (now characterized and analysed[2987]) which in each case was converted into the more stable polymorph, m.p. 178–180°, by seeding with material supplied from the original specimen.[894]

min.$^{-1}$) for deaminations in an excess of 5*M*-potassium hydroxide at 80°: 2-amino- (10.3), 4-amino- (6.7), 2-methylamino- (3.4), 4-methylamino- (3.3), 2-amino-4,6-dimethyl- (2.4), 4-amino-2,6-dimethyl- (1.3), 2-amino-5-bromo- (94) and 4-amino-5-bromo-pyrimidine (155).[2993] It is clear that electron-withdrawal (by bromine) facilitates deamination, and electron-release (by methyl) retards the process. The original paper[2993] should be consulted for interesting details and conclusions: for example, the rate of hydrolytic deamination in concentrated alkali shows an exponential catalytic dependence on the concentration of hydroxide.

The use of nitrous acid in replacing a primary amino by an hydroxy group is exemplified in the formation of 2-hydroxy-4,6-dimethylpyrimidine (70%),[2601] 4-t-butyl-2-hydroxypyrimidine (30%),[2602] 2-hydroxy-4,6-diphenylpyrimidine (diazotization in concentrated sulphuric acid),[2556] 2-benzylthio-4-chloro-6-hydroxypyrimidine (99% using acetic acid and sodium nitrite),[2589] 4-chloro-6-hydroxy-2-methylthiopyrimidine (85%),[2589] 4-hydroxy-6-methoxy-2-methylthiopyrimidine (90%),[2589] 1-benzyl(or methyl)uracil (from 4-amino-1-benzyl-1,2-dihydro-2-oxopyrimidine or its 1-methyl-homologue),[2337] 2-hydroxy-5-nitropyrimidine,[2431] and 4-chloro-6-hydroxypyrimidine (63%).[2777] See also ref. 2488.

E. From Alkoxypyrimidines (*H* 232)

A methoxy group is often easier to hydrolyse to an hydroxy group than is a chloro substituent in a similar molecular environment. This is evident (neglecting the mild effect of each group on the other) in the hydrolysis (82%) of 4-amino-6-chloro-2-methoxy- (**5**, R = Me) to 4-amino-6-chloro-2-hydroxy-pyrimidine (**5**, R = H) by 10% alkali,[2837] of 4-chloro-2,6-dimethoxy- to 4-chloro-2,6-dihydroxy-pyrimidine (*ca.* 60%) by hydrochloric acid,[2165, 2243, 2611, 2645] and of 4-bromo(or iodo)-2,6-dimethoxy- to the corresponding dihydroxypyrimidines (52%, 73%) by hydrobromic acid or by sodium iodide in dimethylformamide, respectively.[2611] The two-step process for converting a chloro- into an hydroxy- *via* an alkoxy-pyrimidine is often preferred to direct hydrolysis, especially in the presence of other groups sensitive to hydrolytic conditions. The process is seen in the conversion of 4-benzylamino-2-chloro- (**6**) into 4-benzylamino-2-ethoxy- (**7**) and thence by warming with concentrated hydrochloric acid for a few minutes into 4-benzylamino-2-hydroxy-pyrimidine (**8**);[2674] in the conversion of 4-amino-2-chloro- into 4-amino-2-methoxy- and thence by acid into 4-amino-2-hydroxy-pyrimidine;[2766] and in similar sequences leading to 2-β-amino-β-carboxyethyl-4-hydroxypyrimidine,[2192] and other hydroxypyrimidines.[2674]

The conversion of alkoxy- into hydroxypyrimidines is further exemplified in the formation of 2-hydroxy- (10% sulphuric acid),[2759] 4-amino-2-hydroxy- (ethereal pyridine hydrochloride),[2677] 2-hydroxy-4-sulphanilylamino- (ethereal pyridine hydrochloride or aq. hydriodic acid),[2677] 2-dimethylamino-4-hydroxy- (hydrochloric acid),[2676] 4-amino-6-dimethylamino-2-hydroxy- (alkali),[2837] 5-benzoyl-2,4-dihydroxy- (hydrochloric acid),[2814] 1,6-dihydro-4-hydroxy-1-methyl-6-thio- (alkali),[2760] 5-bromo-2,4-dihydroxy- (hydrogen bromide in aq. acetic acid),[2994] and 2,4-dihydroxy-6-methyl-5-nitro-pyrimidine (hydrochloric acid; rates roughly determined at 50, 75, and 100°);[2437] also in the formation of 1-methyluracil[2995] (**10**, R = H) and its 5-fluoro derivative[2572, 3226] (**10**, R = F) from 1,6-dihydro-4-methoxy-1-methyl-6-oxopyrimidine (**9**, R = H) and its 5-fluoro derivative (**9**, R = F), respectively, by acid, and in other examples.[2996, 3002]

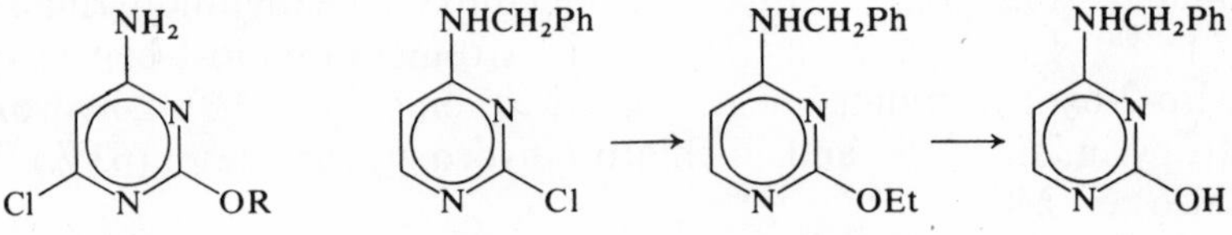

(**5**) (**6**) (**7**) (**8**)

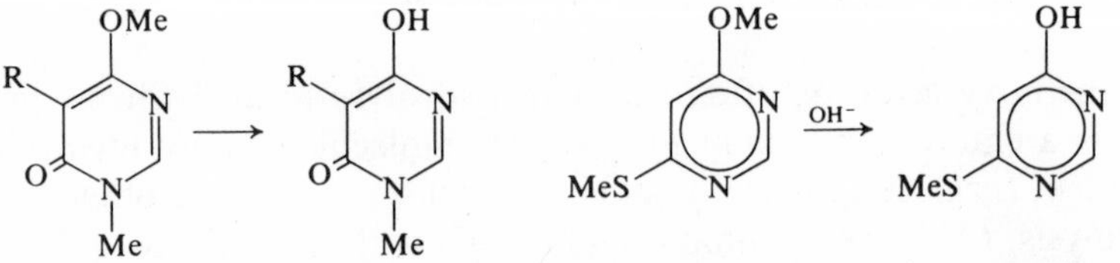

(**9**) (**10**) (**11**) (**12**)

Selective dealkylation of 4,6-dibenzyloxypyrimidine and its 2- or 5-methyl(or phenyl)derivatives is possible with hydrogen chloride in acetonitrile to give 4-benzyloxy-6-hydroxypyrimidine and appropriate derivatives.[3001, 3767] The relative ease of hydrolysing a methoxy and a methylthio group in the same molecule (again neglecting mutual electronic effects) is indicated by the alkaline hydrolysis of 4-methoxy-6-methylthiopyrimidine (**11**) giving a single product, 4-hydroxy-6-methylthiopyrimidine (**12**).[2997]

F. From Mercaptopyrimidines and Related Derivatives (*H* 233)

Mercaptopyrimidines continue to be used widely as sources of hydroxypyrimidines *via* known indirect routes.

(1) *By S-Alkylation and Hydrolysis of the Alkylthiopyrimidines* (*H* 233)

The alkylation step is treated later [Ch. VIII, Sect. 1.D(2)]; the second step is exemplified in the acid hydrolysis* to the corresponding hydroxy compound of 4-dimethylamino-2-methylthiopyrimidine,[2745] 2-benzylthio-4-morpholinopyrimidine,[2910] 4-hydroxy-5-β-hydroxyethyl-2-methylthiopyrimidine,[2202] 4-hydroxy-5-methoxy-2-methylthiopyrimidine,[2586] 1-benzyl-1,6-dihydro-4-methyl-2-methylthio-6-oxopyrimidine,[2985] and 1-benzyloxy-1,4-dihydro-2-methylthio-4-oxopyrimidine (to 1-hydroxyuracil).[2964] Like alkoxy groups, alkylthio groups are often hydrolysed more easily than a chloro group in the same molecule: 4-amino-6-chloro-2-hydroxypyrimidine (**14**) is the sole product (70–90%) from acid hydrolysis of the corresponding 2-methylthiopyrimidine (**13**),[2475, 2777] and only 6-chloro-3-methyluracil (**15**, R = OH) is formed from 4-chloro-1,6-dihydro-1-methyl-2-methylthio-6-oxopyrimidine (**15**, R = SMe) in acid.[2440]

The chloroacetic acid version of this process (*H* 233) still proves useful: 1- and 3-methyluracil have been made in good yield from the corresponding 2-thiouracils with aq. chloroacetic acid,[2195, 2311] and so has 2,4-dihydroxy-5-*p*-methoxybenzylpyrimidine;[2213] the process for making cytosine from thiocytosine (4-amino-2-mercaptopyrimidine) (*H* 234) with chloroacetic acid has been modernized;[2915] other cytosine derivatives so formed include 4-amino-1-benzyl-1,2-dihydro-2-oxopyrimidine,[2337] its 1-methyl homologue,[2337] 2-hydroxy-4-methylaminopyrimidine,[2998] and several 4-amino-5-aryl-2-hydroxypyrimidines (**16**, R = OH) *via* isolated 2-carboxymethylthio intermediates (**16**,

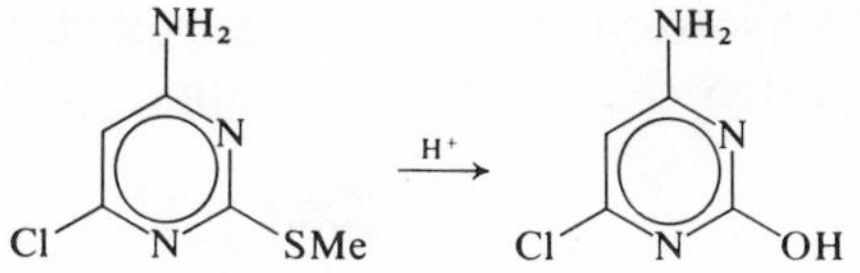

(**13**) (**14**) 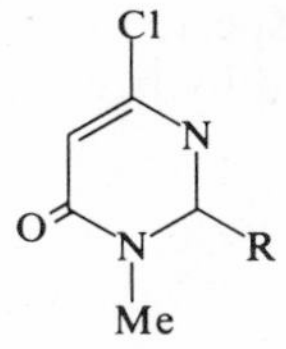 (**15**)

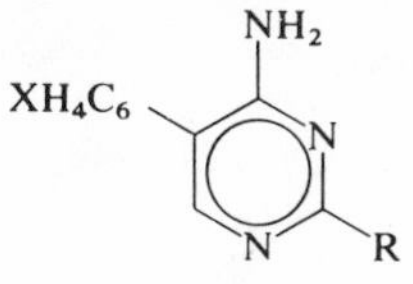

(**16**) 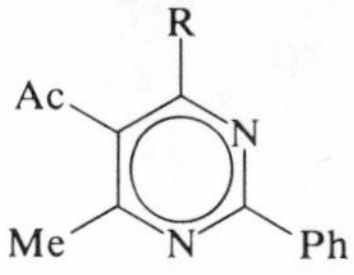 (**17**) 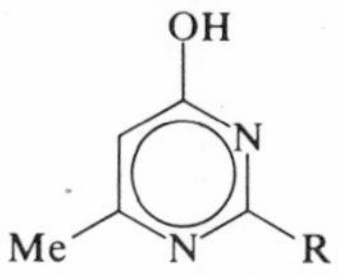 (**18**)

* Colleagues are apt to complain if the fume-hood is ineffective.

$R = SCH_2CO_2H$);[2274, 2893, 2983] 4,6-diethyl-5-*o*-hydroxyphenyl-2-hydroxypyrimidine and related compounds have been made similarly *via* isolated 2-carboxymethylthiopyrimidines.[2893]

(2) *By Oxidation to Sulphinic or Sulphonic Acid and Hydrolysis* (*H* 234)

This oxidative route to hydroxypyrimidines has been little used recently. However, 5-acetyl-4-mercapto-6-methyl-2-phenylpyrimidine (**17**, R = SH) has been converted satisfactorily into its hydroxy analogue (**17**, R = OH) by hydrogen peroxide in aq. ethanolic alkali,[2314] and 4-hydroxy-2-mercapto-6-methylpyrimidine (**18**, R = SH) has been oxidized with permanganate to 4-hydroxy-6-methyl-2-sulphopyrimidine (**18**, $R = SO_3H$; isolated in 57% yield as the potassium salt) which underwent facile hydrolysis in acid to 2,4-dihydroxy-6-methylpyrimidine (**18**, R = OH).[2999] The successful oxidation of nontautomeric thiopyrimidines to oxopyrimidine cannot proceed *via* a sulphonic or sulphinic acid, and is discussed later [Ch. X, Sect. 1.A(6)].

(3) *By S-Alkylation, Oxidation to a Sulphone or Sulphoxide, and Hydrolysis* (*H* 236)

The first two steps in this useful process are discussed in Ch. VIII, Sects. 1.D(2) and 5.A, respectively. The hydrolytic step is exemplified by the conversion of 2-methylsulphonyl- (**19**, R = H) and 4-methylsulphinyl-pyrimidine (**20**) by cold alkali into 2- and 4-hydroxypyrimidine, respectively,[2619] of 5-hydroxy(or methoxy)-2-methylsulphonyl- into 2,5-dihydroxy-(or 2-hydroxy-5-methoxy)-pyrimidine by alkali,[3000]

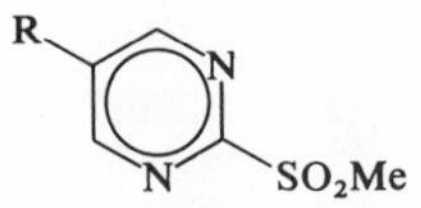

R N N SO_2Me

(19)

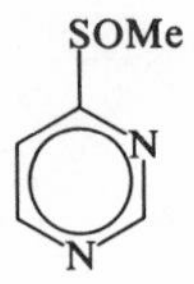

SOMe N N

(20)

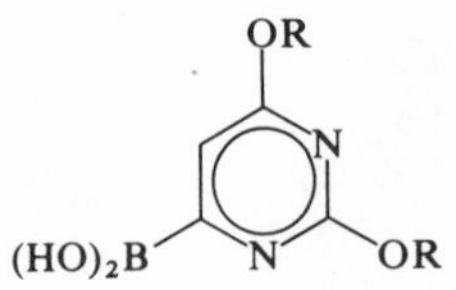

OR N N $(HO)_2B$ OR

(21)

$N{=}CHNMe_2$ N N $Me_2NHC{=}N$ SCN

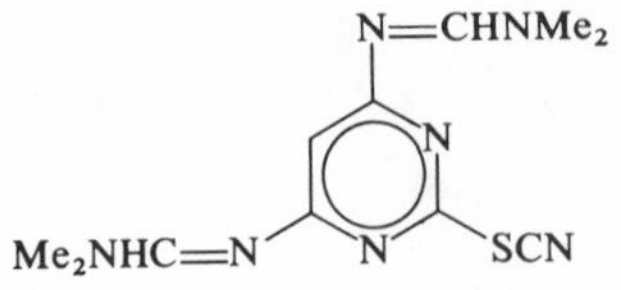

(22)

of 5-benzyloxy-2-methylsulphonyl- into 5-benzyloxy-2-hydroxy-pyrimidine by alkali,[2585] of 4-ethoxy-2-ethylsulphonyl- into 4-ethoxy-2-hydroxy-5-methylpyrimidine,[2582] of 5-fluoro-2-methylsulphonylpyrimidine (**19**, R = F) and its 4-methyl derivative by alkali into 5-fluoro-2-hydroxypyrimidine and its 4-methyl derivative, respectively,[2679] and of 4,6-dichloro-2-methylsulphonyl- into 4,6-dichloro-2-hydroxy-pyrimidine by brief treatment with 'slightly warm' alkali.[2165] 4-β-Hydroxyphenethylamino-2-methylthiopyrimidine and hydrogen peroxide in acetic acid eventually give the corresponding 2-hydroxypyrimidine in small yield.[2597] Although the preparation of 2-hydroxy-4-methoxypyrimidine has been claimed (*H* 236) by the peroxide oxidation of 2-ethylthio-4-methoxypyrimidine, the work was unrepeatable;[3399] the authentic material has been made now from 1,2-dihydro-4-methoxy-2-oxo-1-tetrahydro-2′-furylpyrimidine and hydrogen chloride, and its structure was confirmed by conversion into cytosine.[3399]

G. By Reductive Cleavage of Benzyloxypyrimidines (*H* 237)

Benzyloxy is unique among alkoxy groups in giving an hydroxy group by hydrogenolysis as well as by hydrolysis. This has been utilized to avoid hydrolytic conditions in converting 2-amino-4-benzyloxy-6-fluoro- into 2-amino-4-fluoro-6-hydroxy-pyrimidine;[2610] in similarly making its isomer, 4-amino-6-fluoro-2-hydroxypyrimidine,[2610] and its analogue, 2-fluoro-4,6-dihydroxypyrimidine;[2737] and in converting 2,4-dibenzyloxypyrimidin-5-ylboronic acid (**21**, R = CH_2Ph) into uracil-5-boronic acid (**21**, R = H).[2923]

H. By Other Methods (*New*)

The rare hydrolysis of a cyano- to an hydroxy-pyrimidine is illustrated by the formation of uracil (43%) on treatment of 2-chloro-4-cyanopyrimidine with boiling alkali;[2954] and of 5-cyano-4-hydroxy- from 4,5-dicyano-2-methylpyrimidine.[2175] 4,6-Bisdimethylaminomethyleneamino-2-thiocyanatopyrimidine (**22**) has been converted into 2,4,6-trihydroxypyrimidine (60%) by acid hydrolysis.[2859]

2. Preparation of 5-Hydroxypyrimidines (*H* 237)

The recent information on synthesis of 5-hydroxypyrimidines is of such a nature that it seems best to list each compound with its method(s) of formation.

5-Hydroxypyrimidine has been made (46%) by refluxing 5-methoxypyrimidine with potassium hydroxide in glycol.[2323] *4,5-Dihydroxypyrimidine* (**23**) was formed (as indicated)* by acid hydrolysis of 5-benzamido-4-hydroxy-, 5-amino-4-hydroxy- (75%), 4,5-diamino- (95%), 5-benzyloxy-4-hydroxy- (68%), 4-amino-5-benzyloxy-, 4-benzamido-5-benzyloxy-, 4-hydroxy-5-methoxy- (73%), 4-mercapto-5-methoxy- (low yield), or 4-amino-5-methoxy-pyrimidine (80%);[2584] also by desulphurization and debenzylation with Raney nickel of 5-benzyloxy-4-hydroxy-2-mercaptopyrimidine.[2584] *4-Amino-5-hydroxypyrimidine* was made (72%) by hydrogenolysis of 4-amino-5-benzyloxypyrimidine.[2584] *2,5-Dihydroxypyrimidine* was made (70%) by brief hydrobromic acid hydrolysis of 5-benzyloxy-2-hydroxypyrimidine.[2585] *5-Hydroxy-2-methylthiopyrimidine* (49%) survived as the product of a long vigorous acid hydrolysis of 5-benzyloxy-2-methylthiopyrimidine![2585] *2-Chloro-5-hydroxypyrimidine* (29%) resulted from treatment of its 2,4-dichloro analogue with zinc dust in water.[2585] *2-Amino-4,5-dihydroxypyrimidine* (80%) came from acid hydrolysis of its 5-benzyloxy analogue.[2210] The isomeric, *4-amino-5,6-dihydroxypyrimidine* was formed (60%) by coupling 4,5-dihydroxypyrimidine with diazotized *p*-chloroaniline and reducing the resulting 4-*p*-chlorophenylazo-5,6-dihydroxypyrimidine; also (12%) by the action of ammonium persulphate on 4-amino-6-hydroxypyrimidine and acid hydrolysis of the resulting pyrimidin-5-yl hydrogen sulphate.[2210] *2,4,5-Trihydroxypyrimidine* (isobarbituric acid; 75%), and *4,5-dihydroxy-2-methylpyrimidine* (71%) were made by acid hydrolysis of their respective 5-benzyl ethers;[2210] the former also by the action of nitrous acid and 5-aminouracil.[3750] *5-Hydroxy-1(or 3)-methyluracil* were made from their 5-methoxy analogues by vigorous acid hydrolysis.[2198] Two new syntheses of *divicine* (2,4-diamino-5,6-dihydroxypyrimidine) are discussed later (Sect. 8.C).

3. Preparation of Extranuclear Hydroxypyrimidines (*H* 241)

All the known methods for making ω-hydroxypyrimidines are represented in recent examples. 5-Hydroxymethylpyrimidines and their derivatives have been reviewed.[3150]

* This satisfying study by J. F. W. McOmie and his colleagues must be uniquely deep for a pyrimidine without direct biological connexions.

A. By Primary Synthesis (*H* 242)

Examples in this category of synthesis have been given in Chs. II and III.

(23)

B. From Amino Derivatives (*H* 242)

The dihydrochloride of 4-amino-5-aminomethyl-2-ethylpyrimidine (**24**, R = NH_2; made by primary synthesis) has been converted into 4-amino-2-ethyl-5-hydroxymethylpyrimidine (**24**, R = OH) in 80% yield by treatment with aq. sodium nitrite at 60°.[2270] 5-Aminomethyl- has been converted into 5-hydroxymethyl-2,4-dimethylpyrimidine (40%).[3007]

C. From Halogeno Derivatives (*H* 243)

Examples of this method have been discussed in Ch. VI, Sect. 7.A.

D. By Reduction of Esters (*H* 243)

This straightforward method is well represented by several lithium aluminium hydride reductions of methyl or ethyl esters in tetrahydrofuran or ether to give the corresponding hydroxymethylpyrimidines. The method was used to reduce 5-ethoxycarbonyl-2-methylpyrimidine (21% yield),[2654] 4-hydroxy-5-methoxycarbonyl-2-methylpyrimidine (32%),[3003] 2-amino-4-methoxycarbonylpyrimidine (10%),[3004] 5-ethoxycarbonyl-2-methyl-4-methylaminopyrimidine (42–70%),[2651, 2653] 4-dimethylamino-5-ethoxycarbonyl-2-methylpyrimidine (74%),[2651, 3005] 4-chloro-5-ethoxycarbonyl-2-methylpyrimidine (49%),[2651] 5-ethoxycarbonyl-4-methoxy-2-methylpyrimidine (**25**)* (9%),[2651] 5-ethoxycarbonyl-2,4-dimethylpyrimidine (54%),[3006, 3007] 5-ethoxycarbonyl-4-hydrazino(or methylamino)-2-methylthiopyrimidine (67% each),[3474] 4-aziridino-5-ethoxycarbonyl-2-methylthiopyrimidine (46%),[2255] 4-amino-5-ethoxycarbonyl-2-methoxypyrimidine (69%),[2205] and other such esters.[3005] 4-Ethoxycarbonylmethyl-2,6-dihydroxy- gave 2,4-dihydroxy-6-β-hydroxyethyl-pyrimidine (62%).[2674]

E. By the Action of Formaldehyde (*H* 243)

The reaction of formaldehyde with activated methyl groups to give β-hydroxyethylpyrimidine[2424] has been discussed in Ch. IV, Sect. 2.C.

F. From Pyrimidine Aldehydes or Ketones (*H* 244)

The direct reduction of a pyrimidine aldehyde to a primary alcohol is exemplified by the hydrogenation of 2-amino-4-formyl-6-hydroxy-

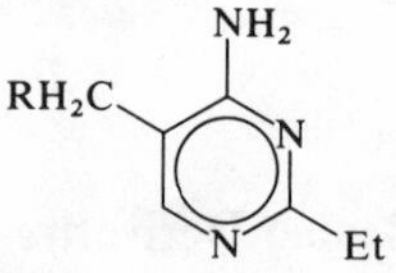

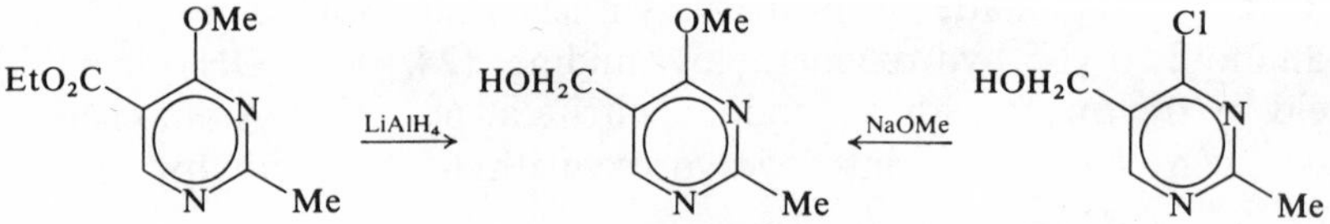

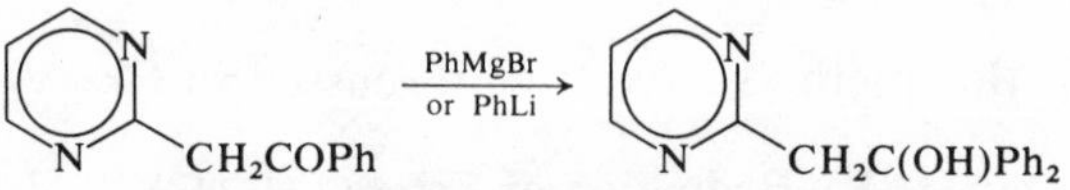

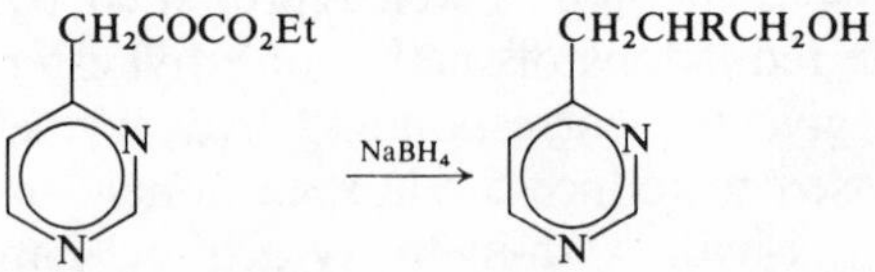

* 5-Hydroxymethyl-4-methoxy-2-methylpyrimidine (**26**) is better made (43%) from its 4-chloro analogue (**27**) and methanolic sodium methoxide. The 4-mercapto and methylthio analogues can only be made by such nucleophilic displacement.[2651]

over palladium to give 2-amino-4-hydroxy-6-hydroxymethyl-5-δ-phenylbutylpyrimidine (80%).[2980] The action of phenyl magnesium bromide on 2-phenacylpyrimidine (**28**) gave 81% of the secondary alcohol, 2-β-hydroxy-ββ-diphenylethylpyrimidine (**29**); replacement of the Grignard reagent by phenyl-lithium (under appropriate conditions) gave only 25% of the same alcohol (**29**).[3008] Reduction of 4-ethoxyoxalylmethylpyrimidine (**30**) with sodium borohydride in methanol gave a separable mixture of 4-βγ-dihydroxypropylpyrimidine (**31**, R = OH) in 33% yield, and 4-γ-hydroxypropylpyrimidine (**31**, R = H) in 14% yield.[3009] 4-Formyl-5-hydroxymethyl- gave 4,5-bis-hydroxymethyl-2-methylthiopyrimidine (62%) using sodium borohydride in methanol.[3443]

G. By Hydroxyalkylation and Other Means (*H* 244)

The examples in this section do not represent an exhaustive survey but simply illustrate the formation of some less usual types of ω-hydroxypyrimidines. Hydroxyalkylation by β-chloro(or bromo)ethanol (ethylene halohydrins) occurs in the conversion of 2-mercapto- into 2-β-hydroxyethylthio-4,6-dimethylpyrimidine (**32**),[3010] of 2-hydroxy-4,6-dimethyl- into 1,2-dihydro-1-β-hydroxyethyl-4,6-dimethyl-2-oxopyrimidine (**33**),[3010] of 2-amino-4-hydroxy-6-methyl- into a separable 1:1 mixture of 2-amino-1,4-dihydro-1-β-hydroxyethyl-6-methyl-4-oxo- and 2-amino-1,6-dihydro-1-β-hydroxyethyl-4-methyl-6-oxo-pyrimidine,[3010] of 6-amino-1-dimethylaminouracil into 4-amino-3-dimethylamino-1,2,3,6-tetrahydro-1-β-hydroxyethyl-2,6-dioxopyrimidine,[2291] and of 2-amino- into 1,2-dihydro-1-β-hydroxyethyl-2-iminopyrimidine (**34**).[2633] Hydroxyethylation by ethylene oxide is illustrated in the transformation of 5-ethylamino-2,4-dihydroxy- (**36**) into 2,4-dihydroxy-5-ethyl-(β-hydroxyethyl)aminopyrimidine (**37**, R = Et) or 5-ethyl-(β-hydroxyethyl)amino-3-β-hydroxyethyluracil (**38**) according to the conditions,[2643] of 2-amino- into 2-β-hydroxyethylaminopyrimidine (**35**),* [3011] of 5-amino- into 5-bis-(β-hydroxyethyl)amino-methyl-2,4-dihydroxypyrimidine,[2647] of 4-amino-1,2-dihydro-1-methyl-2-oxo- (**39**) into 1,2,3,4-tetrahydro-3-β-hydroxyethyl-4-imino-1-methyl-2-oxo- (**40**)† and its 4-β-hydroxyethylimino analogue (**41**),[3012] and in other cases.[2336, 2643] Alkylation with ethylene carbonate similarly changed 2,4-dihydroxypyrimidine into a separable mixture of 1- and 3-β-hydroxyethyluracil and 1,2,3,4-tetrahydro-1,3-bis-β-hydroxyethyl-2,4-

* This has been made unambiguously from 2-chloropyrimidine and ethanolamine;[2642] also by rearrangement of 1,2-dihydro-1-β-hydroxyethyl-2-iminopyrimidine (**34**).[2633]

† This initial product also underwent Dimroth rearrangement in alkaline media to give 1,2-dihydro-4-β-hydroxyethylamino-1-methyl-2-oxopyrimidine (**42**) which reacted with ethylene oxide to give the second product (**41**) by an alternative route.[3012]

dioxopyrimidine.[3013] The formation of an ω-hydroxypyrimidine by alkylation with an agent bearing a protected (or potential) alcoholic hydroxy group is exemplified in the reaction of 2,4-dihydroxypyrimidine with δ-chlorobutyl *p*-nitrobenzoate to give 1-δ-*p*-nitrobenzoyloxybutyluracil (**43**) which underwent deacylation by boiling with methanolic butylamine to give 1-δ-hydroxybutyluracil.[2547]

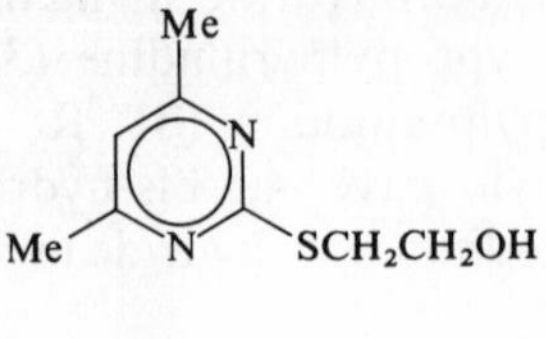

Me, N, Me, N, SCH_2CH_2OH **(32)**; Me, N, O, Me, N, CH_2CH_2OH **(33)**

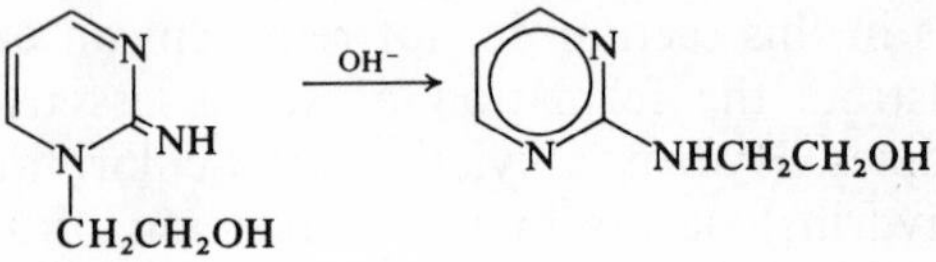

N, NH, N, CH_2CH_2OH **(34)** $\xrightarrow{OH^-}$ N, N, $NHCH_2CH_2OH$ **(35)**

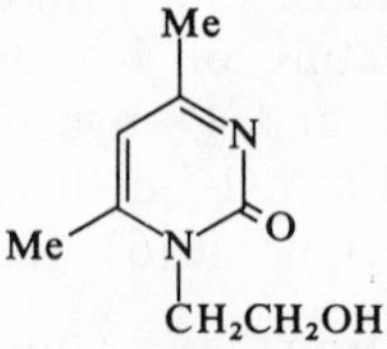

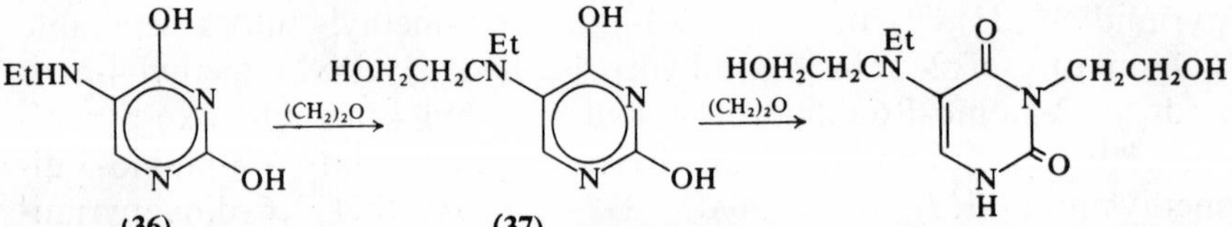

OH, EtHN, N, N, OH **(36)** $\xrightarrow{(CH_2)_2O}$ OH, Et, HOH_2CH_2CN, N, N, OH **(37)** $\xrightarrow{(CH_2)_2O}$ Et, O, HOH_2CH_2CN, N, CH_2CH_2OH, N, H, O **(38)**

Aminolysis of chloro, methoxy, methylthio, and such like pyrimidines by ethanolamine and other related amines gives rise to ω-hydroxylated pyrimidines. The process is illustrated by the formation of 4-amino-6-β-hydroxyethylamino-5-nitropyrimidine (83%) from the 6-chloro analogue,[2724] and by many other examples in Ch. VI and elsewhere.

An unusual but logical reaction sequence is that of 5-hydrazinocarbonylmethyl-2-methylpyrimidine (**44**) with nitrous acid (twice) to give 5-hydroxymethyl-2-methylpyrimidine (**45**) in 31% yield.[2654]

4. Preparation of Alkoxy- and Aryloxy-Pyrimidines (*H* 245)

The preparation of alkoxypyrimidines by the Principal Synthesis has been covered in Ch. II; their preparation from halogenopyrimidines in Ch. VI, Sects. 5.D, 6.B, and 7.A.

Other methods have been used but little.

Both 2- and 4-methylsulphonylpyrimidine reacted with sodium alkoxide in an appropriate alcohol to give 2-ethoxy- and 4-butoxypyrimidine, respectively;[2619] 4-methylsulphinylpyrimidine similarly gave 4-propoxypyrimidine;[2619] 5-methoxy-2-methylsulphonyl- gave 2,5-dimethoxy-pyrimidine;[3000] and 4-amino-5-methoxy-2-methylsulphonyl- gave 4-amino-2,5-dimethoxy-pyrimidine in 60% yield.[2205]

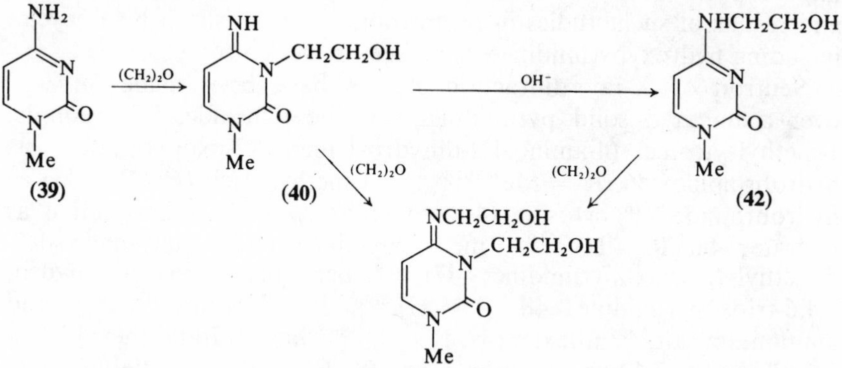

NH_2 … N … O … N … Me **(39)** —$(CH_2)_2O$→ NH … N–CH_2CH_2OH … O … N … Me **(40)** —OH^-→ $NHCH_2CH_2OH$ … N … O … N … Me **(42)**

(40) —$(CH_2)_2O$→ **(41)** ←$(CH_2)_2O$— **(42)**

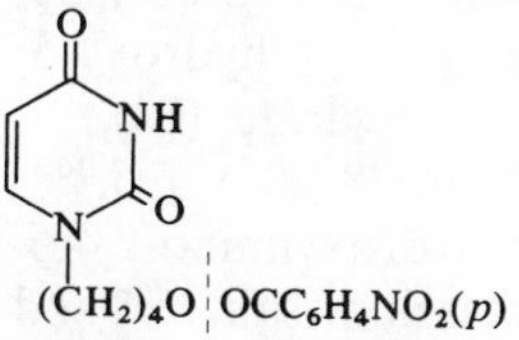

NCH_2CH_2OH … N–CH_2CH_2OH … O … N … Me **(41)**

O … NH … O … N … $(CH_2)_4O$ ⁞ $OCC_6H_4NO_2(p)$ **(43)**

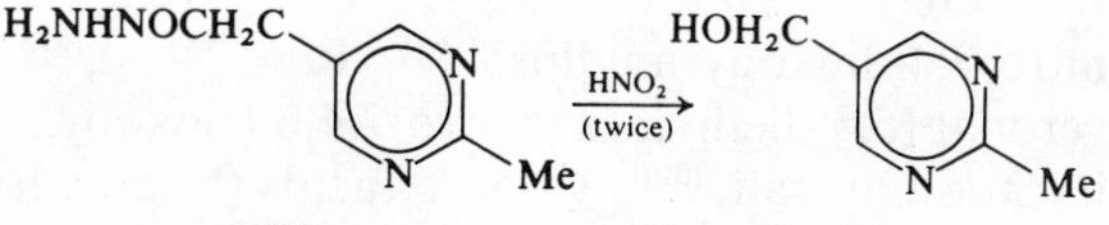

$H_2NHNOCH_2C$ … N … N … Me **(44)** —HNO_2 (twice)→ HOH_2C … N … N … Me **(45)**

O-Alkylation is represented in the formation of 5-benzyloxymethyl-2,4-dihydroxypyrimidine,[2847, 3014] and of 4-methoxy-2,5-dimethylpyrimidine (accompanied by 1,6-dihydro-1,2,5-trimethyl-6-oxopyrimidine* which was the major product of methylating 4-hydroxy-2,5-dimethylpyrimidine with diazomethane).[2818] Diazomethane methylation of 2-*p*-chlorophenyl-4-ethoxycarbonylmethyl-6-hydroxypyrimidine was unusual in giving a greater amount of the 6-methoxy analogue (67% yield) than of the 1,6-dihydro-1-methyl-6-oxo analogue (3–15%).[3481]

Simple alkoxypyrimidines are often made from their chloro derivatives. Thus appropriate 5-alkoxy-2-amino-4-chloropyrimidines gave 2-amino-5-methoxy(ethoxy, propoxy, butoxy, or s-butoxy)pyrimidine in good yield by hydrogenolysis (Pd/H_2) or in poorer yield by the action of zinc dust and alkali.[2212]

* The structure was based on p.m.r. spectra.[2818]

5. The Fine Structure of Hydroxypyrimidines (*H* 249)

The fine structure of hydroxypyrimidines is of interest both in the solid state and in solution. Such information on the molecular geometry and the site of protonation in solid naturally-occurring pyrimidines is obviously relevant to the structure of nucleic acids and their synthetic analogues; similar information on biologically-active molecules is important in connexion with their mode of action. The methods for, and results of such studies in prototropic tautomerism of heterocycles including hydroxypyrimidines, have been reviewed to 1963.[3036]

Neutron or X-ray diffraction studies have been made for well over a hundred solid pyrimidines.[3015] These include, for example, 1-methylcytosine (4-amino-1,2-dihydro-1-methyl-2-oxopyrimidine as hydrobromide; **46**, R = Me),[3016, 3017] 1-methyluracil (**60**, R = H) as hydrobromide,[3173] cytosine (4-amino-1,2-dihydro-2-oxopyrimidine as hydrate; **46**, R = H),[3018] 1-methylthymine* (1,2,3,4-tetrahydro-1,5-dimethyl-2,4-dioxopyrimidine; **47**),[3019] barbituric acid (hexahydro-2,4,6-trioxopyrimidine; **48**, R = H),[3020] its dihydrate[3021, 3174] and ammonium salt,[3022] alloxan (**49**, R = O),[3023] its 'hydrate' (hexahydro-5,5-dihydroxy-2,4,6-trioxopyrimidine; **48**, R = OH),[3024] dialuric acid (1,2,3,4-tetrahydro-5,6-dihydroxy-2,4-dioxopyrimidine as hydrate; **50**, R = OH),[3025, 3026] dilituric acid (1,2,3,4-tetrahydro-6-hydroxy-5-nitro-2,4-dioxopyrimidine; **50**, R = NO_2),[3027] its dihydrate (**51**),[3028] veronal (5,5-diethylhexahydro-2,4,6-trioxopyrimidine; **48**, R = Et),[3029] its sodium salt,[3030] violuric acid (hexahydro-5-hydroxyimino-2,4,6-trioxopyrimidine; **49**, R = NOH),[3031, 3032] its potassium salt,[3033] and 4-amino-5-pent-2′-en-4′-on-2′-ylaminopyrimidine (**52**).[3034] Much of this data has been reviewed.[3035]

A number of 'hydroxypyrimidines' have been examined by a variety of physical methods to find what is the predominant tautomeric form(s) of each in aqueous or other solution. The first examination of *2-hydroxypyrimidine* in dimethyl sulphoxide by p.m.r. spectra indicated equivalence of H-4 and H-6: therefore, either the accepted formulation (*H* 482) as 1,2-dihydro-2-oxopyrimidine (**53**) based on ionization and ultra-violet spectra was in error, or the equivalence of H-4 and H-6 simply indicated a relatively fast *N*-proton exchange in the oxo-form (**53**).[2410] The latter explanation proved correct and would be accepted now as axiomatic.[3036, 3037] Similar p.m.r. examination of *4-hydroxypyrimidine* in dimethyl sulphoxide and in water confirmed the predominance (*H* 482) of 1,6-dihydro-6-oxopyrimidine (**54**) with a con-

* Measurements were made[3019] on the stable prisms rather than the metastable needles which first appear[669] from aqueous solution.

tribution from 1,4-dihydro-4-oxopyrimidine (**55**) in each solution.[3038] An early p.m.r. study of *cytosine* seemed to indicate a zwitterionic form (**56**) for the neutral molecule, and a cation (**57**).[3039] However, careful pK_a and ultra-violet spectral comparisons with methylated model compounds indicated conclusively that the predominating neutral molecule in water or dimethyl sulphoxide was 4-amino-1,2-dihydro-2-oxopyrimidine (**46**, R = H),[2745, 2987] and the cation was in the form (**58**);[2987, 3037, 3040] the p.m.r. results were rationalized.[2987, 3037, 3269] Although the predominantly dioxo nature of uracil, thymine, and their simple *C*-, *N*-, and *O*-derivatives has not been doubted for many years,[3036] a variety of methods have been used in recent years to confirm the following structures: *uracil* (**59**, R = H),[3039, 3041, 3042] its anion(s),[3042] and its cation;[3043] *thymine* (**59**, R = Me),[3039,3041,3044] and its anion(s);[3044,3045] *1-methyluracil* (**60**, R = H),[3042, 3046] and its anion;[3042] *1-* and *3-methylthymine* (**47** and **61**, R = Me) and their anions,[3044] *1,3-dimethyl-* and *1,3,6-trimethyluracil* probably as the cations (**62**, R = H or Me),[3047] *2-ethoxy-4-* and *4-ethoxy-2-hydroxypyrimidine* which now appear to exist in non-aqueous solution as the oxo forms (**63**) and (**64**, R = H), respectively,[3046, *cf.* 1634] *4-ethoxy-2-hydroxy-5-methylpyrimidine* as the oxopyrimidine (**64**, R = Me) and its anion,[3044] *5-halogenouracils* (**59**, R = Br, Cl, F, or I) and the anion(s) of the fluoro derivative,[3039, 3041, 3045, 3048, 3270] *5-bromo-1-methyluracil* which is essentially the dioxo form (**60**, R = Br) but has a larger contribution from the hydroxy form (**65**) than do other uracil derivatives,[3036, 3046] *5-nitrouracil* (**59**, R = NO_2),[824, 3041] *2-thiouracil* as 1,2,3,4-tetrahydro-4-oxo-2-thiopyrimidine;[1767, 2410] also *1-methylcytosine* cation* (**46**, R = Me; protonated at N-3),[3049, 3050] *isocytosine* as

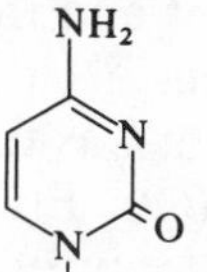

(**46**) 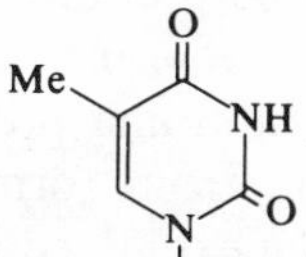 (**47**) 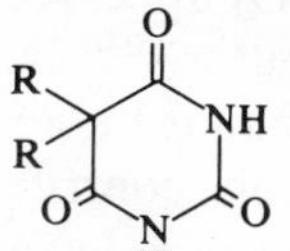 (**48**) 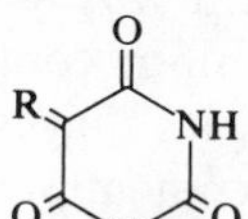 (**49**) 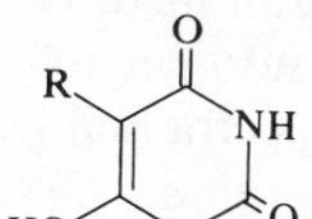 (**50**)

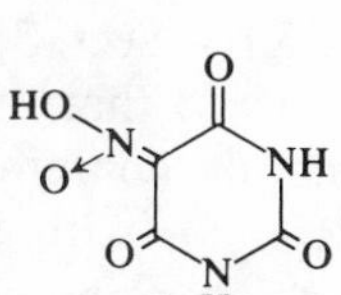

(**51**) 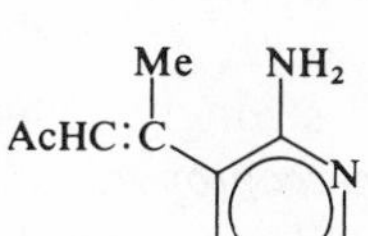 (**52**) 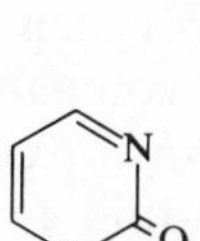 (**53**) 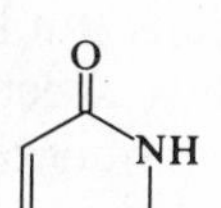 (**54**) 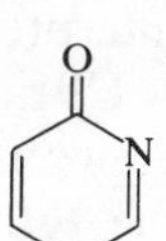 (**55**)

* As judged by ^{15}N n.m.r. spectroscopy using labelled pyrimidines; the spectra for other such pyrimidines are also recorded.[3049]

a mixture of the oxo forms (**66**) and (**67**),[2776] *pseudo-cytosine* (4-amino-6-hydroxypyrimidine) as the amino-oxo form (**68**),[2776] and other such pyrimidines.[3041, 3270, 3271]

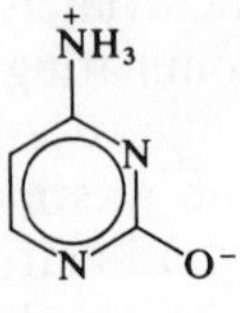

(**56**) 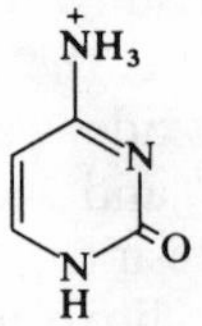 (**57**) 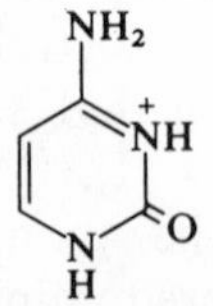 (**58**) 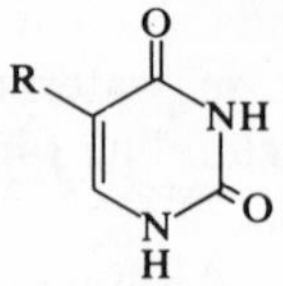 (**59**) 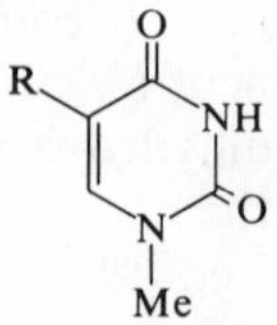 (**60**)

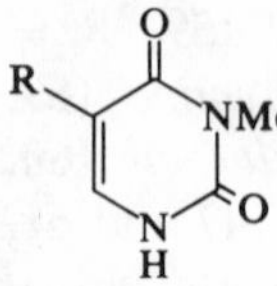

(**61**) 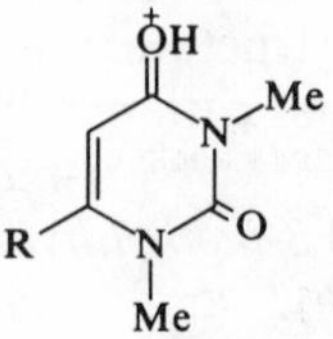 (**62**) 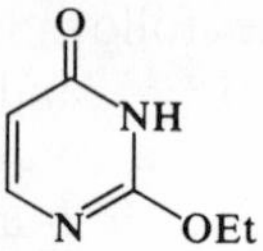 (**63**) 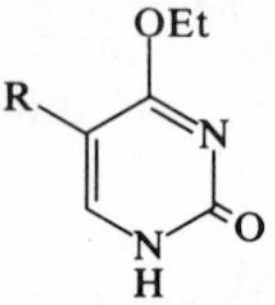 (**64**) 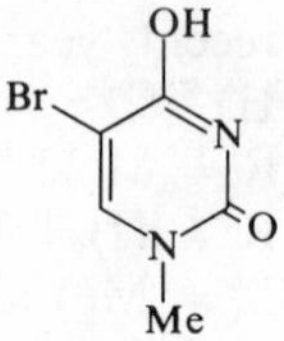 (**65**)

Early infra-red measurements suggested[1451, 1747] that solid *4,6-dihydroxypyrimidine* was largely 1,6-dihydro-4-hydroxy-6-oxopyrimidine (**69**). However, ultra-violet spectral comparisons with methylated model compounds suggested that an aqueous solution essentially consisted of 1,4,5,6-tetrahydro-4,6-dioxopyrimidine (**70**, R = O) with some 1,4-dihydro-6-hydroxy-4-oxopyrimidine (**71**);[2997] in other hands[3051] similar comparisons suggested a major contribution by the hydroxy-oxo form (**69**) and a minor one from form (**70**, R = O). When the p.m.r. spectra of aq. dimethyl sulphoxide solutions were examined, the form (**70**, R = O) appeared to be predominant and the tautomer (**69**) a minor contributor,[3038] but in yet other hands[3052] the spectra and basicities suggested that the main tautomer present was the betaine (**72**) with an appreciable contribution from form (**69**). Elsewhere,[3053–3056] these and similar data have been explained in terms of a major contribution from form (**69**) in dimethyl sulphoxide solution, but from the 'bipolar-ionic' form (**73**) in D_2O solution. None of these interpretations can be considered final yet. (See also Ch. XIII, Sect. 3).

Ultra-violet spectral comparisons strongly suggest[2242, 2760] that an aqueous solution of *4-hydroxy-6-mercaptopyrimidine* mainly contains the tetrahydro-4-oxo-6-thio forms, (**74**) and/or (**70**, R = S), with a possible contribution from 1,6-dihydro-4-hydroxy-6-thiopyrimidine (**75**); the same is true of 5-n-alkyl-4-hydroxy-6-mercaptopyrimidines.[2242] Alternative explanations of the data in terms of charge-separated molecules may be anticipated!

Association of 4-hydroxypyrimidines in solution by hydrogen bonding has been suggested on the basis of p.m.r. data,[3172] and so has 3,4-hydration in 2-hydroxypyrimidine and some of its derivatives.[3459]

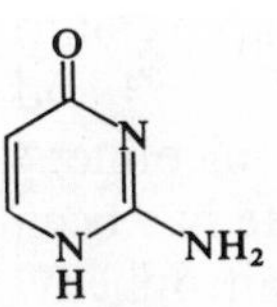

(66) 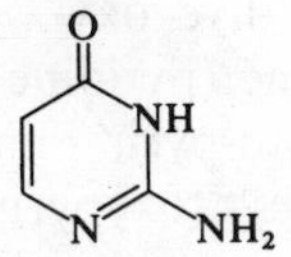 (67) 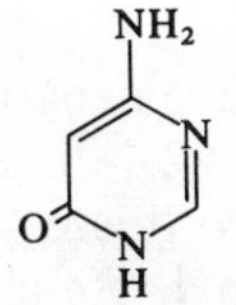 (68) 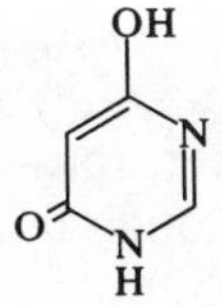 (69) 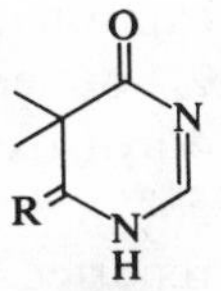 (70)

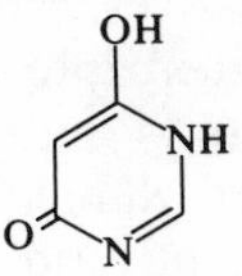

(71) 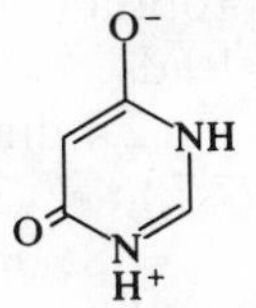 (72) 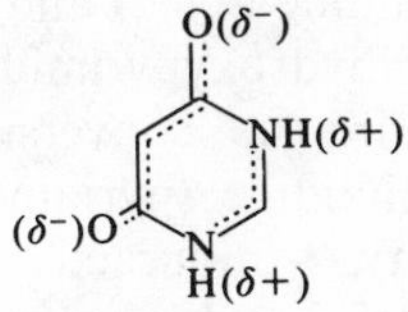 (73) 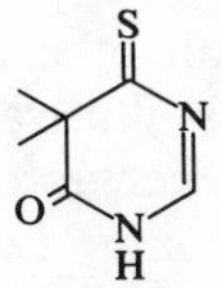 (74) 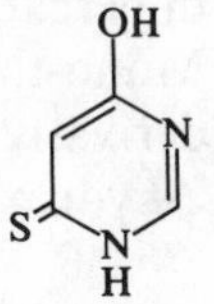 (75)

The dipole moment of 2-methoxypyrimidine has been measured in benzene (2.20 D) and compared with those of 2-aminopyrimidines, methoxytriazines, methoxypyridines, aminotriazines, and aminopyridines.[3084]

6. Reactions of Hydroxypyrimidines (*H* 250)

Despite their great importance, hydroxypyrimidines undergo relatively few types of reaction.

A. Conversion into Halogenopyrimidines (*H* 250)

The formation of halogeno- from hydroxy-pyrimidines has been discussed in Ch. VI. It is safe to forecast a dramatic increase in the use of thionyl chloride + dimethylformamide (*cf*. Ch. VI, Sect. 1.C) for such metathesis of 2-, 4-, and 6-hydroxypyrimidines.

B. Conversion into Mercaptopyrimidines (*H* 251)

It is now clear that the thiation of 2-, 4-, and 6-hydroxypyrimidines by phosphorus pentasulphide* can be carried out more effectively in

* Better results are sometimes obtained with a good quality phosphorus pentasulphide than with a technical grade. A process for purifying the reagent has been outlined.[2165]

pyridine or a homologue than in xylene or tetralin. The process is exemplified in the conversion of 4-hydroxy- into 4-mercapto-pyrimidine (pyridine: 58%),[2642] 4-hydroxy- into 4-mercapto-2,6-dimethylpyrimidine (pyridine: 65%),[2746] 5-bromo-2-hydroxy- into 5-bromo-2-mercapto-pyrimidine (pyridine: 25%),[2746] 4-hydroxy-2-mercapto- into 2,4-dimercapto-5,6-tetramethylenepyrimidine (pyridine: 67%),[2183] and 4-hydroxy-2-mercapto- into 2,4-dimercapto-pyrimidine (α-picoline:* 56%).[2915] Although failure to thiate 4,6-dihydroxypyrimidine has been reported (*H* 252), its 5-phenyl, 5-*o*-chlorophenyl, and 5-*p*-nitrophenyl derivatives undergo the process successfully in pyridine to give 4,6-dimercapto-5-phenylpyrimidine (74%) and appropriate derivatives.[2522] Mono-thiation of a 2,4-dihydroxypyrimidine leads to a 4-mercapto-derivative as illustrated in the conversion of 2,4-dihydroxy- into 2-hydroxy-4-mercapto-pyrimidine (pyridine: 77%),[3057] 3-methyluracil (**76**, R = O) into 3-methyl-4-thiouracil (**76**, R = S; in slightly aq. pyridine: *ca.* 30%),[2641] 2,4-dihydroxy- into 2-hydroxy-4-mercapto-5-methylpyrimidine (pyridine: 83%),[2581] and 5-fluoro-2,4-dihydroxy- into 5-fluoro-2-hydroxy-4-mercapto-pyrimidine (pyridine: 76%).

Thiation of an hydroxy group in the presence of amino groups is now considered generally possible (*cf. H* 251): thus, thiation of the corresponding hydroxypyrimidine gave 5-amino-4-mercapto-6-methyl- (pyridine: 51%),[2227] 4,5-diamino-6-mercapto- (pyridine: 50%; β-picoline: 68%; 5-ethyl-2-methylpyridine: 41%),[3058, 3059] 4,5-diamino-6-mercapto-2-methyl- (triethylamine: 51%; β-picoline: 49%; 2,6-dimethylpyridine: 29%; pyridine: 0%),[3058, 3060] 2,4,5-triamino-6-mercapto- (β-picoline: *ca.* 45%),[3061] 4,5-diamino-6-mercapto-2-phenyl- (β-picoline: 51%),[2284] 4-amino-6-mercapto-2-methyl- (triethylamine: 50%; β-picoline, 50%; 2,6-dimethylpyridine: 27%),[3060] 2-amino-5-anilino-4-mercapto- (slightly aq. pyridine: ?%),[2567] 4-amino-2,6-dimercapto- (from the dihydroxy analogue in β-picoline: 43%; from 6-hydroxy-2-mercapto analogue in β-picoline: 23%),[3060] and 5-amino-2-hydroxy-4-mercapto-pyrimidine (pyridine: 50%);[2775] also other such compounds.[2567, 3467] Thiation of 5-acetamido-4-amino-6-hydroxy-2-methylpyrimidine (**77**) in pyridine was accompanied by cyclization to yield a separable mixture of 4-amino-2,6-dimethylthiazolo[5,4-*d*]-pyrimidine (**78**; 81%) and 6-mercapto-2,8-dimethylpurine (**79**; 12%); other such pyrimidines behaved similarly.[3060]

The conversion of necessarily oxopyrimidines into thiopyrimidines is illustrated in the formation of 1,6-dihydro-1-methyl-6-thiopyrimidine (**80**, R = H; in pyridine: 71%),[2173] its 4-methoxy derivative (**80**,

* The reaction is complete[2915] in 15 min. instead of the 8 hr. needed in xylene.[356] Pyridine or β-picoline is less satisfactory.

R = OMe; in pyridine: *ca.* 35%),[2760] 2-amino-1,6-dihydro-1-methyl-6-thiopyrimidine (**81**, R = H; in pyridine: *ca.* 36%),[2776] and its 2-dimethylamino homologue (**81**, R = Me; in pyridine: *ca.* 30%).[2776]

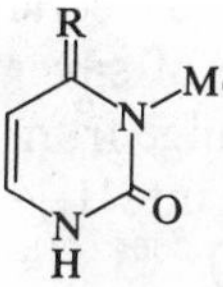

(**76**) 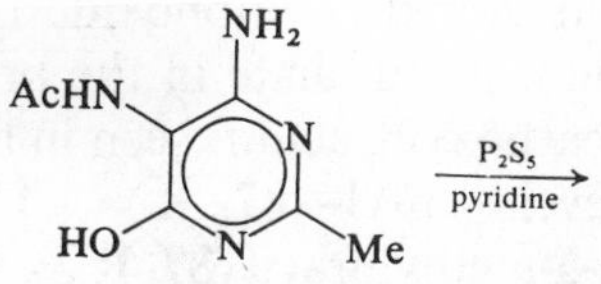 (**77**) $\xrightarrow[\text{pyridine}]{P_2S_5}$ (**78**) + (**79**)

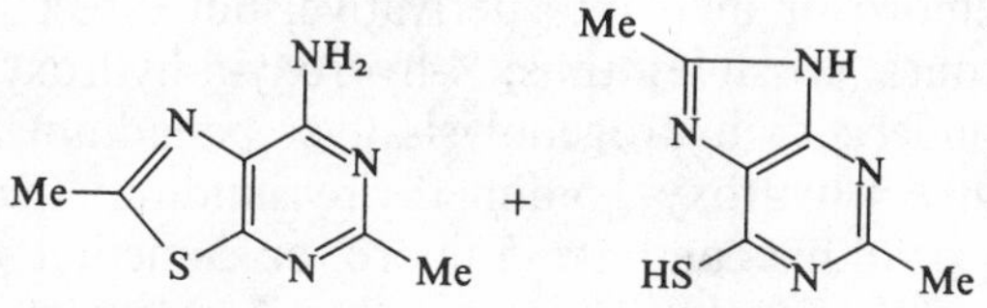

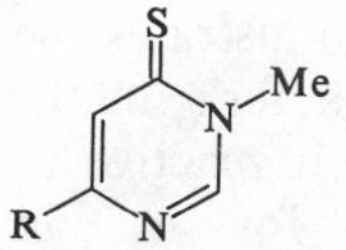

(**80**) 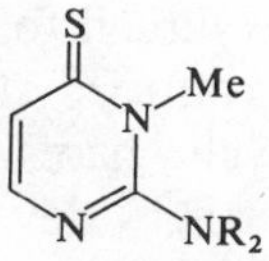 (**81**)

C. Other Reactions of Hydroxypyrimidines (*H* 252)

O-Acylation of 5- and extranuclear-hydroxypyrimidines occurs fairly readily. Thus treatment of 4-amino-5-hydroxypyrimidine (**82**) with benzoyl chloride in pyridine at 70° gives 4-benzamido-5-hydroxypyrimidine (**83**; 31%) and 4-benzamido-5-benzoyloxypyrimidine (**84**; 23%),[2584] and 2,5-dihydroxypyrimidine gives 5-benzoyloxy- or 5-acetoxy-2-hydroxypyrimidine by boiling in benzoyl chloride or acetic anhydride, respectively;[2585] acylation of alcoholic-hydroxy groups is illustrated in the formation of 5-β-acetoxyethyl-2,4-dihydroxypyrimidine (acetic anhydride/pyridine: 83%),[2202] 5-β-acetoxycyclopentyl-1,2,3,4-tetrahydro-1,3-dimethyl-2,4-dioxopyrimidine (**85**; by acetic anhydride/pyridine: 89%),[2952] 5-acetoxymethyl-4-amino-2-methylthiopyrimidine (**86**; R = H; by limited acetic anhydride in ethyl acetate: 94%) which was converted into 4-acetamido-5-acetoxymethyl-2-methylthiopyrimidine (**86**, R = Ac) by treatment with an excess of anhydride,[3062] 4-amino-2-methylthio-5-propionoxymethylpyrimidine,[3062] and related compounds.[3062] It must be accepted with caution that the ease of acylation is in the order: ω-OH > 2-, 4-, or 6-NH_2 > 5-OH.

The *O*- and *N*-alkylation of hydroxypyrimidines is discussed elsewhere: Ch. VII, Sect. 4.C and Ch. X, Sect. 1.A(3), respectively.

The oxidation of an hydroxymethyl- to a formyl-pyrimidine is represented by the conversion of 2,4-dihydroxy-5-hydroxymethyl-pyrimidine into 5-formyl-2,4-dihydroxypyrimidine by manganese dioxide in water (40–60%) or in dimethyl sulphoxide (85%); by ceric sulphate (70%); or by potassium persulphate in the presence of silver ion (90%).[3063] Oxidation to a carboxylic acid is seen in the change from 1-γ-hydroxy-δ-hydroxymethylcyclopentyl- (**87**, R = CH_2OH) into 1-γ-carboxy-δ-hydroxycyclopentyl-5-methyluracil (**87**, R = CO_2H).[3064] The removal of hydroxy groups is generally done *via* the corresponding chloro or mercapto derivative, but direct reductive removal is sometimes possible: thus, 4-hydroxy-5-hydroxymethyl-2-methylpyrimidine undergoes hydrogenolysis over palladium in ethanolic acetic acid to give 4-hydroxy-2,5-dimethylpyrimidine* (*ca.* 55%);[3003] 2,4-dihydroxypyrimidine and its 5-fluoro or 5-methyl derivatives are reduced by sodium amalgam/acetic acid to 2-hydroxypyrimidine and its 5-fluoro or 5-methyl derivative, respectively, in small yield.[3065, 3066] The hydrogenolysis of 1-hydroxyuracil to uracil illustrates the reductive removal of an *N*-hydroxy group.[2200] L. Birkofer's method[3067, 3068] for *N*-acylation of heterocyclic bases *via* their reactive trimethylsilyl derivatives has been further developed for use in nucleoside syntheses.[3069–3077, 3212] Such a process is illustrated simply by the reaction of 2,4-dihydroxypyrimidine (**88**, R = H) and chlorotrimethylsilane in

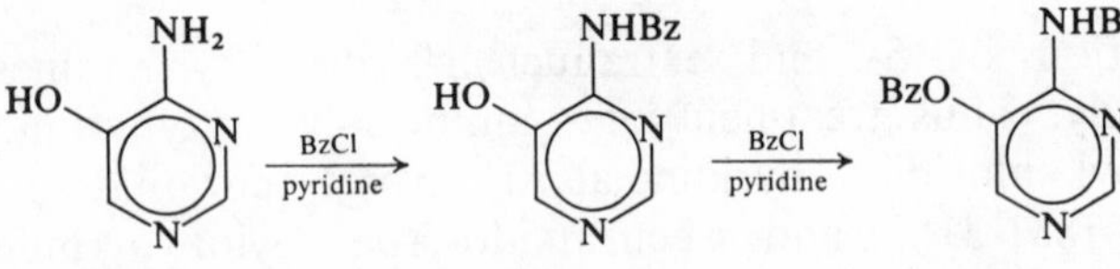

NH_2 HO N N —BzCl, pyridine→ NHBz HO N N —BzCl, pyridine→ NHBz BzO N N

(82) **(83)** **(84)**

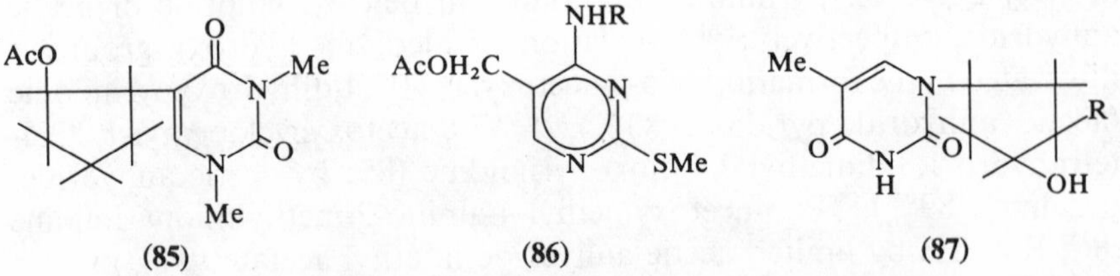

AcO O Me N O N Me — NHR AcOH$_2$C N N SMe — Me N O O N H R OH

(85) **(86)** **(87)**

dioxan containing triethylamine to give 2,4-bistrimethylsiloxypyrimidine (**89**, R = H).[3070, 3071] The silicon-containing groups protect the molecule from subsequent *O*-alkylation and appear to activate it towards *N*-alkylation by an alkyl halide or appropriate acylated sugar halide. Thus on treating the compound (**89**, R = H) with one mole of

* Further reduction by sodium/ethanol gives 1,3-diamino-2-methylpropane (50%).[3003]

methyl iodide* a molecule of iodotrimethylsilane is eliminated with the formation of the (unisolated) oxopyrimidine (**90**, R = H) which on mild hydrolysis yields 1-methyluracil (**91**, R = H);[3075] when an acylated sugar halide has been used, alkaline hydrolysis is needed not only to complete the last step but also to deacylate the sugar hydroxy groups.[3070, 3071] Other trimethylsiloxypyrimidines described include,

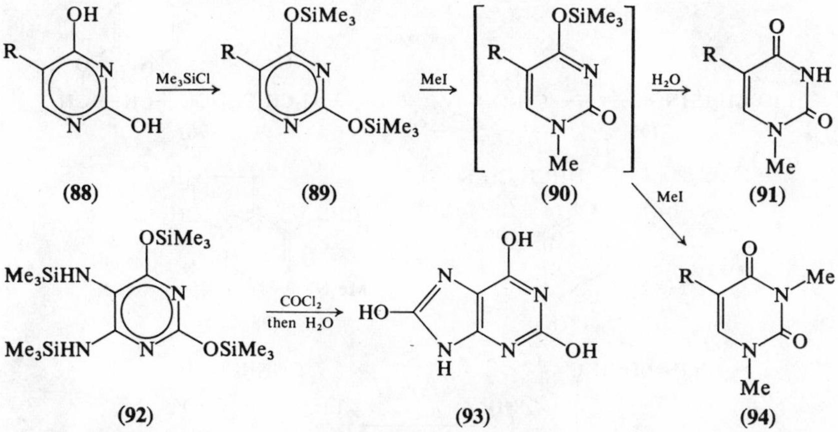

for example, 2,4-bistrimethylsiloxy-5,6-bistrimethylsilylaminopyrimidine (**92**),† which was made in 94% yield by treating 4,5-diamino-2,6-dihydroxypyrimidine with 'hexamethyldisilazan' (bistrimethylsilylamine: $Me_3SiNHSiMe_3$);[3068] 2-trimethylsiloxy-4-trimethylsilylamino-pyrimidine [Me_3SiCl + NEt_3: 69%;[3070, 3071] $(Me_3Si)_2NH$: 90%];[3074] 4-ethoxy-2-trimethylsiloxypyrimidine (85%);[3074] 2-trimethylsiloxy-5-trimethylsiloxymethyl-4-trimethylsilylaminopyrimidine (95%);[3074] and 5-bromo(chloro, fluoro, or iodo)-2,4-bistrimethylsiloxypyrimidine (**89**; R = Br, Cl, F, or I) from the corresponding uracils (**88**; R = Br, Cl, F, or I).[3075] The whole subject was reviewed in 1965.[3079]

Several extranuclear-hydroxy derivatives of pyrimidine undergo cyclization reactions. Thus 5-acetyl-1-α-ethoxycarbonylmethylcarbamoyl-β-hydroxyethyluracil (**95**) in pyridine containing methane- or toluene-sulphonyl chloride gives 6-acetyl-3-ethoxycarbonylmethylcarbamoyl-2,3-dihydro-7-oxo-oxazolo[3,2-*a*]pyrimidine (**96**) which may be used to make other pyrimidines: see Ch. III, Sect. 5.E(6).[2318]

* An excess of alkylating agent leads eventually to an *NN'*-dialkyluracil: e.g., 5-methyl-2,4-bistrimethylsiloxypyrimidine (**89**, R = Me)[3071, 3074] and dimethyl sulphate give eventually 1,5-dimethyluracil (**91**, R = Me; 16%) and 1,2,3,4-tetrahydro-1,3,5-trimethyl-2,4-dioxopyrimidine (**94**, R = Me; 34%).[3076]

† Treatment with phosgene followed by mild hydrolysis gave uric acid (**93**).[3068]

5-Hydroxymethyluracil and its 6-methyl or 6-phenyl derivatives have been shown[3523] to react with phenols or naphthols in the presence of a little acid to give 5-benzyl- or 5-naphthylmethyl-uracils. For example, 2,4-dihydroxy-5-*p*-hydroxybenzylpyrimidine was thus made in 95% yield, and eight other analogues in similarly good yield.[3523]

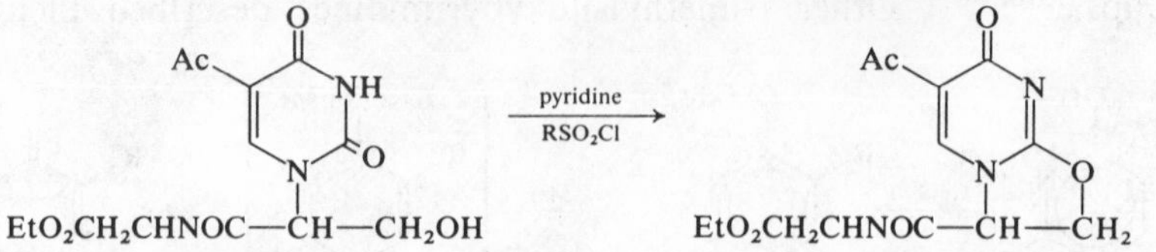

(95) → pyridine, RSO_2Cl → (96)

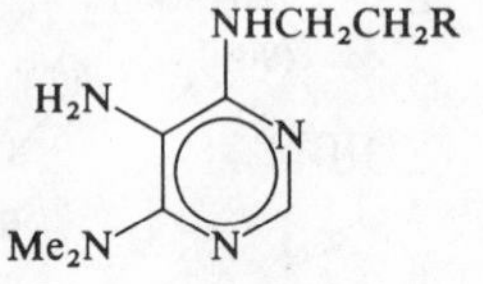

(97) 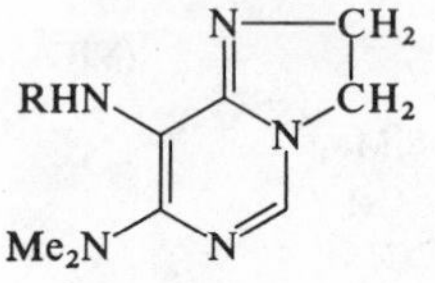 (98)

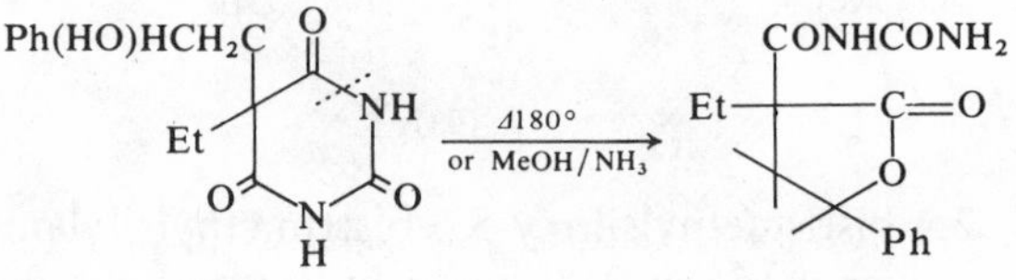

(99) → Δ180° or $MeOH/NH_3$ → (100)

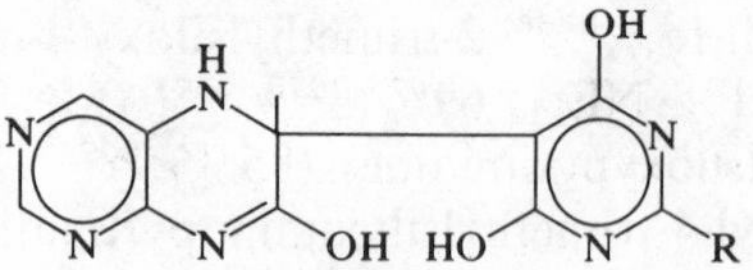

(101)

In attempting the synthesis of a purine by heating 5-amino-4-dimethylamino-6-β-hydroxyethylaminopyrimidine (**97**, R = OH) in formamide, J. H. Lister[3080] isolated 6-dimethylamino-7-formamido-1,3a,5-triazaindane (**98**, R = CHO); it was synthesized unambiguously as were analogues (Ch. VI, Sect. 3.B) by treating the alcohol (**97**, R = OH) with thionyl chloride and allowing the unisolated ω-chloro derivative (**97**, R = Cl) to cyclize into 7-amino-6-dimethylamino-1,3a,5-triazaindane (**98**, R = H) which was then formylated to the compound (**98**, R = CHO).[3080] An ω-hydroxy group is involved in the interesting rearrangement of 5-ethyl-5-β-hydroxyphenethylbarbituric acid (**99**) into α-ethyl-γ-phenyl-α-ureidocarbonylbutyrolactone (**100**).[3081]

4,6-Dihydroxy- and 2,4,6-trihydroxy-pyrimidine act as Michael reagents, e.g., towards 7-hydroxypteridine to which they add at the 5:6-bond giving 6-4′,6′-dihydroxypyrimidin-5′-yl-5,6-dihydro-7-hydroxypteridine (**101**, R = H) and its 2′,4′,6′-trihydroxy-analogue (**101**, R = OH), respectively;[3082] also towards other pteridines.[3245]

A remarkable analytical compilation, *Komplexchemische Identifizierung von Pyrimidinderivaten im Mikromass-stab*,[3083] lists the appearance in ultra-violet light and the colour reactions for a variety of pyrimidines. Reagents used include sodium hydroxide, nitrous acid, alkaline cobalt sulphate, picric acid, ferrous ion, and other more specialized test solutions.[3083]

(**101a**) $\xrightarrow{OH^-}$ Et(Pr^i)C(CONHCONH$_2$)(CO$_2$H) $\xrightarrow{-CO_2}$ Et(Pr^i)CHCONHCONH$_2$ $\xrightarrow{-CO_2,\ -NH_3}$ Et(Pr^i)CHCONH$_2$ (**101b**)

Silver nitrate sprays have been used successfully in biochemistry to distinguish purine from pyrimidine spots on paper chromatograms.[3179]

In an effort to develop a more sensitive identification procedure for barbiturates in tissues, the silver salts of all commonly used barbiturates have been prepared in crystalline form.[3166] Some were suitable for X-ray diffraction identification.[3166] Another process of possible application to analysis is the hydrolysis of 5,5-disubstituted-barbituric acids by aq. ammonia at 200° during 10 min.[3167] The resulting amides are formed in good yield (*ca.* 80%) and are readily identified by sharp melting points between 50° and 120°. The steps involved are exemplified by the conversion of 5-ethyl-5-isopropylbarbituric acid (**101a**) into α-ethyl-β-methylbutyramide (**101b**).[3167]

7. Reactions of Alkoxy- and Aryloxy-pyrimidines (*H* 254)

A. Conversion into Hydroxy Derivatives (*H* 254)

The mechanism of the acid-catalysed hydrolysis of 2-methoxypyrimidine has been studied using ^{18}O material.[2759, 4085] The conversion of an *N*-alkoxy into an *N*-hydroxy group is exemplified in the treatment of 1- or 3-benzyloxyuracil with hydrogen bromide in acetic acid to give 1- or 3-hydroxyuracil [uracil-1(or 3)-*N*-oxide],[2200] and in a similar conversion of 6-amino-1-benzyloxyuracil (**102**, R = CH_2Ph) into 6-amino-1-hydroxyuracil (**102**, R = H).[2199] Other aspects are discussed fully in Sects. 1.E, 1.G, and 2.

B. Conversion into Halogeno Derivatives (*H* 254)

This has been treated briefly in Ch. VI, Sect. 3.C.

C. Aminolysis of Alkoxypyrimidines (*H* 255)

The rates for butylaminolysis of simple alkoxypyrimidines have been measured under preparative conditions.[2697] The apparent first-order rate constants indicate that 2-methoxy- reacts more slowly than 4-methoxypyrimidine by a factor of ten; that the higher alkoxypyrimidines undergo aminolysis much more slowly than do the methoxy homologues (for example, 2-methoxypyrimidine reacts with butylamine 80 times faster than does 2-isopropoxypyrimidine); and that 2-chloropyrimidine is aminolysed some 40,000 times faster than is 2-methoxypyrimidine.[2697] The last fact should not discourage use of methoxypyrimidines as intermediates in the preparation of aminopyrimidines because an appropriate temperature and time of reaction will overcome the initial disadvantage. Thus 4-methoxypyrimidine and butylamine at 150° for 3 hr. give a 75% yield of pure 4-butylaminopyrimidine.[2697] However, it must be remembered that a small amount of rearrangement to an *N*-methyloxopyrimidine will occur during aminolysis of simple methoxypyrimidines;[2630, 2697] no practical difficulty from this source has been reported.

Aminolysis of alkoxypyrimidines is illustrated as a synthetic tool by the following: 4-ethoxy-2-hydroxypyrimidine and methanolic methylamine or dimethylamine give 2-hydroxy-4-methylamino- or 4-dimethylamino-2-hydroxy-pyrimidine (93%), respectively;[3086, 3087] 4-ethoxy-1,2-dihydro- or 1,2-dihydro-4-methoxy-1-methyl-2-oxopyrimidine similarly give 1,2-dihydro-4-methylamino- and 4-dimethylamino-1,2-dihydro-1-methyl-2-oxopyrimidine;[2987, 3087] 1,2-dihydro-4-methoxy-2-oxo-1-tetrahydropyran-2′-ylpyrimidine (**103**, R = H) and aq. ammonia or methylamine at *ca.* 100° give 4-amino-1,2-dihydro- (**104**, R = R′ = H) or 1,2-dihydro-4-methylamino-2-oxo-1-tetrahydropyran-2′-ylpyrimidine (**104**; R = H, R′ = Me), respectively;[2996] the methoxy

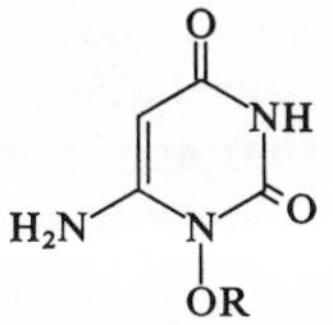

(102)

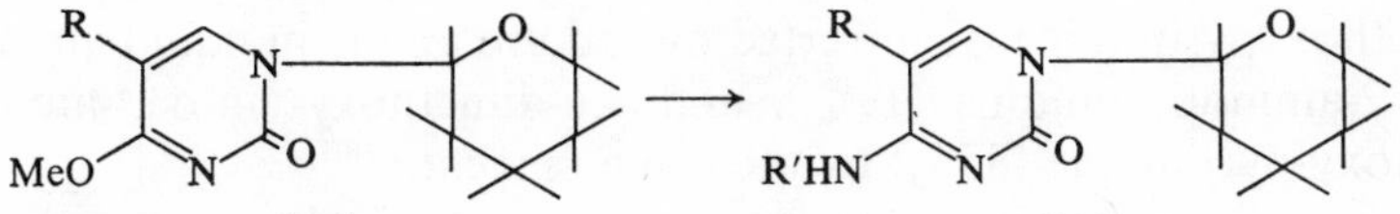

(103) (104)

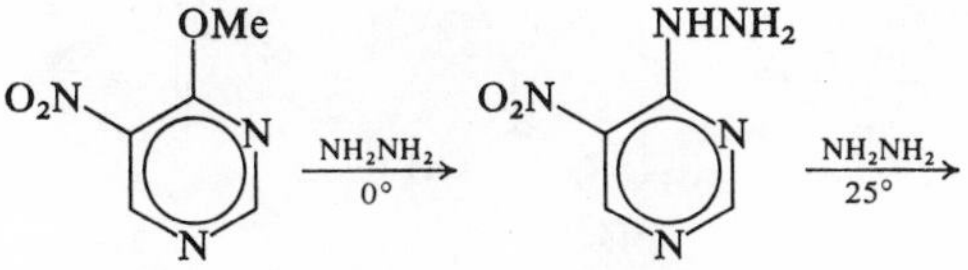

(105) (106)

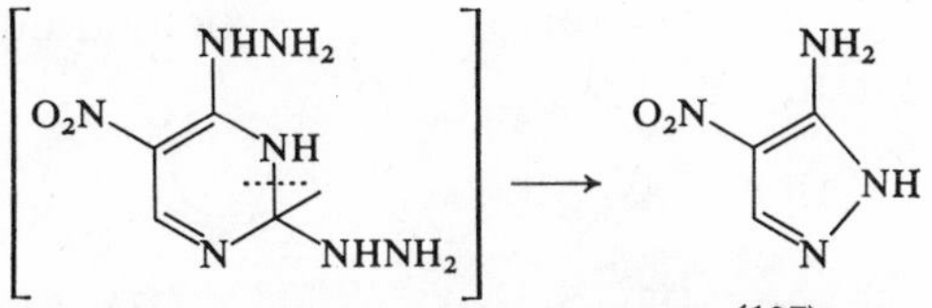

(107)

derivatives (**103**; R = Br, I, or Me) similarly give 4-amino-5-bromo-1,2-dihydro-2-oxo-1-tetrahydropyran-2′-ylpyrimidine (**104**; R = Br, R′ = H) and the analogues (**104**; R = I or Me, R′ = H);[2996] 5-amino-4-mercapto-6-methoxypyrimidine with refluxing hydrazine hydrate gives 5-amino-4-hydrazino-6-mercaptopyrimidine (89%);[2227] and 4-methoxy- gives 4-hydrazino-6-methyl-2-methylthiopyrimidine.[2452] Other examples are known.[3399]

The powerful activation of methoxy groups by a 5-nitro substituent is shown by the conversion of 4-methoxy-5-nitropyrimidine into 4-t-butylamino-5-nitropyrimidine (90%) within 3 hr. by refluxing methanolic t-butylamine,[2562] a reagent with but 0.1% of the reactivity of an n-alkylamine;[2694] of 4,6-dimethoxy- into 4,6-dihydrazino-5-nitropyrimidine at 25°;[2562] and of 4-methoxy- (**105**) into 4-hydrazino-* (**106**) or 4-α-methylhydrazino-5-nitropyrimidine at $<0°$.[2857, 2858] 4-Hydrazino-2-methyl-5-nitropyrimidine was made similarly.[3312]

* Above 25°, a second molecule of hydrazine adds to the 2:3-bond of compound (**106**) with eventual loss of the (N-1 + C-2) fragment of the pyrimidine to give 3-amino-4-nitropyrazole (**107**; 60%) in which the 3-amino group has been derived from N-3 of the pyrimidine.[2857, 2858]

4-Ethoxy-2-hydroxypyrimidine (**108**) reacts with methanolic hydroxylamine (3 days at room temperature) to give *ca.* 50% yield of 2-hydroxy-4-hydroxyaminopyrimidine (**109**) and >25% of its addition product 4,5-dihydro-2-hydroxy-4,6-bishydroxyaminopyrimidine (**110**) which reverts with hydrochloric acid to the pyrimidine (**109**);[3089, 3090] 1,2-dihydro-4-hydroxyamino-1-methyl-2-oxopyrimidine (**111**, R = H) and its 1,5-dimethyl homologue (**111**, R = Me) were prepared similarly.[3090] 2-Methoxypyrimidine is converted by sodium amide in liquid ammonia into 2-aminopyrimidine (19%), and 4,6-dimethoxy- into 4-amino-6-methoxy-pyrimidine (85%) by the same reagent.[2765]

OEt NHOH NHOH NHOH
NH_2OH NH_2OH H^+
R N OH HOHN O Me
(**108**) (**109**) (**110**) (**111**)

OH $OCH_2CH{:}CH_2$ O
$H_2C{:}HCH_2C$ $CH_2CH{:}CH_2$
Me Me Me
(**112**) (**113**) (**114**)

Me Me
NH_2 OH
H^+
O N NH_2 N H N NH_2
(**115**) (**116**)

4,6-Diethoxypyrimidine undergoes an odd reaction with aq. hydrazine hydrate at 210°: 3-methyl-1,2,4-triazole is formed; similarly, 4,6-diethoxy-5-methylpyrimidine gives 3-ethyl-1,2,4-triazole, thus indicating that fission of the 4:5-bond is involved in the mechanism.[3203]

D. Rearrangement of Alkoxy- or Aryloxy-pyrimidines (*H* 256)

The thermal rearrangement of alkoxy- to *N*-alkyloxo-pyrimidines is discussed later in Chap. X, Sect. 1.A(4); for convenience, the *ortho*-Claisen rearrangement of alkenyloxypyrimidines, e.g., the formation of 5-allyl-4-hydroxy-2-methylpyrimidine (**112**) (as well as 1-allyl-1,6-

dihydro-2-methyl-6-oxopyrimidine, **114**) from 4-allyloxy-2-methylpyrimidine (**113**)[2555] is discussed in the same section.

A fascinating rearrangement of *o*-aminophenoxy- to *o*-hydroxyanilino-pyrimidines takes place under acidic conditions. Thus, for example, 2-amino-4-*o*-aminophenoxy- (**115**) gives 2-amino-4-*o*-hydroxyanilino-6-methylpyrimidine (**116**) in methanolic hydrochloric acid;[3091] the rearrangement is not confined to pyrimidines.[3091]

8. Some Naturally Occurring Hydroxypyrimidines (*H* 256)

The subject of pyrimidine nucleosides and nucleotides cannot be discussed in this *Supplement*. Indeed, because of the rapid growth in such research,* even specialized reviews tend to become outdated more rapidly than in other fields. For this reason 'progress' books are perhaps the most valuable in this field. Recent introductory or general books include those of T. L. V. Ulbricht (two),[3093, 3094] D. W. Hutchinson,[3095] D. O. Jordan,[3096] E. Chargaff,[3097] and A. Michelson;[3098] more specialized reference or progress books include those written or edited by R. F. Steiner and R. F. Beins,[3099] G. L. Cantoni and D. R. Davies,[3100] W. W. Zorbach and R. S. Tipson,[3101] R. S. Scharffenberg and R. E. Beltz,[3102] and J. N. Davidson and W. E. Cohn.[3103]

A. and B. Uracil and Thymine (*H* 256 and 258)

The methods for preparing uracil and thymine have been reviewed in detail.[2201] An interesting preparation of thymine, albeit in small yield, involves aromatization of its 5,6-dihydro derivative with palladium in refluxing quinoline.[3104] Details for preparing the stable [5-^{2}H]uracil, [6-^{2}H]uracil, [5,6-$^2H_2$]uracil, [3-^{15}N]uracil, [5,6-$^2H_2$]dihydrouracil, [α-^{2}H]thymine, and [6-^{2}H]thymine have been published;[2196] exchange deuteration was used to make the 'unstable' [1,3-$^2H_2$]uracil and [1,3-$^2H_2$]thymine.[2196] Uracil, pyrimidine, cytosine, 4,6-dichloropyrimidine, and 4-chloro-6-hydroxypyrimidine have been tritiated by $^1H^3HO$ ('HTO') in the presence of aluminium chloride using ethylene dichloride as the solvent.[3105] Studies of isotope distribution in tritiated

* In his fascinating review of chemical publications,[3092] R. S. Cahn has pointed out that the number of papers on nucleic acids is doubling each 18 months compared with a doubling-period of *ca.* 8 years for chemistry as a whole.

uracil, uridine, and cytidine have indicated a marked preference for the 5-position.[3178] The preparation of [2,6-$^{14}C_2$]thymine has been carried out by condensing [^{14}C]urea[3169] with [^{14}CN]α-cyanopropionic acid and reductively deaminating the resulting [2,4-$^{14}C_2$]6-aminothymine.[2306]

The reaction of uracil with aq. formaldehyde has been studied in some depth by physical means: apparently 'the acid imino-group of uracil reacts with one molecule of formaldehyde', but 'paper chromatography of the uracil-formaldehyde solutions failed to show the presence of any spots additional to those of the reactants'.[3106, *cf.* 1168]

Uracil and thymine are converted into their respective 1,3-dibenzoyl derivatives by benzoyl chloride in dioxan containing pyridine; under less vigorous conditions the 3-benzoyl derivatives are isolated.[3170] The di- can be converted into the mono-benzoyl derivatives by treatment with ethanolic chloroform or ethanolic hydrogen chloride.[3170]

Treatment of uracil and thymine with hydrazine hydrate gives 3-hydroxypyrazole (3-pyrazolone) and 3-hydroxy-5-methylpyrazole, respectively, with loss of urea in each case.[3203, 3204] The pyrimidines react similarly with methylhydrazine to give 5-hydroxy-1-methylpyrazole (*via* its isolated 3-ureido derivative) and 5-hydroxy-1,4-dimethylpyrazole, respectively;[3205, 3206] with 1,2-dimethylhydrazine they give 2,3-dihydro-1,2-dimethyl-3-oxopyrazole (*via* its isolated 5-ureido derivative) and 2,3-dihydro-1,2,4-trimethyl-3-oxopyrazole (*via* its isolated 5-ureido derivative), respectively.[3205, 3206] The action of hydrazines on uridine and related nucleosides and nucleotides has been well studied[3204, 3207–3210] (see also a selected bibliography[3204]).

The mass spectra have been measured of pyrimidine, uracil, thymine, 6-methyluracil, 1,3-dimethyluracil, 5,6-dihydrouracil, 5,6-dihydrothymine, 5-hydroxymethyluracil, other pyrimidines such as cytosine, and the corresponding deuterated compounds. Molecular ions were observed for all compounds and fragmentation patterns characteristic of the position and nature of the substituents emerged.[2428] The ^{15}N n.m.r. spectra of appropriately labelled uracil and related pyrimidines have been studied.[3049] Some interesting observations and calculations have been made on the ultra-violet spectral shifts occasioned by the oxygen of uracil being replaced partly or fully by sulphur or selenium.[3107] Dipole moments have been measured in dioxane for 1,3-dimethyluracil (3.9 D) and 5-bromo-3-methyluracil (4.5 D);[3108] they have been calculated for uracil (4.05 D), 1,3-dimethyluracil (3.7 D), 3-methylthymine (3.5 D), 5-bromo-3-methyluracil (4.5 D), and other compounds.[3108, 3109] Resonance energies for two possible tautomeric forms of uracil and thymine have been calculated,[3110] and a detailed interpretation of the ultra-violet spectra of uracil and other bases has been attempted[3111] following 'a semi-empirical self-consistent-field calculation for the bases after an appropriate optimization of the integral values using reference compounds'. The spectra of uracil, thymine, and other common pyrimidines have been examined in dimethyl sulphoxide containing chloranil, bromanil, and *p*-benzoquinone: most exhibited bands which were concluded to be charge transfer in origin.[3168]

Results continue to pour out from experiments on the irradiation of uracil, thymine, and their derivatives. Three categories are recognizable: the 5:6-hydration of uracil and its derivatives in solution by light or γ-irradiation; the oxidation of uracil and especially thymine during irradiation; and the formation of dimers on irradiation of uracil or thymine in an ice matrix, or of thymine in aqueous solution.*

The photohydration of uracil derivatives had been quite well studied by 1960 (see *H* 257). Since then, the 4,5-dihydro-2,4,6-trihydroxypyrimidine (**118**) resulting from photohydration of uracil (**117**) has been identified with a sample made unambiguously by hydrogenation of 5,5-dibromo-4,5-dihydro-2,4,6-trihydroxypyrimidine (**119**);[3112] a study has been made of the kinetics for photohydration of 1,3-dimethyluracil to hexahydro-4-hydroxy-1,3-dimethyl-2,6-dioxopyrimidine (**120**) and for the reverse reaction in acid or alkali;[3113] the last-mentioned pyrimidine (**120**) has been identified as the product of γ-irradiation of 1,3-dimethyluracil solution in the absence of air;[3114] the dehydration in aqueous base of the photohydrates of 1-methyl- and 1,6-dimethyluracil has been studied kinetically to establish a mechanism for the reactions; and a 'zwitterion intermediate' (**121**)† has been suggested[3113] to account for the kinetics and other phenomena associated with photohydration of uracil derivatives. Irradiation of uracil and 1,3-dimethyluracil in isopropanol gave the respective 5,6-dihydro derivatives on a preparative scale.[3773]

Although it is resistant to photo-hydration, thymine (**122**) undergoes photo-oxidation during ultra-violet irradiation of its aqueous solution in the presence of air.[3116] The products, 2,4-dihydroxy-5-hydroxymethylpyrimidine (**123**), 5-formyl-2,4-dihydroxypyrimidine (**124**), and uracil (**126**) were identified by chromatographic comparison with authentic specimens; although no 5-carboxyuracil (**125**) could be found, it was assumed to be an intermediate as shown in the sequence (**122** → **126**).[3116] This assumption was strengthened by the identification

* A survey of the quantum efficiencies of ultra-violet photolysis of a range of substituted pyrimidine nucleosides has been made without reference to the products.[3171]

† The term and formulation cannot be intended literally. A polarized intermediate with δ^+ and δ^- charges would be more realistic.

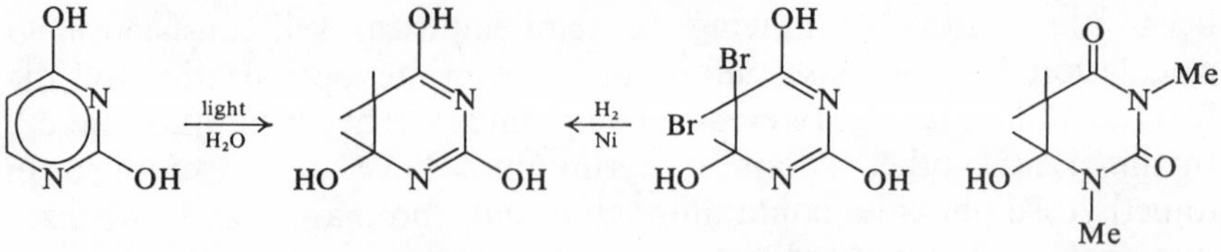

of 5-carboxy-1,2,3,4-tetrahydro-1,3-dimethyl-2,4-dioxopyrimidine (**128**) among the photo-oxidation products from 1,3-dimethylthymine (**127**).[3117] S. Y. Wang and R. Alcantara have themselves discussed the possible biological importance of their elegant work described above,[3118] and they have suggested 'dithymine peroxide' (**129**) as the primary oxidation product of thymine.[3118, *cf.* 3124] The formation of different types of peroxide and other products from aqueous solutions of thymine and other pyrimidines under the influence of γ-radiation in the presence of air has been studied by D. Barszcz and D. Shugar,[3119] who have also discussed earlier work[3114, 3120–3122] in this area;* uracil in the solid state is more stable to γ-rays than is thymine but both give dimers (*v.i.*) and thymine gives 2,4-dihydroxy-5-hydroxymethylpyrimidine also.[3123]

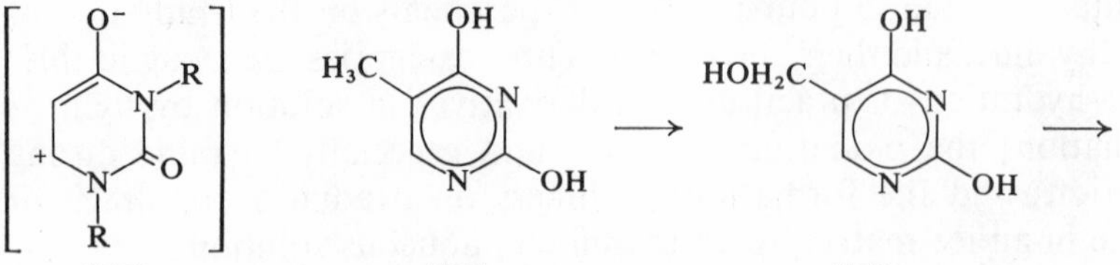

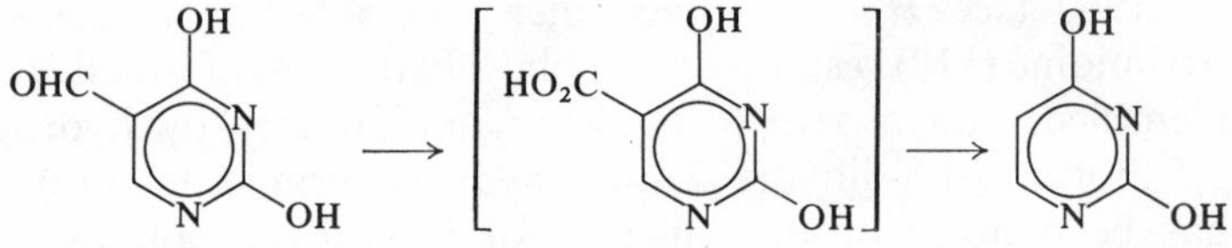

The discovery by R. Beukens and his colleagues[3126–3128] of the photo-dimerization of thymine and uracil has been called[3125] 'a new chapter of radiation chemistry and biology'. Ultra-violet irradiation of a frozen aqueous solution of thymine gives a dimer,[3126–3133] at first tentatively formulated[3127] in the *anti*-configuration (**131**) without

* The same workers have shown that brominated-thymine (5-bromo-4,5-dihydro-2,4,6-trihydroxy-5-methylpyrimidine; **130**) gives uracil (**117**), thymine (**122**) and 2,4-dihydroxy-5-hydroxymethyluracil (**123**) among the products of ultra-violet irradiation.[2636]

regard to the *cis*- or *trans*-relationship of the pyrimidine rings. However it was always admitted,[3127, 3133] and later proven,[3134] that only a *syn*-configuration could result directly from photo-dimerization of adjacent thymine residues on deoxyribonucleic acid, and it later became apparent[3132, 3135] that the photo-dimer of thymine was, in fact, the *cis-syn*-isomer (**132**, R = H). This was proven conclusively by G. M. Blackburn and R. J. H. Davies[3136, 3137, 3760] and confirmed elsewhere.[3138, 3139] Other such dimers are formed by irradiation of frozen aqueous solutions or solid films of uracil, which appears to yield two interconvertible dimers broadly analogous to that from thymine;[3140–3143, 3147, 3784] from 1,3-dimethyluracil which gives four distinct isomeric dimers, yet unformulated;[3144] and from 1,3-dimethylthymine which initially was known to give two isomeric dimers,[3131, 3133] one proven to be the *cis-syn*-compound (**132**, R = Me),[3138, 3139] and the other probably the *cis-anti*-isomer.[3139] More recently, all four possible isomers have been isolated and identified.[3460, 3779] Dimers are also produced similarly from 5-bromouracil, cytosine, orotic acid, and other pyrimidines.[3125, 3143, 3502]

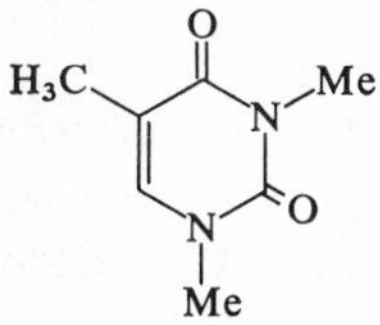

(127)

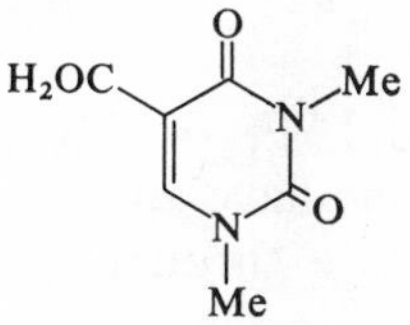

(128)

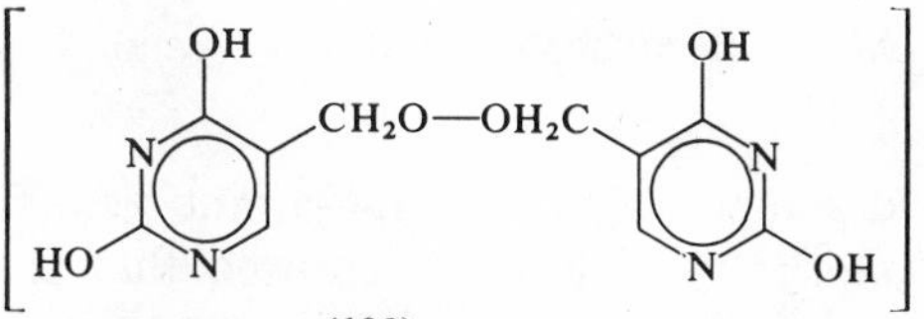

(129)

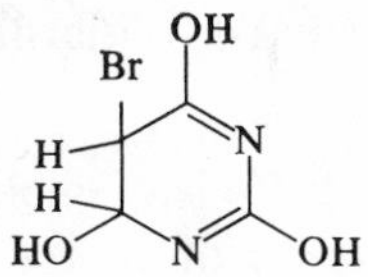

(130)

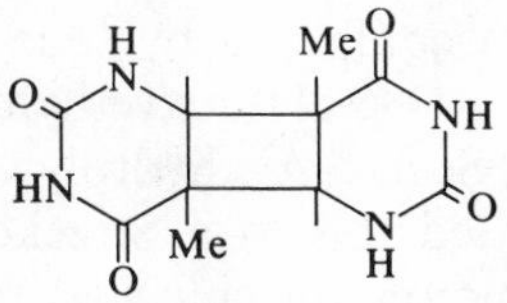

(131)

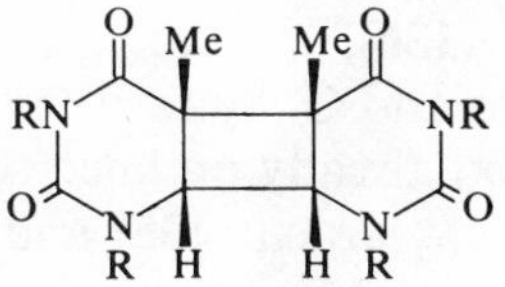

(132)

Why must the aqueous solution of thymine be frozen during photo-dimerization; and why indeed does the dimer revert* to thymine when

* The kinetics of photo-reversion have been studied.[3145]

similarly irradiated in unfrozen aqueous media? These questions have been discussed quite fully by S. Y. Wang:[3135, 3146] it seems that the formation of crystalline ice forces the solute into solid aggregates in which the molecules of thymine monohydrate are suitably arranged for dimerization during irradiation. If other frozen solvents are used, no dimerization takes place. When the dimer is suspended in (liquid) water and irradiated, the equilibrium (or steady state) is reversed because the monomeric molecules are distributed at random in the solution as they are formed, and hence have little tendency to dimerize again.

Most of the adverse biological effects produced by ultra-violet irradiation of tissues are currently blamed on the formation of dimer from adjacent thymine residues in a polynucleotide strand. Hence it is natural that the present emphasis is on photo-dimerization in such biopolymers.[3134, 3148, 3149] A review of the photochemistry of nucleic acids is available in Polish.[3088]

C. Divicine: 2,4-Diamino-5,6-dihydroxypyrimidine (*H* 259)

Details of the J. F. W. McOmie and J. H. Chesterfield synthesis (*H* 259) of divicine have appeared now.[2210] Identification with natural material was made through the diacetate (**133**) derived from each source.[2210]

D. Willardiine: 1-β-Amino-β-carboxyethyluracil (*New*)

In 1959 R. Gmelin isolated a new nonprotein L-α-amino acid from the seeds of *Acacia willardiana*,[3151] and later he showed that it was present also in some other species of Acacia.[3152] The new acid, willardiine, was formulated as the L-isomer of 1-β-amino-β-carboxyethyl-pyrimidine (**136**),[3151] and this has been confirmed by four syntheses: J. H. Dewar and G. Shaw[2338, 3153] made 1-ββ-diethoxyethyluracil (**134**) by a method already outlined (Ch. III, Sect. 5.A), hydrolysed it to the 1-formylmethyluracil (**135**) and submitted this to a Strecker reaction (potassium cyanide, ammonia, and ammonium chloride) to give DL-willardiine (**136**) which they resolved; the L-isomer proved identical with natural material.[2338, 3153] A new and better preparation of the key intermediate (**134**) was later devised;[3154] it consisted simply of alkylating uracil with β-bromo-αα-diethoxyethane. In a third synthesis[3155] β-ethoxyacryloyl isocyanate (EtOCH:CHCONCO) and methyl β-amino-α-*p*-toluenesulphonamidopropionate were used to give the intermediate (**137**) which underwent cyclization to the pyrimidine

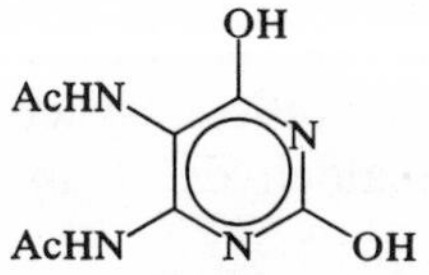

(**133**)

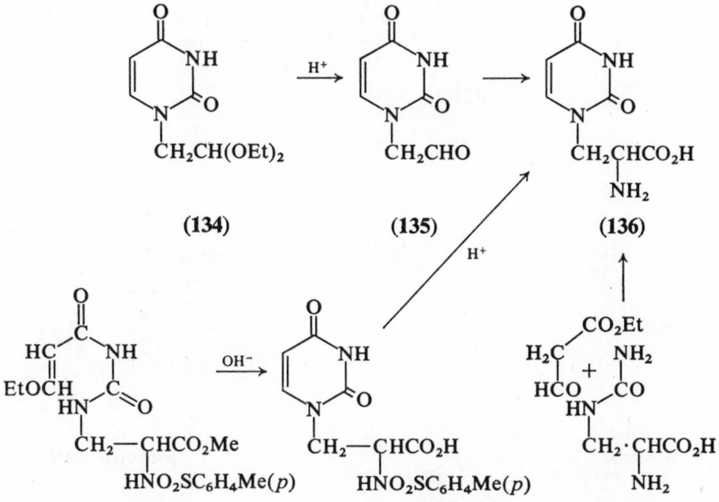

(**134**) (**135**) (**136**)

(**137**) (**138**) (**139**)

(**138**) in alkali; deacylation in acidic media gave willardiine (**136**).[3155] The fourth method involved a Principal Synthesis using ethyl formylacetate and *N*-β-amino-β-carboxyethylurea (**139**) to give willardiine directly.[2310]

9. The Alloxan Group of Pyrimidines (*H* 260)

The chemistry of the alloxan group has been rather neglected of recent years, but some preparations and reactions of alloxan (**141**) have been reported.

Chlorination ($HCl + H_2O_2$) of 5-nitrobarbituric acid gives 5-chloro-5-nitrobarbituric acid (**140**) which on heating gives alloxan (**141**) as hydrate in 63% overall yield;[2635] 5-chlorobarbituric acid (**142**) and nitric acid gives alloxan (**141**) in 76% yield.[2635] The preparation from urea and diethyl malonate of barbituric acid, its conversion into

5-benzylidene barbituric acid (**143**), and the final oxidation (CrO_3) to alloxan (**141**) have been reinvestigated to utilize ^{14}C-urea effectively (50% overall) in producing 2-^{14}C-alloxan.[3156]

The crystal structure of alloxan has been measured[3023] and comments have been made on some theoretical aspects of the data.[3157]

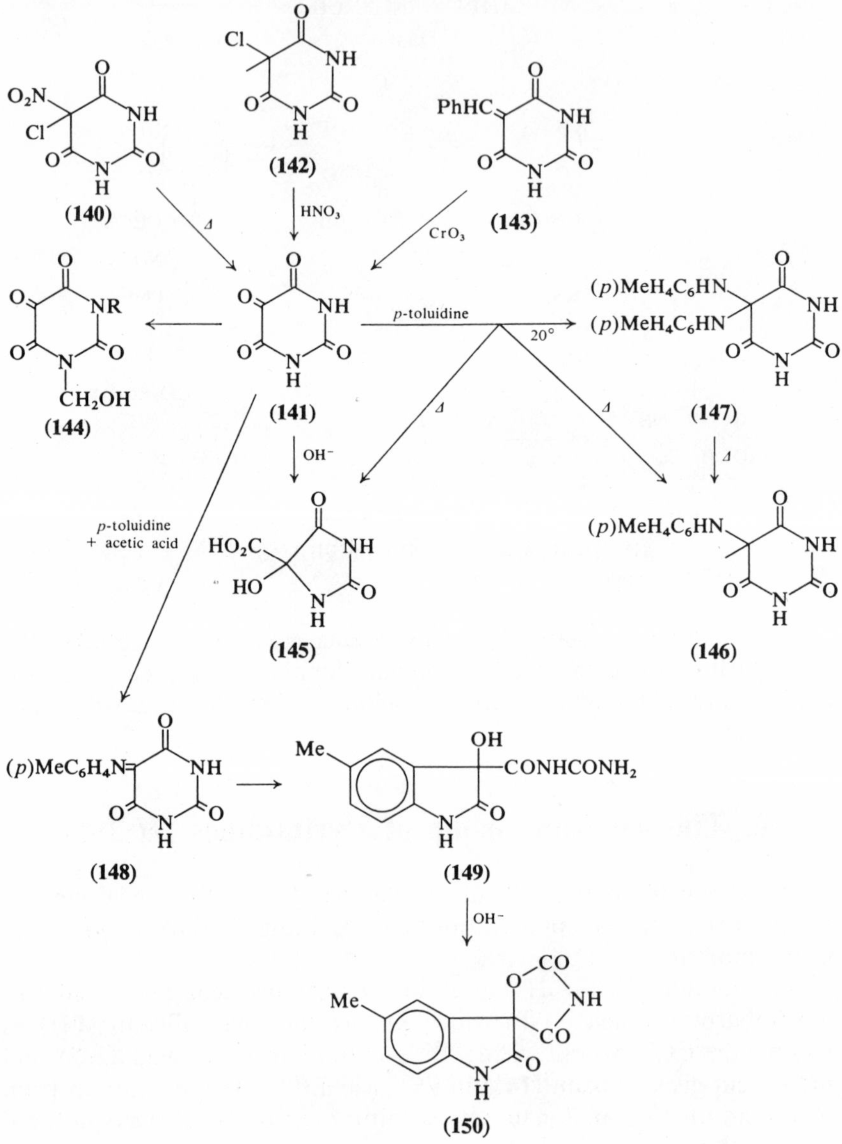

When treated with formaldehyde in mildly acidic media, alloxan yields 1-hydroxymethyl- (**144**, R = H) or 1,3-bis-hydroxymethyl-alloxan (**144**, R = CH_2OH) according to the amount of aldehyde;[3158] these compounds are of potential use as crease-resisting agents in textiles.[3158] The benzilic acid rearrangement of alloxan into alloxanic acid (**145**) has been investigated kinetically[3159] and also by using tagged alloxan derivatives.[3160] The course of the rearrangement is far less simple than previously thought.[1045, 3161] When ethanolic alloxan hydrate is boiled with *p*-toluidine, the toluidine salt of alloxanic acid (**145**) is isolated in 58% yield and *p*-toluidinobarbituric acid in 15% yield;[3162] the same reactants in the cold give 5,5-ditoluidinobarbituric acid (**147**) which on boiling in ethanol gives the above mixture (**145** + **146**).[3162] In contrast, anhydrous alloxan and *p*-toluidine in acetic acid give 5-*p*-tolyliminobarbituric acid (**148**) which gradually gives the indole (**149**) and thence, in alkali, the spiro compound (**150**).[3163] Other examples of these reactions are known.[3162, 3163]

Alloxan forms a 5-semicarbazone (**151**, R = O) and thiosemicarbazone (**151**, R = S).[3164] In alkali, each of these undergoes cyclization to the (unisolated) penta-azanaphthalene (**152**, R = OH or SH) which undergoes fission of the pyrimidine ring to give 6-carboxy-3,5-dihydroxy-1,2,4-triazine or its 3-mercapto analogue, respectively, in good yield.[3164] When *S*-ethylthiosemicarbazide is condensed with alloxan in acetic acid, 3-ethylthio-6,8-dihydroxy-1,2,4,5,7-penta-azanaphthalene (**152**, R = SEt) is isolated and may be hydrolysed first to

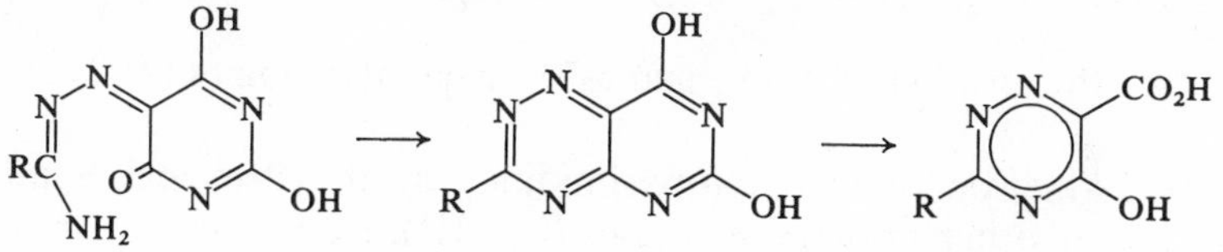

(**151**) (**152**) (**153**)

the corresponding 3,6,8-trihydroxy compound and thence to the triazine (**153**, R = OH).[3164] A similar condensation of alloxan with aminoguanidine gives the derivative (**151**, R = NH_2), cyclized in ammonia to 3-amino-6,8-dihydroxy-1,2,4,5,7-penta-azanaphthalene **152**, R = NH_2).[2388, 2529, 3165]

CHAPTER VIII

Sulphur-containing Pyrimidines (*H* 272)

Although alkylthiopyrimidines are still used widely as intermediates, the derived sulphones are much better leaving groups. Therefore it is often wise to oxidize a thioether to a sulphone prior to nucleophilic replacement. The general availability of Raney nickel for removing mercapto and alkylthio groups in favour of hydrogen, coupled with improved methods for thiation of hydroxypyrimidines with phosphorus pentasulphide (Ch. VII, Sect. 6.B), have made the removal of an hydroxy group from pyrimidine often easier *via* a mercapto- than *via* a chloro-pyrimidine.

1. The Mercaptopyrimidines (*H* 272)

A. Preparation of 2-, 4-, and 6-Mercaptopyrimidines (*H* 272)

The formation of mercaptopyrimidines by the Principal Synthesis (Ch. II), by other primary syntheses (Ch. III), from chloropyrimidines (Ch. VI, Sect. 5.G), and by thiation of hydroxypyrimidines (Ch. VII, Sect. 6.B) have been discussed already.

The conversion of an alkylthio- into a mercaptopyrimidine (*H* 274) is seen in the treatment of 2,4,5-triamino-6-benzylthiopyrimidine (**1**, R = CH_2Ph) in liquid ammonia with sodium to give 2,4,5-triamino-6-mercaptopyrimidine (**1**, R = H; 20%);[3061] also in the hydrolysis of 1-benzyl-2-benzylthio-1,4-dihydro-4-oxopyrimidine by hot hydrochloric acid for 15 min. to give 1-benzyl-2-thiouracil (86%).[3187]

B. Preparation of 5-Mercaptopyrimidines (*H* 276)

2-Amino-5-mercaptopyrimidine (**3**, R = NH_2) has been made by reducing di(2-aminopyrimidin-5-yl) disulphide (**2**, R = NH_2) with alkaline sodium dithionite; it was isolated (81%) after *S*-methylation.[2304] 5-Mercapto-2-phenylpyrimidine (**3**, R = Ph) was made and isolated similarly (91%) from the corresponding disulphide (**2**, R = Ph).[2304] An odd reaction is that of the sulphide, 1,2,3,4-tetrahydro-6-hydroxy-1,3-dimethyl-2,4-dioxo-5-(1,2,3,4-tetrahydro-6-hydroxy-1,3-dimethyl-2,4-dioxopyrimidin-5-ylthio)pyrimidine (**4**), with acetic anhydride to give 5-acetylthio-1,2,3,4-tetrahydro-6-hydroxy-1,3-dimethyl-2,4-dioxopyrimidine (**5**; 71%).[2304]

D. Reactions of 2-, 4-, and 6-Mercaptopyrimidines (*H* 277)

The fine structure of mercaptopyrimidines (previously mentioned *H* 277) is discussed in a new section (1.F) below.

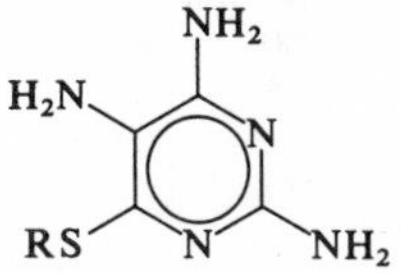

(**1**) 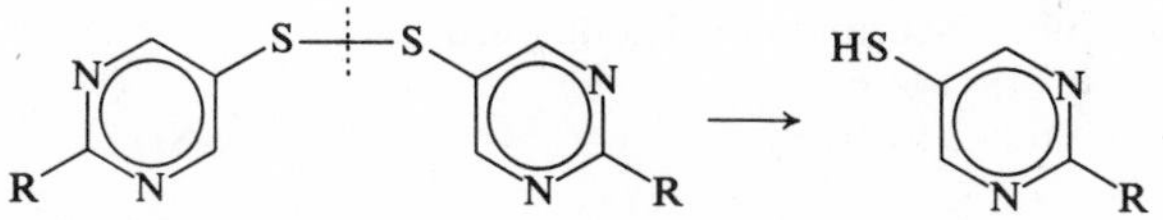 (**2**) (**3**)

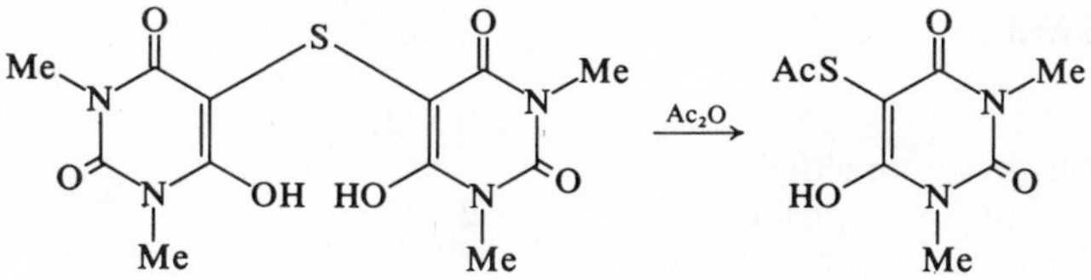

Ac_2O

(**4**) (**5**)

Ni

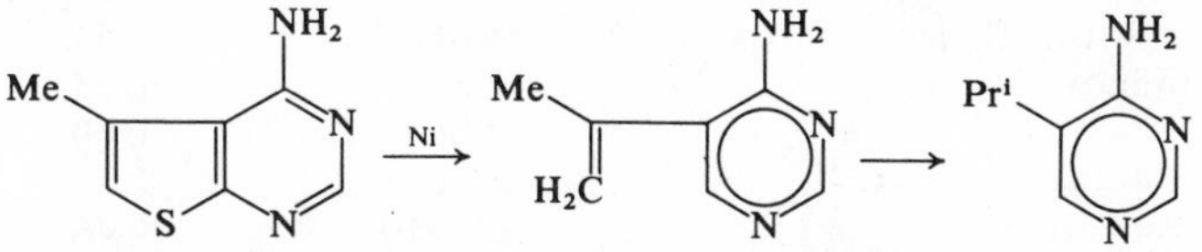

(**6**) (**7**) (**8**)

(1) *Removal of Mercapto Groups* (*H* 277)

The recent use of Raney nickel in desulphurization of 2-, 4-, and 6-mercapto- and alkylthio-pyrimidines is summarized in Table XIII. In addition, several thieno[2,3-*d*]pyrimidines have been desulphurized to yield pyrimidines:[2531, 2898] 4-amino-5-methylthieno[2,3-*d*]pyrimidine (**6**) gave a mixture of 4-amino-5-isopropenyl- (**7**) and 4-amino-5-isopropyl-pyrimidine (**8**), which was hydrogenated to give only the latter product (**8**);[2898] 4-amino-5,6-tetramethylenethieno[2,3-*d*]pyrimidine gave a mixture of 4-amino-5-cyclohexenyl- and 4-amino-5-cyclohexyl-pyrimidine;[2898] and other similar examples are known.[2531, 2898]

An unusual oxidative removal of a mercapto group is seen in the formation of 5-ethoxycarbonylmethyl-4-hydroxypyrimidine (64%) by

TABLE XIII. Some Raney Nickel[a] Desulphurizations (*H* 279)

Pyrimidine produced	Derivative desulphurized	Yield (%)	Ref.
2-Amino-5-γ-anilinopropyl-4-methyl-	6-SH[b]	77	3190
4-Amino-5-benzamido-	6-SH	*ca.* 20	3186
4-Amino-6-chloro-	2-SMe[b]	63	2777, 3192
2-Amino-5-*p*-chlorophenyl-4-methyl-	6-SH[b]	72	2816
2-Amino-1,4-dihydro-4-imino-1-methyl-[c]	6-SMe[c]	64	2288
5-Amino-1,4-dihydro-4 imino-1-methyl-	2-SMe	28	2288
4-Amino-6-dimethylamino-	2-SMe[b]	60	2433
4-Amino-5-formamido-1,6-dihydro-1-methyl-6-oxo-	2-SMe	*ca.* 75	3188
4-Amino-2-hydroxy-	6-SH[b]	43	2777, 3192
	6-SMe[b]	36	2777, 3192
4-Amino-6-hydroxy-	2-SH[b]	90, 75	3192, 3457
	2-SMe[b]	79	3192
5-Amino-4-hydroxy-6-methyl-	2-SH	73	2227
4-Amino-6-methyl-5-methylamino-	2-SMe	33	2675
5-Amino-4-methyl-6-methylamino-	2-SH	—	2675
2-Amino-4,5-pentamethylene-	6-SH[b]	38	3192
2-Amino-4,5-tetramethylene-	6-SH[b]	67	3192
2-Amino-4,5-trimethylene-	6-SH[b]	51	3192
4-Benzylamino-	2-SH[b]	*ca.* 33	3185
	$2\text{-SCH}_2\text{CO}_2\text{H}$[b]	*ca.* 60	3185
1-Benzyl-1,4-dihydro-4-oxo-	$2\text{-SCH}_2\text{Ph}$	—	3187
5-Bromo-4-hydroxy-	2-SMe	46	2191
4-t-Butyl-6-hydroxy-	2-SH	94	2574
4-Carboxy-6-hydroxy-	2-SH	48	2214
4-Chloro-6-hydroxy-	2-SMe[b]	51	2777, 3192
5-Chloro-4-hydroxy-	2-SMe	46	2191
2,4-Diamino-	6-SH[b]	65	3192
4,5-Diamino-	6-SH[b]	77	3192
	2,6-diSH	84	3189
	2-SH	64	3457
4,6-Diamino-	2-SH[b]	83,[d] 70[e]	2296
	2-SMe[b]	63	3192
	$2\text{-SCH}_2\text{CH}_2\text{NEt}_2$	*ca.* 30	3191

Pyrimidine produced	Derivative desulphurized	Yield (%)	Ref.
4,5-Diamino-3,6-dihydro-3-methyl-6-oxo-	2-SH	*ca.* 65	2492
4,5-Diamino-6-ethoxycarbonyl-	2-SH[f]	80	3199
4,5-Diamino-2-phenyl	6-SH	60	2284
4,6-Diamino-5-sulphamoyl	2-SMe	31	2364
4-Diethoxymethyl-6-methyl-	2-SMe[b]	60	2301
1,2-Dihydro-1,6-dimethyl-2-oxo-	4-SMe	30	2630
1,6-Dihydro-1,5-dimethyl-6-oxo-	2-SEt[b]	*ca.* 25	2630
4,6-Dihydroxy-5-β-hydroxyethyl-	2-SH	40	2638
5-β-(1,3-Dioxolan-2-yl)ethyl-4-hydroxy-6-methyl-	2-SH	86	3190
5-Ethoxycarbonyl-2-methyl	4-SH	30	2654
2-Ethoxy-5-fluoro-	4-SH	—	3530
1-Ethyl-1,4-dihydro-4-oxo-	2-SH	52	3187
5-Fluoro-4-hydroxy-	2-SH	56	2191
5-Fluoro-2-methoxy-	4-SH	—	3530
4-Heptylamino-	2-SH[b]	*ca.* 40	3185
4-Hexylamino-	2-SH[b]	*ca.* 30	3185
4-Hydroxy-	2-SH	70	3457
4-Hydroxy-5-isoamyl-6-methyl-	2-SH[b]	54	2816
4-Hydroxy-5-methoxy-6-methyl-	2-SH	—	3243
4-Hydroxy-6-methyl-	2-SH	95	2574
4-Hydroxy-5-methyl-6-phenyl-	2-SH	67	2241
4-Hydroxy-5,6-pentamethylene-	2-SH[b]	67	3192
4-β-Hydroxyphenethylamino-	2-SMe[b]	56	2597
4-Hydroxy-6-phenyl-	2-SH	73, 92	2519, 2574
4-Hydroxy-5,6-tetramethylene-	2-SH[b]	49	3192
4-Tetrahydrofurfurylamino-	2-SH[b]	*ca.* 15	3185
4,5-Tetramethylene-	2,6-diSH[b]	38	3192
2,4,5-Triamino-	6-SMe[b]	76	3192
4,5,6-Trismethylamino-	2-SMe	—	2454
Unsubstituted	2-SH	27[g]	3457

[a] In some examples a commercially available sponge nickel catalyst[3192] was used.
[b] In ethanol; others are in aqueous media.
[c] As hydrochloride.
[d] In an autoclave at 120°.
[e] Under reflux.
[f] In 'diglyme'.
[g] As mercuric chloride complex.

treating its 2-mercapto derivative with lead tetra-acetate and hydrogen peroxide.[2577]

A new reductive desulphurization of 4,5-diamino-2-mercaptopyrimidine with nickel boride in aqueous suspension at 200° gave 4,5-diaminopyrimidine (45%; *cf.* Raney nickel, 65%). The reagent failed with three other mercaptopyrimidines, but was moderately successful in other series.[3457]

(2) S-*Alkylation of Mercaptopyrimidines* (*H* 282)

Recent examples of this important reaction are contained in Table XIIIa.

TABLE XIIIa. Examples of *S*-Alkylation of Mercaptopyrimidines (*New*)

Pyrimidine produced	Reagents[a]	Yield (%)	Ref.
5-Acetyl-2,4-dimethyl-6-methylthio-	NaOH/MeI	86	2905
5-Acetyl-4-methyl-6-methylthio-2-phenyl-	NaOH/MeI	95	2905
4-Allylthio-2,5,6-triamino-	KOH/CH_2:$CHCH_2Cl$[b]	56	2778
5-Amino-4-benzylidene-hydrazino-6-methylthio-	NaOH/MeI[c]	91	2227
2-Amino-4-benzylthio-	NaOH/$PhCH_2Cl$	—[d]	2165
4-Amino-6-benzylthio-	NaOH/$PhCH_2Cl$	—[d]	2165
2-Amino-4-benzylthio-5-bromo-6-methyl-	NaOH/$PhCH_2Cl$	—[d]	2165
2-Amino-4-benzylthio-6-carboxy-	NaOH/$PhCH_2Cl$	69	2214
2-Amino-4-benzylthio-6-methyl-	NaOH/$PhCH_2Cl$	—[d], 61	2165, 2842
2-Amino-4,6-bisbenzylthio-	NaOH/$PhCH_2Cl$	—[d]	2165
2-Amino-4,6-bisbenzylthio-5-phenyl-	NaOH/$PhCH_2Cl$	—[d]	2165
2-Amino-4,6-bis-*o*-chloro-benzylthio-5-phenyl-	NaOH/(*o*)$ClC_6H_4CH_2Cl$	—[d]	2165
2-Amino-4,6-bis-2′,4′-dichlorobenzylthio-	NaOH/(2,4)$Cl_2C_6H_3CH_2Cl$	—[d]	2165
4-Amino-2,6-bis-2′,4′-dichlorobenzylthio-	NaOH/(2,4)$Cl_2C_6H_3CH_2Cl$	—[d]	2165
2-Amino-4,6-bis-2′,4′-dichlorobenzylthio-5-phenyl-	NaOH/(2,4)$Cl_2C_6H_3CH_2Cl$	—[d]	2165
2-Amino-4,6-bisethylthio-	NaOH/EtI	—[d]	2165
2-Amino-4,6-bismethylthio-	NaOH/MeI	—[d]	2165
4-Amino-2,6-bismethylthio-	NaOH/MeI	—[d]	2165
5-Amino-4,6-bismethylthio-	NaOH/MeI	—[d]	2165
2-Amino-4,6-bismethylthio-5-phenyl-	NaOH/MeI	—[d]	2165
2-Amino-4,6-bispropylthio-	NaOH/PrI	—[d]	2165
2-Amino-5-bromo-4-*o*-chlorobenzylthio-6-methyl-	NaOH/(*o*)$ClC_6H_4CH_2Cl$	—[d]	2165
2-Amino-5-bromo-4-methyl-6-methylthio-	NaOH/MeI	—[d]	2165

Pyrimidine produced	Reagents[a]	Yield (%)	Ref.
2-Amino-5-bromo-4-methyl-6-*p*-nitrobenzylthio-	NaOH/(*p*)$O_2NC_6H_4CH_2Cl$	—[d]	2165
2-Amino-5-bromo-4-methyl-6-propylthio-	NaOH/PrI	—[d]	2165
2-Amino-4-butylthio-6-hydroxy-	NaOH/BuI	—[d]	2165
4-Amino-2-butylthio-6-hydroxy-	NaOEt/BuI[e]	—	3216
2-Amino-4-butylthio-6-methyl-	NaOH/BuI	—[d]	2165
2-Amino-4-carboxy-6-*o*-chlorobenzylthio-	NaOH/(*o*)$ClC_6H_4CH_2Cl$	68	2214
2-Amino-4-carboxy-6-methylthio-	NaOH/MeI	69	2214
4-Amino-2-carboxymethylthio-5-*p*-chlorophenyl-6-ethyl-	$ClCH_2CO_2Na$	50	2983
2-Amino-4-*o*-chlorobenzylthio-6-methyl-	NaOH/(*o*)$ClC_6H_4CH_2Cl$	—[d]	2165
2-Amino-4-chloro-6-2′,4′-dichlorobenzylthio-	NaOH/(2,4)$Cl_2C_6H_3CH_2Cl$	—[d]	2165
2-Amino-4-chloro-6-ethylthio-	NaOH/EtI	—[d]	2165
2-Amino-4-chloro-6-methylthio-	NaOH/MeI	—[d]	2165
2-Amino-4-chloro-6-propylthio-	NaOH/PrI	—[d]	2165
2-Amino-4-cyclohexylthio(?)[f] -6-hydroxy-	NaOH/$(CH_2)_5CHCl$(?)	—[d]	2165
2-Amino-4-2′,4′-dichlorobenzylthio-	NaOH/(2,4)$Cl_2C_6H_3CH_2Cl$	—[d]	2165
4-Amino-6-2′,4′-dichlorobenzylthio-	NaOH/(2,4)$Cl_2C_6H_3CH_2Cl$	—[d]	2165
2-Amino-4-2′,4′-dichlorobenzylthio-6-methyl-	NaOH/(2,4)$Cl_2C_6H_3CH_2Cl$	—[d]	2165
4-Amino-6-ethoxycarbonyl-2-methylthio-5-nitro-	NaOH/MeI	—	3458
2-Amino-4-ethylthio-	NaOH/EtI	—[d]	2165
4-Amino-6-ethylthio-	NaOH/EtI	—[d]	2165
2-Amino-4-ethylthio-6-hydroxy-	NaOH/EtI	—[d]	2165
4-Amino-2-ethylthio-6-hydroxy-	NaOEt/EtI[e]	—	3216
2-Amino-4-ethylthio-6-methyl-	NaOH/EtI	—[d]	2165
4-Amino-6-hydroxy-2-isopropylthio-	NaOEt/Pr^iI[e]	—	3216
2-Amino-4-hydroxy-6-methylthio-	NaOH/MeI	—[d]	2165

continued

TABLE XIIIa (*continued*).

Pyrimidine produced	Reagents[a]	Yield (%)	Ref.
4-Amino-2-hydroxy-6-methylthio-	NaOH/MeI	—[d]	2165
	$NaOH/Me_2SO_4$	*ca.* 100	2777
4-Amino-6-hydroxy-2-methylthio-	KOH/Me_2SO_4	*ca.* 80	2475
	NaOEt/MeI[e]	—	3216
2-Amino-4-hydroxy-6-propylthio-	NaOH/PrI	—[d]	2165
4-Amino-6-hydroxy-2-propylthio-	NaOEt/PrI[e]	—	3216
5-Amino-2-methyl-4-methylamino-6-methylthio-	NaOH/MeI	*ca.* 90	2675
2-Amino-4-methyl-6-methylthio-	NaOH/MeI	—[d]	2165
5-Amino-4-methyl-6-methylthio-	NaOH/MeI	72	2227
2-Amino-4-methyl-6-*p*-nitrobenzylthio-	NaOH/(*p*)$O_2NC_6H_4CH_2Cl$	—[d]	2165
2-Amino-4-methylthio-	NaOH/MeI	—[d]	2165, 2776
4-Amino-6-methylthio-	NaOH/MeI	—[d]	2165
4-Amino-6-*p*-nitrobenzylthio-	NaOH/(*p*)$O_2NC_6H_4CH_2Cl$	—[d]	2165
4-Amylamino-2-carboxymethylthio-	$ClCH_2CO_2H$	*ca.* 50	3185
4-Benzylamino-2-carboxymethylthio-	$ClCH_2CO_2H$	*ca.* 50	3185
2-Benzylthio-4-bis-(β-hydroxyethyl)amino-	$NaOH/PhCH_2Cl$[g]	86	2910
2-Benzylthio-4-carboxy-6-hydroxy-	$NaOH/PhCH_2Cl$	62	2214
4-Benzylthio-2,6-dihydroxy-	$NaOH/PhCH_2Cl$	—[d]	2165
4-Benzylthio-6-hydroxy-	$NaOH/PhCH_2Cl$	—[d]	2165
2-Benzylthio-4-morpholino-	$NaOH/PhCH_2Cl$[g]	82	2910
4,6-Bisbenzylthio-5-bromo-	$NaOH/PhCH_2Cl$	—[d]	2165
4,6-Bisbenzylthio-5-chloro-	$NaOH/PhCH_2Cl$	—[d]	2165
2,4-Bisbenzylthio-6-methyl-	$NaOH/PhCH_2Cl$	—[d]	2165
4,6-Biscarboxymethylthio-5-*o*-chlorophenyl-	$NaOMe/EtO_2CCH_2Cl$[h]	37	2522
4,6-Biscarboxymethylthio-5-*p*-chlorophenyl-	$NaOMe/EtO_2CCH_2Cl$[h]	76	2522
4,6-Biscarboxymethylthio-5-phenyl-	$NaOMe/EtO_2CCH_2Cl$[h]	67	2522
2,4-Bis-2′,4′-dichlorobenzylthio-	NaOH/(2,4)$Cl_2C_6H_3CH_2Cl$	—[d]	2165
2,4-Bis-2′,4′-dichlorobenzylthio-6-methyl-	NaOH/(2,4)$Cl_2C_6HC_3H_2Cl$	—[d]	2165
4-Bis-(β-hydroxyethyl)amino-2-carboxymethylthio-	$ClCH_2CO_2H$	*ca.* 90	2648

Pyrimidine produced	Reagents[a]	Yield (%)	Ref.
4,6-Bismethylthio-	NaOH/MeI	—[d]	2165
5-Bromo-4,6-bis-2′,4′-dichlorobenzylthio-	NaOH/(2,4)$Cl_2C_6H_3CH_2Cl$	—[d]	2165
5-Bromo-4,6-bismethylthio-	NaOH/MeI	—[d]	2165
5-Bromo-4,6-bis-*p*-nitrobenzylthio-	NaOH/(*p*)$O_2NC_6H_4CH_2Cl$	—[d]	2165
5-Bromo-4,6-bispropylthio-	NaOH/PrI	—[d]	2165
5-Bromo-2-methylthio-	$NaHCO_3$/MeI	*ca.* 90	2746
5-Butyl-4-carboxymethylthio-6-chloro-	NaOMe/MeO_2CCH_2Cl[h]	87	2522
2-Butyl-4-hydroxy-6-methylthio-	NaOH/MeI	93	2242
5-Butyl-4-hydroxy-6-methylthio-	NaOH/MeI	98	2242
2-Butylthio-4-carboxy-6-hydroxy-	NaOH/BuI	53	2214
4-Carboxy-2,6-bismethylthio-	NaOH/MeI	49	2214
5-Carboxy-2,4-bismethylthio-	NaOH/MeI	—[d]	2165
4-Carboxy-2-2′,4′-dichlorobenzylthio-6-hydroxy-	NaOH/2,4-di$ClC_6H_3CH_2Cl$	72	2214
4-Carboxy-2-ethylthio-6-hydroxy-	NaOH/EtI	79	2214
4-Carboxy-2-hydroxy-6-methylthio-	NaOH/MeI	46	2214
4-Carboxy-6-hydroxy-2-propylthio-	NaOH/PrI	42	2214
4-Carboxymethylthio-6-chloro-5-*o*-chlorophenyl-	NaOMe/MeO_2CCH_2Cl[h]	62	2522
4-Carboxymethylthio-6-chloro-5-*p*-chlorophenyl-	NaOMe/MeO_2CCH_2Cl[h]	71	2522
4-Carboxymethylthio-6-chloro-5-phenyl-	NaOMe/MeO_2CCH_2Cl[h]	72	2522
2-Carboxymethylthio-4,6-dihydroxy-5-phenyl-	NaOMe/MeO_2CCH_2Cl[h]	*ca.* 10	2522
2-*o*-Chlorobenzylthio-4-dimethylamino-5-methyl-	NaOH/(*o*)$ClC_6H_4CH_2Cl$[b]	46	2583
4-*o*-Chlorobenzylthio-2-hydroxy-	NaOH/(*o*)$ClC_6H_4CH_2Cl$	—[d]	2165
2-*o*-Chlorobenzylthio-4-hydroxy-5-methyl-	NaOH/(*o*)$ClC_6H_4CH_2Cl$[b]	94	2583
5-Chloro-4,6-bis-2′,4′-dichlorobenzylthio-	NaOH/(2,4)$Cl_2C_6H_3CH_2Cl$	—[d]	2165
5-Chloro-4,6-bisethylthio-	NaOH/EtI	—[d]	2165
5-Chloro-4,6-bismethylthio-	NaOH/MeI	—[d]	2165
4-Chloro-2-dimethylamino-6-methylthio-	NaOH/MeI	*ca.* 85	2776

continued

TABLE XIIIa (*continued*).

Pyrimidine produced	Reagents[a]	Yield (%)	Ref.
2,4-Diamino-6-benzylthio-	$K_2CO_3/PhCH_2Cl$[i]	*ca.* 90	2778
4,6-Diamino-2-benzylthio-	$PhCH_2Cl$[j]	—[k]	2296
2,5-Diamino-4-benzylthio-6-β-hydroxyethylamino-	$KOH/PhCH_2Cl$[g]	85	2504
2,4-Diamino-6-methylthio-	KOH/MeI	85	2288, 2778
4,6-Diamino-2-methylthio-	$NaOH/Me_2SO_4$	*ca.* 80	2433
4,5-Diamino-6-methylthio-2-phenyl-	NaOH/MeI	40	2284
2,4-Diamino-6-propylthio-	KOH/PrI	*ca.* 90	2778
4-2′,4′-Dichlorobenzylthio-2-hydroxy-	$NaOH/(2,4)Cl_2C_6H_3CH_2Cl$	—[d]	2165
4-2′,4′-Dichlorobenzylthio-6-hydroxy-	$NaOH/(2,4)Cl_2C_6H_3CH_2Cl$	—[d]	2165
4-2′,4′-Dichlorobenzylthio-6-methyl-2-methylthio-	—[l]	—[d]	2165
1,4-Dihydro-1-methyl-6-methylthio-4-oxo-	NaOH/MeI	*ca.* 55	2760
1,6-Dihydro-1-methyl-4-methylthio-6-oxo-	NaOH/MeI	*ca.* 90	2760
4,6-Dihydroxy-5-β-hydroxyethyl-2-methylthio-	$NaHCO_3/Me_2SO_4$	61	2638
4,6-Dihydroxy-2-methylthio-	$NaOH/Me_2SO_4$[m]	*ca.* 70	2165
	KOH/MeI	*ca.* 30	2454
5-3′,4′-Dimethoxybenzyl-4-hydroxy-2-methylthio-	$NaOH/Me_2SO_4$	—	2213
2-Dimethylamino-4-methylthio-	NaOH/MeI	*ca.* 80	2776
4-Dimethylamino-2-methylthio-	NaOH/MeI	*ca.* 70	2745
4-Dimethylamino-6-methylthio-	KOH/MeI	*ca.* 70	2676
2,4-Dimethyl-6-methylthio-	NaOH/MeI	*ca.* 85	2746
5-Ethoxycarbonyl-4-methyl-6-methylthio-2-phenyl-	NaOH/MeI	92	2905
4-Ethyl-6-hydroxy-2-methylthio-	KOH/Me_2SO_4	70	2753
4-Ethylthio-	NaOH/EtI	*ca.* 70	2697
4-Hydroxy-2,5-bismethylthio-	NaOMe/MeI	88	3765
4-Hydroxy-2,6-bismethylthio-	NaOH/MeI	—[d]	2165
2-β-Hydroxyethylthio-4,6-dimethyl-	$NaOH/HOCH_2CH_2Cl$[n]	73	3010
4-Hydroxy-5-2′-hydroxycyclopentyl-2-methylthio-	NaOH/MeI	82	2952
4-Hydroxy-5-methyl-6-methylthio-	NaOH/MeI	77	2242

Pyrimidine produced	Reagents[a]	Yield (%)	Ref.
2-Hydroxy-4-methylthio-	NaOH/MeI	63	3215
4-Hydroxy-6-methylthio-	NaOH/MeI	—[d]	2165
4-Hydroxy-2-methylthio-6-phenyl-	KOH/Me_2SO_4	67	2753
4-β-Hydroxyphenethylamino-2-methylthio-	$NaOH/Me_2SO_4$	88	2597
2-Isopropylthio-	$NaHCO_3/Pr^iI$	*ca.* 50	2697
4-Methoxy-6-methylthio-	NaOH/MeI	*ca.* 75	2760
4-Methylamino-6-methylthio-	NaOH/MeI	*ca.* 80	2781
4-Methyl-2,6-bismethylthio-	NaOH/MeI	—[d]	2165
4-Methyl-2-methylthio-	$NaHCO_3/Me_2SO_4$	*ca.* 70	2746
2-Methylthio-	$NaHCO_3/Me_2SO_4$[o]	83	2512
2-Methylthio-5-nitro-	$NaHCO_3/MeI$	0[p]	2746
	CH_2N_2 (ether)	0[p]	2746
2,4,5-Triamino-6-benzylthio-	$K_2CO_3/PhCH_2Cl$[i]	76	2778
2,4,5-Triamino-6-butylthio-	KOH/BuI[b]	47	2778
2,4,5-Triamino-6-*p*-chlorobenzylthio-	$K_2CO_3/(p)ClC_6H_4CH_2Cl$[i]	83	2778
2,4,5-Triamino-6-ethylthio-	KOH/EtI	52	2778
2,4,5-Triamino-6-methylthio-	KOH/MeI	67	2778
2,4,5-Triamino-6-propylthio-	KOH/PrI[g]	50	2778
2,4,6-Triscarboxymethylthio-5-phenyl-	$NaOMe/MeO_2CCH_2Cl$[h]	42	2522
2,4,6-Tris-*o*-chlorobenzylthio-	$NaOH/(o)ClC_6H_4CH_2Cl$	—[d]	2165
2,4,6-Tris-2′,4′-dichlorobenzylthio-	$NaOH/(2,4)Cl_2C_6H_3CH_2Cl$	—[d]	2165
2,4,6-Trismethylthio-	NaOH/MeI	—[d]	2165

[a] Aqueous media unless otherwise indicated.
[b] In aq. dioxan.
[c] With a little sodium dithionite.
[d] 'Yields of recrystallized products usually 80–95%.'[2165]
[e] In ethanol(?).
[f] This substituent is recorded[2165] as 'n-$C_6H_{11}S$'.
[g] In aq. ethanol.
[h] Alkylated in methanol and resulting ester saponified by aq. alkali.
[i] In dimethylformamide.
[j] In refluxing ethanol.
[k] Isolated as the 5-nitroso derivative in 95% yield.
[l] Route not indicated.
[m] An excess of methylating agent gave 1,6-dihydro-4-hydroxy-1-methyl-2-methylthio-6-oxopyrimidine (33%).[2440]
[n] In aq. propanol.
[o] This improved yield resulted from saturation of the aqueous media with sodium chloride prior to extraction of the product.
[p] Failure reported, but required product made otherwise.

The use of an $\alpha\omega$-dihalogenoalkane in place of a simple alkyl halide for *S*-alkylating a mercaptopyrimidine permits two pyrimidine rings to be linked by an —S—$(CH_2)_n$—S— chain. Thus 2-mercaptopyrimidine and ethylene dibromide in a 2:1 ratio gave >90% of 2-β-pyrimidin-2′-ylthioethylthiopyrimidine; higher homologues were formed by using appropriate dihalogenoalkanes.[3229] 2-Mercapto-4,6-dimethylpyrimidine similarly gave 2-β-4′,6′-dimethylpyrimidin-2′-ylthioethylthio-4,6-dimethylpyrimidine and its homologues.[3229]

(3) *Mercapto- to Hydroxy-pyrimidines* (*H* 284)

This change has been discussed in Ch. VII, Sect. 1.F.

(5) *Mercapto- to Amino-pyrimidines* (*H* 284)

Mercaptopyrimidines are usually converted into alkylthio- or even alkylsulphonyl-pyrimidines prior to aminolysis. However 4-mercaptopyrimidines often undergo aminolysis directly, and the process can be especially useful with 2,4-dimercaptopyrimidines in which the 4-mercapto group is replaced preferentially.

Such preferential aminolyses are exemplified by the conversion of 5-amino-2,4-dimercapto- (**9**, R = NH_2) into 5-amino-2-mercapto-4-methylamino-pyrimidine (**10**; R = NH_2, R′ = Me) in 65% yield by aq. methylamine at 130°;[2458] of 1-benzyl-2,4-dithiouracil (**11**, R = Ph) into 4-amino-1-benzyl-1,2-dihydro-2-thiopyrimidine (**12**, R = Ph) by aq. ammonia at 120°;[2337] of 1-methyl-2,4-dithiouracil (**11**, R = H) into 4-amino-1,2-dihydro-1-methyl-2-thiopyrimidine (**12**, R = H);[2337] of 2,4-dimercaptopyrimidine (**9**, R = H) into 4-furfurylamino-,[3217] 4-amino- (**10**, R = R′ = H),[2915] 4-butylamino- (**10**, R = H, R′ = Bu),[3185] 4-isoamylamino-,[3185] 4-hexylamino-,[3185] 4-heptylamino-,[3185] 4-benzylamino-,[3185, 3217] 4-bis-(β-hydroxyethyl)amino-,[2648, 2910] and 4-β-hydroxyphenethylamino-2-mercaptopyrimidine (86%);[2597] of 2,4-dimercapto- into 4-hydrazino-2-mercapto-6-methylpyrimidine (80%) by refluxing ethanolic hydrazine hydrate;[3218] and of 2,4-dimercapto-6-methylpyrimidine with ethylenediamine into 2-mercapto-4-β-(2′-mercaptopyrimidin-4′-ylamino)ethylaminopyrimidine (**13**) in 71% yield.[2710] 2,4-Diselenylpyrimidine (**14**, R = SeH) behaves as its dithio analogue in reacting with ammonia at 100° to give a 50% yield of 4-amino-2-selenylpyrimidine (**14**, R = NH_2);[2779] 4-Amino-5-methyl-2-selenylpyrimidine is formed similarly.[2779]

Other aminolyses of 4-mercaptopyrimidines furnished 4-amino-2-hydroxy-,[3057] 2-hydroxy-4-methylamino-,*[2984] 4-ethyl-6-hydrazino-

* Attempts to aminolyse 3-methyl-4-thiouracil (**15**, R = SH) with dimethylamine did not give the expected 4-dimethylamino-2,3-dihydro-3-methyl-2-oxopyrimidine (**15**,

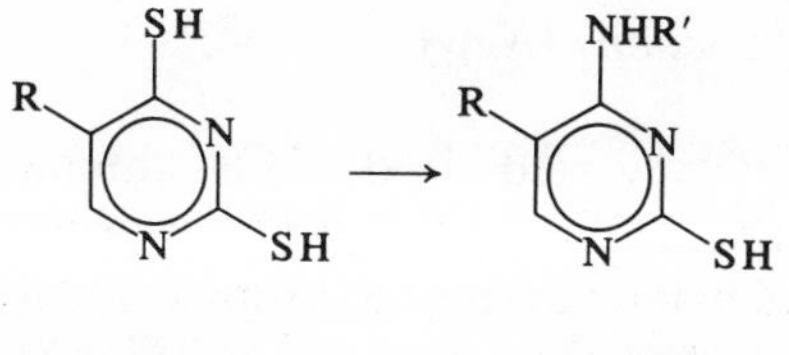

2-hydroxy-,[2753] 4-hydrazino-2-hydroxy-6-phenyl(or propyl)-,[2753] 2-hydroxy-4-(*N*-β-hydroxyethyl-*N*-methylamino)- (**16**),[2641] 4-γ-hydroxypropylamino-2-hydroxy-,[2641] and 4-amino-2-hydroxy-5-2′-hydroxycyclopentyl-pyrimidine.[2952]

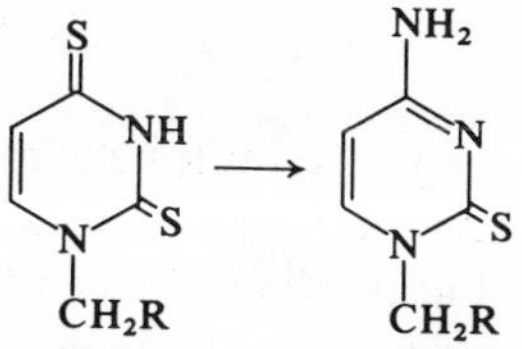

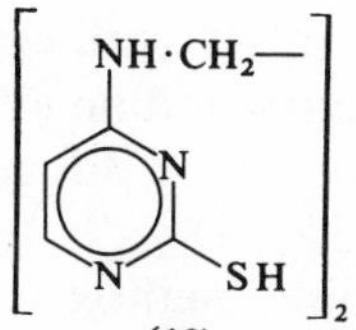

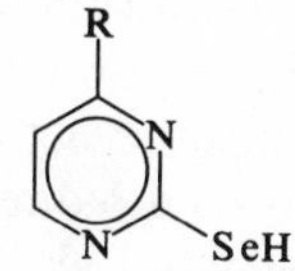

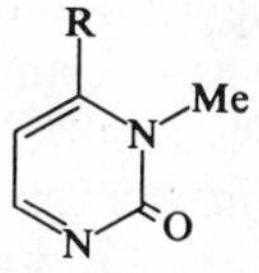

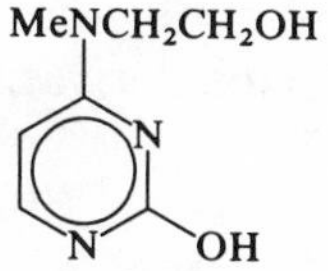

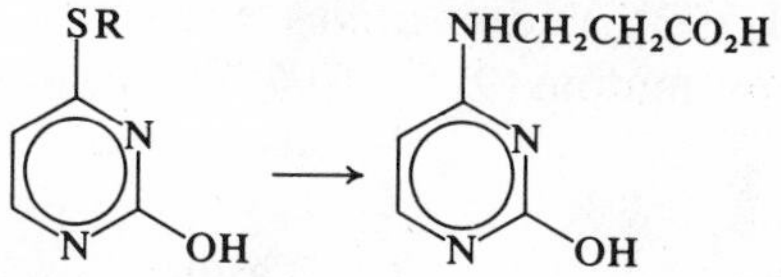

The preparation of 4-β-carboxyethylamino-2-hydroxypyrimidine (**18**) from both 2-hydroxy-4-mercaptopyrimidine (**17**, R = H) and its *S*-methyl derivative (**17**, R = Me) furnishes a direct preparative comparison of the ease of aminolysis of mercapto and methylthio compounds: for comparable yields, using an aqueous solution of the sodium salt of β-alanine at 100°, the mercaptopyrimidine required 34 hr. but the methylthiopyrimidine only 2 hr.[2581]

The unusual aminolysis of a 2-mercaptopyrimidine is seen in the conversion of 5-hexyl-4-hydroxy-2-mercapto-6-methylpyrimidine into the 2-hydrazino analogue (63%) by refluxing propanolic hydrazine hydrate in 6 hr.[2225]

(7) *Oxidation of Mercaptopyrimidines* (*H* 286)

The oxidation of mercaptopyrimidines to disulphides and sulphonic acids is treated in Sects. 3 and 4.A, respectively.

R = NMe_2); instead, ring fission and degradation gave *NN*-dimethylurea, *trans*-β-dimethylaminothioacrylic acid methylamide ($Me_2NCH{:}CHCSNHMe$) and a third unidentified product.[3219] A mechanism involving nucleophilic attack of dimethylamine at C-6 was postulated.[3219]

(8) *Other Reactions* (*New*)

Mercaptopyrimidines may be converted into their thiocyanato analogues by treatment with cyanogen bromide; yields appear to be much better than those obtained from nucleophilic replacement of a chloro group by thiocyanate ion (Ch. VI, Sect. 5.H).[2783] Thus an aqueous solution of the sodium salt of 2-amino-4-mercapto-5,6-dimethylpyrimidine (**19**, R = H) stirred for 5 min. at 10° with cyanogen bromide gave 2-amino-4,5-dimethyl-6-thiocyanatopyrimidine (**19**, R = CN, 86%; 50%*).[2783] Similar reactions afforded 2-thiocyanatopyrimidine (82%; 25%*) and its 4,6-dimethyl (75%; 18%*), 4-amino-5-cyano (57%), 5-carboxy-4-methyl (15%), 4-amino-5-ethoxycarbonyl (62%), 4,6-diamino (54%), 4-amino-6-hydroxy (53%), and 4-amino-6-methyl (52%) derivatives;[2783] also 4-methyl-2-methylthio-6-thiocyanatopyrimidine (91%; 61%*) and more complicated derivatives.[2783]

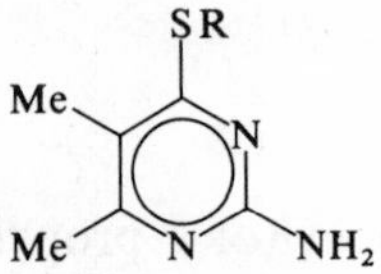

(19) 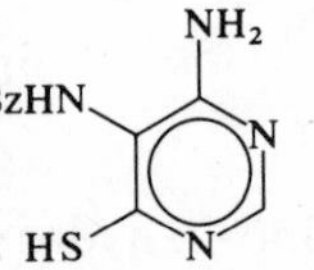 (20) 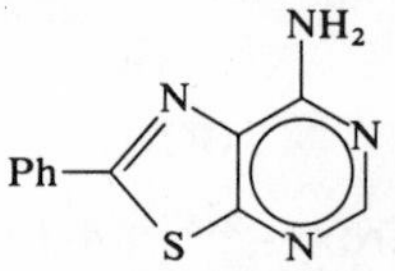 (21)

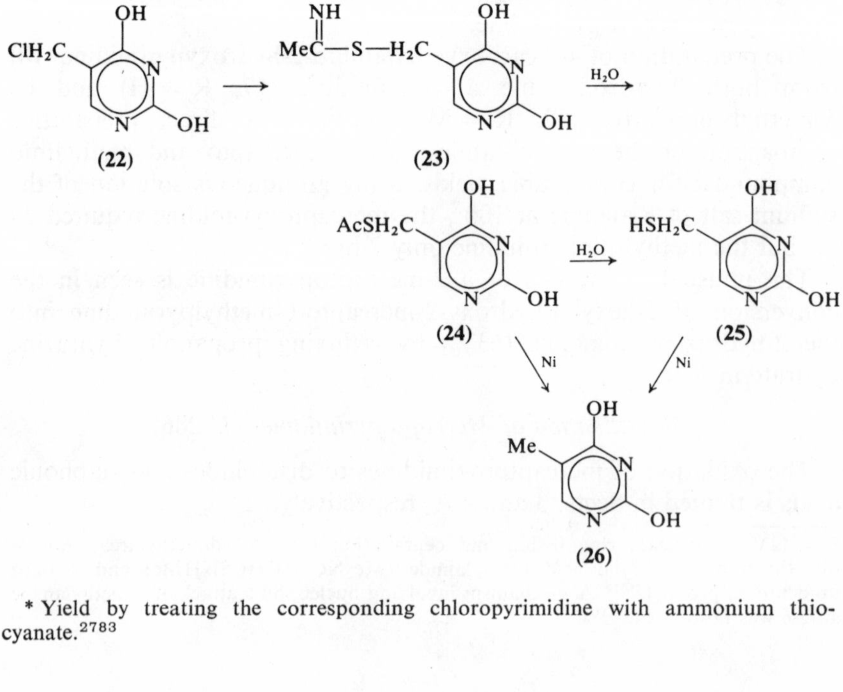

(22) → (23) $\xrightarrow{H_2O}$

(24) $\xrightarrow{H_2O}$ (25)

(24) $\xrightarrow{Ni}$ (26) $\xleftarrow{Ni}$ (25)

* Yield by treating the corresponding chloropyrimidine with ammonium thiocyanate.[2783]

Mercapto groups can be involved in the cyclization of appropriate pyrimidines to bicyclic systems. For example, 4-amino-5-benzamido-6-mercaptopyrimidine (**20**) may be dehydrated to 7-amino-2-phenyl-thiazolo[5,4-*d*]pyrimidine (**21**),[3186] and 5-amino-2-chloro-4-mercapto-6-trifluoromethylpyrimidine reacts with triethyl orthoformate to give 5-chloro-7-trifluoromethylthiazolo[5,4-*d*]pyrimidine in good yield;[2735] other such cyclizations are known.[2735]

E. Reactions of 5- and ω-Mercaptopyrimidines (*H* 286)

Apart from their oxidation (Sects. 3 and 4.A), few reactions of such mercaptopyrimidines are described. However, 2,4-dihydroxy-5-mercaptomethylpyrimidine (**25**) can be desulphurized by Raney nickel to give thymine (**26**; 72%) or alkylated by methyl or ethyl iodide and alkali to give the 5-methylthiomethyl derivative or its ethylthiomethyl homologue, respectively.[2650] 5-Acetylthiomethyl-2,4-dihydroxypyrimidine (**24**), prepared indirectly from chloromethyluracil (**22**) by treatment with thioacetamide and subsequent hydrolysis of the resulting 5-acetimidoylthiomethyluracil (**23**), may be hydrolysed to 5-mercaptomethyluracil (**25**) or desulphurized directly to thymine (**26**).[2650] 2,4-Dihydroxy-5-mercaptopyrimidine may be *S*-alkylated to its methylthio, propylthio, and other homologues;[3002] the same 5-mercaptopyrimidine (2 moles) reacts with ethylene dibromide (1 mole) in boiling aq. sodium hydroxide (2 moles) to give 5-β-2′,4′-dihydroxypyrimidin-5′-ylthio-ethylthio-2,4-dihydroxypyrimidine, in which the two pyrimidine molecules are joined by an —S—CH_2CH_2—S— bridge.[3230] The higher homologues (up to six methylene groups in the bridge) have been made by using appropriate αω-dihalogenoalkanes.[3230]

Ring closures can also occur with ω-mercaptoalkylpyrimidines. For example, 2-amino-4-hydroxy-5-β-mercaptoethylpyrimidine (**27**) is dehydrated by polyphosphoric acid to give 96% of 2-amino-5,6-dihydro-

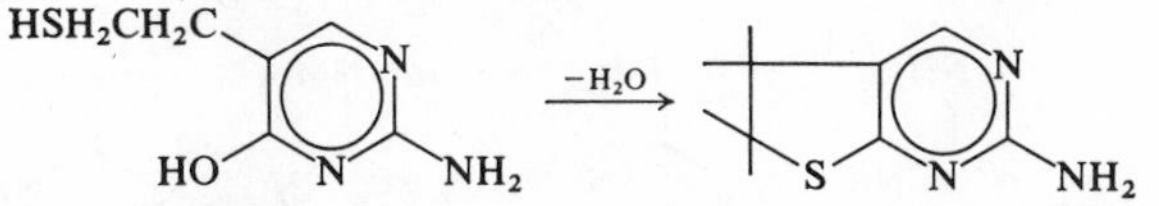

HSH_2CH_2C … HO, N, N, NH_2 $\xrightarrow{-H_2O}$ S, N, N, NH_2

(**27**) (**28**)

Me, $HS(CH_2)_2H_2C$ … HO, N, N, OH $\xrightarrow{-H_2O}$ Me, S, N, N, OH

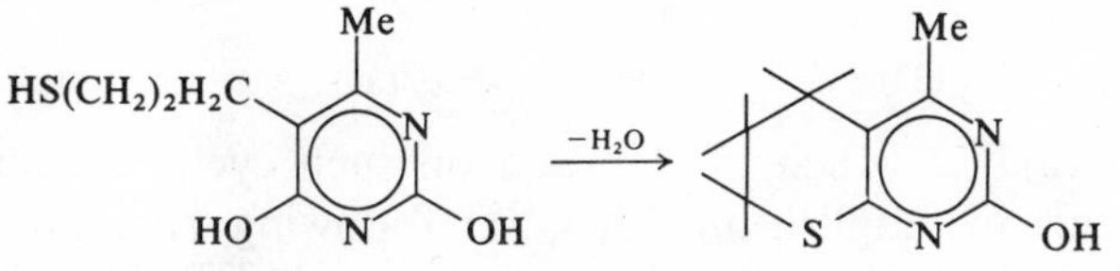

(**29**) (**30**)

thieno[2,3-*d*]pyrimidine (**28**);[2908] and in a similar way, 2,4-dihydroxy-5-γ-mercaptopropyl-6-methylpyrimidine (**29**) gives 6,7-dihydro-2-hydroxy-4-methyl-5*H*-thiopyrano[2,3-*d*]pyrimidine (**30**).[2908] Other such examples are described.[2908]

F. Fine Structure of Mercaptopyrimidines (*New*)

The fine structure of mercaptopyrimidines which carry an hydroxy substituent, e.g., 4-hydroxy-6-mercaptopyrimidine,[2242, 2760] have been discussed in Ch. VII, Sect. 6. Available data on the structure of other mercaptopyrimidines is summarized here.

A. Albert and G. B. Barlin[2173] have compared the ultra-violet spectra and ionization constants of 2- and 4-mercaptopyrimidine with those of their *N*- and *S*-methyl derivatives. They concluded that in aqueous solution, 2-mercaptopyrimidine existed mainly as 1,2-dihydro-2-thiopyrimidine (**32**) with a minute contribution of the thiol (**31**); and that 4-mercaptopyrimidine existed as an approximately 2:1 mixture of 1,6-dihydro-6-thio- (**33**) and 1,4-dihydro-4-thio-pyrimidine (**35**), with the thiol (**34**) contributing very little indeed.[2173] In addition, comparisons of ultra-violet spectra with appropriate *N*- and *S*-methylated derivatives have revealed[2776] that thiocytosine in aqueous media may well be a mixture of 4-amino-1,2(and 2,3)dihydro-2-thiopyrimidine (**36** and **37**); 4-dimethylamino-2-mercaptopyrimidine exists mainly as 4-dimethylamino-1,2-dihydro-2-thiopyrimidine (**38**); 4-amino-6-mercaptopyrimidine is mainly 4-amino-1,6-dihydro-6-thiopyrimidine (**39**); and thioisocytosine is almost certainly 2-amino-1,6-dihydro-6-thio-

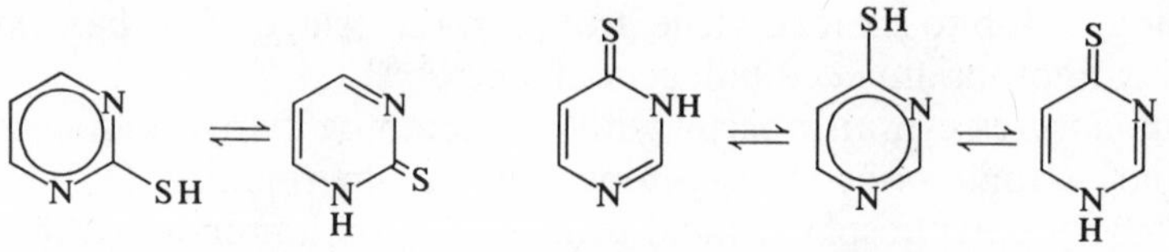

(31) (32) (33) (34) (35)

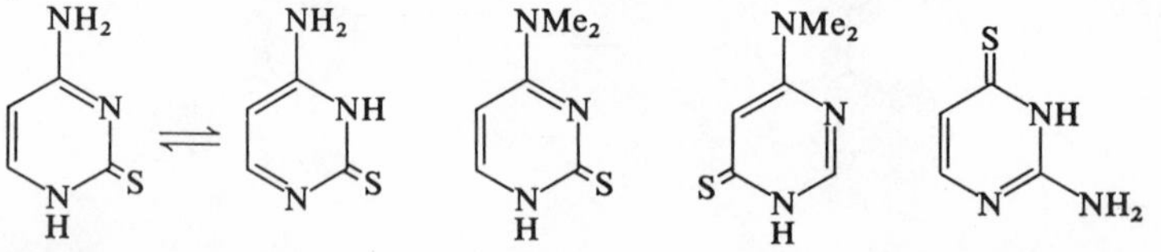

(36) (37) (38) (39) (40)

pyrimidine (**40**).[2776] Much of the data on such cyclic thioamides has been reviewed in English in 1964,[3220] following spectral work on 2-thiouracil derivatives and related compounds.[3221] Dihydro-2-thiouracils have also been examined similarly.[3222] The crystal and molecular structures of 2,4-dimercaptopyrimidine and 2,4-diselenylpyrimidine have been determined by X-ray diffraction.[3223, 3224] Relevant spectra are available.[3225]

2. The Thioethers: Alkylthio- and Arylthiopyrimidines (*H* 286)

A. Preparation of 2-, 4-, and 6-Alkylthiopyrimidines (*H* 286)

All the usual ways of making alkylthiopyrimidines have been discussed already: the Principal Synthesis in Ch. II; *S*-alkylation of mercaptopyrimidines in Sect. 1.D(2); and sodium alkylmercaptide and chloropyrimidines in Ch. VI, Sect. 5.F. In addition, 4-cyano-5-ethoxycarbonyl-2-methylthiopyrimidine (**41**, R = CN) has been converted into 32% of 5-ethoxycarbonyl-2,4-bismethylthiopyrimidine (**41**, R = SMe) by sodium cyanide(!) in hot dimethyl sulphoxide;[2599] and 5-ethoxycarbonyl-2-methylthiopyrimidin-4-yl trimethylammonium chloride (**41**, R = $N^+Me_3Cl^-$) into the same product (**41**, R = SMe) in small yield by potassium cyanide in acetamide![2599]

B. Preparation of 5-Alkylthiopyrimidines (*H* 288)

5-Methylthiopyrimidine (70%) may be made by treating 5-bromopyrimidine with ethanolic sodium methylmercaptide.[2619]

C. Reactions of Alkylthiopyrimidines (*H* 288)

Most of the reactions of alkylthiopyrimidines are discussed elsewhere: reductive removal of alkylthio groups by Raney nickel in Table XIII; hydrolyses in Ch. VII, Sect. 1.F; dealkylation to mercaptopyrimidines in Sect. 1.A; and oxidation to sulphoxides and sulphones in Sect. 5.A.

The aminolysis of alkylthiopyrimidines (*H* 289) has been studied by measuring rates of reaction in simple examples.[2697, 2746] Like the corresponding chloropyrimidines, 4-methylthio- and 4-ethylthio-pyrimidine underwent butylaminolysis more rapily than their respective 2-isomers at several temperatures. However, at 150° 4-methylthiopyrimidine behaved exceptionally in reacting more slowly than its 2-isomer because the plots of rate constant *versus* temperature crossed *ca.* 160°, within the range of temperatures studied.[2697] Each ethylthiopyrimidine reacted more slowly than the corresponding methylthio homologue.[2697] The usual substituent effects were evident: electron-release by *C*-methyl groups progressively decreased the rate of aminolysis, but electron-withdrawal by a bromo or nitro substituent increased the rate of reaction very markedly.[2746] This was quite impressive on a preparative scale: 4-methyl-2-methylthiopyrimidine (**42**) heated with butylamine at 180° for 65 hr. gave a 42% yield of 2-butylamino-4-methylpyrimidine (**43**) but 2-methylthio-5-nitropyrimidine (**44**) and butylamine at 20° for 1 min. gave *ca.* 95% yield of 2-butylamino-5-nitropyrimidine (**45**); both figures were in fair agreement with calculations using predetermined rate constants.[2746]

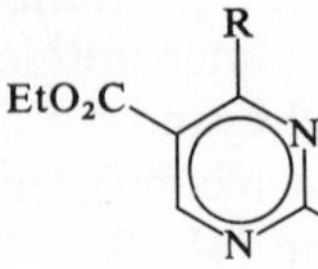

R, EtO_2C, N, N, SMe — (**41**)

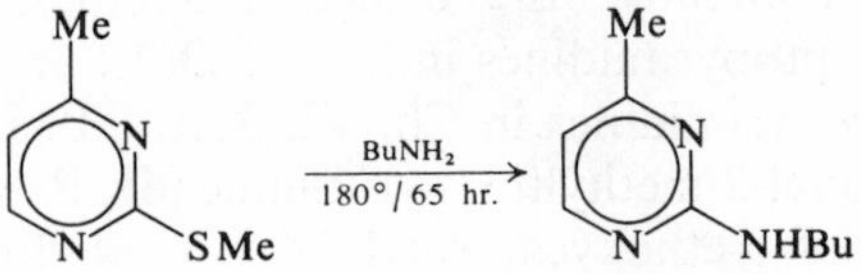

Me, N, N, SMe — (**42**) $\xrightarrow[180°/65\ hr.]{BuNH_2}$ Me, N, N, NHBu — (**43**)

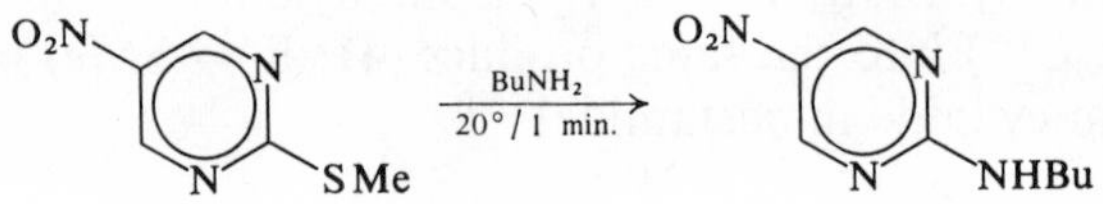

O_2N, N, N, SMe — (**44**) $\xrightarrow[20°/1\ min.]{BuNH_2}$ O_2N, N, N, NHBu — (**45**)

The process is further exemplified in the conversion of 2-hydroxy-4-methylthio- into 4-amino-2-hydroxy-pyrimidine (ethanolic ammonia at 150°);[3227] of 4-amino-6-hydroxy-2-methylthio- (**46**, R = H) into 4-amino-2-dimethylamino-6-hydroxy-pyrimidine (**47**; dimethylammonium acetate at 160°),[3228] into 4-hydroxy-2,6-bismethylaminopyrimidine (transamination as well as replacement using methylammonium acetate at 160°),[3228] or into 4-amino-2-anilino(or *p*-chloroanilino)-6-hydroxypyrimidine (aniline or *p*-chloroaniline at 160°);[2430] of 2-hydroxy-5-methyl-4-methylthio- into 4-benzylamino-2-hydroxy-5-methyl-pyrimidine (85%; benzylamine in refluxing butanol for 10 hr.);[2674] of 2-hydroxy-4-methylthio- (**48**) into 4-amino-2-hydroxy-pyrimidine (**49**, R = H; *ca.* 100%; methanolic ammonia at 25° for a week),[3057] into 4-β-carboxyethylamino-2-hydroxypyrimidine (**49**, R = $CH_2CH_2CO_2H$; 90%),[2581] or into 2-hydroxy-4-hydroxyaminopyrimidine (**49**, R = OH; refluxing ethanolic hydroxylamine hydrochloride);[3215] of 4,6-dihydroxy-2-methylthiopyrimidine into 2-γ-(dipropylamino)propylamino-4,6-dihydroxypyrimidine and its homologues (at 170°);[2741] of 2-hydroxy-4-methylthiopyrimidine into 2-hydroxy-4-3′-methylbut-2-′-enylaminopyrimidine (95% as hemi-sulphate);[3471] of 4-carboxy-6-hydroxy-5-methyl-2-methylthio- into 2-amino-4-carboxy-6-hydroxy-5-methyl-pyrimidine*(89%);[3183] of 4-amino-5-carbamoyl-2-ethylthio- into 4-amino-5-carbamoyl-2-β-hydroxyethylamino-pyrimidine;[2273] of 4-ethyl-6-hydroxy-2-methylthio- into 4-ethyl-2-hydrazino-6-hydroxy-pyrimidine (74%; refluxing ethanolic hydrazine hydrate for 5 hr.);[2452] and of 4-hydroxy-6-methyl-2-methylthio- into 2-hydrazino-4-hydroxy-6-methyl-pyrimidine.[2580]

The aminolytic displacement of a 2-methylthio group is assisted considerably by a 5-nitroso substituent. Thus while 4-amino-6-hydroxy-2-methylthiopyrimidine (**46**, R = H) needed amine at 160° for aminolysis (see above), the 5-nitroso derivative (**46**, R = NO) reacted with aq. morpholine even at room temperature to give 4-amino-6-hydroxy-2-morpholino-5-nitrosopyrimidine (87%).[2473] Other amines reacted with similar ease to give 4-amino-6-hydroxy-5-nitroso-2-piperidinopyrimidine, (**50**; 75%), 2,4-diamino-6-hydroxy-5-nitroso-pyrimidine (also made by nitrosation), and 4-amino-6-hydroxy-2-hydroxyamino(or β-hydroxyethylamino)-5-nitrosopyrimidine.[2473] Likewise, 4,6-diamino-2-methylthio-5-nitrosopyrimidine reacted with hot aq. dimethylamine in a few minutes to give 4,6-diamino-2-dimethylamino-5-nitrosopyrimidine (85%), and other amines gave analogous products.[2473]

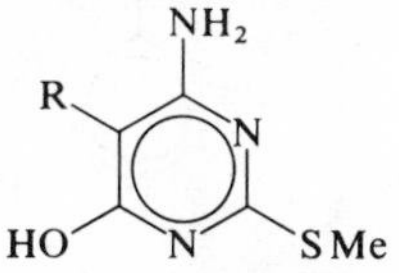

NH_2 R HO N N SMe
(**46**)

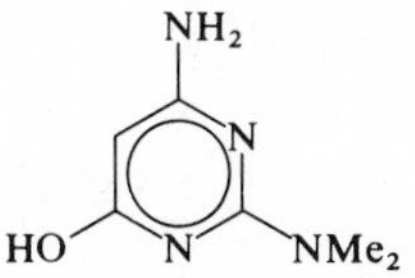

NH_2 HO N N NMe_2
(**47**)

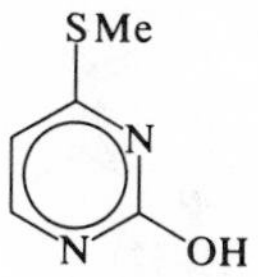

SMe N N OH
(**48**)

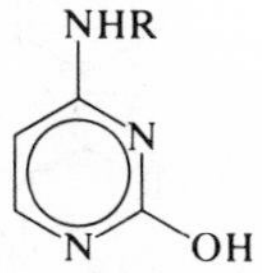

NHR N N OH
(**49**)

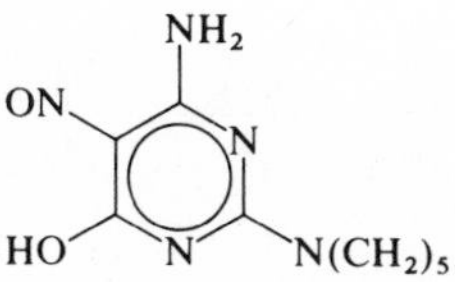

NH_2 ON HO N N $N(CH_2)_5$
(**50**)

3. Dipyrimidinyl Disulphides and Sulphides (*H* 291)

Disulphides are generally made by oxidation of the corresponding mercaptopyrimidines. This process is illustrated in the formation of di(5-acetyl-4-methyl-2-phenylpyrimidin-6-yl) disulphide (**51**, R = Ac)

* Five 5-alkyl homologues were made similarly.[3183]

using bromine as oxidant,[2314] di(5-ethoxycarbonyl-4-methyl-2-phenylpyrimidin-6-yl) disulphide* (**51**, R = CO_2Et) using bromine[2314] or air,[2313] di(2,4-diaminopyrimidin-6-yl) disulphide (80%) using iodine or (in unstated yield) by using nitrous acid,[2503] di(4-anilino-6-hydroxypyrimidin-5-yl) disulphide using iodine or air,[2899] di(5-acetyl-4-methyl-2-undecylpyrimidin-6-yl) disulphide (92–98%) using nitrous acid or iodine,[2905] di(4-aminopyrimidin-6-yl) disulphide (64–95%) using aq. nitrous acid or ethanolic isoamyl nitrite,[2777] and elsewhere.[2650].

Di[2-amino(or phenyl)pyrimidin-5-yl] disulphide may be made by the Principal Synthesis, which also serves to make di(2-aminopyrimidin-5-yl) sulphide.[2304] Other sulphides have been made in several ways. Thus di(4-methyl-6-phenylpyrimidin-2-yl) sulphide (**53**) was made (94%) by boiling an ethanolic mixture of 2-chloro- and 2-mercapto-4-methyl-6-phenylpyrimidine (**52**), or (in 11% yield) by treating the

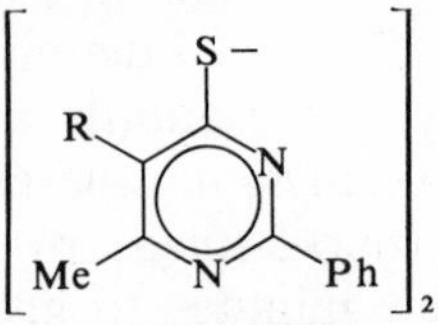

(**51**)

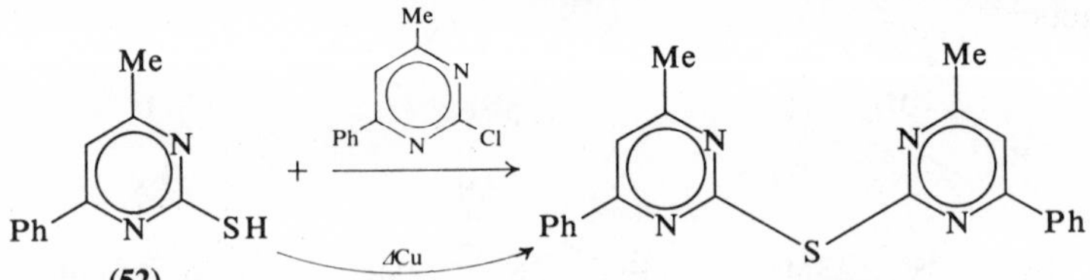

(**52**) (**53**)

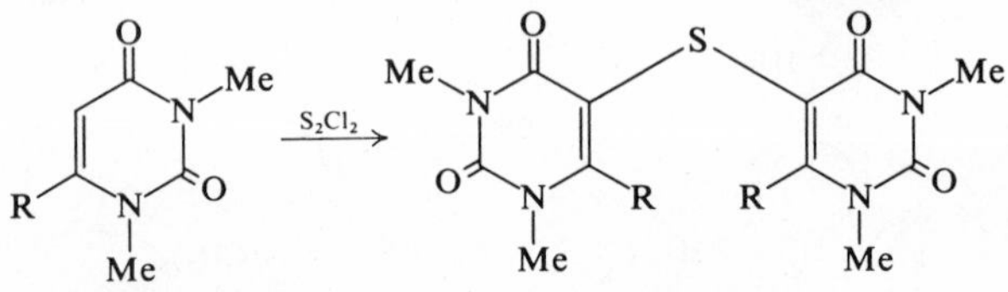

(**54**) (**55**)

mercaptopyrimidine (**52**) with copper bronze in refluxing *p*-cymene.[2603] Similarly, 5-chloromethyl-2,4-dihydroxy- and 2,4-dihydroxy-5-mercaptomethyl-pyrimidine combined to give di(2,4-dihydroxypyrimidin-5-ylmethyl) sulphide in excellent yield.[2650] 1,2,3,4-Tetrahydro-6-hydroxy-1,3-dimethyl-2,4-dioxopyrimidine (**54**, R = OH) reacted with sulphur chloride (S_2Cl_2) in acetic acid to give di(1,2,3,4-tetra-

* The melting point was recorded as 137° at first[2314] but as 175° later.[2313]

hydro-6-hydroxy-1,3-dimethyl-2,4-dioxopyrimidin-5-yl) sulphide (**55**, R = OH; 94%);[2304] thionyl chloride gave the same product (**55**, R = OH) in 60% yield.[2304] Sulphur chloride was used also to make di(1,3-diethyl-1,2,3,4-tetrahydro-6-hydroxy-2,4-dioxopyrimidin-5-yl) sulphide (93%); di(4-chloro-1,2,3,6-tetrahydro-1,3-dimethyl-2,6-dioxopyrimidin-5-yl) sulphide (**55**, R = Cl; 50%); its 4-anilino (**55**, R = NHPh) and 4-amino (**55**, R = NH_2) analogues; and di(2,4-dihydroxy-6-methylpyrimidin-5-yl) sulphide.[2304]

4,6-Dichloro-5-*p*-nitrophenylpyrimidine treated with 2 moles of potassium hydrogen sulphide gave, not the expected dimercapto analogue, but di(4-mercapto-5-*p*-nitrophenylpyrimidin-6-yl) sulphide.[2522] The reaction of thiourea with 2-amino-4-chloro-6-methylpyrimidine appears to be influenced by the solvent: in absolute ethanol, 2-amino-4-mercapto-6-methylpyrimidine (54%) was formed, but in 80% ethanol, di(2-amino-4-methylpyrimidin-6-yl) sulphide was the only product.[2842]

Few reactions of pyrimidine sulphides or disulphides have been reported recently, apart from a few oxidations and the reduction of di(4-anilino-6-hydroxypyrimidin-5-yl) disulphide by sodium borohydride to 4-anilino-6-hydroxy-5-mercaptopyrimidine.[2899] However some good precedents for the pyrimidine series are provided in the excellent purine paper of I. L. Doerr, I. Wempen, D. A. Clarke, and J. J. Fox.[3231]

4. Pyrimidine Sulphonic Acids and Related Compounds (*H* 295)

A. Preparation (*H* 295)

Oxidative methods for making sulphopyrimidines are represented recently in the treatment of 4-amino-2-mercapto-5-phenylpyrimidine (**56**) with alkaline peroxide to give 4-amino-5-phenyl-2-sulphopyrimidine (**57**; 46%),[2274] and of 4-hydroxy-2-mercapto-6-methylpyrimidine with alkaline permanganate to give 4-hydroxy-6-methyl-2-sulphopyrimidine (57%).[2999]

The direct sulphonation of 2-aminopyrimidine with chlorosulphonic acid has been modified[3232] in detail, thereby improving the melting point of the resulting 2-amino-5-sulphopyrimidine (**58**) by 20°, and the yield from 28%[1474] to 58%.[3232] Chlorosulphonic acid has been used also to prepare 4,6-diamino-2-hydroxy(or methylthio)-, 2,4-diamino-6-chloro(or hydroxy)-, 2-amino-4-hydroxy-, and 4-amino-6-hydroxy-

2-methyl-5-sulphopyrimidine, all in moderate yield;[2364, 2588] in contrast, the same reagent with 4-amino-1,2,3,6-tetrahydro-1,3-dimethyl-2,6-dioxopyrimidine (**59**, R = H) gave the 5-chlorosulphonyl derivative (**59**, R = SO_2Cl),[2364] and with 4-benzyl-5-bromo-2,6-dihydroxypyrimidine gave 5-bromo-4-*p*-chlorosulphonylbenzyl-2,6-dihydroxypyrimidine (**60**).[2655]

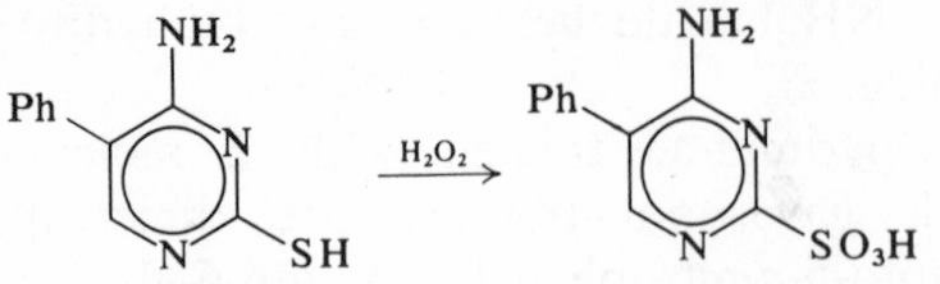

(56) (57) 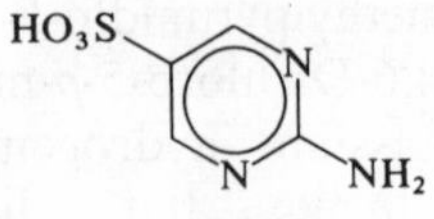 (58)

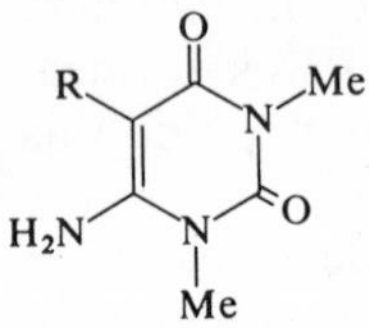

(59) 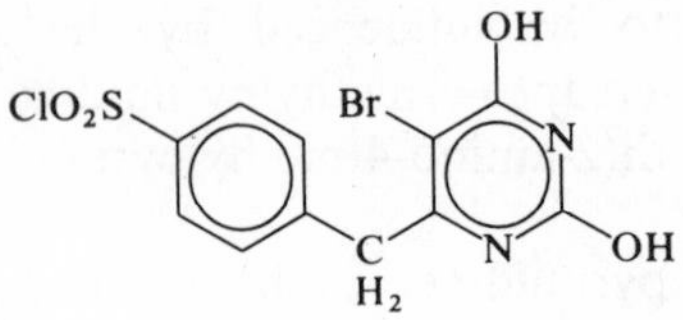 (60) 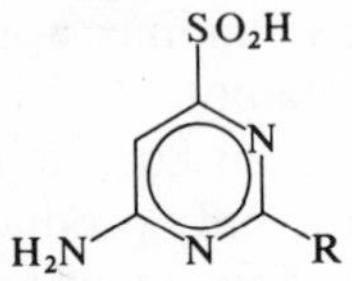 (61)

ω-Sulphonation has been achieved indirectly by condensing 4-chloro-5-ethoxycarbonyl-2-methylthiopyrimidine with taurine to give 5-ethoxycarbonyl-2-methylthio-4-β-sulphoethylaminopyrimidine.[2744]

Alkaline hydrogen peroxide has been used to oxidize appropriate mercaptopyrimidines to the sulphinic acids: 4,5-diamino-6-hydroxy-2-sulphino- (*ca.* 65%),[2478] 2,4-diamino-6-sulphino- (**61**, R = NH_2; 45%),[2503] and 4-amino-6-sulphino-pyrimidine (**61**, R = H; 49%).[2777]

B. Reactions (*H* 297)

The lability of pyrimidine sulphonic and sulphinic acids under hydrolytic conditions is illustrated by the liberation of sulphur dioxide from the potassium salt of 4-hydroxy-6-methyl-2-sulphopyrimidine (**65**, R = SO_2H) in water at 60° to give 2,4-dihydroxy-6-methylpyrimidine (**65**, R = OH),[2999] and by the conversion of 4,5-diamino-6-hydroxy-2-sulphinopyrimidine (**63**) into 4,5-diamino-6-hydroxypyrimidine (**62**) or 4,5-diamino-2,6-dihydroxypyrimidine (**64**) by ethanolic or aq. hydrogen chloride, respectively.[2478] The sulpho group of 4-hydroxy-6-methyl-2-sulphopyrimidine (**65**, R = SO_3H) was displaced rapidly by hydrazine hydrate at 80° to give 2-hydrazino-4-hydroxy-6-methylpyrimidine (**65**, R = $NHNH_2$).[2999] A further pointer (*cf. H* 297) that sulpho may prove to be a useful leaving group is the conversion of 2,4-dimethoxy-

6-sulphopyrimidine by sodium sulphanilamide into 2,4-dimethoxy-6-sulphanilylaminopyrimidine.[2771]

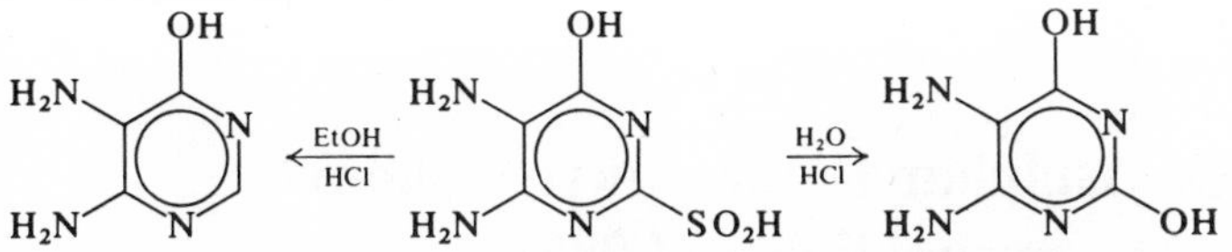

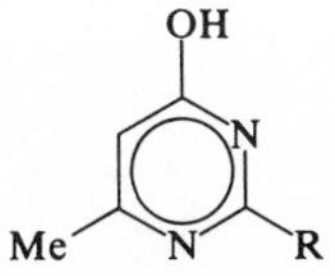

(62) (63) (64) (65)

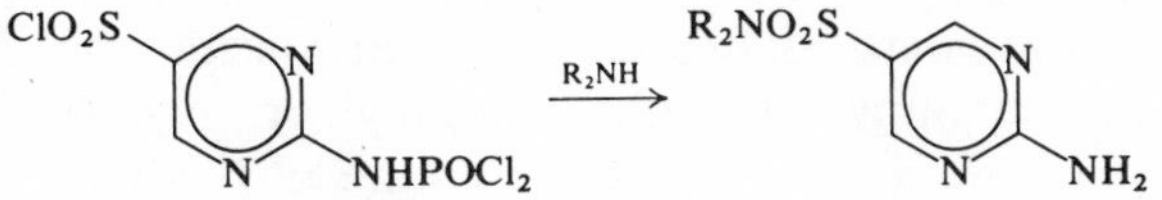

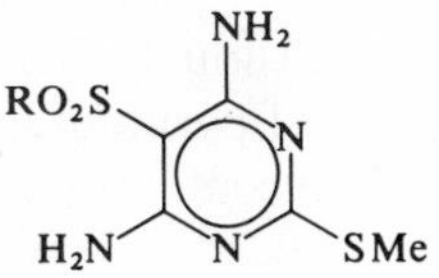

(66) (67) (68)

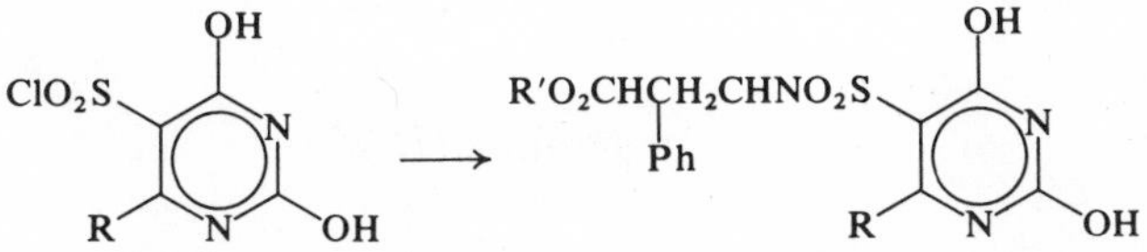

(69) (70)

Other recently recorded reactions involve modification of the sulpho group without fission of the C—S bond. Thus the sodium salt of 2-amino-5-sulphopyrimidine (**58**) reacted with phosphorus pentachloride to give, when poured into ice, 5-chlorosulphonyl-2-dichlorophosphinylaminopyrimidine (**66**).[3232] This reacted with aq. ammonia or an appropriate amine to give 2-amino-5-sulphamoyl- (**67**, R = H), 2-amino-5-dimethylsulphamoyl- (**67**, R = Me), and 2-amino-5-butylsulphamoyl-pyrimidine, as well as a dozen analogues.[3232] In a similar way 4,6-diamino-2-methylthio-5-sulphopyrimidine (**68**, R = OH) was converted into the 5-chlorosulphonyl derivative (**68**, R = Cl) by phosphoryl chloride and then allowed to react with ammonia to give the 5-sulphamoyl derivative (**68**, R = NH_2);[2364] 4-amino-5-chlorosulphonyl-1,2,3,6-tetrahydro-1,3-dimethyl-2,6-dioxopyrimidine (**59**, R = SO_2Cl) reacted with ethylamine to give the 5-ethylsulphamoyl analogue (**59**, R = SO_2NHEt) or with methanol to give the ester, i.e., the methoxysulphonyl derivative (**59**, R = SO_3Me);[2364] the chlorosulphonyl derivative (**60**) with ammonia gave 5-bromo-2,4-dihydroxy-6-*p*-sulphamoylbenzylpyrimidine;[2655] and 5-chlorosulphonyl-2,4-dihydroxypyrimidine (**69**, R = H) and its 6-methyl derivative (**69**, R = Me) have been converted by appropriate β-amino acids and esters (in the presence of triethylamine) into 5-β-carboxy(or ethoxycarbonyl)-β-phenylethylsulphamoyl-2,4-dihydroxypyrimidine (**70**, R = H, R′ = H or Et), their 6-methyl derivatives (**70**; R = Me, R′ = H or Et), and other analogues.[3233]

5. Alkylsulphinyl- and Alkylsulphonyl-pyrimidines (*H* 298)

Although there are very few pyrimidine sulphoxides known yet, the use of pyrimidine sulphones as intermediates has increased considerably in recent years. Both alkylsulphinyl and alkylsulphonyl are equally excellent as leaving groups, and in this respect the nature of the alkyl (or aryl) part appears to have little effect.[2619, 2668] Relevant data are more numerous at present in other heterocyclic series.[2668–2672, 3236, 3237]

A. Preparation (*H* 298)

The sulphoxides are rather more difficult to make and purify than the sulphones. 2-Methylthiopyrimidine (**71**) has been oxidized with periodate to give 2-methylsulphinylpyrimidine (**72**; *ca.* 50%); 4-methylthiopyrimidine by *m*-chloroperbenzoic acid in chloroform to give 4-methylsulphinylpyrimidine (*ca.* 70%); and the same reagent was used to make 5-methylsulphinyl- and 2-phenylsulphinyl-pyrimidine, both of which required chromatographic purification.[2619] Both fuming nitric acid and hydrogen peroxide in acetic acid have been used to make 2,4-dihydroxy-5-methylsulphinylpyrimidine, its ethyl-, and its amyl-sulphinyl homologues in 40–60% yield.[3002] Peracetic acid in aq. acetic was used to oxidize 4-amino-5-carboxymethylthio-1-ethyl-1,2,3,6-tetrahydro-2,6-dioxo-3-propylpyrimidine to its 5-carboxymethylsulphinyl derivative (**74**; 75%);[2363] and peracetic acid in 1:1 acetone/acetic acid oxidized 2,4-dihydroxy-5-methylthiopyrimidine to the corresponding sulphoxide in 83% yield.[3765]

Most sulphones have been made by oxidation of the corresponding thioethers. Thus 2-methylthiopyrimidine (**71**) was converted into 2-methylsulphonylpyrimidine (**73**) by chlorine, possibly *via* 2-methylsulphinylpyrimidine (**72**) which has been oxidized to the sulphone (**73**) in good yield by *m*-chloroperbenzoic acid.[2619] 2-Phenylthio- (**76**) was oxidized similarly to 2-phenylsulphonyl-pyrimidine (**77**),[2619] and other examples of oxidative procedures are given in Table XIIIb.

TABLE XIIIb. Pyrimidine Sulphones Made by Oxidation of Alkyl- or Arylthiopyrimidines (*New*)

Pyrimidine produced	Reagent[a]	Yield (%)	Ref.
5-Acetoxymethyl-4-amino-2-methylsulphonyl-	Cl_2	53	3062
4-Amino-5-carbamoyl-2-methylsulphonyl-	Cl_2	29	3062
4-Amino-5-carboxymethylsulphonyl-1-ethyl-1,2,3,6-tetrahydro-2,6-dioxo-3-propyl-	AcO_2H	76	2363
4-Amino-5-cyano-2-methylsulphonyl-	Cl_2	67	3062
4-Amino-5-methoxy-2-methylsulphonyl-	Cl_2	80	2205
5-Amylsulphonyl-2,4-dihydroxy-	$KMnO_4$	40	3002
5-Benzyloxy-2-methylsulphonyl	H_2O_2	79	3000
	Cl_2	—[b]	2585
4-Benzylsulphonyl-2,6-dimethoxy-	H_2O_2/HCO_2H	73	2774
4-*o*-Bromoanilino-5-ethoxycarbonyl-2-methylsulphonyl-	Cl_2	92	3062
5-Bromo-4-chloro-2-methylsulphonyl-	Cl_2	—	2255
5-Bromo-4-methyl-2-methylsulphonyl-	Cl_2	—	2255
4-*o*-Chloroanilino-5-ethoxycarbonyl-2-methylsulphonyl-	Cl_2	87	3062
4-Chloro-5-ethoxycarbonyl-2-methylsulphonyl-	Cl_2	94	3062
2-Chloro-4-methyl-6-phenylsulphonyl-	$H_2O_2/AcOH$	61	2742
4-Chloro-6-methylsulphonyl-	Cl_2	72	2749
2-Chloro-4-methyl-6-*o*-tolylsulphonyl-	$H_2O_2/AcOH$	61	2742
2-Chloro-4-methyl-6-*p*-tolylsulphonyl-	$H_2O_2/AcOH$	62	2742
4,6-Dichloro-2-methylsulphonyl-	Cl_2	*ca.* 50	2165
	H_2O_2/Ac_2O	—	3238
2,4-Dihydroxy-5-methylsulphonyl-	$KMnO_4$	45	3002
	AcO_2H	88	3765
2,4-Dimethoxy-6-phenylsulphonyl-	AcO_2H	85	2771
4,6-Dimethoxy-2-phenylsulphonyl-	AcO_2H	64	2771
4-Ethoxy-2-ethylsulphonyl-5-methyl-	Cl_2	97	2582
5-Ethylsulphonyl-2,4-dihydroxy-	$KMnO_4$	54	3002
4-Ethylsulphonyl-2,6-dimethoxy-	AcO_2H	52	2771
5-Fluoro-4-methyl-2-methylsulphonyl-	Cl_2	31	2679
5-Fluoro-2-methylsulphonyl-	Cl_2	60	2679
4-Hydroxy-5-methoxy-2-methylsulphonyl-	Cl_2	50	2205
4-Hydroxy-2-methyl-6-methylsulphonyl-	AcO_2H	90	3505
4-Hydroxy-6-methylsulphonyl-	Cl_2	83	2749
5-Hydroxy-2-methylsulphonyl-	H_2O_2	69	3000
4-Iodo-6-methylsulphonyl-	Cl_2	33	2609
4-Methoxy-6-methylsulphonyl-	Cl_2	23	2682
5-Methoxy-2-methylsulphonyl-	Cl_2	88	2198

continued

TABLE XIIIb (*continued*).

Pyrimidine produced	Reagent[a]	Yield (%)	Ref.
4-Methyl-2,6-bisphenylsulphonyl-	H_2O_2/AcOH	62	2742
4-Methyl-2,6-bis-*p*-tolylsulphonyl-	H_2O_2/AcOH	65	2742
2-Methylsulphonyl-	Cl_2	*ca.* 60	2619
4-Methylsulphonyl-	*m*-$ClC_6H_4CO_3H$	*ca.* 50	2619
5-Methylsulphonyl-	*m*-$ClC_6H_4CO_3H$	55	2619
2-Methylsulphonyl-4-phenyl-	AcO_2H	—	3239
2-Methylsulphonyl-4-trifluoromethyl-5,6-trimethylene-[c]	AcO_2H	—	2235
2-Phenylsulphonyl-	Cl_2	*ca.* 80	2619

[a] The indicated reagents have been used under a variety of conditions. For example, chlorine was used in methanol, water, hydrochloric acid, alkali, as preformed sodium hypochlorite, or in other ways.
[b] Not characterized but hydrolysed to 5-benzyloxy-2-hydroxypyrimidine (65%).
[c] And many analogues.

An alternative approach to sulphones is illustrated by the treatment of 2-chloropyrimidine (**75**) with sodium benzenesulphinate to give 2-phenylsulphonylpyrimidine (**77**) directly;[2619] 2,4-dimethoxy-6-phenylsulphonylpyrimidine (35%) and 4-*p*-acetamidophenylsulphonyl-2,6-dimethoxypyrimidine (40%) were made similarly using potassium benzenesulphinate and its *p*-acetamido derivative respectively with 4-chloro-2,6-dimethoxypyrimidine.[2771]

The Principal Synthesis has been used to make two 5-sulphones: ethyl α-ethoxymethylene-α-methylsulphonylacetate and acetamidine or propionamidine gave 4-hydroxy-2-methyl-5-methylsulphonylpyrimidine and its 2-ethyl analogue, respectively, both in moderate yield.[3506]

Di(2,4-dihydroxypyrimidin-5-ylmethyl) sulphone was made by peroxide oxidation of the corresponding sulphide.[2650]

B. Reactions (*H* 300)

The conversion of pyrimidine sulphones and sulphoxides into hydroxypyrimidines [Ch. VII, Sect. 1.F(3)] and into alkoxypyrimidines (Ch. VII, Sect. 4) has already been discussed.

The 2- and 4-sulphones and sulphoxides of pyrimidine undergo rapid aminolysis. The rates for amylaminolysis and cyclohexylaminolysis of 2- (**73**) and 4-methylsulphonylpyrimidine, 2- (**72**) and 4-methylsulphinylpyrimidine, and 2-phenylsulphonylpyrimidine (**77**) in dimethylsulphoxide have been measured and are appreciably greater than

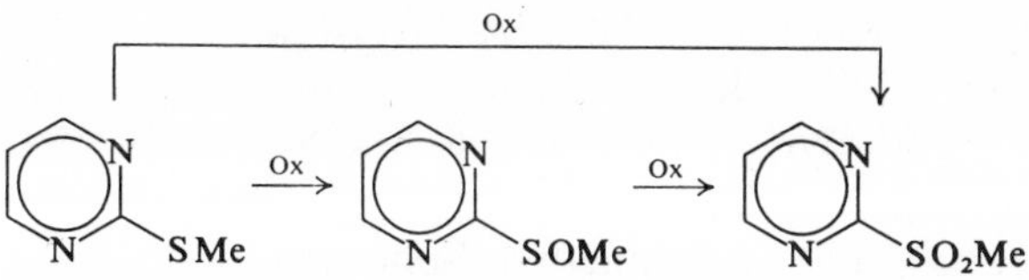

Ox

Ox Ox

SMe SOMe SO_2Me

(71) (72) (73)

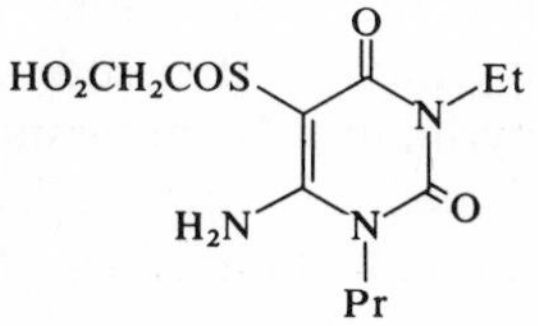

HO_2CH_2COS O Et N N O H_2N Pr

(74)

$PhSO_2Na$

PhSNa Cl_2

Cl SPh SO_2Ph

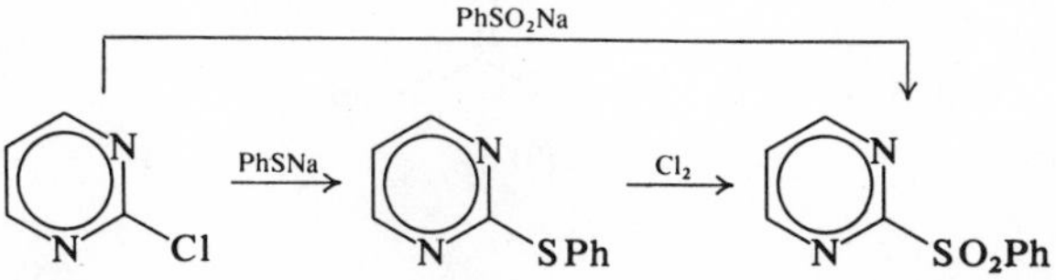

(75) (76) (77)

the rates for the corresponding aminolyses of 2-chloropyrimidine (**75**).[2619] Preparative scale aminolyses are illustrated by the conversion of 2-methylsulphonyl- into 2-cyclohexylamino-, 2-amylamino- (60%), or 2-hydrazino-pyrimidine (45%);[2619] of 2-methylsulphinyl- into 2-amylamino-pyrimidine;[2619] of 4-methylsulphonyl- into 4-amylamino-pyrimidine;[2619] of 4-amino-5-ethoxycarbonyl-2-ethylsulphonyl- into 4-amino-2-aziridino-5-ethoxycarbonyl-pyrimidine (**78**; 44%);[2205] of 4,6-dichloro-2-methylsulphonyl- into 2-aziridino-4,6-dichloro-pyrimidine (*ca.* 50%);[2255] of 5-acetoxymethyl-4-amino-2-methylsulphonyl-pyrimidine (**79**) by appropriate amines (with concomitant deacylation) into 2,4-diamino- (**80**, R = H) 4-amino-2-ethylamino- (**80**, R = Et) and 4-amino-2-propylamino-5-hydroxymethylpyrimidine (**80**, R = Pr);[3062] and of 4-*o*-bromo(or chloro)anilino-5-ethoxycarbonyl-2-methylsulphonyl- into 2-amino-4-*o*-bromo(or chloro)anilino-5-ethoxycarbonyl-pyrimidine.[3062]

Several pyrimidine sulphones have been used to make ‘sulphanilamido-’ (i.e., sulphanilylamino-) pyrimidines by displacement of the alkylsulphonyl group by sodium sulphanilamide in dimethylsulphoxide or other suitable medium. For example, 4-methoxy-6-methylsulphonylpyrimidine (**81**, R = H) rapidly gave 4-methoxy-6-sulphanilylaminopyrimidine (**82**, R = H) in good yield;[2682] 4-benzyl(or ethyl)-

sulphonyl-2,6-dimethoxy-, 2,4-dimethoxy-6-methylsulphonyl- (**81**, R = OMe) or 2,6-dimethoxy-6-phenylsulphonyl- gave 2,4-dimethoxy-6-sulphanilylamino-pyrimidine (**82**, R = OMe);[2609, 2771, 2774] 5-methoxy-2-methylsulphonyl- gave 5-methoxy-2-sulphanilylamino-pyrimidine (78%);[2198] and 4,6-dimethyl-2-phenylsulphonyl- gave 4,6-dimethyl-2-sulphanilylamino-pyrimidine (97%).[2771]

Other reactions include the conversions of 2-methylsulphonylpyrimidine into 2-azidopyrimidine by sodium azide in dimethylformamide, into 2-cyanopyrimidine by potassium cyanide in dimethylformamide, and into 2-methylthiopyrimidine by reduction with hydriodic acid;[2619] also the reduction of 2-methylsulphinylpyrimidine

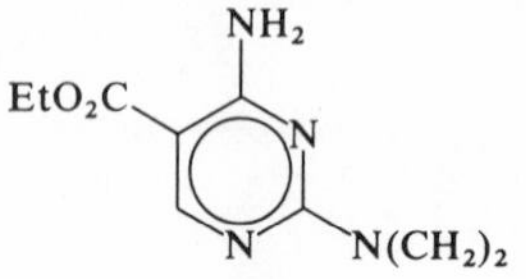

(78) 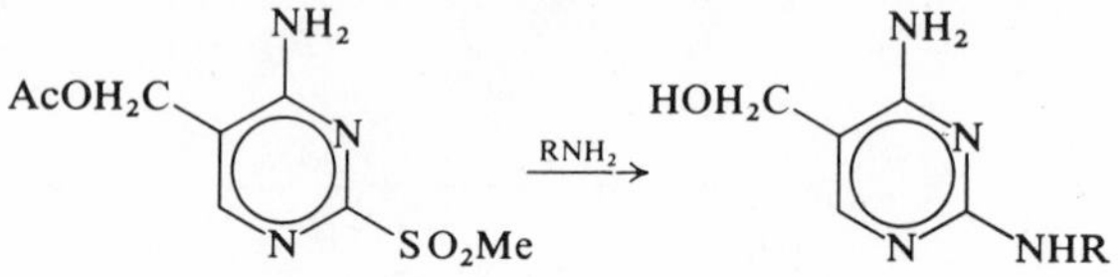 **(79)** **(80)**

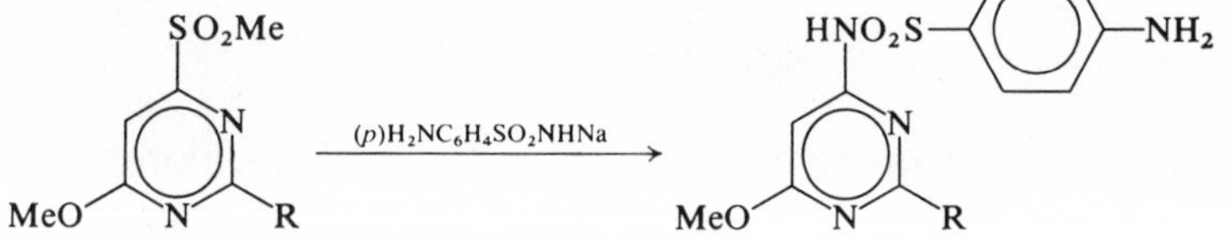

(81) **(82)**

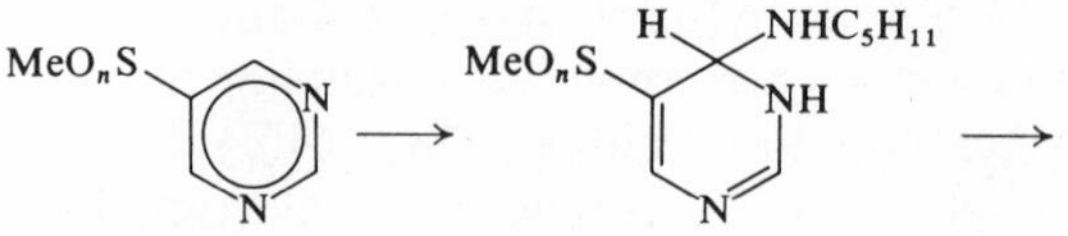

(83) **(84)**

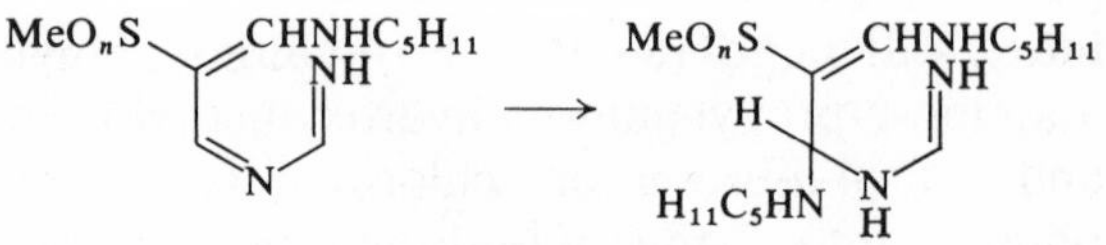

(85) **(86)**

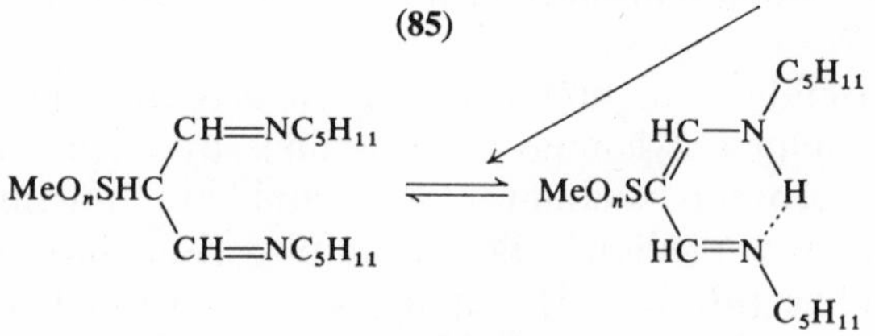

(87)

by hydriodic acid to an unidentified product (2-methylthiopyrimidine?).[2619]

When attempts were made to amylaminolyse 5-methylsulphonylpyrimidine (**83**, $n = 2$) or the corresponding sulphoxide (**83**, $n = 1$), a reaction more facile than the expected displacement occurred with loss of ammonia.[3240] The products, 1,3-diamylimino-2-methylsulphonylpropane (**87**, $n = 2$) and its analogue (**87**, $n = 1$), probably arose *via* the intermediates (**84**) to (**86**). In such behaviour, and in their covalent hydration as cations,[3240] the sulphone and sulphoxide resembled 5-nitropyrimidine, in which acute localization of π-electrons also caused facile ring fission and covalent hydration.[2688] In a similar way, benzylamine gave 1,3-dibenzylimino-2-methylsulphonyl(and methylsulphinyl)propane.[3240]

CHAPTER IX

The Aminopyrimidines (*H* 306)

1. Preparation of 2-, 4-, and 6-Aminopyrimidines (*H* 306)

All the important methods of making such aminopyrimidines have been discussed already: by primary syntheses in Chs. II and III; from halogenopyrimidines in Ch. VI, Sects. 5.B and 5.C; from alkoxypyrimidines in Ch. VII, Sect. 7.C; and from mercapto-, alkylthio-, sulpho-, alkylsulphonyl-, or alkylsulphinyl-pyrimidines in appropriate sections [1.D(5); 2.C; 4.B; and 5.B] of Ch. VIII. The Dimroth Rearrangement (Ch. X, Sect. 2.B) must not be overlooked as a means of preparing some alkylaminopyrimidines. A convenient method has been reported for the hitherto difficult process of converting the salts of 4,5,6-triaminopyrimidines into their free bases. It has been applied to 4,5,6-triaminopyrimidine (92%) and its 2-methyl (80%), 2-methylthio (64%), 2-amino (58%), and other derivatives.[2933]

H. By Other Methods Including Transamination (*H* 310)

The device of directly replacing one type of amino group by another (exchange amination; transamination; *umaminierung*) has been used quite widely in the pyrimidine series of recent years. The process seems to be more effective with 4- or 6- than with 2-amino (or substituted amino) groups.

The first simple examples were described in 1960 by C. W. Whitehead and J. J. Traverso,[2153] who heated the hydrochloride of 4-amino-2,6-dimethylpyrimidine (**1**) with a 10% excess of amine at 170° for 20 hr. to obtain the 4-butylamino (**2**; 91%), 4-anilino (53%), 4-cyclohexylamino (75%), 4-benzylamino (71%), 4-heptylamino (70%), and 4-3′4′-dimethylphenethylamino (69%) analogues.[2153] Rather similarly, 4-amino-2-hydroxypyrimidine and a molar proportion of amine hydrochloride, to which was added a little of the free amine, gave 4-anilino- (64%), 4-benzylamino- (79%), and 4-phenethylamino-2-hydroxypyrimidine (63%);[2153] and 4-amino-6-hydroxypyrimidine (**3**, R = H) or its 2-methyl derivative (**3**, R = Me) gave 4-anilino- (**4**, R = H, R′ = Ph; 13%) and 4-benzylamino-6-hydroxypyrimidine (**4**, R = H, R′ = CH_2Ph; 37%), their 2-methyl derivatives (**4**, R = Me, R′ = Ph or CH_2Ph; 57%, 60%), 4-hydroxy-6-*p*-methylbenzylaminopyrimidine (60%), and 4-hydroxy-2-methyl-6-phenethylaminopyrimidine (65%).[2153] In addition, 4-amino-6-octylaminopyrimidine (**5**) and benzylamine hydrochloride gave 4,6-bisbenzylaminopyrimidine (**6**; <25%), but attempts to transaminate 4-amino-6-hydroxypyrimidine with heptylamine or benzylamino as free bases were unsuccessful.[2153] Similar processes using the hydrochloride or acetate salt of the appropriate amine converted 2-amino- into 2-benzylamino-4,6-dimethylpyrimidine in small yield;[2153] 4-amino- into 4-anilino- (50%),[2414] 4-benzylamino- (75%), or 4-furfurylamino-1,2,3,6-tetrahydro-1,3-dimethyl-2,6-dioxopyrimidine (65%) or into analogues;[2153] 4-amino- into 4-cyclohexyl-

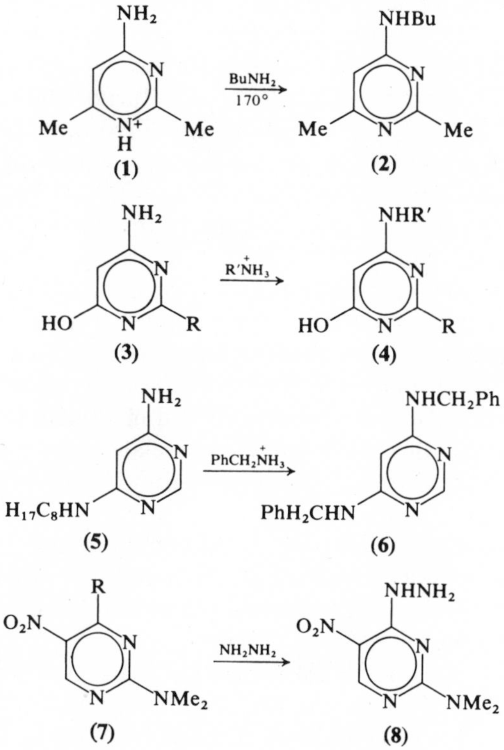

amino(or β-benzylhydrazino)-2,6-dihydroxypyrimidine;[2838, 3254] 4-amino-6-hydroxy-2-methylthio- into 4-hydroxy-2,6-bismethylamino- (*ca.* 60%)[3228, 3253] or 2,4-bis-*p*-chloroanilino-6-hydroxy-pyrimidine (60%);[2430] 2,4-bisdimethylamino- (**7**, R = NMe_2) into 2-dimethylamino-4-hydrazino-5-nitropyrimidine (**8**; 30%) by warming with hydrazine hydrate for a few minutes, or (for comparison) from 2-dimethylamino-4-chloro-5-nitropyrimidine (**7**, R = Cl) and hydrazine;[2712] 2,4-diamino-6-hydroxy- into 2-amino-4-hydroxy-6-methylamino- (*ca.* 50%) or 2-amino-4-benzylamino(or furfurylamino)-6-hydroxy-pyrimidine (23%, 30%);[3228] 4-amino-2,6-dihydroxy- into 4-furfurylamino-2,6-dihydroxy- (55%) or 2,4-dihydroxy-6-methylamino(or β-hydroxyethylamino)-pyrimidine (*ca.* 75%, 50%);[3228] 4-amino-2-dimethylamino-6-hydroxy- into 2-dimethylamino-4-hydroxy-6-methylamino-pyrimidine (44%);[3228] and 4-hydroxy-2,6-bismethylamino- into 4-amino-6-hydroxy-2-methylamino-pyrimidine (24%).[3228]

Some more complicated examples include conversion of 4-amino- (**9**, R = H) into 4-$\beta\beta$-di-isobutylhydrazino-1,2,3,6-tetrahydro-1,3-dimethyl-2,6-dioxopyrimidine (**9**, R = $NBu^i{}_2$) or related compounds;[3254] 4-amino- into 4-β-benzylhydrazino-3-dimethylamino-1,2,3,6-tetrahydro-1-methyl-2,6-dioxopyrimidine or related compounds;[3254] 6-amino- into 6-anilino(or benzylamino)-1-dimethylaminouracil or analogue;[2246] 4-amino- into 4-benzylamino(or *p*-chloroanilino)-1,2,3,6-tetrahydro-1-methyl-2,6-dioxo-3-piperidinopyrimidine;[2246] cytidine into its 2′,4′-dinitrophenylhydrazino analogue by ethanolic dinitrophenylhydrazine and a little hydrogen chloride at 22° during 24 hr.;[3255] and other such examples.[2246, 2430, 2776, 3254]

Cytosine (**10**, R = R′ = H) behaves in an interesting way with hydroxylamine at pH 7 to give an (unisolated) addition product (**11**, R = H) which undergoes additional transamination by hydroxylamine to give 4,5-dihydro-2-hydroxy-4,6-bishydroxyaminopyrimidine (formulated as the tautomer: **12**, R = H). This reversibly eliminates the original hydroxylamine molecule in acid to give 2-hydroxy-4-hydroxyaminopyrimidine (**13**, R = R′ = H).[3256–3258] 1-Methylcytosine (**10**, R = Me, R′ = H) behaves similarly with hydroxylamine to give 1,2-dihydro-4-hydroxyamino-1-methyl-2-oxopyrimidine (**13**, R = Me, R′ = H) *via* the intermediates (**11** and **12**; R = Me).[3258] In contrast, 5- or 6-substitution appears to interfere with addition of hydroxylamine and a direct transamination occurs:[3090] e.g., the hydrochloride of 5-methylcytosine (**10**, R = H, R′ = Me) and aq. hydroxylamine give 2-hydroxy-4-hydroxyamino-5-methylpyrimidine (**13**, R = H, R′ = Me; 70%);[3090] 5-hydroxymethylcytosine (**10**, R = H, R′ = CH_2OH) behaves similarly, as do related compounds.[3090]

Change of one type of amino group for another can be brought about without transamination. Such processes are illustrated in the conversion of 6-benzylamino-1-methyluracil (**14**, R = CH_2Ph) into 6-amino-1-methyluracil (**14**, R = H) by hydrogenation over a palladium catalyst;[2560] of 2-hydrazinopyrimidine or 4-hydrazino-2,6-dimethylpyrimidine into the corresponding aminopyrimidine (70%, 50%) by boiling with Raney nickel in ethanol;[3259] of 2,4-dimethoxy-6-α-methylhydrazinopyrimidine (**15**, R = Me) into 2,4-dimethoxy-6-methylaminopyrimidine (**16**, R = Me; 91%) by a similar procedure;[2770, 3259] of 4-hydrazino-2,6-dimethoxypyrimidine (**15**, R = H) into 4-amino-2,6-dimethoxypyrimidine (**16**, R = H; *ca.* 90%) by hydrogenation over Raney nickel (but not palladium) or by boiling with a massive amount of Raney nickel in ethanol;[3259] and of 4-benzylideneamino-, 4-ethylideneamino-, 4-β-benzoylhydrazino-, or 4-β-acetylhydrazino-2,6-dimethoxypyrimidine into 4-amino-2,6-dimethoxypyrimidine (**16**, R = H; 30–55%) by Raney nickel in ethanol.[3259]

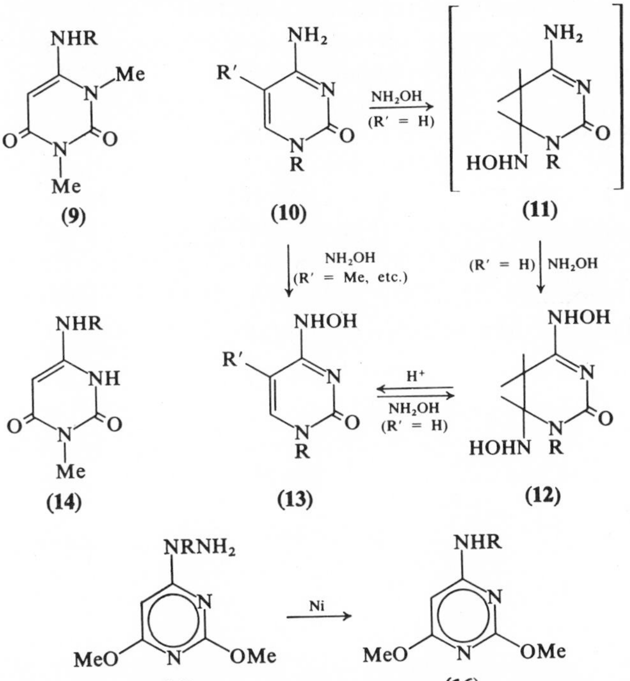

An interesting conversion of 2-amino-4-diethoxycarbonylmethyl- into 2,4-diamino-6-methyl-5-nitropyrimidine (88%) by heating in aq. ammonia has been reported briefly.[2790]

The modification of amino groups by indirect alkylation (involving Dimroth rearrangement) or direct alkylation must not be overlooked. The former is discussed in Ch. X, Sect. 2.B., and the latter is illustrated in the reaction of 2-aminopyrimidine with styrene oxide to give 2-β-hydroxyphenethylaminopyrimidine.[2597]

2. Preparation of *N*- and 5-Aminopyrimidines (*H* 313)

Most known *N*-aminopyrimidines have been made by the primary syntheses. For example, diethyl malonate and *N*-piperidinourea ('1,1-pentamethylenesemicarbazide') gave 1-piperidinobarbituric acid;[2246] ethyl ββ-diethoxypropionate (**17**) and benzaldehyde semicarbazone (**18**), on refluxing in ethanolic sodium ethoxide, gave 1-benzylideneaminouracil (**19**), which underwent acidic hydrolysis to 1-aminouracil (**20**);[3260] β-methoxycarbonylaminoacryloyl chloride (**21**) and benzylidene hydrazine ($PhCH:NNH_2$) gave the acyclic intermediate (**22**, R = —N:CHPh) which underwent cyclization to 3-benzylideneaminouracil (**23**, R = —N:CHPh) yielding 3-aminouracil (**24**) on hydrolysis;[3261] alternatively the same acid chloride (**21**) was treated

H
C(OEt)₂
H_2C
EtO_2C
+
HN—N=CHPh
H_2N C O
NaOEt →
N:CHPh
N
HO N O
H⁺ →
NH_2
N
HO N O

(**17**) (**18**) (**19**) (**20**)

NH
OC CO_2Me
Cl
RNH_2 →
NH
OC CO_2Me
NHR
NaOEt →
N
O N OH
R
H⁺ →
N
O N OH
NH_2

(**21**) (**22**) (**23**) (**24**)

with benzyloxycarbonylhydrazine ($PhCH_2OCONHNH_2$) to give the intermediate (**22**, R = $—HNCO_2CH_2Ph$), 3-benzyloxycarbonylaminouracil (**23**, R = $—HNCO_2CH_2Ph$), and finally 3-aminouracil (**24**);[3261] and other Principal Syntheses are outlined in Ch. II.

An indirect method of making *N*-aminouracils has been reported:[3260, 3261] uracil reacted with hydroxylamine-*O*-sulphonic acid to give a mixture of 1- and 3-aminouracils (**20**, **24**) isolated as their benzylidene derivatives, e.g., (**19**); in addition, some 1,3-diaminouracil was isolated as 1,3-bisbenzylideneamino-1,2,3,4-tetrahydro-2,4-dioxopyrimidine.

Nearly all 5-aminopyrimidines have been made by reduction of the corresponding nitro, nitroso, or azo derivative (Ch. V), but a few by nucleophilic displacement of 5-halogeno derivatives (Ch. VI, Sect. 6.A) and by other less important methods. A method of some potential for making 5-alkylaminopyrimidines from 5-aminopyrimidines has been outlined by C. Temple, R. L. McKee, and J. A. Montgomery:[2752] 5-amino-4,6-dichloropyrimidine (**25**) was acylated by trifluoroacetic anhydride at 20° to give the 5-trifluoroacetamido derivative (**26**), which underwent easy methylation by methyl iodide in dimethylformamide containing potassium carbonate at 20°. The resulting 4,6-dichloro-5-*N*-methyltrifluoroacetamidopyrimidine (**27**) was hydrolysed by aq. triethylamine at 20° to 4,6-dichloro-5-methylaminopyrimidine (**28**) in *ca.* 60% overall yield;[2752] deacylation by hydrazine or methylhydrazine gave 4-chloro-6-hydrazino-5-methylaminopyrimidine (**29**, R = H) and

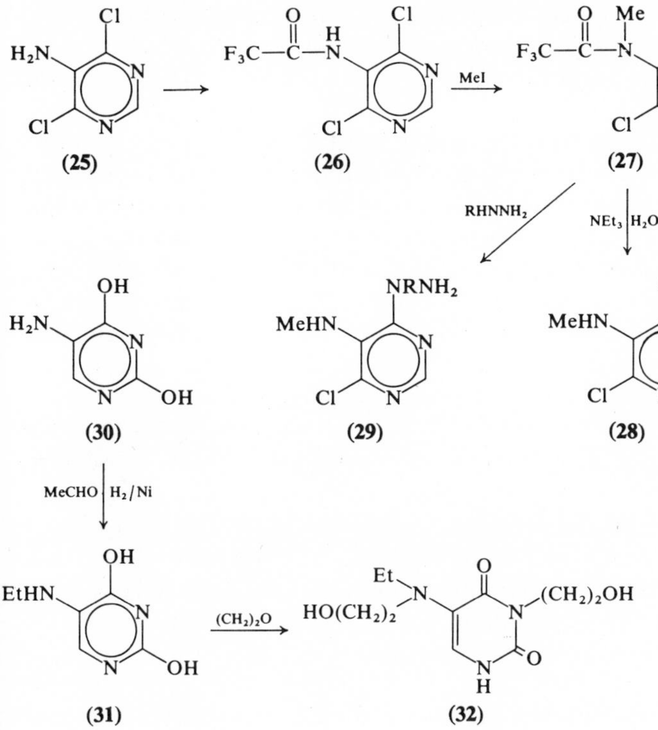

(**25**) → (**26**) —MeI→ (**27**)

(**27**) —RHNNH₂→ (**29**); (**27**) —NEt₃ | H₂O→ (**28**)

(**30**) —MeCHO | H₂/Ni→ (**31**) —(CH₂)₂O→ (**32**)

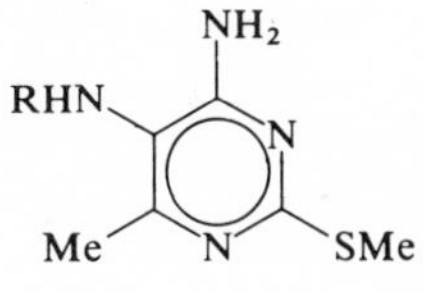

(**33**)

its 6-α-methylhydrazino homologue (**29**, R = Me). In a rather similar way, 4,5-diamino-1,2,3,6-tetrahydro-1,3-dimethyl-2,6-dioxopyrimidine was acylated to its 5-*p*-toluenesulphonamido derivative. This was methylated to the 5-*N*-methyl-*p*-toluenesulphonamido analogue and then deacylated by sodium in liquid ammonia to give 4-amino-1,2,3,6-tetrahydro-1,3-dimethyl-5-methylamino-2,6-dioxopyrimidine in good yield.[3289] 5-Aminopyrimidine has been made from 5-carbamoyl- or 5-hydrazinocarbonyl-pyrimidine with sodium hypochlorite (61%) or nitrous acid (58%), respectively.[2323]

Reductive alkylation of a 5-amino group has been used to convert 5-amino- (**30**) into 5-ethylamino-2,4-dihydroxypyrimidine (**31**) by treatment with acetaldehyde and hydrogenation over Raney nickel;[2643] the ethylamino derivative (**31**) was further alkylated by ethylene oxide to give 3-β-hydroxyethyl-5-*N*-ethyl-*N*-β-hydroxyethylaminouracil (**32**).[2643] The use of ethylene oxide is further illustrated in the conversion of 5-β-hydroxyethylamino- into 5-bis(β-hydroxyethyl)amino-1-methyluracil.[2643] Reductive alkylation has been used to advantage by T. Sugimoto and S. Matsuura in Nagoya.[3475] Thus 4,5-diamino-6-hydroxypyrimidine and ethanolic acetaldehyde were hydrogenated over Raney nickel to give 4-amino-5-ethylamino(or diethylamino)-6-hydroxypyrimidine according to the relative amount of acetaldehyde. Similarly, 4-amino-5-benzylamino-6-hydroxypyrimidine was made using benzaldehyde; subsequent ethylation (acetaldehyde) gave 4-amino-5-*N*-benzyl-*N*-ethylamino-6-hydroxypyrimidine which on debenzylation (Pd/H_2) gave the above 4-amino-5-ethylamino-6-hydroxypyrimidine again. The same methods furnished, e.g., 4,5-bisethylamino-, 5-diethylamino-4-ethylamino-, 4-amino-5-dimethylamino-, and other such 6-hydroxypyrimidines. Similar procedures converted 4,5-diamino-2-hydroxypyrimidine into its 5-ethylamino or 5-diethylamino analogues; 4,5-diaminopyrimidine into its 5-benzylamino or 5-dibenzylamino analogues; and 2,4,5-triamino-6-hydroxypyrimidine into its 5-benzylamino and thence into other such derivatives, e.g., 2,4-diamino-6-hydroxy-5-propylaminopyrimidine (by propylation and subsequent debenzylation).[3475]

The process of alkylation *via* a formamido derivative (*H* 315) has been used to convert 4,5-diamino- (**33**, R = H) into 4-amino-5-formamido-6-methyl-2-methylthiopyrimidine (**33**, R = CHO) and thence with lithium aluminium hydride into 4-amino-6-methyl-5-methylamino-2-methylthiopyrimidine (**33**, R = Me);[2675] also to make 4,5,6-trismethylaminopyrimidine by a lithium aluminium hydride reduction of 5-formamido-4,6-bismethylamine-2-methylthiopyrimidine*

* Direct reduction of 5-formamido-4,6-bismethylaminopyrimidine failed, probably for lack of solubility in appropriate solvents.[2454]

and desulphurization of the crude product with Raney nickel.[2454] The above triamine was also made by direct methylation of 5-amino-4,6-bismethylaminopyrimidine with methyl iodide.[2454] Direct arylations of 5-aminopyrimidines are seen in the reaction of 5-amino-2,4-dihydroxypyrimidine with 2,4-dinitrofluorobenzene or 4-fluoro-3-nitrobenzoic acid to give 5-2′,4′-dinitroanilino- or 5-4′-carboxy-2′-nitroanilino-2,4-dihydroxypyrimidine, respectively.[2567]

3. Preparation of Extranuclear Aminopyrimidines (*H* 316)

A. By Reduction of a Nitrile (*H* 316)

The well-known, but often unsatisfactory, transformation of a cyano- into an aminomethyl-pyrimidine has been used in preparing 2-amino-5-γ-aminopropyl- (75%) from 2-amino-5-β-cyanoethyl-4-hydroxy-6-methylpyrimidine by hydrogenation over a platinum oxide catalyst in acidic aq. ethanol;[3262] 4,6-diamino-5-aminomethyl-2-methylpyrimidine (**34**; 55%; W-7 Raney nickel*);[2681] 5-aminomethyl-2-methyl-4,6-bismethylaminopyrimidine;[2681] 5-aminomethyl-4,6-dimethoxy-2-methylpyrimidine;[2681] and 5-aminomethyl-4-dimethylamino(or methoxy)-2-methylpyrimidine (70%; 80%).[2681] 5-Cyano-4,6-bisdimethylamino-2-methylpyrimidine failed to reduce to its 5-aminomethyl analogue and 5-cyano-2-methylpyrimidine gave only a tetrahydro derivative of the expected product.[2681]

C. By the Mannich Reaction (*H* 318)

Mannich reactions involving activated methyl groups or the 5-position in pyrimidines have been known for many years but only recently has such an attack on a ring nitrogen atom been recorded. Thus 5-ethyl-5-phenylbarbituric acid (**37**), formaldehyde, and piperidine gave the 1-piperidinomethyl derivative (**38**; 71%).[3263] Similar reactions gave 5-cyclohexyl-1-morpholinomethyl-3,5-dimethyl- (95%), 5-ethyl-

* When 'normal grade Raney nickel' was used in methanolic ammonia to reduce 4,6-diamino-5-cyano-2-methylpyrimidine (**35**), the secondary base 4,6-diamino-5-4′,6′-diamino-2′-methylpyrimidin-5′-ylmethylaminomethyl-2-methylpyrimidine (**36**) resulted.[2681]

1-morpholinomethyl-3-methyl-5-phenyl- (65%), 5-ethyl-1,5-bispiperidinomethyl-, and 5-ethyl-1,5-bismorpholinomethyl-3-methyl-barbituric acid (86%).[3263, 3264] The structures of the last two compounds followed from removal of their 1-substituents by hydrogenolysis or treatment with ethereal hydrogen chloride.[3264]

D. By Primary Syntheses (*H* 318)

This unimportant way to make ω-amino derivatives has been illustrated recently by the condensation of *C*-phthalimidoacetamidine with ethyl acetoacetate to give 4-hydroxy-6-methyl-2-phthalimidomethylpyrimidine (**39**), converted by successive treatment with hydrazine and aq. hydrobromic acid into 2-aminomethyl-4-hydroxy-6-methylpyrimidine.[2875]

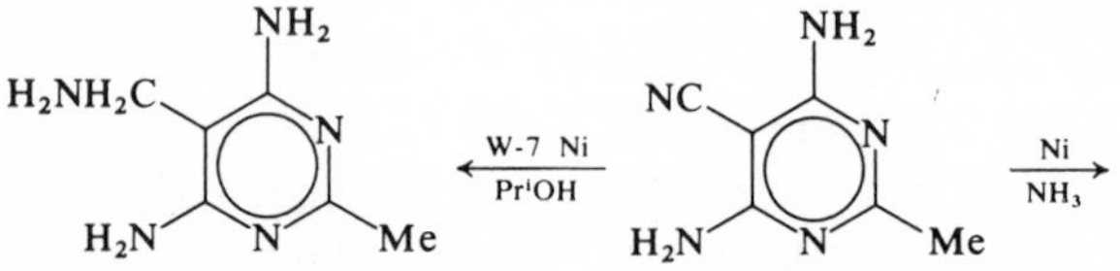

NH_2 H_2NH_2C N H_2N N Me **(34)** ← W-7 Ni / Pr^iOH — NH_2 NC N H_2N N Me **(35)** — Ni / NH_3 →

NH_2 H NH_2 N CH_2NH_2C N Me N NH_2 NH_2 N Me **(36)**

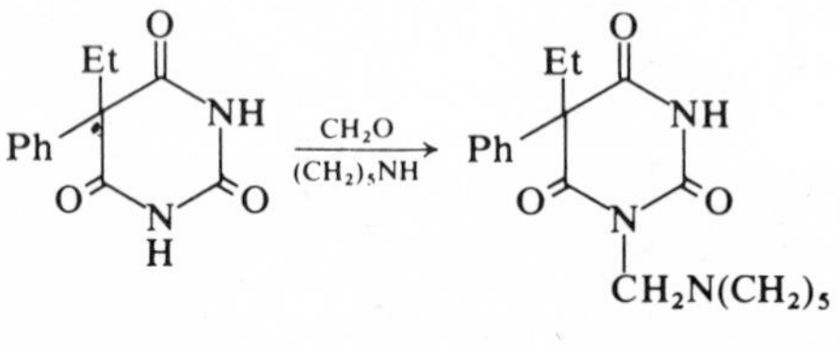

O Et NH Ph O N O H **(37)** — CH_2O / $(CH_2)_5NH$ → O Et NH Ph O N O $CH_2N(CH_2)_5$ **(38)**

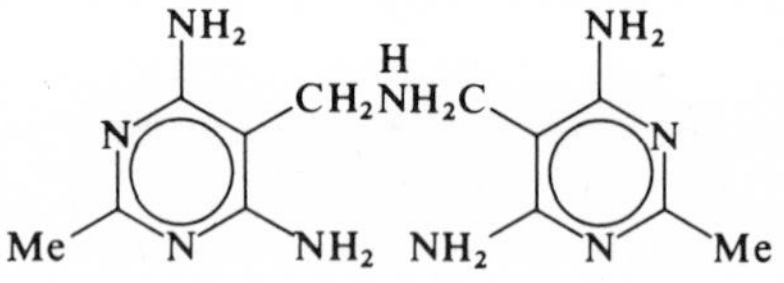

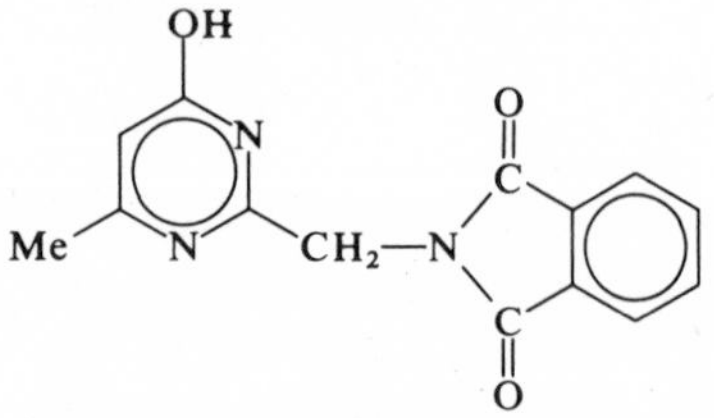

OH N O C Me N CH_2—N C O **(39)**

E. By Other Routes (*H* 319)

The aminolysis of ω-halogenopyrimidines has been discussed (Ch. VI, Sect. 7). The reduction of an ω-nitro to an ω-amino derivative is seen in the formation of 2,4-diamino-6-*p*-aminophenethyl-5-δ-phenylbutylpyrimidine (92%) by hydrogenation of its *p*-nitro analogue.[2981] ω-Aminopyrimidines have been made occasionally *via* the corresponding acylamino derivative, especially in rather complicated systems.[2290, 2652] Another rather specialized reaction is illustrated by the treatment of 4,5-diamino-6-chloropyrimidine with dimethylformamide and phosphoryl chloride to give 4-chloro-5,6-bis(dimethylaminomethyleneamino)pyrimidine (**40**). An aminomethylene or related group may be introduced directly into the 5-position rather similarly:[3265] 1,2,3,4-tetrahydro-6-hydroxy-1,3-dimethyl-2,4-dioxopyrimidine (**41**) gave its 5-aminomethylene derivative (**42**, R = H; 90%) by either heating with formamide at 200° or treatment with methanolic thioformamide containing triethylamine at 40°; *N*-methylformamide gave the analogous 5-methylaminomethylene derivative; acetamide or propionamide gave the 5-α-aminoethylidene and 5-α-aminopropylidene derivatives, respectively; and formanilide or thioformanilide gave the 5-anilinomethylene derivative.[3265] The dimethylaminomethylene derivative (**42**, R = Me) has been made in 80% yield by treating dimethylbarbituric acid (**41**) with phosgene and dimethylformamide;[3266] the same reagent (dimethylaminomethylene chloride hydrochloride) converted 4,6-dihydroxypyrimidine into its 5-dimethylaminomethylene derivative (**43**, R = H),[2594, 2732] which was easily hydrolysed to 5-formyl-4,6-dihydroxypyrimidine.[2594, 2732] 5-Dimethylaminomethylene-4,5-dihydro-6-hydroxy-4-oxo-2-phenyl(or methyl)pyrimidine (**43**, R = Ph or Me) were made similarly and easily gave the corresponding 5-formyl derivatives.[2594]

Pyrimidine aldehydes and ketones sometimes have furnished convenient intermediates for ω-amino derivatives. Thus 4-ethoxalylmethyl-2,6-diethoxypyrimidine (**44**) was converted into its oxime (2,4-diethoxy-6-β-ethoxycarbonyl-β-hydroxyiminoethylpyrimidine; **45**) and then reduced by stannous chloride (with incidental hydrolysis) to 4-β-amino-β-carboxyethyl-2,6-dihydroxypyrimidine (**46**);[2423] the isomeric 2-β-amino-β-carboxyethyl-4,6-dihydroxypyrimidine was made satisfactorily by hydrogenation of the appropriate oxime over Raney nickel, but use of palladium led to isolation of a dihydro derivative of the product.[2423] Similar oximes were reduced by sodium amalgam to give 4-β-amino-β-carboxyethyl-5-ethyl-6-hydroxy-2-mercaptopyrimidine (**47**; 52%) and homologues[3267] and stannous chloride was used to make 2-amino-4-β-amino-β-carboxyethylpyrimidine similarly.[2955] Equally effective intermediates were provided by Schiff bases of pyrimidine aldehydes: 5-formyl-4-hydroxy-6-methyl-2-piperidinopyrimidine and methylamine in dioxane gave 4-hydroxy-6-methyl-5-methyliminomethyl-2-piperidinopyrimidine which was hydrogenated (Pt/acetic acid) to give the corresponding 5-methylaminomethyl derivative (isolated as its *N*-nitroso derivative);[2490] 5-β-formylethyl-4-hydroxy-6-methylpyrimidine (**48**, R = H) was allowed to react with aniline and the crude Schiff base reduced by sodium borohydride to give 5-γ-

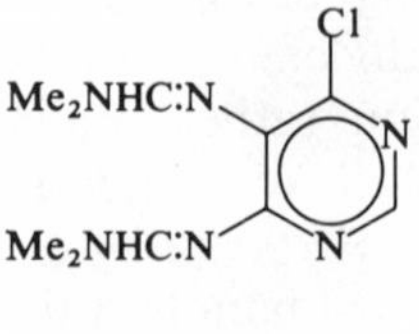

(40) 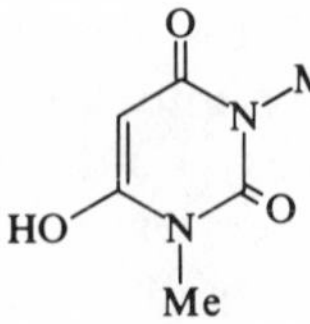 **(41)**

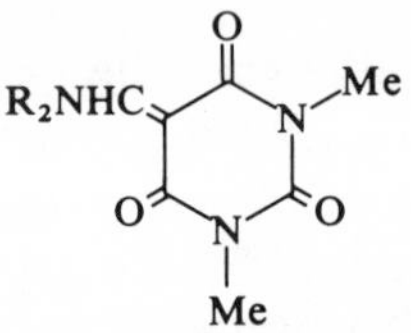

(42) 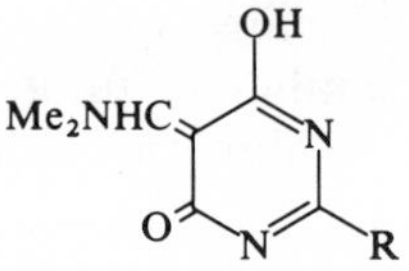 **(43)**

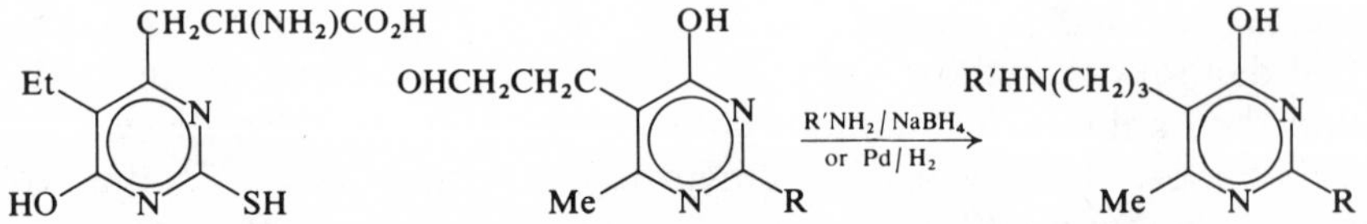

(44) → **(45)** $\xrightarrow{SnCl_2}$ **(46)**

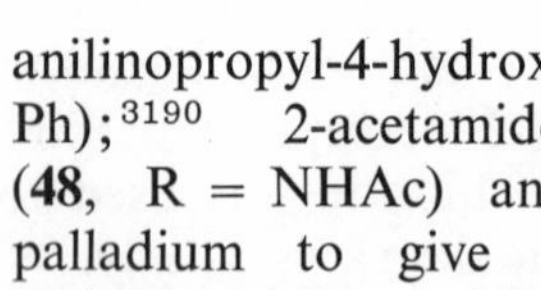

(47) 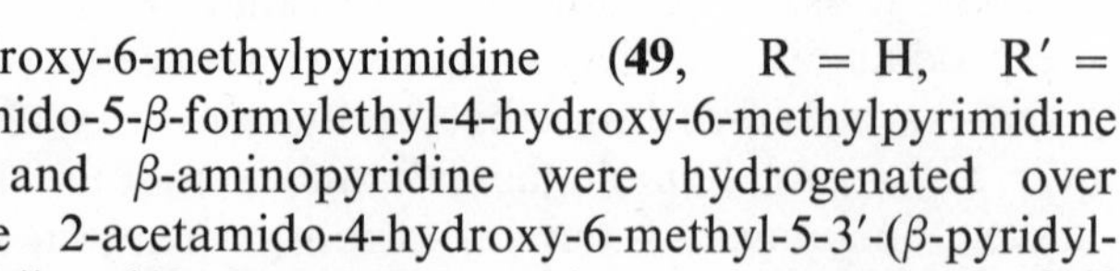 **(48)** $\xrightarrow[\text{or Pd/H}_2]{R'NH_2/NaBH_4}$ **(49)**

anilinopropyl-4-hydroxy-6-methylpyrimidine (**49**, R = H, R′ = Ph);[3190] 2-acetamido-5-β-formylethyl-4-hydroxy-6-methylpyrimidine (**48**, R = NHAc) and β-aminopyridine were hydrogenated over palladium to give 2-acetamido-4-hydroxy-6-methyl-5-3′-(β-pyridylamino)propylpyrimidine (**49**, R = NHAc, R′ = β-pyridyl; 98%);[3268] and many other examples have been recorded using borohydride or catalytic reduction.[2566, 3268]

4. Properties and Structure of Aminopyrimidines (*H* 320)

The predominance of tautomer (**50**) in a dimethyl sulphoxide solution of 2-aminopyrimidine has been confirmed by p.m.r. spectra,[2401] and other aminopyrimidines have also been examined.[2607] The mass spectra of some twenty aminopyrimidines have been measured and fragmentation patterns characteristic of the nature and position of the substituents have been recorded.[2629, 3272] The molecular association of 2-aminopyrimidine in various solvents has been studied by infra-red measurements.[3273] An interesting correlation has been observed in pyridine, the diazines, and their amino derivatives: on protonation, the magnitude and direction of $\lambda_{max.}$ displacement in the ultra-violet spectra depends on the distance between the nitrogen atoms.[2408] Some SCF–MO calculations for aminopyrimidines have been carried out and

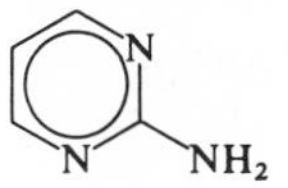

(**50**) 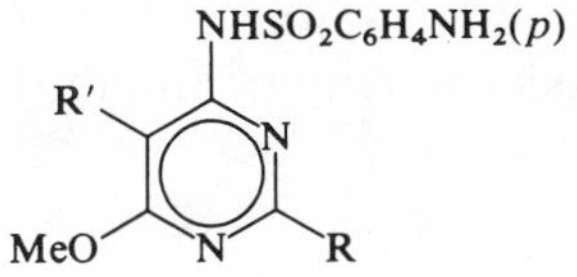 (**51**) 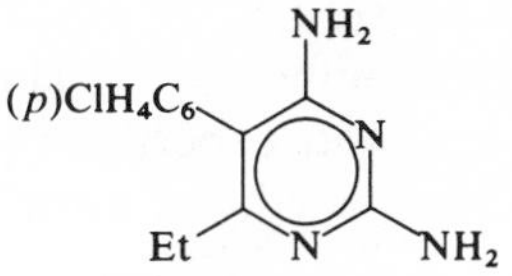 (**52**)

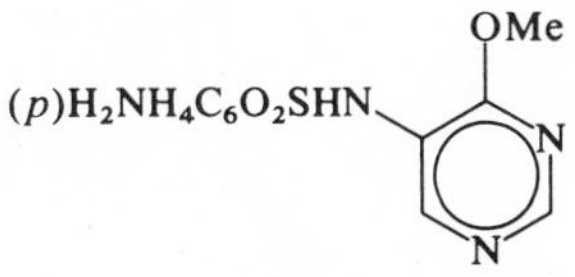

(**53**)  (**54**)

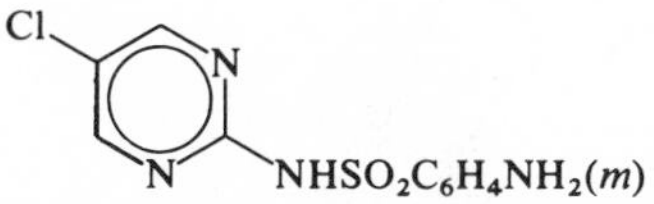

(**55**) 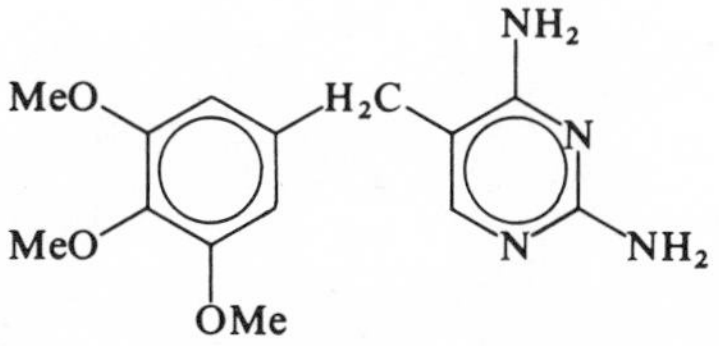 (**56**)

satisfactory agreements between experimental and theoretical singlet-singlet transition energies were reported.[3274] With the help of LCAO–MO calculations the π-electronic structures of 2- and 4-aminopyrimidine have been formulated.[3335] In some 2-aminopyrimidines, the calculated electron density at the amino group and the chemical shifts of the amino protons have been shown to bear a linear relationship: this is also true in other systems.[3757]

Several aminopyrimidine derivatives have antimalarial properties.[3275] Perhaps the most promising is *sulformetoxine* (4,5-dimethoxy-6-sulphanilylaminopyrimidine; **51**, R = H, R′ = OMe) which stays in the body for a long time; others include *pyrimethamine* (2,4-diamino-5-*p*-chlorophenyl-6-ethylpyrimidine, Daraprim; **52**) which is still widely used in prophylaxis, *sulfalene* (4-methoxy-5-sulphanilylaminopyrimidine; **53**), *sulfadimethoxine* (2,4-dimethoxy-6-sulphanilylaminopyrimidine; **51**, R = OMe, R′ = H), *sulfadiazine* (2-sulphanilylaminopyrimidine; **54**), *metachloridine* (2-*m*-aminobenzenesulphonamido-5-chloropyrimidine; **55**), and *trimethoprim* (2,4-diamino-5-3′,4′,5′-trimethoxybenzylpyrimidine; **56**) still classified[3275] as an experimental compound. The quinoline derivative, *chloroquine*, seems still to be the drug of choice for treatment of malaria, but some of the above pyrimidines are particularly useful in cases of resistance.

The tuberculostatic activity of hydrazinopyrimidines have been discussed,[3276] and some pyrimidinyl amino acids have been reported as antibacterials.[3277] The biological activity of aminopyrimidines has been reviewed expertly by C. C. Cheng,[2553] but the following compounds seem worthy of mention here: *trimethoprim* (**56**) and its 3′,4′-dimethoxy

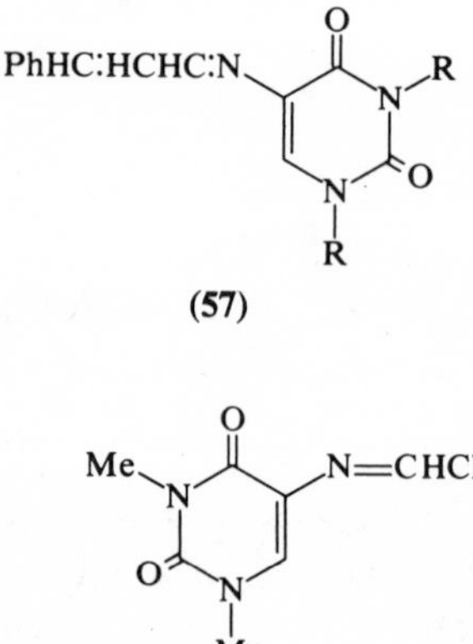

(57) 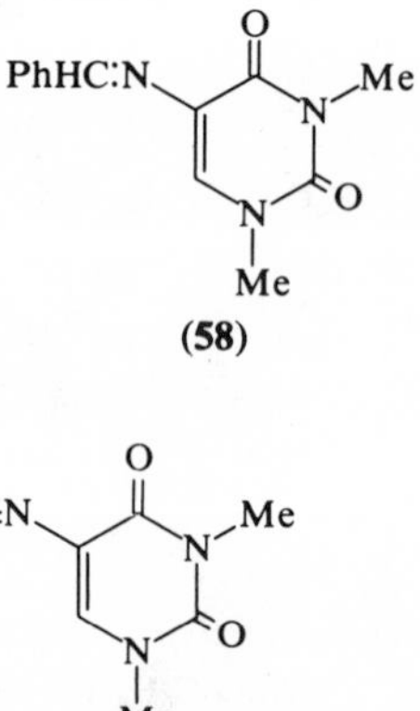 (58)

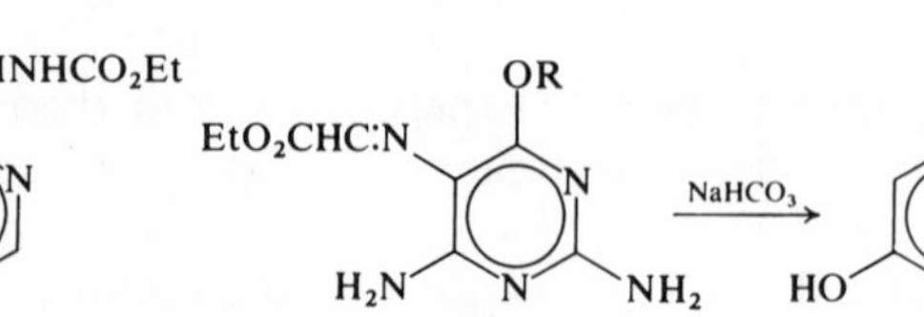

(59)

(60) (61) (62)

analogue, *diaveridine*, are both effective synergists to sulphonamides, the former in human bacterial infections,[3484] and the latter in coccidial infections;[3485] *sulphaorthodimethoxine* (Fanasil, 4,5-dimethoxy-6-sulphanilylaminopyrimidine) has a half-life in the body of about 200 hr. and is used in the treatment of chronic infections such as bronchitis;[3486] and *glymidine* (Glycodiazine, 2-benzenesulphonamido-5-β-methoxyethoxypyrimidine) has shown promise as a hypoglycaemic drug in diabetes.[3487]

5. Reactions of Aminopyrimidines (*H* 321)

The replacement of amino groupings by the hydroxy group has been discussed in Ch. VII, Sects. 1.D, 2, and 3.B; the preparation of chloro- and bromopyrimidines from aminopyrimidines in Ch. VI, Sects. 1.C, 1.D, and 3.E.

C. Formation of Schiff Bases (*H* 321)

Although no simple Schiff bases derived from 2-, 4-, 6-, or ω-aminopyrimidines appear to be known, those from 5-aminopyrimidines are well represented. Thus 5-amino- gave 5-cinnamylideneamino-2,4-dihydroxypyrimidine (**57**, R = H, 93%) by stirring with cinnamaldehyde in hydrochloric acid;[3278] 5-amino- gave 5-cinnamylideneamino-1,2,3,4-tetrahydro-1,3-dimethyl-2,4-dioxopyrimidine (**57**, R = Me; 75%) by refluxing with cinnamaldehyde in hydrochloric acid, or its 5-benzylideneamino analogue (**58**, 76%) by using benzaldehyde without acid;[3278] the same 5-aminopyrimidine with malondialdehyde or its nitro derivative gave the double Schiff base, 1,2,3,4-tetrahydro-1,3-dimethyl-2,4-dioxo-5-γ-(1′,2′,3′,4′-tetrahydro-1′,3′-dimethyl-2′,4′-dioxopyrimidin-5′-ylimino)propylideneaminopyrimidine (**59**, R = H) and its β-nitro derivative (**59**, R = NO_2), respectively;[3278] 4,5-diamino- gave 4-amino-5-benzylideneamino-pyrimidine by warming with benzaldehyde in dilute acetic acid;[1391] 4-amino-5-benzylideneamino-2-methylpyrimidine resulted from heating 4,5-diamino-2-methylpyrimidine with benzaldehyde and piperidine;[2562] 5,6-diamino- gave 6-amino-5-benzylideneamino-3-methyluracil by addition of benzaldehyde, and the Schiff base was subsequently reduced to the corresponding 5-benzylaminopyrimidine by sodium amalgam;[2489] 2,4,5-triamino-6-phenylpyrimidine with *p*-nitrocinnamaldehyde in ethanolic acetic acid gave 2,4-diamino-5-*p*-nitrocinnamylideneamino-6-phenylpyrimidine (93%), subsequently reduced with sodium borohydride to the corresponding 5-*p*-nitrocinnamylamino- derivative;[2944] 5-amino-4-chloro-6-β-ethoxycarbonylhydrazino- gave 4-chloro-6-ethoxycarbonylhydrazino-5-isopropylideneamino-pyrimidine (**60**; 59%) by stirring in acetone with sodium bicarbonate;[3279] 2,4,5-triamino- gave 2,4-diamino-5-ethoxycarbonylmethyleneamino-6-methoxypyrimidine (**61**; R = Me)* by shaking with ethyl ethoxyglycollate;[3280] and the 6-benzyloxy homologue (**61**, R = CH_2Ph)* was made similarly.[3280] Additional examples include 4-amino-5-$\beta\beta$-diethoxyethylideneamino-2-hydroxypyrimidine (not purified),[2948] 2,4-diamino-5-3′-hydroxy-1′-methylbut-2′-enylideneamino-6-methylthiopyrimidine,[2487] and others.[3281]

Formamide can behave as an aldehyde with aminopyrimidines to give a Schiff base: 4-amino-5-ethoxycarbonyl-2-mercaptopyrimidine and formamide at 120° gave 4-aminomethyleneamino-5-ethoxycarbonyl-2-mercaptopyrimidine (**63**);[2275] similarly, 4-amino-6-hydroxypyrimidine with dimethylformamide (containing phosphoryl chloride) gave 4-chloro-6-dimethylaminomethyleneaminopyrimidine (**64**, R = H), and 4-amino-2,6-dichloropyrimidine gave the 2,4-dichloro analogue (**64**, R = Cl).[2593] In contrast, aldehydes can attack an activated 5-position in preference to an amino group, e.g., 4-amino-2,6-dihydroxypyrimidine and formaldehyde gave 4-amino-5-4′-amino-2′,6′-dihydroxypyrimidin-5′-ylmethyl-2,6-dihydroxypyrimidine (**65**), and many homologues were made similarly;[2926] alternatively, aldehydes may react abnormally with aminopyrimidines as exemplified in the formation of 2-hydroxymethylaminopyrimidine (**66**) or 2-pyrimidin-2′-ylaminomethylaminopyrimidine (**67**) from 2-aminopyrimidine and formaldehyde according to conditions,[3282] and in the conversion of 2-amino- into 2-bis(hydroxymethyl)amino-4-β-(5′-nitro-2′-furyl)vinylpyrimidine (**68**).[3283]

The pyrimidinylhydrazones, i.e., the products from hydrazinopyrimidines and aldehydes or ketones, are formally akin to Schiff bases. They are made under the usual conditions for hydrazone formation and are illustrated by 4-benzylidenehydrazino(or ethylidenehydrazino)-2,6-dimethoxypyrimidine,[3259] 4-benzylidenehydrazino-6-methyl-2-methylthiopyrimidine (**69**),[2938] 4-cyclohexylidenehydrazinopyrimidine,[3284] 4-α-methylpropylidenehydrazinopyrimidine,[3284] 5-amino-4-benzylidenehydrazino-6-chloropyrimidine,† [2673] 2,4-dibenzyl-

* Warming with sodium bicarbonate caused cyclization to the 2-amino-4-alkoxy-7-hydroxypteridine (**62**).[3280]

† Note the preferential attack by benzaldehyde on the 4-hydrazino rather than on the 5-amino group.

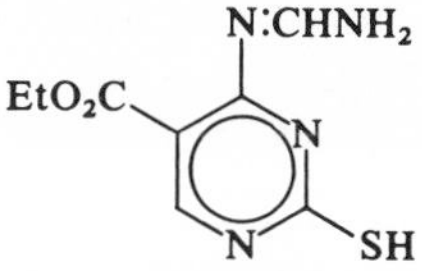

(63) 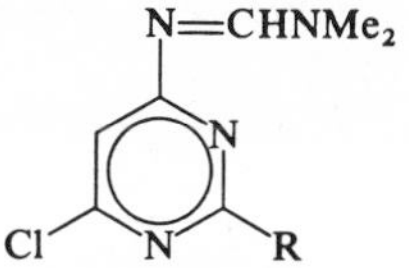 (64) 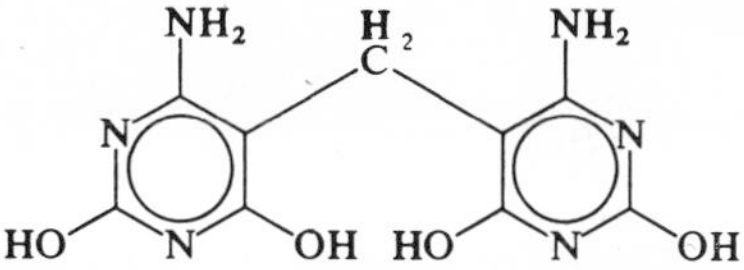 (65)

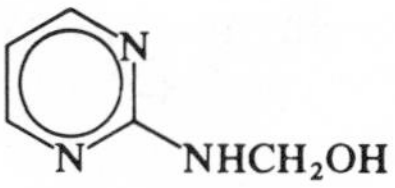

(66)  (67)

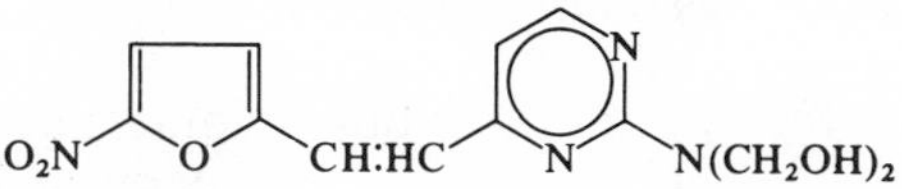

(68)

oxy-6-β-ethoxycarbonyl-α-methylethylidenehydrazinopyrimidine (**70**) which was cyclized by alkali to the pyrazole (**71**),[2558] and other such compounds.[2452] Acid hydrazides also undergo such reactions with aldehydes and ketones as in the conversion of 5-hydrazinocarbonyl- into 5-isopropylidenehydrazinocarbonyl-4-methyl-2-phenylpyrimidine (**72**) by acetone, or into 5-4′-hydroxy-3′-methoxybenzylidenehydrazino-carbonyl-4-methyl-2-phenylpyrimidine by vanillin.[3285] 4-Chloro-6-methyl-2-5′-nitrofurfurylidenehydrazinopyrimidine was made by stirring the 2-benzylidenehydrazino analogue with 5-nitrofurfural in cold aq. hydrochloric acid.[3477]

D. Acyl Derivatives of Aminopyrimidines (*H* 324)

The acylation of aminopyrimidines is such a well-known process that recent examples can be dismissed quite briefly in the subsections that follow.

(1) *Acetylation* (*H* 324)

Hot acetic anhydride was used to convert appropriate aminopyrimidines into 4-acetamido-6-amino- (87%),[3286] 4,6-diacetamido- (78%),[3286] 4-acetamido-6-anilino(or piperidino)- (35%, 54%),[3286] 2-hydroxy-4-*N*-methylacetamido- (*ca.* 90%),[2984] 2-acetamido-5-*p*-chlorophenyl-4-hydroxy-6-methyl- (70%),[2816] 4-acetamido-2-amino-6-methoxy-5-nitroso- (66%),[3291] 4-dimethylamino-6-*N*-methylacetamido-5-nitro- (77%),[3291] 2,4-diacetamido-6-hydroxy-5-isoamyl- (81%),[2816] 2-diacetylamino-4-methyl- (*ca.* 50%),[2955] 2-acetamido-4-hydroxy-5-nitro- (95%),[2464] 2,4-diacetamido-5,6-dihydroxy-* (51%),[2210] 4-acetamido-5-hydroxy-,[2584] 2-acetamido-4-hydroxy-6-methyl- (58%),[2710] 5-acetamido-2,4-dimethoxy- (91%),[2897] 2-acetamido-5-γ-acetamidopropyl-4-hydroxy-6-methyl-,[3262] and 4-acetamido-5-ethoxycarbonyl-2-methyl-pyrimidine;[2698] also into 6-acetamido-1-methyluracil,[2831] *N*-4-acetylcytidine,[3287] 4-β-acetylhydrazino-2,6-dimethoxypyrimidine (55%; using acetyl chloride),[3259] and 4-β-acetylhydrazino-5-amino-6-chloropyrimidine (48%).[2752]

(2) *Formylation* (*H* 325)

The process of formylating aminopyrimidines, generally with hot 90–100% formic acid, is illustrated in the preparation of 5-formamido-4,6-bismethylamino- (*ca.* 90%),[2454] 4-amino-5-formamido-6-methyl-2-methylthio-† (**73**, R = H),[2675] 4-amino-6-methyl-5-*N*-methylformamido-2-methylthio-† (**73**, R = Me),[2675] 5-formamido-2,4-dihydroxy-6-methylamino-,[2470] and 4-amino-2,6-dichloro-5-formamido-pyrimidine,‡ [3288] as well as other simple derivatives,[2190, 2675, 3253, 3288, 3291] and 6-amino-5-formamido-1-hydroxyuracil.[2316]

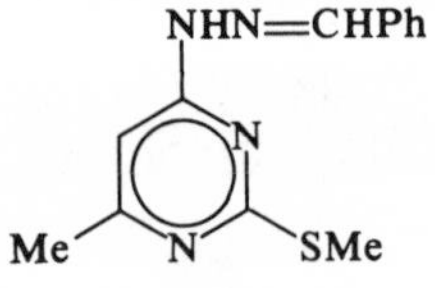

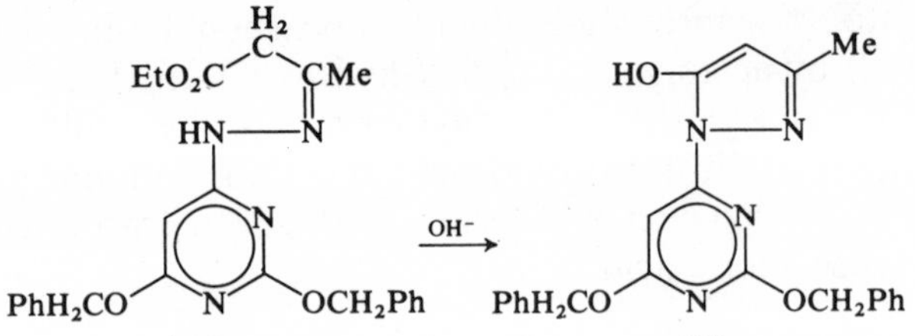

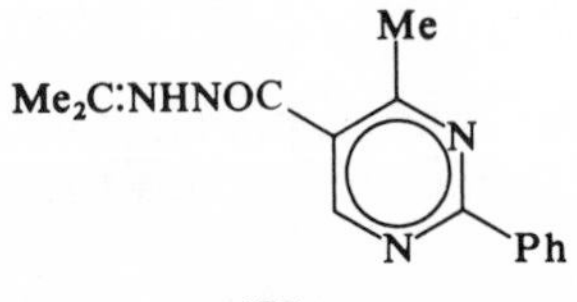

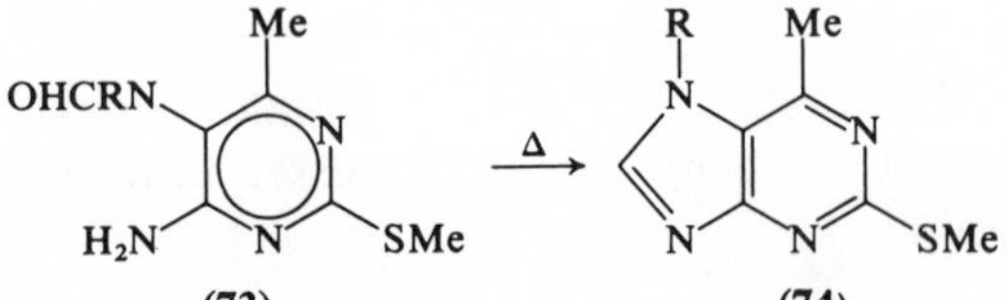

(69) **(70)** **(71)** **(72)** **(73)** **(74)**

* An uncharacterized triacetate(?) was formed but on dissolution in aq. ammonia and reprecipitation with acid it gave the diacetate.

† Like most other 4-amino-5-formamidopyrimidines, this was subsequently cyclized to the purine (**74**).[2675]

‡ Methylation of this and other analogues gave 5-*N*-methylformamidopyrimidines which underwent cyclization to 7-methylpurines, thus constituting a new and useful route to these difficult compounds.[3288]

Hydrazinopyrimidines may be easily formylated at the β-position to give, e.g., 5-amino-4-chloro-6-β-formylhydrazino- (57% by refluxing in butyl formate),[2752] 2-β-formylhydrazino-5-hexyl-4-hydroxy-6-methyl- (88% by formic acid at 60°),[2225] and 2-ethylthio-4-β-formylhydrazino-6-methyl-pyrimidine (*ca.* 50% by formic acid);[3218] also 6-β-formylhydrazino-3-methyluracil (*ca.* 60% by formic acetic anhydride),[2755] its 5-nitro derivative (*ca.* 70%),[2755] 6-β-formyl-α-methylhydrazino-3-methyluracil (*ca.* 65%),[2440] and such like.[2439, 2440]

Treatment of an aminopyrimidine with an orthoester can give the enol-ether of a formamido derivative. Such compounds are 4-amino-5-ethoxymethyleneaminopyrimidine (**75**, R = R′ = H; *ca.* 55% by triethyl orthoformate),[2144] its 6-chloro derivative (**75**, R = H, R′ = Cl; 68%),[3290] 4-amino-6-chloro-5-α-ethoxypropylideneaminopyrimidine (**75**, R = Et, R′ = Cl; 34%) and 4-chloro-5,6-bis-α-ethoxypropylideneaminopyrimidine (63%) by using triethyl orthopropionate,[2144] and the like.[2144] Hydrazinopyrimidines react similarly to give, for example, 4-β-ethoxymethylenehydrazino- and 4-β-(α-ethoxyethylidene)hydrazino-5-nitropyrimidine, and their 2-methyl derivatives.[3312]

(3) *Benzoylation and Other Acylations* (*H* 327)

Most benzoylations have been done using benzoyl chloride (or a derivative) in pyridine as illustrated in the preparation of 4-amino-5-benzamido-,[3186] 4,5-dibenzamido-,[3186] 4-benzamido-2,6-dimethoxy-,[2662] 4-*p*-bromobenzamido-2,6-diethoxy-,[2662] 4-benzoylhydrazino-2,6-dimethoxy-,[3259] and many other substituted-benzamidopyrimidines.[2662, 2663, 3292] However, Schottan–Baumann conditions have been used to make 4-amino-5-benzamido-6-methyl- (41%), 4,6-diamino-5-benzamido- (36%), and other such pyrimidines.[3186] Esters may also be used as acylating agents: e.g., 4-amino-2,6-diethoxypyrimidine and phenyl salicylate in α-methylnaphthalene at 200° gave 2,4-diethoxy-6-*o*-hydroxybenzamidopyrimidine.[2662] Both *O*- and *N*-benzoylation can occur together, as in the formation of 4-benzamido-5-benzoyloxy- (**76**, R = Bz; 23%) and 4-benzamido-5-hydroxy- (**76**, R = H; 31%) from 4-amino-5-hydroxy-pyrimidine and benzoyl chloride in pyridine;[2584] hydroxyaminopyrimidines undergo *O*-benzoylation under similar conditions to give, for example, 4-amino-2-*N*-benzoyloxyanilino-5-cyanopyrimidine (**77**, R = Bz) from the 2-*N*-hydroxyanilino analogue (**77**, R = H).[2330]

Acylation by unusual agents produced 2,4-dimethoxy-6-3′-pyridylcarbonylamino- (nicotinoyl chloride in pyridine),[2662] 5-bromo-4-

phthaloylglycylamino- (**78**; by phthaloylglycyl chloride in pyridine),[2722] 4-amino-1,3-dibenzyl-5-glycolloylamino-1,2,3,6-tetrahydro-2,6-dioxo- (by glycollic acid or its ethyl ester),[3293] 2-acetoacetylamino- (80%, by diketene),[3294] 4-acetoacetylamino-6-chloro- (diketene),[3294] 4-amino-1,2,3,6-tetrahydro-1,3-dimethyl-2,6-dioxo-5-valerylamino- (by valeric acid at 130°),[3295] 2-phthalimido- (65%, by phthalic anhydride at 140°),[3347] and 4-amino-5-*β*-butenoylamino-1,2,3,6-tetrahydro-1,3- dimethyl-2,6-dioxo-pyrimidine (by vinylacetic acid at 130°),[3295] as well as other related compounds.[2190, 2662, 3295–3297] Halogenated acylating agents were used to produce 4-amino-5-trichloroacetamido- (trichloroacetyl chloride in acetone),[1391] 4-amino-5-trifluoroacetamido- (trifluoroacetic anhydride under reflux),[1391] 4-amino-2-hydroxy-5-trifluoroacetamido- (by refluxing trifluoroacetic acid),[2661] 2-amino-5-*γ*-bromoacetamidopropyl-4-hydroxy-6-methyl-* (by *p*-nitrophenyl bromoacetate in aq. acetone),[2751] 4-amino-2,6-dihydroxy-5-trichloroacetamido- (by trichloroacetic acid at 100°),[3298] 4-amino-5-dichloroacetamido-6-hydroxy- (by dichloroacetic acid at 120°),[3298] 2,4-diamino-5-*γ*-*p*-bromoacetamidophenylpropylamino-6-phenyl- (by bromoacetic anhydride),[2944] and other pyrimidines.[2981, 3298, 3299] The 'acylation' of aminopyrimidines by (inorganic) phosphorus halides is illustrated in

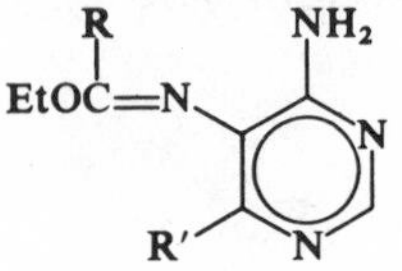

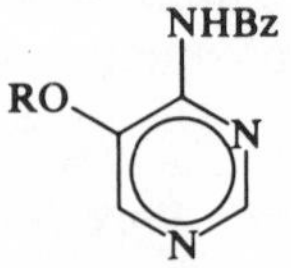

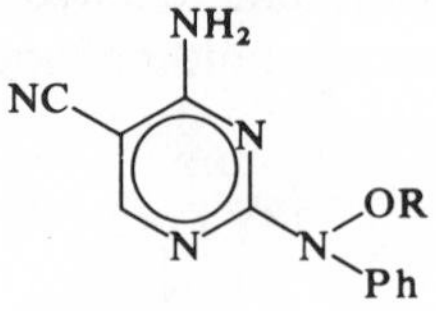

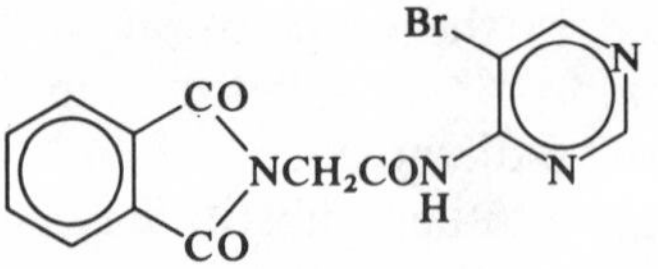

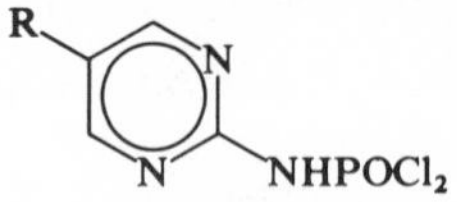

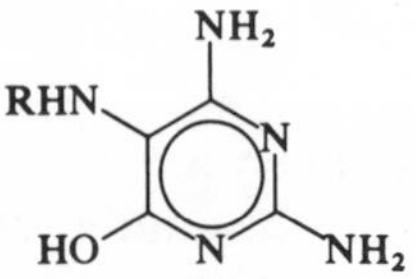

(**75**) (**76**) (**77**)

(**78**) (**79**)

(**80**)

* The selective bromoacylation of *ω*-amino- in polyaminopyrimidines has been studied and the technique must be varied with the relative basic strengths and reactivity of the amino groups.[2751]

the formation of 2-dichlorophosphinylaminopyrimidine (**79**, R = H; 90% by using phosphoryl chloride),[3301] 4-chloro-2-dichlorophosphinylamino-6-methylpyrimidine (by phosphoryl chloride),[3300] 5-bromo-2-dichlorophosphinylaminopyrimidine (**79**, R = Br),[3300] 5-chlorosulphonyl-2-dichlorophosphinylaminopyrimidine (**79**, R = SO_2Cl; by phosphorus pentachloride on 2-amino-5-sulphopyrimidine, followed by water: 71% yield),[3232] and 4-chloro-2-dichlorophosphinylamino-5-nitropyrimidine.[2462] The *P*-chloro substituents in these easily underwent aminolysis, e.g., by ethyleneimine, to give phosphoric triamides of interest as potential anti-tumor agents.

(4) *Other Ways of Preparing Acylaminopyrimidines* (*H* 328)

Examples of acylaminopyrimidines made by the Principal Synthesis are in Ch. II, and of 5-formamidopyrimidines, made by reducing a 5-nitropyrimidine with a metal plus formic acid, in Ch. V. Transacylation was used to convert 2,4-diamino-5-formamido-6-hydroxy- (**80**, R = CHO) into 2,4-diamino-6-hydroxy-5-hydroxyoxalylamino-pyrimidine (**80**, R = $COCO_2H$) by refluxing in aq. oxalic acid;[3302] the process was also involved in the change of 5-acetamido-4,6-dichloro- into 4-acetylhydrazino-5-amino-6-chloro-pyrimidine (*ca.* 45%) by aq. hydrazine,[2466] and in related reactions.[2752] The hydrogenation of 5-cyano-2,4-dimethylpyrimidine over Raney nickel in acetic anhydride gave 5-acetamidomethyl-2,4-dimethylpyrimidine (40%).[3007]

(5) *Deacylation of Acylaminopyrimidines* (*H* 329)

Most such deacylations have been done recently under acidic conditions. Thus 2-amino-5-formamido- gave 2,5-diamino-4-hydroxy-6-methylaminopyrimidine (*ca.* 70% by 15% methanolic hydrogen chloride),[2483] 2,4-diamino-5-formamido- gave 2,4,5-triamino-1,6-dihydro-1-methyl-6-oxopyrimidine (55% by aq. sulphuric acid,[3303] or *ca.* 85% by methanolic hydrogen chloride[2493]), 5-acetamido-2-amino- gave 2,5-diamino-4-hydroxy-6-methylpyrimidine (83% by 6*N*-hydrochloric acid),[2303] 4-amino-5-formamidomethyl- gave 4-amino-5-aminomethyl-pyrimidine (by 10% hydrochloric acid),[2190] 4-amino-5-formamido- gave 4-amino-5-amino-methyl-2-propylpyrimidine (by 14% ethanolic hydrogen chloride at 130°),[2190] and 4,6-dichloro-5-*N*-methyltrifluoroacetamido- gave 4-chloro-6-hydroxy-5-methylamino-pyrimidine (76% by methanolic hydrogen chloride).[2752] The deacylation of

2-hydroxy-4-*N*-methylacetamidopyrimidine was followed spectrally at room temperature using *N*-hydrochloric acid or *N*-sodium hydroxide (complete in 10 hr. at 25°), but the product was not isolated.[2984] A convenient deacylation technique, e.g., in the presence of easily-hydrolysed alkoxy groups, would be to warm with the appropriate alcoholic sodium alkoxide (*cf.* the deacylation of 2,4-diacetamido-pteridine).[3304]

(6) *Arylsulphonamidopyrimidines* (*H* 329)

Arylsulphonamidopyrimidines may be made by three chief methods. The use of primary synthesis, e.g., in making 2-benzenesulphonamido-5-phenylpyrimidine,[3310] has been discussed in Ch. II. The second route involves nucleophilic displacement of a halogeno, sulphone, trialkyl-ammonio, or other such group by sodio sulphanilamide or a related compound. Some of these reactions have been discussed under the appropriate leaving groups, e.g., in Ch. VI, Sect. 5.H (displacement of chlorine), and in Ch. VIII, Sect. 5.B (displacement of alkylsulphonyl). The displacement of a trialkylammonio group is typified in the conversion of 5-chloro-4-methoxypyrimidin-2-yl trimethyl ammonium chloride (**81**) into 5-chloro-4-methoxy-2-sulphanilylaminopyrimidine (**82**; 79%) by warming with sodio sulphanilamide in acetamide at 100°;[2571] also in the analogous formation of 5-methoxy-, 5-methoxy-2-methyl-, 5-methoxy-2-methylthio-, and analogous 4-sulphanilyl-aminopyrimidines.[2586]

The third method, direct acylation, is generally done by treating the aminopyrimidine with a benzenesulphonyl chloride in pyridine. In this way have been made: 2-benzenesulphonamidopyrimidine,[3305] 2,4-dimethoxy-6-*p*-nitrobenzenesulphonamidopyrimidine (60%),[2774] 4-amino-6-hydroxy-2-methyl-5-*p*-toluenesulphonamidopyrimidine,[3060] 2-butyl-4,5-dimethyl-6-sulphanilylaminopyrimidine (using *p*-acet-amidobenzenesulphonyl chloride in methylene chloride and trimethyl-amine followed by removal of the acetyl group by methanolic hydro-chloric acid),[2349] 2-dimethylamino-4-*p*-nitrobenzenesulphonamidopyri-midine (49%),[2837] 1-allyl-6-amino-5-benzenesulphonamido-3-ethyl-1,2,-3,4-tetrahydro-2,4-dioxopyrimidine (93%),[3306] 4-amino-5-ethane-sulphonamido-1,2,3,6-tetrahydro-1-methyl-3-β-methylallyl-2,6-dioxo-pyrimidine (58%),[3306] 4-methoxy-6-*p*-nitrobenzenesulphonamido-pyrimidine (79%),[2446] 4-methyl-6-2′(3′ or 4′)-pyridyl-2-sulphanil-ylaminopyrimidine (*via* the acetamido analogue),[2186] 4-*p*-acetamido-

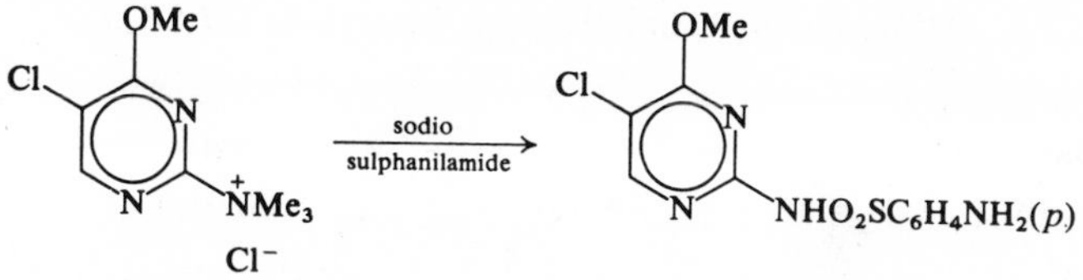

OMe Cl N N $\overset{+}{N}Me_3$ Cl^- **(81)** sodio sulphanilamide → OMe Cl N N $NHO_2SC_6H_4NH_2(p)$ **(82)**

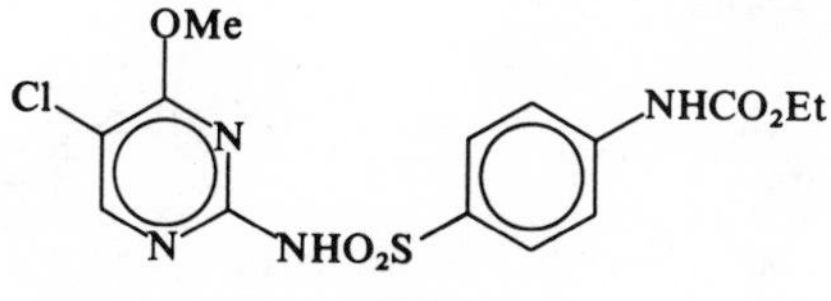

OMe Cl N N NHO_2S $NHCO_2Et$ **(83)**

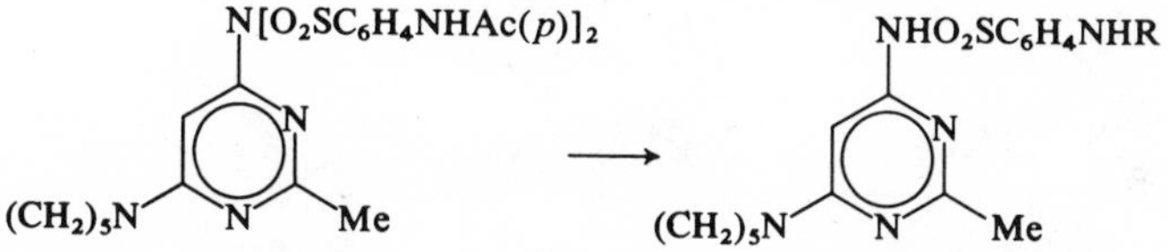

$N[O_2SC_6H_4NHAc(p)]_2$ N $(CH_2)_5N$ N Me **(84)** → $NHO_2SC_6H_4NHR$ N $(CH_2)_5N$ N Me **(85)**

benzenesulphonamido-6-methoxy-2-methylthiopyrimidine,[3216] 5-chloro-2-*p*-ethoxycarbonylaminobenzenesulphonamido-4-methoxypyrimidine* (**83**),[2571] 2,4-dimethoxy-6-*p*-ethoxycarbonylaminobenzenesulphonamidopyrimidine* (88%),[2280] 4-*p*-ethoxycarbonylaminobenzenesulphonamido-2-hydroxypyrimidine* (46%),[2677] 2-ethoxy-4-*p*-nitrobenzenesulphonamidopyrimidine (53% using trimethylamine; 25% using pyridine),[3307] and others.[2212, 2280, 2524, 2677 2678, 2837, 2886, 3216, 3306–3309, 3311]

Sometimes such acylation leads to a diacylated derivative but usually this can be partly hydrolysed under mild conditions to give the mono-acylated derivative. Thus 4-bis-(*p*-ethoxycarbonylaminobenzenesulphonyl)amino-2-methoxy-5,6-dimethylpyrimidine in alkali gave 2-methoxy-4,5-dimethyl-6-sulphanilylaminopyrimidine;[2280] 4-bis-(*p*-nitrobenzenesulphonyl)amino-5,6-dimethoxypyrimidine in methanolic alkali gave 4,5-dimethoxy-6-*p*-nitrobenzenesulphonamidopyrimidine (*ca.* 85%);[2524] and 4-bis-(*p*-acetamidobenzenesulphonyl)amino-2-methyl-6-piperidinopyrimidine (**84**) in alkali gave 2-methyl-4-piperidino-6-sulphanilylaminopyrimidine (**85**, R = H; 75%), or in dilute ammonia gave 4-*p*-acetamidobenzenesulphonamido-2-methyl-6-piperidinopyrimidine (**85**, R = Ac; 50%).[3308]

* Alkaline hydrolysis gave the corresponding sulphanilylaminopyrimidine.

E. Diazotization and Related Reactions (*H* 331)

The use of nitrous acid to convert amino- into hydroxy-pyrimidines has been discussed in Ch. VII, Sects. 1.D and 3.B; the conversion of amino- into halogeno-pyrimidines in Ch. VI, Sects. 1.C, 1.D, and 3.E; and the formation of (*N*-nitrosoamino)pyrimidines from alkylamino-pyrimidines in Ch. V, Sect. 2.D.

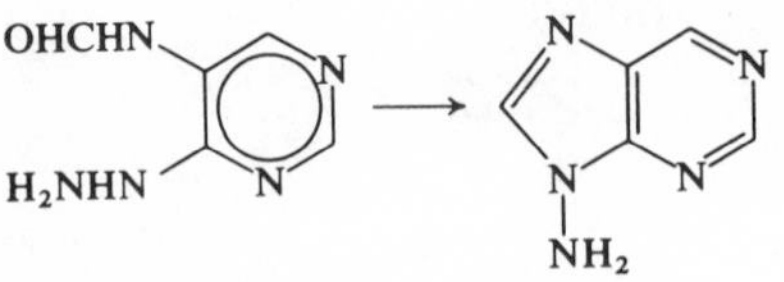

OHCHN, H_2NHN (**86**) → NH_2 (**87**) H_2N, EtOHC:NHN (**88**) → HN, H (**89**)

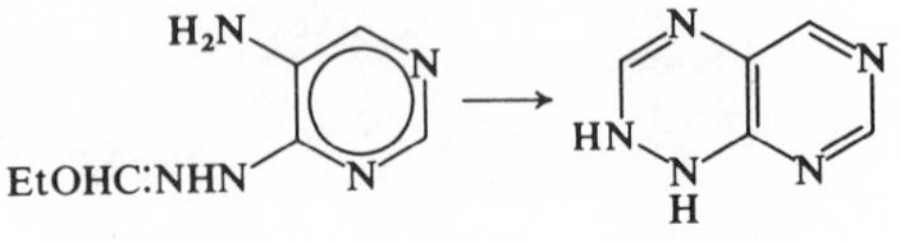

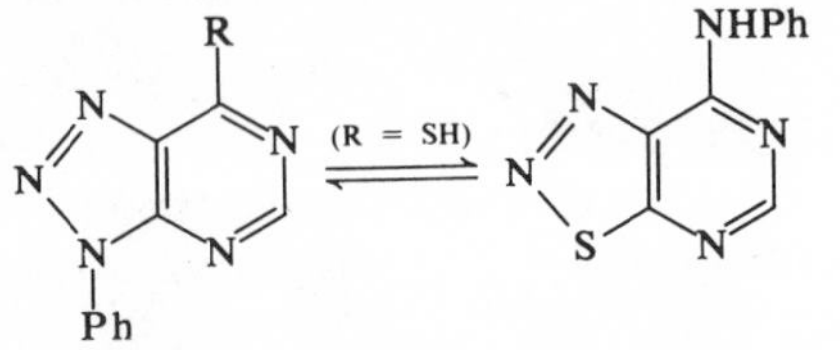

R, Ph (**90**) ⇌ (R = SH) NHPh, S (**91**)   O, CH_2Ph (**92**)

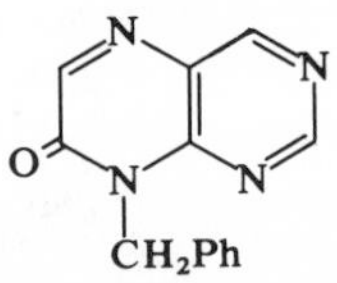

F. Bicyclic Heterocycles from Aminopyrimidines (*H* 333)

A few typical cyclizations involving amino- or diamino-pyrimidines as starting materials are given below, more to draw attention to their importance as intermediates than in any effort to be systematic or complete.

Purines are still generally made from aminopyrimidines[3198, 3313] as illustrated in random examples[2144, 2243, 2661, 2675, 2724, 2982, 3058, 3327] or the following less usual reactions: 4,5-diamino-6-hydroxypyrimidine and acetamidine hydrochloride (with or without sodium acetate) gave 6-hydroxy-8-methylpurine, and the reaction was shown to be quite general for 8-substituted purines;[3314, 3315] 4,5-diaminopyrimidine condensed with carbon disulphide in dimethylformamide or pyridine to give 8-mercaptopurine (*ca.* 95%),[3117, 3316] again a general reaction; 5-amino-4-hydrazinopyrimidine and formic acid gave (*via* 5-formamido-4-hydrazinopyrimidine; **86**?) 9-aminopurine (**87**),[2673] but reduction of the nitro group in 4-β-ethoxymethylenehydrazino-5-nitropyrimidine to give the enol ether (**88**) of 5-amino-4-formylhydrazinopyrimidine was followed by cyclization to 1,2-dihydropyrimido[5,4-*e*]-*as*-triazine

(**89**);[3312, cf. 3214] 4,5-diamino-6-methylaminopyrimidine with formamide gave 6-amino-9-methylpurine, but with triethyl orthoformate/acetic anhydride gave 6-methylaminopurine, an apparently general phenomenon in such condensations;[3318] and the recent use of 4-amino-5-nitrosopyrimidines[2485, 2486, 2838, 3319–3321] or related compounds in several new reactions leading to purines.[2950, 2951, 3291, 3322]

8-*Azapurines* (1,2,3,5,7-penta-azaindenes) can generally be made from 4,5-diaminopyrimidines with nitrous acid (although an alternative approach from triazoles is proving fruitful[2530, 3249–3252]). The use of aminopyrimidines is illustrated in the formation of 4-chloro-,[2568] 4,6-dihydroxy-,[2489] 4-chloro-1-phenyl-* (**90**, R = Cl),[2887] and other 1,2,3,5,7-penta-azaindenes.[2489, 2568, 2887] 4-Amino-5-phenylazopyrimidines gave 2-phenylpenta-azaindenes on oxidation.[3785]

Pteridines are made generally from a 4,5-diaminopyrimidine and an α-dicarbonyl compound.[3323, 3324, 3478] When the latter is unsymmetrical two products can result. Typical examples are the condensation of tetra-aminopyrimidine with hydroxymethylglyoxal to give either 2,4-diamino-6-hydroxymethylpteridine[3325] or its 7-hydroxymethyl isomer[3304] according to conditions; the 6-benzyl-7-hydroxy- or 7-benzyl-6-hydroxy-pteridines obtained from phenylpyruvic acid and diaminopyrimidines according to the other substituents in the latter;[3326] and the (unambiguous) formation of 8-benzyl-7,8-dihydro-7-oxopteridine (**92**) from 5-amino-4-benzylaminopyrimidine and ethyl ethoxyglycollate.[2675] Pteridines are also made by the Timmis synthesis from 4-amino-5-nitrosopyrimidines [see Ch. V, Sect. 2.B(2)], and by other routes from aminopyrimidines.[3332, 3334]

Thiazolo[5,4-d]*pyrimidines* may be made by two routes from aminopyrimidines. The first is typified by the condensation of 5-amino-4-mercapto-2-methyl-6-methylaminopyrimidine (**93**) with formic acid to give 5-methyl-7-methylaminothiazolo[5,4-*d*]pyrimidine (**94**);[2675] the second in the treatment of 2,4-dichloro-5-nitro-6-trifluoromethylpyrimidine with potassium thiocyanate in acetic acid, followed by heating with aniline and reduction with iron, to give 2-amino-5-anilino-7-trifluoromethylthiazolo[5,4-*d*]pyrimidine.[2735]

Other systems derived from aminopyrimidines are illustrated in the preparation of 1,3-diaza-1,3-dihydro-7-dimethylamino-2-phenyl-2-phospholo[4,5-*d*]pyrimidine-2-oxide (**95**) from 4,5-diamino-6-dimethylaminopyrimidine and benzenephosphonicdiamide;[2725, 3328] 2,4-dihydroxy-

* Treatment with thiourea gave the 4-mercapto-analogue (**90**, R = SH) which is in equilibrium with 4-anilino-1-thia-2,3,5,7-tetra-azaindene (**91**). The kinetics and preparative aspects of this equilibrium (the 'Christmas rearrangement') have been studied.[2887, 3252]

pyrimido[4,5-*d*]pyrimidine (**96**; 85%) from 4-amino-5-bromo-2,6-dihydroxypyrimidine and 2 molecules of formamide;[2628] 3-amino-4,6-diphenyl-1*H*-pyrazolo[3,4-*d*]pyrimidine (**97**) by heating 5-cyano-4-hydrazino-2,6-diphenylpyrimidine;[2370] 5,7-dimethyl-2-phenylimidazo-[1,2-*a*]pyrimidine (**98**) from 2-amino-4,6-dimethylpyrimidine and ω-bromoacetophenone;[3329] and 2-hydroxy-4-methylpyrimido[4,5-*b*]-diazepine (**99**) from 4,5-diaminopyrimidine and ethyl acetoacetate.[3330]

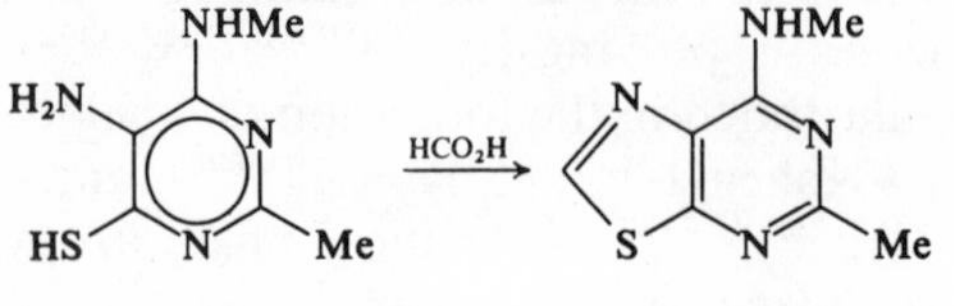

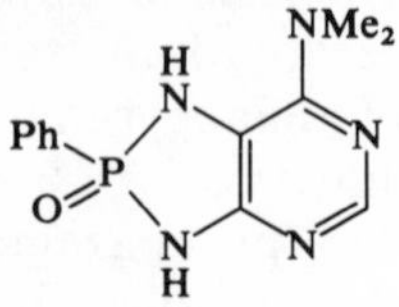

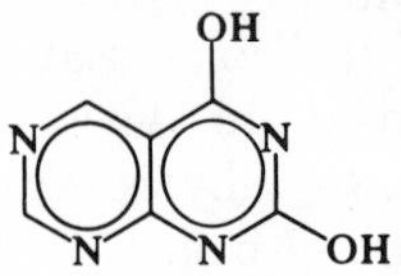

G. Other Reactions (*H* 335)

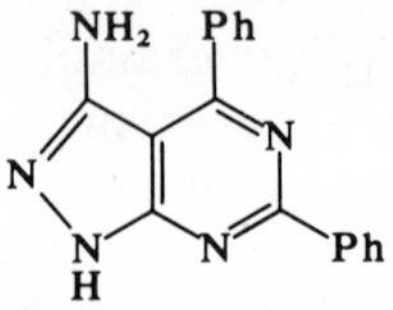

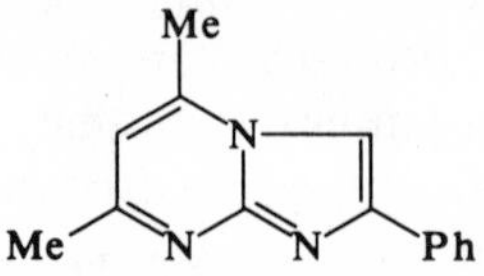

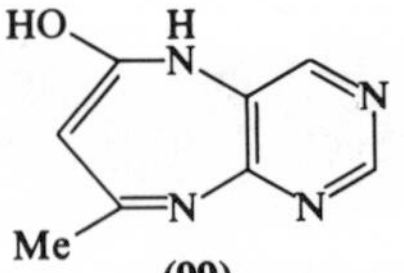

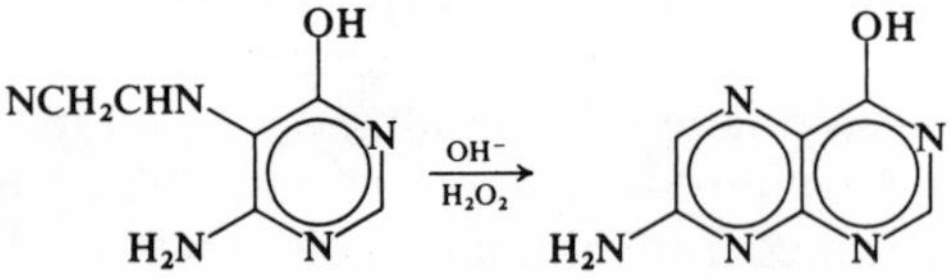

The reductive removal of an amino group has been claimed.[2306] Thus 'by catalytic hydrogenation of 4-aminouracil, uracil is obtained, (and) in the same way we obtained thymine by catalytic hydrogenation of 4-aminothymine'. However, the supporting experimental details appear to involve hydrogenation (Pt/H_2) of the acyclic cyanoacetyl- or α-cyanopropionyl-urea, isomeric with aminouracil and aminothymine, respectively.[2306] 4-Amino-2-hydroxypyrimidine underwent hydrogenation (Pt/H_2) to 1,4,5,6-tetrahydro-2-hydroxypyrimidine (identified as its picrate).[3227, 3336]

When tetra-aminopyrimidine hydrochloride* in methanolic acetic

* The commonly available but rather insoluble sulphate was converted into this soluble salt which is reported to be a mixture of di- and tri-hydrochloride.[3332]

acid was treated first with sodium cyanide and then with acetaldehyde, 2,4,6-triamino-5-α-cyanoethylaminopyrimidine (*ca.* 90%) resulted;[3332] 4-amino-5-cyanomethylamino-6-hydroxypyrimidine (**100**) and other such derivatives were made similarly.[2492, 3332] Subsequent oxidative cyclization gave 2,4,7-triamino-6-methyl-, 7-amino-4-hydroxy- (**101**), and other pteridines.[2492, 3332]

Ultra-violet irradiation of 4-amino-2,6-dimethylpyrimidine has been studied in some detail.[3337, 3338] The primary (isolated) product was 3-cyano-2,4-di-iminopentane, HN:CMeCH(CH)CMe:NH, almost certainly in the hydrogen-bonded form (**102**) according to very convincing spectral data. Its further break down was followed, and the likely mechanisms were discussed.[3337, 3338]

The oxidative removal of a hydrazino group in favour of hydrogen has been discussed in Ch. VI, Sect. 5.A, and the reduction of a hydrazino- or α-methylhydrazino- to an amino- or methylamino-pyrimidine, respectively, in Sect. 1.H of this Chapter. Hydrazinopyrimidines may be converted into the corresponding azido derivatives with nitrous acid: thus, 4,6-dihydrazinopyrimidine and aq. nitrous acid at 25° gave 4,6-diazidopyrimidine (77%).[2754] However, such azidopyrimidines have been shown to be in equilibrium with tetrazolopyrimidines, and this

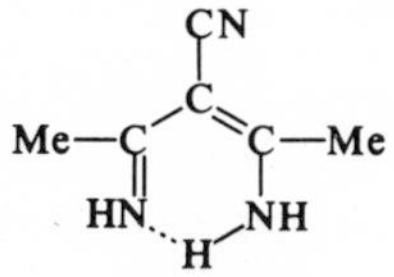

(102) 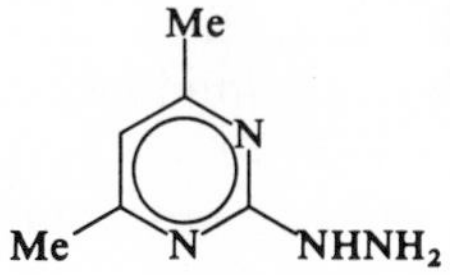 (103) 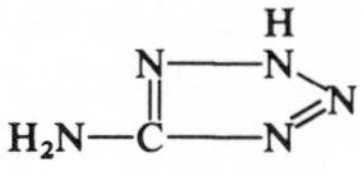 (104)

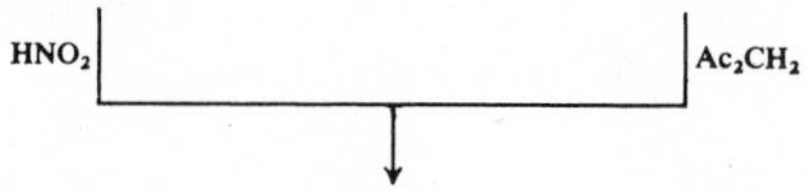

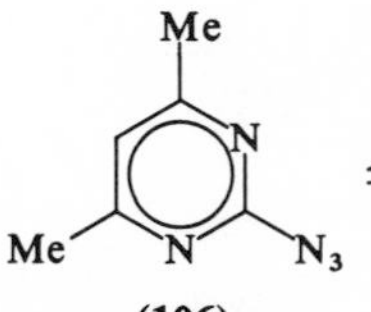

HNO_2 Ac_2CH_2

(106) 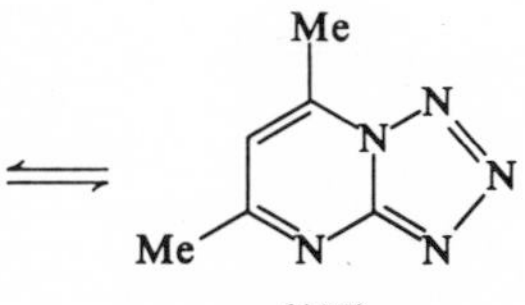 ⇌ (105)

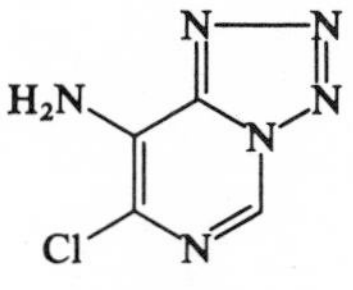

(107) 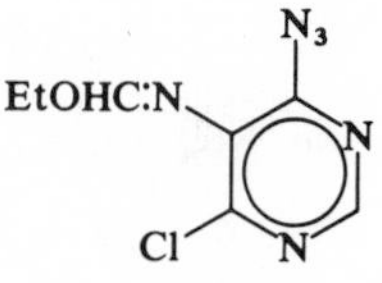 (108)

aspect has been studied in detail by J. A. Montgomery and his colleagues. Thus the reaction of 5-aminotetrazole (**104**) with acetylacetone and the reaction of 2-hydrazino-4,6-dimethylpyrimidine (**103**) with nitrous acid both led to the same product, formulated as 5,7-dimethyltetrazolo[1,5-*a*]pyrimidine (**105**).[2188, 3339]

The equilibrium was examined by infra-red and p.m.r. spectra: in the solid state or in dimethyl sulphoxide solution, only the bicyclic form (**105**) could be detected; in trifluoroacetic acid, only the monocyclic form (**106**) was present; and in other solvents, a mixture of both was evident.[3340] Similar results were obtained in the equilibrium, 4-azidopyrimidine $\rightleftharpoons$ tetrazolo[1,5-*c*]pyrimidine,[2466] and in related systems.[2466, 3341–3343] The effect of substitution on such equilibria was clearly seen in the azido derivatives from 5-amino-4-chloro-6-hydrazino- and 4-chloro-5-ethoxymethyleneamino-6-hydrazino-pyrimidine. That from the 5-amino compound was in the bicyclic form (**107**), but that from the enol ether was in the monocyclic form, 4-azido-6-chloro-5-ethoxymethyleneaminopyrimidine (**108**).[2934]

Some interesting reactions have been recorded by W. Klötzer for *N*-1(or 3)-amino groups in pyrimidines. Removal: 1-amino-uracil (**110**, R = H) and nitrous acid gave uracil (**109**, R = H; 88%),[3260] 1-amino-1,2,3,4-tetrahydro-3-methyl-2,4-dioxopyrimidine (**110**, R = Me) similarly gave 3-methyluracil (**109**, R = Me),[3260] and 1-amino-1,2,3,6-tetrahydro-3-methyl-2,6-dioxopyrimidine gave 1-methyluracil.[3261] Schiff bases: 1-aminouracil reacted with an appropriate aldehyde or ketone to give 1-α-methylbenzylideneamino- (**111**), 1-5′-chloro-2′-hydroxy-

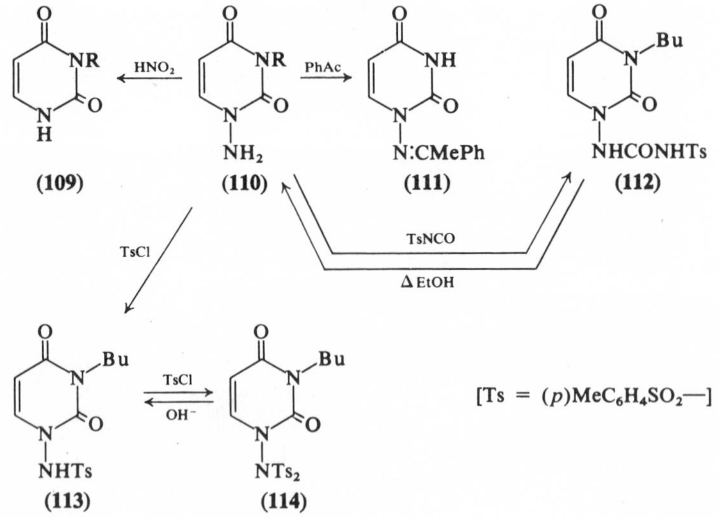

[Ts = $(p)MeC_6H_4SO_2$—]

benzylideneamino-, 1-5′-nitrofurfurylideneamino-, and other such uracil derivatives.* [3260]

Acylation: 1-amino-3-butyl-1,2,3,4-tetrahydro-2,4-dioxopyrimidine (**110**, R = Bu) and *p*-toluenesulphonyl chloride in pyridine gave a separable mixture of 1-butyl-1,2,3,6-tetrahydro-2,6-dioxo-3-*p*-toluenesulphonamidopyrimidine (**113**) and 1-bis-(*p*-toluenesulphonyl)amino-3-butyl-1,2,3,4-tetrahydro-2,4-dioxopyrimidine (**114**), which was converted into the monoacyl derivative (**113**) by mild alkaline treatment.[3260] Acylureide: The above 1-amino-3-butyl derivative (**110**, R = Bu) and *p*-toluenesulphonyl isocyanate gave 1-butyl-1,2,3,6-tetrahydro-2,6-dioxo-3-*p*-toluenesulphonylureidopyrimidine (**112**) which was reconverted into the original pyrimidine on prolonged boiling in ethanol.[3260]

6. Urethanes (Alkoxycarbonylaminopyrimidines) (*H* 336)

Urethanes have been made recently from the corresponding aminopyrimidines with ethyl chloroformate or from the acid azide with ethanol.

The first route was used to make 2-ethoxycarbonylaminopyrimidine (**115**) by adding ethyl chloroformate to a solution of 2-aminopyrimidine in pyridine at $<10°$;[3345] a similar procedure converted 5-amino- into 5-ethoxycarbonylamino-6-methyl-4-*p*-nitrobenzylidenehydrazinopyrimidine (79%).[2227] In place of pyridine, aq. alkali or simply hot benzene, has been used by E. Dyer and her colleagues as a medium in which to use alkyl chloroformates for preparing 2-methoxy(ethoxy, phenoxy, or amyloxy)carbonylaminopyrimidine, the corresponding 4-isomers, 5-ethoxycarbonylamino-2,4-dihydroxypyrimidine, and related compounds.[3346] Appropriate alkyl chlorothioformates in ethanol or acetone were used to make 2-(ethylthio)carbonylaminopyrimidine (**116**; 55%), 5-(butylthio)carbonylamino-2,4-dihydroxypyrimidine (61%), 2,4-bis-[(ethylthio)carbonylamino]pyrimidine, the 4,6-isomer (**117**, X = S), and many of their homologues.[3347]

The bis-thiourethane (**117**, X = S) was converted by ethanolic triethylamine, containing a little mercuric chloride, into its oxygen analogue, 4,6-bis-(ethoxycarbonylamino)pyrimidine (**117**, X = O; 83%); this was made also by treating 4,6-diaminopyrimidine with ethyl chloroformate as above (25% yield) or with ethyl pyrocarbonate

* Several Schiff bases of 1-amino-5,6-dihydrouracil are described, but they were made by primary syntheses.[3344]

[$(EtOCO)_2$] in ethanol (95% yield).[3347] Ethyl pyrocarbonate also gave gratifying yields in the several similar reactions in which it was used.[3347]

The second route to urethanes is much less convenient but has been used to convert 4-ethoxycarbonylpyrimidine, *via* its amide and azide, into 4-ethoxy(or methoxy)carbonylaminopyrimidine;[3346] and 4-carboxy-2,6-dihydroxypyrimidine (*via* its butyl ester, hydrazide, and azide) into 4-ethoxy(amyloxy, benzyloxy, or butoxy)carbonylamino-2,6-dihydroxypyrimidine.* [3348]

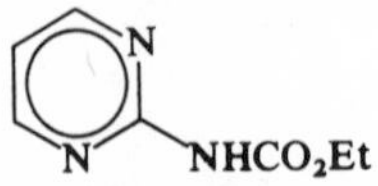

(115) 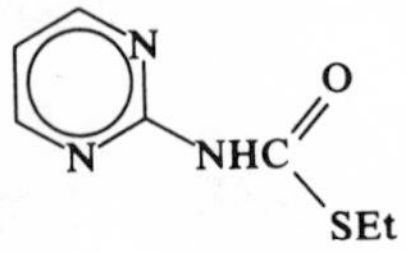 (116) 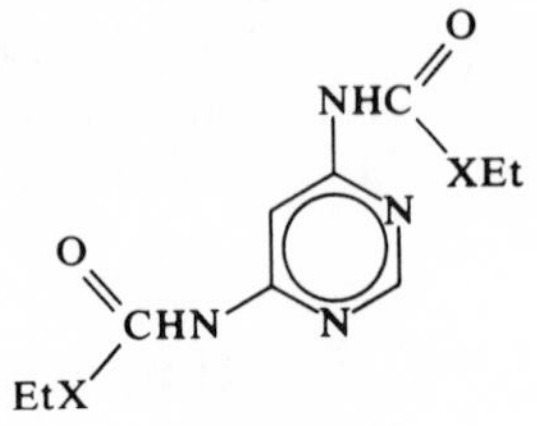 (117)

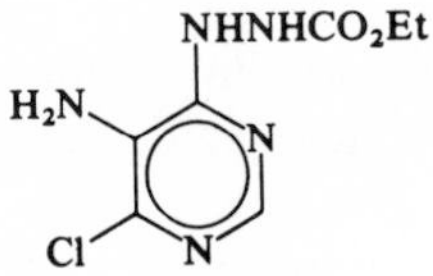

(118) 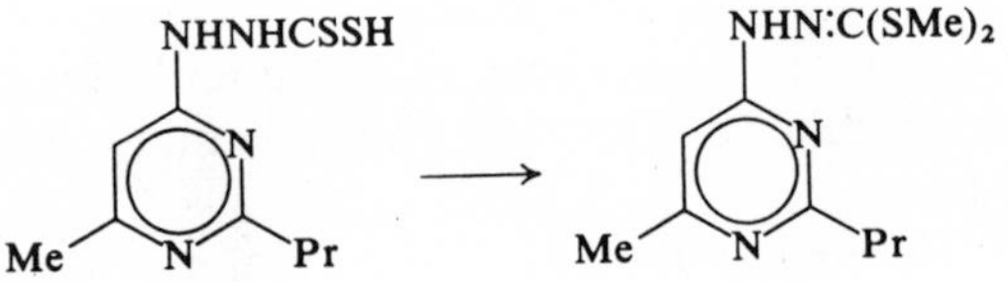 (119) (120)

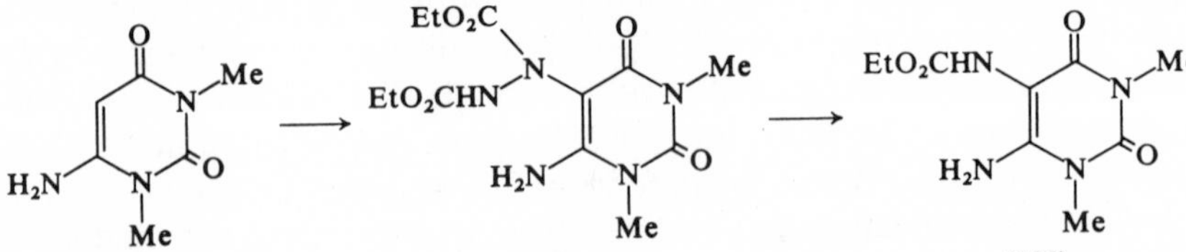

(121) (122) (123)

β-Alkoxycarbonylhydrazino derivatives are known also. Thus 5-amino-4-chloro-6-hydrazinopyrimidine reacted with ethyl chloroformate in dioxane at 25° to give 5-amino-4-chloro-6-β-ethoxycarbonylhydrazinopyrimidine (**118**; 70%).[3279] The benzyloxy homologue was made similarly.[3772]

A related derivative was made by first treating 4-hydrazino-6-methyl-2-propylpyrimidine with carbon bisulphide in aq. potassium carbonate to give 4-dithiocarboxyhydrazino-6-methyl-2-propylpyrimidine (**119**), which was dimethylated by methyl iodide in aq. potassium carbonate

* Hydrogenolysis of the benzyl urethane (Pd/H_2) gave 4-amino-2,6-dihydroxypyrimidine.[3348]

to give 4-bis(methylthio)methylenehydrazino-6-methyl-2-propylpyrimidine (**120**).[3349]

A new general method for introducing an $\alpha\beta$-diethoxycarbonyl hydrazino group into the unoccupied 5-position of an aminopyrimidine has appeared.[3350] Thus treatment of 4-amino-1,2,3,6-tetrahydro-1,3-dimethyl-2,6-dioxopyrimidine (**121**) with diethyl azodicarboxylate ($EtO_2C \cdot N{:}N \cdot CO_2Et$) gave the 5-$\alpha\beta$-diethoxycarbonylhydrazino derivative (**122**; 77%) which on refluxing with Raney nickel in ethanol gave the corresponding 5-ethoxycarbonylaminopyrimidine (**123**).[3350] These and other such pyrimidines were shown to be valuable intermediates for purines and penta-azanaphthalene systems.[3350]

7. Ureidopyrimidines (*H* 339)

The usual method for making ureido- and thioureido-pyrimidines is to treat the corresponding amino derivative with an isocyanate or isothiocyanate. This process has produced 2- and 4-phenylureido-,[3346] 2-hydroxy-4-phenylureido-,[3346] 4-amino-6-ethyl(phenyl, or *p*-methoxyphenyl)ureido-,[3286] 4-allylthioureido-6-amino-[3286] 4-butylthioureido-2,6-dimethyl-,[2791] 4-methylthioureido-2,6-dimethyl-,[2791] 4-amino-6-butyl(phenyl, or *o*-carboxyphenyl)ureido-,[3286] 5-ethoxycarbonyl-2-mercapto-4-phenylureido- (**124**),[2275] 4,6-diamino-5-ethylthioureido-2-methylthio-,[2743] 4-amino-5-ethylureido-1,2,3,6-tetrahydro-1-methyl-3-β-methylallyl-2,6-dioxo- (75%),[2309] 1-allyl-6-amino-3-ethyl-5-ethylureido-1,2,3,4-tetrahydro-2,4-dioxo- (82%),[2309] 1-allyl-6-amino-5-butylthioureido-3-ethyl-1,2,3,4-tetrahydro-2,4-dioxo-* (80%),[2309] and other such pyrimidines.[2309] In a similar way, 4-amino-2,6-dimethylpyrimidine (**125**) and benzoyl isothiocyanate gave 4-benzoylthioureido-2,6-dimethylpyrimidine (**126**) which in alkali gave 2,4-dimethyl-6-thioureidopyrimidine (**127**), also produced directly from the amine (**125**) and acidified potassium thiocyanate.[3351]

Some related semicarbazidopyrimidines have been made from hydrazinopyrimidines: 4-hydrazino-6-methyl-2-propyl- gave 4-methyl-2-propyl-6-semicarbazido-pyrimidine (*ca.* 80%) on treatment with acidified sodium cyanate solution;[2791] 2,4-dimethyl-6-methylsemicarbazido- (**128**), 4-methyl-6-methylsemicarbazido-2-propyl-, 4-methyl-6-t-octylsemicarbazido-2-propyl-, and other such pyrimidines were made using an appropriate alkyl isocyanate in an organic solvent.[2791]

* Converted into the corresponding 5-butylureido derivative (75%) by alkaline hydrogen peroxide; other similar transformations were also successful.[2309]

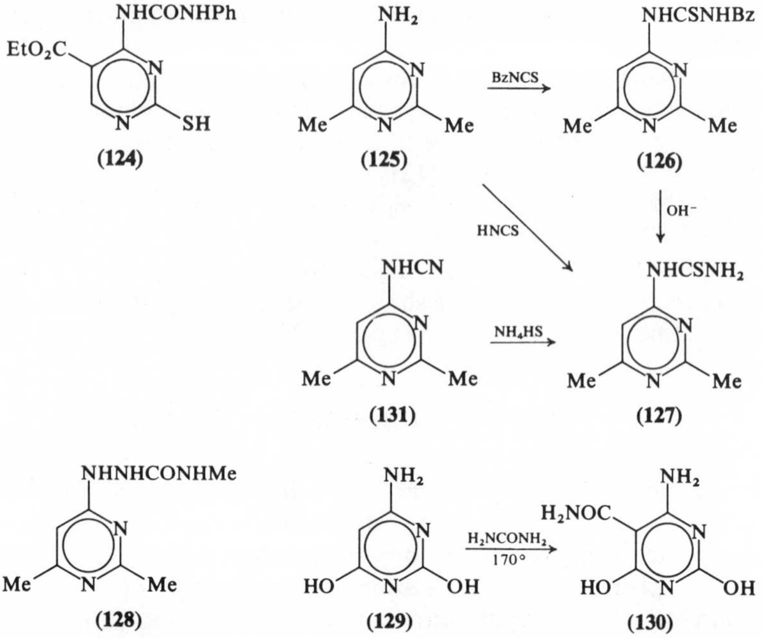

(124) (125) (126) (127) (128) (129) (130) (131)

A less important way to make ureidopyrimidines is illustrated in the reaction of aq. ammonium hydrogen sulphide with 4-cyanoamino-2,6-dimethylpyrimidine (**131**) to give 2,4-dimethyl-6-thioureidopyrimidine* (**127**);[2791] and in the acidic hydrolysis of 4-cyanoamino-6-hydroxy-2-methylpyrimidine to 4-hydroxy-2-methyl-6-ureidopyrimidine (and thence in alkali to 4-amino-6-hydroxy-2-methylpyrimidine).[2222]

Two compounds, formed by fusion of 4-amino-2,6-dihydroxy- (**129**) and 2,4-diamino-6-hydroxy-pyrimidine with urea, were first described[3228] as 2,4-dihydroxy- and 2-amino-4-hydroxy-6-ureidopyrimidine. However they later proved to be isomeric amides, 4-amino-5-carbamoyl-2,6-dihydroxy- (**130**) and 2,4-diamino-5-carbamoyl-6-hydroxy-pyrimidine, respectively, formed by introduction of carbamoyl groups into the vacant 5-positions during fusion.[3352]

8. Other Substituted-amino-pyrimidines (*H* 341)

A. Nitroamines (*H* 341)

The formation of some nitroaminopyrimidines has been discussed in Ch. V, Sect. 1.A(2).

* The given m.p. is 18° higher than that recorded for the same material made by another route.[3351]

D. Cyanoamines (*H* 343)

Some examples of the primary synthesis of cyanoaminopyrimidines have been given in Ch. II. They may also be made from a chloropyrimidine: thus 4-chloro- gave 4-cyanoamino-2,6-dimethylpyrimidine by prolonged refluxing with ethanolic sodium cyanamide;[2791] and 4-chloro- gave 4-cyanoamino-6-methyl-2-propylpyrimidine by heating with sodium cyanamide in dimethylformamide.[2791]

The conversion of cyanoamines into ureides, etc., has been mentioned in Sect. 7 above.

E. Guanidinopyrimidines (*H* 344)

Guanidinopyrimidines have been made by the Principal Synthesis (see Ch. II) and by aminolysis of chloropyrimidines: thus 4-amino-6-chloro- gave 4-amino-6-guanidino-5-nitropyrimidine (**132**; 78%) by refluxing with ethanolic guanidine.[2164] Treatment of the guanidine (**132**) with alkali gave 4-amino-5-nitro-6-ureidopyrimidine (**133**) and 2,4,6-triamino-5-nitropyrimidine (**134**), the latter by fission of the 1:2- or 2:3-bond, loss of C-2 as formic acid, and formation of a ring involving the guanidino group. The ureide (**133**) underwent a similar rearrangement to give 4,6-diamino-2-hydroxy-5-nitropyrimidine (**135**), which was accompanied by the hydrolytic product 4,6-diamino-5-nitropyrimidine (**136**). Acid conditions caused the guanidine (**132**) to undergo rather similar reactions; during these experiments, the aliphatic rearrangement intermediate, 3-guanidino-3-imino-2-nitropropionamide (**137**), was isolated.[2164]

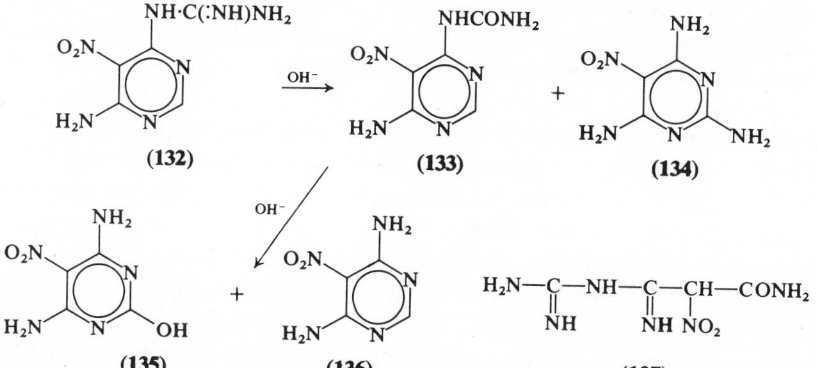

F. Hydroxylamines (*H* 344)

Aminolysis of chloropyrimidines and such like by hydroxylamine has been discussed in appropriate sections of Chs. VI and VIII.

G. Trimethyl Pyrimidinylammonium Chlorides (*H* 345)

Such quaternary pyrimidines have evoked more interest in recent years because the trimethylammonio group has been recognized as a good leaving group in nucleophilic displacements. However, extreme conditions must be avoided or the group may split out methyl halide spontaneously to give a dimethylamino group, useless for displacement.

The following examples illustrate the preparation and use of 2-, 4-, or 6-ammoniopyrimidines. 4-Chloro-2-methylthiopyrimidine and trimethylamine in anhydrous benzene gave trimethyl 2-methylthiopyrimidin-4-ylammonium chloride (**138**; *ca.* 90%).[2262, 2745, 2954] This reacted with potassium cyanide in acetamide to give 4-cyano-2-methylthiopyrimidine (**142**; 44%),[2954] and with benzyloxyurea to give *inter alia* an unstable product (**139**?) which lost HNCO to give 4-benzyloxyamino-2-methylthiopyrimidine (**140**), converted by acidic hydrolysis into 2-hydroxy-4-hydroxyaminopyrimidine (**141**).[2262] 2-Ethylthio-5-fluoropyrimidin-4-yl trimethylammonium chloride (50%) was made from its 4-chloro analogue as above;[2209, 2515] treatment with sodio sulphanilamide in molten acetamide gave 2-ethylthio-5-fluoro-4-sulphanilylaminopyrimidine (28%).[2209] 5-Fluoro-2-methoxypyrimidin-4-yl trimethylammonium chloride was made and allowed to react in a similar way.[2209] Trimethyl pyrimidin-2-ylammonium chloride was made from 2-chloropyrimidine in 90% yield.[2557] 4-Chloro-5-ethoxycarbonyl-2-methylthiopyrimidine and trimethylamine gave 5-ethoxycarbonyl-2-methylthiopyrimidin-4-yl trimethylammonium chloride (96%) which failed to give the corresponding 4-cyano derivative by heating with potassium cyanide in acetamide or in methanol; instead 4-dimethylamino-5-ethoxycarbonyl- (5%) and 4-methoxy-5-methoxycarbonyl-2-methylthiopyrimidine* (79%), respectively, were isolated.[2599] In contrast, trimethyl 4-methyl-6(or 2)-phenylpyrimidin-2(or 6)-ylammonium chloride (prepared in the usual way in 78% and 52% yield, respectively) did react with potassium cyanide in acetamide to give 2(or 4)-cyano-4(or 6)-methyl-6(or 2)-phenylpyrimidine (54%, 68%).[2603] 2,4-Dimethoxypyrimidin-6-yl trimethylammonium chloride (**143**, R = N^+Me_3; *H* 345) reacted with aq. sodium azide and with potassium

* Note an unexpected change in the ester grouping.

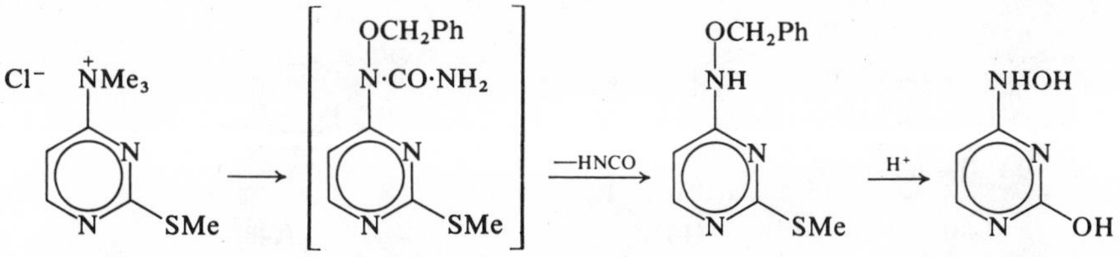

fluoride in glycol to give 4-azido- (**143**, R = N_3; 76%) and 4-fluoro-2,6-dimethoxypyrimidine (**143**, R = F; 42%), respectively;[2611] also with sodio sulphanilamide to give 2,4-dimethoxy-6-sulphanilylamino-pyrimidine (**143**, R = $(p)H_2NC_6H_4SO_2NH$; 43%).[2682] 2-Chloro-4,6-diphenylpyrimidine gave trimethyl 4,6-diphenylpyrimidin-2-ylammonium chloride (>85%) which in turn gave 2-cyano-4,6-diphenylpyrimidine (96%).[3354] Similar trimethylated amines have been used to make sulphonamides conveniently,[2516, 3353] and a pyridinium substituent has been used as a leaving group:[2718] e.g., 2-amino-4-chloro-6-hydroxy-5-nitropyrimidine condensed with pyridine in dimethylformamide to give 2-amino-4-hydroxy-5-nitropyrimidin-6-ylpyridinium chloride (**144**) which gave 2-amino-4,6-dihydroxy(or 4-hydroxy-6-methoxy)-5-nitropyrimidine (**145**, R = H or Me) by warming in water or methanol, respectively;[2718] the preparation and use of other such pyridinium derivatives are described.[2718]

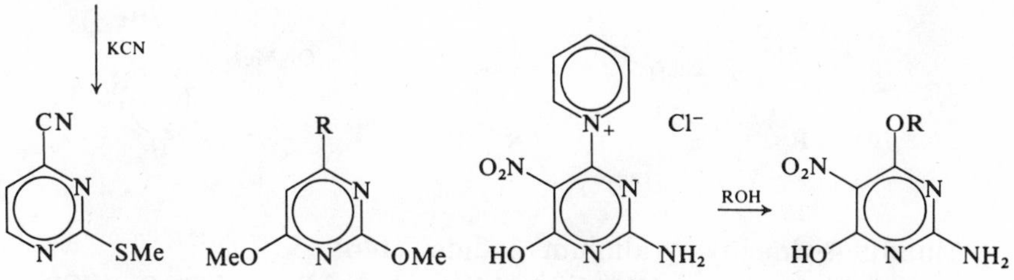

ω-Quaternary-amino substituents are also effective as leaving groups. The formation and reactions of these ω-ammoniopyrimidines are seen in the following sequences: 5-chloromethyl-2,4-dihydroxypyrimidine and trimethylamine gave 2,4-dihydroxypyrimidin-5-ylmethyl trimethylammonium chloride (*ca.* 75%);[2640] 4-amino-5-bromomethyl-2-trifluoromethylpyrimidine and ethanolic trimethylamine gave 4-amino-2-trifluoromethylpyrimidin-5-ylmethyl trimethylammonium bromide (82%) which, with sodium sulphanilamide in acetamide, gave a small yield of 4-amino-5-sulphanilylaminomethyl-2-trifluoromethylpyrimidine;* [2516] and 4-methylpyrimidine with iodine and pyridine gave

* This is much better made from its 5-*p*-nitrobenzenesulphonamidomethyl analogue.[2516]

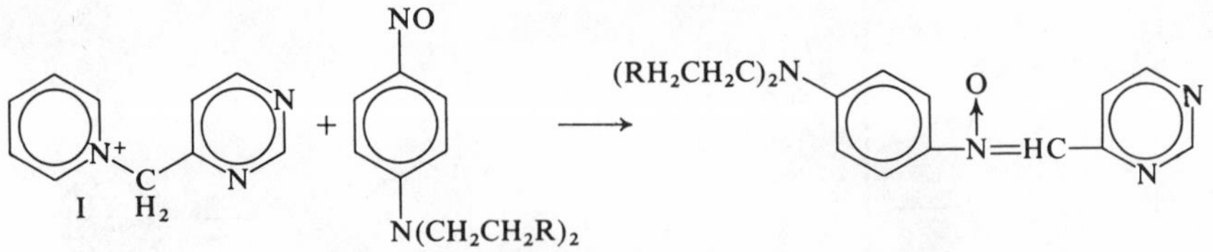

(146) (147) (148)

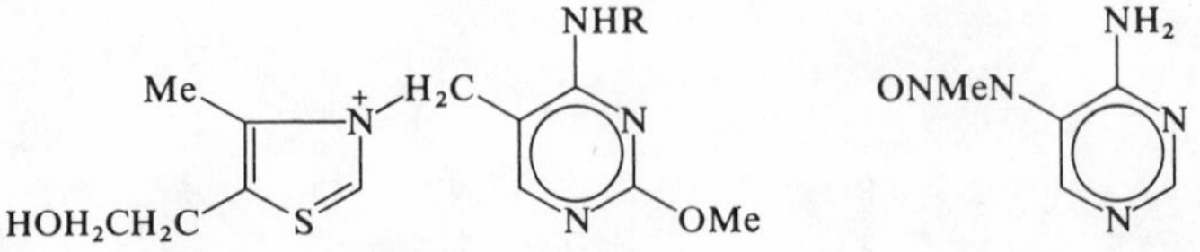

(149) (150)

pyrimidin-4-ylmethylpyridinium iodide (**146**; 58%),[2834, 3009] which reacted with *p*-nitroso-*NN*-bis-(β-chloroethyl)aniline (**147**, R = Cl) or its β-hydroxyethyl analogue (**147**, R = OH) to give the nitrones (**148**, R = Cl or OH).[2834] Thiamine* (Vitamin B_1; **149**, R = H) is a member of this group of compounds. Its 4-methylamino homologue (**149**, R = Me) was made by quaternizing 5-β-hydroxyethyl-4-methylthiazole with 5-chloromethyl-2-methyl-4-methylaminopyrimidine.[2654]

H. Nitrosoamines (*New*)

Nitrosoamines derived from alkylaminopyrimidines are exemplified in 4-amino-5-methylnitrosoaminopyrimidine (**150**) formed from 4-amino-5-methylaminopyrimidine and ethanolic isoamyl nitrite;[2491] it failed to undergo hydrogenation to 4-amino-5-α-methylhydrazino-pyrimidine.[1391]

9. Some Naturally Occurring Aminopyrimidines (*H* 345)

A. Cytosine: 4-Amino-2-hydroxypyrimidine (*H* 346)

The best synthetic route to cytosine, involving the sequence (**151** → **152** → **153** → **156**), was reported by R. S. Karlinskaya and N. V.

* Much of the recent work on thiamine, its homologues, and related compounds with Vitamin B_1 activity has been done in Japan. T. Matsukawa and S. Yurugi have contributed an excellent review of this work in English.[3355] An interesting synthesis of labelled thiamine has been reported.[3356]

Khromov-Borisov in 1957;[2766] it has been discussed (Ch. VI, Sect. 2.D) and as a by-product yielded isocytosine (**157**) *via* its *O*-methyl derivative (**154**).[2766] Another good route (**158** → **159** → **160** → **161** → **156**) has been modified recently.[2915] A third route, of particular use for making isotopically labelled cytosine,[184] has been modified to improve the yield of the intermediate (**155**) and of cytosine (**156**) resulting from its condensation with urea.[2272] A fourth route involving desulphurization by Raney nickel of 4-amino-2-hydroxy-6-mercapto(or methylthio)-pyrimidine should not be overlooked.[2777] For a fifth route[3465] see Ch. X, Sect. 1.A(4.a).

The polarographic behaviour of cytosine has been studied;[3357, 3358] its fine structure has been discussed in Ch. VII, Sect. 5. Cytosine appears to be photochemically hydrated or otherwise affected by irradiation but the subject is still rather obscure; available information has been summarized.[3088, 3119, 3146, 3361] Some SCF–MO calculations have been made[3359] for cytosine and its basic strength has been discussed in theoretical terms;[3360] so too has its tautomerism.[3466]

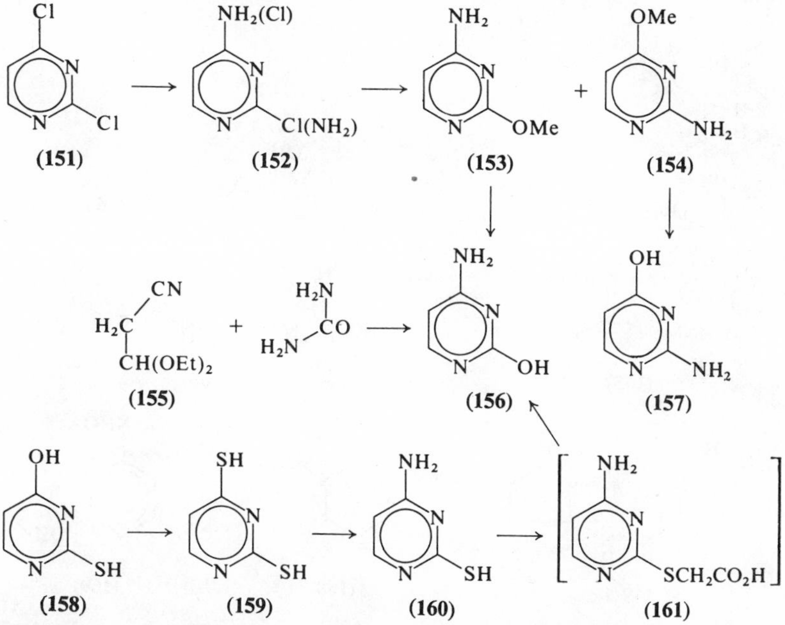

The degradation of cytosine by permanganate oxidation has been studied:[2337] the ammonia, oxalate, formate, urea, and biuret formed were explained by postulating an initial 'cytosine glycol' (**162**, R = NH_2) which was subsequently degraded directly to biuret and indirectly *via* (**162**, R = OH) to urea, etc.[2337] The action of hydroxylamine on cytosine to give 2-hydroxy-4-hydroxyaminopyrimidine has been discussed above (Sect. 1.H). Hydrazine also reacts in a rather complicated way with cytosine:[3204–3206] thus hydrazine hydrate at 80° gave 4-hydrazino-2-hydroxypyrimidine (**163**, R = H), 3-aminopyrazole (**164**), and urea; at 90° it gave the pyrazole (**164**) and 2-hdryoxy-4-β-2′-hydroxypyrimidin-4′-ylhydrazinopyrimidine (**166**); and aq. hydrazine as mono-cation (pH 6) at 80° gave the pyrimidine (**163**, R = H) and 3-β-pyrazol-3′-ylhydrazinopyrazole (**167**).[3204, 3206] In contrast, aq. methylhydrazine mono-cation at 80° gave only 2-hydroxy-4-α-methylhydrazinopyrimidine (**163**, R = Me) and aq. *NN*′-dimethylhydrazine under similar conditions only 4-αβ-dimethylhydrazino-2-hydroxypyrimidine (**165**).[3206] The mechanisms of these transformations have been discussed.[3204, 3206] Cytosine was first oxidized to its 3-*N*-oxide (**168**) by

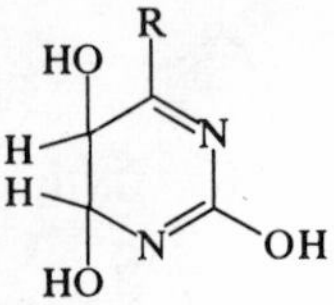

(**162**) 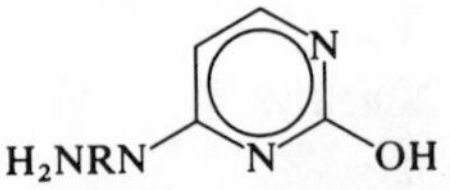 (**163**) 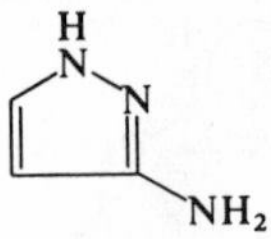 (**164**)

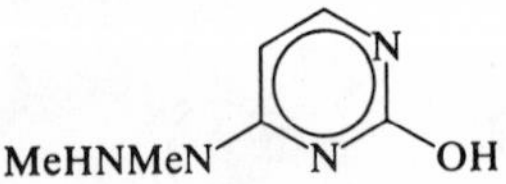

(**165**) 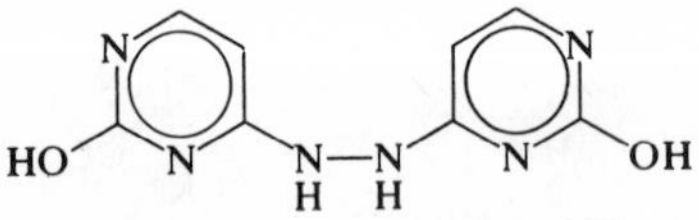 (**166**)

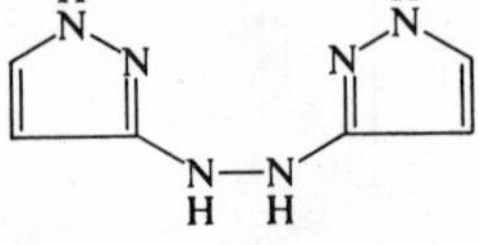

(**167**) 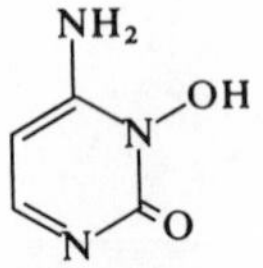 (**168**) 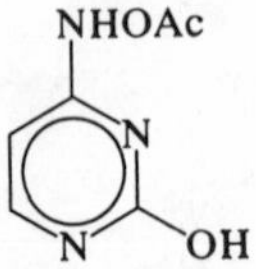 (**169**)

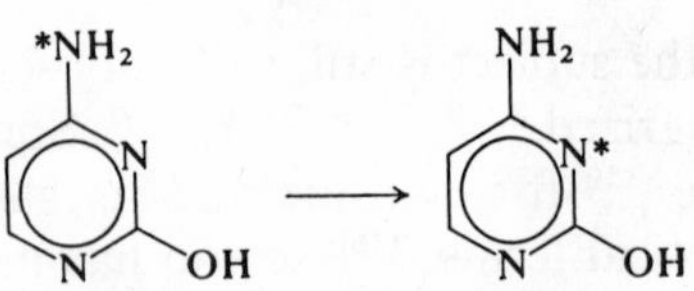

(**170**) → (**171**)

perphthalic acid[3362] but *m*-chloroperbenzoic acid proved a better reagent on a preparative scale.[3215] The structure was proven by unambiguous syntheses of cytosine-1(and 3)-*N*-oxides.[2262] The 3-*N*-oxide was stable to acid and alkali but rearranged during acetylation in acetic anhydride/acetic acid to give 4-acetoxyamino-2-hydroxypyrimidine (**169**).[3215] This rearrangement was akin to the transformation under comparable conditions of 2-hydroxy-4-methylaminopyrimidine into 4-amino-2,3-dihydro-3-methyl-2-oxopyrimidine,[2984] [NH_2-^{15}N]-cytosine (**170**) into [3-^{15}N]cytosine (**171**),* [3227] and such like.[3363] Alkylation of cytosine is discussed in Ch. X, Sect. 1.A(3.e).

D. Lathyrine or Tingitanin: 2-Amino-4-β-amino-β-carboxyethylpyrimidine (*New*)

Lathyrine was isolated from the seeds of the Tangier pea (*Lathyrus tingitanus*) in 1961.[3364, 3365] Its structure (**175**, R = NH_2) rested at first on spectral data,[3366] but later it was confirmed by oxidation to 2-amino-4-carboxypyrimidine,[2955] and by four distinct syntheses: the first (as judged by the submission date of the paper) involved the sequence (**172**, R = NAc_2) → (**173**, R = NH_2) → (**174**, R = NH_2) → (**175**, R = NH_2);[2955] the second and third syntheses were alike, involving the sequences (**172**, R = OMe or SMe) → (**173**, R = OMe or SMe) → (**174**, R = OMe or SMe) → (**175**, R = OMe or SMe) → (**175**, R = NH_2);[2832] and in the fourth synthesis 2-amino-4-methylpyrimidine (**172**, R = NH_2) was converted by several steps into its chloro derivative (**176**) which was condensed with ethyl α-acetamido-α-cyanoacetate to give the pyrimidine (**177**), converted into the product (**175**, R = NH_2) by treatment with acid.[3004]

E. Other Aminopyrimidines Derived from Natural Sources (*New*)

Acidic hydrolysis of thiamine (Vitamin B_1) yields *toxopyrimidine* (pyramin: 4-amino-5-hydroxymethyl-2-methylpyrimidine; **178**, R = H),[3367] which produces death by convulsions in rats and mice.[3368] It and its analogues, e.g., the 4-methylamino homologue (**178**, R = Me)[2817] or 4-amino-2-ethyl-5-hydroxymethylpyrimidine,[2270] have many and varied biological effects such as a Vitamin B_6 antagonism in man. The subject has been reviewed well.[2553]

Among the pyrimidine antibiotics, *bacimethrin*, 4-amino-5-hydroxymethyl-2-methoxypyrimidine (**179**), is the simplest.[3370–3372] It was isolated from *Bacillus megatherium* and is active against a variety of

* The equilibrium mixture contained about 50% of each cytosine, present before acidic hydrolysis as its acetyl derivative.[3227]

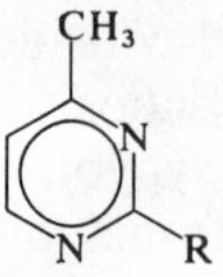

CH_3, N, N, R **(172)** — CH_2COCO_2Et, N, N, R **(173)** — $CH_2C(:NOH)CO_2Et$, N, N, R **(174)** — $CH_2CH(NH_2)CO_2H$, N, N, R **(175)**

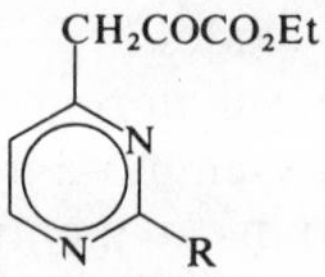

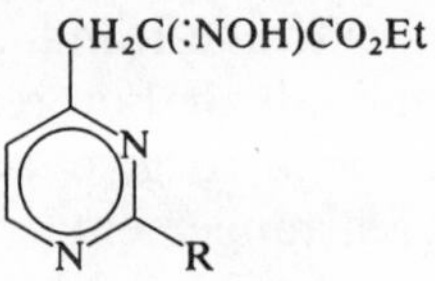

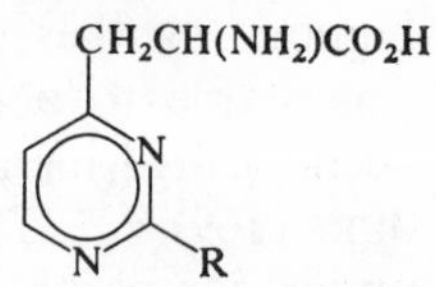

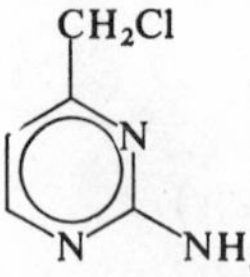

CH_2Cl, N, N, NH_2 **(176)** — CN, CH_2C—CO_2Et, NHAc, N, N, NH_2 **(177)** — NHR, HOH_2C, N, N, Me **(178)** — NH_2, HOH_2C, N, N, OMe **(179)**

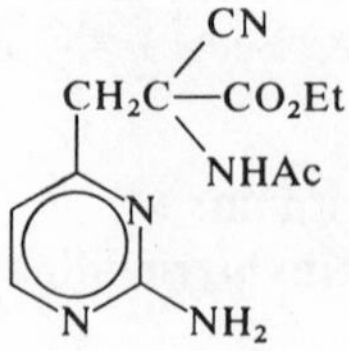

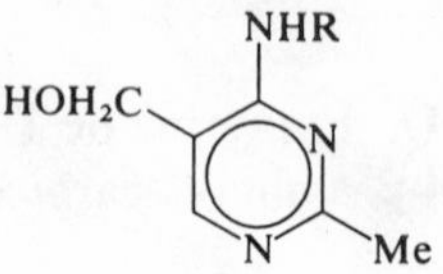

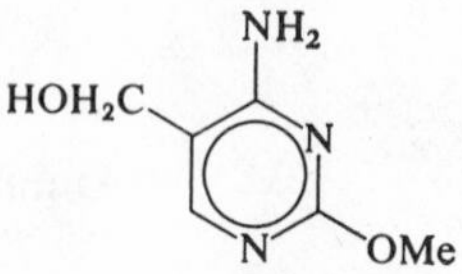

organisms.[2553] It has been synthesized from 4-amino-5-ethoxycarbonyl-2-ethylthiopyrimidine by treatment with sodium methoxide followed by reduction of the ester grouping.[2205, 3369] Other pyrimidine antibiotics are rather complicated and include *amicetin*;[3373] *gougerotin*, the postulated structure[3374] of which has now been revised;[2553, 3375] *blasticidin S*, an antiphytopathogenic fungal agent of confirmed structure;[3376–3378] and *albomycin*, which has 3-methylcytosine as part of its structure.[2998]

CHAPTER X

The *N*-Alkylated Pyrimidines and the Pyrimidine-*N*-Oxides (*H* 356)

The advance of knowledge within the topics covered in this chapter has been very uneven in recent years. Thus the only important contributions to *N*-alkylated oxopyrimidines have been studies of their formation from alkoxypyrimidines in the presence of alkyl halides (Hilbert–Johnson conditions) or by simple thermal rearrangement; and the start[2762] of a long-awaited[3379] systematic study of the alkylation of hydroxypyrimidines. Many 1(or 3)-alkylated iminopyrimidines have been made in connexion with a study[2855] of their Dimroth rearrangement into alkylaminopyrimidines. A few rather important pyrimidine-*N*-oxides have been made, in particular by W. Klötzer (see below).

1. The Oxopyrimidines (*H* 357)

Although the oxo group in pyrimidines is nearly always associated with an *N*-alkyl group, several examples, in which this has been replaced by an *N*-acyl or related group, are included below.

A. Preparation of Oxopyrimidines (*H* 357)

The formation of oxopyrimidines by primary syntheses (*H* 357, 358) has been covered in Chs. II and III. Typical examples are the condensation of *N*-methylurea with 1,1,3,3-tetraethoxy-2-methylpropane to give 1,2-dihydro-1,5-dimethyl-2-oxopyrimidine (**1**),[2630] and the cyclization of *N*-methyl-*N*′-β-cyano-β-ethoxycarbonylvinylurea to 4-amino-5-ethoxycarbonyl-2,3-dihydro-3-methyl-2-oxopyrimidine (**2**).[2987]

(3) *By Alkylation of Hydroxypyrimidines* (*H* 359)

The formation of alkoxypyrimidines (generally as by-products) during *N*-alkylation of hydroxypyrimidines has been discussed (Ch. VII, Sects. 4 and 6.C).

(a) *Cases with One Hydroxy Group* (*H* 359). The potassium salt of 2-hydroxypyrimidine and benzyl bromide or isopropyl iodide in dimethylformamide at 60° gave 1-benzyl-1,2-dihydro- (53%) or 1,2-dihydro-1-isopropyl-2-oxopyrimidine (26%), respectively;[2762] use of ethanolic alkali as a medium gave a lower yield of the benzyl derivative[2626] and rather poor yields of 1-ethyl-1,2-dihydro-2-oxo- and 1,2-dihydro-2-oxo-1-propyl-pyrimidine.* [2511] The prolonged action of an excess of methyl iodide on the sodium salt of 2-hydroxypyrimidine gave 1,2-dihydro-1,3-dimethyl-2-oxopyrimidinium iodide (**3**; 42%).[2762]

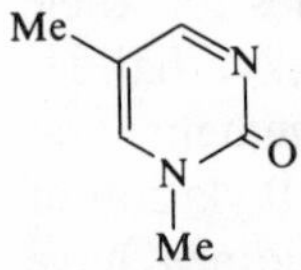

(1) 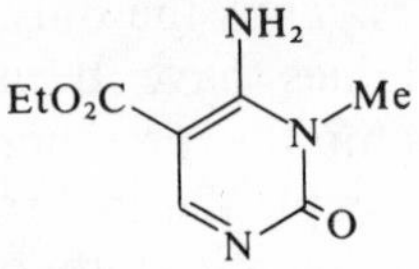 (2) 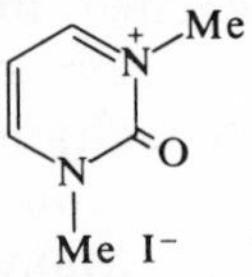 (3) 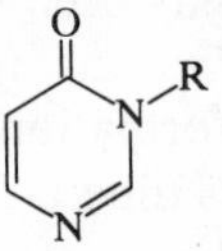 (4)

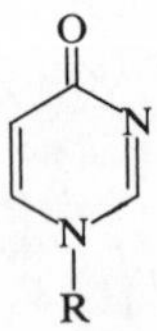

(5) 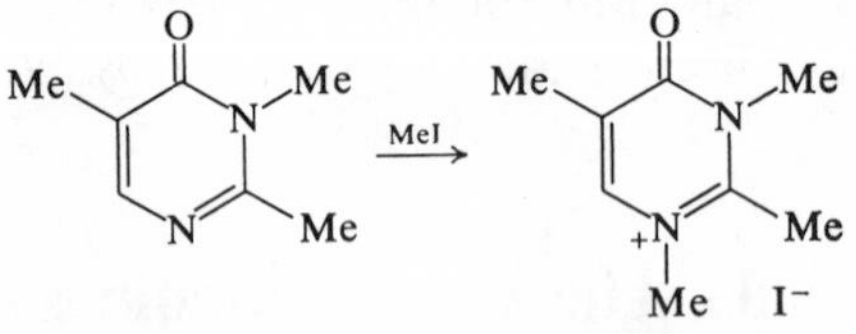 (6) (7)

2-Hydroxy-4-methylpyrimidine with methyl iodide/sodium methoxide gave only 1,2-dihydro-1,4-dimethyl-2-oxopyrimidine, of established structure.[2746] 2-Hydroxy-4,6-dimethylpyrimidine reacted with β-chloroethanol/alkali to give 1,2-dihydro-1-β-hydroxyethyl-4,6-dimethyl-2-oxopyrimidine (57%);[3010] the 1-β-hydroxypropyl homologue (38%) was made similarly.[3010]

4-Hydroxypyrimidine underwent methylation by methyl iodide in hot ethanolic alkali to give 1,6-dihydro-1-methyl-6-oxopyrimidine (**4**, R = Me; 25%) and the 1,4-dihydro-1-methyl-4-oxo isomer† (**5**,

* These and related compounds are probably best made by the Principal Synthesis.[2511]

† A claim[2512] to have prepared the isomer (**5**, R = Me) in 47% yield using dimethyl sulphate/methanolic sodium methoxide appears to be in error: the melting points of the product and its picrate[2512] correspond exactly to those of 1,6-dihydro-1-methyl-6-oxopyrimidine (**4**, R = Me) and its picrate.[273, 3187]

R = Me; 4%);[3187] ethyl iodide similarly gave a mixture of 1-ethyl-1,4(and 1,6)-dihydro-4(and 6)-oxopyrimidine (**5** and **4**, R = Et) with the latter predominating, but benzyl chloride gave a 1:1 mixture of 1-benzyl-1,4(and 1,6)-dihydro-4(and 6)-oxopyrimidine (**5** and **4**, R = CH_2Ph).[3187] Diazomethane and 4-hydroxy-2,5-dimethylpyrimidine gave 1,6-dihydro-1,2,5-trimethyl-6-oxopyrimidine (**6**; 58%) as the major product, as judged by the p.m.r. spectrum; this was quaternized by prolonged treatment with methyl iodide to give 1,6-dihydro-1,2,3,5-tetramethyl-6-oxopyrimidinium iodide (**7**; 80%).[2818] The structures of 1,6-dihydro-1,4-dimethyl-6-oxopyrimidine,[64] and its 1,5-dimethyl isomer, prepared by methylation of 4-hydroxy-6(or 5)-methylpyrimidine,[64, 2630] have been confirmed by p.m.r. spectra and other means.[2630] The mercury salts of 2- and 4-hydroxypyrimidine were allowed to react with appropriate sugar halides to give 1,2-dihydro-2-oxo-1-β-D-ribofuranosyl-, 1-β-D-glucopyranosyl-1,6-dihydro-6-oxo-, and 1,6-dihydro-6-oxo-1-β-D-ribofuranosyl-pyrimidine.[3380] 4-Benzyloxy-6-hydroxy-5-methylpyrimidine and diazomethane gave the methoxy derivative (30%) and 1,6-dihydro-6-oxo derivative (58%).[3767]

(b) *Cases with Two Hydroxy Groups* (*H* 360). Uracil and 2,4-dihydroxy-6-methylpyrimidine have been satisfactorily mono-*N*-methylated for the first time by boiling their salts with methyl iodide in a solvent of low polarity (toluene or dioxane): the yields of 1-methyl-(**8**, R = H) and 1,6-dimethyl-uracil (**8**, R = Me) were 40–50%.[3381] The more common dimethylation in polar solvents was illustrated in the formation of 5-ethyl-1,2,3,4-tetrahydro-1,3-dimethyl-2,4-dioxo-6-propylpyrimidine,[3766] 1,2,3,4-tetrahydro-1,3,5-trimethyl-2,4-dioxopyrimidine (**9**; 91%),[3076] and 1,2,3,4-tetrahydro-5-2′-hydroxycyclopentyl(or hydroxymethyl)1,3-dimethyl-2,4-dioxopyrimidine.[2952, 3762] Providing the alkyl residue is large enough (?), 1-monoalkylation of uracils appears to be feasible, e.g., by using an alkyl halide in dimethyl sulphoxide containing potassium carbonate: this technique gave 1-*p*-nitrobenzyl- (41%), 1-but-3′-enyl- (37%), 1-δ-methoxybutyl- (49%), 1-ethoxycarbonylmethyl- (**10**; 35%), and other similar derivatives of uracil.[2547, 3382] Another procedure is exemplified in the preparation of 1-dimethylaminoethyluracil (40%) by boiling uracil in dioxan with sodium hydride and then adding dimethylaminoethyl chloride; other such derivatives were made similarly.[3191] 1-Ethoxy(and methoxy)-carbonyluracil were made by boiling uracil in benzene containing the alkyl chloroformate;[3346] 1-(ethylthio)carbonyluracil from uracil and *S*-ethyl chlorothioformate under Schotten–Baumann conditions;[3347] 1-amino-1,2,3,6-tetrahydro-3-methyl-2,6-dioxopyrimidine from 3-aminouracil with diazomethane;[3261] 1-benzylideneamino-1,2,3,4-tetrahydro-3-methyl-2,4-dioxopyrimidine from the sodium salt of 1-benzylideneaminouracil and methyl iodide;[3260] and alkylations of more complicated systems are also represented.[3383–3386] *N*-Alkylation *via* *O*-trimethylsilyl derivatives[3076] has been discussed in Ch. VII, Sect. 6.C.

1-Acetyl-5-benzyloxymethyluracil (**11**) was prepared 'in a high yield' by acetylation with acetic anhydride at 140°,[3014] and 1,3-dibenzoyl-1,2,3,4-tetrahydro-2,4-dioxopyrimidine (**12**, R = Bz) was formed on treatment of uracil with benzoyl chloride in anhydrous dioxane and pyridine, but even column chromatography in ethanolic chloroform removed the 1-acyl group to give 3-benzoyluracil* (**12**, R = H; 92%).[3170]

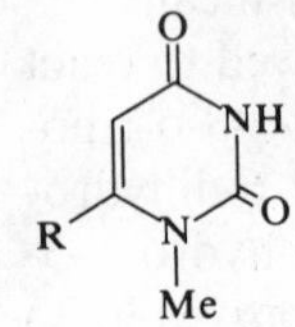

(**8**) R; NH; O; N–Me

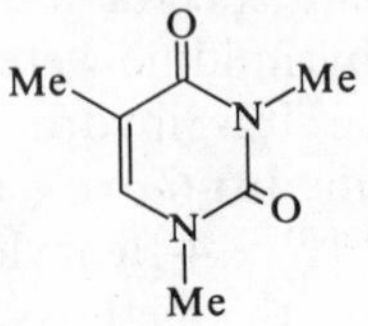

(**9**) Me; N–Me; O; N–Me

(**10**) NH; O; N–CH_2CO_2Et

(**11**) PhH_2COH_2C; NH; O; N–Ac

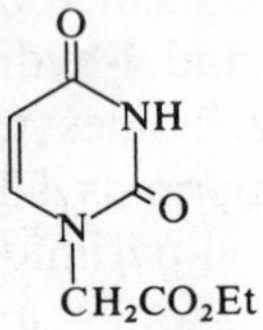

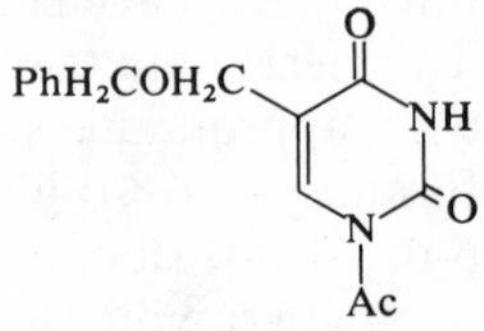

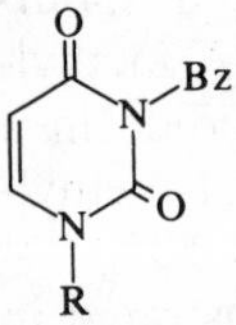

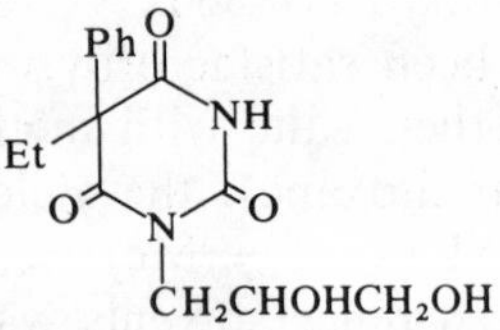

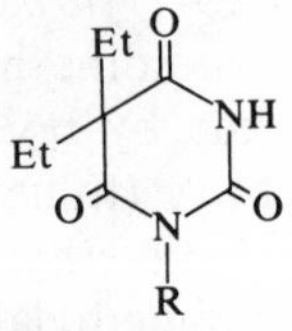

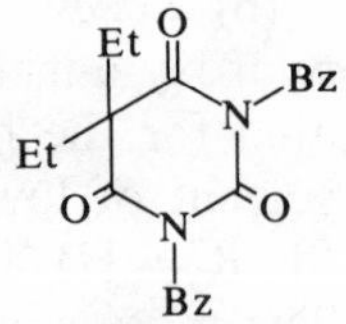

(**12**) N–Bz; O; N–R

(**13**) Ph; Et; NH; O; N–$CH_2CHOHCH_2OH$

(**14**) Et; Et; NH; O; N–R

(**15**) Et; Et; N–Bz; O; N–Bz

Treatment of the sodium salt of uracil with ethylene carbonate in dimethylformamide gave a mixture of 1-β-hydroxyethyluracil and 1,3-bis-β-hydroxyethyluracil. These were separable directly by chromatography;[3770, 3771] acetylation of the mixture gave the corresponding acetoxyethyl derivatives which were separated by conventional procedures.[3770] The hydroxyethyl derivatives were converted by thionyl chloride into 1-β-chloroethyl- and 1,3-bis-β-chloroethyl-uracil, respectively;[3770, 3771] the former lost hydrogen chloride to potassium t-butoxide in dimethyl sulphoxide giving 1-vinyluracil (60%),[3770] which was also made (10%) by pyrolysis of 1-acetoxyethyluracil.[3770] Direct vinylation of uracil by a mixture of sulphuric acid, vinyl acetate, and mercuric acetate failed to give the same product, but similar treatment of 3-methyluracil (made by an improved method[3770]) or 4-ethoxy-2-hydroxypyrimidine furnished the respective 1-vinyl derivatives satisfactorily.[3770]

* The orientation was proven by methylation and subsequent deacylation to 1-methyluracil (**8**, R = H).[3170]

(c) *Cases with Three Hydroxy Groups* (*H* 362). Alkylation of barbiturates is represented in the formation of 1-$\beta\gamma$-dihydroxypropyl-5-ethyl-5-phenylbarbituric acid (**13**), 1-γ-butoxy-β-carbamoylpropyl-5-ethyl-5-phenylbarbituric acid, and such like.[3387]

The acylation of 5,5-diethylbarbituric acid (**14**, R = H) and related pyrimidines has been studied by J. Bojarski and his colleagues in Kraków. By heating the silver salts of barbiturates with benzoyl chloride, *p*-nitrobenzoyl chloride, or *o*-chlorobenzoyl chloride in benzene or pyridine, mono- or di-acyl derivatives such as 1-benzoyl-(**14**, R = Bz) or 1,3-dibenzoyl-5,5-diethylbarbituric acid (**15**) and 5-ethyl-1-*p*-nitrobenzoyl- or 5-ethyl-1,3-bis-*p*-nitrobenzoyl-5-phenylbarbituric acid were formed according to conditions;[2448, 3388–3390, 3393, 3394] similar acetylations gave, e.g., 1,3-diacetyl-5,5-diethylbarbituric acid.[3391] The mechanism of such diacylations is not as simple as might be thought: experimentally based schemes have been suggested.[2448] The alkaline degradation of these acylated barbiturates to (acyclic) urea derivatives has been studied.[3392, 3395]

(d) *In the Presence of Nitro Groups* (*H* 363). As anticipated (*H* 363) the product from 2-hydroxy-5-nitropyrimidine and methyl iodide was not 2-methoxy-5-nitropyrimidine but 1,2-dihydro-1-methyl-5-nitro-2-oxopyrimidine (**16**).[2431, 3483] This was proven by obtaining the same nitropyrimidine (**16**) by two other routes: rearrangement of authentic 2-methoxy-5-nitropyrimidine[2431, 2630, 3483] and nitration of 1,2-dihydro-1-methyl-2-oxopyrimidine.[2431, 3483] Similarly, 2-hydroxy-5-nitropyrimidine, alcoholic potassium hydroxide, and ethyl iodide gave 1-ethyl-1,2-dihydro-5-nitro-2-oxopyrimidine (27%);[2762] in contrast, the use of isopropyl iodide with dimethylformamide as solvent gave a 1:1 mixture of 2-isopropoxy-5-nitro- and 1,2-dihydro-1-isopropyl-5-nitro-2-oxo-pyrimidine.[2762] Methylation of 2,4-dihydroxy-6-methyl-5-nitropyrimidine by dimethyl sulphate/alkali gave 1,2,3,4-tetrahydro-1,3,6-trimethyl-5-nitro-2,4-dioxopyrimidine.[2427]

(e) *In the Presence of Amino Groups* (*H* 364). Some earlier confusion (*H* 346, 364), about the products formed by methylation of cytosine (**18**, R = H) appears to be resolved. Use of methyl iodide/ethanolic alkali gave only one isolable product, 4-amino-1,2-dihydro-1-methyl-2-oxopyrimidine (1-methylcytosine; **19**, R = H; *ca.* 15%), the structure of which was proven by comparison with material made unambiguously, and confirmed by deamination with nitrous acid to 1-methyluracil.[2337, *cf.* 668, 669] In contrast, the use of dimethyl sulphate in dimethylformamide to methylate cytosine led to a mixture, separable with difficulty, of 1,2,3,4-tetrahydro-4-imino-1,3-dimethyl-2-oxopyrimidine and 4-amino-2,3-dihydro-3-methyl-2-oxopyrimidine (3-methylcytosine; **17**, R = H).[2168, 2641] The latter compound was also made by decarboxylation of its 5-carboxy derivative (**17**, R = CO_2H) *in vacuo*,[2987] although earlier attempts to do this at atmospheric pressure had given a rearranged product, 2-hydroxy-4-methylaminopyrimidine (**18**, R = Me);[274, 280] 3-methylcytosine was also made by methylation of cytidine followed by degradation with perchloric acid.[2168] Alkaline hydrolysis of 3-methylcytosine gave 3-methyluracil.[2168] 4-Amino-2-hydroxy-5-hydroxymethylpyrimidine underwent similar monomethylation to 4-amino-2,3-dihydro-5-hydroxymethyl-3-methyl-2-oxopyrimidine (**17**, R = CH_2OH), identified by analysis and the close resemblance of its ultra-violet spectra with those of 3-methylcytosine.[3396]

Alkylation of cytosine in the presence of base is further illustrated in the reaction of acrylonitrile with cytosine in aq. pyridine to give only 4-amino-1-β-cyanoethyl-1,2-dihydro-2-oxopyrimidine.[3363] In a similar way, the cytosine derivative, 4-amino-6-chloro-2-hydroxypyrimidine,

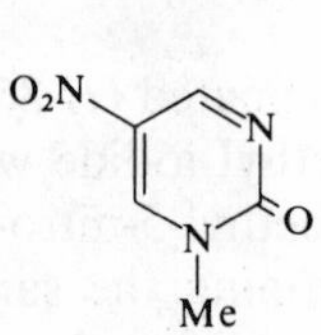

(16) 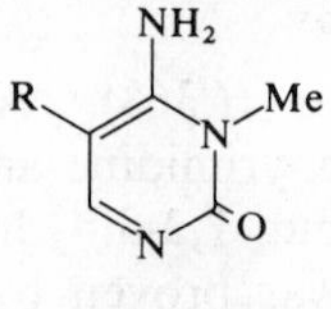 (17) 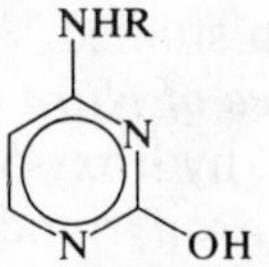 (18) 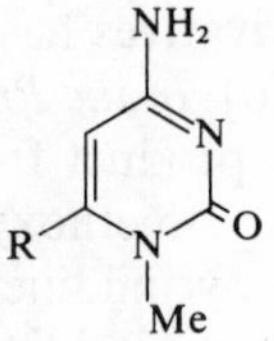 (19)

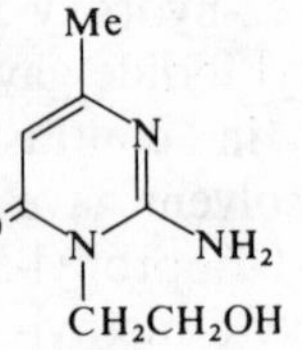

(20) 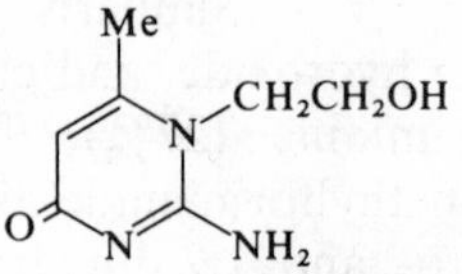 (21) 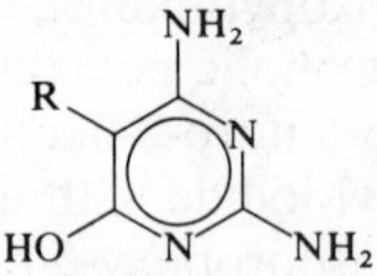 (22) 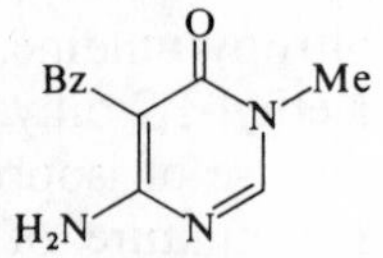 (23)

was methylated by dimethyl sulphate/alkali at the 1-position to give 4-amino-6-chloro-1,2-dihydro-1-methyl-2-oxopyrimidine (**19**, R = Cl), identified by alkaline hydrolysis to 6-amino-3-methyluracil (**19**, R = OH);[2475] 4,6-diamino-2-hydroxypyrimidine similarly gave 4,6-diamino-1,2-dihydro-1-methyl-2-oxopyrimidine (**19**, R = NH_2) without ambiguity.[2481] 4-Dimethylamino-2-hydroxypyrimidine gave a separable mixture of 4-dimethylamino-1,2(and 2,3)-dihydro-1(and 3)-methyl-2-oxopyrimidine.[2676] Benzylation of cytosine (**18**, R = H) by benzyl chloride/ethanolic alkali gave only 4-amino-1-benzyl-1,2-dihydro-2-oxopyrimidine (*ca.* 80%).[2337]

As with the methylation of isocytosine (*H* 365),[2288] β-hydroxyethylation of 2-amino-4-hydroxy-6-methylpyrimidine gave a separable *ca.* 1:1 mixture of two isomers: 2-amino-1,6(and 1,4)-dihydro-1-β-hydroxyethyl-4(and 6)-methyl-6(and 4)-oxopyrimidine (**20** and **21**), distinguished by comparison of their ultra-violet spectra with those of 1- and 3-methylisocytosine.[3010] Related compounds also prepared by methylation include 2(and 4)-amino-5-formamido-1,6-dihydro-1-methyl-4(and 2)-methylamino-6-oxopyrimidine,[2483, 3253] 2,4-diamino-5-formamido-1,6-dihydro-1-methyl-6-oxopyrimidine,[3253] and related compounds;[3253] exceptionally, 2,4-diamino-6-hydroxypyrimidine (**22**, R = H) and methyl iodide/alkali gave only its 5-methyl derivative (**22**, R = Me), proven in structure by an unambiguous synthesis.[2288] 2-Dimethylamino-4-hydroxypyrimidine with diazomethane or with methyl iodide/methanolic sodium methoxide gave a separable mixture of 2-dimethylamino-4-methoxypyrimidine and 2-dimethylamino-1,6-dihydro-1-methyl-6-oxopyrimidine in which the former predominated.[2676]

As with 4-amino-6-hydroxypyrimidine, methylation of 4-amino-5-benzoyl-6-hydroxypyrimidine occurred on the nitrogen adjacent to the hydroxy group to give 4-amino-5-benzoyl-1,6-dihydro-1-methyl-6-oxopyrimidine (**23**).[2389] Other methylations of amino-hydroxypyrimidines are described.[3397, 3398]

(f) *In the Presence of Alkylthio Groups* (*H* 368). Methylation of the sodium salt of 2-hydroxy-4-methylthiopyrimidine with methyl iodide in dimethylformamide gave mainly 1,2-dihydro-1-methyl-4-methylthio-2-oxopyrimidine, but higher (more bulky) alkyl halides gave more and more *O*-alkylation:[2762] thus ethyl iodide gave 81% of 1-ethyl-1,2-dihydro-4-methylthio-2-oxo- and only 14% of 2-ethoxy-4-methylthiopyrimidine (**25**, R = Et); propyl bromide gave 75% of 1,2-dihydro-4-methylthio-2-oxo-1-propyl- (**24**, R = Pr) and 18% of 4-methylthio-2-propoxy-pyrimidine (**25**, R = Pr); and isopropyl iodide gave only 40% of 1,2-dihydro-1-isopropyl-4-methylthio-2-oxo- (**24**, R = Pr^i) but 56% of 2-isopropoxy-4-methylthiopyrimidine (**25**, R = Pr^i).[2762] 3-Alkylated isomers, e.g. (**26**), were not formed.

The dimethylation[273] of 4-hydroxy-2-mercaptopyrimidine to give 1,4(and 1,6)-dihydro-1-methyl-2-methylthio-4(and 6)-oxopyrimidine has been repeated and a modified procedure for separation of the isomers has been outlined.[3187] Similarly 4-hydroxy-5-methoxy-2-methylthiopyrimidine gave about equal amounts of two monomethyl derivatives, 1,4(and 1,6)-dihydro-5-methoxy-1-methyl-2-methylthio-4-(and 6-)oxopyrimidine, distinguished by vigorous hydrolysis to 1- and 3-methylisobarbituric acid, respectively.[2198] In contrast, benzylation of 2-benzylthio-4-hydroxypyrimidine gave a single product, 1-benzyl-

2-benzylthio-1,4-dihydro-4-oxopyrimidine (**27**);*[3187] and from benzylation of 4-hydroxy-6-methyl-2-methylthiopyrimidine only 1-benzyl-1,6-dihydro-4-methyl-2-methylthio-6-oxopyrimidine (**28**)* was isolated.[2985] 1,6-Dihydro-5-methoxy-4-methoxymethyl-1-methyl-2-methylthio-6-oxopyrimidine has been made by methylation.[2198] The action of allyl bromide/ethanolic alkali on 4-hydroxy-2-methylthiopyrimidine gave 4-allyloxy-2-methylthio- (15%), 1-allyl-1,6-dihydro-2-methylthio-6-oxo- (26%), and 1-allyl-1,4-dihydro-2-methylthio-4-oxo-pyrimidine (8%);[2555] the same substrate with acrylonitrile in pyridine gave only 1-cyanoethyl-1,6-dihydro-2-methylthio-6-oxopyrimidine.[3363]

(g) *In the Presence of Other Groups* (*H* 369). Alkylations in this category are few: 5-bromo-2-hydroxypyrimidine and diazomethane gave 5-bromo-1,2-dihydro-1-methyl-2-oxopyrimidine (**29**, R = Me);[2746] 4-benzyl-5-bromo-2,6-dihydroxypyrimidine and *m*-nitrobenzyl chloride or chloroacetonitrile in dimethyl sulphoxide containing potassium carbonate gave 6-benzyl-5-bromo-3-*m*-nitrobenzyluracil (10%; separated from its isomer by thin layer chromatography) or 6-benzyl-5-bromo-3-cyanomethyluracil (9%; separated similarly from its isomer and dialkylated analogue), respectively;[2655] and 5-bromo-2-hydroxy- with ethyl iodide in ethanolic alkali gave 5-bromo-1-ethyl-1,2-dihydro-2-oxo-pyrimidine (**29**, R = Et; *ca.* 60%).[2630]

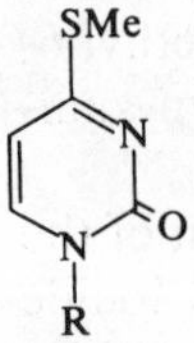

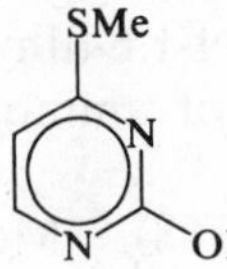

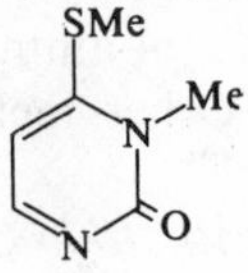

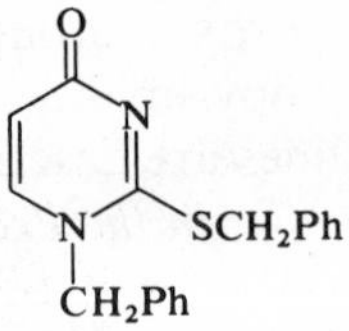

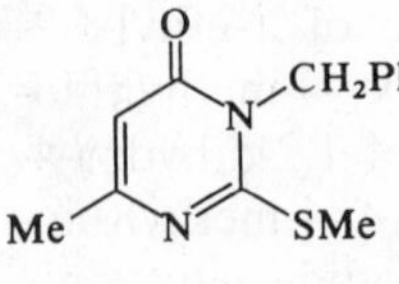

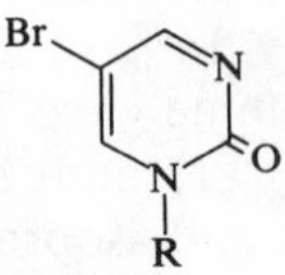

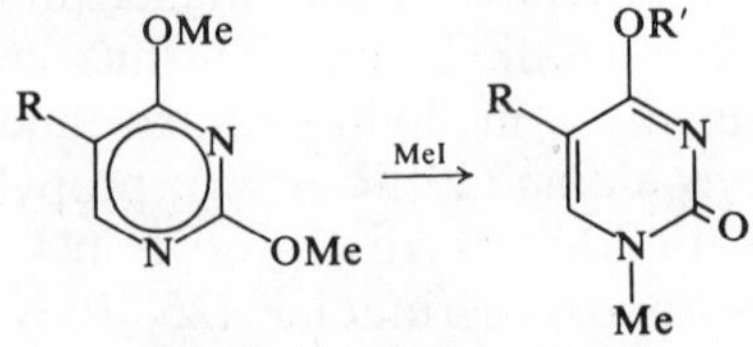

(4) *From Alkoxypyrimidines* (*H* 371)

The conversion of alkoxypyrimidines into *N*-alkyldihydro-oxopyrimidines may be brought about by an alkyl halide or through

* It will be observed that the benzylated products (**27** and **28**) are of opposite configurations. However, both structures appear to be quite firmly based, the first by de-

rearrangement by simple heating. The reactions proceed by different mechanisms, and although both were used first in the pyrimidine series (*H* 371) by G. E. Hilbert and T. B. Johnson, the term 'Hilbert–Johnson reaction' is now generally reserved for that involving an alkyl (or sugar) halide.

(a) *By the Hilbert–Johnson Reaction* (*H* 371). As defined above, this reaction is seen in the conversion of 5-bromo-2,4-dimethoxy- (**30**, R = Br) by methyl iodide into 5-bromo-1,2-dihydro-4-methoxy-1-methyl-2-oxo-pyrimidine (**31**, R = Br, R′ = Me), confirmed in structure by hydrolysis to 5-bromo-1-methyluracil (**31**, R = Br, R′ = H);[3046] of 5-fluoro-2,4-dimethoxy- (**30**, R = F) by methyl iodide at 65° for 18 hr. or 20° for 10 days, into 5-fluoro-1,2-dihydro-4-methoxy-1-methyl-2-oxo-pyrimidine (**31**, R = F, R′ = OMe) in 70–80% yield;[2572, 3226] of 2,4-diethoxy-5-fluoro- (**32**) into 4-ethoxy-5-fluoro-1,2-dihydro-1-methyl-2-oxo-pyrimidine (**33**; 77%), confirmed in structure by acidic hydrolysis to 5-fluoro-1-methyluracil (**31**, R = F, R′ = H);[2748] of 2,4-dibenzyloxy-, by benzyl chloride and methyl iodide at 20°, into 1-benzyl-4-benzyloxy-1,2-dihydro-2-oxo- (35%) and 4-benzyloxy-1,2-dihydro-1-methyl-2-oxo-pyrimidine (**31**, R = H, R′ = CH_2Ph; 65%), respectively;[2829] of 5-acetamido-2,4-dimethoxy- (**30**, R = NHAc) into 5-acetamido-1,2-dihydro-4-methoxy-1-methyl-2-oxo-pyrimidine (**31**, R = NHAc, R′ = Me);[2897] of 2,4-dimethoxy-5-nitro- (**30**, R = NO_2) into 13% of 1,2-dihydro-4-methoxy-1-methyl-5-nitro-2-oxo-pyrimidine (**31**, R = NO_2, R′ = Me);[2897] of 5-benzyloxymethyl-2,4-dimethoxy- into 5-benzyloxymethyl-1,2-dihydro-4-methoxy-1-methyl-2-oxo-pyrimidine;[2847] of the 2,4-dimethoxypyrimidines (**30**, R = H, Br, I, or Me) by 2-chlorotetrahydro-pyran (or furan) into 1,2-dihydro-4-methoxy-2-oxo-1-tetrahydropyran-2′-yl(or tetrahydro-2′-furyl)pyrimidine and their 5-bromo, iodo, and methyl derivatives, respectively;[2996, 3399] of 4-methoxy-6-methylaminopyrimidine (**34**, R = H) into 1,4-dihydro-6-methoxy-1-methyl-4-methyliminopyrimidine hydriodide (**35**, R = H) and thence by elimination of methyl iodide into 1,6-dihydro-1-methyl-4-methylamino-6-oxopyrimidine (**39**, R = H);[2781] of 4-dimethylamino-6-methoxypyrimidine (**34**, R = Me) into 4-dimethylamino-1,6-dihydro-1-methyl-6-oxopyrimidine (**39**, R = Me) without isolating the intermediate (**35**, R = Me);[2676] of 4,6-dimethoxypyrimidine into 1,6-dihydro-4-methoxy-1-methyl-6-oxopyrimidine* (**36**; 70%);[2997] of 4-dimethylamino-2-methoxypyrimidine into

sulphurization to 1-benzyl-1,4-dihydro-4-oxopyrimidine[3187] and the second by hydrolysis o 3-benzyl-6-methyluracil.[2985]

* A minor isomeric product, tentatively formulated[2997] as 1,4-dihydro-6-methoxy-1-methyl-4-oxopyrimidine, has been shown to be the betaine (**38**) or one of its dimers arising *via* the (isolated) methiodide (**37**).[3001, 3055, 3056, 3768] The whole position has been summarized in some detail.[3768]

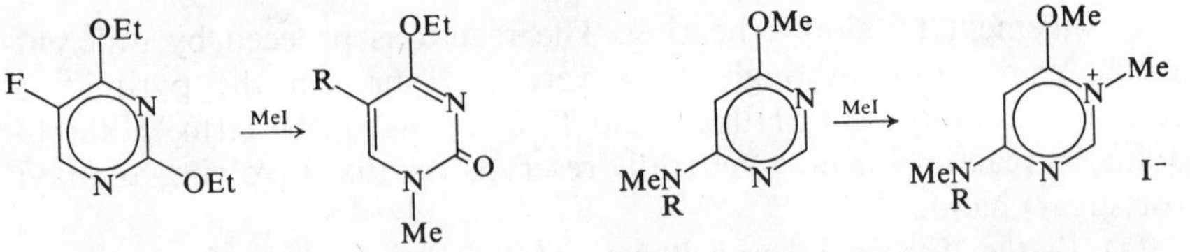

4-dimethylamino-1,2-dihydro-1-methyl-2-oxopyrimidine (**46**, X = O; *ca.* 80%);[2676] of 2-dimethylamino-4-methoxypyrimidine into 2-dimethylamino-1,6-dihydro-1,3-dimethyl-6-oxopyrimidinium iodide, the methiodide of the expected product;*[2676] of 2,4-diethoxypyrimidine by ethyl bromoacetate into 4-ethoxy-1-ethoxycarbonylmethyl-1,2-dihydro-2-oxopyrimidine, isolated as the derived 1-carboxymethyl analogue (80%);[3382] of 2,4-diallyloxypyrimidine by methyl (or allyl) iodide into 4-allyloxy-1,2-dihydro-1-methyl-2-oxopyrimidine (90%) or 1-allyl-4-allyloxy-1,2-dihydro-2-oxopyrimidine (96%), respectively;[2555] and similar cases.[3382, 3768] 5-Bromo-2,4-dimethoxypyrimidine reacted with sodium iodide in acetonylacetone at 100° to give 5-bromo-1,2,3,4-tetrahydro-1,3-dimethyl-2,4-dioxopyrimidine (75%); with acetic acid also present, the product was 5-bromo-1-methyluracil (60%).[2994]

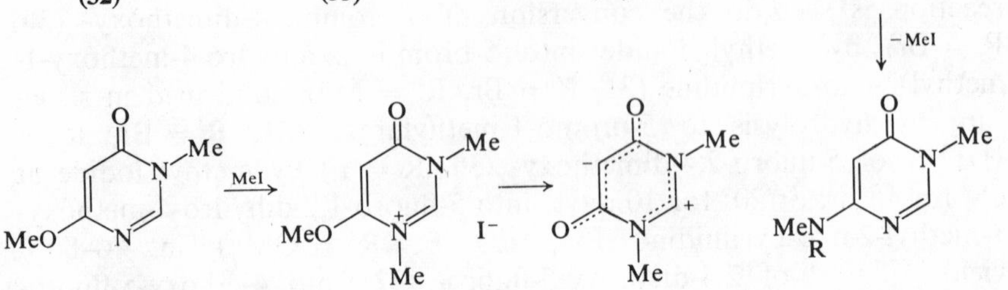

The reaction can also be brought about by an acyl halide: 2,4-dimethoxypyrimidine and acetyl chloride gave 1-acetyl-1,2-dihydro-4-methoxy-2-oxopyrimidine, which underwent ammonolysis and deacylation with ammonia to give 4-amino-2-hydroxypyrimidine; 4-benzylamino-2-hydroxy-5-methylpyrimidine and other analogues were made similarly.[3465]

The Hilbert–Johnson reaction is still used widely to make nucleosides, which are outside the scope of this monograph (*H* 256). However, the process has been reviewed recently and well by J. Pliml and M.

* This methiodide underwent two interesting reactions: on sublimation it lost methyl iodide to give 2-dimethylamino-1,6-dihydro-1-methyl-6-oxopyrimidine (*ca.* 80%); on dissolution in water it lost dimethylamine to give 1,2,3,4-tetrahydro-1,3-dimethyl-2,4-dioxopyrimidine (*ca.* 60%).[2676]

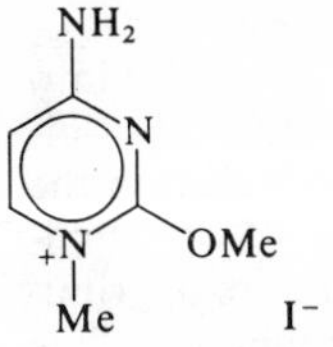

Prystaš,[3400] and optimum conditions or limitations may be confirmed easily from typical papers.[2572, 2829, 2847, 2897, 3001, 3226, 3401–3407, 3767]

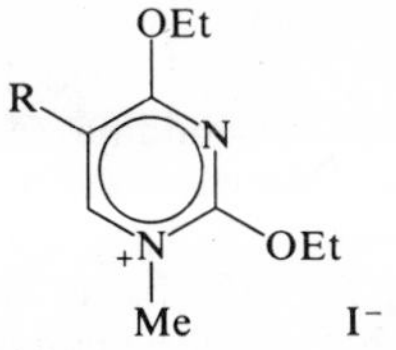

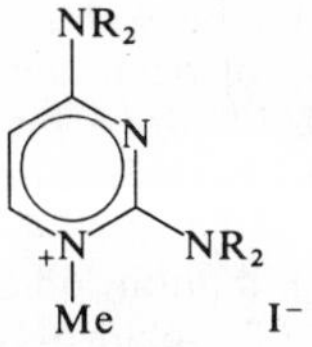

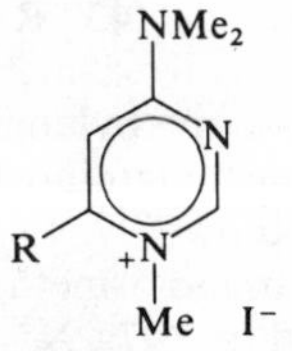

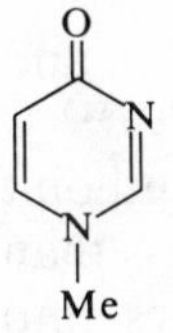

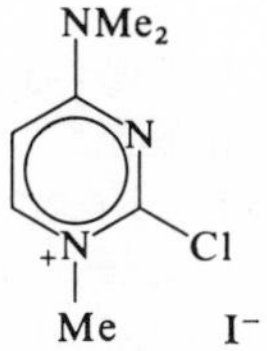

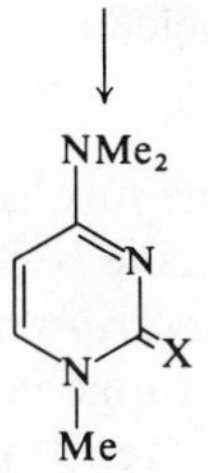

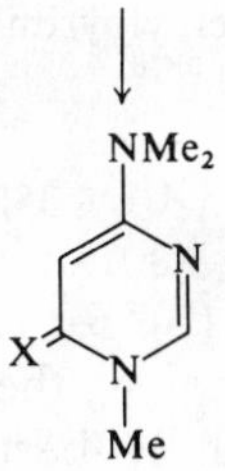

It is clear from the original work of G. Hilbert[668] and from the 'transalkylations' evident in some of the above examples and in each nucleoside synthesis, that a quaternary alkiodide is involved as an intermediate. Indeed, such compounds (**40**; **35**; and **41**, R = H or Me) have been isolated during the reaction with methyl iodide of 4-amino-2-methoxypyrimidine,[668] 4-methoxy-6-methylaminopyrimidine (**34**, R = H),[2781] 2,4-diethoxypyrimidine,[3408] and its 5-methyl derivative,[3408] respectively; each intermediate gave the expected product under appropriate conditions. The quaternary salt (**41**, R = H) showed the anticipated p.m.r. spectrum and it was confirmed in structure by its ready reaction with methanolic ammonia to give the known[2288] 2,4-diamino-1-methylpyrimidinium iodide (**42**, R = H), i.e., 2(4)-amino-1,4(1,2)-dihydro-4(2)-imino-1-methylpyrimidine hydriodide.[3408]

The usual leaving groups are activated in such quaternary salts. Thus 4-dimethylamino-1-methyl-6-methylthiopyrimidinium iodide (**43**, R = SMe) and ethanolic sodium hydrogen sulphide gave 4-dimethylamino-1,6-dihydro-1-methyl-6-thiopyrimidine (**47**, X = S);[2676] likewise 4-amino-2-methylthio-1-tri-*O*-benzoyl-D-ribosylpyrimidinium chloride and hydrogen sulphide in pyridine gave tri-*O*-benzoyl-2-thiocytidine and thence the known[3409] 2-thiocytidine.[3408] These reactions might be considered as *de facto* extensions of the Hilbert–Johnson reaction.

Other replacements are illustrated in the conversion of 2,4-bisdimethylamino-1-methylpyrimidinium iodide (**42**, R = Me) by sodium hydrogen sulphide into 4-dimethylamino-1,2-dihydro-1-methyl-2-thiopyrimidine (**46**, X = S); of the iodide (**43**, R = SMe) by sodium hydroxide at 25° into 4-dimethylamino-1,6-dihydro-1-methyl-6-oxopyrimidine (**47**, X = O); of the iodide (**42**, R = Me) into 4-dimethylamino-1,2-dihydro-1-methyl-2-oxopyrimidine (**46**, X = O); of 4-dimethylamino-1-methylpyrimidinium iodide (**43**, R = H) by warm alkali into the known[273] 1,4-dihydro-1-methyl-4-oxopyrimidine (**44**); of 4-chloro-6-dimethylamino-3-methylpyrimidinium iodide (**43**, R = Cl) by alkali into the oxopyrimidine (**47**, X = O); of 2-chloro-4-dimethylamino-1-methylpyrimidinium iodide (**45**) into the oxopyrimidine (**46**, X = O); and of 4,6-bisdimethylamino-1-methylpyrimidinium iodide (**43**, R = NMe_2) into the oxopyrimidine (**47**, X = O).[2676]

Other aspects of the mechanism of the Hilbert–Johnson reaction and of its modifications have been discussed in some detail.[3400, 3410, 3411]

(b) *By Thermal Rearrangement* (*H* 371). From a preparative point of view, thermal rearrangement is generally a less effective process than the Hilbert–Johnson reaction for converting alkoxypyrimidines into *N*-alkyldihydro-oxopyrimidines. For example, from the rearrangement of 2-methoxypyrimidine (**48**, R = H) in the presence of triethylamine at 160° for 4 hr., only 10% of pure 1,2-dihydro-1-methyl-2-oxopyrimidine (**49**, R = H) could be isolated, although the reaction was *ca.* 50% complete;[2511] 4-methoxy-5-nitropyrimidine gave only 16% of the purified 1,6-dihydro-1-methyl-5-nitro-6-oxopyrimidine after heating at 98° for a week;[2630] and an even lower yield of 1,6-di-

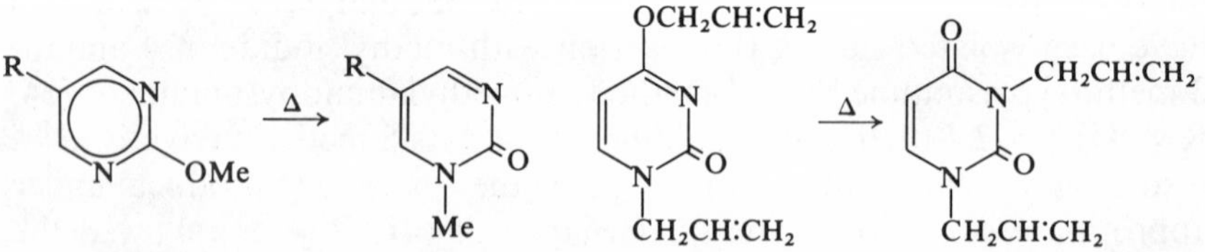

(48) (49) (50) (51)

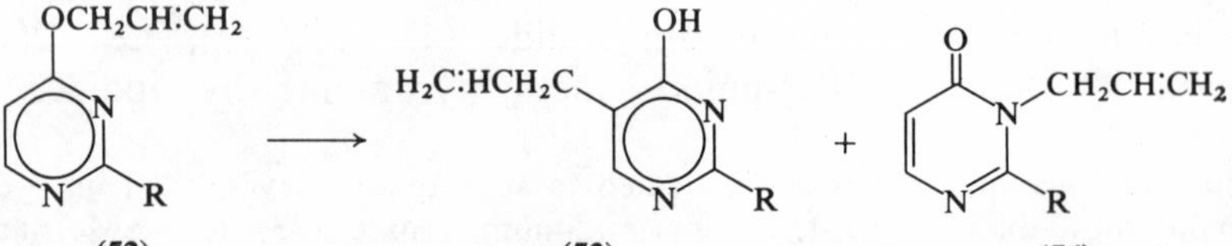

(52) (53) (54)

hydro-1,5-dimethyl-6-oxopyrimidine was obtained by heating 4-methoxy-5-methylpyrimidine.[2630] However, reasonable yields can be obtained sometimes as in the conversion of 1-allyl-4-allyloxy-1,2-dihydro-2-oxopyrimidine (**50**) into 1,3-diallyl-1,2,3,4-tetrahydro-2,4-dioxopyrimidine (**51**; 62%) by heating at 240° for 2 hr.,[2555] although in using such unsaturated groups there is a further complication: 4-allyl-oxypyrimidines often undergo *ortho*-Claisen rearrangement to give more of the 5-allyl-4-hydroxy isomer than of the *N*-alkyl-4-oxo isomer.[3500] Thus 4-allyloxy-2-phenylpyrimidine (**52**, R = Ph) on heating in the presence of *NN*-diethyl-*m*-toluidine at 250° for 16 hr. gave 36% of 5-allyl-4-hydroxy-2-phenylpyrimidine (**53**, R = Ph; isolated) and 14% of 1-allyl-1,6-dihydro-6-oxo-2-phenylpyrimidine (**54**, R = Ph; gas chromatographic and spectral identification);[2555] 4-allyloxy-2-methyl(or methylthio or benzylthio)pyrimidine (**52**, R = Me, MeS, or $PhCH_2S$) behaved similarly;[2194, 2206, 2555] and 4-allyloxy-2-aminopyrimidine gave at least 5-allyl-2-amino-4-hydroxypyrimidine.[2194] 2-Benzylthio-4-but-2′-enyloxy- and 4-but-2′-enyloxy-2-methylthio-pyrimidine underwent *ortho*-Claisen rearrangement with inversion of the alkenyl group to give 2-benzylthio-4-hydroxy- and 4-hydroxy-2-methylthio-5-α-methylallylpyrimidine, respectively, apparently unaccompanied by *N*-alkenyl isomers.[2206] A 'cross-over' rearrangement with a mixture of 4-but-2′-enyloxy-2-methylthio- and 4-allyloxy-2-benzylthio-pyrimidine confirmed the intramolecular nature of such *ortho*-Claisen rearrangements.[2206] 2-Substituted-4-allylamino- and 2-substituted-4-allylthio-pyrimidine did not undergo rearrangement.[2555]

Despite its limited preparative interest the thermal rearrangement of simple alkoxypyrimidines has been studied recently by rate measurements. The rearrangements of 2- and 4-methoxypyrimidine were followed spectrometrically and proved to be first-order reactions which were accelerated by tertiary bases whose efficiencies varied according to their basic strengths.[2511, 2697] Of the higher 2-alkoxypyrimidines, only the ethoxy-, isopropoxy-, and s-butoxy homologues showed any measurable rearrangement, and only in the presence of base and above 200°.[2511] The *C*-methyl derivatives rearranged more slowly than the parent alkoxypyrimidines, but the 5-bromo- and especially the 5-nitro derivatives did so much more quickly.[2630] For example, at 190° in the presence of a 5-molar proportion of triethylamine, the first-order rate constants (10^6k, sec^{-1}) were: 2-methoxy- (25), 2-methoxy-4-methyl- (3.4), 2-methoxy-5-methyl- (7.4), 5-bromo-2-methoxy- (200), and 2-methoxy-5-nitro-pyrimidine (50,000); 4-methoxy- (14), 4-methoxy-2-methyl- (3.0), 4-methoxy-6-methyl- (3.0), 4-methoxy-5-methyl- (7.1), 5-bromo-4-methoxy- (6,000), and 4-methoxy-5-nitro-pyrimidine (30,000).[2630] These rate constants have an evident qualitative relation to the electron-withdrawal or electron-release of the substituents, and this was placed on a more quantitative basis by rate measurements for the rearrangements of 2-methoxy-5-*p*-substituted-phenylpyrimidines: the plot of log *k versus* modified Hammetts *para-sigma* constants for *para*-substituted (NMe_2, Me, H, OMe, Cl, Br, and NO_2) phenyl groups was pleasingly linear.[3412]

A free-radical mechanism for thermal rearrangement was precluded by the very minor changes in rate following addition of benzoyl peroxide or benzoquinone.[2511] An intramolecular mechanism,[2511] suggested by the fact that alkyl-migration always occurred to the α- rather than the γ-nitrogen in pyrimidines (*cf.* the *ortho*-Claisen rearrangement), was also proven wrong:[3412] a 'cross-over' experiment using 4-methoxypyridine* and 2-ethoxy-5-nitropyrimidine gave all four possible rearranged products, thus showing that the mechanism was at least intermolecular,[3412] in line with a similar conclusion[3413] in the pyridine series.

(5) *By Hydrolysis of Iminopyrimidines* (*H* 373)

Most of the pyrimidine imines made recently bore an alkyl group on the adjacent ring nitrogen with a view to Dimroth rearrangement. However, during rearrangement in alkali, some hydrolysis to the corresponding oxopyrimidine generally occurred, and this became a useful reaction when the rearrangement was sluggish. Thus 4-dimethylamino-1,2-dihydro-2-imino-1-methylpyrimidine (**55**, X = NH) in warm alkali gave only a little 4-dimethylamino-2-methylaminopyrimidine (**56**) but much 4-dimethylamino-1,2-dihydro-1-methyl-2-oxopyrimidine (**55**, X = O);[2627] 1,2-dihydro-2-imino-1,4-dimethylpyrimidine in alkali at 25° gave 1,2-dihydro-1,4-dimethyl-2-oxopyrimidine (55%);[2602] 4-t-butyl-1,2-dihydro-2-imino-1-methylpyrimidine similarly gave 4-t-butyl-1,2-dihydro-1-methyl-2-oxopyrimidine (40%);[2602] and 4-dimethylamino-1,6-dihydro-1-methyl-6-methyliminopyrimidine in alkali gave the 6-oxo analogue.[2776]

(6) *From Thiopyrimidines* (*New*)

The conversion of a mercapto- into an hydroxy-pyrimidine is commonplace but the apparently similar conversion of a nontautomeric *N*-alkylated-thiopyrimidine into the corresponding oxopyrimidine is quite unusual, mainly because hydrolytic procedures are inapplicable. However, 4-ethyl-1,2,3,6-tetrahydro-5-methyl-2-oxo-1,3-diphenyl-6-thiopyrimidine (**57**, X = S) and hydrogen peroxide gave 4-ethyl-1,2,3,6-tetrahydro-5-methyl-2,4-dioxo-1,3-diphenylpyrimidine (**57**, X = O; 80%);[2339] 1,2,3,4-tetrahydro-2,4-dioxo-1,3,6-triphenylpyrimidine and

* 4-Methoxypyridine undergoes thermal rearrangement into 1,4-dihydro-1-methyl-4-oxopyridine quite easily.[3414] The rate is comparable with that for the rearrangement of 2-ethoxy-5-nitropyrimidine.[3412]

other analogues were made similarly;[2339] and 5-ethoxycarbonyl-1,4-dihydro-6-methyl-1,2-diphenyl-4-thiopyrimidine (**58**, X = S) was converted by mercuric acetate in acetic acid to the corresponding oxopyrimidine (**58**, X = O; 34%).[2313]

B. Reactions of Oxopyrimidines (*H* 375)

The conversion of oxo- into thiopyrimidines by phosphorus pentasulphide has been discussed in Ch. VII, Sect. 6.B. Oxopyrimidines have not been converted directly into the corresponding imino derivatives but a related conversion into a so-called 'pyrimidone hydrazone' has been done indirectly: 1,6-dihydro-1-methyl-6-oxopyrimidine* (**59**) and triethyloxonium fluoroborate (Et_3OBF_4) gave the quaternary salt (**60**) which was condensed with formylhydrazine; the crude product (**61**; R = CHO) was deacylated by acid to give 4-hydrazono-3,4-dihydro-3-methylpyrimidine (**61**, R = H), converted into other derivatives for analysis.[2512] An isomer, 2-hydrazono-1,2-dihydro-1-methylpyrimidine

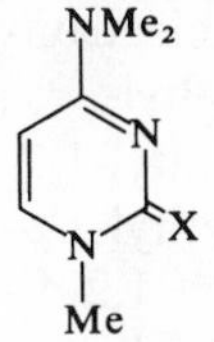

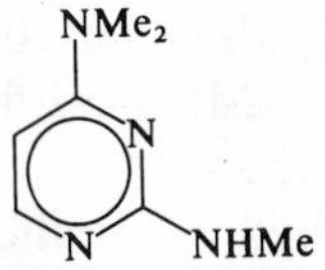

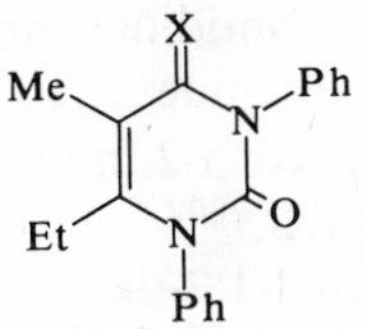

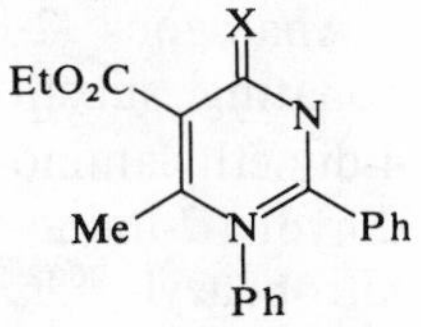

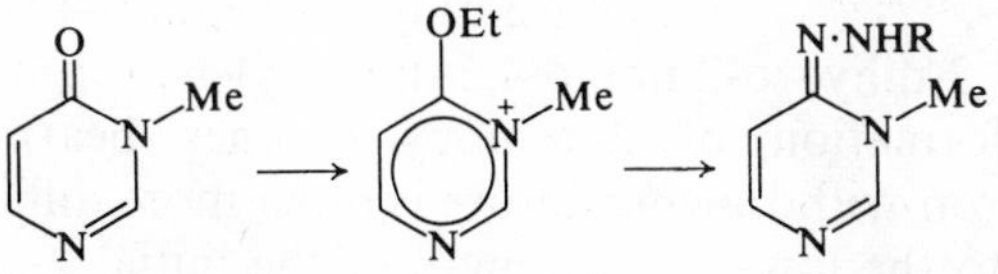

(**62**), was made by quaternizing 2-methylthiopyrimidine with methyl iodide, treating the product with benzoylhydrazine, and then deacylating.[2512]

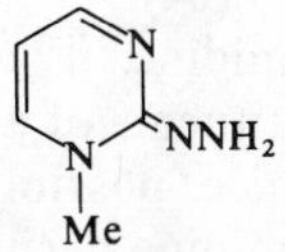

The *C*-methyl-protons of 1,2-dihydro-1,4-dimethyl-2-oxopyrimidine and related oxopyrimidine undergo facile deuterium exchange which apparently is catalysed both by hydrogen ion and by hydroxyl ion.[2861] 4-Trideuteriomethyl-1,2-dihydro-1-methyl-2-oxopyrimidine was prepared thus in a pure state.[2861]

* As pointed out before, this compound was described in error[2512] as the 1,4-dihydro-1-methyl-4-oxo isomer. Its derivatives were therefore similarly misformulated.

2. The Iminopyrimidines (*H* 377)

A. Preparation of Iminopyrimidines (*H* 377)

The treatment of a 2- or 4-aminopyrimidine with an alkyl halide almost invariably leads to alkylation at a ring nitrogen atom with the formation of an alkyldihydro-iminopyrimidine. Thus 2-aminopyrimidine and methyl iodide in boiling ethanol gave 1,2-dihydro-2-imino-1-methylpyrimidine hydriodide (**63**) in good yield within 6 hr[2986, 3282] instead of the 5 days required by an earlier procedure.[1177] Other simple 1,2-dihydro-2-iminopyrimidine hydrohalides, made rather similarly, include the 1-ethyl (EtI/EtOH, 24 hr. reflux),[2627, 2986] 1-propyl (hydrobromide from PrBr/PrOH or $MeOCH_2CH_2OH$;[2633, 2986] hydriodide[2633]), 1-butyl (BuI, 100°, 45%;[2627] BuI/EtOH, reflux, 70%[2986]), 1-heptyl,[2627] 1-dodecyl,[2986] 1-benzyl,[2633, 2986] 1-*p*-methoxybenzyl,[2986] 1-*p*-nitrobenzyl,[2633] 1-allyl,[2986] 1-β-hydroxyethyl ($BrCH_2CH_2OH$/EtOH, 100°, 94%),[2633] 1-prop-2′-ynyl,[3415] and 1-isopropyl[2633] derivatives.

Analogous 2-iminopyrimidines with *C*-alkyl or other electron-donating substituents include 1,2-dihydro-2-imino-1,4-dimethyl-,[2602] 4-dimethylamino-1,2-dihydro-2-imino-1-methyl-,[2627] 4-t-butyl-1,2-dihydro-2-imino-1-methyl-,[2602] 1,2-dihydro-2-imino-4-methyl-1-*p*-nitrobenzyl-,[2602] 5-ethyl-1,2-dihydro-2-imino-1-methyl-,[2627] 1,2-dihydro-2-imino-1,4,6-trimethyl-,[2627] 1,2-dihydro-2-imino-1,4,6-trimethyl-5-propyl(or prop-2′-ynyl or prop-1′-ynyl)-,[2700] 5-allyl-1,2-dihydro-2-imino-1,4,6-trimethyl-,[2700] 1-allyl-1,2-dihydro-2-imino-4,6-dimethyl- (**64**),[2700] and 1,2-dihydro-2-imino-4,6-dimethyl-1-prop-2′-ynylpyrimidine.[2700] The formation of 2-iminopyrimidines bearing *C*-substituents of an electron-withdrawing nature is often more difficult in practice, possibly due to the low basic strength of the initial 2-aminopyrimidine derivative. However, such alkylations have given 5-bromo-1,2-dihydro-2-imino-1-methyl (or 1,4,6-trimethyl)- (MeI/110°),[2626, 2627] 5-bromo-1-ethyl-1,2-dihydro-2-imino- (EtI/130°),[2626] 5-chloro-1,2-dihydro-2-imino-1-methyl-,[2633] 1,2-dihydro-2-imino-5-iodo-1-methyl-,[2633] 1-ethyl-4(or 6)-chloro-1,2-dihydro-2-imino-6(or 4)-methyl-,[2709] 5-bromo-1,2-dihydro-2-imino-1,4-dimethyl-,[2602] 5-bromo-4-t-butyl-1,2-dihydro-2-imino-1-methyl-,[2602] 5-carbamoyl(or cyano)-1,2-dihydro-2-imino-1-methyl- ($MeI/MeOCH_2CH_2OH$, 98°, 24 hr.),[2376] 5-carbamoyl(or cyano)-1,2-dihydro-2-imino-1,4,6-trimethyl-,[2633] 5-carbamoyl(or cyano*)-1,2-dihydro-2-imino-1,4-dimethyl-,[2602] and

* Not purified.

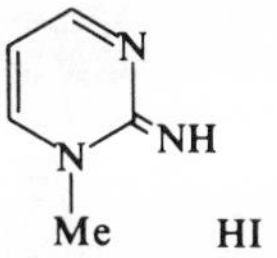

(63) 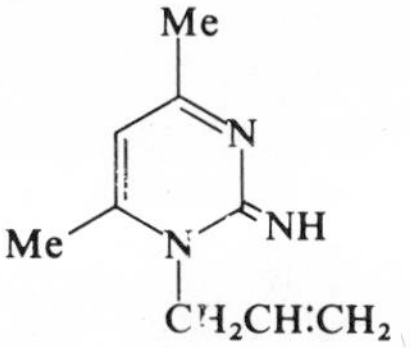 (64) 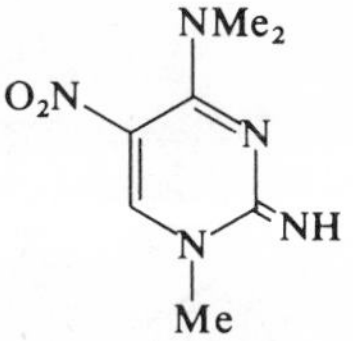 (65) 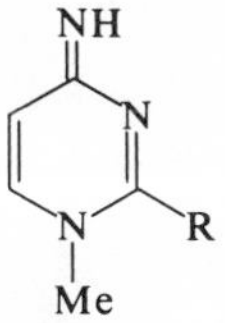 (66)

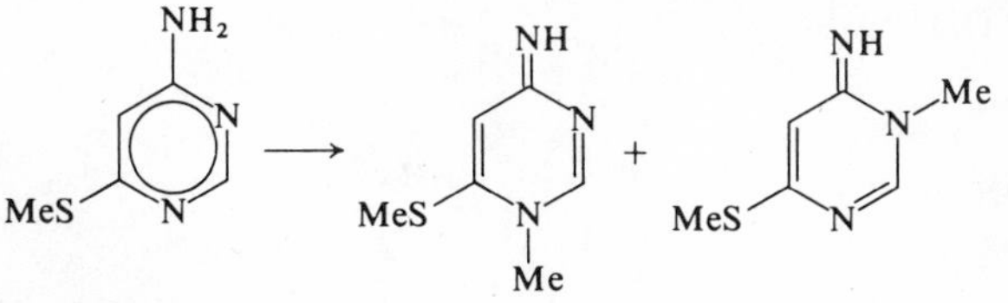

(67) (68) (69) 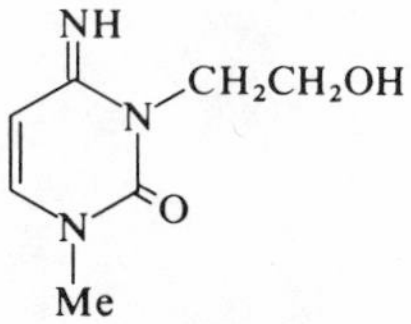 (70)

4-dimethylamino-1,2-dihydro-2-imino-1-methyl-5-nitro-pyrimidine (**65**).[2626]

4-Iminopyrimidines (hydrohalides) are represented by 1,4-dihydro-4-imino-1-methyl-2-methylthiopyrimidine (**66**, R = SMe; 91%; MeI/MeOH);[2288] its 2-methoxy analogue (**66**, R = OMe);[2745] the separable 1:1-mixture of 1,4-dihydro-4-imino-1-methyl-6-methylthio- (**68**) and 1,6-dihydro-6-imino-1-methyl-4-methylthio-pyrimidine (**69**) made from 4-amino-6-methylthiopyrimidine (**67**) with methyl iodide in methanol;[2781] 1,6-dihydro-6-imino-4-methoxy-1-methylpyrimidine, the only product from similar treatment of 4-amino-6-methoxypyrimidine;[2781] its 4-chloro-analogue;[2781] 1,2,3,4-tetrahydro-3-β-hydroxyethyl-4-imino-1-methyl-2-oxopyrimidine, prepared as free base (**70**) by the action of methanolic ethylene oxide on 4-amino-1,2-dihydro-1-methyl-2-oxopyrimidine;[3012] 5-amino-1-carbamoylmethyl-1,4(or 1,6)-dihydro-4(or 6)-iminopyrimidine (67%) from 4,5-diaminopyrimidine and ethanolic iodoacetamide;[2722] and the betaine of its 1-carboxymethyl analogue.[2722]

Imines produced by alkylation of other diamino- and triamino-pyrimidines are typified by 2-amino-1,4-dihydro-4-imino-1-methyl-pyrimidine* (**72**; or tautomer), from 2,4-diaminopyrimidine (**71**);[2288] the separable mixture of 2-amino-1,4-dihydro-4-imino-1-methyl-6-methylthio- (**73**) and its isomer, 2-amino-1,6-dihydro-6-imino-1-methyl-4-methylthio-pyrimidine* (**74**) from 2,4-diamino-6-methylthiopyrimi-

* The position of the methyl group in the products (**72**–**74**) was proven by converting the known 2,4-diamino-3,6-dihydro-3-methyl-6-oxopyrimidine (**80**) into 2-amino-4-chloro-1,6-dihydro-6-imino-1-methylpyrimidine (**79**) and thence into the methylthiopyrimidine (**74**); desulphurization of its isomer (**73**) gave the imine (**72**).[2288]

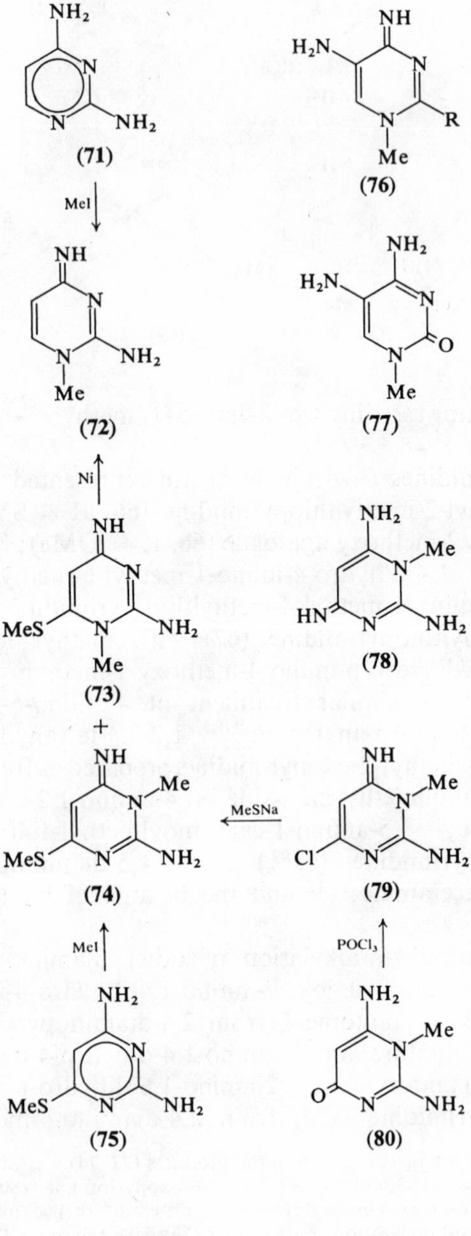

NH2
N
N NH2
(71)

MeI

NH
N
N NH2
Me
(72)

Ni

NH
N
MeS N NH2
Me
(73)

+

NH
N Me
MeS N NH2
(74)

MeI

NH2
N
MeS N NH2
(75)

NH
H2N N
N R
Me
(76)

NH2
H2N N
N O
Me
(77)

NH2
N Me
HN N NH2
(78)

NH
N Me
Cl N NH2
(79)

MeSNa

POCl3

NH2
N Me
O N NH2
(80)

dine (**75**);[2288] 1,4-dihydro-4-imino-1-methyl-2-methylaminopyrimidine;[2288] 5-amino-1,4-dihydro-4-imino-1-methylpyrimidine* (**76**, R = H) from 4,5-diaminopyrimidine;[2288] 5-amino-1,4-dihydro-4-imino-1-methyl-2-methylthiopyrimidine* (**76**, R = SMe);[2288] 2,4-diamino-3,6-dihydro-6-imino-3-methylpyrimidine (**78** or a tautomer);[2288] 4,5-diamino-3,6-dihydro-6-imino-1-methylpyrimidine (or tautomer);[2454] 4,5-diamino-1,2-dihydro-2-imino-1-methylpyrimidine† (or tautomer);[2454] and the 1-methylated derivatives of most of the possible methylamino and dimethylamino analogues of 2,4-diaminopyrimidine, with or without a 5- or 6-halogeno substituent.[2624]

Pyrimidines bearing an alkylimino group are well known. Their hydriodides are made from alkylaminopyrimidines by further alkylation on a ring nitrogen atom. Thus 2-methylaminopyrimidine with methyl iodide gave 1,2-dihydro-1-methyl-2-methyliminopyrimidine (**81**, R = H; 90%),[2627] and further examples are: 2-ethyl(or butyl)imino-1,2-dihydro-1-methyl-,[2627, 3416] 1-ethyl(or butyl)-1,2-dihydro-2-methylimino,[2627, 3416] 5-bromo-1,2-dihydro-1-methyl-2-methylimino- (**81**, R = Br),[2627] 2-benzylimino-1,2-dihydro-1-methyl(or isopropyl)-,[2626] 1-benzyl-1,2-dihydro-2-methyl(or isopropyl)imino-,[2626] 1-benzyl-2-benzylimino-1,2-dihydro-,[2626] 2-allylimino-1,2-dihydro-1-propyl- (**82**),[3416] 1-allyl-1,2-dihydro-2-propylimino-,[3416] 2-dodecylimino-1,2-dihydro-1-methyl-,[2986] 4-benzylimino-1,4(or 3,4)-dihydro-1,2,6(or 2,3,6)-trimethyl-,[2153] 1,4-dihydro-1-methyl-4-methylimino-6-methylthio-,[2781] 4-chloro-3,6-dihydro-3-methyl-6-methylimino-,[2781] 1,2,3,4-tetrahydro-3-β-hydroxyethyl-4-β-hydroxyethylimino-1-methyl-2-oxo- (**83**),[3012] 5-cyano-1,2-dihydro-1-methyl-2-methylimino- (**81**, R = CN; 81% using MeI/$MeOCH_2CH_2OH$ at 98° for 24 hr.),[2376] and 5-cyano-1,2-dihydro-1,4,6-trimethyl-2-methylimino-pyrimidine.[2633]

B. Reactions of Iminopyrimidines; the Dimroth Rearrangement (*H* 379)

The hydrolysis of imino- to oxopyrimidines has been discussed above in Sect. 1.A(5). The deuteration of the *C*-methyl groups in 1,2-dihydro-2-imino-1,4-dimethyl(and 1,4,6-trimethyl)pyrimidine has been studied; it was both acid and base catalysed.[2861]

The term Dimroth rearrangement was coined[2627] in 1963 to cover an

* For structural proof the imine (**76**, R = H) was prepared by desulphurizing its derivative (**76**, R = SMe), which was hydrolysed also to the known 4,5-diamino-1,2-dihydro-1-methyl-2-oxopyrimidine (**77**).[2288]

† Proven in structure by hydrolysis to the known 4,5-diamino-1,2-dihydro-1-methyl-2-oxopyrimidine.[2454]

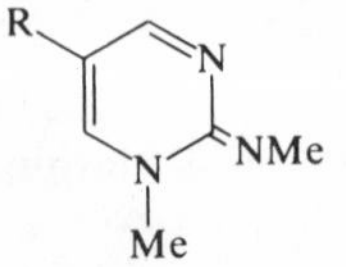

(81) 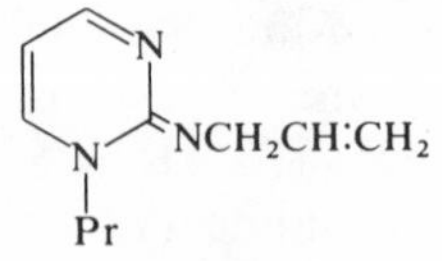 (82) 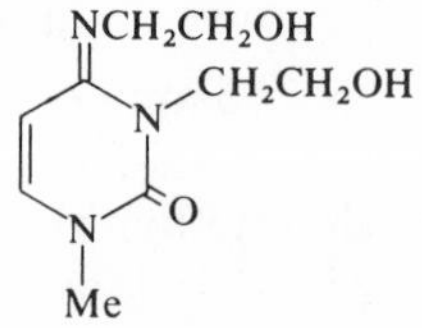 (83)

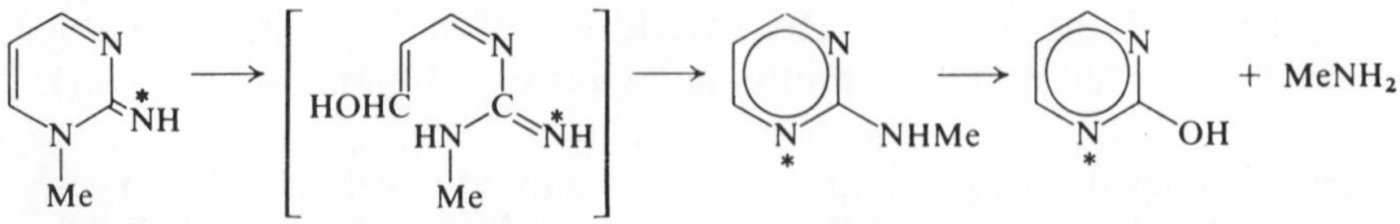

(84) → [(85)] → (86) → (87) + $MeNH_2$

isomerization proceeding by ring-fission and subsequent recyclization whereby a ring-nitrogen and its attached substituent exchanged places with an imino (or potential imino) group in the α-position. It is simply illustrated by the rearrangement of 1,2-dihydro-2-imino-1-methylpyrimidine (**84**) to 2-methylaminopyrimidine (**86**) in aq. alkali at room temperature.[2627] Since that date a detailed study has been made of the isomerization in the pyridine, pyrimidine, and related series; the whole subject has been reviewed recently.[2855] Early pyrimidine examples have been discussed (*H* 379) and more recent studies in the series are summarized briefly below.

The gross mechanism of the Dimroth rearrangement was proven independently in two laboratories.[2154, 2986] Aminolysis of 2-chloropyrimidine by [^{15}N]-ammonia gave 2-aminopyrimidine isotopically labelled on the extracyclic nitrogen. Its methylated derivative (**84**) was rearranged in alkali. The resulting methylaminopyrimidine (**86**) was hydrolysed to 2-hydroxypyrimidine (**87**) and methylamine. Only the former showed ^{15}N-enrichment on mass-spectral examination, indicating that the extracyclic nitrogen in the imine (**84**) had been incorporated into the ring during rearrangement. This inferred an acyclic intermediate such as the guanidine (**85**). A similar mechanism has been inferred for the exchange of nuclear- and extranuclear-nitrogen atoms in 2-aminopyridine, 2-amino-3-(and 5-)nitropyridine, 2-aminoquinoline, and 3-aminoisoquinoline, etc.[3468–3470]

The first rate studies on the Dimroth rearrangement were done on known[494, 2493, 3417] pteridines. These indicated two successive reactions (ring fission and reclosure) which were slowed by electron-donating substituents and accelerated by electron-withdrawing substituents.[2374, 3418–3420] This immediately explained existing qualitative data (*H* 379) and some puzzling failures of compounds in the pyrimidine series[2168, 2215, 2288] to undergo rearrangement. A spectral study of the rearrangement of 1,2-dihydro-2-imino-1-methylpyrimidine (**84**) indicated that ring-fission was the rate-determining step ($t_{1/2}$ = 114 min. at 25° and pH 14), and that no build-up of the intermediate (**85**) occurred.[2374, 2627] However, by carrying out the reaction in the presence of hydroxylamine, the intermediate was trapped and characterized as its oxime;[2375] in addition, a hydrolytic by-product, malondialdehyde, was identified,[2375] and a little hydrolytic de-methylamination of the intermediate was shown to occur under some conditions to give 2-hydroxypyrimidine on recyclization.[3421] This picture was general for the rearrangement of pyrimidine imines.

The effect of added electron-donating substituents on the rate of rearrangement is summarized in Table XVa.[2376, 2627, 2633] The powerful

TABLE XVa. The Effect of Electron-donating *C*-Substituents on the Rate of Dimroth Rearrangement at pH 14 and 25° (*New*)

Parent pyrimidine	Substituents	$t_{1/2}$ (min.)
1,2-Dihydro-2-imino-1-methyl-	—	114
	5-Et	196
	4,6-Me_2	166
	4,6-Me_2-5-Pr	178
	4-NMe_2	2000
	4,6-$(NH_2)_2$	∞
5-Carbamoyl-1,2-dihydro-2-imino-1-methyl	—	7.5
	4,6-Me_2	26

donation by the dimethylamino group in 4-dimethylamino-1,2-dihydro-2-imino-1-methylpyrimidine (**88**, X = NH) so slowed rearrangement to the diamine (**89**) that the competing hydrolysis to the oxopyrimidine (**88**, X = O) accounted for most of the reactant.[2627]

The hastening effect of electron-withdrawing substituents on rearrangement is summarized in Table XVb.[2374–2376, 2626, 2627, 2633, 2700, 3418, 3422] Of particular note is the vast effect of the nitro group: each such example underwent instant isomerization in alkali. A mildly exceptional rapidity in the rearrangement of 1,2-dihydro-2-imino-5-iodo-1-methylpyrimidine (**90**, R = I), *vis à vis* its 5-bromo- and 5-chloro analogues (**90**, R = Br or Cl), has been reported without explanation;[3423] the exceptional behaviour of the nitrile (**90**, R = CN) is discussed below. As might be expected, 1,2-dihydro-2-imino-1,4,6-trimethyl-5-prop-1′-ynylpyrimidine (**91**, R = C⋮CMe), in which the

TABLE XVb. The Effect of Electron-withdrawing 5-Substituents on the Rate of Dimroth Rearrangement at pH 14 and 25° (*New*)

Parent pyrimidine	5-Substituent	$t_{1/2}$ (min.)
1,2-Dihydro-2-imino-1-methyl-	—	114
	Br	39[a]
	Cl	49[a]
	I	31[a]
	$CONH_2$	7.5
		14[a]
	CN	—[b]
1,2-Dihydro-2-imino-1,4,6-trimethyl-	—	166
	Pr[c]	178
	$CH_2CH{:}CH_2$	157
	$CH_2C{\vdots}CH$	109
	$C{\vdots}CMe$	64
	$CONH_2$	26
	CN	0.1
		0.2[a]
4-Amino-1,6-dihydro-6-imino-1-methyl-	—	15
	NO_2	<0.1
4-Dimethylamino-1,2-dihydro-2-imino-1-methyl-	—	2000
	NO_2	<0.1

[a] At 20°.
[b] Abnormal product; see text.
[c] For comparison with allyl.

triple bond was conjugate with the ring, rearranged more quickly than the 5-prop-2′-ynyl isomer (**91**, R = $CH_2C{\vdots}CH$).[2700]

The effect of variation in the *N*-1-alkyl group of 1-alkyl-1,2-dihydro-2-iminopyrimidine on its rate of rearrangement is evident from Table XVc.[2374, 2627, 2633, 2700, 3423] Thus the 1-methyl derivative underwent rearrangement more slowly than the higher homologues, probably on account of some steric factor since there is little difference in the electronic properties of alkyl groups. Contrariwise, the 1-allyl and 1-β-hydroxyethyl derivatives (**92**, R = $CH_2CH{:}CH_2$ or CH_2CH_2OH) rearranged more quickly than the sterically similar propyl derivative (**92**, R = Pr) on account of the relative electronic properties of the groups; greater electron-withdrawal in the prop-2′-ynyl derivative (**92**, R = $CH_2C{\vdots}CH$) caused even more rapid rearrangement.* The same rationalization applied to the rearrangement of the 1-*p*-nitrobenzyl relative to the 1-benzyl derivative.

* Although rearrangement of the imine (**92**, R = $CH_2C{\vdots}CH$) occurred normally in aqueous media to give mainly the propynylaminopyrimidine (**93**), in alcoholic media preferential cyclization took place to give 2-methyl-1,3a,7-triazaindene (**94**).[2700, 3415]

TABLE XVc. The Effect of Variation in the *N*-1-Substituent on the Rate of Dimroth Rearrangement at pH 14 and 25° (*New*)

Parent pyrimidine	*N*-1-Substituent	$t_{1/2}$ (min.)
1,2-Dihydro-2-imino-	Me	114
	Et	63
	Pr	55
	Bu	58
	n-C_7H_{15}	57
	$CH_2CH{:}CH_2$	33
	$CH_2C{:}CH$	5
	CH_2CH_2OH	36
	CH_2Ph	23
	$CH_2C_6H_4NO_2(p)$	8
1,2-Dihydro-2-imino-4,6-dimethyl-	Me	166
	$CH_2CH{:}CH_2$	133
	$CH_2C{:}CH$	22
5-Bromo-1,2-dihydro-2-imino-	Me	39[a]
	Et	38[a]

[a] At pH 12.2 and 20°.

1,2-Dihydro-2-imino-1,4-dimethylpyrimidine (**95**, X = NH) rearranged so abnormally slowly that the product contained only 10% of 4-methyl-2-methylaminopyrimidine along with 80% of the oxopyrimidine (**95**, X = O) and 10% of 2-hydroxy-4-methylpyrimidine.[2602] 4-t-Butyl-1,2-dihydro-2-imino-1-methylpyrimidine and related derivatives behaved similarly,[2602] and no satisfactory explanation has emerged. Another fascinating abnormality occurred[2376] with 5-cyano-1,2-dihydro-2-imino-1-methylpyrimidine (**96**) which in alkali rapidly gave the

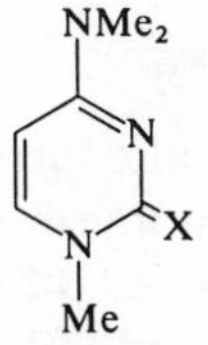

(**88**) 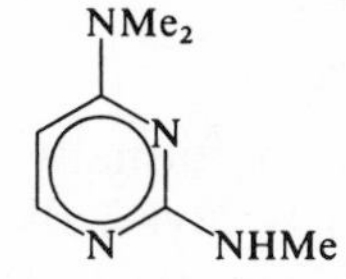 (**89**) 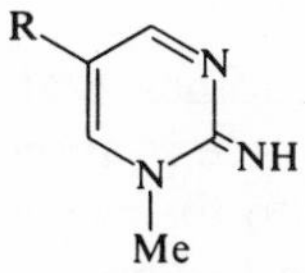 (**90**) 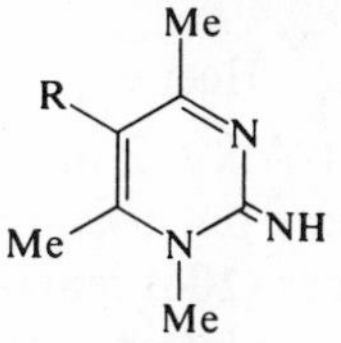 (**91**)

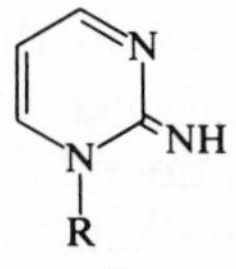

(**92**) 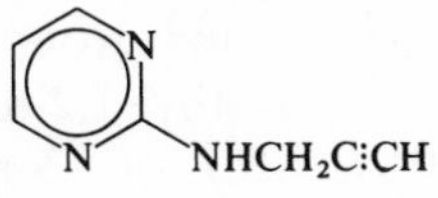 (**93**) 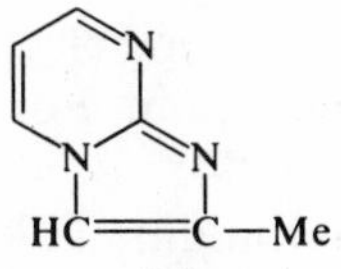 (**94**) 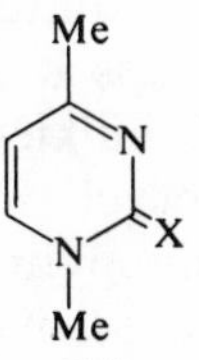 (**95**)

(characterized) intermediate (**97**). In acid this was reconverted into the parent imine salt; in aq. ammonia (pH 9) it gave the normal rearranged product, 5-cyano-2-methylaminopyrimidine (**100**); and in alkali (pH 14) it gave 4-amino-5-formyl-2,3-dihydro-2-imino-3-methylpyrimidine (**98**) by an alternative ring-closure involving the cyano group. The formylpyrimidine (**98**) underwent a slow second Dimroth rearrangement to 5-carbamoyl-2-methylaminopyrimidine (**103**), presumably *via* the (unisolated) intermediate (**99**).[2376] The methylimine (**101**) behaved similarly in giving the abnormal formylpyrimidine (**102**) but, of its nature, failed to give products corresponding to the amines (**100**) or (**103**).[2376] The above abnormal cyclization probably occurred because the formyl group in the intermediate (**97**) was deactivated by hydration in strong alkali. This hypothesis was strengthened by the completely normal rearrangement of 5-cyano-1,2-dihydro-2-imino-1,4,6-trimethylpyrimidine, involving a ketonic intermediate not prone to such hydration.[2633]

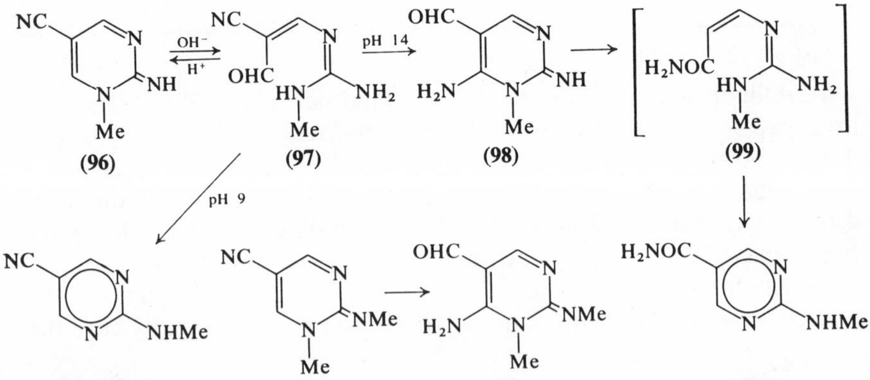

1-Alkyl-2-iminopyrimidines rearrange into formally aromatic products but this driving force is absent when 1-alkyl-2-alkyliminopyrimidines (**104**) rearrange into their isomers (**106**) *via* the intermediate (**105**). In the latter case an equilibrium mixture of the isomers (**104**) and (**106**) was attained, independent of which isomer was used as starting material, and controlled by the steric and electronic factors associated with the groups R and R′.[2626, 3416] Thus equilibrium favoured the isomer with the bulkier and/or more electron-withdrawing group attached to the exocyclic nitrogen atom. For example, 1-ethyl-1,2-dihydro-2-methyliminopyrimidine (**104**; R = Et, R′ = Me) or 2-ethylimino-1,2-dihydro-1-methylpyrimidine (**106**; R = Et, R′ = Me) gave in alkali the same equilibrium mixture containing 2 parts of the former and 3 parts of the latter (steric control); and 1-allyl-1,2-dihydro-2-propylimino-pyrimidine (**104**; R = $CH_2CH{:}CH_2$, R′ = Pr) or 2-allylimino-1,2-dihydro-1-propylpyrimidine (**106**; R = $CH_2CH{:}CH_2$, R′ = Pr) attained an equilibrium mixture of 1 part of the propylimino derivative and 3 parts of the allyliminopyrimidine (electronic control).[2626, 3416]

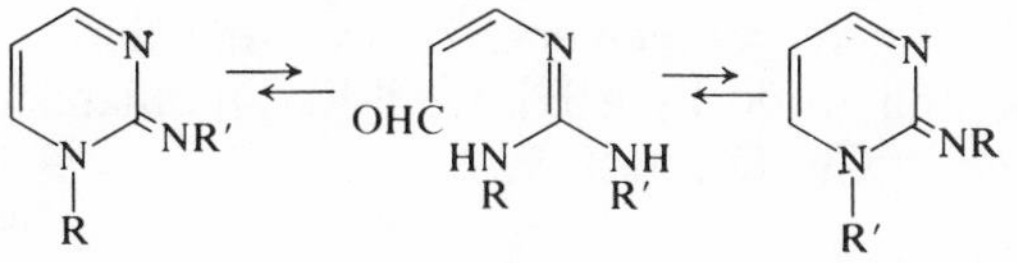

(**104**) (**105**) (**106**)

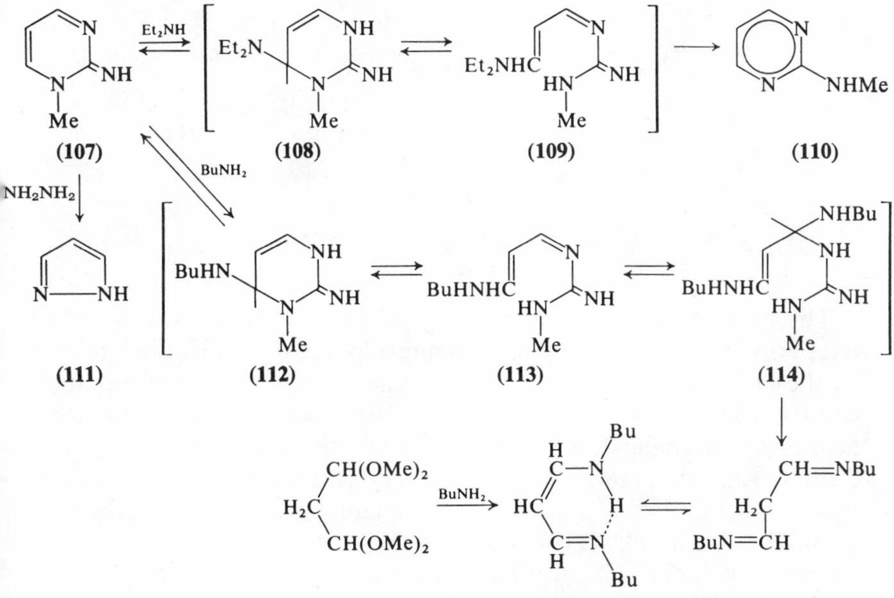

(**107**) (**108**) (**109**) (**110**)

(**111**) (**112**) (**113**) (**114**)

(**115**) (**116**) (**117**)

The mechanism of the Dimroth rearrangement has been discussed in terms of the kinetics of its several steps,[2375] and it has been shown that water plays an essential role under normal conditions.[2375, 3423] However, anhydrous secondary amines also provide a satisfactory medium for rearrangement. Thus when the hydriodide of 1,2-dihydro-2-imino-1-methylpyrimidine (**107**) was warmed with diethylamine, the adduct (**108**) underwent ring opening to the acyclic intermediate (**109**), which recyclized to 2-methylaminopyrimidine (**110**; 60%).[3240] The imine (**107**) was unaffected by triethylamine,[3240] but with the primary amine, butylamine, an unexpected reaction took place leading to 1,3-bis-butyliminopropane (**117**) probably *via* the intermediates (**112**–**114**).[3240, cf. 3219] The product was stabilized in the cyclic form (**116**) and was made unambiguously from 1,1,3,3-tetramethoxypropane. (**115**)[3240] The imine (**107**) reacted rather similarly with hydrazine to give pyrazole (**111**; 60%).[3240]

Some interesting rearrangements of cytosine derivatives that take place in acetic anhydride by a Dimroth-like mechanism have already been mentioned in Ch. IX, Sect. 9.A.

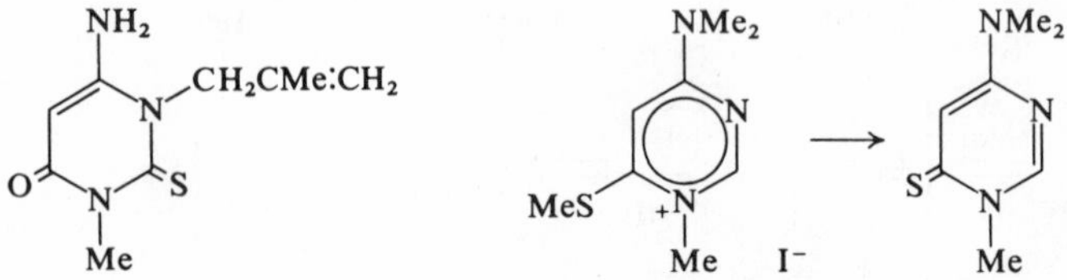

(**118**) (**119**) (**120**)

3. The *N*-Methylated Thiopyrimidines (*H* 381)

These compounds may be made by thiation[2173] of the corresponding oxopyrimidines with phosphorus pentasulphide (Ch. VII, Sect. 6.B) or by the Principal Synthesis (Ch. II). The latter route is exemplified in the condensation of *N*-methyl-*N'*-β-methylallylthiourea with cyanoacetic acid to give 4-amino-1,2,3,6-tetrahydro-1-methyl-3-β-methylallyl-6-oxo-2-thiopyrimidine (**118**).[2309] Other primary syntheses (Ch. III) have also been used,[2242, 2313, 2339] and 4-dimethylamino-1-methyl-6-methylthio-pyrimidinium iodide (**119**) has been converted by ethanolic sodium hydrogen sulphide into 4-dimethylamino-1,6-dihydro-1-methyl-6-thio-pyrimidine (**120**).[2676]

4. The Pyrimidine-*N*-Oxides (*H* 382)

The peroxide oxidation 4-methylpyrimidine gives a separable mixture of the *N*-1- and *N*-3-oxides (Ch. IV, Sect. 2.C), and their dipole moments and p.m.r. spectra have been compared with pyrimidine- and 5-methylpyrimidine-*N*-oxide;[2418] the dipole moments of 4-ethoxy-6-methylpyrimidine and its *N*-3-oxide have been compared with those of related pyridazines.[3424]

2,4-Diamino-6-2′,4′-dichlorophenoxypyrimidine and related compounds have been oxidized to the corresponding 3-*N*-oxides by *m*-chloroperbenzoic acid.[3508]

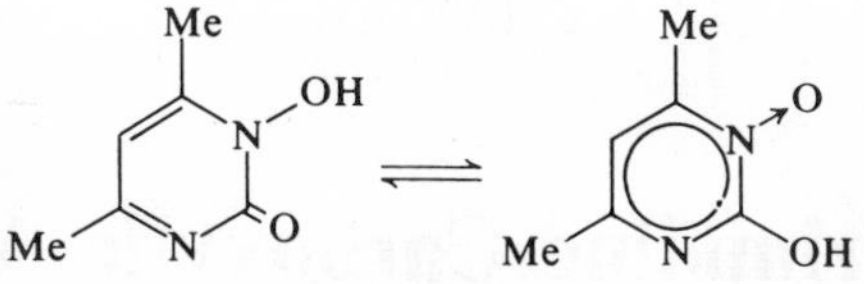

(121)

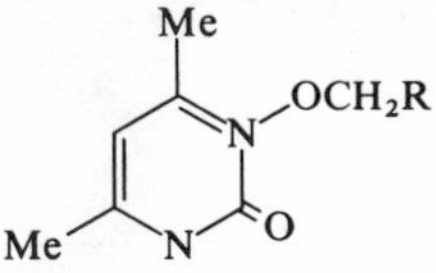

(122) 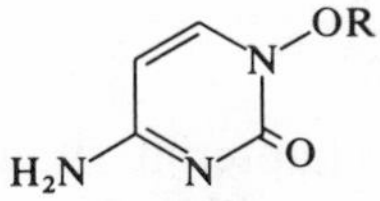 (123)

The Principal Synthesis has been used to make *N*-oxides directly or *via* the corresponding *N*-alkoxy derivatives. These processes are illustrated in the condensation of *N*-hydroxyurea with acetylacetone to give 1,2-dihydro-1-hydroxy-4,6-dimethyl-2-oxopyrimidine (i.e., 2-hydroxy-4,6-dimethylpyrimidine-*N*-oxide; *ca.* 10%);[2869] of *N*-benzyloxy-(or methoxy)urea with acetylacetone to give 1-benzyloxy-1,2-dihydro-4,6-dimethyl-2-oxopyrimidine (**122**, R = Ph) and the methoxy analogue (**122**, R = H), respectively, each of which gave the *N*-oxide (**121**) on acidic hydrolysis;[2869] and of *N*-benzyloxyurea with αβ-dibromopropionitrile to give 4-amino-1-benzyloxy-1,2-dihydro-2-oxopyrimidine (**123**, R = CH_2Ph) which was hydrogenated to the corresponding *N*-hydroxy compound (cytosine-1-oxide; **123**, R = H).[2262] Other primary syntheses have been used to make *N*-oxides: *N*-(β-amino-β-hydroxyiminopropionyl)urea [$H_2NC(:NOH)CH_2CONHCONH_2$] in refluxing dimethylformamide gave 4-amino-2,6-dihydroxypyrimidine-3-oxide (12%),[2316] and other examples[2199, 2200, 2262, 2964] have been discussed in Ch. III. 1-Benzyloxy-5-fluorouracil has been made recently by a Principal Synthesis from ethyl α-fluoro-α-methoxymethyleneacetate with *N*-benzyloxyurea and also by another primary synthesis from β-benzyloxyamino-α-fluoroacrylamide with oxalyl dichloride;[3464] hydrogenolysis gave the *N*-oxide.[3464]

4-Amino-5-fluoro-2-hydroxypyrimidine-3-*N*-oxide and acetic anhydride gave 4-acetoxyamino-5-fluoro-2-hydroxypyrimidine, also prepared by selective aminolysis of 2,4-dichloro-5-fluoropyrimidine with hydroxylamine, hydrolysis, and acetylation.[3464]

CHAPTER XI

The Pyrimidine Carboxylic Acids and Related Derivatives (*H* 389)

In using this chapter it should be remembered that a given interconversion reaction will be discussed only at the first logical opportunity. For example, the formation of esters from the corresponding carboxylic acids is dealt with as a reaction of acids rather than a preparative method for esters, simply because acids are discussed prior to esters.

1. The Carboxypyrimidines (*H* 389)

A. Preparation of Carboxypyrimidines (*H* 389)

Examples of the formation of carboxypyrimidines by primary syntheses have been given in Chs. II and III.

(2) *By Hydrolysis of Esters, Amides, and Nitriles* (*H* 390)

This useful route from esters to carboxylic acids is illustrated in the hydrolysis of 5-ethoxycarbonyl- to 5-carboxy-pyrimidine (88%; 2*N*-sodium hydroxide at 50°);[2680] 1-benzyloxy-5-ethoxycarbonyl- (**1**) to 1-benzyloxy-5-carboxy-uracil (**2**; *ca.* 70%; 2*N*-sodium hydroxide at 100°) or to 5-carboxy-1-hydroxyuracil (**3**; 53%; hydrobromic acid/acetic acid at 100° for 14 hr.);[2199] 4,5-diamino-6-ethoxycarbonyl- to 4,5-diamino-6-carboxy-pyrimidine (*ca.* 85%; *N*-sodium carbonate at 100°);[3199] 2-amino-5-ethoxycarbonyl- to 2-amino-5-carboxy-pyrimidine (*ca.* 40%; boiling aq. ethanolic potassium hydroxide),[2278] 4-dimethylamino-5-ethoxycarbonyl- to 5-carboxy-4-dimethylamino-2-methylpyrimidine (*ca.* 55%; *N*-potassium hydroxide);[2698] 1-dimethylamino-5-ethoxycarbonyl- to 5-carboxy-1-dimethylamino-uracil (30%; hydrochloric acid);[2870] 5-chloro-4-ethoxycarbonyl- to 4-carboxy-5-chloro-6-hydroxy-2-methylpyrimidine (61%; 5*N*-sodium hydroxide at 20°);[2191] 4-amino-5-ethoxycarbonyl- to 4-amino-5-carboxy-2,3-dihydro-3-methyl-2-oxopyrimidine (**4**; R = NH_2; 70%; *N*-sodium hydroxide*);[2987] and other such examples.[2954, 3425] Some ω-carboxypyrimidines produced from the corresponding ethyl esters include 5-carboxymethyl-,[2577] 4-carboxymethylthio-5,6-tetramethylene-,[2183] 2,4-diamino-5-*p*-carboxyphenyl-6-methyl- (from the methyl ester),[2282] and such like pyrimidines.[2978, 3426]

The formation of carboxylic acids from amides or nitriles is seen in the hydrolysis of 2-cyano- to 2-carboxy-pyrimidine (70%; boiling 2*N*-sodium hydroxide);[2607] 4-cyano- to 4-carboxy-2-methylthiopyrimi-

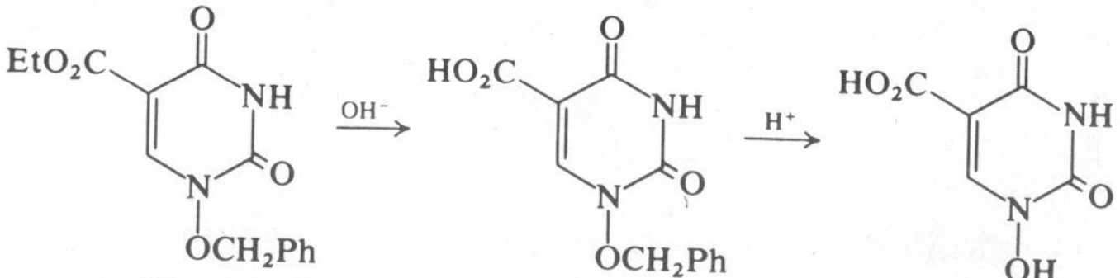

(1) (2) (3)

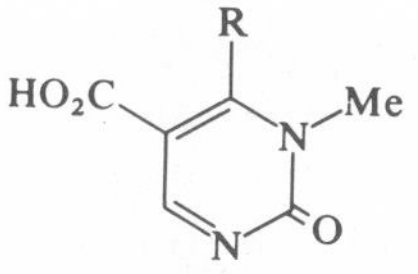

(4)

dine (43%; boiling 2*N*-sodium hydroxide for 2 hr.†);[2608] 4-chloro-6-methoxycarbonyl- to 4-amino-6-carbamoyl- (not isolated; aq. ammonia at 130°) and thence to 4-amino-6-carboxy-pyrimidine (37% overall; 2.5*N*-sodium hydroxide);[2954] 2,4-diamino-5-*p*-cyanophenyl- to 2,4-diamino-5-*p*-carboxyphenyl-6-ethyl(and methyl)pyrimidine (quantitative yield; potassium hydroxide in refluxing glycol);[2282] 2-amino-5-cyano- to 2-amino-5-carboxy-pyrimidine (*ca.* 30%; 2*N*-potassium hydroxide at 100°);[2278] 2-amino-5-β-cyanoethyl- to 2-amino-5-β-carboxyethyl-4-hydroxy-6-methylpyrimidine (86%; 6*N*-hydrochloric acid under reflux);[2232] and 4-amino-5-phenylcarbamoyl- to 4-amino-5-carboxy-pyrimidine (62%; 2*N*-sodium hydroxide).[2359] Attempts to

* Hydrolysis with 2.5*N*-sodium hydroxide gave also some 5-carboxy-3-methyluracil (**4**, R = OH).[280, 2987]

† Longer boiling gave 4-carboxy-2-hydroxypyrimidine (58%).[2954]

hydrolyse 2-amino-5-cyano(or carbamoyl)-4,6-dimethylpyrimidine to the corresponding acid failed.[2633]

(3) *By Oxidative Procedures* (*H* 391)

The oxidation of alkyl- and alkenyl-pyrimidines to carboxypyrimidines and such like has been mentioned in Ch. IV, Sect. 2.C. Ozonization has also been used to convert 4-acetamido-6-β-2′-furylvinyl- (**5**) into 4-acetamido-6-carboxy-2-methylpyrimidine (**6**);[2451] 4-amino-6-methyl-2-β-5′-nitro-2′-furylvinylpyrimidine into its 2-carboxy analogue;[2451] and other vinylpyrimidines into carboxylic acids which were decarboxylated without purification.[2451]

(4) *By Other Methods* (*H* 392)

The conversion of trichloro- or tribromomethyl groups into carboxy groups by the use of silver nitrate[2579, 2806] has been mentioned in Ch. VI, Sect. 7. In addition, 2,4-dihydroxy-5-trifluoromethylpyrimidine has been converted by warm sodium hydroxide, or even sodium bicarbonate, into 5-carboxy-2,4-dihydroxypyrimidine.[2666, 2667]

Some pyrimidine-5-carboxylic acids have been made by M. Mehta and his colleagues[2519] from 5-bromopyrimidines by halogen–metal interconversion with butyl-lithium followed by treatment with carbon dioxide; chloro groups were unaffected by such treatment. Thus 5-bromo- (**7**) gave the intermediate (**8**) and thence 5-carboxy-4-chloro-6-methoxypyrimidine (**12**; 53%); 5-bromo- gave 5-carboxy-2-methoxy-4,6-dimethylpyrimidine; and other such acids were made similarly in 50–90% yield.[2519]

The formation of ω-carboxypyrimidines by treating a pyrimidine with a reagent bearing a carboxy group is exemplified in the alkylation of uracil with chloroacetic acid to give 1-carboxymethyluracil (**9**; 70%),[3427] or with acrylic acid in liquid ammonia (or acrylonitrile in sodium hydroxide) to give 1-β-carboxyethyluracil (**10**; 85%);[3363, 3428] in the aminolyses of appropriate chloropyrimidines with amino acids to give 2-amino-4-carboxymethylamino-6-methylpyrimidine[2970] and related compounds (see Ch. VI, Sect. 7);[2719, 2720, 2832, 2971, 2972] and in the aminolysis of 1,2-dihydro-1-methyl-4-methylthio-2-oxopyrimidine with sodium β-alanate to give 4-β-carboxyethylamino-1,2-dihydro-1-methyl-2-oxopyrimidine (**11**; *ca.* 60%).[3363]

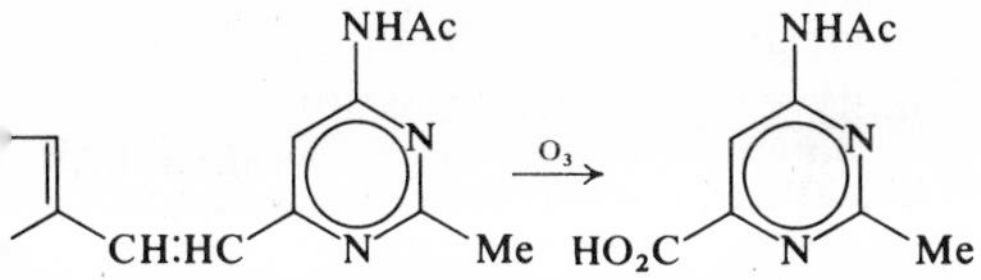

(5) (6) 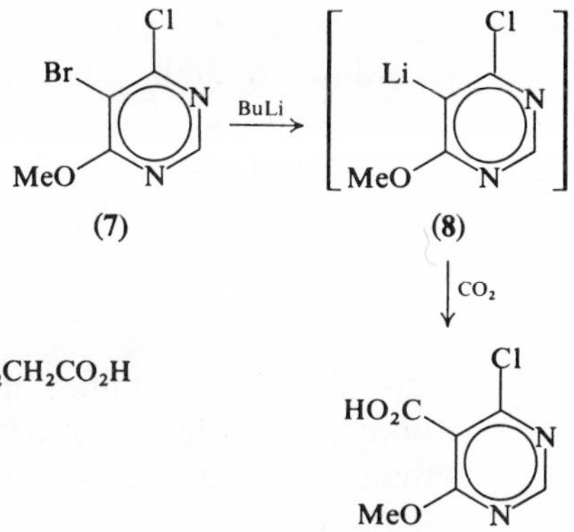 (7) (8) (9) 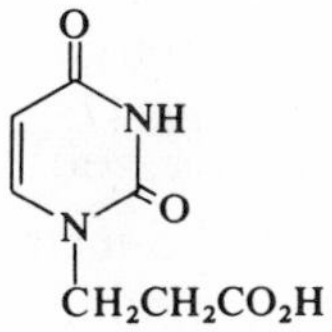 (10) 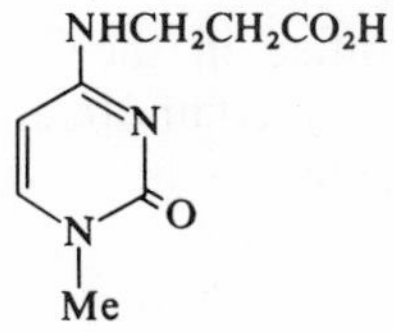 (11) (12)

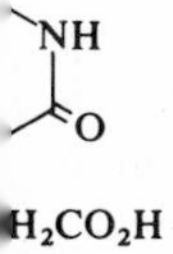

B. Reactions of Carboxypyrimidines (*H* 393)

The infra-red spectra of 36 pyrimidinecarboxylic acids have been summarized.[3429]

(1) *Decarboxylation* (*H* 393)

The process is illustrated simply in the formation of 5-bromo-2,4-dihydroxypyrimidine from its 6-carboxy derivative by heating 0.5 g. portions at 285° for 5 min.;[2196] of 2-phenylpyrimidine from its 4-carboxy derivative;[2603] of 5-ethyl(or butyl)-2,4-dihydroxypyrimidine (70%, 80%) and similar compounds by heating the 6-carboxy derivatives in refluxing quinoline with copper powder;[2347] of 5-benzoyl-2-phenylpyrimidine (91%) by heating the 4,6-dicarboxy derivative in acetic anhydride/acetic acid at 120°;[2372] of 5-bromo-2,4-dihydroxy-6-methyl- from 5-bromo-4-carboxymethyl-2,6-dihydroxy-pyrimidine by refluxing in *N*-ethylaniline (76% yield), morpholine (33% yield), or piperidine (38% yield);[2880] of 4-acetamido-2-methylpyrimidine from its 6-carboxy derivative;[2451] of 5-β-benzyloxycarbonylamino-β-carboxyethylpyrimidine (**13**; R = H) from the ββ-dicarboxy analogue (**13**, R = CO_2H; 63%) simply by acidifying the dipotassium salt;[2579] and other such examples.[2451, 2786, 3430, 3522] Simple decarboxylation of 4-amino-5-carboxy-2,3-dihydro-3-methyl-2-oxopyrimidine (**14**) was possible *in vacuo* at 250°,[2987] although at atmospheric pressure the reaction was accompanied by rearrangement to give 2-hydroxy-4-methylamino-pyrimidine (*H* 380). An exceptionally easy decarboxylation occurred during attempted recrystallization of 4-carboxy-2-*p*-chlorophenyl-6-hydroxypyrimidine from ethanol.[4381]

(2) *Esterification* (*H* 393)

Esterification is typified in the following preparations: 4-methoxy-carbonylpyrimidine by methanol/sulphuric acid in 47% yield,[2806] by methanol/hydrogen chloride in 52–86% yield,[2806, 3009] or by diazo-methane in a yield 'sensiblement quantitatif';[2806] 5-methoxycarbonyl-pyrimidine and 5-bromo-2-methoxycarbonylpyrimidine by diazo-methane;[2806] 2-methoxycarbonylpyrimidine (54%) by heating the silver salt of the acid with methyl iodide;[2607] 5-bromo-4-ethoxycarbonyl-2-phenylpyrimidine (73%) by ethanol/sulphuric acid;[2603] 4-butoxy-(isobutoxy or ethoxy)carbonyl-2,6-dihydroxypyrimidine (78%, 98%) and related esters by using the appropriate alcohol with sulphuric acid;[3431, 3432, 3434, 3436] 5-butoxycarbonyl-2,4-dihydroxypyrimidine (similarly);[3431] 4-hydroxy-6-methoxycarbonyl-2-methylthiopyrimidine (**15**, R = H; 61%) by methanolic hydrogen chloride;* [2214] 5-fluoro-4-

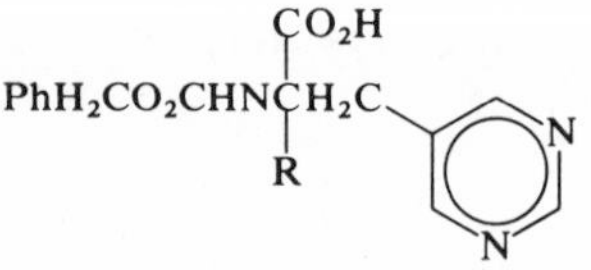

(13)

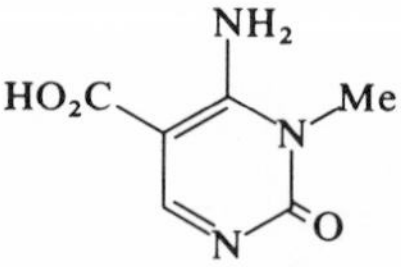

(14)

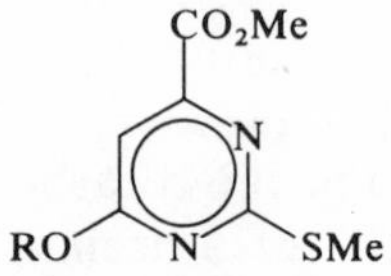

(15)

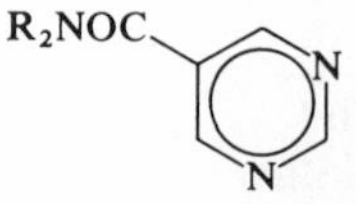

(16)

methoxycarbonyl-2,6-dihydroxypyrimidine (84%) using methanolic sulphuric acid;[3433] 2,4-diamino-6-ethyl-5-*p*-methoxycarbonylphenyl-pyrimidine (93%) by methanolic hydrogen chloride;[2282] and others.[2917, 3009, 3427, 3435]

* An interesting by-product, 4-methoxy-6-methoxycarbonyl-2-methylthiopyrimidine (**15**, R = Me; 17%), was also formed.[2214]

(3) *Formation of Acid Chlorides* (*H* 394)

Although chlorocarbonylpyrimidines continue to be made as intermediates, few have been purified and analysed. Thus 5-carboxypyrimidine and thionyl chloride gave 5-chlorocarbonylpyrimidine which was converted by ethereal ammonia into 5-carbamoylpyrimidine (**16**, R = H) in 67% overall yield, or by ethereal diethylamine into 5-diethylcarbamoylpyrimidine (**16**, R = Et);[2680] 2-amino-5-β-carboxyethyl- with thionyl chloride/pyridine gave 2-amino-5-β-chlorocarbonylethyl-4-hydroxy-6-methylpyrimidine (71%) which was characterized as 2-amino-4-hydroxy-6-methyl-5-β-phenylcarbamoylethylpyrimidine;[2232] 5-carboxy- was converted to 5-carbamoyl-2-methoxyl-4,6-dimethylpyrimidine *via* the acid chloride (made using thionyl chloride/triethylamine);[2519] and other such sequences are recorded, some using thionyl chloride/dimethylformamide as reagent.[2519, 3436, 3437]

Two methods for the preparation of the acid chloride of orotic acid, 4-chlorocarbonyl-2,6-dihydroxypyrimidine (**17**), have been reported. The first involved treatment of orotic acid monohydrate with an excess of thionyl chloride and a little pyridine. It gave a crude acid chloride, which 'was satisfactory for esterification without purification,' i.e., gave pure 2,4-dihydroxy-6-methoxycarbonylpyrimidine in excellent yield on refluxing with methanol.[2591] The second method consisted of boiling anhydrous orotic acid in benzene with thionyl chloride and dimethylformamide. It gave an apparently pure acid chloride (96%; with reasonable elemental analyses) which was converted into 4-carbamoyl-2,4-dihydroxypyrimidine in 95% yield.[3438] However, it has been claimed[3432] that the crude product obtained by the first procedure was 'largely the anhydride' (orotic anhydride; analysis given), and that the product from the second procedure 'was invariably a mixture of orotyl chloride, orotic anhydride, and orotic acid'. Whatever the truth, the so-called orotyl chloride was satisfactorily converted into a number of amides such as 2,4-dihydroxy-6-β-hydroxyethylcarbamoylpyrimidine,[3438] 4-carboxymethylcarbamoyl-2,6-dihydroxypyrimidine (**18**, R = H),[3438] and 2,4-dihydroxy-6-methoxycarbonylmethylcarbamoylpyrimidine (**18**, R = Me).[3439]

Orotic acid and phosphoryl chloride containing phosphorus pentachloride gave 2,4-dichloro-6-chlorocarbonylpyrimidine (64%), which in cold water gave 4-carboxy-2,6-dichloropyrimidine, or in methanolic sodium methoxide gave 2,4-dimethoxy-6-methoxycarbonylpyrimidine.[2591]

(4) *Other Reactions* (*H* 395)

Orotic acid has been converted directly (?) into the amide, 4-butylcarbamoyl-2,6-dihydroxypyrimidine (*ca.* 80%) by salt formation with butylamine followed by treatment (dehydration?) with a little phos-

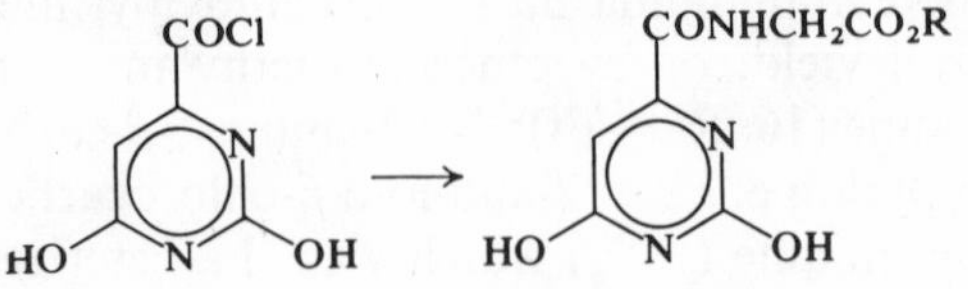

(17) → (18)

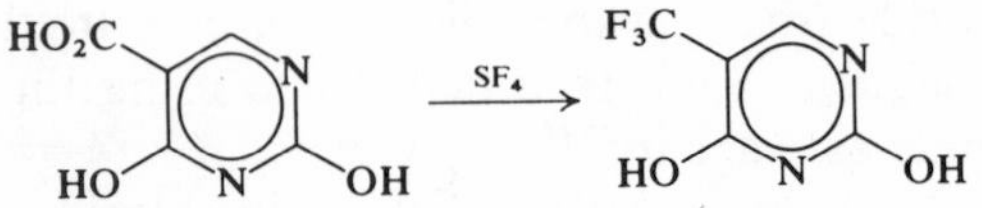

(19) —SF_4→ (20)

phoryl chloride in toluene.[3440] The 4- and 5-ethylcarbamoyl-, 4- and 5-diethylcarbamoyl-, and other such 2,6-dihydroxypyrimidines were made similarly.[3440]

5-Carboxy-2,6-dihydroxypyrimidine (**19**) has been converted into 2,4-dihydroxy-5-trifluoromethylpyrimidine (**20**; 77%) by sulphur tetrafluoride at room temperature.[2664]

2. Alkoxycarbonylpyrimidines (Esters) (*H* 395)

A. Preparation of Esters (*H* 395)

The esterification of carboxylic acids and the conversion of acid chlorides into esters[3436, 3437] have been discussed in Sects. 1.B(2) and (3) above. The formation of esters in the Principal Synthesis has been covered in Ch. II.

Transalkylation of an ester is possible sometimes. Thus 2-*p*-chlorophenyl-4-methoxy-6-methoxycarbonylmethylpyrimidine was converted into the corresponding benzyloxycarbonylmethyl compound by boiling in toluene with benzyl alcohol, aluminium foil, and some mercuric chloride; or into the isopropoxycarbonylmethyl compound by aluminium isopropoxide in boiling toluene.[3481]

The formation of an ester from an amide is illustrated in the treatment of 4-carbamoylmethyl-2-*p*-chlorophenylpyrimidine with hot methanolic hydrogen chloride to give (in 20 min.) the 4-methoxycarbonylmethyl analogue;[3481] the same ester was formed, also, from the corresponding nitrile by the action of hydrogen chloride in cold benzene-methanol followed by the addition of water.[3481]

The direct introduction of ester groupings into a *C*-methyl group is possible sometimes. Thus 4- or 5-*p*-chlorophenyl-2-methylpyrimidine and an equal weight of sodium hydride were refluxed with diethyl carbonate. Work-up under anhydrous conditions gave 4- or 5-*p*-chlorophenyl-2-diethoxycarbonylmethylpyrimidine (83% and 70%, respectively).[3481] When water was used during working-up, hydrolysis and (mono)decarboxylation ensued; such a process gave 4-carboxymethyl-6-*p*-chlorophenyl-2-ethoxypyrimidine.[3481]

The addition to a pyrimidine of a fragment already bearing an ester grouping is exemplified in the following conversions. 4-Methylpyrimidine with butyl glyoxylate in pyridine at 100° for 3 days gave 4-β-butoxycarbonylvinylpyrimidine (**21**);[3009] 4-chloro-6-ββ-diethoxyvinyl-5-nitropyrimidine (**22**) was hydrolysed by cold aq. ethanol to give 4-chloro-6-ethoxycarbonylmethyl-5-nitropyrimidine (**23**, R = Cl) which was not isolated but converted into the 4-amino analogue (**23**, R = NH_2);[2785] 4-hydroxy-2-mercapto- and ethyl chloroacetate gave 2-

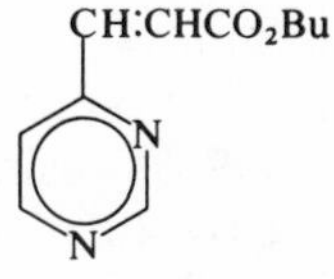

(21) 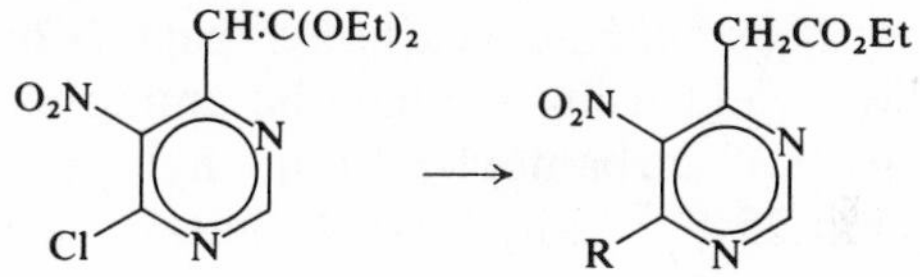 **(22)** **(23)**

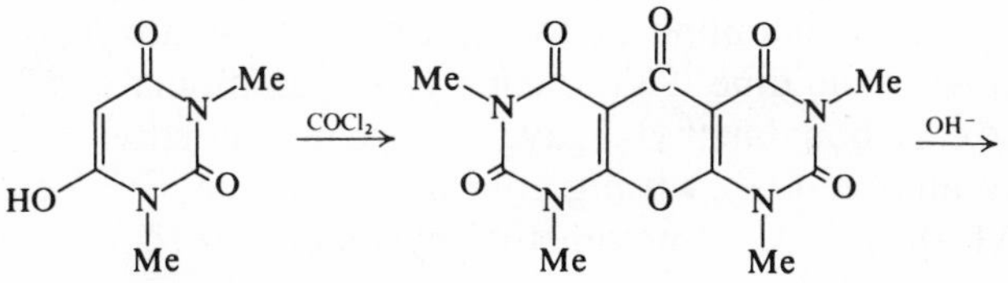

(24) **(25)**

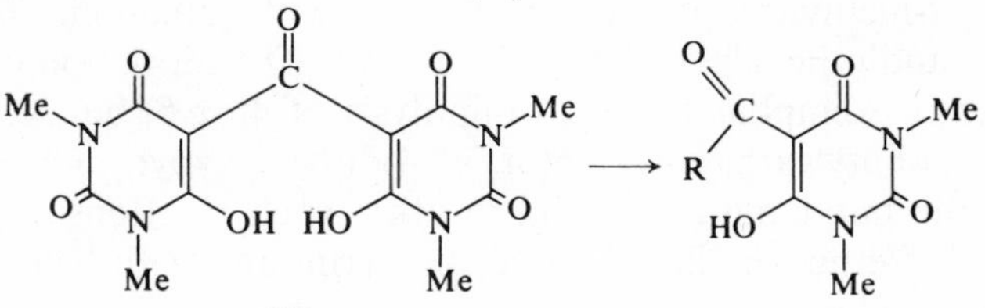

(26) **(27)**

ethoxycarbonylmethylthio - 4 - hydroxy - 5,6 - tetramethylenepyrimidine (74%);[2183] uracil and methyl acrylate gave 1-methoxycarbonylmethyluracil (30%);[3428] 2-amino-4-formyl- gave 2-amino-4-β-ethoxycarbonylvinyl-6-hydroxy-5-δ-phenylbutylpyrimidine on treatment with ethoxycarbonylmethylene triphenyl phosphorane;[2978] and other related reactions.[2192, 2237, 2290, 2310, 2652, 2788, 2827, 2860, 3277]

A rather specialized route to 1,2,3,4-tetrahydro-6-hydroxy-5-methoxycarbonyl-1,3-dimethyl-2,4-dioxopyrimidine (**27**, R = OMe) and the corresponding amide (**27**, R = NH_2) or anilide (**27**, R = NHPh) is provided by the sequence (**24**) → (**27**). The first step is catalysed by tetraethylurea and the last step is brought about by boiling methanol, ethanolic ammonia, or ethanolic aniline according to the desired product.[3266]

B. Reactions of Esters (*H* 396)

The hydrolysis and reduction of esters to give carboxy- and hydroxymethyl-pyrimidines, respectively, have been discussed in Sect. 1.A(2) above and in Ch. VII, Sect. 3.D. 4-Methoxycarbonylpyrimidine has been reduced to 4-formylpyrimidine (15%) with lithium aluminium hydride in tetrahydrofuran at −70°.[3009]

The formation of amides from esters is illustrated in the following examples: 4- or 5-methoxycarbonyl- gave 4- or 5-carbamoyl-pyrimidine in excellent yield using methanolic ammonia at 25°;[2806] 5-ethoxycarbonyl- gave 5-carbamoyl-2,4-dimethylpyrimidine (99%) using aq. ammonia at 25°;[3007] 2-hydroxy-4-methoxycarbonyl- gave 4-carbamoyl-2-hydroxy-6-methyl(or propyl)pyrimidine (89%) using aq. ammonia;[2184] 4-ethoxycarbonyl- gave 4-phenylcarbamoyl-pyrimidine (42%) in boiling aniline/aniline hydrochloride;[2427] 5-ethoxycarbonyl- gave 5-carbamoyl-pyrimidine (98%) using liquid ammonia at 60°;[2323] 4-chloro-5-ethoxycarbonylmethyl- gave 5-carbamoylmethyl-4-chloro-2-methylthiopyrimidine (79%) using cold aq. ammonia;[2744] 5-ethoxycarbonylmethyl- gave 5-carbamoylmethyl-pyrimidine (87%) in ethanolic ammonia;[2577] 4-chloro-6-methoxycarbonyl- gave 4-carbamoyl-6-chloro-2-methylthiopyrimidine (66%) with ethanolic ammonia at 0°;[2214] and others.[2175, 2184, 2282, 2522, 3427] The formation of substituted amides is exemplified in the aminolysis of 4(or 5)-butoxycarbonyl- to 4(or 5)-butylcarbamoyl-2,6(or 2,4)dihydroxypyrimidine (refluxing ethanolic butylamine),[3431] and in other such reactions.[2214, 3432] Treatment of 5-ethoxycarbonyl-4-hydroxypyrimidine with benzylamine gave acyclic products.[3442]

Hydrazides are usually made from esters with ethanolic hydrazine hydrate. Thus were made 5-hydrazinocarbonylpyrimidine (54%),[2323] 5-hydrazinocarbonylmethylpyrimidine (93%),[2577] 2-*NN*-dimethylguanidino-5-hydrazinocarbonylmethyl-4-hydroxypyrimidine,[2208] 4(or 5)-hydrazinocarbonyl-2,6(or 2,4)-dihydroxypyrimidine,[3431] 4-ethyl-6-hydrazinocarbonyl-2-hydroxypyrimidine,[2184] 3-hydrazinocarbonylmethyluracil,[3427] and analogues.[2184, 2208, 2282, 2522]

An apparently direct removal of an ester group occurred on heating 4-amino-5-ethoxycarbonyl-2-mercaptopyrimidine in formamide for 2 days: 4-amino-2-mercaptopyrimidine resulted in *ca.* 70% yield.[2275]

3. Carbamoylpyrimidines (Amides) and Related Compounds (Hydrazides and Azides) (*H* 397)

A. Preparation of Amides (*H* 397)

The conversion of esters into amides has been discussed above in Sect. 2.B, and acid chlorides into amides has been mentioned in Sect. 1.A(3) and is further exemplified in the reaction of 4-chlorocarbonyl-2,6-dihydroxypyrimidine with amines to give 4-benzylcarbamoyl-, 4-butylcarbamoyl-, 4-2′-pyridylcarbamoyl- (**28**), and other 2,6-dihydroxypyrimidines,[3441] as well as in other such reactions.[2232, 2680, 3436, 3437] A few amides have been made by primary syntheses (Chs. II and III).

(3) *By Controlled Hydrolysis of Nitriles* (*H* 398)

The conditions for making amides from nitriles are illustrated in the preparation of 4-carbamoyl-2-chloro- (47%; sulphuric acid at 25°),[2954] 2-amino-5-carbamoyl-4-methyl(or 4,6-dimethyl)- (75%, 62%; sulphuric acid at 25° and 100°, respectively),[2602, 2633] 4-amino-5-carbamoyl-2-methyl- (*ca.* 90%; by boiling with an aqueous suspension of Amberlite IRA-400 resin),[2698] 4-amino-5-carbamoyl-2-methylamino-* (*ca.* 70%; sulphuric acid at 30°),[2273] and 4-amino-5-carbamoyl-2-methylthiopyrimidine (80%; refluxing 0.1*N*-sodium hydroxide).[3062] 5-Acetoxymethyl-4-cyano-2-methylthiopyrimidine in methanolic ammonia at 20° gave 4-carbamoyl-5-hydroxymethyl-2-methylthiopyrimidine (63%) and 4-carboxy-5-hydroxymethyl-2-methylthiopyrimidine lactone (7%).[3443]

* Seven analogues bearing different 2-substituents were made similarly.[2273]

No successes with the Radziszewski reaction (hydrogen peroxide/sodium hydroxide) have been reported recently.

Treatment of 2-, 4-, or 5-cyanopyrimidine with ethanolic ammonium hydrogen sulphide gave 2-, 4-, or 5-thiocarbamoylpyrimidine (**29**, X = S) in 80–90% yield.[2806] Each was converted by refluxing ethanolic hydroxylamine into the corresponding hydroxyamidinopyrimidine (**29**, R = NOH; 23–92% yield).[2806]

(4) *By Other Means* (*H* 399)

An amide group was introduced into the vacant 5-position of 4-amino-2,6-dihydroxy- and 2,4-diamino-6-hydroxy-pyrimidine on fusion with urea.[3352] The resulting 5-carbamoyl derivatives (**30**, R = OH or NH_2) were first described as ureido derivatives.[3228]

4-Methylpyrimidine, sulphur, and boiling aniline gave 4-*N*-phenyl-(thiocarbamoyl)pyrimidine (**31**, X = S; 57%), which was converted into the 4-phenylcarbamoylpyrimidine (**31**, X = O; 70%) by heating with selenium dioxide in dioxan.[2427] Analogues of the thioanilide and anilide were made similarly.[2427]

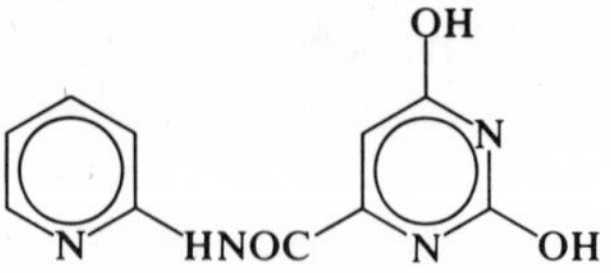

(28) 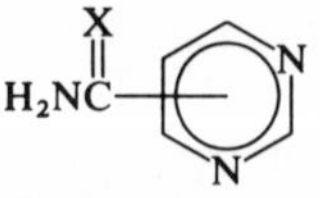 **(29)**

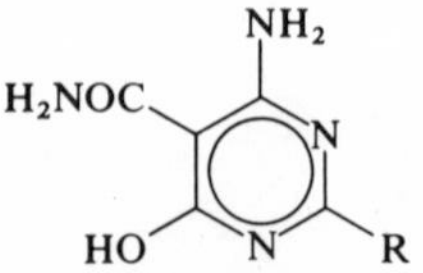

(30) 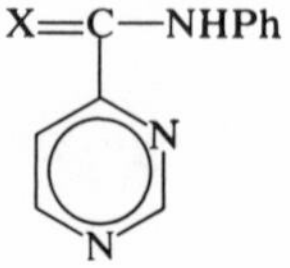 **(31)**

When 2- or 4-cyanopyrimidine was warmed with an ether-benzene solution of diethylamino magnesium bromide (Et_2NMgBr; made from ethyl magnesium bromide and diethylamine), 2- or 4-diethylcarbamoylpyrimidine (*ca.* 20%) and 2- or 4-carbamoylpyrimidine (*ca.* 35%) were isolated.[2806]

B. Reactions of Amides (*H* 400)

The hydrolysis of amides to carboxylic acids has been mentioned in Sect. 1.A(2) above; and the Hofmann and related reactions leading to aminopyrimidines have been exemplified in appropriate sections of Ch. IX.

The dehydration of amides to nitriles has been used to make 2-, 4-, or 5-cyanopyrimidine (70–80%) by refluxing the appropriate amide in phosphoryl chloride,[2806] 5-cyanopyrimidine (77%) by dry distillation of the amide with phosphorus pentoxide at 250°,[2680] 5-cyano-2,4-dimethylpyrimidine (49%) by refluxing the amide in xylene containing phosphoryl chloride,[3007] 4,5-dicyano-2-methylpyrimidine (57%) from the corresponding diamide by the same method,[2175] and 5-cyano-2-methylpyrimidine (44%) by adding the amide to pyridine and phosphoryl chloride (exothermic).[2681]

RNH_2 ; $CONHNH_2$ (32) $\xrightarrow{HNO_2}$ CON_3 (33) $\xrightarrow{ROH}$ $NHCO_2R$ (34); (33) $\xrightarrow{RNH_2}$ CONHR (35); (32) → CONHN:CRMe (36); (32) → CONHNHC(:O)R (37); (34) $\xrightarrow{H_2\ (R\ =\ CH_2Ph)}$ NH_2 (38); each on a 2,6-dihydroxypyrimidine ring (HO, N, OH)

C. Preparation and Reactions of Hydrazides and Azides (*H* 400)

The preparation of acid hydrazides from esters has been discussed in Sect. 2.B above. The conversion of hydrazides into azides is illustrated in the preparation of 4-azidocarbonyl-2,6-dihydroxypyrimidine (**33**; 56%) from the hydrazide (**32**) with nitrous acid.[3431] The azide (**33**) has been converted by appropriate alcohols into the urethanes, 4-ethoxycarbonylamino- (**34**, R = Et) 4-butoxycarbonylamino- (**34**, R = Bu), and other such derivatives of 2,6-dihydroxypyrimidine.[3348] The same azide did not rearrange to the corresponding isocyanate in boiling toluene, was hydrolysed to orotic acid in boiling water, and gave with amines the appropriate *N*-alkylamides, e.g., 4-ethylcarbamoyl- (**35**,

R = Et), 4-benzylcarbamoyl- (**35**, R = CH_2Ph), and other such 2,6-dihydroxypyrimidines.[3444]

Some of the possible reactions of acid hydrazides are illustrated by the conversion of 4-hydrazinocarbonyl-2,6-dihydroxypyrimidine (**32**) into 4-β-ethylidenehydrazinocarbonyl- (**36**, R = H; by acetaldehyde), 4-β-isopropylidenehydrazinocarbonyl- (**36**, R = Me; by acetone), 4-β-(α-carboxyethylidene)hydrazinocarbonyl- (**36**, R = CO_2H; by pyruvic acid), 4-β-formylhydrazinocarbonyl- (**37**, R = H; by formic acid), and 4-β-*p*-toluenesulphonylhydrazinocarbonyl-2,6-dihydroxypyrimidine (**37**, R = *p*-$MeC_6H_4SO_2$; by *p*-toluenesulphonyl chloride).[3444] The benzyloxycarbonylamino derivative (**34**, R = CH_2Ph) underwent hydrogenolysis to 4-amino-2,6-dihydroxypyrimidine (**38**).[3348]

4. Pyrimidine Nitriles (*H* 401)

The preparation of nitriles by primary syntheses (Chs. II and III) and by dehydration of amides (Sect. 5.B, above) have been covered already.

(3) *By Other Means of Preparation* (*H* 404)

As mentioned in Ch. VI, cuprous cyanide has been used to convert a 4-iodo- to a 4-cyano-pyrimidine,[2608] and 5-bromo- into 5-cyanopyrimidine.[2607] Several other 5-cyanopyrimidines[2376, 2602, 2633] and an ω-cyanopyrimidine[2980] have been made also by direct replacement of a halogeno substituent. However this route is seldom satisfactory and it is often better to proceed *via* a sulpho or trimethylammonio intermediate. An illustration of the first indirect route was the reaction of 2-chloropyrimidine with sodium sulphite followed by potassium cyanide to give 21% of 2-cyanopyrimidine;[2806] and of the second route, the reaction of 2-chloropyrimidine with trimethylamine to give crude trimethyl pyrimidin-2-ylammonium chloride (**39**), which reacted with potassium cyanide in acetamide, to give 2-cyanopyrimidine (*ca.* 40%).[2607, 3445] Other examples of such reactions have been discussed (Ch. IX, Sect. 8.G).

The oximes of pyrimidine aldehydes, which may be made either from the aldehydes or by nitrosation of suitable methylpyrimidines, can be dehydrated to give pyrimidine nitriles. Thus 2-hydroxy-4-hydroxyiminomethylpyrimidine (**40**, R = H) in phosphoryl chloride gave 2-chloro-4-cyanopyrimidine (**41**, R = H; 60%);[2954] its 6-phenyl

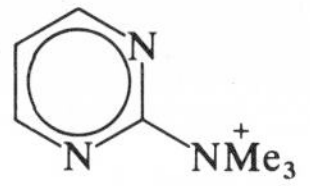

CH:NOH
N
Cl^-
N
$\overset{+}{N}Me_3$
N
$POCl_3$
R
N
OH
(39) 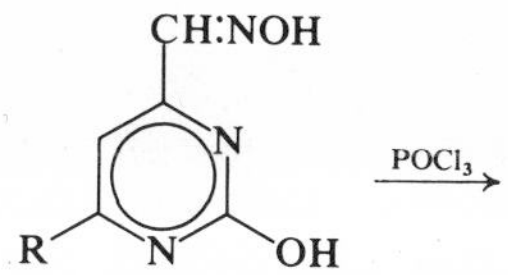
(40)

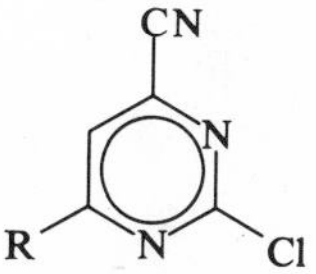

CN
CH:NOH
EtO_2C
N
N
R
N
Cl
N
SMe
(41)
(42) 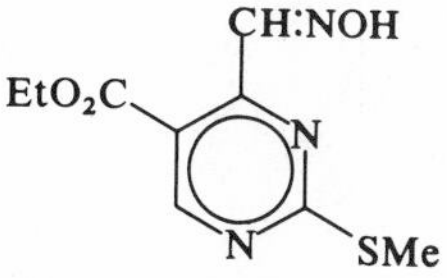

derivative (**41**, R = Ph; 50%) was made similarly;[2592] and 5-ethoxy-carbonyl-4-hydroxyiminomethyl-2-methylthiopyrimidine (**42**) and refluxing acetic anhydride gave 4-cyano-5-ethoxycarbonyl-2-methylthio-pyrimidine (90%).[2599]

B. Reactions of Cyanopyrimidines (*H* 405)

Most of the reactions of nitriles have already been discussed (see *H* 405 for references to appropriate sections).

The 2-, 4-, and 5-iminoethers of pyrimidine (**44**; 2-, 4-, and 5-*C*-ethoxy-*C*-iminomethylpyrimidine) have been made from appropriate nitriles (**43**) in the usual way. The 2- and 5-isomers were converted into the corresponding amidinopyrimidines (**45**),[2806] and were also used to make *N*-substituted amidines by warming with aromatic amines in the presence of aluminium chloride. Some *N*-substituted amidines were made directly from 2-, 4-, or 5-cyanopyrimidine by warming with an aromatic amine in the presence of aluminium chloride. Typical products were 2-phenylamidino- (67%), 2-β-naphthylamidino- (63%), 4-*p*-tolylamidino- (67%), 4-benzylamidino- (36%), 5-*p*-methoxyphenyl-amidino- (20%), and 5-thiazol-2′-ylamidino-pyrimidine (**46**; 10%).[2806] 2-Cyanopyrimidine reacted with ammonium benzenesulphonate at 240° to give the benzenesulphonate salt of 2-amidinopyrimidine (42%);[2603] a similar route gave 2-amidino-4,6-dimethyl-, 2-amidino-4-methyl-6-phenyl-, and 4-amidino-6-methyl-2-phenyl-pyrimidine, a useful general method.[2603]

4-Cyanopyrimidine was converted by methyl magnesium iodide into 4-acetylpyrimidine (20%) but 2-cyanopyrimidine failed to undergo a similar reaction.[2806] Some miscellaneous reactions of nitriles are

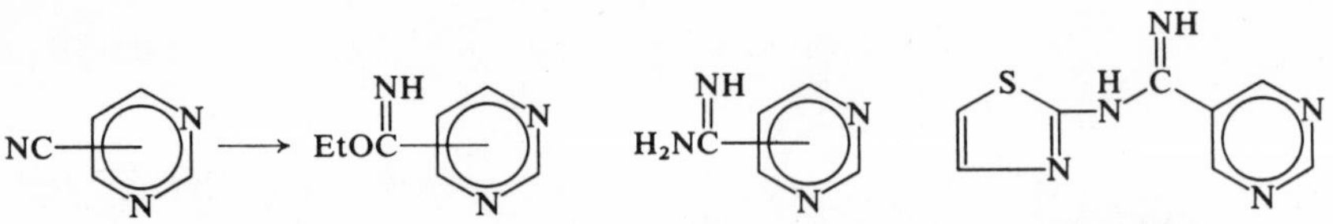

(43) (44) (45) (46)

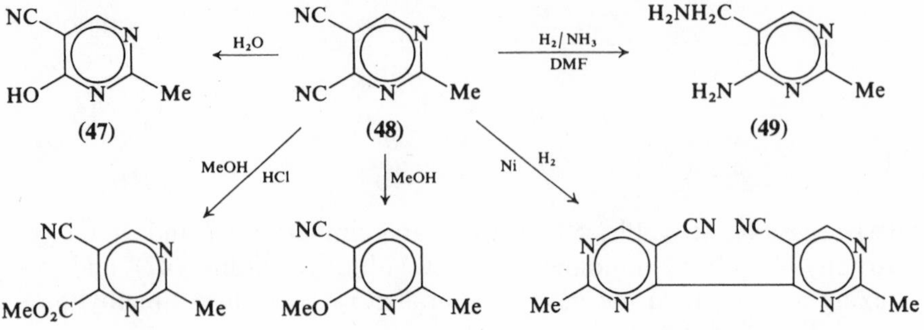

(47) (48) (49)

(50) (51) (52)

illustrated in the following transformations of 4,5-dicyano-2-methylpyrimidine (**48**): hot water gave 5-cyano-4-hydroxy-2-methylpyrimidine (**47**); hydrogenation in dimethylformamide containing ammonia gave 4-amino-5-aminomethyl-2-methylpyrimidine (**49**); methanolic hydrogen chloride gave 5-cyano-4-methoxycarbonyl-2-methylpyrimidine (**50**); refluxing methanol gave 5-cyano-4-methoxy-2-methylpyrimidine (**51**); and hydrogenation over nickel in dimethylformamide gave 5-cyano-4-5′-cyano-2′-methylpyrimidin-4′-yl-2-methylpyrimidine (**52**).[2175] Attempted reduction of 5-acetoxymethyl-4-cyano-2-methylthiopyrimidine with chromous acetate caused reductive cleavage at the CN group to give 5-hydroxymethyl-2-methylthiopyrimidine.[3443]

4-Amino-5-cyanopyrimidine has been converted into 4-amino-5-formylpyrimidine by hydrogenation in acidic media over Pd/C and subsequent treatment with aq. ammonia; the yield was 49%.[3764] 5-Formyl-4-hydroxypyrimidine (60%) was made similarly from 5-cyano-4-hydroxypyrimidine.[3764]

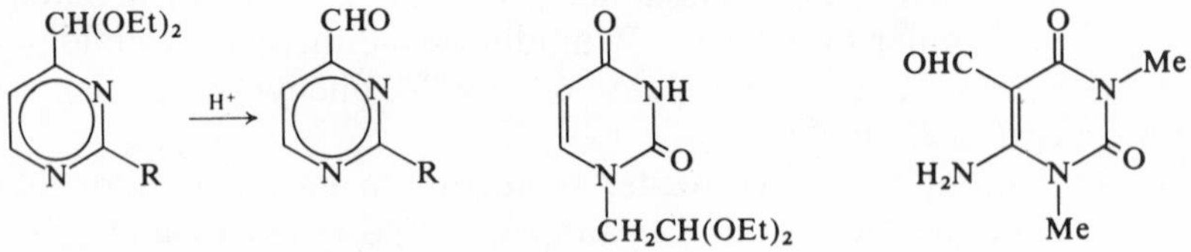

(53) (54) (55) (56)

5. Pyrimidine Aldehydes and Their Acetals (*H* 406)

A. Preparation of Formylpyrimidines (*H* 406)

(1) *By Primary Syntheses* (*H* 406)

The formation of acetals by primary syntheses has been covered in Chs. II and III. Unmasking of the aldehyde group is subsequently done by acid hydrolysis. Thus formamidine acetate and α-diethoxyacetyl-β-dimethylaminoethylene [$Me_2NCH{:}CHCOC(OEt)_2$] gave 4-diethoxymethylpyrimidine (**53**, R = H; 74%) which in aq. sulphuric acid at 60° gave 4-formylpyrimidine (**54**, R = H; 47%).[2301] In a similar way, 4-diethoxymethyl-2-phenylpyrimidine (**53**, R = Ph) was prepared and then converted into 4-formyl-2-phenylpyrimidine (**54**, R = Ph) using aq. ethanolic hydrochloric acid;[2301] the same general route provided 4-formyl-6-methylpyrimidine, its 2-methylthio derivative, 4-formyl-2-methylthiopyrimidine (**54**, R = SMe), 4-formyl-2-methylpyrimidine, and other such derivatives.[2301] 1-ββ-Diethoxyethyluracil (**55**), prepared by primary synthesis, was converted into 1-formylmethyluracil by warming with dilute hydrochloric acid; the product was isolated as a hydrate and further characterized as 1-β-2′,4′-dinitrophenylhydrazonoethyluracil.[2338]

Diethylacetals were also used as intermediates to give 2-acetamido-5-β-formylethyl-4-hydroxy-6-methyl(or phenyl)pyrimidine (by boiling in water),[2176, 2220] 2-acetamido-4-formyl-6-hydroxy-5-δ-phenylbutylpyrimidine (by 98% formic acid at 100°),[2979] 5-β-formylethyl-4-hydroxy-6-methylpyrimidine (by boiling water),[3190] 2,4-diamino-5-*p*-chlorophenyl-6-formylpyrimidine (by hot aq. hydrochloric acid),[2981] 2,4-diamino(or diacetamido)-6-formyl-5-δ-phenylbutylpyrimidine (by aq. ethanolic hydrochloric acid or 98% formic acid, respectively),[2981] 4-formyl-6-hydroxy-2-mercaptopyrimidine (by aq. sulphuric acid),[2219] and ten related compounds.[2219]

(2) *By Direct and Indirect* C-*Formylation* (*H* 408)

The 5-formylation of 4-amino-1,2,3,6-tetrahydro-1,3-dimethyl-2,6-dioxopyrimidine by formic acetic anhydride (*H* 408) has been confirmed and extended: diethyl malonyl dichloride gave the same 5-formyl derivative (**56**) as did the anhydride! The mechanism is complicated.[2361]

The Reimer–Tiemann reaction has been used to make 5-formyl-

2,4-dihydroxypyrimidine (45%),[3193] and also a number of other 5-formyl derivatives all isolated only as their hydrazones.[3446, 3447]

The Vilsmeier reaction has been applied successfully to several alkoxy- or amino-pyrimidines. Thus the dimethylformamide-phosphoryl chloride adduct reacted with solutions of 2,4,6-trimethoxy-, 2-dimethylamino-4,6-dimethyl-, and 4-chloro-2-dimethylamino-6-methoxypyrimidine in dimethylformamide to give the corresponding 5-formyl derivatives in up to 88% yield.[2815] When an hydroxy group was present, it was usually converted into a chloro substituent at the same time: 2-amino-4,6-dihydroxy- gave 2-amino-4,6-dichloro-5-formylpyrimidine (**57**; 28%);[2593] and 4-amino-2,6-dichloro-5-formyl-, 4-amino-6-chloro-5-formyl-2-methylthio-, and 4,6-dichloro-5-formyl-pyrimidine were obtained in this way.[2593] Another complication was the occasional attack of an amino group in lieu of formylation: 4-amino-2,6-dichloro-(or dimethoxy)pyrimidine gave 2,4-dichloro(or dimethoxy)-6-dimethylaminomethyleneaminopyrimidine (**58**, R = Cl or OMe).[2593] 4,6-Dichloropyrimidine was unaffected by dimethylformamide-phosphoryl chloride.[2593]

The reaction has also been carried out in two stages using the reagent phosgene-dimethylformamide: 4,6-dihydroxypyrimidine gave the hydrochloride of 5-dimethylaminomethylene-4,5-dihydro-6-hydroxy-4-oxo- (**59**, R = H) which with water gave 5-formyl-4,6-dihydroxypyrimidine (**60**, R = H).[2594, 2732] The 2-methyl and 2-phenyl derivatives (**60**, R = Me or Ph) were made similarly.[2594]

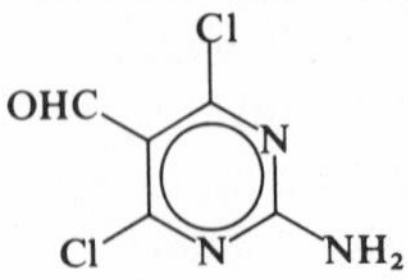

(**57**) 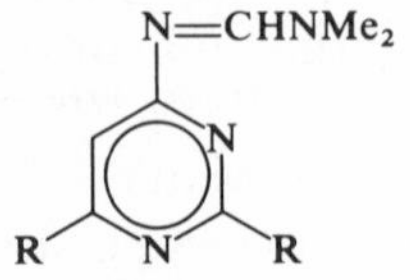 (**58**) 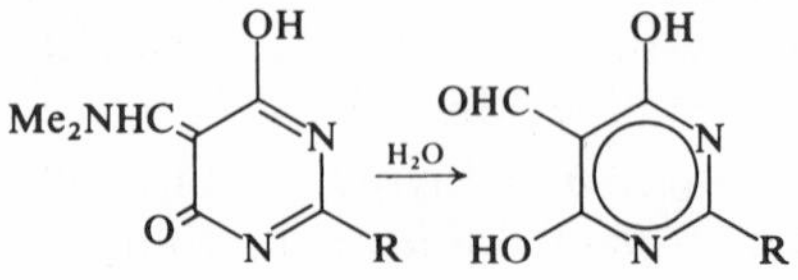 (**59**) $\xrightarrow{H_2O}$ (**60**)

(**61**) $\xrightarrow{SeO_2}$ (**62**) (**63**) $\xrightarrow{Ox}$ (**64**)

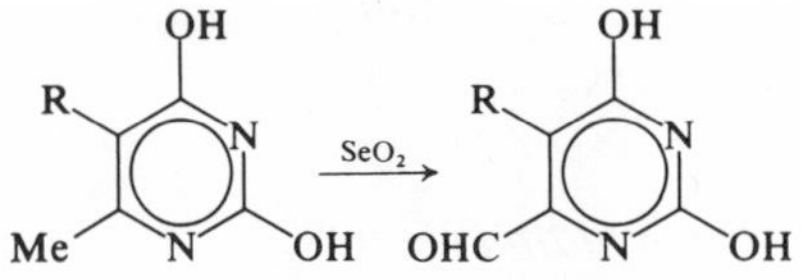

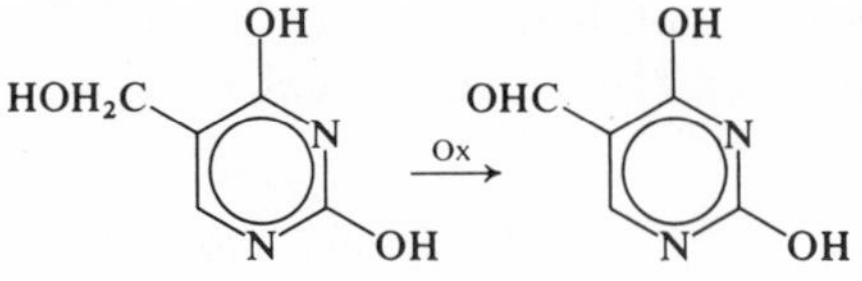

(4) *By Oxidative or Reductive Processes* (*H* 411)

Selenium dioxide has been used to convert 2,4-dihydroxy-6-methylpyrimidine (**61**, R = H) and its 5,6-dimethyl homologue (**61**, R = Me) into 4-formyl-2,6-dihydroxypyrimidine (**62**, R = H; 58%) and its 5-methyl derivative (**62**, R = Me; 94%), respectively.[2947] 2,4-Dihydroxy-5-hydroxymethylpyrimidine (**63**) has been oxidized to 5-formyl-2,4-dihydroxypyrimidine (**64**) by sodium dichromate in acetic acid (34% yield),[3116] manganese dioxide in boiling water (40–60% yield),[3063] manganese dioxide in dimethylsulphoxide at 100° (85% yield),[3063] ceric sulphate (70% yield),[3063] and aq. potassium persulphate containing a little silver ion (70–90% yield).[3063]

A significant reduction is that of 4-methoxycarbonylpyrimidine by lithium aluminium hydride in tetrahydrofuran at −70° to give 4-formylpyrimidine in 15% yield.[3009]

(5) *By Other Means* (*H* 412)

5-Bromo-2-formylpyrimidine (**66**) has been made by reducing 5-bromo-2-tribromomethylpyrimidine with tin and hydrochloric acid in acetone, and treating the resulting 2-dibromomethyl analogue (**65**) with silver nitrate.[2575]

The conversion of a methylpyrimidine into the corresponding hydroxyiminomethylpyrimidine (oxime of the aldehyde) has been discussed in Ch. IV, Sect. 2.C. A typical example of this method is the reaction of 4-methylpyrimidine (**67**) with amyl nitrite/sodium hydride

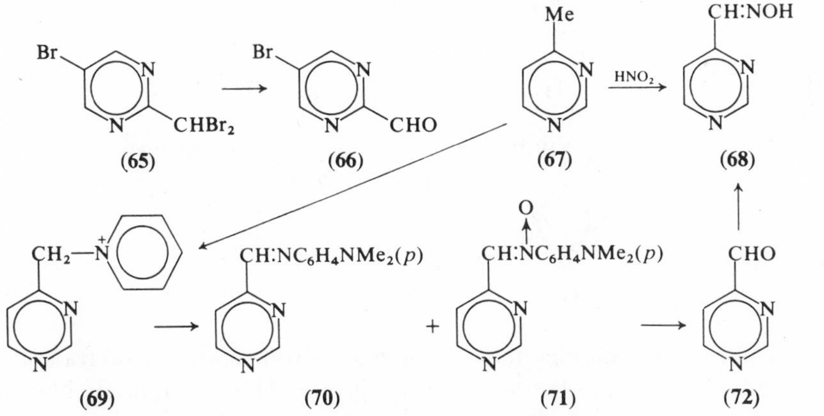

to give 4-hydroxyiminomethylpyrimidine (**68**).[3242] The same transformation has been done by treating 4-methylpyrimidine (**67**) with iodine/pyridine to give *N*-pyrimidin-4′-ylmethylpyridinium iodide (**69**); this with *p*-nitroso-*NN*-dimethylaniline gave a mixture of 4-*p*-dimethylaminophenyliminomethylpyrimidine (**70**) and the corresponding nitrone

(**71**), both of which gave 4-formylpyrimidine (**72**) on acidic hydrolysis; the crude product was treated with hydroxylamine and isolated as the oxime (**68**).[3242]

An apparent (rather than real) displacement of a 5-nitroso group by a formyl group has been recorded: 4-amino-3-benzyloxy-1,2,3,6-tetrahydro-1-methyl-5-nitroso-2,6-dioxopyrimidine (**73**, R = NO) was reduced by Raney nickel/formic acid to give the 5-formyl analogue (**73**, R = CHO) of proven structure; a mechanism has been suggested.[3397]

An anomalous preparation of 2,4-dihydroxy-5-hydroxyiminomethylpyrimidine by treatment of 5-chloromethyluracil with hydroxylamine has been described;[2820] in a similar way, 5-chloromethyl- or 5-mercaptomethyl-uracil and phenylhydrazine gave 2,4-dihydroxy-5-phenylhydrazonomethylpyrimidine (**74**; *ca.* 40%).[2820]

4-Amino-5-formyl-2,3-dihydro-2-imino-3-methylpyrimidine was formed by an abnormal Dimroth rearrangement of 5-cyano-1,2-dihydro-2-imino-1-methylpyrimidine.[2376] The mechanism has been discussed in Ch. X, Sect. 2.B.

4-Amino-5-formyl- and 5-formyl-4-hydroxy-pyrimidine have been made from the 5-cyano analogues by hydrogenation.[3764]

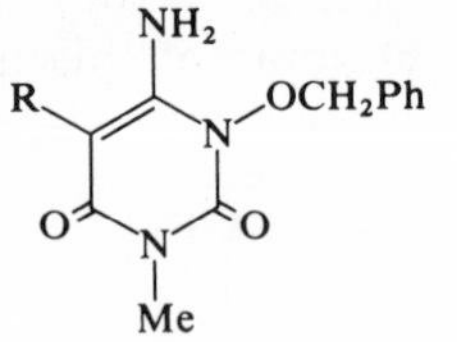

(**73**)

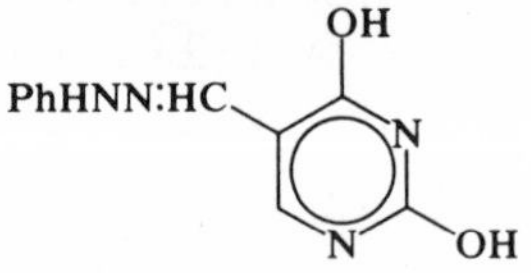

(**74**)

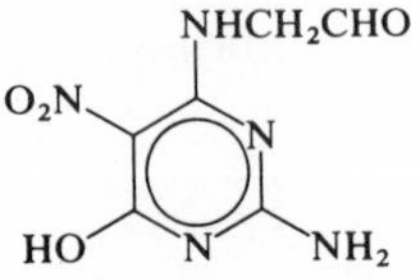

(**75**)

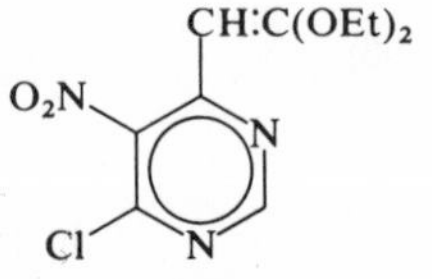

(**76**)

ω-Formylpyrimidines have been made from addition of fragments already bearing a potential aldehydo group. Thus 2-amino-4-chloro-6-hydroxy-5-nitropyrimidine and aminoacetaldehyde diethylacetal gave 2-amino-4-ββ-diethoxyethylamino-6-hydroxy-5-nitropyrimidine (67%) which was hydrolysed to 2-amino-4-formylmethylamino-6-hydroxy-5-nitropyrimidine (**75**; 86%) by acid.[2715] A similar route gave 4-ββ-diethoxyethylamino- and thence 4-formylmethylamino-2-hydroxy-5-nitropyrimidine;[2948] 4,5-diamino-2-hydroxypyrimidine treated with glyoxal and subsequently hydrogenated in ethanol over Raney nickel gave 4-amino-5-$\beta\beta$-diethoxyethylamino-2-hydroxypyrimidine.[2948]

When 4,6-dichloro-5-nitropyrimidine was allowed to react with ketene diethylacetal, 4-chloro-6-$\beta\beta$-diethoxyvinyl-5-nitropyrimidine (**76**) was formed. This is not the acetal of an aldehyde, but the equivalent of an ortho ester: aq. ethanol gave 4-chloro-6-ethoxycarbonylmethyl-5-nitropyrimidine, isolated after ammonolysis to the 4-amino analogue.[2785]

B. Reactions of Formylpyrimidines (*H* 413)

(1) *Oxidation and Reduction* (*H* 413)

There are no recent examples of the oxidation of aldehydes to acids; the reduction of aldehydes to hydroxymethylpyrimidines has been discussed (Ch. VII, Sect. 3.F).

(2) *Formation of the Usual Aldehyde Derivatives* (*H* 413)

The aldoximes, 5-hydroxyiminomethyl-2,4,6-trimethoxy-, 2-dimethylamino-5-hydroxyiminomethyl-4,6-dimethyl-, and 4-chloro-2-dimethylamino-5-hydroxyiminomethyl-6-methoxy-pyrimidine, are typical examples of the direct formation of such derivatives from the corresponding aldehydes.[2815] 2-Amino-4-hydroxy-6-β-hydroxyiminoethylamino-5-nitropyrimidine was made from the corresponding formylmethylamino derivative,[2715] and 2-ethylthio-4-hydroxy-6-hydroxyiminomethylpyrimidine (**78**) was made from the acetal, 4-diethoxymethyl-2-ethylthio-6-hydroxypyrimidine (**77**).[2611]

Some typical hydrazones prepared from the corresponding aldehydes (or their derivatives) include 2,4-dihydroxy-6-phenylhydrazonomethyl- (**79**; 74%),[2947] 2,4-dihydroxy-5-phenylhydrazonomethyl-,[3449] 5-2′,4′-dinitrophenylhydrazonomethyl-4-hydroxy-2-mercapto-,[3446] 5-dimethylhydrazonomethyl-4,6-dihydroxy-2-mercapto-,[3446] 2,4,6-trihydroxy-5-*p*-nitrophenylhydrazonomethyl-,[3447] 4,6-dihydroxy-5-phenylhydrazonomethyl-,[2594] 4-*p*-chlorophenylhydrazonomethyl-5-ethyl-6-hydroxy-2-mercapto-,[2219] and 4-hydrazonomethyl-2,6-dihydroxy-pyrimidine.[3448] Other examples are contained in the above references. Other types of hydrazone are exemplified in 6-benzenesulphonylhydrazonomethyl-3-methyl-2-thiouracil (**80**),[2219] 2,4-dihydroxy-6-α-naphthylmethylenehydrazonomethylpyrimidine,[3448] and 5-formylhydrazonome-

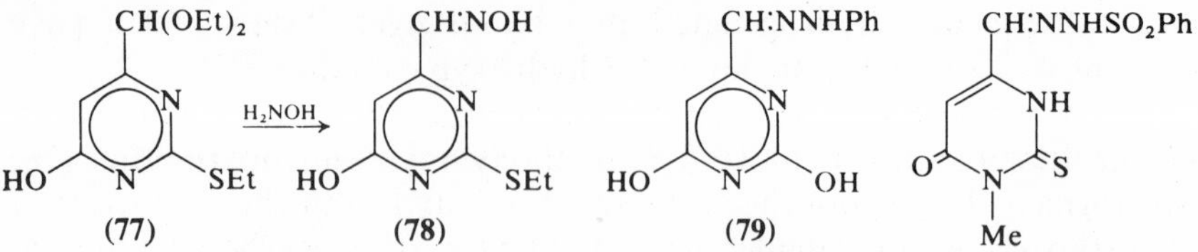

$CH(OEt)_2$ N HO N SEt (77) $\xrightarrow{H_2NOH}$ CH:NOH N HO N SEt (78) CH:NNHPh N HO N OH (79) $CH:NNHSO_2Ph$ NH O N S Me (80)

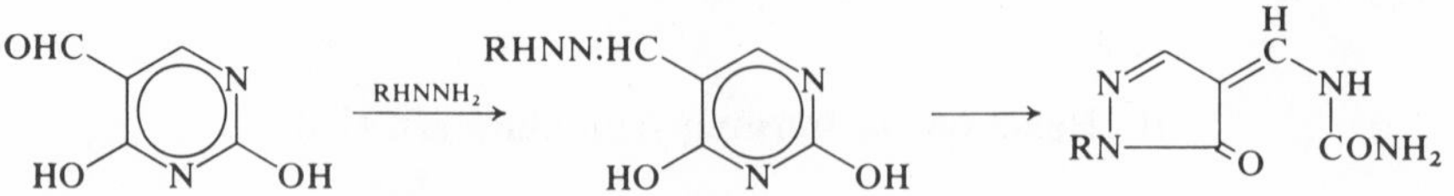

OHC N HO N OH (81) $\xrightarrow{RHNNH_2}$ RHNN:HC N HO N OH (82) ⟶ H C N NH RN O $CONH_2$ (83)

thyl-2,4-dihydroxypyrimidine.[3449] When 5-formyl-2,6-dihydroxypyrimidine (**81**) was treated with hydrazine or methylhydrazine under exceptionally gentle conditions, 5-hydrazonomethyl-2,4-dihydroxy- (**82**, R = H) and 2,4-dihydroxy-5-methylhydrazonomethyl-pyrimidine (**82**, R = Me), respectively, were isolated in small yield;[3449] under more normal conditions, the hydrazones rearranged to yield 4,5-dihydro-5-oxo-4-ureidomethylenepyrazole (**83**, R = H) and its 1-methyl derivative (**83**, R = Me), respectively.[3449]

Examples of semicarbazones are 4-hydroxy-2-mercapto-6-semicarbazonomethylpyrimidine (**84**, X = O),[2219] its thiosemicarbazonomethyl analogue (**84**, X = S),[2219] 5-ethyl-4-hydroxy-2-mercapto-6-phenylsemicarbazonomethylpyrimidine,[2219] 2,4-dihydroxy-6-thiosemicarbazonomethylpyrimidine,[2947] and such like.[3446]

The type of Schiff base made from a pyrimidine aldehyde and an amine (generally aromatic) are well represented, e.g., by 5-*p*-fluorophenyliminomethyl-2,4,6-trihydroxypyrimidine (**85**),[3447] its 2-mercapto analogue,[3446] 4-allyl(or benzyl)-iminomethyl-6-hydroxy-2-mercaptopyrimidine,[2219] and 2-amino-4-mercapto-6-methyl-5-γ-phenyliminopropylpyrimidine;*[2566] the last mentioned anil was reduced by methanolic sodium borohydride to give 2-amino-5-γ-anilinopropyl-4-mercapto-6-methylpyrimidine.[2566] Similar reductions of such anils have furnished 2-amino-4-hydroxy-6-methyl-5-γ-*m*-trifluoromethylanilinopropylpyrimidine,[3268] its 5-γ-butylaminopropyl analogue,[3450] and other such compounds.[3268]

The recovery of an aldehyde from its Schiff-base is illustrated in the reaction of 2,4-diamino-6-*p*-bromoanilino-5-*p*-bromophenylimino-

* This compound actually has a bicyclic structure[2566] but can be thought of as an anil for practical purposes.

methylpyrimidine with refluxing 0.1*N*-hydrochloric acid to give the 5-formyl analogue (61%).[3461]

(3) *Other Reactions* (*H* 414)

An aldehyde may be converted into the corresponding nitrile *via* its oxime: e.g., 4-diethoxymethyl-2-ethylthio-6-hydroxy- gave 2-ethylthio-4-hydroxy-6-hydroxyiminomethyl- and thence (using acetic anhydride as dehydrating agent) 4-cyano-2-ethylthio-6-hydroxy-pyrimidine.[2611]

An aldehyde may serve as starting material to make a dihalogenomethyl derivative: 5-formyl- gave 5-difluoromethyl-2,4-dihydroxy-pyrimidine (**86**) by prolonged treatment with sulphur tetrafluoride.[2664]

The 'pyrimidoin', 4-$\alpha\beta$-dihydroxy-β-pyrimidin-4'-ylvinylpyrimidine (enolic form; **87**) was made in 60% yield by the action of potassium cyanide on the hydrate of 4-formylpyrimidine.[2301]

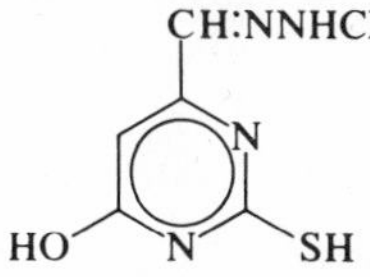

(84)

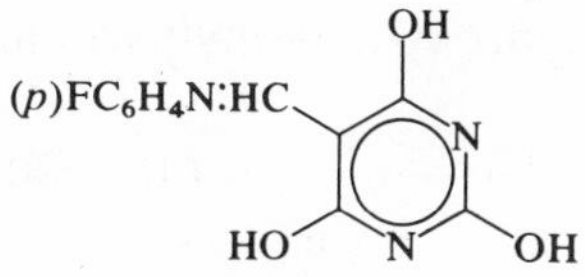

(85)

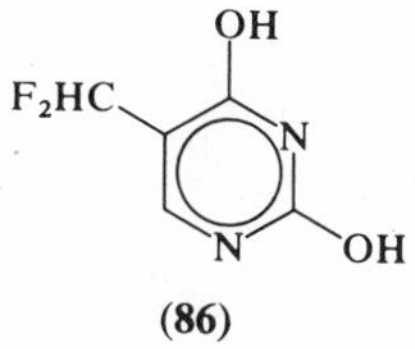

(86)

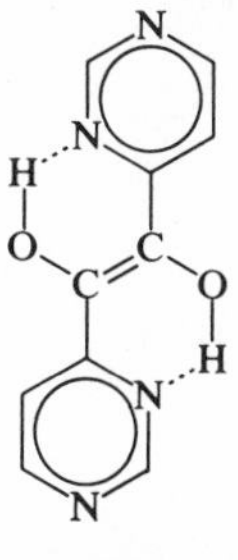

(87)

6. Pyrimidine Ketones and Derivatives (*H* 415)

A. Preparation of C-Acylpyrimidines (*H* 415)

The formation of pyrimidine ketones by primary syntheses will be found in Chs. II and III.

Direct *C*-acylation of 1,2,3,4-tetrahydro-6-hydroxy-1,3-dimethyl-2,4-dioxopyrimidine to give the 5-acetyl derivative (**88**) has been carried out with acetic anhydride (*H* 416) or less directly (*via* a bicyclic intermediate) using malonic acid/acetic anhydride.[2361] Other *C*-acylations are exemplified in the reaction of 2,4-diethoxy-6-methylpyrimidine with potassium ethoxide/diethyl oxalate to give 4-ethoxalylmethyl-2,6-diethoxypyrimidine (**89**, R = H; 73%),[2423] and in the analogous formation of its 5-nitro derivative (**89**, R = NO_2; 91%) and other such compounds.[2223]

Oxidative methods for making ketones are poorly represented: 5-benzyl-4,6-dimethyl-2-phenylpyrimidine (**90**) in boiling aq. potassium permanganate eventually gave 53% of 5-benzoyl-4,6-dicarboxy-2-phenylpyrimidine (**91**) which decarboxylated to 5-benzoyl-2-phenylpyrimidine (**92**) in boiling acetic anhydride.[2372] 2,4-Diethoxy-5-α-hydroxybenzylpyrimidine underwent oxidation by chromium trioxide in pyridine to give 5-benzoyl-2,4-diethoxypyrimidine (60%);[2814] the *p*-fluoro and *p*-methyl derivatives were made similarly.[2814] Ozonization of 4-hydroxy-6-methyl-2-α-5′-nitrofurfurylidene-ethylpyrimidine gave 2-acetyl-4-hydroxy-6-methylpyrimidine.[2451]

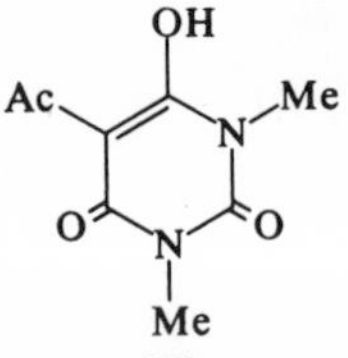

OH, Ac, Me, O, O, N, N, Me
(**88**)

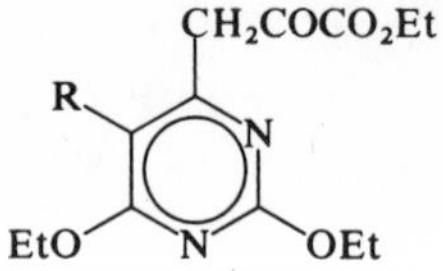

CH_2COCO_2Et, R, N, EtO, N, OEt
(**89**)

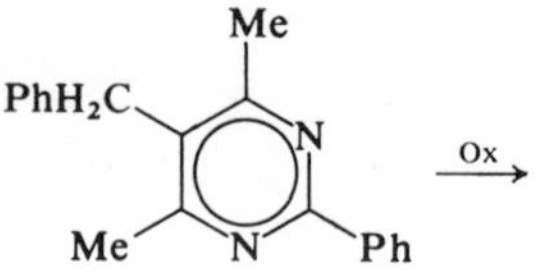

Me, PhH_2C, N, Me, N, Ph, Ox →
(**90**)

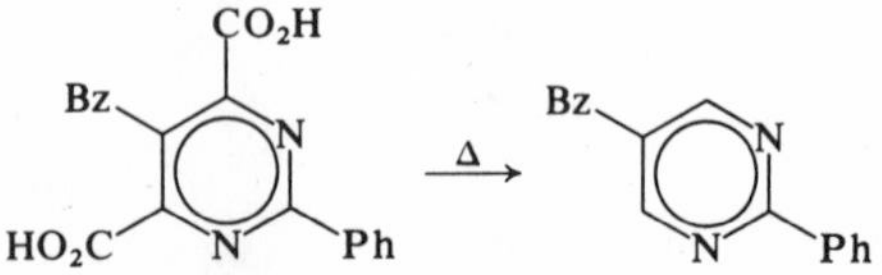

CO_2H, Bz, N, HO_2C, N, Ph
(**91**)

Δ →

Bz, N, N, Ph
(**92**)

4-Cyanopyrimidine was converted into 4-acetylpyrimidine (20%) by methyl magnesium iodide,[2806] but 2-cyanopyrimidine behaved differently.[2806] 4-Acetonylamino-2,6-dihydroxy-5-nitropyrimidine was made from the 4-chloro analogue with aminoacetone,[2456] 2-aminopyrimidine and diketene gave 2-acetoacetylaminopyrimidine (80%), and its 4-methyl derivative was made similarly.[3294]

B. Reactions of C-Acylpyrimidines (*H* 417)

It appears that most pyrimidine ketones must be used for biological testing or simply bottled, because few reactions are recorded.

5-Benzoyl-2,4-dihydroxypyrimidine was reduced by aq. potassium borohydride to give 2,4-dihydroxy-5-α-hydroxybenzylpyrimidine (40%).[2814]

Oximes include 4-β-ethoxycarbonyl-β-hydroxyiminoethyl-6-methoxy-,[2919] 5-α-hydroxyiminobenzyl-2-phenyl-,[2372] 4,6-dibenzyloxy-2-β-ethoxycarbonyl-β-hydroxyiminoethyl-,[2423] and 4-β-carboxy-β-hydroxyiminoethyl-5-ethyl-6-hydroxy-2-mercapto-pyrimidine.[3267] Reduction of the last mentioned oxime with sodium amalgam gave the 4-β-amino-β-carboxyethyl analogue;[3267] other such reductions are described.[2423, 2919, 3267] Hydrazones described include 4-β-hydrazonopropylamino-2,6-dihydroxy-5-nitropyrimidine,[2456] 4-hydroxy-6-methyl-2-α-*p*-nitrophenylhydrazonoethylpyrimidine,[2451] and a number of unanalysed 2,4-dinitrophenylhydrazones.[2317]

7. The Isocyanato-, Thiocyanato-, and Isothiocyanatopyrimidine Family; the Nitrile Oxides (*H* 418)

Apart from the new member of this family, the nitrile oxides, little new information has been reported.

The formation of thiocyanatopyrimidines from halogenopyrimidines has been discussed in Ch. VI, Sect. 5.H; the route from mercaptopyrimidines with cyanogen bromide in Ch. VIII, Sect. 1.D(8).

1,4,6-Triamino-1,2-dihydro-2-thiopyrimidine (**93**; prepared by primary synthesis) in dimethylformamide with phosphoryl chloride underwent rearrangement involving the participation of doubly bonded sulphur as a neighbouring group to give eventually 4,6-bisdimethylaminomethyleneamino-2-thiocyanatopyrimidine (**94**);[2859, 3451] with acetic acid this gave 4,6-diformamido-2-thiocyanatopyrimidine.[2859] The mechanism of rearrangement has been discussed and other reactions of the product recorded.[2859]

D. Pyrimidine Nitrile Oxides (*New*)

The following nitrile oxides (R—C≡N → O) are among the first heterocyclic members to be prepared;[2815] the flanking groups were designed to inhibit dimerization to which nitrile oxides are prone.[3452] The oxime, 5-hydroxyiminomethyl-2,4,6-trimethoxypyrimidine (**95**), was dehydrogenated by alkaline hypobromite to give 2,4,6-trimethoxypyrimidine-5-nitrile oxide (**96**; 19%); 4-chloro-2-dimethylamino-6-

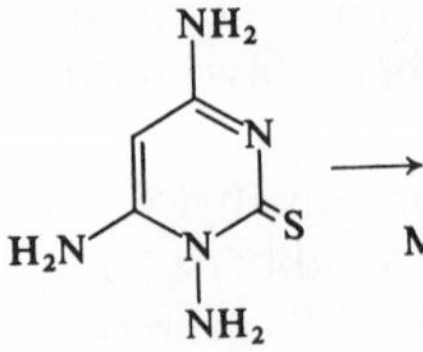

methoxypyrimidine-5-nitrile oxide (22%) and 2-dimethylamino-4,6-dimethylpyrimidine-5-nitrile oxide (34%) were made similarly.[2815] The last-mentioned oxide was warmed with aniline to give the 5-α-anilino-hydroxyiminomethyl analogue (**97**).[2815]

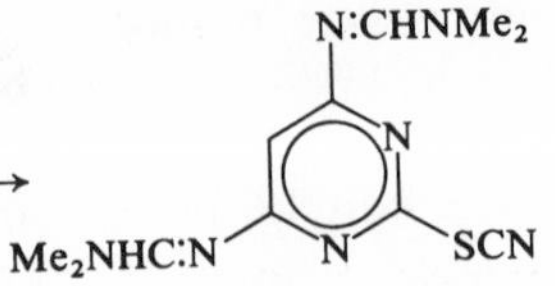

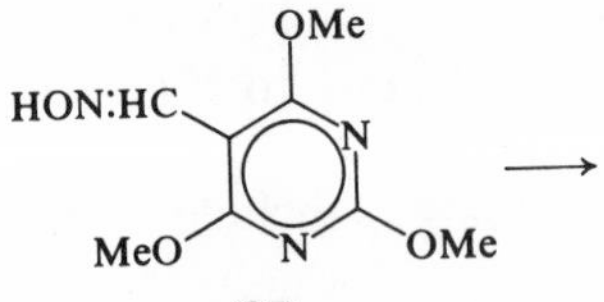

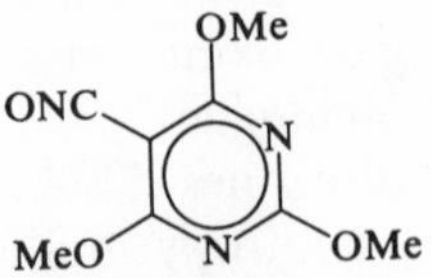

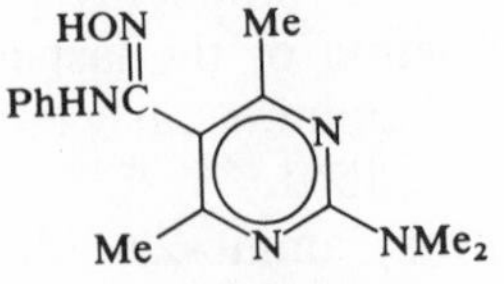

A better reagent with which to dehydrogenate aldoximes appears to be *N*-bromosuccinimide in dimethylformamide containing sodium methoxide.[3518] With its help, 2,4,6-trimethoxy- (**96**; 80%) and 4,6-dichloro-2-dimethylamino-pyrimidine-5-nitrile oxide (73%) were made from the corresponding 5-hydroxyiminomethyl derivatives.[3518]

8. Orotic Acid (*H* 422)

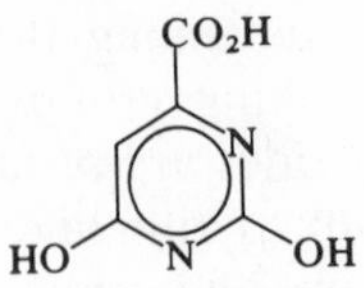

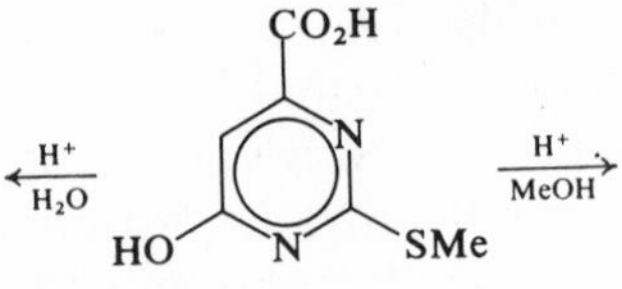

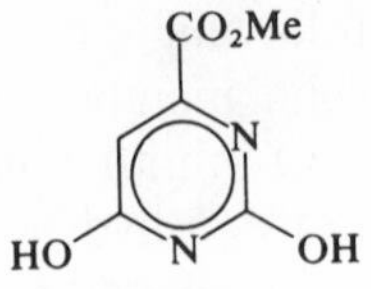

A convenient large-scale preparation of orotic acid (4-carboxy-2,6-dihydroxypyrimidine; **98**) has been reported:[2214] diethyl oxalacetate was condensed with *S*-methylthiourea in aq. sodium hydroxide to give 4-carboxy-6-hydroxy-2-methylthiopyrimidine (**99**; 63%), which in refluxing 2*N*-hydrochloric acid gave orotic acid (**98**; 79%);[2214] the same intermediate (**99**) with methanolic hydrogen chloride gave the ester, 2,4-dihydroxy-6-methoxycarbonylpyrimidine (**100**; 71%).[2214] A modified synthesis of orotic acid from monoethyl oxalacetate and urea has been published,[2216] along with procedures for esterification and formation of other derivatives.[2216] There is also an interesting route to orotic acid from the monoureide of maleic acid,[3453] and some variations on the classical synthesis of orotic acid derivatives.[2346]

The metal complexes of 5-substituted orotic acids have been studied[3454, 3456] with a view to 'elucidating the effect of metal ions on the double-stranded helical structure of nucleic acids'. The amine and metal salts of orotic acid and its derivatives have also been studied.[3434, 3455]

CHAPTER XII

The Reduced Pyrimidines* (*H* 430)

The statement (*H* 431) that there are nine hydrogenated derivatives (**1**–**9**) of pyrimidine, five dihydro, three tetrahydro, and one hexahydro, does not take into account the complications arising from the mobile nature of hydrogen atoms attached to nitrogen or from the possibilities for existence of different ring conformations of cyclic compounds. Thus the 1,4- and 1,6-dihydropyrimidines (**2** and **3**) will be in tautomeric equilibrium with each other, as will the 1,2,3,4- and 1,2,5,6-tetrahydropyrimidines (**6** and **7**), respectively. In addition, since the tetrahydro and hexahydro compounds are formally derivatives of cyclohexene or cyclohexane in which CH or CH_2 groups have been replaced by N or NH groups, isomerism resulting from chair-boat or other conformational interconversions is at least a theoretical possibility. As the name suggests, hydropyrimidines may be, and some actually are, formed by the addition of hydrogen to the pyrimidine nucleus. However, in a large number of compounds the pyrimidine nucleus is inert to attack by a variety of reducing agents, or else the hydropyrimidine produced is unstable under the reaction conditions and only decomposition products can be isolated. Other preparative methods thus become necessary and these fall broadly into two groups. The smaller group consists of the addition of reagents other than hydrogen to a pyrimidine nucleus; the much larger consists of the wide variety of synthetic methods (Chs. II and III) used in the preparation of the pyrimidine nucleus itself modified to produce this same nucleus at a different level of reduction.

This chapter is organized into four sections dealing with the preparation, the physical properties and structure, the chemical properties, and the uses of hydropyrimidines respectively. Each section deals in order with the di-, the tetra-, and finally the hexa-hydropyrimidines.

In our discussion of their preparation, we closely adhere to the

* By R. F. Evans, Department of Chemistry, University of Queensland, St. Lucia, Australia, 4067.

earlier scheme (*H* 431) and discuss first synthetic methods which involve the building up of the pyrimidine ring and then proceed to discuss the addition of hydrogen and other reagents to the pyrimidine nucleus. Following an earlier suggestion (*H* 430), nomenclature involving imaginary hydrogen atoms is avoided when possible: for example, the so-called hexahydropyrimidin-2-one (**10**) is regarded as a tetrahydro derivative since it is actually the more stable tautomer of 1,4,5,6-tetrahydro-2-hydroxypyrimidine (**11**).

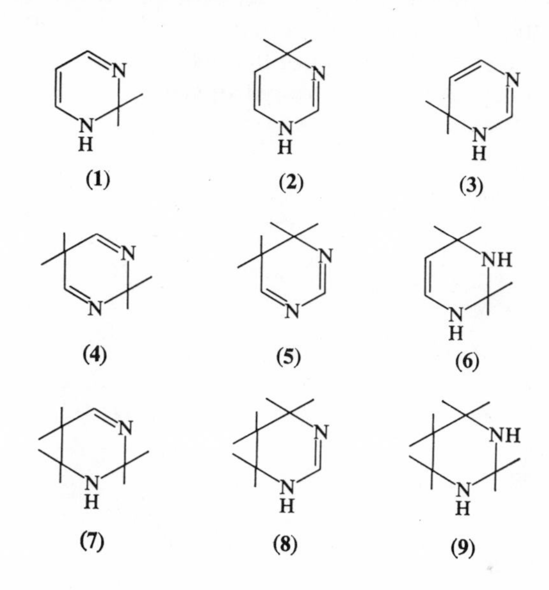

1. Preparation of Dihydropyrimidines (*H* 431)

A. By a Modified Principal Synthesis (*H* 431)

The condensation of a three-carbon entity with an N—C—N compound is exemplified by the reaction of 3-ethoxy-2-methoxymethylene-propionitrile[2276] (**12**, R = CN) or a related ester[2277, 2279, 3533] (**12**, R = CO_2Et) with urea or its *N*-methyl or *N*-phenyl derivatives to yield a 5,6-dihydropyrimidine derivative (**13**). Similarly, the reaction of benzamidine with an $\alpha\beta$-unsaturated ketone like mesityl oxide affords a tautomeric mixture of the 1,4- and 1,6-dihydro compounds (**14** and **15**),[3534] but thiourea affords only 10% of the desired dihydropyrimidine (**16**), the major product being a 2-imino-dihydro-1,3-thiazine (**17**).[3535]

B. From an Aminopropane Derivative and an Isocyanate (*H* 434)

This reaction was recently exemplified by the condensation[3536] of a β-aminopropionic ester with potassium cyanate to give an intermediate ureido compound which cyclizes easily to a 5,6-dihydropyrimidine (**18**). An analogous reaction starting with benzoyl isothiocyanate and ethyl β-aminopropionate led to 5,6-dihydro-2-thiouracil.[3537] Again, condensations of β-aminopropionitrile with phenyl (and other) isothiocyanates give thioureas, converted by methyl iodide into isothiouronium salts (**19**). These cyclize in anhydrous ammonia to aminodihydropyrimidines (**20**).[3538] The dihydropyrimidine (**21**) was made by cyclization of a ureide.[3539]

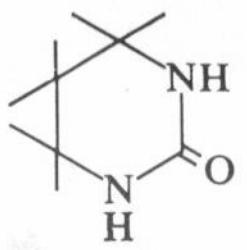

NH O N H ⇌ N OH N H

(**10**) 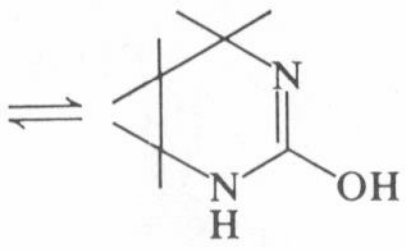 (**11**)

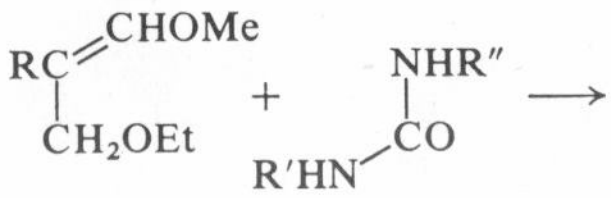

RC=CHOMe, CH$_2$OEt + NHR″, CO, R′HN → R, R″, N, O, N, R′

(**12**) 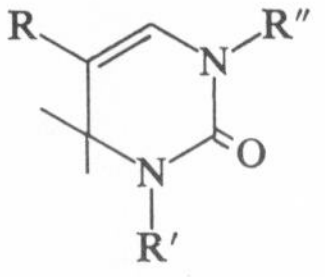 (**13**)

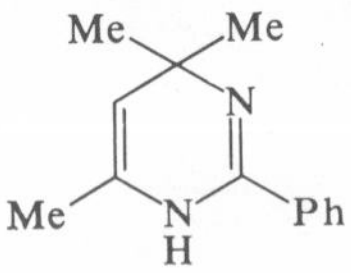

Me Me N Me N H Ph ⇌ Me Me NH Me N Ph

(**14**) 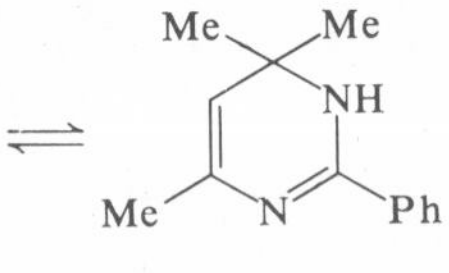 (**15**)

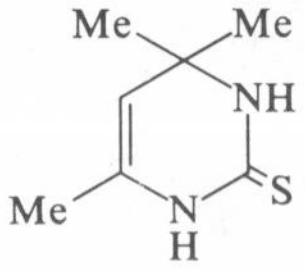

Me Me NH Me N H S

(**16**)

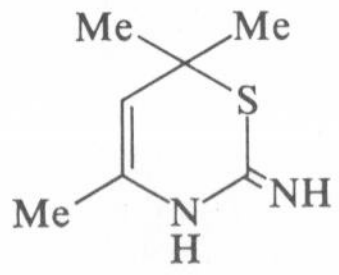

Me Me S Me N H NH

(**17**)

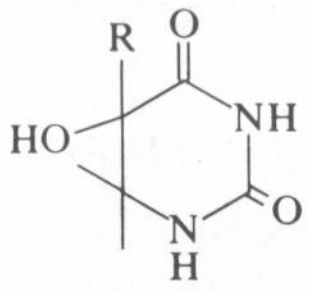

R O HO NH N H O

(**18**)

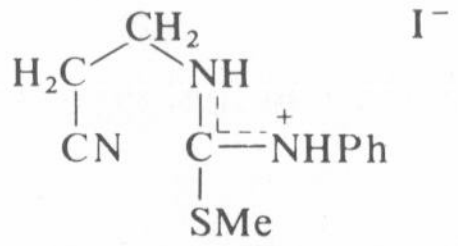

CH$_2$ H$_2$C NH CN C=NHPh SMe I$^-$ →

(**19**)

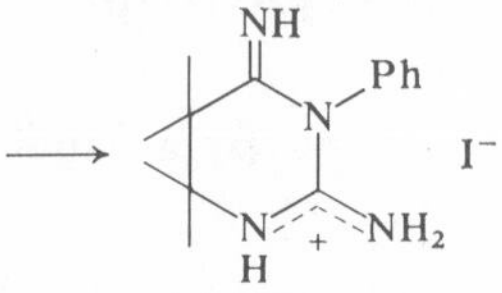

NH Ph N I$^-$ N H + NH$_2$

(**20**)

NH$_2$ N O N CH$_2$CH$_2$CN

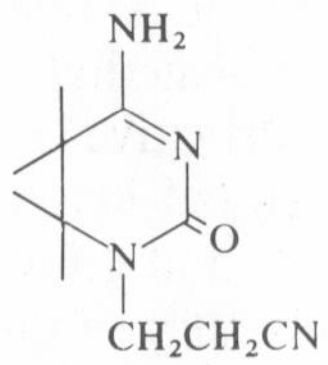

(**21**)

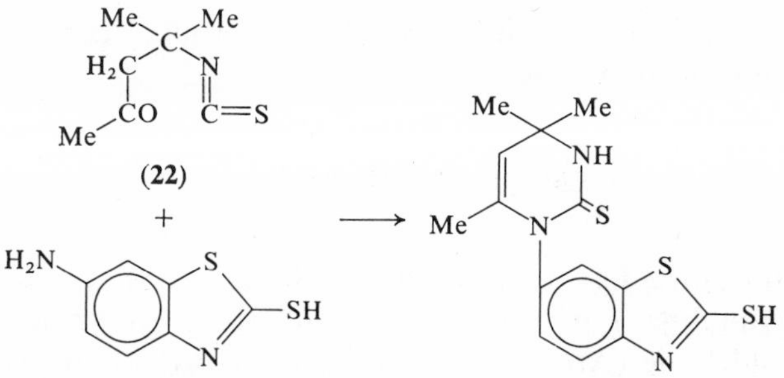

C. From an Isocyanatopropane Derivative and an Amine (*H* 436)

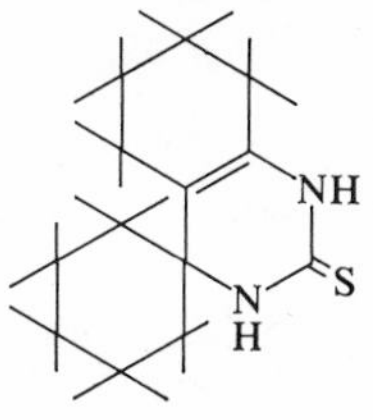

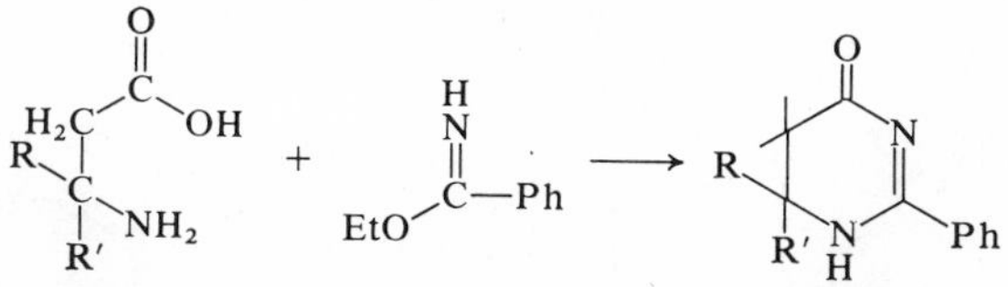

Condensation of 2-isothiocyanato-2-methylpentan-4-one (**22**) with the aminobenzothiazole (**23**) affords a 3,4-dihydropyrimidine derivative (**24**).[3540] A similar type of isothiocyanato intermediate is probably involved in the condensation of cyclohexanone with ammonium thiocyanate or thiourea to give the tricyclic compound (**25**); other ketones also undergo this condensation.[3541]

D. From an Aminopropane Derivative and *O*-Methylurea (*H* 438)

Since ethyl benzimidate (**27**) is, in a sense, a relative of *O*-methylurea, its reaction with the β-aminopropionic acids (**26**, R = R′ = Me; or

R = H, R′ = Ph) is included here.[3542] The product is an appropriate 5,6-dihydro derivative (**28**).

F. The Biginelli Reaction (*H* 440)

This procedure has been used to obtain the fluorine-containing dihydropyrimidine (**29**, Ar = $C_6H_4F(p)$) from the reaction of an aromatic aldehyde (ArCHO) with urea and ethyl acetoacetate.[3543] In another example, an intermediate has been isolated and separately condensed with acetophenone to give the dihydrohydroxy-pyrimidine (**30**).[3544]

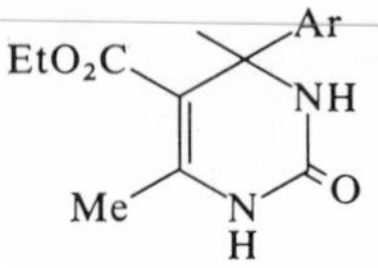

(**29**) 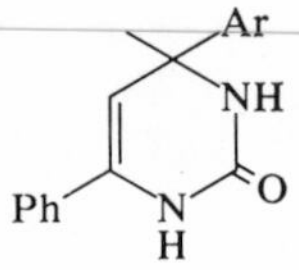 (**30**)

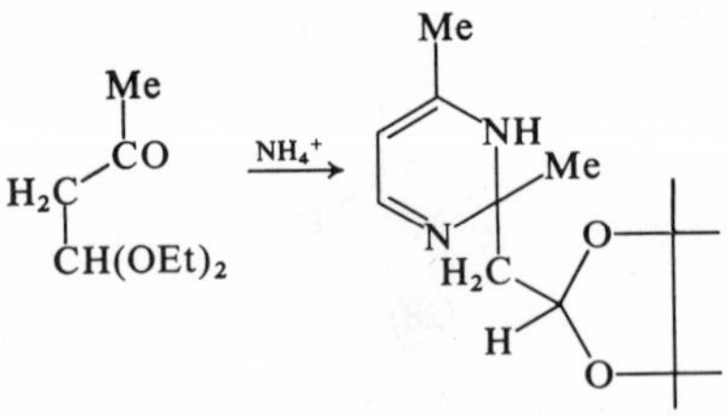

(**31**) (**32**) 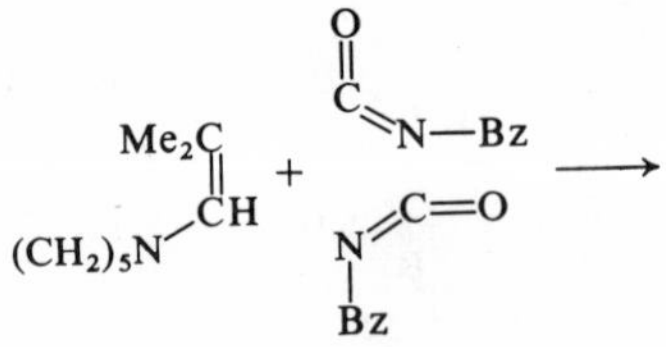 (**33**) (**34**)

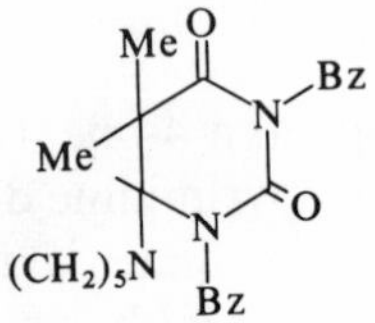

(**35**) 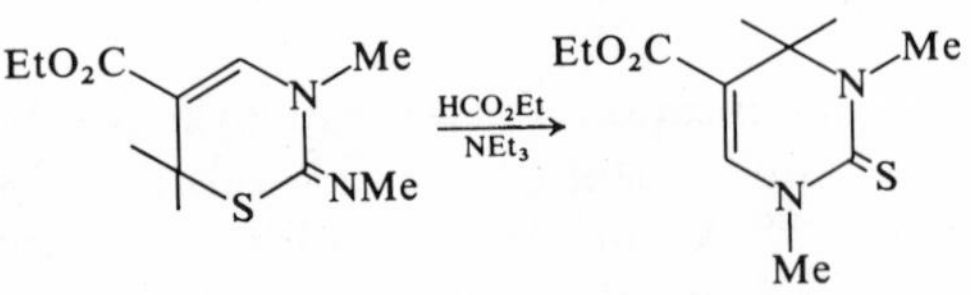 (**36**) (**37**)

G. Other Syntheses (*H* 442)

No details are available for the curious condensation of 3-oxobutanal acetal (**31**) with ammonia and in the presence of ammonium nitrate to give the 1,2-dihydropyrimidine (**32**);[3545] nor for the condensation between an acyl isocyanate (**34**) and an enamine (**33**) to afford the 5,6-dihydropyrimidine (**35**).[3546]

It has been mentioned earlier that some dihydropyrimidine syntheses proceed predominantly by a second route involving a different heterocyclic system. However, such systems, e.g., those 1,3-thiazines like (**36**), can be rearranged to dihydropyrimidines (**37**).[2277, 2279, 3533, 3547, 3548]

H. By Reduction of Pyrimidines (*H* 442)

Reductive methods have been classified according to the active entities involved.[3549] Hydrogen atoms are implicated or thought to be implicated in (*a*) catalytic reduction, (*b*) electrochemical reduction at a low-overvoltage electrode, (*c*) photochemical reduction, (*d*) reduction involving complex metal hydrides derived from Group IVB metals, (*e*) diimide reductions, and (*f*) homogeneous reductions by dihydro derivatives of Group VIII metals. Hydrogen is transferred as hydride ion in (*g*) reduction involving complex metal hydrides containing boron or aluminium, (*h*) homogeneous hydrogenation involving certain monohydrido complexes of Group VIII metals, and (*i*) hydrogen transfer from one substrate to another. Finally, hydrogen is transferred as protons in (*j*) reduction brought about by dissolving metals, (*k*) electrochemical reduction at a high-overvoltage electrode, and (*l*) reduction, the first stage of which is attack by an anion.

However, it is symptomatic of the state of hydropyrimidine chemistry that only a few of the possible reduction types have been used systematically for pyrimidine compounds. Thus although catalytic reduction of pyrimidine and its simple alkyl derivatives in the presence of palladized charcoal gives 1,4,5,6-tetrahydro compounds as final products,[3550] dihydro derivatives can be detected as intermediates, e.g., in the case of 2-aminopyrimidine,[3551] or actually isolated if addition of the second molecule of hydrogen is slower than the first, e.g., with 4,6-dimethylpyrimidine which affords a salt of the 1,4(1,6)-dihydro compound (**38**).[3550] This is probably due to the steric effect of the two methyl groups.

Use of a different catalyst (Pt) with 2-hydroxypyrimidine or of its *N*-methyl derivative and interruption of the reduction at the dihydro stage led to isolation of the corresponding 3,6-dihydro derivatives (**39**, R = H or Me).[2988] These and analogous compounds were also prepared from the Raney nickel desulphurization[3552] of the 5,6-dihydro-4-thiouracil compounds (**40**, R = H or Me, R′ = H; or R = H,

R′ = Me). Similarly, dihydro-2-thiouracil compounds could be dethiated to dihydropyrimidines with different bond arrangements so that (**41**, R = SH) gave (**41**, R = H), regarded as the stable tautomer of a 5,6-dihydro compound; and (**42**, R = SH) the 5,6-dihydro derivatives (**42**, R = H, R′ = H or Me). Dethiation and reduction occurs

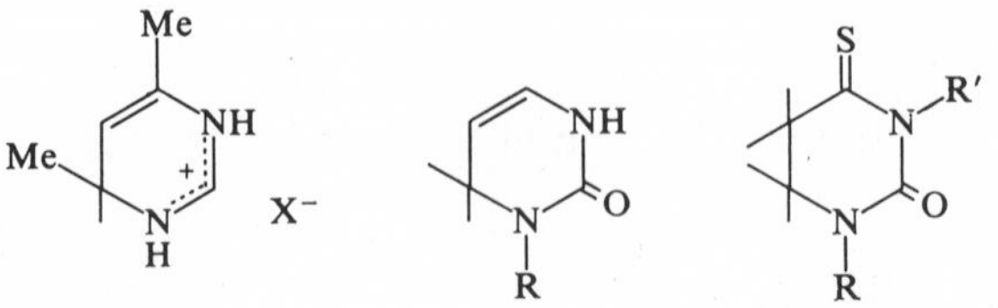

(38) (39) (40)

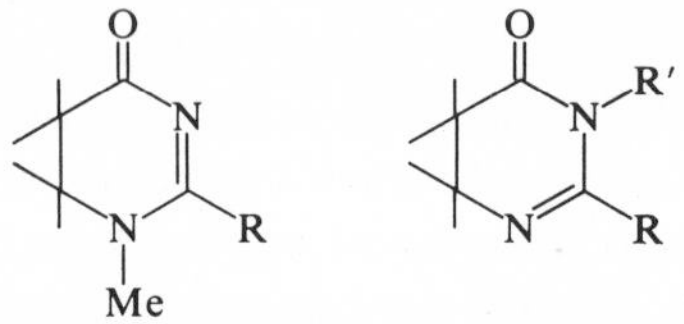

(41) (42)

with the nontautomerizing thio compound (**43**) giving the important medicinal Butazolidin, 5-butylhexahydro-4,6-dioxo-1,3-diphenylpyrimidine.[3553]

The isolated double bond in the predominating tautomeric form of biologically important uracil (**44**) and its derivatives is readily reducible. Deuterogenation,[2196] effected in the presence of platinum, is claimed to produce 5,6-dihydrouracil-5,6d_2 (**45**) although the possibility of hydrogen/deuterium scrambling does not appear to have been considered.

Stereospecific addition of hydrogen to the 5,6-double bond of thymidine was noted in the rhodium-on-alumina catalyzed reduction, to give the S(−) isomer (**46**) only.[3555]

Active metal hydride reduction of heterocyclic ring systems is temperature dependent, the ring being reduced before reducible substituents at high temperatures, the reverse being true at lower temperatures. It has been shown[3474] that a variety of ethyl 4-substituted-2-methylthiopyrimidine-5-carboxylates are reduced to the 1,6-dihydro compounds by lithium aluminium hydride and other active metal hydrides like sodium or lithium borohydride at room temperature. Substituents like cyano, hydroxyimino, chloro, methyl, or hydroxy were present.

Dihydropyrimidines of unknown configuration result from the lithium aluminium hydride reduction of a number of 5-ethoxycarbonyl-, 5-carbamoyl-, or 5-cyano-pyrimidines,[2681] while it is likely that a 1,6-dihydro compound rather than a tetrahydro derivative, is formed

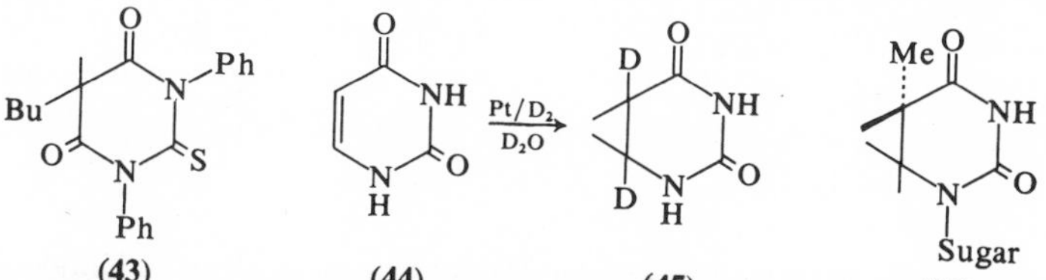

(43) **(44)** **(45)** **(46)**

with 5-ethoxycarbonyl-2,4-dimethylpyrimidine.[3007] From the limited data available it seems that ring reduction is facilitated by the presence of an electron-withdrawing group in the 4- or 5-positions. Steric hindrance seems to play its part too, an unsubstituted position in the above series of compounds obviously aiding reduction. Even if 1,4-reduction had occurred, it would be expected that complete tautomerization to the more stable conjugated system of 1,6-dihydro compounds would follow. Sodium borohydride does not reduce pyrimidine nor its 4,6-dimethyl derivative but will attack the corresponding methiodides (affording unidentified substances), the positive charge now making the electron-deficient nucleus even more so, obviously beyond a critical level. The more powerful lithium aluminium hydride (in refluxing ether) reduces pyrimidine readily (possibly to the 1,2,3,4-tetrahydro stage[3550]), but leaves 4,6-dimethylpyrimidine substantially unchanged.[3556] Aqueous potassium borohydride does not reduce uracil, but reduces its 1,3-dimethyl derivative to the dihydro stage.[3551] This is probably bound up with the protection of the nucleus against nucleophilic hydride attack by conversion into the negatively charged anion. It is intriguing to note however, that derivatives of uracil, e.g., uridine, if simultaneously irradiated with ultra-violet light will suffer reduction at the 5,6-double bond by sodium borohydride.[3557] It is claimed that the 5,6-double bond assumes some free radical character. Thymidine is reduced but only further transformation products of the 5,6-dihydro intermediate can be isolated.[3555] With the 6-carboxy-5-fluoro- derivative of uracil (5-fluoro-orotic acid) both reduction of the 5,6-double bond and hydrogenolysis of the C—F bond occurred.[3558] Irradiation of uracil in isopropanol gives 5,6-dihydrouracil.[3773]

Other methods of reduction have been little used for production of dihydropyrimidines. Zinc dust in glacial acetic acid reduces 5-methyl-2-phenylpyrimidine to an unstable 1,6-dihydro derivative (**47**).[3473]

Electrolytic reduction of 2-hydroxy-4,6-diphenylpyrimidine gives a dihydro derivative (e.g., **30**, Ar = Ph).[3544] Polarographic reduction of cytosine and some 5-substituted derivatives (**48**) led to reduction of the 3,4-double bond and hydrogenolytic splitting off of the amino group to give the reduced compound (**49**).[3559] This confirmed some earlier work on the subject.[3358, 3560] Hydrogenolytic fission[3561] of an imine group is involved in the reduction at a lead cathode of the appropriate barbituric acid derivative to the 2,5-dihydro compound (**50**).

I. Addition of Reagents Other than Hydrogen to a Double Bond (*New*)

t-Butyl magnesium chloride adds across the 3,4-bond of pyrimidine to give the 4-t-butyl-3,4-dihydro compound (**51**);[3551] lithium n-butyl, on the other hand, underwent addition to 4,6-dimethoxypyrimidine to yield a 2,5-dihydro compound (**52**).[2519]

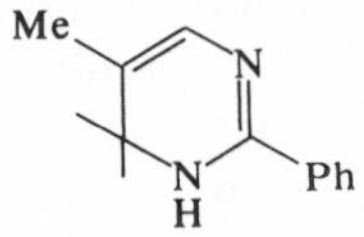

(**47**) 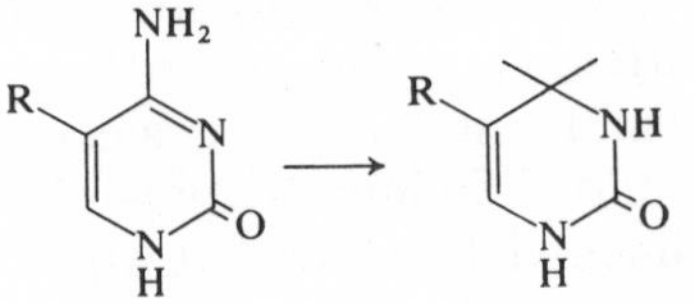 (**48**) (**49**) 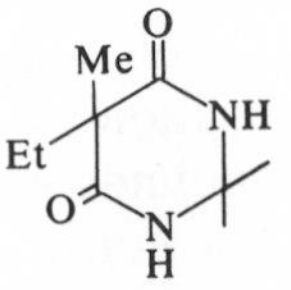 (**50**)

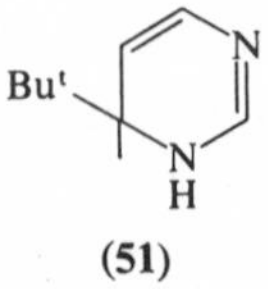

(**51**) 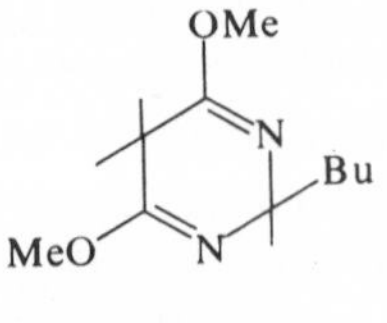 (**52**) 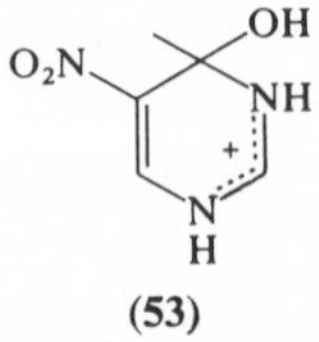 (**53**) 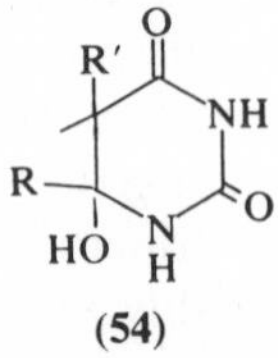 (**54**)

Covalent hydration of nitrogenous heterocycles has hitherto been thought to be confined to di- and poly-azanaphthalenes. Recently, however, the rapid acid catalyzed deuteration of 2-hydroxypyrimidine has been explained in terms of the presence of a small percentage of a hydrate, 1,6-dihydro-2,6-dihydroxypyrimidine.[3459] More spectacularly, 5-nitro-, 5-methylsulphonyl-, and 5-methylsulphinyl-pyrimidine in aqueous acid are completely converted to the hydrated cations, e.g., (**53**).[2688, 3240]

Hydration of the 5,6-double bond of uracil derivatives is effected in aqueous solution under the stimulus of ultra-violet radiation, so that 5-fluorouracil yields dl-5-fluoro-5,6-dihydro-6-hydroxyuracil (**54**, R = H, R′ = F),[3562] and uracil-5,6-t_2 yields a 5,6-dihydro-6-hydroxyuracil-

5,6-t_2 (**54**, R = R′ = t).[3563] γ-Ray irradiation of cytosine affords a dihydrohydroxycytosine[3564, 3565] while ultra-violet irradiation of aq. cytidylic acids leads to a similar dihydro-4-hydroxy derivative.[3361]

Bromine water attacks orotic acid so that the net result is the addition of hypobromous acid across the 5,6-double bond.[3566]

2. Preparation of Tetrahydropyrimidines (*H* 445)

Of the three possibilities, the 1,4,5,6-tetrahydro system is undoubtedly the one most intensively investigated and the one most readily obtained.

A. From 1,3-Diaminopropane or a Derivative (*H* 445)

The condensation of 1,3-diaminopropanes with a carboxylic acid or a molecule related thereto continues to be one of the reactions exhaustively used to prepare 1,4,5,6-tetrahydropyrimidine derivatives. Thus 1,3-diaminopropane (**55**, R = H) reacts with hydrocyanic acid (**56**, R′ = H) with elimination of ammonia affording the parent 1,4,5,6-tetrahydropyrimidine (**57**, R′ = H).[3567, 3568] The diamino compound is usually used as the free base, but sometimes as a salt, e.g., the mono(or di)-*p*-toluenesulphonate.[3569, 3570] In the latter case, there must be equilibrium with a species containing an uncharged NH_2 group, since the condensation depends upon nucleophilic attack of such a group upon the cationoid carbon atom of the nitrile group. A protonated amino group would have no unshared pair of electrons and thus no nucleophilic character. Variations in the structure of the diamine component seem to be small in number, being confined to *N*- and *C*-alkyl or aryl[3567–3569] substituted compounds and to 1,3-diaminopropan-2-ol.[3569] However, variations in the nitrile component (**56**) are legion. Thus R′ may be substituted benzyl,[3569] substituted α-naphthyl,[3572] 9-xanthenylmethyl, its 10-thia analogue,[3573] or 3-t-butylaminopropyl.[3571] Some less usual R′ substituents have been trifluoromethyl,[3574] polyvinyl[3575] (affording a polyvinyl-tetrahydropyrimidine), amino, substituted-amino,[3570] and cyano[3576] leading to a bi-1,4,5,6-tetrahydropyrimidin-2-yl. When R′ is bromo, the product is not the 2-bromo derivative (**57**; R′ = Br) but the salt of the 2-amino compound.[3577–3580] Thus with (**55**, R = $C_{12}H_{25}$) the intermediate bromoamidine (**58**) evidently prefers to eliminate hydrogen bromide (and not ammonia) which is retained by salt formation and the product is the hydrobromide of the 2-amino compound (**57**, R = $C_{12}H_{25}$, R′ = NH_2). The first reaction of a 1,3-diaminopropane with carboxylic acids must be one of neutralization. Vigorous conditions will be necessary subsequently to convert the amine salt into an amide, e.g., (**59**) which can be dehydrated to the tetrahydropyrimidine (**57**). Consequently, it is preferable to use an acid derivative which affords the amide directly. The acid derivative of choice is usually an ethyl ester,[3581, 3582] and often the amide.[3576] Thus oxalic acid has been heated with the diamine (**55**, R = $C_{12}H_{25}$) to give the tetrahydropyrimidinecarboxylic acid (**57**, R = $C_{12}H_{25}$, R′ = CO_2H).[3583] Phenylacetic acid condenses with 3-β-phenethylaminopropylamine to afford 1,4,5,6-tetrahydro-1-β-phenethyl-2-phenylpyrimidine (**57**, R = $PhCH_2CH_2$, R′ = Ph).[3581]

Ortho esters may be used instead of ordinary esters so that trimethylenediamine gives the corresponding 2-substituted 1,4,5,6-tetrahydropyrimidine (**57**, R = H, R′ = Me, Et, Ph) with ethyl ortho acetate, propionate, and benzoate.[3584] The R′ group (in the ester component) has been varied between wide limits and can be trifluoromethyl,[3585] ethoxymethyl,[3576, 3586] diethoxymethyl,[3586] and diethylaminomethyl.[3587] However, in the case of ethyl acetoacetate the side chain becomes deacetylated and the product from the condensation with 3-ethylaminopropylamine is 1-ethyl-1,4,5,6-tetrahydro-2-methylpyrimidine.[3588]

The intermediate amide (**59**, R′ = Me) has also been produced *in situ* when the acetyl derivative of the product (**60**) from the addition of a primary amine to acrylonitrile has been reduced with hydrogen and Raney nickel. Under the reaction conditions, cyclization of the intermediate acetamide occurs and the product consists largely of the 1-alkyl-1,4,5,6-tetrahydro-2-methylpyrimidine (**57**, R = Me, Et, Pr, Bu, $C_{12}H_{25}$; R′ = Me).[3554, 3747]

Yet another method of general applicability involves the condensation of a 1,3-diaminopropane derivative (**61**, R = R′ = H or Me; or R = H, R′ = Me) with an amidine salt to give good yields of the reduced pyrimidine salt (**62**; R″ = H, Me, Bu^t, Ph).[2957] The anion X can be acetate, halide, or picrate and has little effect on the reaction. The nitrogenous groups of the amidine are eliminated so that 1,3-bismethylaminopropane and formamidine hydrochloride afford the 1,3-dimethylpyrimidine salt and ammonia only. DL-αγ-Diaminobutyric acid, with one protonated amino group, yielded the zwitterion, 1,4,5,6-tetrahydropyrimidine-4-carboxylic acid. At room temperature, guanidine hydrochloride forms an addition complex with trimethylenediamine in which the double bond is still present and

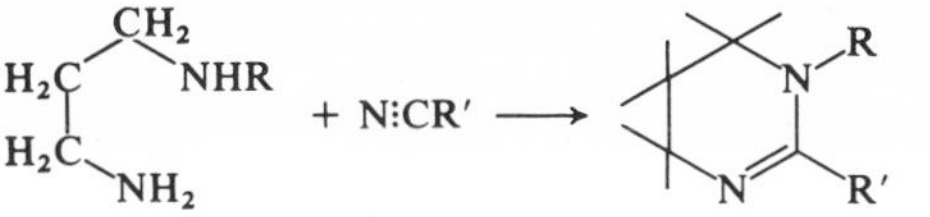

(55) + N:CR′ (56) ⟶ (57) 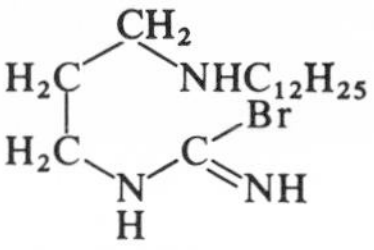 (58)

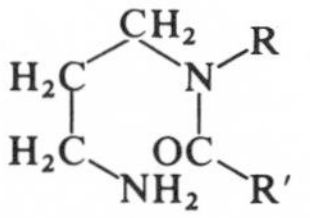

(59) 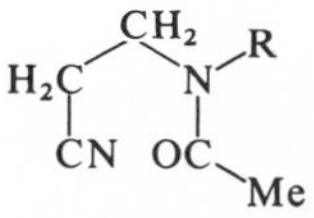 (60)

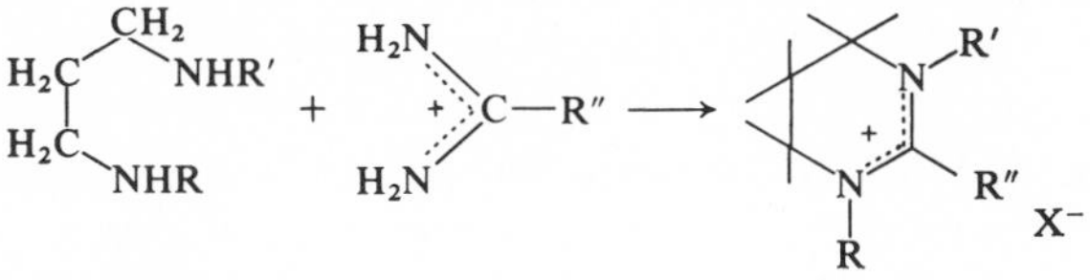

(61) + (H$_2$N)$_2$C$^+$—R″ ⟶ (62) X$^-$

which is easily broken up by picric acid. However, at 140°, elimination of ammonia occurs and 2-amino-1,4,5,6-tetrahydropyrimidine can be isolated.

1,3,5-Triazine can serve as a source of formamidine and the observation that it reacts with trimethylenediamine to give the parent 1,4,5,6-tetrahydropyrimidine[1561] has been confirmed.[2162] However, the substance originally claimed to be 1,4,5,6-tetrahydropyrimidine picrate[1561] is probably ammonium picrate.

During the condensation of 1,3-diaminopropan-2-ol with acetamidine hydrochloride, ammonia is evolved, but no hydropyrimidine can be isolated. Yet reaction of this diamine with ethyl formate, ethyl acetate, or even stearic acid proceeds as expected giving 1,4,5,6-tetrahydropyrimidines with hydrogen, methyl, or heptadecyl attached to C-2.[3589,3590]

Variations on the diamine-amidine theme include condensations of diamines with imidates ($EtO \cdot CR{=}N^+H_2Cl^-$; R = $ClCH_2$, $HOCH_2$, etc.),[3591, 3592] with a salt of an *O*-alkylurea,[3593] or with isothiouronium salts[3594] to give 2-substituted-1,4,5,6-tetrahydropyrimidines (**62**, R = R′ = H, R″ = $ClCH_2$, etc.). With the urea derivatives, an alcohol or a mercaptan is eliminated in addition to ammonia, and the product is a 2-aminotetrahydro derivative.

The condensations seem to involve reaction between one end of the diamine component and the amidine or other component[3314] so that a 3-amidino- or 3-guanidino-propylamine intermediate results. This rapidly cyclized to the tetrahydropyrimidine, as observed with 1-guanidino-3-dodecylaminopropane.[3595]

A guanidinopropane derivative in which cyclization to a tetrahydropyrimidine does not involve the elimination of ammonia is β-guanidinopropionaldehyde diethyl acetal (**63**). Refluxing a solution containing the acetal in the presence of a trace of acid causes hydrolysis to the corresponding aldehyde which immediately cyclizes to 2-amino-1,4,5,6-tetrahydro-4-hydroxypyrimidine hydrochloride.[3596]

Carbonic acid derivatives continue to be used as the source of the carbon atom required to complete the pyrimidine ring starting with 1,3-diaminopropanes. Thus diethyl carbonate reacts with various 2-alkyl derivatives of 1,3-diaminopropane to give 5-substituted-1,4,5,6-tetrahydro-2-hydroxypyrimidines (**64**, R = H, R′ = alkyl; or R = R′ = alkyl).[3597]

Phosgene has been used similarly,[3598, 3599, 3634] but in contrast to the diamine-amidine reaction, phosgene reacts even if one of the amino groups in the 1,3-diamine is tertiary. Thus 1-anilino-3-dimethylaminopropane (**65**) is converted into hexahydro-1-methyl-2-oxo-3-phenylpyrimidine (**67**) possibly by loss of methyl chloride from the quaternary salt (**66**).[3599]

However, urea and *N*-alkyl-ureas appear to be the most popular reagents for this task of supplying one carbon atom to complete the tetrahydropyrimidine ring,[3600–3603] the products being derivatives of 1,4,5,6-tetrahydro-2-hydroxypyrimidine. Thiourea may be used to

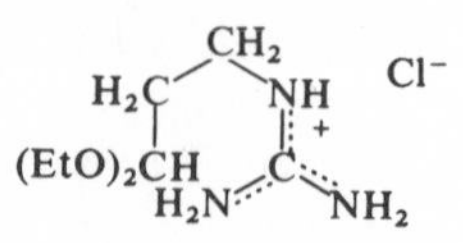

(**63**) 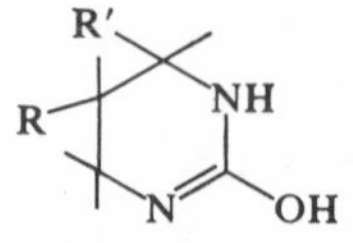 (**64**)

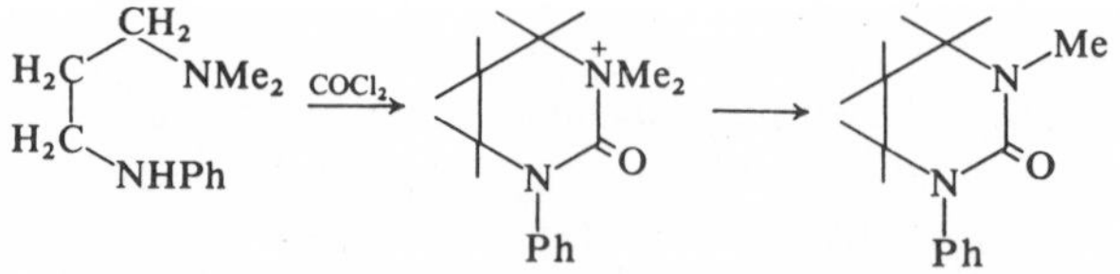

(**65**) (**66**) (**67**)

produce the 2-mercapto analogues[3604] although 1,4,5,6-tetrahydro-2-mercaptopyrimidine itself is readily obtainable from the reaction of 1,3-diaminopropane with carbon disulphide,[3605] or from the attempted dehydrogenation of 1,4,5,6-tetrahydropyrimidine with sulphur, which results in the insertion of a sulphur atom at C-2.[2162] Carbon diselenide behaves like carbon disulphide and reacts with 1,3-diaminopropane to give 1,4,5,6-tetrahydro-2-selenylpyrimidine.[3606] Both sulphur and selenium compounds are formed by the intramolecular cyclization of an intermediate 3-aminopropyl isothio(or seleno)cyanate

$$[H_2NCH_2CH_2CH_2N{:}C{:}S(\text{or Se})].$$

The reaction of benzyl isocyanate dichloride ($PhCH_2N{:}CCl_2$) with 1,3-diaminopropane forms a salt of 2-benzylamino-1,4,5,6-tetrahydropyrimidine probably again by an internal nucleophilic displacement in the intermediate carbodiimide ($H_2NCH_2CH_2CH_2N{:}C{:}NCH_2Ph$).[3607]

On at least one occasion, it has been proved that an *N*-alkylurea may also supply one or both nitrogen atoms in addition to C-2 to complete the tetrahydropyrimidine ring, e.g., it so reacts with both 1-amino-3-hydroxy- and 1,3-dihydroxy-propane.[3602]

Related methods of obtaining 1,4,5,6-tetrahydro-2-hydroxypyrimidines include the alkaline degradation of the diurethane (**68**), perhaps through a monourethane, to give the 4,6-diphenyl derivative (**69**),[3604] and the reduction of cyanoethyl isocyanurate (**70**). In the latter, the nitrile group is reduced to CH_2NH_2, and on heating this product momentarily depolymerizes to γ-aminopropyl isocyanate ($H_2NCH_2CH_2CH_2NCO$) which cyclizes to 2-hydroxy-1,4,5,6-tetrahydropyrimidine (**64**, R =′R′ = H).[3608]

Finally, in this section may be mentioned the condensation of benzaldehyde with the formal 1,3-diaminopropane derivative, 3-aminopropylamidoxime (**71**). The product may have one of three structures (**72**), two of which are tetrahydropyrimidines, or it may be a tautomeric mixture.[3609]

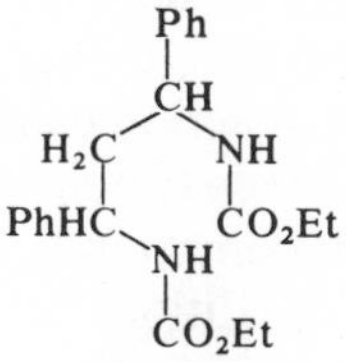

Ph, CH, H_2C, NH, PhHC, CO_2Et, NH, CO_2Et

(**68**)

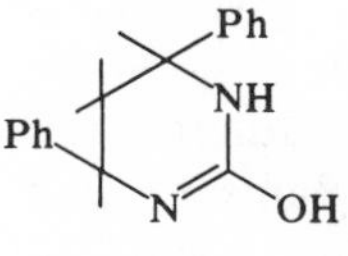

Ph, NH, Ph, N, OH

(**69**)

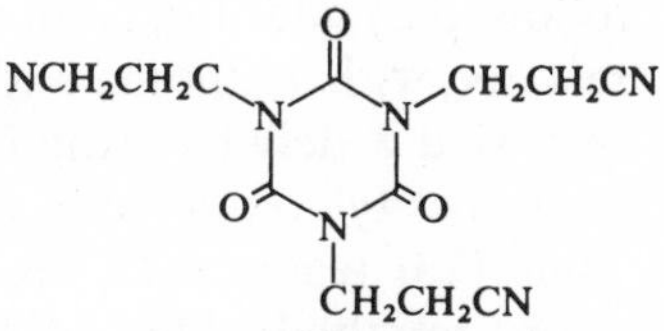

NCH_2CH_2C, O, N, N, CH_2CH_2CN, O, N, O, CH_2CH_2CN

(**70**)

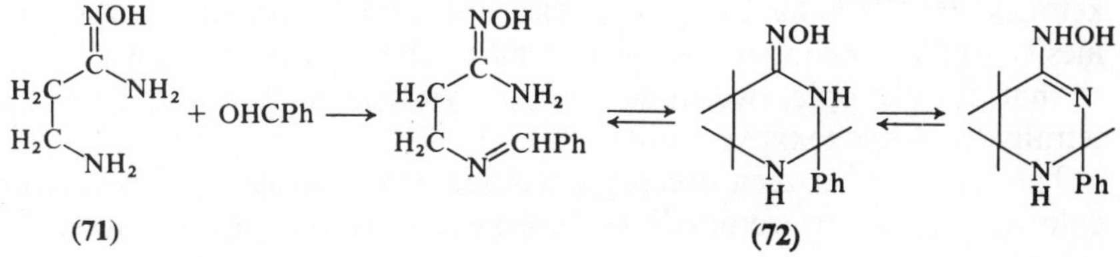

NOH, C, H_2C, NH_2, H_2C, NH_2 + OHCPh ⟶ NOH, C, H_2C, NH_2, H_2C, N=CHPh ⇌ NOH, NH, N H, Ph ⇌ NHOH, N, N H, Ph

(**71**)

(**72**)

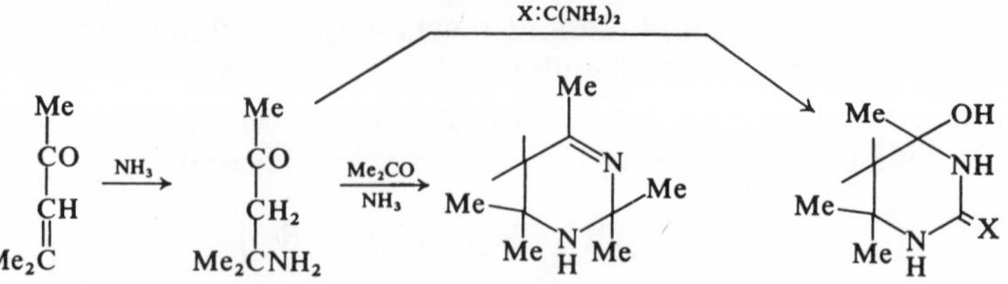

B. From Carbonyl Compounds and Ammonia, Amines, or Ureas (*H* 448)

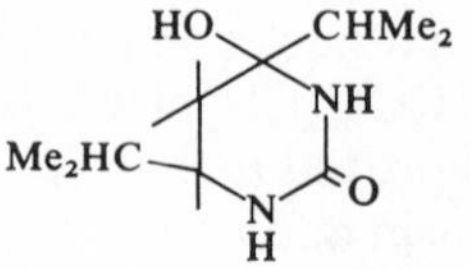

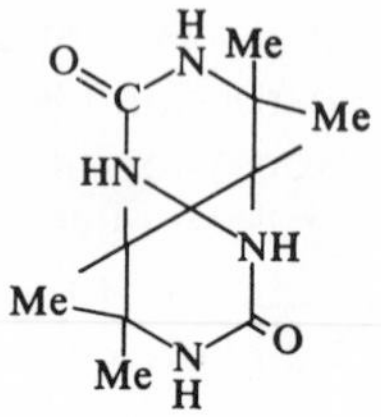

1,2,5,6-Tetrahydro-2,2,4,6,6-pentamethylpyrimidine (**75**) is obtained in excellent yield when acetone and liquid ammonia, with the catalysts calcium and ammonium chloride, are kept at 20° for 24 hr.[1578] It forms a crystalline monohydrate, spectroscopic examination of which indicates that it simply contains water of crystallization rather than being formed by covalent hydration of the C=N bond in (**75**).[3610]

The mechanism of the reaction is seen as involving the initial condensation of two molecules of acetone to give mesityl oxide (**73**) followed by addition of ammonia to give diacetonamine (**74**) and further reaction with ammonia and acetone to give (**75**). This basic reaction can be varied widely by using *inter alia* different amino compounds to react with mesityl oxide and other αβ-unsaturated carbonyl compounds. Thus (**73**) reacts with urea or thiourea under alkaline conditions to afford tetrahydro derivatives of a 4-hydroxypyrimidine (**76**, X = O or S).[3611, 3612] Acrolein[3613, 3614] and other αβ-unsaturated aldehydes and ketones[2119, 3615] behave similarly with urea and *N*-alkyl-ureas. As with mesityl oxide, precursors may be used in the condensation with urea so that, for example, two moles of isobutyraldehyde and urea form the tetrahydro-4-hydroxypyrimidine (**77**).[3616]

The reaction between urea and acetone in the presence of hydrogen chloride affords 'triacetonediurea' assigned structure (**78**).[3617] Since the

$PhCH(NHCONH_2)_2$

(79) (80) 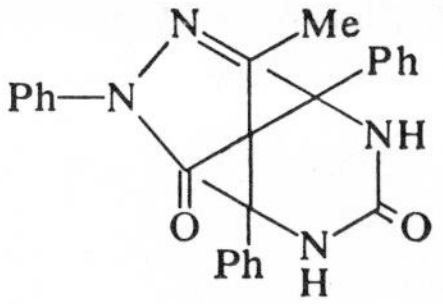 (81)

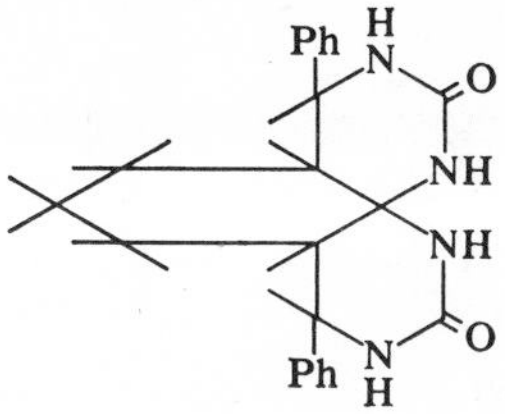

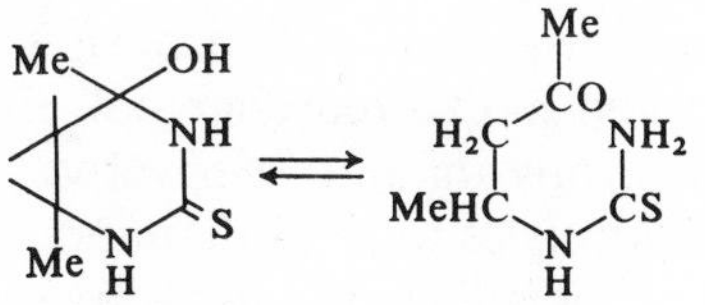

(82) (83)

same compound is obtained if phorone ($Me_2C:CHCOCH:CMe_2$) is used as starting material,[3617, 3618] the spiro structure is to be preferred to the alternative reduced pyrimidopyrimidine structure.[3619]

Spiro structures also result when αα-bisureidotoluene (**79**; an intermediate in the Biginelli reaction) condenses with cyclic ketones or other carbonyl compounds. Cyclohexanone gives (**80**)[3620] while an appropriate methyl-*N*-phenylpyrazolone affords (**81**).[3621]

Again, an αβ-unsaturated ketone such as 1-acetylpropene (AcCH:CHMe) may react with thiocyanic acid prior to reaction with ammonia or an amine and this leads again to a tetrahydro-4-hydroxy compound (**82**).[3622, 3623] Under certain circumstances this exhibits tautomerism with an open chain form, the *N*-ketoalkylthiourea (**83**).[3622]

C. From 1,3-Dihalogenopropane Derivatives (*H* 449)

The condensation of 1,3-dihalogenopropanes with 2,6-dichloro-*N*-phenylbenzamidine (**84**) to afford 2-2′,6′-dichlorophenyl-1,4,5,6-tetrahydro-1-phenylpyrimidine (**85**) is the only recent example of this type of reaction described.[3624]

D. By Reduction of Pyrimidines or Dihydropyrimidines (*H* 450)

A number of catalysts have been used to prepare tetrahydropyrimidines by reduction methods but palladized charcoal seems to be the

$$H_2C(CH_2X)CH_2X + PhN{=}C(NH_2)C_6H_3Cl_2 \longrightarrow \text{(85)}\ X^-$$

(84) (85)

one most widely used to date.[2162, 2685, 2686, 3625] A variety of pyrimidines (**86**; Table XVd) absorb two moles of hydrogen when reduced in acidic media with hydrogen and palladized charcoal. Each product is a salt of the 1,4,5,6-tetrahydropyrimidine derivative (**87**), confirmed in structure by alkaline hydrolysis to a 1,3-diaminopropane (**88**). Aqueous inorganic acid or glacial acetic acid is used save that with 2-methoxypyrimidine, an exactly equivalent quantity of methanolic acetic acid must be used in order to avoid the ready splitting of the ether linkage. The acid is required to protonate both the pyrimidine starting material (since it was likely that the pyrimidinium cation and not the neutral molecule was the species reduced) and also the product, which as neutral molecule would poison the catalyst; moreover, the products (**87**) are salts containing the resonating amidinium system which are more stable than the free bases. The parent 1,4,5,6-tetrahydropyrimidine hydrochloride is obtained both from the reduction of pyrimidine in aq. hydrochloric acid and from the reductive dehalogenation of 2,4-dichloropyrimidine in water.

Some pyrimidine compounds do not follow this 1,4,5,6-reduction pattern: 5-aminopyrimidine upon reduction in acidic solution absorbs only one mole of hydrogen. This is probably added to the 1,6-bond to afford the salt of the enamine (**89**, R = NH_2) which is rapidly hydrolysed (**89**, R = OH) and tautomerized to the keto compound (**90**). The presence of a positive charge elsewhere in the molecule amplified the tendency of the carbonyl grouping to hydrate, so that the ultimate product is a salt of the gem diol (**91**).[2686] 5-Hydroxypyrimidine also absorbs only one mole of hydrogen upon reduction and finally affords the same product (**91**) after a similar sequence of 1,6-addition of hydrogen (**89**, R = OH) followed by tautomerization and hydration.[3625]

4-Aminopyrimidine forms an unstable 1,2,5,6-tetrahydro derivative (**92**) on reduction in an acidic medium.[2685] Only the three decomposition fragments, an acrylamidinium salt, an ammonium salt, and formaldehyde, can be detected or isolated.[2685, 2686] Catalytic reduction of 5-acetamidopyrimidine in hot acetic anhydride at 100° yields the diacetyl derivative of a 1,2,3,4-tetrahydro compound (**93**).[3625] A

TABLE XVd. Catalytic Hydrogenation (Pd/C) of Pyrimidines (*New*)

(86) →[H_2/Pd/C, HX] (87) →[OH^-] (88)

R	R′	R″	R‴	X
H	H	H	H	Cl
Me	Me	H	Me	Cl
NH_2	H	H	H	Cl, AcO
NHAc	H	H	H	Cl, AcO
NHBz	H	H	H	Cl
NH_2	Me	H	Me	Cl
H	H	NHAc	H	Cl
OMe	H	H	H	AcO
H	H	OMe	H	Br
OH	H	H	H	Cl

1,4,5,6-tetrahydro derivative (**94**) results from the palladium catalyzed hydrogenation of the 3,4(or 4,5)-dihydro-2-hydroxy-4,6-diphenylpyrimidine.[3544]

Platinum resembles palladium in its catalytic activity and 2-amino-4-methylpyrimidine and its 4-β-amino-β-carboxyethyl analogue are reduced to salts of the 1,4,5,6-tetrahydro stage.[2955] Both 2-hydroxypyrimidine and its 1,4-dihydro derivative afford 1,4,5,6-tetrahydro-2-hydroxypyrimidine upon platinum catalyzed hydrogenation in ethanol solution.[2988] 2-Hydroxy-4,6-dimethylpyrimidine and its *N*-methyl derivative afforded 1,4,5,6-tetrahydro compounds under similar conditions.[3556]

Rhodium was a better catalyst than platinum for the reduction of 4-hydroxypyrimidine or its 3-methyl derivative to the 1,2,5,6-tetrahydro derivatives (**95**, R = H or Me) in ethanol solution,[2988] although in acetic acid platinum could cope with the job in the case of the 4-hydroxy-2-methyl(or 2,6-dimethyl)pyrimidine.[3626] The 2,5-dimethyl compound gave a reduction product which could be hydrogenolysed to an open chain amide.

Both platinum and rhodium catalysts are too active during reduction of cytosine (**96**, R = H)[3627] and cytidine (**96**, R = a sugar),[3628]

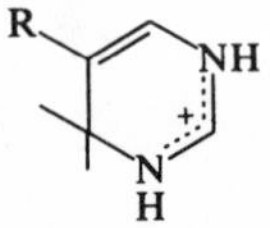

R, NH, +, N, H (**89**)

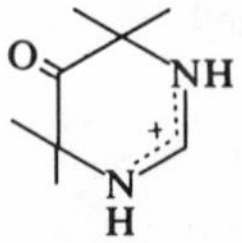

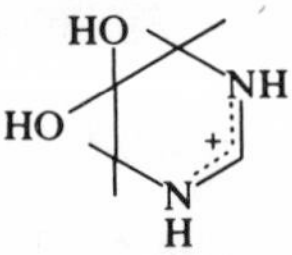

O, NH, +, N, H (**90**)

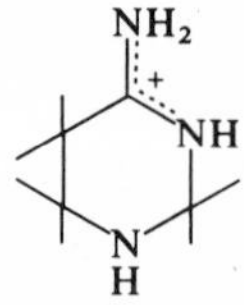

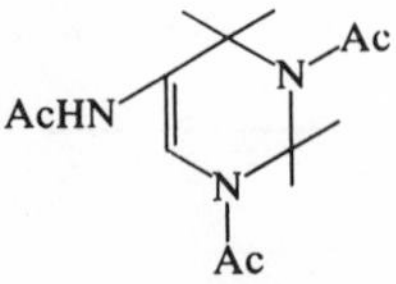

HO, HO, NH, +, N, H (**91**)

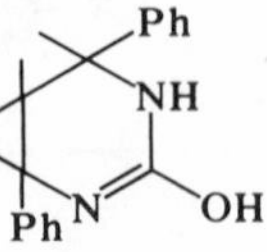

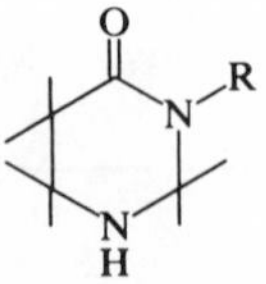

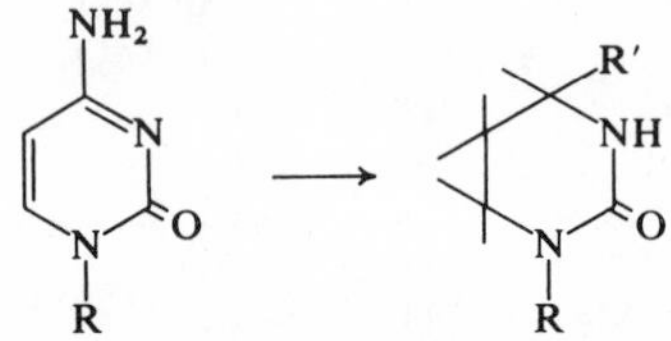

NH_2, +, NH, N, H (**92**)

AcHN, N, Ac, N, Ac (**93**)

Ph, NH, Ph, N, OH (**94**)

O, N, R, N, H (**95**)

NH_2, N, O, N, R (**96**) → R′, NH, O, N, R (**97**)

R, NH, N, H (**98**)

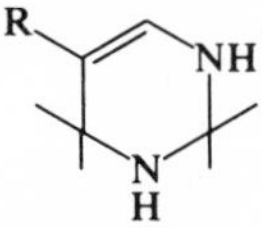

respectively, since hydrogenolysis of the 4-amino group in (**97**, R′ = NH_2) accompanies reduction to give the 1,4,5,6-tetrahydro derivative (**97**, R′ = H). Some of the product (**97**, R = a sugar; R′ = NH_2) was isolated (prior to deamination) in the cytidine case.

Raney nickel may have catalyzed the reduction of 1,4-dihydro-1-methyl-4-oxopyrimidine to a 1,2,5,6-tetrahydro derivative.[3552]

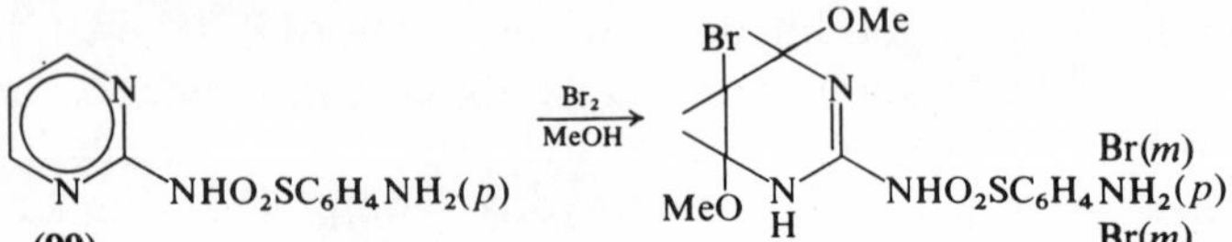

N, N, $NHO_2SC_6H_4NH_2(p)$ (**99**) $\xrightarrow[MeOH]{Br_2}$ Br, OMe, N, MeO, N, H, $NHO_2SC_6H_4NH_2(p)$, Br(m), Br(m) (**100**)

$N\cdot CO_2C_6H_4Cl(p)$ (**101**) $\xrightarrow[200°]{p\text{-anisidine}}$ CH_2, H_2C, $NHC_6H_4OMe(p)$, H_2C, $NHCO_2C_6H_4Cl(p)$ (**102**) → N, $C_6H_4OMe(p)$, N, H, O (**103**)

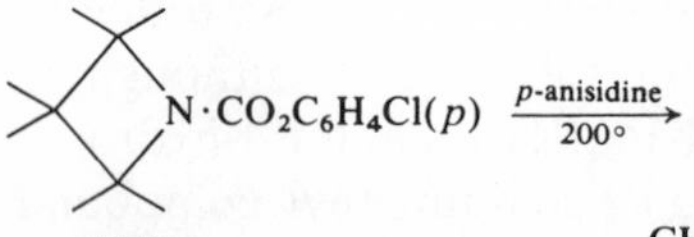

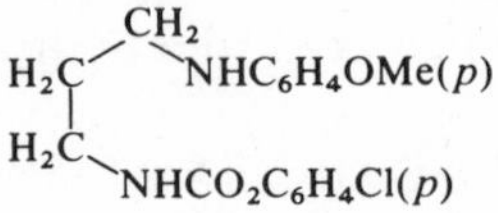

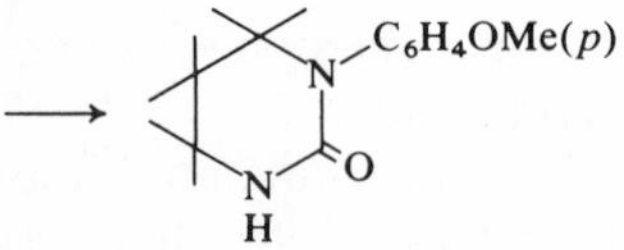

Pyrimidine is reduced by a boiling ethereal solution of lithium aluminium hydride, probably to an unstable 1,2,3,4-tetrahydropyrimidine (**98**, R = H).[3551] Identifiable decomposition products are formaldehyde and ammonia. Both the amide and the ester groups in the 5-ethoxycarbonyl-1,4(3,4)-dihydro-2-hydroxypyrimidine were reduced by a tetrahydrofuran solution of lithium aluminium hydride to the 1,2,3,4-tetrahydro compound (**98**, R = CH_2OH).[3629]

E. By Addition of Reagents Other than Hydrogen to the Pyrimidine Nucleus (*New*)

Bromination of the sulphanilylaminopyrimidine (**99**) in methanol solution introduces three bromine atoms into the molecule. One of these enters the 5-position of the pyrimidine nucleus, rendering it so deficient of electrons that both 4- and 6-positions are subjected to nucleophilic attack by solvent molecules. The product ultimately isolated is one in which two molecules of methanol have added to the pyrimidine ring giving a derivative of 1,4,5,6-tetrahydropyrimidine (**100**).[3630]

The conversion of pyrimidine into its 2,4-di-t-butyl derivative with lithium-t-butyl must proceed through a 1,2,3,4-tetrahydro derivative.[3631]

F. From Other Heterocyclic Substances (*New*)

The ring expansion of the azetidine derivative (**101**) is brought about by heating with a primary aromatic amine, especially one with an electron-donating substituent in the ring, such as *p*-anisidine. Nucleophilic attack takes place on the carbon atom next to the nitrogen atom of the azetidine ring, and the intermediate carbamic acid ester (**102**) recyclizes with expulsion of *p*-chlorophenol to form 1,4,5,6-tetrahydro-2-hydroxy-1-*p*-methoxyphenylpyrimidine (**103**).[3632]

3. Preparation of Hexahydropyrimidines (*H* 452)

These compounds are mostly prepared from 1,3-diaminopropanes with an aldehyde or ketone to supply C-2. Reference to Table XVe indicates how the scope of this basic reaction has been widened to take in representative combinations of 1,3-diaminopropane, its *N* and *NN'*-alkyl,[3610, 3633, 3634] aralkyl,[3635–3641] sulphonyl,[3642] and even nitro[3642] substituted compounds with formaldehyde, aliphatic or aromatic aldehydes, and ketones.

TABLE XVe. Preparation of Hexahydropyrimidines from Aldehydes or Ketones with 1,3-Diaminopropane Derivatives

$$H_2C(CH_2CH_2NHR')... \quad \begin{matrix} CH_2 \\ H_2C \quad NHR' \\ H_2C \quad NHR \end{matrix} + O{=}C\begin{matrix} R'' \\ R''' \end{matrix} \longrightarrow \textbf{(104)}$$

(104)

R	R′	R″	R‴	Ref.
H	H	H	H	3610
H	H	H	Me	3610
H	H	Me	Me	3610
H	H	Et	Et	3610
H	Me	H	H	3610
H	Me	Me	Me[a]	3610
H	Bu	H	H	3610
H	Bu^t	H	H	3610
H	Bu^t	H	Ph[a]	3610
H	Bu^t	Me	Me[a]	3610
Me	Me	H	H	3610
Me	Me	H	Me	3633
Me	Me	H	Et	3633
Me	Me	H	Pr^i	3633
Me	Me	H	Ph	3633
Bu	Bu	H	H	3610, 3634
Bu	Bu	H	Me	3634
Bu^t	Bu^t	H	H	3610
p-$Me_2NC_6H_4CH_2$	*p*-$Me_2NC_6H_4CH_2$	H	Pr^i	3635
NO_2	NO_2	H	H	3641
$PhSO_2$	$PhSO_2$	H	H	3642
—[b]	—[b]	H	—[c]	3636

[a] The product is the open chain tautomer, $R'HNCH_2CH_2CH_2N{:}CR''R'''$.
[b] 2′-Hydroxy-3′-methoxybenzyl.
[c] 3′,4′-Dimethoxyphenyl.

The parent hexahydropyrimidine (**104**, R = R′ = R″ = R‴ = H) results from the condensation of aq. 1,3-diaminopropane monohydrochloride with formaldehyde.[3610] The reaction may be visualized as proceeding by Schiff base formation at the unprotonated amino group followed by complete tautomerization to the cyclic hexahydropyrimidine salt. Basification of the mixture yields free hexahydropyrimidine (**104**) which, contrary to past belief, exists completely in the cyclic form. If unprotonated, trimethylenediamine and formaldehyde condense to give a tetramer (**105**) consisting of four hexahydropyrimidine rings, joined by CH_2 groups through their nitrogen atoms.[3643] 1,3-Diaminopropane was used in the mono protonated form for condensation with acetaldehyde, acetone, or diethyl ketone, but its derivatives were used as the uncharged neutral species.

3-Dimethylaminopropylamine alone reacted with formaldehyde to give the hexahydro-2,4,6-trisdimethylaminopropyl-1,3,5-triazine. However p.m.r. evidence showed that an aqueous solution of the diamine monohydrochloride was converted by formaldehyde into the hexahydropyrimidine quaternary salt (**106**).[3644]

When an *N*-alkyl-1,3-diaminopropane was condensed with acetone or benzaldehyde, strain in the molecule destabilized the cyclic structure so that the product was the open chain tautomer formulated in the footnote to Table XVe.[3610]

The condensation of nitroparaffins with formaldehyde in the presence of primary amines affords hexahydropyrimidines, again probably through the intermediacy of a 1,3-diaminopropane. Thus whether the hexahydropyrimidine (**108**) or the intermediate 1,3-diaminopropane (**107**) is isolated from the reaction of *p*-toluidine with nitroethane and

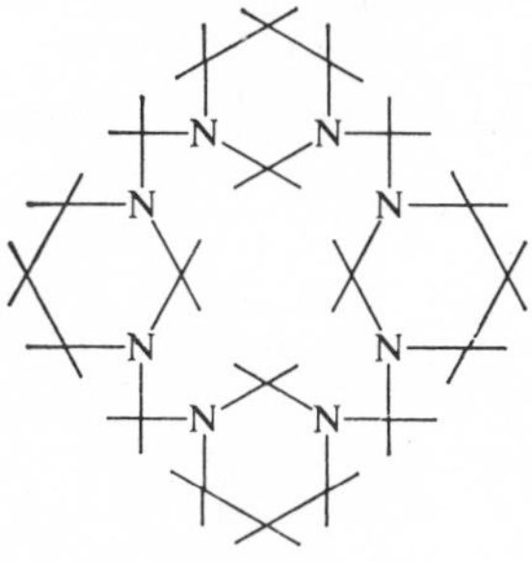

N N N N N N N N

(**105**)

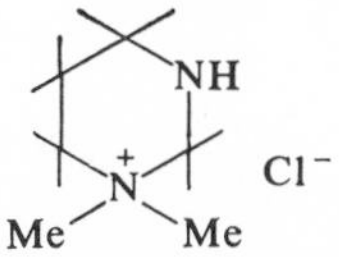

NH N⁺ Me Me Cl⁻

(**106**)

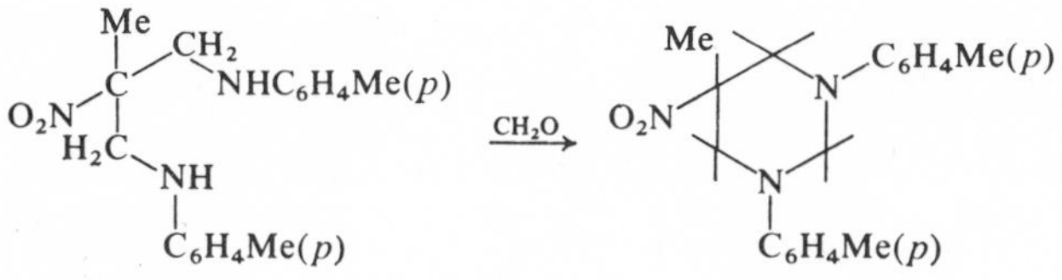

Me CH_2 C $NHC_6H_4Me(p)$ O_2N H_2C NH $C_6H_4Me(p)$

(**107**)

$\xrightarrow{CH_2O}$

Me O_2N N $C_6H_4Me(p)$ N $C_6H_4Me(p)$

(**108**)

formaldehyde depends solely on whether the last reagent is present in excess or not.[3645] Although the condensation is carried out in acetone solution, no trace of the 2,2-dimethyl analogue of (**108**) is reported. The precursor of the 1,3-diaminopropane is probably the 2-nitropropan-1,3-diol derivative. This can be prepared separately and will undergo condensation with primary amines and formaldehyde. Thus 2,2-dinitropropan-1,3-diol condenses with $\gamma\gamma\gamma$-trinitropropylamine and formaldehyde to give a 1,3-bis-$\gamma\gamma\gamma$-trinitropropyl-5,5-dinitrohexahydropyrimidine.[3646] Other examples are known.[3647–3650]

If the aliphatic nitro compound contains only one hydrogen atom activated by a nitro group, condensation with a di(primary)amine and formaldehyde still affords a hexahydropyrimidine derivative. Trimethylenediamine, formaldehyde, and β-nitropropane give hexahydro-1,3-bis-β-nitroisobutylpyrimidine (**109**),[3651] through the intermediacy of β-nitroisobutyl alcohol [$Me_2C(NO_2)CH_2OH$].

Compounds with —CH_2— activated by other electron-withdrawing groups (CO, SO_2) can also take part in the reaction with primary amines and formaldehyde to yield hexahydropyrimidines. Ethyl methyl ketone affords 2-acetyl-2-methylpropan-1,3-diol [$AcC(CH_2OH)_2Me$], converted by primary amines (RNH_2) into the hexahydro-pyrimidine (**110**).[3652] The cyclic sulphone (**111**) yields a spiro hexahydropyrimidine (**112**).[3653]

A novel reaction is the displacement of dimethylamine from dimethylaminodimethoxymethane [$MeOCH(NMe_2)OMe$] by 1,3-bis-ethylaminopropane, followed by expulsion of methoxide ion to give hexahydro-2-methoxy-1,3-dimethylpyrimidine (**113**).[3654] The 2-cyano analogue is made similarly using α-dimethylamino-α-methoxyacetonitrile as starting material.[3655]

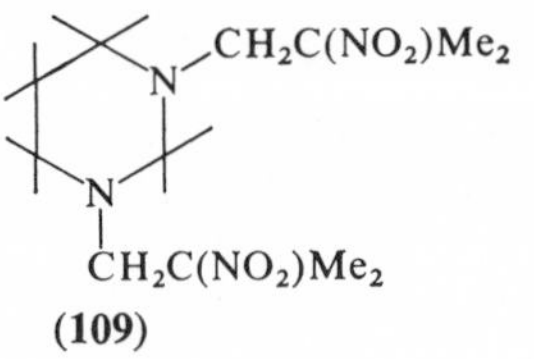

$CH_2C(NO_2)Me_2$ N N $CH_2C(NO_2)Me_2$

(**109**)

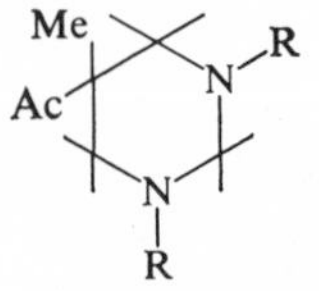

Me Ac N R N R

(**110**)

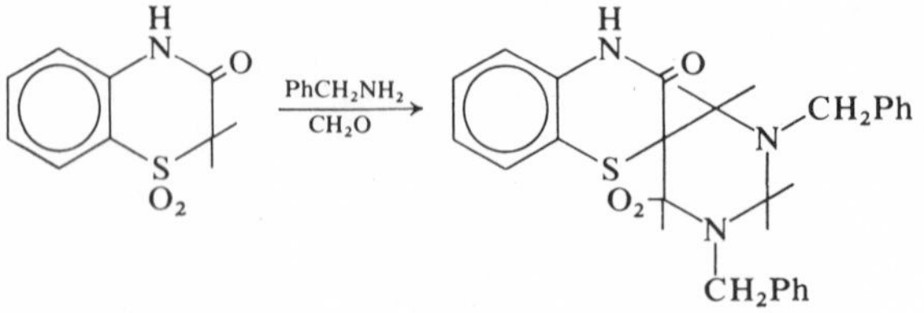

H N O S O_2 $\xrightarrow[CH_2O]{PhCH_2NH_2}$ H N O S O_2 N CH_2Ph N CH_2Ph

(**111**) (**112**)

N Me OMe N Me

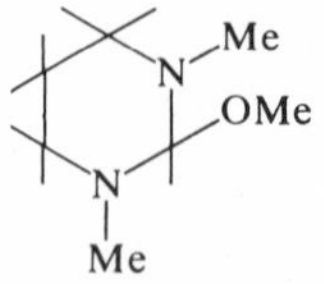

(**113**)

The report[714] that 2-amino-4,6-dichloropyrimidine was reduced in acidic solution to give 2-aminohexahydropyrimidine is erroneous.[3551] The original product was characterized as a picrate with a melting point suspiciously close to that of the picrate of 2-amino-1,4,5,6-tetrahydropyrimidine, and repetition of the reduction followed by direct comparison with an authentic sample confirmed the identity. Consequently the only authenticated case of the production of a hexahydropyrimidine by a reduction process is the reduction of the C=N group in 1,2,5,6-tetrahydro-2,2,4,6,6-pentamethylpyrimidine by sodium in alcohol.[1578] In this case there is no need to interrupt the powerful amidinium resonance of a 1,4,5,6-tetrahydro compound and the reduction proceeds in a different manner, electron and proton being successively added to the double bond.

4. Reactions of Reduced Pyrimidines (*H* 454)

A. Reductions (*H* 454)

The catalytic reduction of several pyrimidine compounds to the 1,4,5,6-tetrahydro stage has been documented above. Although dihydro compounds must be involved as intermediates, only in one case has this intermediate been isolated and characterized as a result of interrupting the reduction after the absorption of one mole of hydrogen. This intermediate is 3,6-dihydro-2-hydroxypyrimidine (**39**, R = H) and it, in turn, is separately reduced in the presence of platinum up to, but not beyond, the resonance stabilized 1,4,5,6-tetrahydro stage (**97**, R = R′ = H).[2988] The same state (**94**) is reached by electrolytic reduction of dihydro-2-hydroxy-4,6-diphenylpyrimidine.

Lithium aluminium hydride reduces the amide carbonyl of the stable tautomer of the 5-ethoxycarbonyl-1,4(1,6)-dihydro-2-hydroxypyrimidine to give ostensibly the 1,2,3,4-tetrahydro derivative (**98**, R = CH_2OH).[3629] Since two tautomeric changes of mobile hydrogen atoms would convert this into the 1,4,5,6-tetrahydro isomer, it would be interesting to investigate the p.m.r. spectrum of the product. The stability of the ring in 1,6-dihydro derivatives of ethyl 2-mercapto-4-substituted-pyrimidinecarboxylates may be more apparent than real. Steric factors affect the reduction of nitrogen heterocycles in which a hydride ion attack at positions adjacent to ring nitrogen atoms is involved.[3656] In the cases examined, the 2- and 4-positions were occupied by substituents capable of hindering the approach of the aluminohydride ion and the remaining position contained an ethoxycarbonyl group with the carbonyl conjugated to the two double bonds of the nucleus. In this event the ester grouping wins out in the competition for hydride ion since it is preferentially reduced.

The lithium aluminium hydride reduction of 5-ethyl-5-phenylbarbituric acid to the tetrahydro derivative (**114**) must surely involve reduction of the amide carbonyl of an intermediate dihydro compound e.g., (**115**).[1540]

Catalytic reduction, however, readily converts 5-ethoxycarbonyl-1,6-dihydro-2-methylpyrimidine into the 1,4,5,6-tetrahydro compound.[923] The catalytic hydrogenolysis of a tetrahydro-4-hydroxypyrimidine has been mentioned.[3626]

The ring in *NN'*-disubstituted-hexahydropyrimidines is stable towards catalytic reduction. Thus the nitro group in many 1,3-dialkylhexahydro-5-nitropyrimidines is reduced to amino using Raney nickel at 70° and hydrogen at 1000 p.s.i. without ring fission under these vigorous conditions.[3657] However, catalytic reduction under mild ambient conditions may cause fission of the ring in certain hexahydropyrimidines where there are no *N*-substituents or only one. These are the cases where steric factors cause strain in the cyclic structure and

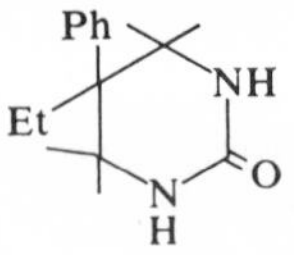

Ph Et NH O N H

(**114**)

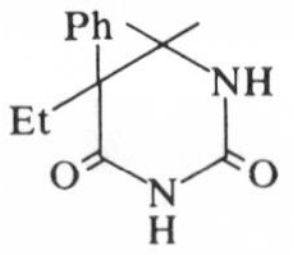

Ph Et NH O O N H

(**115**)

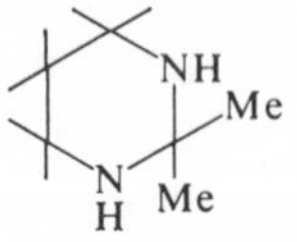

NH Me N H Me

(**116**)

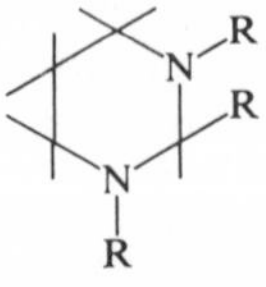

N R R N R

(**117**)

OR Me NH O N Sugar

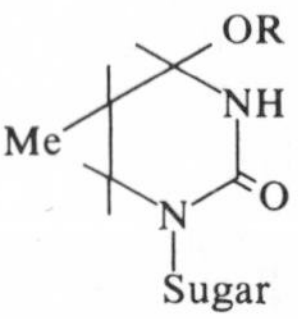

(**118**)

give rise to a dynamic equilibrium with an open chain tautomer containing a C:N-bond. The latter group is very readily reduced, causing the continual disappearance of the open chain form and the displacement of the equilibrium so that the final result is the apparent hydrogenolysis of the hexahydropyrimidine ring. Thus hexahydro-2,2-dimethylpyrimidine (**116**) is readily converted into 3-isopropylaminopropylamine ($H_2NCH_2CH_2CH_2NHPr^i$) on shaking with hydrogen and Adams catalyst under ordinary catalytic conditions because the cyclic compound is in equilibrium with 3-isopropylideaminopropylamine ($H_2NCH_2CH_2CH_2N{:}CMe_2$) the double bond of which is the group being readily reduced.[3610]

Sodium borohydride is equally proficient at this task when some of the open chain tautomer is present in an equilibrium mixture. However, it has also proved proficient when no open chain tautomer is (or can be) present. A small yield of 3-methylaminopropylamine may be obtained from hexahydropyrimidine itself, and much better yields of the corresponding diamine from 1,2,3-trialkylhexahydropyrimidines (**117**).[3610, 3633, 3634] A suggested mechanism involves the reduction of an immonium ion produced by interaction of one nitrogen atom's lone pair of electrons with the alcohol solvent molecules.

Sodium borohydride also causes ring fission at the amide bond of dihydrouracils, 5,6-dihydrouracil itself giving γ-ureidopropanol ($NH_2CONHCH_2CH_2CH_2OH$) with methanolic sodium borohydride.[3557] Dihydrothymidine gives the analogous compound together with the *O*-methyl ether (**118**, R = Me) of the intermediate carbinolamide (**118**, R = H). This was probably formed *via* the tautomeric open chain aldehyde by loss of methanol from its dimethyl acetal.[3555]

B. Oxidation (*H* 455)

The limited evidence available reveals that 1,6-dihydropyrimidines exhibit the whole range of stability towards oxidising or dehydrogenating agents. Thus atmospheric oxygen attacks 1,6-dihydro-5-methyl-2-phenylpyrimidine (**47**) so readily that it cannot be obtained free from the oxidation product, 5-methyl-2-phenylpyrimidine.[3473] 1,6-Dihydro-2-methylthio compounds, on the other hand, seem perfectly stable towards oxygen, but can be aromatized to the pyrimidine compound with a quinone.[3474] With the 2-oxo compounds (**13**, R = H, R′ = Me; and *vice versa*) refluxing in a dioxane solution with 2,3-dichloro-5,6-dicyanoquinone is required for dehydrogenation to the corresponding pyrimidine. Potassium permanganate in acetone was also an effective oxidizer for 4-t-butyl-3,4-dihydropyrimidine to 4-t-butylpyrimidine.[3551] γ-Irradiation of dihydrouracil and its derivatives is claimed to bring about dehydrogenation to the uracil.[3676]

The attempted dehydrogenation of 1,4,5,6-tetrahydropyrimidine with sulphur led to insertion of sulphur in the molecule and formation of 1,4,5,6-tetrahydro-2-mercaptopyrimidine.[2162] Palladized charcoal, on the other hand, led to dehydrogenative coupling and formation of bi-(1,4,5,6-tetrahydropyrimidin-2-yl) (**119**).[2162] There thus appears to be a reluctance on the part of the 1,4,5,6-tetrahydro system to dehydrogenate. If a blocking group, e.g., an alkyl group, is placed at C-2 instead of hydrogen, dehydrogenation of the ring occurs, but only under vigorous conditions. Thus a stream of acetic acid and 1,3-diaminopropane vapours, passed over platinum-on-alumina at 400°, forms 1,4,5,6-tetrahydro-2-methylpyrimidine *in situ* which then dehydrogenates to 2-methylpyrimidine.[3677] On the other hand, the chromic acid/pyridine oxidation of 1,4,5,6-tetrahydro-isocytosine to the dihydro compound (**42**; R = NH_2, R′ = H) involves the easier conversion of a secondary alcohol group to carbonyl.[3596]

Of the hexahydropyrimidines, only the *NN′*-dimethyl compound is recorded as reacting with an oxidizing agent. Thus chloranil affords a variety of coloured solutions from which only the hydrolytic decomposition product, 1,3-bismethylaminopropane, can be isolated.[3610]

Hexahydro-2-methylpyrimidine (**120**) is unique in the compounds examined so far. When shaken with platinum and hydrogen under ambient conditions more hydrogen is evolved and the 1,4,5,6-tetrahydro compound (**121**) can be isolated. Thus under reducing conditions, an oxidation has occurred.[3610] It may be that in the presence of platinum catalyst, an equilibrium between hexahydro and tetrahydro derivatives is set up and only in the 2-methyl derivative does the equilibrium favour the tetrahydro species.

C. Nitrosation and Nitration (*H* 455)

The *N*-nitrosation of dihydropyrimidines seems not to have been attempted. The reduction product of 4-hydroxypyrimidine yielded only a mono-nitroso compound upon treatment with aq. nitrous acid.[2988] The more basic secondary amino group was nitrosated to give 1,2,5,6-tetrahydro-4-hydroxy-1-nitrosopyrimidine which from the infra-red data quoted for KBr discs, appeared to exist as the oxo-tautomer (**122**, R = NO). The strong band at 3389 cm^{-1} in the reduced molecule disappeared upon nitrosation and therefore must refer to νNH of the secondary amino group.

With hexahydropyrimidines, two types of behaviour emerge with nitrous acid. If the hexahydropyrimidine exhibits tautomerism, the molecule reacts in the tautomeric open chain form. This must contain an aliphatic primary amino group which is deaminated by nitrous acid. If the hexahydropyrimidine does not exhibit tautomerism, *N*-nitroso derivatives result. Thus hexahydropyrimidine and its 2-methyl derivative readily give *NN′*-dinitroso derivatives, but hexahydro-2,2-dimethylpyrimidine undergoes decomposition.[3610]

Direct nitration of hydrogenated pyrimidines has been attempted in even fewer cases. A mixture of acetic anhydride and nitric acid converts 1,4,5,6-tetrahydro-2,5-dihydroxypyrimidine (**123**, R = H) into the *NN'*-dinitro 5-nitrate ester (**123**, R = NO_2). Hexahydro-*NN'*-dinitro-

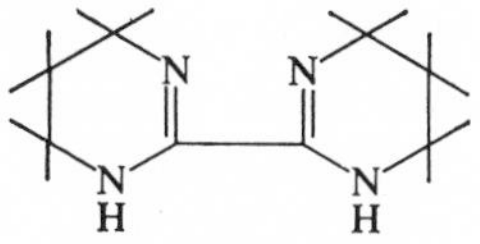

(119) 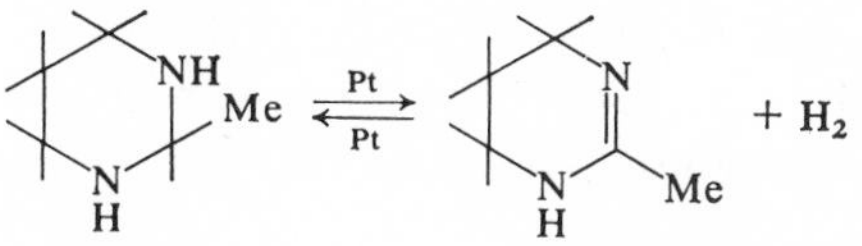 (120) (121)

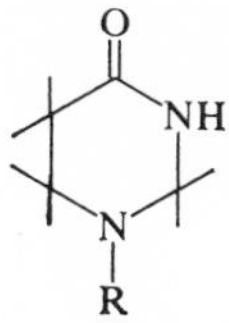

(122) 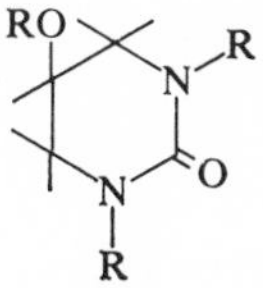 (123) 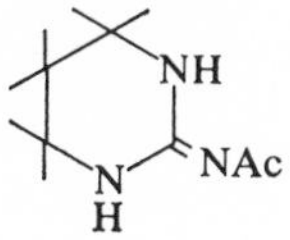 (124) 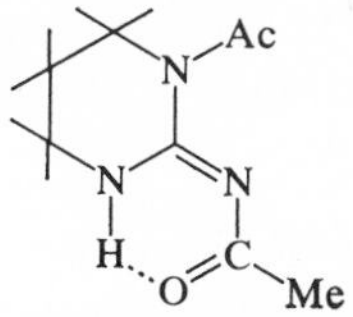 (125)

pyrimidine arises from condensation of appropriately substituted 1,3-diaminopropane with formaldehyde[3642] but not by direct nitration. Similar remarks apply to other 5-nitro derivatives.[3647]

D. Acylation (*H* 456)

Acylation readily occurs if the reduced pyrimidine contains an OH or NH group. Acetic anhydride acetylated 1,6-dihydro-2-methylthiopyrimidines at N-1.[3474]

Benzoylation of 1,4,5,6-tetrahydropyrimidine affords 1,3-bisbenzamidopropane, so that the unprotected 2-position is a point of weakness in the ring. With a group other than hydrogen at C-2 this complication does not arise. Thus 1,4,5,6-tetrahydro-2-phenylpyrimidine forms an *N*-*p*-toluenesulphonyl derivative[3658] without destruction of the reduced pyrimidine ring. *N*-Acylation is preferred to *O*-acylation when the 2-hydroxy analogue is treated with a dialkylmalonyl dichloride.[3659] 2-Amino-1,4,5,6-tetrahydropyrimidine reacts with ethyl acetate to afford a mono-acetylimino compound (**124**) judged to exist in this tautomeric form because of a similarity of its p.m.r. spectrum to that of a related imidazolidine derivative.[3660] Formation of a hydrogen-bonded unsymmetrical diacetyl derivative (**125**) is brought about when the acetate salt is heated with acetic anhydride.

The sodio derivatives of 2-aminotetrahydropyrimidines acylate in conjunction with acid chlorides.[3661]

Hexahydropyrimidines will be readily acylated provided that NH groups are present and that none of the open chain tautomer is present. Hexahydropyrimidine and its 2-methyl derivative readily form *NN'*-dibenzoyl or *p*-toluenesulphonyl derivatives. The 2,2-dimethyl derivative affords the corresponding derivatives of 1,3-diaminopropane through hydrolytic fission of the azomethine linkage in the tautomeric open chain form.[3610] Phosgene reacts with hexahydropyrimidine to give a dicarbamoyl chloride.[3662]

E. Halogenation (*H* 457)

The *N*-halogenation of 5,6-dihydro-6-methyluracil in the presence of sodium hydroxide has been observed.[3663]

F. Metatheses of Mercapto and Hydroxy Derivatives (*H* 457)

During the period under review, the metathetical reactions of sulphur compounds attracted patent attention, especially those belonging to the tetrahydro series. Alkyl substitution of 1,4,5,6-tetrahydro-2-mercaptopyrimidine took place at the sulphur atom with alkyl bromides,[3664] with benzyl chloride,[3665, 3666] and with 2-chloromethylthiophene.[3667, 3668] The 2-methylthio derivatives have been extensively investigated because of the ready displacement of the thioether group by amines.[3669] Thus 1,4,5,6-tetrahydro-2-methylthiopyrimidine (**126**) with hexylamine yields the 2-hexylamino analogue, preferably isolated as a *p*-toluenesulphonate salt.[3670] The amino group of glycine also acts as a nucleophile, but the resulting 2-carboxymethylamino-1,4,5,6-tetrahydropyrimidine (**127**) cyclizes to the reduced triazaindene (**128**).[3671] Phenyl isothiocyanate has been used to build on a second ring to the 2-alkylthiotetrahydro compound (**126**)[3666] to form the reduced tetraazanaphthalene (**129**).

G. *N*-Alkylation (*H* 458)

The addition of the NH groups in the molecule (**130**, R = H) across the activated double bond of two acrylonitrile molecules may be regarded as a dialkylation[3672] giving the product (**130**, R = CH_2CH_2CN).

The hydrogen of the NH group in 1,4,5,6-tetrahydropyrimidine has acidic properties and may be displaced by potassium. The potassio salt may then react with acetylene to give 1,4,5,6-tetrahydro-1-vinylpyrimidine,[3673] one starting point for polyvinyltetrahydropyrimidine.[3674]

Free 1,4,5,6-tetrahydropyrimidine is a strong nucleophile and reacts rapidly with methyl iodide, forming initially the *N*-monomethiodide. This must lose hydrogen iodide to unchanged tetrahydropyrimidine and the resulting base, tetrahydro-1-methylpyrimidine, reacts further with methyl iodide to give the *NN'*-dimethyl derivative.

2-Amino-1,4,5,6-tetrahydropyrimidines alkylate on a ring nitrogen atom.[3675] 1,3-Dimethylhexahydropyrimidine reacts rapidly with one equivalent of methyl iodide to give the monomethiodide **(131)**.[3610] Coulombic repulsion between the two positive charges stops formation of a dimethiodide in addition to the adverse axial methyl-methyl interactions that would be present in such a molecule. Moreover, steric strains in the molecule are responsible for the fact that, though 1,3-di-t-butyl and di-n-butyl hexahydropyrimidine react with methyl iodide, the methiodide breaks up into unknown substances.[3551]

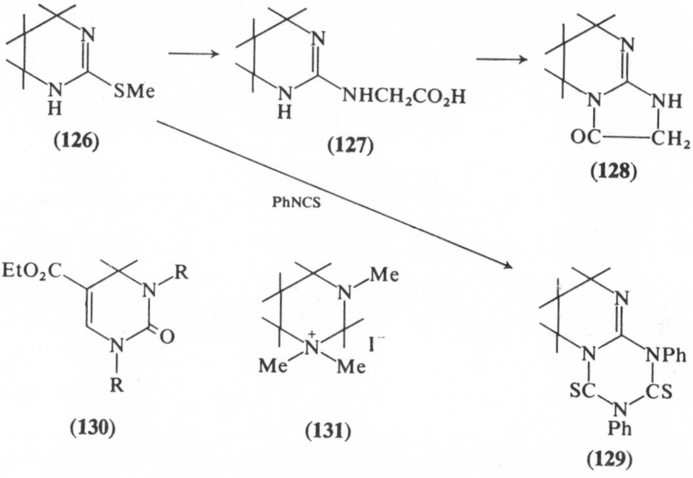

SMe NHCH$_2$CO$_2$H NH OC CH$_2$ PhNCS

(126) **(127)** **(128)**

EtO$_2$C R O R Me I$^-$ Me Me NPh SC CS Ph

(130) **(131)** **(129)**

H. Other Reactions (*H* 459)

(1) *Nonhydrogenolytic Ring Opening* (*New*)

Reduced derivatives of 2- and 4-hydroxypyrimidines are actually cyclic ureas or amides so that it is not surprising for these to hydrolyse with destruction of the reduced ring. Thus 1,2,5,6-tetrahydro-4-hydroxy-2-methylpyrimidine (**132**) affords β-aminopropionic acid on treatment with alkali.[3626] The acidic hydrolysis of 2-butyl-2,5-dihydro-4,6-dimethoxypyrimidine (**133**) to dimethyl malonate, ammonia, and valeraldehyde may also be noted in this connection.[2519] 1,4,5,6-Tetrahydropyrimidine (**134**, R = H) is unstable when dissolved in water, even at room temperature: 15 min. suffice to bring about hydrolytic splitting of the ring to 3-formamidopropylamine ($H_2NCH_2CH_2CH_2 \cdot NHCHO$). When the hydrogen at C-2 is replaced by other groups, hydrolytic splitting becomes more difficult. Thus the aqueous solution of the 2-phenyl compound (**134**, R = Ph) is stable at room temperature for 24 hr. The 2-amino compound (**134**, R = NH_2), on boiling as an aqueous solution for 45 min., gave a mixture of the ureide ($H_2NCH_2CH_2CH_2NH.CONH_2$) together with a further hydrolysis product, 1,3-diaminopropane. Alkali attacked both rings in 1,4,5,6-tetrahydro-2-1′,4′,5′,6′-tetrahydropyrimidin-2′-ylpyrimidine giving oxalic acid and 1,3-diaminopropane.[2162, 2957]

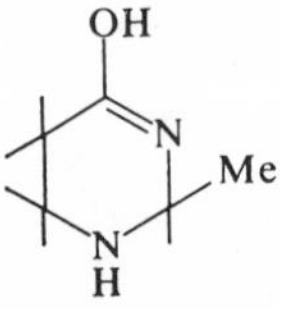

(132) 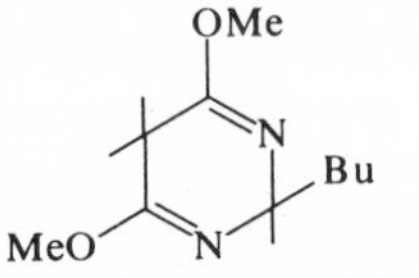 **(133)** 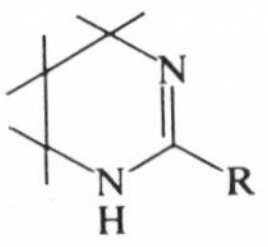 **(134)** 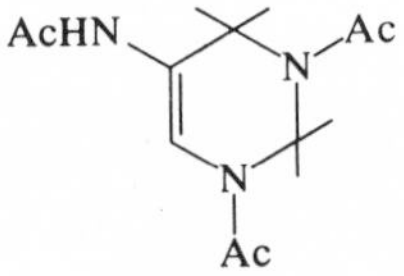 **(135)**

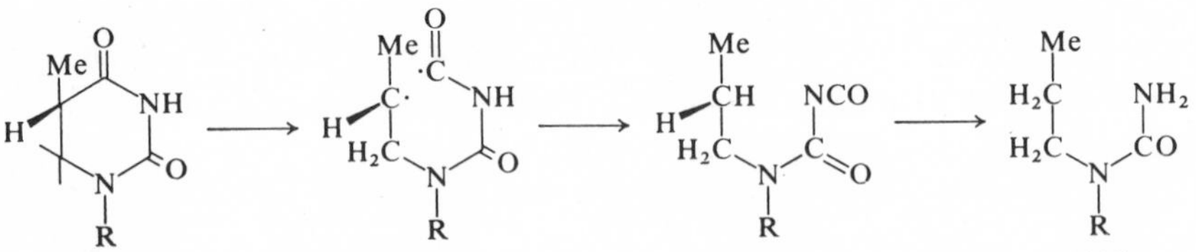

(136) → **(137)** → **(138)** → **(139)**

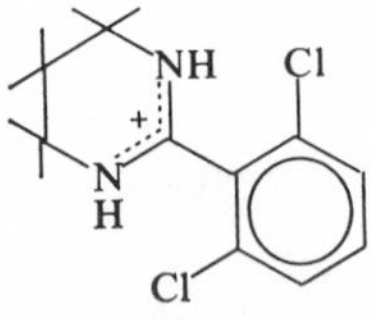

(140)

A weak point in the ring of 1,4,5,6-tetrahydropyrimidine is undoubtedly C-2. Thus α-naphthyl isocyanate gives the bis-α-naphthylureido derivative of 1,3-diaminopropane, and attempted benzoylation with benzoyl chloride gives 1,3-dibenzamidopropane.

Hexahydropyrimidines are rapidly hydrolysed in aqueous mineral acid back to the diaminopropanes and the aldehyde or ketone.[3610] In fact, any hydropyrimidine with two hydrogen atoms, one or two alkyl groups, and a hydrogen atom attached to C-2 has a weakness at this point and is rapidly hydrolysed in aqueous acid to the carbonyl compound and another open chain compound. A case in the dihydropyrimidine series has been mentioned above. The reduction of 5-acetamidopyrimidine in hot acetic anhydride to give the 1,3,5-triacetyl-1,2,3,4-tetrahydro compound (**135**) is accompanied by hydrolytic ring fission of some of the product to the propene [(AcHNCH$_2$C(NHAc):CHNHAc].[3625]

Ring fission of a different kind occurs with the selective photolysis of dihydrothymidine (**136**, R = a sugar). The diradical (**137**) produced by fission of the 4,5-bond rearranges to the isocyanate (**138**). The fate of this depends upon the solvent: water gives the amide (**139**) accompanied by decarboxylation.[3678]

The conversion of 1,4,5,6-tetrahydro-1-methylpyrimidine hydrochloride into the 2-amino compound with cyanogen chloride probably involves ring opening followed by recyclization.[3679]

The tetrahydropyrimidine ring of the herbicide (**140**) is disrupted in the soil and the hindered 2,6-dichlorobenzonitrile remains.[3705]

(2) *Reaction with Aldehydes* (*New*)

The reactions that have been investigated seem to fall into two categories depending upon whether the site of reaction is a ring NH group or an activated side chain methyl group. Only the latter seems to have been observed with dihydropyrimidines. Thus the 4-methyl group on 1,6-dihydro-2-hydroxy-4,6,6-trimethylpyrimidine (**141**, R = Me) is activated by the double bond of the ring and condenses with benzaldehyde[3680] to give the benzylidene derivative (**141**, R = CH:CHPh) or with formaldehyde and dimethylamine to give the Mannich base (**141**, R = CH$_2$CH$_2$NMe$_2$).[3681] In a similar fashion the 2-methyl group in 1,4,5,6-tetrahydro-1,2-dimethylpyrimidine condenses with thiophene-2-aldehyde. In this case the intermediate aldol may be isolated; this rapidly loses water, especially in the presence of stannous fluoride to give the *trans* condensation product (**142**).[3682]

With trimethyleneurea, the stable tautomeric form of 1,4,5,6-tetrahydro-2-hydroxypyrimidine, the amidic NH groups retain sufficient nucleophilic power to react with formaldehyde to give bishydroxymethyl derivatives.[3600, 3603, 3613] Condensation between the aldehyde group of 5-nitrofurfural[3683] and the *N*-amino (?) derivative of trimethylene urea is possible.

Hexahydropyrimidine with two replaceable NH groups readily condenses with formaldehyde, the product depending upon the ratio of the reagents. Hexahydro-1,3-bishydroxymethylpyrimidine results from a condensation with an excess of formaldehyde.[3684] Under other circumstances it is possible to isolate the cyclic substance (**105**) formed by

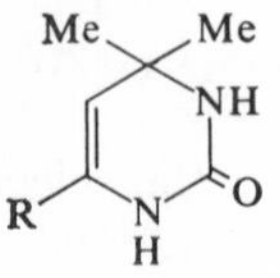

(**141**) 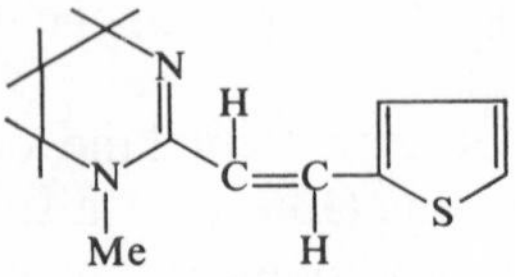 (**142**)

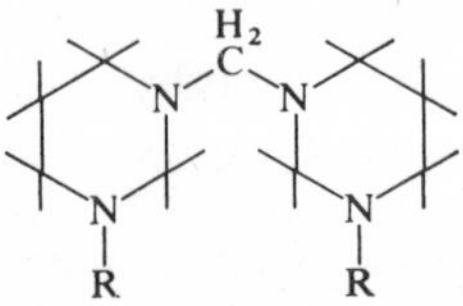

(**143**) 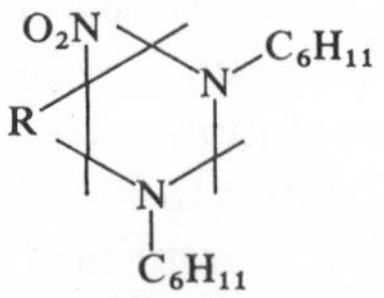 (**144**)

the condensation of 4 molecules of hexahydropyrimidine with 4 of formaldehyde.[3610] 1-Alkyl and other hexahydropyrimidines react with formaldehyde so that a molecule of water is eliminated between the formaldehyde and the NH groups of two molecules with formation of a methane derivative (**143**).

Finally formaldehyde is displaced from the hexahydropyrimidine derivative (**144**, R = CH_2OH) upon attack by an aryl diazonium chloride,[3685] to give the azo compound (**144**, R = N:NAr).

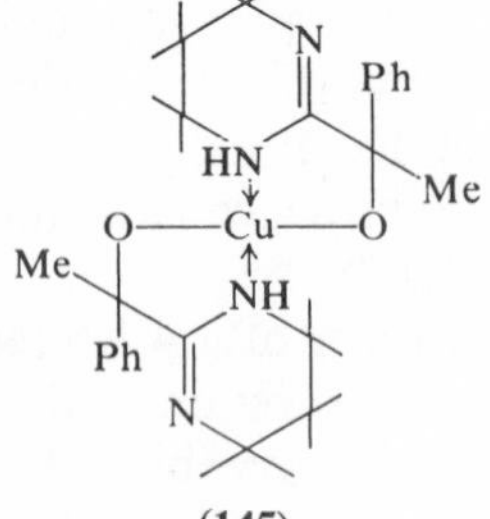

(**145**) 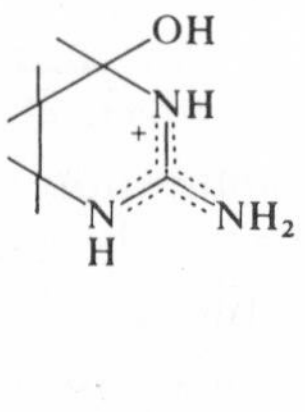 (**146**) 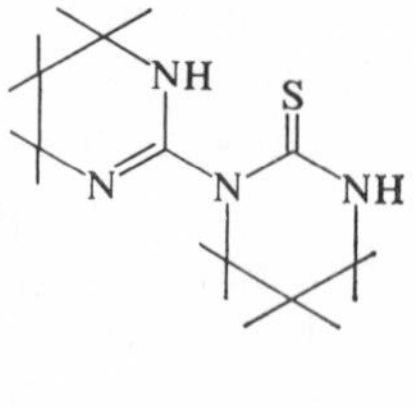 (**147**)

(3) *Miscellaneous Reactions* (*New*)

Trimethyleneurea and its thio analogue are catalysts for the maleic → fumaric acid isomerization in aqueous solution.[3686, 3687] Trimethyleneurea is a relatively weak ligand and forms complexes with metals like zinc, cobalt, or (ferric) iron in which the bonding is between the oxygen and metal atoms.[3688] 2:1 Square planar metal complexes, e.g., (**145**) are also formed between (cupric) copper and a number of α-hydroxy-1,4,5,6-tetrahydropyrimidines.[3689]

The hydroxy group in 2-heptadecyl-1,4,5,6-tetrahydro-5-hydroxypyrimidine can be sulphonated with chlorosulphonic acid to give a useful wetting agent.[3590]

The 4-hydroxyl group in the tetrahydro compound (**146**) is readily replaced by an ethoxy or methoxy group on treatment with the corresponding alcohol. Dehydration accompanied by acetylation of the exocyclic amino group occur with hot (135°) acetic anhydride. A 4,5-carbon-carbon and not a 3,4-carbon-nitrogen double bond is then formed.[3596]

1,4,5,6-Tetrahydro-2-mercaptopyrimidine gives *N*-mono- and di-acetyl products when treated with acetic anhydride. Acetyl chloride and acetic acid, however, give a product (**147**) isolated as a hydrochloride which appears to be formed by the displacement of thioacetic acid from an intermediate *S*-acetyl derivative.[3690] Diketene adds to other cyclic isothioureas, e.g., 2-*p*-bromobenzylthio-1,4,5,6-tetrahydro-1-methylpyrimidine to afford *cis*- and *trans*-crotonyl ureas, e.g., (**148**), in which the *S*-alkyl group has migrated.[3691]

Treatment of an aqueous solution of dihydrouracil with γ-rays causes hydroxylation of the CH_2 groups, with formation of the 6-hydroxy derivative[3692] or the 5,6-dihydroxy dihydrouracil.[3693]

5. Properties of Reduced Pyrimidines (*New*)

A. Ionization (*New*)

With the exception of a few tetrahydropyrimidines (see Table XVf), there does not seem to have been a systematic investigation of the basic or acidic properties of reduced pyrimidines. As befits their cyclic amidine or guanidine structure, the 1,4,5,6-tetrahydro compounds are highly basic, and in the case of the 2-amino compound, outstandingly basic even with an electron-withdrawing group at position 2. Thus benzoylation of a pyridine solution of 2-amino-1,4,5,6-tetrahydropyrimidine gives the hydrochloride of 2-benzamido-1,4,5,6-tetrahydropyrimidine because the latter is still one hundred times stronger as a base than pyridine. The pK_a of 1,4,5,6-tetrahydro-2-phenylpyrimidine is 0.2 units lower than that of the parent molecule so that, contrary to expectation, the resonance interaction between the phenyl ring and the hydropyrimidine ring is smaller in the cation than in the neutral molecule. This suggests that in the solvated cation, the angle between the plane of the two rings may be larger than in the free base.

TABLE XVf. Some pK_a Values at 20° for Reduced Pyrimidines (*New*)

	pK_a	Ref.
1,4,5,6-Tetrahydropyrimidines		
Unsubstituted	13.0	2162
2-Amino-	14.1	2686
2-Acetamido-	8.34	2686
2-Benzamido-	7.12	2686
2-Benzyl-	13.0	2957
2-Phenyl-	12.8	2957
2-Methoxy-	10.79	3625
5-Methoxy-	11.72	3625
5-Amino-	>12; 5.89	3625
5,5-Dihydroxy-	10.03	2686
5-Hydroxy-	10.9	3589
5-Hydroxy-2-methyl-	11.4	3589
2-Amino-4-hydroxy-	11.8	3596
2-Mercapto-	-0.93	3694
1,2,5,6-Tetrahydropyrimidines		
2,2,4,6,6-Pentamethyl-	8.11	3610
Hexahydropyrimidines		
Unsubstituted	9.75	3610
1,3-Dimethyl-	8.40	3610

The introduction of a 5-methoxy group, at a distance of two carbon atoms from the basic centre of 1,4,5,6-tetrahydropyrimidine, lowers the pK_a value by 1.8 units, compared with the drop of 2.2 units when a 2-methoxy group is inserted. In the latter case the inductive electron-withdrawing effect of the methoxy group outweighs the electron-releasing mesomeric effect. Introduction of a 5-hydroxyl group causes a drop in the pK_a value, although the second decrease is not as great as the first.

Acylation of the 2-amino-1,4,5,6-tetrahydropyrimidine causes a large decrease of 6–7 units in the pK_a value (14.1) but the guanidinium resonance ensures, as pointed out earlier, that the acylated compound still has a high pK_a value. The importance of guanidinium-type resonance is illustrated by the high pK_a values for 2-amino-4-hydroxy derivatives of hydropyrimidines which are formally analogous to the aldehyde ammonias, themselves weak bases.

1,4,5,6-Tetrahydro-2-mercaptopyrimidine has a very low basic pK_a, in agreement with its actual structure being that of a cyclic thiourea.

The pK_a value found for hexahydropyrimidine is close to that predicted[3748] on the basis that it has the cyclic structure and is a typical secondary amine (11.15) which contains a ring (ΔpK_a 0.2), two groups which have equal probabilities of accepting a proton (ΔpK_a 0.3) and an alkylamine group both one ($-\Delta$pK_a 1.7) and three carbon atoms ($-\Delta$pK_a 0.45) away from the site of protonation. The pK_a of hexahydro-1,3-dimethylpyrimidine is also in agreement with expectation, tertiary aliphatic amines being about 0.7 units weaker than the corresponding secondary amines.[3748] The pK_a of the 1,2,5,6-tetrahydropentamethylpyrimidine agreed with its being a cyclic amine whose basicity was considerably weakened by its contiguity with an electron-withdrawing C=N group.

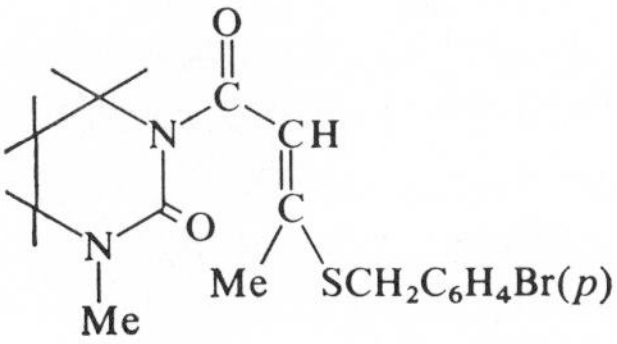

(148) 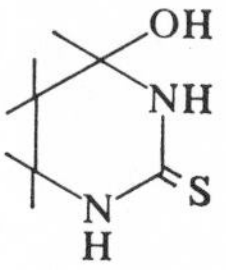 (149)

B. Ring-Chain Tautomerism (*New*)

The possibility of ring-chain tautomerism in the product from the condensation of 3-aminopropylamidoxime and benzaldehyde has been mentioned above but no definitive spectroscopic evidence has been produced. The tetrahydro thiouracil (**149**) is claimed to have the cyclic structure in the solid state but to be in tautomeric equilibrium with an open chain isomer ($OHC.CH_2CH_2NHCSNH_2$) in solution.[3695]

As long ago as 1951,[3696] it was shown that when 1,4,5,6-tetrahydro-2-alkylamino-1,1-dimethylpyrimidinium chloride was dissolved in chloroform, the infra-red spectrum of the solution exhibited a band at

2130 cm^{-1}. This was believed to be due to the carbodiimide bond of the open chain tautomeric form ($RN:C:NCH_2CH_2CH_2N^+HMe_2$).

Ring-chain tautomerism in hexahydropyrimidines with at least one NH group in the ring becomes evident when the cyclic structure is partially destabilized by introduction of two alkyl groups at C-2.[3610] Thus hexahydropyrimidine exists completely as the cyclic form and has no absorption in the double bond region of the infra-red spectrum. However, its 2,2-dimethyl derivative must be a tautomeric mixture in which the open chain form gives rise to the C=N stretching vibration at 1670 cm^{-1} in the infra-red spectrum. If one of the NH groups is replaced by *N*-t-butyl, the tautomeric equilibrium shifts in favour of the open chain form. The infra-red spectrum of 1-t-butylhexahydro-2,2-dimethylpyrimidine (or its 2,2-diethyl or 2-phenyl analogue) exhibits a weak ν(NH) band near 3300 cm^{-1} while the ν(C=N) band around 1650 cm^{-1} is now the strongest band. Thus these three compounds should be regarded properly as the open-chain tautomeric 1-alkylidene-amino-3-t-butylaminopropanes.

C. Nuclear Magnetic Resonance (*New*)

Although a systematic investigation is lacking, there are some interesting data which have been used in diagnosis of structure. Thus acetylation of a 1,6- or 1,4-dihydropyrimidine causes the hydrogens attached to the carbon atom next to the *N*-acetyl group to appear *ca.* 0.5τ downfield from those in the unacetylated material and so could be used to distinguish between the two original isomers. In the p.m.r. spectra of 2-oxo(and thio)-1,2,3,4-tetrahydropyrimidines and their *N*-alkyl or *N*-acetyl derivatives, the NH of the 1,4-dihydro compound appears at a lower field than does the NH of a 1,6-dihydropyrimidine.[2277, 2279, 3474]

In the 2,5-dihydropyrimidine derivative (**133**), strong coupling (J = 5 c./s.) extended over the five bonds between H-2 and H-5. The doublet due to the 5-methylene hydrogen atoms collapses to a singlet when the H-2 methine quintuplet peaks are saturated in double resonance experiments. This would be consistent with the 2,5-dihydro-pyrimidine ring being time-averaged planar.[2519]

The protons of the 5-methylene group in 1,2,5,6-tetrahydro-2,2,4,6,6-pentamethylpyrimidine at room temperature give rise to a singlet which is due in this case to rapid oscillation of the molecule between two conformations (**150**) so that only a time-averaged signal is seen.[3610] Similarly the trimethylene chain in 1,4,5,6-tetrahydropyrimidine salts

TABLE XVg. The p.m.r. Spectra of Hexahydropyrimidine and Some *N*-Substituted Derivatives Compared with Trimethylenediamine and Its Salts (τ)

$H_2NCH_2CH_2CH_2NH_2$	β-H_2	$\alpha + \alpha'$-H_2
—[a]	8.23–8.73	7.28, 7.38, 7.50
/HCl[a]	7.94–8.32	6.97, 7.10, 7.30
/2HCl[a]	7.70–8.20	6.75, 6.87, 7.00

R	R′	5-H_2	4 + 6-H_2	2-H_2
H	H[a]	7.87–8.83	6.97, 7.06, 7.15	6.12
H	H.HCl[a]	8.00–8.20	6.75, 6.87, 7.00	5.81
Me	Me[b]	7.95–8.50	7.56, 7.64, 7.75	7.03
Me	Me.MeI[a]		6.51, 6.61, 6.71	6.16
			7.21, 7.30, 7.40	
H	Bu^n [c]	Under Bu^n peaks	7.19–7.60	6.80
H	Bu^t [c]	8.25–8.60	7.21, 7.31, 7.39	6.54
Bu^n	Bu^n [c]			6.97
Bu^t	Bu^t [c]	8.42–8.64	7.84, 7.90, 7.96	7.37
PhCO	PhCO[b]	8.00–8.38	6.12, 6.21, 6.29	4.85
Tosyl	Tosyl[b]	8.53–9.02	7.72, 6.82, 6.91	5.31
NO	NO[b]	7.82–8.33	5.40, 5.49, 5.59	3.50
			5.42, 5.52, 5.62	4.07
			5.96, 6.07, 6.17, 6.27	4.58

[a] In D_2O.
[b] In $CDCl_3$.
[c] In CCl_4.

must at room temperature be conformationally flexible. The 4- and 6-methylene groups give rise to superimposed triplets (τ 6.58) while the 5-methylene gives rise to a quintet (τ 8.1).[3551] The spectrum of 1,3-diaminopropane and its cations consist of simple spin multiplets because free rotation about C—C and C—N bonds averages the various conformations of the molecule. The four protons of both methylene groups next to the nitrogen atoms give rise to a triplet (τ 7.1) while the two protons of the central methylene group of the three carbon atom chain afford a multiplet, usually a quintet or sextet (τ 8.2).

If the absorption peaks due to the *N*-substituents are ignored, hexahydropyrimidine and its *N*-substituted derivatives exhibit in their p.m.r. spectra the gross absorption pattern of a trimethylene diamine with an additional singlet for 2-H_2 falling in the range observed in analogues, e.g., hexahydro-1,3,5-triazines.[3702] These p.m.r. measurements refer to temperatures slightly greater than ambient, which is above the temperature where ring inversion between two equivalent

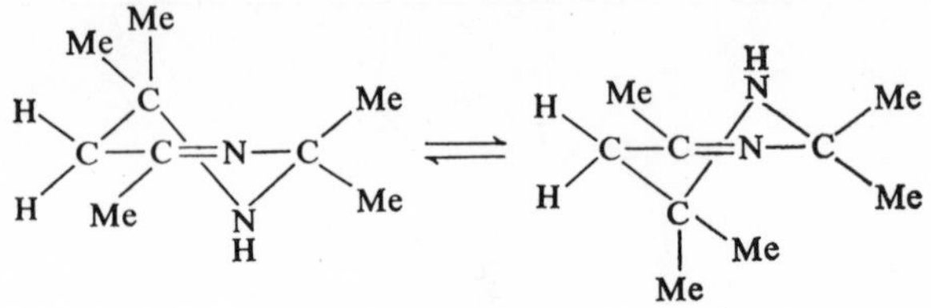

(150)

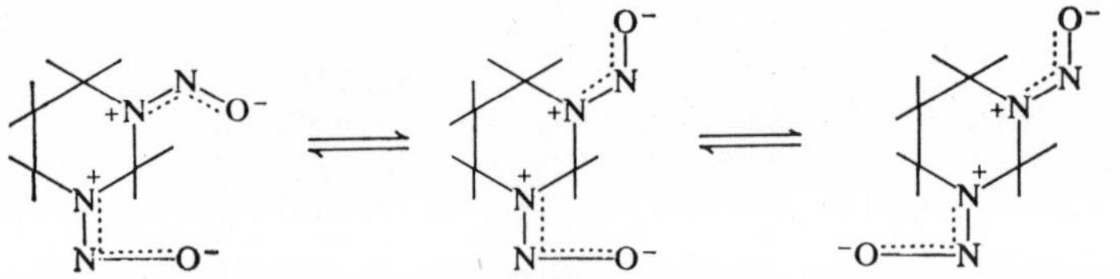

(151)

chair forms is slow enough to cause differentiation between axial and equatorial hydrogen atoms. Otherwise this methylene group situated between two nitrogen atoms should give rise to an AB quartet (Table XVg).

The *NN'*-dinitroso compound affords a complicated spectrum when examined in a number of solvents. The phenomenon is associated with the restriction of rotation about the N—N bond of *N*-nitrosamines which causes hexahydro-1,3-dinitrosopyrimidine to exist, at room temperature, in three molecular configurations **(151)** in the ratio of 6:16:21, respectively. On heating to 160° in *o*-dibromobenzene, the spectrum is transformed into three broad peaks centred at *ca.* τ 4, 6, and 8, respectively. When the solution is cooled to room temperature, the original spectrum reappears.

D. Conformation (*New*)

In derivatives of dihydrouracil, spectroscopic data of various types have been interpreted in terms of a half-chair model for the dihydropyrimidine ring (**152**), in which the —HN—CO—NH—CO— part of the molecule is approximately planar.[3697–3700]

Thus in 5-bromodihydrouracil, its 5-methyl and its 5-methyl-6-hydroxy derivatives, the absolute values of the coupling constants between H-1 and each H-6 were interpreted in terms of one molecular conformation in which the bromine atom adopted an axial position (**152**, X = Br, Y = H or Me). Both hydroxyl and methyl groups, if

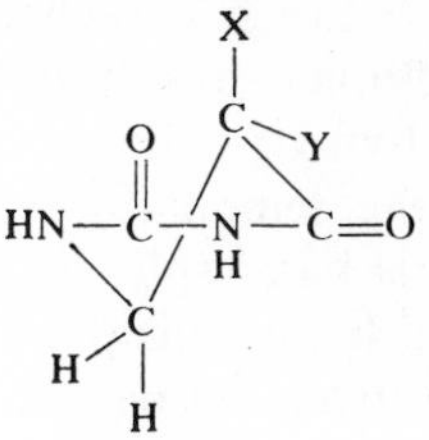

(**152**)

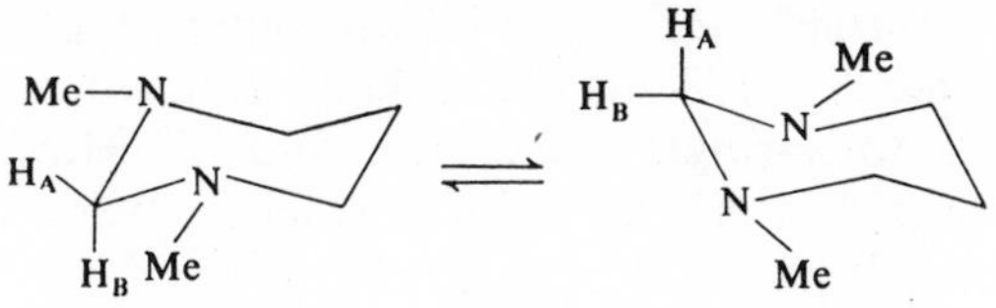

(**153**)

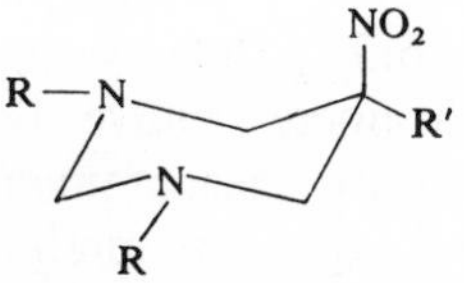

(**154**)

present in the molecule, adopted equatorial positions in one conformationally distinct isomer, although it was claimed that the ring in both 5-methyl- and 6-methyldihydrouracil now had the shape of a deformed half-chair. In dihydrouracil itself rapid interconversion between two possible half-chair conformers took place so that the coupling constant represented only a mean value.

Comparison of ultra-violet and pK_a data for various 1-aryl derivatives of dihydro-2-thiouracil and for the open chain analogue *N*-acetyl-*N'*-phenylthiourea led to the conclusion that the ring had almost a planar conformation. No p.m.r. data were available to check this conclusion.[3222]

At the tetrahydro level, X-ray data show that crystals of 1,4,5,6-tetrahydro-2-mercaptopyrimidine exist in the more stable tautomeric form, trimethylenethiourea.[3701] The ring approximates to a chair conformation with the —N—C(=S)—N— system planar.

In the protonated species of other 1,4,5,6-tetrahydropyrimidines,[3551] the p.m.r. spectra at room temperature indicate that two conformers are rapidly interconverting. The distinction between equatorial and axial type protons is lost and simple-type spin multiplets result. Thus 1,4,5,6-tetrahydropyrimidine hydrochloride in D_2O affords a 4-proton triplet (τ 6.58) due to the methylene protons of the two —HN—CH_2— groups and a two proton quintet (τ 8.01), respectively.

The p.m.r. spectrum of 1,2,5,6-tetrahydro-2,2,4,6,6-pentamethylpyrimidine has a single peak for 5-H_2, again showing that at room temperature distinction between axial and equatorial protons is lost because of rapid equilibration between two conformers (**150**).[3610] In the hexahydropyrimidine series, p.m.r. spectra measured at room temperature are explicable in terms of rapid equilibration between two chair conformers (**153**) and again simple-type spin multiplets result (see Table XVg). Thus the 2-H_2 of hexahydro-1,3-dimethylpyrimidine give rise to a single peak at τ 7.03.[3610] However, if the temperature is lowered, a critical point is reached, below which this band splits into an AB quartet (J_{AB} = 8.9 c./s.).[3702, 3703, 3749] At and below this point, the coalescence temperature ($T_C = -29 \pm 4°$), ring inversion between two equivalent chair forms has become slow on the p.m.r. time scale. However, nitrogen inversion between axial and equatorial position is still fast enough to yield only a time-averaged signal for *N*-Me. Thermodynamic activation parameters have been determined from these variable p.m.r. temperature measurements, both for hexahydro-1,3-dimethylpyrimidine and also its methiodide.[3703] It is found that quaternization raises the free energy of activation for the chair-chair interconversions considerably.

The dipole moment is another physical property which has been linked up with conformation in the case of certain hexahydropyrimidines. 1,3,5-Trialkylhexahydro-5-nitropyrimidines (**154**) have high dipole moments because each compound exists in a chair conformation with the nitro group in the axial position.[3648]

E. Mass Spectra (*New*)

Many mass-spectral data are available for the hexahydropyrimidines but not for other hydropyrimidines.

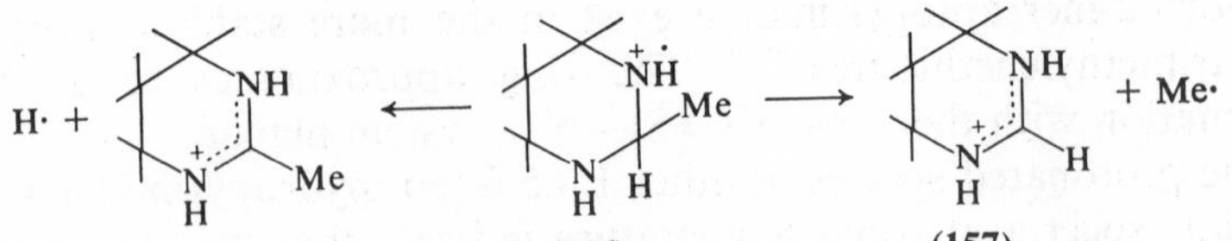

H· + (**155**) ⟵ (**156**) ⟶ (**157**) + Me·

Table XVh represents a favoured pathway for the fragmentation of hexahydropyrimidine itself,[3704] in which observed metastable peaks are indicated by an asterisk. Of the four patterns indicated, the first on the left is of interest since by loss of a hydrogen atom the molecular ion gives the resonance stabilized cyclic amidinium ion, already noted as responsible for the highly basic nature of 1,4,5,6-tetrahydropyrimidine. The stability of the amidinium ion is indicated by its intensity being greater than that of the parent molecular ion.

TABLE XVh. Cracking Pattern for Hexahydropyrimidine (*New*)

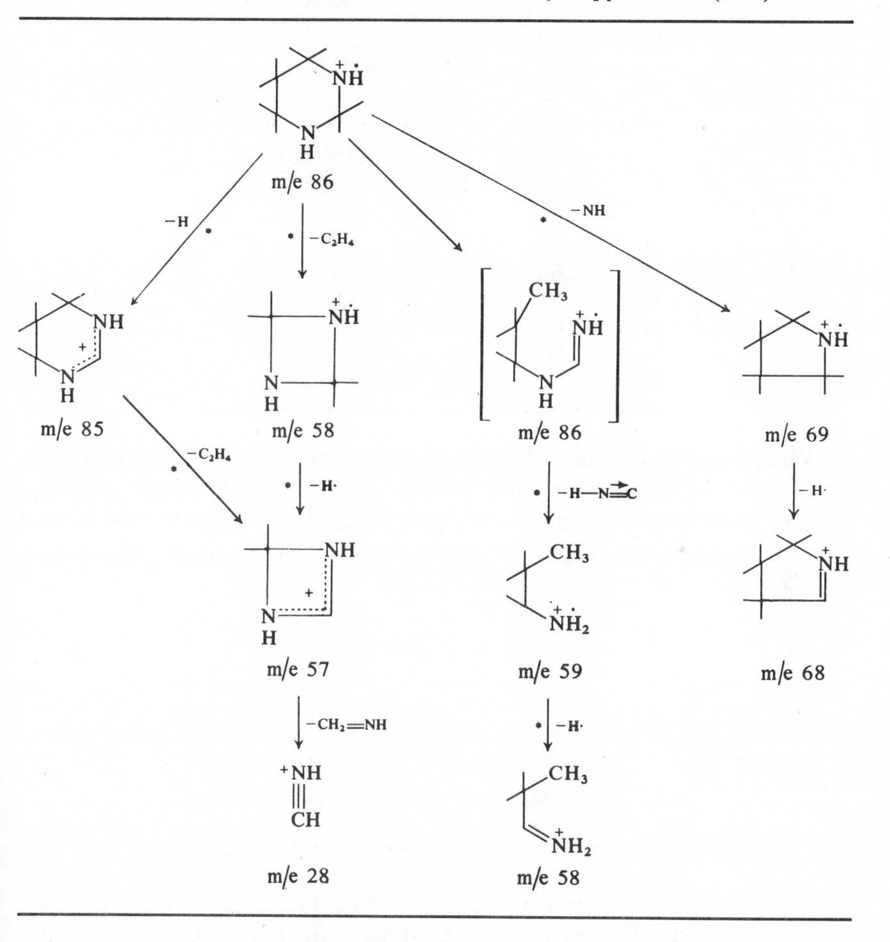

Other alkylated hexahydropyrimidines exhibit this outstanding feature in the mass spectrum whereby a group is ejected from C-2 to give a resonance stabilized entity which gives rise to an ion peak of greater intensity than the molecular ion. Table XVi shows a selection of data in which this is true for the ejection of a hydrogen atom or a methyl radical from the molecular ion $M^{+\cdot}$ of a variety of alkylated hexahydropyrimidines.[3551, 3610, 3704]

TABLE XVi. Cracking Pattern of Alkylhexahydropyrimidines

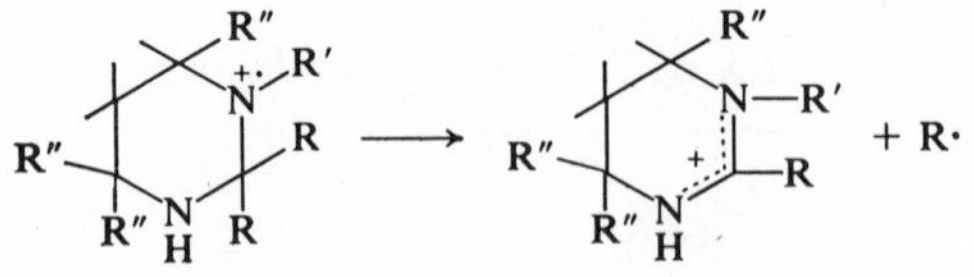

R	R′	R″	$\frac{\%[M - R]^+}{\%M^{+\cdot}}$
H	H	H	7.5
H	Bu	H	4.5
H	H	Me	7.5
H	Pr^i	Me	1.9
Me	H	H	11.5
Me	H	Me	30

When there is a choice of ejection of one of two free radicals to give two different amidinium ions of comparable stability, the cracking pattern involving the more stable and hence less energetic free radical predominates. Thus with the molecular ion of hexahydro-2-methyl-

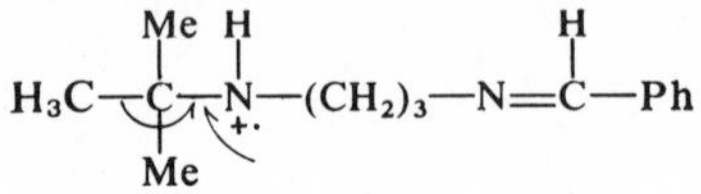

$$H_3C{-}C(Me)_2{-}\overset{+\cdot}{N}H{-}(CH_2)_3{-}N{=}CH{-}Ph$$

(158)

$$H_3C\cdot + Me_2C{=}\overset{+}{N}H{-}(CH_2)_3{-}N{=}CH{-}Ph$$

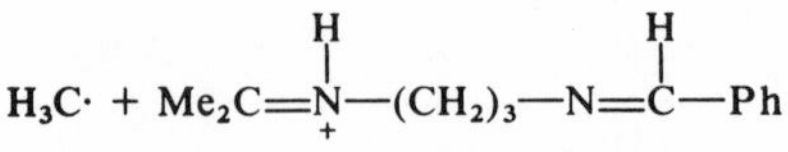

(159)

pyrimidine (**156**), the preferred pathway by a factor of over 3 is ejection of a methyl radical to give the 1,4,5,6-tetrahydropyrimidinium ion (**157**) rather than loss of a hydrogen atom to give the 2-methyl analogue (**155**). The ratio of peak heights is $(M\text{-}15)^+ : (M\text{-}1)^+ : M^{+\cdot} = 18:5:1$.

Those hexahydropyrimidines in which a small quantity of open-chain tautomer is present in the liquid phase give cracking patterns explicable solely in terms of the cyclic structure. This contrasts with those hexahydropyrimidines which are mainly open chain compounds: the mass spectra arise from the open chain structure and no peak corresponding to a cyclic amidinium ion is observed. Thus in 1-benzylideneamino-3-t-butylaminopropane (the stable open-chain tautomer of 1-t-butylhexahydro-2-phenylpyrimidine) the molecular ion peak corresponding to (**158**) is small compared with that due to its decomposition product (**159**), formed by the expulsion of a methyl radical from the t-butyl group.

In Table XVj some of the favoured pathways for the fragmentation of the 1,4,5,6-tetrahydropyrimidine molecular ion are represented. The

TABLE XVj. Cracking Pattern for a Tetrahydropyrimidine (*New*)

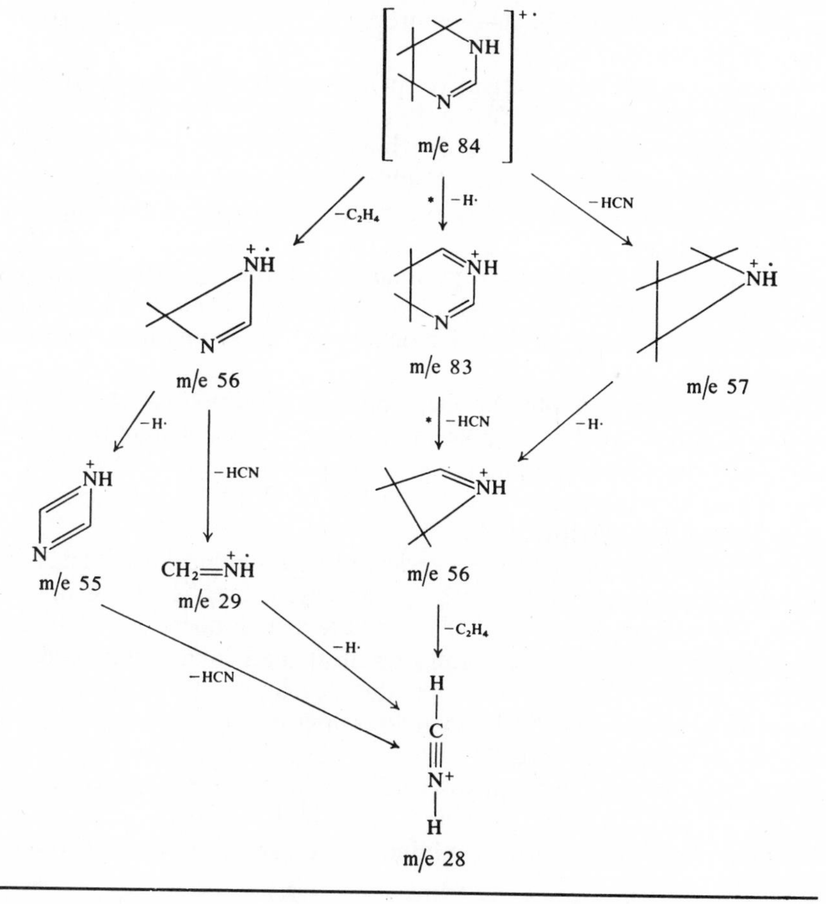

NH
N
m/e 84

$-C_2H_4$ | * $-H\cdot$ | $-HCN$

$\overset{+\cdot}{NH}$
N
m/e 56

$\overset{+}{N}H$
N
m/e 83

$\overset{+\cdot}{NH}$
m/e 57

$-H\cdot$ | $-HCN$ | * $-HCN$ | $-H\cdot$

$\overset{+}{N}H$
N
m/e 55

$CH_2{=}\overset{+\cdot}{NH}$
m/e 29

$\overset{+}{N}H$
m/e 56

$-HCN$ | $-H\cdot$ | $-C_2H_4$

H
|
C
|||
N^+
|
H
m/e 28

base peak is that of the molecular ion itself in contrast with the hexahydropyrimidines, but again ejection of neutral molecules like ethylene or hydrogen cyanide leads to the same protonated form of hydrogen cyanide.

F. Applications of Hydropyrimidines (*New*)

In view of pyrimidine being one of the building blocks for nucleic acids, it is not surprising that hydropyrimidines have many biological activities. Trimethyleneurea (1,4,5,6-tetrahydro-2-hydroxypyrimidine) and its *NN'*-dinitro derivative are fungicides;[3706–3708] 2-alkylthio-[3664, 3709] or 2-phenylmethylthio-tetrahydropyrimidines[3667, 3668] act likewise. The first class of compound also has a radioprotective effect in mice.[3710] 1-Phenyl-2-2',6'-dichlorophenyl-1,4,5,6-tetrahydropyrimidine cation is toxic to plants.[3705]

Various *N*-substituted derivatives of 2-amino-1,4,5,6-tetrahydropyrimidine are either bactericides or fungicides,[3669, 3675, 3711, 3712] as are also 1,2-dialkyl-1,4,5,6-tetrahydropyrimidines.[3713–3715]

So far as man is concerned, *N*-alkyl-*N'*-*p*-toluenesulphonyl derivatives of trimethyleneurea[3716] have a use as hypoglycemic agents; 1,4,5,6-tetrahydro-2-phenylthiopyrimidine is a diuretic;[3665] a 2-*p*-aminophenyl-1,4,5,6-tetrahydropyrimidine affords protection against sun burn;[3717] and 2-amino-1,4,5,6-tetrahydropyrimidine has been investigated for the preferential blocking of the sympathetic nervous system.[3718]

The 1-methoxy-2-naphthoic ester of 2-hydroxymethyl-1,4,5,6-tetrahydropyrimidine may be used as an analgesic and local anaesthetic.[3719] The drug, oxyphencyclimine, is used for ulcer treatment[3745] and so is the α-phenylcyclohexane glycollic ester of 1,4,5,6-tetrahydro-2-hydroxymethyl-1-methylpyrimidine.

Hydropyrimidines also have a variety of nonbiological uses. Thus the *NN'*-bishydroxymethyl derivative of trimethyleneurea is a cross-linking agent for cotton fabrics;[3720–3725] trimethylenethiourea[3726] stabilizes polyoxymethylene polymers towards heat and oxidation; and its *N*-lower-alkyl derivatives improve the development of the positive image in photography.[3727] Polyvinyltetrahydropyrimidine has a use in the treatment of cellulose;[3674, 3728] the *NN'*-dinitro derivative of the nitrate ester of 1,4,5,6-tetrahydro-2,5-dihydroxypyrimidine has found use as an explosive;[3729] and the t-butyl hydroperoxide adduct with 2-amino-1,4,5,6-tetrahydropyrimidine is used in an adhesive formulation.[3744]

An *N*-decyl hydrobromide salt mixed with cobalt phthalocyanine is used in ink production,[3730] and a 2-carbamato derivative improves the fastness of dyes.[3731] Various 2-alkyl-1-hexadecyl-1,4,5,6-tetrahydropyrimidine succinamates have been added to petrols as detergents and anti-stalling substances.[3732, 3733]

Hexahydropyrimidines show a similarly wide spectrum of uses. Hexahydro-1,3-dinitropyrimidine is a plant growth regulator;[3734] various 1,3,5-trialkyl-5-amino or 5-nitro compounds exhibit bactericidal properties[3652, 3735–3739] and are used on account of these properties in adhesives, in petroleum lubricants, and in treatment of flood water! The sedative drug G-TRIL is a hexahydropyrimidine derivative[3740] and a 1,3-dibenzyl-2-*p*-bischloroethylaminophenylhexahydropyrimidine has attracted attention for its activity against Walker carcinoma.[3638]

Hexahydro-1,3-bishydroxymethylpyrimidine is claimed to be the best cross-linking agent for cotton.[3684] The salt of 2-ethyl-2-heptylhexahydro-1-isopropylpyrimidine with a long chain 2-sulphocarboxylic acid enhances the detergent properties of lubricating oils,[3741] and 5-ethylhexahydro-1,3-diisopropyl-5-nitropyrimidine stabilizes such hydrocarbon oils.[3650] 1,3-Nitroalkyl derivatives of 5,5-dinitrohexahydropyrimidines are used as explosives.[3647, 3742] Various *NN'*-disubstituted hexahydropyrimidines mixed with resorcinol improve adhesion between textiles and rubber.[3743]

CHAPTER XIII

The Ionization and Spectra of Pyrimidines (*H* 464)

Although many ionization constants and electronic spectra of pyrimidines have been published, no new principles have emerged recently. This supplementary chapter therefore consists of a list of new pK_a values and a new section (written by T. J. Batterham) on nuclear magnetic resonance spectra, a field which has become relevant to pyrimidines only since 1960. The number of available data on the mass spectra of pyrimidines is too small to warrant review yet.

1. The Ionization of Pyrimidines (*H* 464)

A modern general discussion of the ionization constants and ultraviolet spectra of pyrimidines and purines was written recently by A. Albert.[3781] An invaluable compilation of all known pK_a values for organic bases was made by D. D. Perrin in 1965;[2937] notes on the probable reliability of each value were included. A useful rule-of-thumb method for forecasting acidic or basic pK_a values has appeared.[3748, 3782]

A supplementary list of known pK_a values for pyrimidines is given in Table XVI.

TABLE XVI. The pK_a Values[a] of Some Pyrimidines[b] in Water (*H* 472)

Pyrimidine	Acidic pK_a	Basic pK_a	Ref.[c]
4-Acetamido- (20°)		2.76	2675
4-Acetamido-5-ethoxycarbonyl-2-methyl- (25°)		1.43	2698
5-Allenyl-4,6-dimethyl-2-methylamino- (20°)		4.90	2700

Pyrimidine	Acidic pK_a	Basic pK_a	Ref.[c]
2-Allylamino- (20°)		3.57	2334
2-Allylamino-4,6-dimethyl- (20°)		5.02	2700
5-Allyl-2-amino-4,6-dimethyl- (20°)		5.15	2700
5-Allyl-2-chloro-4,6-dimethyl- (20°)		−0.46	2700
1-Allyl-1,2-dihydro-2-imino-4,6-dimethyl- (20°)		11.16	2700
5-Allyl-1,2-dihydro-2-imino-1,4,6-trimethyl- (20°)		11.99	2700
1-Allyl-1,2-dihydro-2-imino- (20°)		10.53	2633
5-Allyl-1,2-dihydro-2-imino-1,4,6-trimethyl-		11.99	2700
1-Allyl-1,2-dihydro-2-propylimino- (20°)		12.25	3416
5-Allyl-4,6-dimethyl-2-methylamino- (20°)		5.36	2700
5-Allyl-2-hydroxy-4,6-dimethyl- (20°)	10.69	3.96	2700
2-Allylimino-1,2-dihydro-1-propyl- (20°)		11.47	3416
2-Amino- (20°)		3.71	*H*, 2334
2-Amino-5-*o*-amino-*p*-chlorophenylthio-4-hydroxy-	8.6	3.2; 1.5	3488
2-Amino-5-*o*-aminophenylthio-4-hydroxy-	8.7	3.6; 2.1	3488
2-Aminobarbituric acid/1-methyl-5-nitro- (20°)	5.55		2718
4-Amino-5-benzylamino-2,6-dihydroxy- (20°)	8.45	4.26	3489
4-Amino-5-benzylamino-6-hydroxy-[d]	9.99	3.74	3475
2-Amino-5-bromo- (20°)		1.95	2627
4-Amino-5-bromo- (20°)		3.97	3490
2-Amino-5-bromo-4-t-butyl- (20°)		2.93	2602
2-Amino-5-bromo-4,6-dimethyl- (20°)		3.35	2626
2-Amino-5-bromo-4-hydroxy-	8.05	2.56	2610
4-Amino-5-bromo-2-hydroxy-	10.33	3.04	2610
2-Amino-5-bromo-4-methyl- (20°)		2.66	2602
2-Amino-4-t-butyl- (20°)		4.63	2602
4-Amino-5-butylamino-6-hydroxy-[d]	9.00	4.73	3475
2-Amino-5-carbamoyl- (20°)		2.06	2376
4-Amino-5-carbamoyl- (20°)		4.18	2376
2-Amino-5-carbamoyl-4,6-dimethyl- (20°)		3.36	2633
2-Amino-5-carbonyl-4-methyl- (20°)		2.65	2602
4-Amino-5-carbamoyl-2-methyl- (25°)		4.97	2698
5-Amino-4-carboxy-2,6-dihydroxy-	2.63; 8.72		3456
4-Amino-5-carboxy-2-methyl-	6.28	2.14	2698, 3425
2-Amino-5-chloro- (20°)		1.73	2633
2-Amino-4-chloro-1,6-dihydro-6-imino-1-methyl- (20°)		9.90	2288
2-Amino-4-chloro-6-hydroxy-	8.06	0.50	2610
2-Amino-5-chloro-4-hydroxy-	7.97	2.53	2610

continued

TABLE XVI (*continued*).

Pyrimidine	Acidic pK_a	Basic pK_a	Ref.[c]
4-Amino-5-chloro-2-hydroxy-	10.36	2.94	2610
4-Amino-6-chloro-2-hydroxy	8.59	3.26	2610
4-Amino-6-chloro-2-methylamino- (20°)		3.81	*H*, 2288
5-Amino-2-chloro-4-methyl-6-methyl amino- (20°)		3.45	2334
5-*o*-Amino-*p*-chlorophenylthio-4-hydroxy-2-piperidino-	8.5	2.8; 1.4	3488
2-Amino-5-cyano- (20°)		0.66; −1.48	2376
4-Amino-5-cyano- (20°)		2.54	2376
2-Amino-5-cyano-4,6-dimethyl- (20°)		1.67	2633
2-Amino-5-cyano-4-methyl- (20°)		1.14	2602
4-Amino-5-cyano-2-methyl- (25°)		3.51	2698
4-Amino-5-cyanomethylamino-3,6-dihydro-3-methyl-6-oxo- (20°)	>12	3.83	2492
4-Amino-5-cyanomethylamino-6-hydroxy- (20°)	9.79		2492
2-Amino-4-ββ-diethoxyethylamino-1,6-dihydro-1-methyl-5-nitro-6-oxo- (20°)	12.75	0.88	2909
2-Amino-4-ββ-diethoxyethylamino-6-hydroxy-5-nitro- (20°)	8.99	0.58	2909
2-Amino-4-*N*-ββ-diethoxyethyl-*N*-methylamino-6-hydroxy-5-nitro- (20°)	8.21	0.73	2909
4-Amino-5-diethylamino-6-hydroxy-[d]	10.38	5.05	3475
2-amino-1,4-dihydro-4-imino-1-methyl- (20°)		12.9	2288
4-Amino-1,6-dihydro-6-imino-1-methyl- (20°)		12.7	*H*, 2626
5-Amino-1,4-dihydro-4-imino-1-methyl-(20°)		$\not<$12.5	2288
2-Amino-1,4-dihydro-4-imino-1-methyl-6-methylthio- (20°)		$\not<$12.0	2288
2-Amino-3,4-dihydro-4-imino-3-methyl-6-methylthio- (20°)		$\not<$11.2	2288
5-Amino-1,4-dihydro-4-imino-1-methyl-2-methylthio- (20°)		$\not<$12.6	2288
4-Amino-1,6-dihydro-6-imino-1-methyl-5-nitro- (20°)		*ca.* 7.6	2626
2-amino-4-hydroxy-6-methoxy-5-nitro- (20°)	7.41		2718
2-amino-1,6-dihydro-1-methyl-4-methyl-amino-5-nitro-6-oxo- (20°)		−0.17	2909, 2927
4-Amino-1,2-dihydro-1-methyl-2-oxo- (25°)		4.55	*H*, 3086
4-Amino-2,3-dihydro-3-methyl-2-oxo- (20°; 25°)	13–14	7.38 or 7.4 or 7.49	2168, 2641, 2987

Pyrimidine	Acidic pK_a	Basic pK_a	Ref.[c]
2-Amino-1,6-dihydro-1-methyl-6-thio-		2.92	2776
2-Amino-4,6-dihydroxy-	7.00	1.27	*H*, 3492
4-Amino-2,6-dihydroxy-		0.80	3492
5-Amino-2,4-dihydroxy-6-β-hydroxyethyl-amino- (20°)			3489
2-Amino-4,6-dihydroxy-5-methyl-	7.44		3492
4-Amino-2,6-dihydroxy-5-methyl-	9.22	0.02	3492
2-Amino-4,6-dimethyl-5-prop-1′-ynyl- (20°)		4.04	2700
2-Amino-4,6-dimethyl-5-propyl- (20°)		5.47	2700
2-Amino-4,6-dimethyl- (20°)		4.99	*H*, 2334
4-Amino-2,6-dimethyl- (20°)		6.98	2334
2-Amino-4-dimethylamino- (20°)		7.96	2627
2-Amino-4-dimethylamino-1,6-dihydro-1-methyl-5-nitro-6-oxo- (20°)		−0.36	2909
4-Amino-5-dimethylamino-6-hydroxy-[d]	10.40	4.18	3475
2-Amino-4-dimethylamino-6-hydroxy-5-nitro- (20°)	8.57	−0.62	2909
5-Amino-4-dimethylamino-6-methylamino- (20°)		5.35	2454
2-Amino-4-dimethylamino-5-nitro- (20°)		3.49	2626
2-Amino-4,6-dimethyl-5-propyl- (20°)		5.47	2700
2-Amino-4,6-dimethyl-5-prop-1′-ynyl- (20°)		4.04	2700
2-Amino-4,6-dimethyl-5-prop-2′-ynyl- (20°)		4.54	2700
4-Amino-5-ethoxycarbonyl-2-hydroxy-	9.85	3.28	3492
4-Amino-5-ethoxycarbonyl-2-mercapto-	7.58	0.6	3492
4-Amino-5-ethoxycarbonyl-2-methyl- (25°)		4.53	2698, 3425
2-Amino-4,6-diphenyl- (20°)		3.78	2627
4-Amino-5-ethylamino-6-hydroxy-[d]	9.84	4.79	3475
5-Amino-4-ethylamino-6-hydroxy-[d]	10.12	3.52	3475
2-Amino-4-fluoro-6-hydroxy-	8.38	<0	2610
2-Amino-5-fluoro-4-hydroxy-	8.04	2.57	2610
4-Amino-5-fluoro-2-hydroxy-	10.87	2.83 or 2.90	2610, 3086
4-Amino-6-fluoro-2-hydroxy-	9.05	1.50 or 1.52	2610, 3270
4-Amino-5-formamido-2-hydroxy- (20°)	10.87	3.55	2661
4-Amino-5-formamido-2-hydroxy-6-methyl- (20°)	11.39	3.87	2661
4-Amino-5-formamido-2-methylamino- (20°)		6.12	2454
4-Amino-5-formamido-2-methylthio- (20°)		3.71	2675
4-Amino-5-formyl-2,3-dihydro-2-imino-3-methyl- (20°)	12.58 ?	8.47	2376, 2633
4-Amino-5-formyl-2,3-dihydro-3-methyl-2-methylimino- (20°)	13.40 ?	8.67	2376, 2633

continued

TABLE XVI (*continued*).

Pyrimidine	Acidic pK_a	Basic pK_a	Ref.[c]
4-Amino-5-formyl-2-methyl-		4.46	2698
2-Amino-4-*C*-formylmethylamino-1,6-dihydro-1-methyl-5-nitro-6-oxo- (20°)		2.45	2909
2-Amino-4-*C*-formylmethylamino-6-hydroxy-5-nitro- (20°)	8.50	2.34	2909
4-Amino-6-hydrazino-2-methyl-5-nitro- (20°)		4.08	2562
4-Amino-6-hydrazino-5-nitro- (20°)		3.70	2562
2-Amino-4-hydroxy- (20°)	9.59	4.00	*H*, 2288
4-Amino-2-hydroxy- (25°)	>13[e] or 12.15	4.8[e] or 4.61 or 4.58	2153, 3086, 3493
4-Amino-6-hydroxy- (20°)	10.05 or 11.3[e]	1.36 or <2.5[e]	827, 2153
4-Amino-6-hydroxy-2,5-dimethyl-	11.41	2.63	3492
4-Amino-6-hydroxy-2-mercapto-	7.25	0.30	3492
2-Amino-4-hydroxy-6-methylamino-5-nitro- (20°)	8.70	−0.37	2909
2-Amino-4-hydroxy-5-*o*-methylamino-phenylthio-	8.7	3.6; 2.1	3488
2-Amino-4-hydroxy-5-nitropyrimidin-6-ylpyridinium chloride	3.16		2718
2-Amino-4-hydroxy-5-phenylthio-	8.52	2.88	3488
2-Amino-4-hydroxy-5-*m*-toluidino-		4.18	2567
4-Amino-2-hydroxy-5-trifluoroacetamido- (20°)	7.58; 12.77	3.19	2661
2-Amino-5-iodo- (20°)		2.23	2633
2-Amino-4-mercapto- (20°)	8.03	2.86	2776
4-Amino-2-mercapto- (20°)	10.58	3.33	*H*, 2776
4-Amino-6-mercapto- (20°)	9.25	−0.24	2776
5-Amino-4-mercapto-2-methyl-6-methyl-amino- (20°)	10.44	2.28	2675
2-Amino-4-methoxy- (20°)		5.53	2776
5-Amino-4-methoxy- (20°)		4.27	2630
4-Amino-2-methyl- (25°)		6.53	2698
4-Amino-2-methylamino- (20°)		7.55	2288
5-Amino-2-methyl-4,6-bismethylamino- (20°)		6.92	2454
4-Amino-6-methyl-5-methylamino- (20°)		6.12	2675
5-Amino-2-methyl-4-methylamino- (20°)		7.45	2675
5-Amino-4-methyl-6-methylamino- (20°)		6.82	2675
4-Amino-6-methyl-5-methylamino-2-methylthio- (20°)		5.41	2675

Pyrimidine	Acidic pK_a	Basic pK_a	Ref.[c]
5-Amino-2-methyl-4-methylamino-6-methylthio- (20°)		6.02	2675
4-Amino-6-methyl-5-methylformamido-(20°)		4.57	2675
4-Amino-6-methyl-5-methylformamido-2-methylthio- (20°)		4.02	2675
4-Amino-2-methyl-5-nitro- (20°)		2.72	2562
4-Amino-5-methylnitrosoamino- (20°)		3.69	3494
2-Amino-4-methylthio- (20°)		4.75	2776
4-Amino-2-methylthio- (20°)		4.91	2776
4-Amino-6-methylthio- (20°)		3.94	2781
4-Amino-5-nitro- (20°)		1.98	2562
4-Amino-5-nitro-2-styryl- (20°)		2.35	2562
5-*o*-aminophenylthio-2,4-dihydroxy-	8.30	2.99	3488
4-Amino-2-trifluoromethyl-		1.39	2217
2-Amylamino- (20°)		4.04	2627
4-Anilino-2,6-dimethyl-		5.8[e]	2153
2-Amino-4-hydroxy- (20°)	9.59 or 10.8[e]	4.00 or 3.9[e]	2153, 2288
4-Anilino-2-hydroxy-	12.9[e]		2153
4-Anilino-6-hydroxy-	10.9[e]		2153
2-Benzamido- (20°)	11.20	1.56	2686
2-Benzylamino- (20°)		3.56	2626
2-Benzylamino-4,6-dimethyl-		3.9[e]	2153
4-Benzylamino-2,6-dimethyl-		6.3[e]	2153
4-Benzylamino-2-hydroxy-	13.5[e]	3.8[e]	2153
4-Benzylamino-6-hydroxy-	11.7[e]		2153
4-Benzylamino-6-hydroxy-2-methyl-	12.2[e]		2153
1-Benzyl-2-benzylimino-1,2-dihydro- (20°)		10.60	2626
2-Benzyl-4,6-dihydrazino-5-nitro- (20°)		4.11	2562
1-Benzyl-1,2-dihydro-2-imino- (20°)		10.16	2633
1-Benzyl-1,2-dihydro-2-isopropylimino-(20°)		11.55	2626
1-Benzyl-1,2-dihydro-2-methylimino- (20°)		11.17	2626
2-Benzyl-4,6-dihydroxy- (20°)	5.78		2562
2-Benzyl-4,6-dihydroxy-5-nitro- (20°)	3.49		2562
2-Benzylimino-1,2-dihydro-1-isopropyl-(20°)		11.21	2626
2-Benzylimino-1,2-dihydro-1-methyl- (20°)		11.17	2626
4-Benzylimino-1,4(or 3,4)-dihydro-1,2,6(or 2,3,6)-trimethyl-		> 14[e]	2153
2-Benzyl-5-nitro- (20°)		0.51[f]	2688
1-Benzyloxy-1,2-dihydro-4,6-dimethyl-2-oxo-		3.1	2869

continued

TABLE XVI (*continued*).

Pyrimidine	Acidic pK_a	Basic pK_a	Ref.[c]
4,5-Bisethylamino-6-hydroxy-[d]	10.05	4.83	3475
2,4-Bisdimethylamino-6-hydroxy-5-nitro- (20°)	8.52	1.59	2909
4,6-Bismethylamino-2-methylthio-5-nitro- (20°)		1.55	2454
4,6-Bismethylamino-5-nitro- (20°)		2.57	2454
4,6-Bismethylthio-		1.56; −5.60	3220
5-Bromo-2-butylamino- (20°)		2.21	2746
5-Bromo-2-t-butylamino- (20°)		2.14	2746
5-Bromo-4-butylamino- (20°)		4.49	2746
5-Bromo-4-t-butylamino- (20°)		4.69	2746
5-Bromo-4-t-butyl-1,2-dihydro-2-imino-1-methyl- (20°)		11.22	2602
5-Bromo-4-carboxy-2,6-dihydroxy- (25°)	2.38; 7.33		3456
5-Bromo-1,2-dihydro-2-imino-1,4-dimethyl- (20°)		10.33	2602
5-Bromo-1,2-dihydro-2-imino-1-methyl- (20°)		9.95 or 10.06 or 10.17 or 10.41[g]	2375, 2627
5-Bromo-1,2-dihydro-2-imino-1,4,6-trimethyl- (20°)		11.01	2375, 2627
5-Bromo-1,2-dihydro-4-methoxy-1-methyl-2-oxo- (24°)		−3.32	3046
5-Bromo-1,2-dihydro-1-methyl-2-methylimino- (20°)		10.67 or 10.77 or 10.42[h] or 11.05[g]	2375, 2627
5-Bromo-1,2-dihydro-1-methyl-2-oxo- (20°)		0.55	2630
5-Bromo-1,6-dihydro-1-methyl-6-oxo- (20°)		0.14	2630
5-Bromo-2,4-dihydroxy- (24°)	7.83 or 8.05	−7.25	3046, 3270
5-Bromo-4,6-dimethyl-2-methylamino- (20°)		3.57	2626
5-Bromo-2-ethoxy- (20°)		−0.42	2630
5-Bromo-2-ethylamino- (20°)		2.10	2626
5-Bromo-1-ethyl-1,2-dihydro-2-imino- (20°)		10.20 or 9.89[h] or 10.52[g]	2375, 2626
5-Bromo-1-ethyl-1,2-dihydro-2-oxo- (20°)		0.62	2630
5-Bromo-2-hydroxy- (20°)	7.36	0.44	2627, 2630
5-Bromo-4-hydroxy- (20°)	7.15	0.43	2630
5-Bromo-2-mercapto- (20°)	5.47	−0.43	2630
5-Bromo-4-mercapto- (20°)	5.60	−0.46	2630

Pyrimidine	Acidic pK_a	Basic pK_a	Ref.[c]
5-Bromo-2-methoxy- (20°)		−0.77	2746
5-Bromo-4-methoxy- (20°)		1.35	2746
5-Bromo-2-methylamino (20°)		2.09	2627
5-Bromo-4-methyl-2-methylamino- (20°)		2.90	2602
5-Bromo-2-methylthio- (20°)		−0.90	2746
5-Bromo-4-methylthio-		1.02	2746
5-Bromo-1,2,3,4-tetrahydro-1,3-dimethyl-2,4-dioxo- (24°)		−6.44	3046
2-Butoxy- (20°)		1.35	2697
2-s-Butoxy- (20°)		1.63	2697
2-Butylamino- (20°)		4.09	2627
2-t-Butylamino- (20°)		4.24	2334
4-t-Butylamino- (20°)		6.58	2630
2-Butylamino-4,6-dimethyl- (20°)		5.28	2334
2-s-Butylamino-4,6-dimethyl- (20°)		5.13	2334
2-t-Butylamino-4,6-dimethyl- (20°)		5.73	2334
4-Butylamino-2,6-dimethyl- (20°)		7.42	2334
4-s-Butylamino-2,6-dimethyl- (20°)		7.61	2334
4-t-Butylamino-2,6-dimethyl- (20°)		7.87	2334
2-Butylamino-4-methyl- (20°)		4.71	2746
2-Butylamino-5-nitro- (20°)		0.06	2746
2-t-Butylamino-5-nitro- (20°)		0.16	2746
4-t-Butylamino-5-nitro- (20°)		2.60	2562
2-Butyl-4-chloro-6-hydroxy- (20°)	8.25	−0.97	2242
5-Butyl-4-chloro-6-hydroxy- (20°)	8.05	−1.10	2242
1-Butyl-1,2-dihydro-2-imino- (20°)		10.93	2627
4-t-Butyl-1,2-dihydro-2-imino-1-methyl- (20°)		11.69	2602
1-Butyl-1,2-dihydro-2-methylimino- (20°)		11.90	3416
4-t-Butyl-1,2-dihydro-1-methyl-2-oxo- (20°)		2.95	2602
1-Butyl-1,2-dihydro-2-oxo- (20°)		2.76	2630
1-s-Butyl-1,2-dihydro-2-oxo- (20°)		3.00	2630
1-t-Butyl-1,2-dihydro-2-oxo- (20°)		3.58	2630
5-Butyl-4,6-dihydroxy- (20°)	6.24	−0.54	2242
4-t-Butyl-2-hydroxy- (20°)	10.24	2.83	2602
2-Butyl-4-hydroxy-6-mercapto- (20°)	5.25	−1.17	2242
5-Butyl-4-hydroxy-6-mercapto- (20°)	5.09	−1.08	2242
2-Butyl-4-hydroxy-6-methylthio- (20°)	9.39	0.48	2242
5-Butyl-4-hydroxy-6-methylthio- (20°)	9.24	0.40	2242
2-Butylimino-1,2-dihydro-1-methyl- (20°)		12.17	3416
5-Carbamoyl-1,2-dihydro-1,4-dimethyl-2-oxo- (20°)		1.76	2602
5-Carbamoyl-1,2-dihydro-2-imino-1,4-dimethyl- (20°)		9.51	2602

continued

TABLE XVI (*continued*).

Pyrimidine	Acidic pK_a	Basic pK_a	Ref.[c]
5-Carbamoyl-1,2-dihydro-2-imino-1-methyl- (20°)		9.13	2376
5-Carbamoyl-1,2-dihydro-2-imino-1,4,6-trimethyl- (20°)		*ca.* 10.2	2633
5-Carbamoyl-4-dimethylamino-2-methyl- (25°)		5.93	2698
5-Carbamoyl-4,6-dimethyl-2-methylamino- (20°)		3.62	2633
5-Carbamoyl-2-methylamino- (20°)		2.05	2376
5-Carbamoyl-4-methyl-2-methylamino- (20°)		2.89	2602
4-Carboxy-2,6-dihydroxy-	2.07; 9.45		*H*, 3456
5-Carboxy-2,4-dihydroxy-	4.16; 8.89		3456
4-Carboxy-2,6-dihydroxy-5-iodo- (25°)	1.88; 7.63		2454
4-Carboxy-2,6-dihydroxy-5-nitro- (25°)	<1.5; 4.94		3456
5-Carboxy-4-dimethylamino-2-methyl- (25°)		1.90	2698
5-Carboxy-2-ethylthio-4-hydroxy- (25°)	6.01; 10.52		3456
1-Carboxymethoxy-1,2-dihydro-4,6-dimethyl-2-oxo-	2.4 or 3.0	3.0 or 2.4	2869
2-Chloro-4-bis(β-hydroxyethyl)amino-		3.70	2910
4-Chloro-1,6-dihydro-6-imino-1-methyl- (20°)		*ca.* 9	2781
5-Chloro-1,2-dihydro-2-imino-1-methyl- (20°)		10.19	2375
4-Chloro-1,6-dihydro-1-methyl-6-methylimino- (20°)		*ca.* 9	2781
4-Chloro-2,6-dihydroxy-	5.67; *ca.* 13		3270
5-Chloro-2,4-dihydroxy-	7.95; *ca.* 13		3270
2-Chloro-4,6-dimethyl-5-propyl- (20°)		−0.34	2700
2-Chloro-4,6-dimethyl- (20°)		−0.68	2700
4-Chloro-6-hydroxy-5-isopropyl- (20°)	8.11	−0.91	2242
4-Chloro-6-hydroxy-5-methyl- (20°)	7.77	−1.11	2242
2-Chloro-4-methylamino- (20°)		2.83	2288
4-Chloro-2-methylamino- (20°)		2.63	2288
5-Chloro-2-methylamino- (20°)		2.04	2633
4-Chloro-2-methyl-6-methylamino-5-nitro- (20°)		<1.0	2675

Pyrimidine	Acidic pK_a	Basic pK_a	Ref.[c]
5-Cyano-1,2-dihydro-1,4,6-trimethyl-2-methylimino- (20°)		9.42	2633
5-Cyano-4-dimethylamino-2-methyl- (25°)		4.37	2698
5-Cyano-4,6-dimethyl-2-methylamino- (20°)		1.69	2633
5-Cyano-2-methylamino- (20°)		0.76	2376
5-Cyanomethylamino-4-hydroxy-6-methylamino- (20°)	9.90	0.66	2492
2-Cyclohexylamino- (20°)		4.04	2619
4-Cyclohexylimino-1,4(or 3,4)-dihydro-1,2,6(or 2,3,6)-trimethyl-[e]		>14	2153
4-Cyclohexylamino-2,6-dihydroxy-5-nitro- (20°)	5.15		2927
2,5-Diamino- (20°)		*ca.* 4.0; *ca.* 1.0	2627
4,5-Diamino- (20°)		6.04; −0.97	3780
4,6-Diamino- (20°)		6.01	*H*, 827
2,4-Diamino-5-benzylamino-6-hydroxy- (20°)	10.89	5.14; 1.21	3489
4,6-Diamino-5-bromo- (20°)		4.22	*H*, 827
4,5-Diamino-6-carboxy-	1.49	7.19	3199
2,4-Diamino-6-chloro- (20°)		3.57	2288
4,5-Diamino-6-chloro-2-styryl- (20°)		2.85	2562
2,4-Diamino-3,6-dihydro-6-imino-3-methyl- (20°)		$\nless 12.7$	2288
4,5-Diamino-1,2-dihydro-2-imino-1-methyl- (20°)		13.66	2454
4,5-Diamino-3,6-dihydro-6-imino-3-methyl- (20°)		12.11	2454
2,4-Diamino-3,6-dihydro-3-methyl-5-nitro-6-oxo- (20°)		2.80	2909
2,4-Diamino-1,6-dihydro-1-methyl-6-oxo-		unknown	*cf. H*
4,5-Diamino-3,6-dihydro-3-methyl-6-oxo- (20°)	>14	3.60	2492
4,5-Diamino-2,6-dihydroxy-		1.7?	3495
4,6-Diamino-2,5-dimethyl-		7.08; 0.23	3492
4,5-Diamino-6-ethoxycarbonyl-		4.82	3199
2,4-Diamino-5-formamido-6-hydroxy-	9.9	2.5	3496
2,4-Diamino-6-hydroxy-	10.83	3.27	*H*, 3492
4,6-Diamino-2-hydroxy-	6.56 or 6.49	11.98	*H*, 2627, 3492
2,4-Diamino-6-hydroxy-5-methyl- (20°)	11.07 or 11.28	3.61 or 3.64	2288, 3492

continued

TABLE XVI (*continued*).

Pyrimidine	Acidic pK_a	Basic pK_a	Ref.[c]
4,6-Diamino-2-hydroxy-5-methyl-	11.78	6.66	3492
2,5-Diamino-4-hydroxy-6-methylamino- (20°)		decomp.	3489
4,5-Diamino-6-hydroxy-2-methylamino- (20°)	10.63	5.44	2288
4,5-Diamino-6-mercapto-	9.33	2.17	3501
4,6-Diamino-2-mercapto-	10.53	5.02	3492
2,4-Diamino-6-methylthio- (20°)		5.46	2288
4,5-Diamino-2-methylthio- (20°)		5.05	2288
2,4-Diamino-5-*N*-methyl-*p*-toluidino-		6.94; −1.49	2567
4,5-Diamino-2-phenethyl- (20°)		6.65	2562
4,5-Diamino-1,2,3,6-tetrahydro-1,3-dimethyl-2,6-dioxo- (20°)		4.44	3489
2,4-Diamino-5-*p*-toluidino-		7.04; −1.47	2567
5-ββ-Dicarbamoyl-α-phenylethyl-4,6-dihydroxy- (20°)	6.18; ≮13.21		2997
4-ββ-Diethoxyethylamino-2,6-dihydroxy-5-nitro- (20°)	4.45; 12.98		2909
4-*N*-ββ-Diethoxyethyl-*N*-methylamino-2,6-dihydroxy-5-nitro- (20°)	4.48; 12.78		2909
4-*N*-ββ-Diethoxyethyl-*N*-methylamino-2-dimethylamino-6-hydroxy-5-nitro- (20°)	8.29	0.60	2909
2-Diethylamino-4,6-dimethyl- (20°)		5.74	2334
5-Diethylamino-4-ethylamino-6-hydroxy-[d]	10.63	4.70	3475
5,5-Diethyl-1,4,5,6-tetrahydro-1,2-dimethyl-4-oxo-6-thio- (20°)		*ca.* −4	2242
5,5-Diethyl-1,4,5,6-tetrahydro-1,2-dimethyl-6-oxo-4-thio- (20°)		−0.42	2242
4,6-Dihydrazino-2-methyl-5-nitro- (20°)		4.45	2562
1,2-Dihydro-1,4-dimethyl-2-oxo- (20°)		3.21	171
1,2-Dihydro-1,6-dimethyl-2-oxo- (20°)		3.38	2630
1,4-Dihydro-1,5-dimethyl-4-oxo- (20°)		2.49	2630
1,6-Dihydro-1,5-dimethyl-6-oxo- (20°)		2.23	2630
1,2-Dihydro-1-hydroxy-4,6-dimethyl-2-oxo-	6.1	2.85	2869
1,2-Dihydro-1-β-hydroxyethyl-2-imino- (20°)		10.52	2633
1,4-Dihydro-6-hydroxy-1-methyl-4-thio- (20°)	5.05	−1.13	2760
1,6-Dihydro-4-hydroxy-1-methyl-6-thio- (20°)	4.38	−2.02	2760
1,2-Dihydro-2-imino-1,4-dimethyl- (20°)		11.10	2602

Pyrimidine	Acidic pK_a	Basic pK_a	Ref.[c]
1,2-Dihydro-2-imino-4,6-dimethyl-1-prop-2′-ynyl- (20°)		10.41	2700
1,2-Dihydro-2-imino-5-iodo-1-methyl- (20°)		10.24	3422
1,4-Dihydro-4-imino-2-methoxy-1-methyl- (20°)		$\nless$ 12.0	2745
1,6-Dihydro-6-imino-4-methoxy-1-methyl- (20°)		*ca.* 11	2781
1,4-Dihydro-4-imino-1-methyl-2-methyl-amino- (20°)		14.0	2288
1,4-Dihydro-4-imino-1-methyl-2-methyl-thio- (20°)		$\nless$ 12.5	2288
1,4-Dihydro-4-imino-1-methyl-6-methyl-thio- (20°)		*ca.* 11	2781
1,2-Dihydro-2-imino-1-*p*-nitrobenzyl- (20°)		9.64	2633
1,2-Dihydro-2-imino-1-propyl- (20°)		10.96	2633
1,2-Dihydro-2-imino-1-propyl-2′-ynyl- (20°)		10.04	2700
1,2-Dihydro-2-imino-1,4,6-trimethyl- (20°)		11.56	2627
1,2-Dihydro-2-imino-1,4,6-trimethyl-5-propyl- (20°)		12.21	2700
1,2-Dihydro-2-imino-1,4,6-trimethyl-5-prop-1′-ynyl- (20°)		10.97	2700
1,2-Dihydro-2-imino-1,4,6-trimethyl-5-prop-2′-ynyl- (20°)		11.55	2700
1,2-Dihydro-1-isobutyl-2-oxo- (20°)		2.75	2630
1,2-Dihydro-1-isopropyl-2-oxo- (20°)		2.93	2630
1,2-Dihydro-1-methoxy-4,6-dimethyl-2-oxo-		3.1	2869
1,6-Dihydro-4-methoxy-1-methyl-6-methylimino- (20°)		*ca.* 11	2781
1,2-Dihydro-4-methoxy-1-methyl-2-oxo- (24°)		0.65	3046
1,6-Dihydro-4-methoxy-1-methyl-6-oxo- (20°)		−0.44	2760
1,6-Dihydro-1-methyl-4-methylamino-6-oxo- (20°)		0.82	2776, 2781
1,6-Dihydro-1-methyl-4-methylamino-6-thio- (20°)		−0.48	2776
1,2-Dihydro-1-methyl-2-methylimino- (20°)		11.74	2626
1,4-Dihydro-1-methyl-4-methylimino-6-methylthio- (20°)		*ca.* 11	2781
1,4-Dihydro-1-methyl-6-methylthio-4-oxo- (20°)		1.77	2760
1,6-Dihydro-1-methyl-4-methylthio-6-oxo- (20°)		0.13	2760
1,2-Dihydro-1-methyl-2-thio- (20°)		1.66	2173
1,4-Dihydro-1-methyl-4-thio- (20°)		1.16	2173

continued

TABLE XVI (*continued*).

Pyrimidine	Acidic pK_a	Basic pK_a	Ref.[c]
1,6-Dihydro-1-methyl-6-thio- (20°)		0.56	2173
1,2-Dihydro-2-oxo-1-propyl- (20°)		2.75	2630
2,4-Dihydroxy- (25°)	9.51 or 9.43 or 9.5 or 9.46; 13.2	−3.38	*H*, 3042, 3046, 3270, 3493, 3497
4,6-Dihydroxy- (20°)		0.26	*H*, 2997
4,6-Dihydroxy-5-α-(4,6-dihydroxy-pyrimidin-5-yl)benzyl- (20°)	4.91; 9.67; >12	1.2	2997
2,4-Dihydroxy-5-iodo-	8.25; *ca.* 13		3270
4,6-Dihydroxy-5-isopropyl- (20°)	6.40	−0.59	2242
4,6-Dihydroxy-2-mercapto-	3.7; 7.89; >13	−4.39; −6.71	3220
2,4-Dihydroxy-6-methoxycarbonyl- (25°)	7.93		3497
2,4-Dihydroxy-5-methyl- (25°)	9.9 or 9.90		*H*, 3042, 3493, 3497
2,4-Dihydroxy-6-methyl- (25°)	9.68		*H*, 3497
4,6-Dihydroxy-2-methyl- (20°)	6.35	0.21	2997
4,6-Dihydroxy-5-methyl- (20°)	6.01	−0.51	2242
2,4-Dihydroxy-6-methylamino-5-nitro- (20°)	5.11; 13.23		2909
2,4-Dihydroxy-6-methylsulphonyl- (25°)	4.68		3497
4,6-Dihydroxy-2-methylthio- (20°)	5.09		2454
4,6-Dihydroxy-2-methylthio-5-nitro- (20°)	2.00		2454
2,4-Dihydroxy-6-methyl-5-*p*-toluidino-	9.57	−0.99	2567
2,4-Dihydroxy-5-nitro- (25°)	5.56		*H*, 3497
2,4-Dihydroxy-5-nitropyrimidin-6-yl-β-picolinium betain		−0.12	2718
2,4-Dihydroxy-5-nitropyrimidin-6-yl-γ-picolinium betain		0.14	2718
2,4-Dihydroxy-5-nitropyrimidin-6-yl-pyridinium betain		−0.80	2718
2,4-Dihydroxy-5-phenylthio-	8.13		3488
4,6-Dihydroxy-2-propyl- (20°)	6.35	0.23	2997
2,4-Dihydroxy-5-*o*-toluidino-	9.12	−1.34	2567
2,4-Dihydroxy-5-*p*-toluidino-	9.12	−0.66	2567
2,4-Dihydroxy-5-trifluoromethyl-	7.35		3498
2,4-Dimercapto-	6.35; 12.57		*H*, 3220
4,6-Dimercapto-	3.60; 9.70	−2.3; −7.2	3220

Pyrimidine	Acidic pK_a	Basic pK_a	Ref.[c]
2,4-Dimercapto-5-*p*-toluidino-	10.9[i]		2567
4,6-Dimethoxy- (20°)		1.49	2997
4,5-Dimethoxy-2-methyl- (20°)		4.11	2698
2-Dimethylamino-		4.16	*H*, 2334
4-Dimethylamino-1,2-dihydro-2-imino-1-methyl- (20°)		13.68	2627
2-Dimethylamino-1,6-dihydro-1-methyl-6-oxo- (20°)		3.49	2776
4-Dimethylamino-1,6-dihydro-1-methyl-6-oxo-		<1	2776
4-Dimethylamino-2,3-dihydro-3-methyl-2-oxo- (20°)		6.02	2676
2-Dimethylamino-1,6-dihydro-1-methyl-6-thio- (20°)		2.00	2776
4-Dimethylamino-1,2-dihydro-1-methyl-2-thio- (20°)		2.62	2776
4-Dimethylamino-1,6-dihydro-1-methyl-6-thio- (20°)		−0.80	2776
2-Dimethylamino-4,6-dihydroxy- (20°)	*ca.* 6.7	1.45	171
4-Dimethylamino-2,6-dihydroxy-5-nitro- (20°)	4.84; 13.11		2909
4-Dimethylamino-2,6-dimethyl- (20°)		7.73	2334
4-Dimethylamino-5-ethoxycarbonyl-2-methyl- (25°)		5.54	2698
2-Dimethylamino-4-ethoxy-5-nitro- (20°)		1.25	2718
2-Dimethylamino-4-hydroxy- (20°)	9.89	3.68	2776
4-Dimethylamino-2-hydroxy-	12.30	4.25	3086
2-Dimethylamino-4-mercapto- (20°)	8.02	2.20	2776
4-Dimethylamino-2-mercapto- (20°)	11.18	2.40	2776
4-Dimethylamino-6-mercapto- (20°)	*ca.* 9.5	−0.27	2776
2-Dimethylamino-4-methoxy- (20°)		5.87	2776
4-Dimethylamino-2-methoxy- (20°)		6.17	2745
4-Dimethylamino-2-methyl- (25°)		7.49	2698
4-Dimethylamino-2-methylamino- (20°)		8.11	2627
2-Dimethylamino-4-methylamino-5-nitro- (20°)		3.70	2927
4-Dimethylamino-6-methylamino-5-nitro- (20°)		2.90	2454
2-Dimethylamino-4-methylthio- (20°)		5.02	2776
4-Dimethylamino-2-methylthio- (20°)		5.73	2776
4-Dimethylamino-6-methylthio- (20°)		4.57	2776
2,4-Dimethyl-6-methylamino- (20°)		7.57	2334
4,6-Dimethyl-2-methylamino- (20°)		5.25	2627
4,6-Dimethyl-2-methylamino-5-propyl- (20°)		5.63	2700

continued

TABLE XVI (*continued*).

Pyrimidine	Acidic pK_a	Basic pK_a	Ref.[c]
4,6-Dimethyl-2-methylamino-5-prop-1′-ynyl- (20°)		4.13	2700
4,6-Dimethyl-2-methylamino-5-prop-2′-ynyl- (20°)		4.92	2700
2,4-Dimethyl-6-methylthio- (20°)		4.05	2746
4,6-Dimethyl-2-methylthio- (20°)		2.13	2746
4,6-Dimethyl-2-prop-2′-ynylamino- (20°)		4.42	2700
4,6-Dimethyl-2-propylamino- (20°)		5.15	2334
2-Ethoxy- (20°)		1.27	2697
1-Ethoxycarbonylmethoxy-1,2-dihydro-4,6-dimethyl-2-oxo-		2.85	2869
4-Ethoxy-5-ethoxycarbonyl-2-methyl- (25°)		2.70	2698
4-Ethoxy-2-hydroxy- (24°)		1.00	*H*, 3046
2-Ethoxy-5-methyl- (20°)		1.75	2630
2-Ethylamino- (25°)		4.03	2627
1-Ethyl-1,2-dihydro-2-imino- (20°)		10.94	2627
5-Ethyl-1,2-dihydro-2-imino-1-methyl- (20°)		11.45	2627
1-Ethyl-1,2-dihydro-2-methylimino- (20°)		11.86	3416
1-Ethyl-1,2-dihydro-6-methyl-2-oxo- (20°)		3.47	2630
1-Ethyl-1,2-dihydro-2-oxo- (20°)		2.65	2630
2-Ethylimino-1,2-dihydro-1-methyl- (20°)		12.02	3416
5-Ethyl-2-methylamino- (20°)		4.31	2627
4-Fluoro-2,6-dihydroxy-	4.03; *ca.* 13		3270
5-Fluoro-2,4-dihydroxy- (25°)	7.98 or 8.04 or 8.15		3086, 3497, 3498
5-Fluoro-2-hydroxy-4-methylamino		2.66	3086
5-Formamido-4,6-bismethylamino- (20°)		5.00	2454
5-Formamido-4,6-bismethylamino-2-methylthio- (20°)		4.17	2454
5-Formamido-2-methyl-4-methylamino-6-methylthio- (20°)		3.92	2675
2-Heptylamino- (20°)		4.07	2627
1-Heptyl-1,2-dihydro-2-imino- (20°)		10.86	2627
4-Heptylimino-1,4(or 3,4)-dihydro-1,2,6(or 2,3,6)-trimethyl-[e]		>14	2153
2-Hexylamino- (20°)		4.18	2630
2-Hydrazino- (20°)		4.55; −0.46	2619
4-Hydrazino-6-methoxy-5-nitro- (20°)		2.77	2562
4-Hydrazino-5-nitro- (20°)		2.63	2857
2-Hydroxy-4,6-dimethyl- (20°)	10.46 or 9.9	3.82 or 3.75	2700, 2869

Pyrimidine	Acidic pK_a	Basic pK_a	Ref.[c]
4-Hydroxy-2,6-dimethyl- (20°)	9.77	3.06	2700
2-Hydroxy-4,6-dimethyl-5-propyl- (20°)	10.84	4.22	2700
2-Hydroxy-4,6-dimethyl-5-propyl-1′-ynyl- (20°)	9.44	2.72	2700
2-Hydroxy-4,6-dimethyl-5-prop-2′-ynyl- (20°)	10.20	3.56	2700
2-β-Hydroxyethylamino- (20°)		3.58	2334
2-Hydroxy-4-β-hydroxyethylamino-		4.64	2910
4-Hydroxy-5-isopropyl-6-mercapto- (20°)	5.12; 12.30	−0.66	2242
4-Hydroxy-2-mercapto-	7.65; >13	−4.16; −6.64	3220
4-Hydroxy-6-mercapto- (20°)	4.33; 10.52	−1.7	*H*, 827, 2760
4-Hydroxy-6-mercapto-5-methyl- (20°)	4.92; 11.61	−1.16	2242
4-Hydroxy-2-mercapto-6-propyl-	8.25; >12	−4.22; −6.54	3220
4-Hydroxy-2-mercapto-5-*p*-toluidino-	7.74		2567
4-Hydroxy-6-methoxy- (20°)	8.47	−0.22	2997
4-Hydroxy-5-methyl- (20°)	9.12	2.34	2630
4-Hydroxy-2-methylamino- (20°)	9.82	3.93	2288
4-Hydroxy-5-methyl-6-methylthio- (20°)	9.10	−0.03	2242
4-Hydroxy-6-methylthio- (20°)	8.47	−0.11; −7.0	*H*, 3220
2-Hydroxy-4-phenethyl-[e]	13.6	4.0	2153
4-Hydroxy-2-piperidino-	9.49	3.57	3488
5-Iodo-2-methylamino- (20°)		2.44	2633
2-Isobutoxy- (20°)		1.37	2697
2-Isobutylamino-4,6-dimethyl- (20°)		5.37	2334
4-Isobutylamino-2,6-dimethyl- (20°)		7.50	2334
2-Isopropoxy- (20°)		1.58	2697
4-Isopropoxy-2-methyl-		4.46	2698
2-Isopropylamino- (20°)		4.05	2626
2-Isopropylamino-4,6-dimethyl-		5.24	2334
2-Mercapto-	7.14 or 7.04	1.35 or 1.40; −8.3	*H*, 2173, 3220
4-Mercapto-	6.87	−0.75; −6.63	*H*, 2173, 3220
4-Mercapto-2,6-dimethyl- (20°)	8.13	1.80	2630
4-Mercapto-6-methoxy- (20°)	7.51	−1.98	2760
4-Mercapto-6-methoxy-5-methyl- (20°)	7.89	−1.59	2242
4-Mercapto-6-methylamino- (20°)	9.64	−0.27	2781

continued

TABLE XVI (*continued*).

Pyrimidine	Acidic pK_a	Basic pK_a	Ref.[c]
2-Methoxy- (20°)		1.05	*H*, 2697
2-Methoxy-4,6-dimethyl- (20°)		2.75	2746
4-Methoxy-2,6-dimethyl- (20°)		4.76	2746
2-Methoxy-5-methyl- (20°)		1.67	2630
4-Methoxy-2-methyl- (25°)		3.98	2698
4-Methoxy-5-methyl- (20°)		3.62	2630
4-Methoxy-2-methylamino- (20°)		5.76	2627
4-Methoxy-6-methylthio- (20°)		1.62	2760
2-Methylamino- (20°)		4.00	*H*, 2627
4-Methylamino-6-methylthio- (20°)		4.42	2781
2-Methyl-4,6-bismethylamino-5-nitro- (20°)		3.43	2454
4-Methyl-2-methylamino- (20°)		4.57	2602
2-Methyl-5-nitro-		1.49[f]	2688
2-Methylnitrosoamino- (20°)		<1	3491
4-Methylnitrosoamino- (20°)		1.62	3491
3-Methyl-5-nitrouracil-6-ylpyridinium betain		−0.39	2718
2-Methyl-4-phenoxy- (25°)		3.17	2698
2-Methylsulphinyl- (20°)		<−3	2619
4-Methylsulphinyl- (20°)		<0	2619
5-Methylsulphinyl- (20°)		0.42[f]	2619
2-Methylsulphonyl- (20°)		<−3	2619
4-Methylsulphonyl- (20°)		<0	2619
5-Methylsulphonyl- (20°)		0.97[f]	2619
2-Methylthio- (20°)		0.59	2173
4-Methylthio- (20°)		2.48	2173
2-Methylthio-5-nitro- (20°)		−2.65	2746
5-Nitro- (20°)		0.72[f]	2688
2-*p*-Nitrobenzylamino- (20°)		3.05	2633
2-Phenylsulphinyl- (20°)		<−3	2619
2-Propoxy- (20°)		1.34	2697
2-Prop-2′-ynylamino- (20°)		2.90	2700
2-Propylamino- (20°)		4.10	2627
1,2,3,4-Tetrahydro-1,3-dimethyl-2,4-dioxo- (24°)		−3.25	3046
1,2,3,4-Tetrahydro-1,3-dimethyl-6-methylamino-5-nitro-2,4-dioxo- (20°)	9.20		2927
1,2,3,4-Tetrahydro-6-hydroxy-1,3-dimethyl-5-nitroso-2,4-dioxo- (23°)	4.72		3499
1,2,3,4-Tetrahydro-4-imino-1,3-dimethyl-2-oxo-(20°)		9.4	2168

Pyrimidine	Acidic pK_a	Basic pK_a	Ref.[c]
2,4,6-Triamino- (20°)		7.18; 1.07	*H*, 3492
2,4,5-Triamino-6-hydroxy-	10.1	5.1; 2.0	3496
2,4,6-Trihydroxy-5-nitro- (23°)	10.25[i]		3499
2,4,6-Trihydroxy-5-nitroso- (23°)	4.41; 9.66		3499
Trimethyl-2-methylthiopyrimidin-4-yl-ammonium chloride		$\nless$11.9	2745
4,5,6-Trismethylamino- (20°)		6.01	2454
Uracil/5-bromo-1-methyl- (24°)	7.84	−6.60	3046
Uracil/6-chloro-3-methyl-	5.84; *ca.* 13		3270
Uracil/6-cyclohexylamino-3-methyl-5-nitro- (20°)	5.46		2927
Uracil/6-dimethylamino-3-methyl-5-nitro- (20°)	4.96		2909
Uracil/6-ethoxycarbonyl-3-methyl- (25°)		8.18	3497
Uracil/1-methyl- (24°; 25°)	9.77 or 9.72	−3.40	*H*, 3042, 3046, 3497
Uracil/3-methyl- (25°)	10.00 or 9.95 or 9.85		3042, 3270, 3497
Uracil/1-methyl-6-methylamino-5-nitro- (20°)	8.50; 11.97		2909
Uracil/3-methyl-6-methylamino-5-nitro- (20°)	5.29		2909
Uracil/1-methyl-5-nitro- (25°)	7.35		3497
Uracil/3-methyl-5-nitro- (25°)	5.70		3497

[a] The second place of decimal is not necessarily significant.
[b] The list contains most of the pyrimidine pK_a values published from 1960 to 1967 inclusive.
[c] *H* in the reference column indicates that earlier (but not necessarily less accurate) values were given in the original table (*H* 472–476).
[d] At 20° ± 2° (T. Sugimoto, personal communication).
[e] In 66% dimethylformamide (not comparable with other values).
[f] Covalently hydrated cation.
[g] At 10°.
[h] At 30°.
[i] Dianion.

3. The Nuclear Magnetic Resonance Spectra of Pyrimidines* (*New*)

The proton magnetic resonance (p.m.r.) spectra of simple non-tautomeric pyrimidines have not been widely studied, due probably to the synthetic difficulties involved in obtaining many of the compounds. A large number of the most interesting pyrimidines are highly substituted and contain so few hydrogen atoms that the spectra obtained are of little use in structural assignments. However, the power of the p.m.r. technique is aptly illustrated by its use in determining the structures of the preferred forms of many tautomeric oxo-, amino-, and thio-pyrimidines, particularly since the recognition of dimethyl sulphoxide and liquid sulphur dioxide as satisfactory solvents for these compounds. With the free bases, broadening is observed of signals from H-2, H-4, and/or H-6 due to the intermediate relaxation times of the adjacent ^{14}N-resonance. Removal of this broadening by heteronuclear decoupling has, to my knowledge, not been reported.

The spectrum of pyrimidine (**1**) was first reported by S. Gronowitz and R. A. Hoffman[2410] who measured chemical shifts at 40 Mc./s. in a number of solvents with respect to either the solvent peak or an external water standard. They established the relative deshielding of the four ring protons, H-2 > H-4 = H-6 > H-5, and by direct measure-

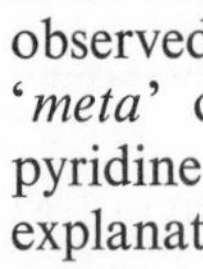

(**1**) 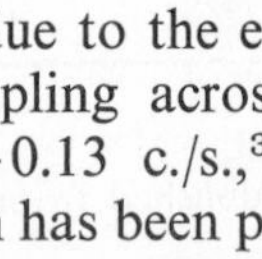 (**2**) 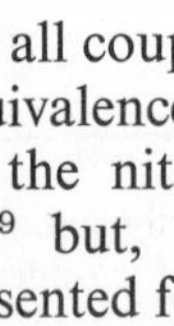 (**3**) 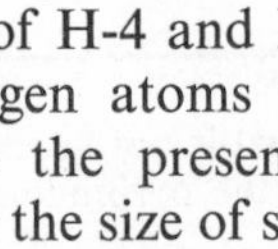 (**4**) 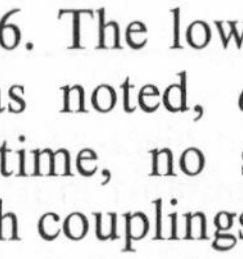 (**5**) 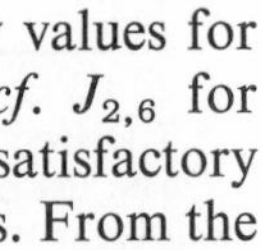 (**6**)

ment obtained values for all coupling constants except $J_{4,6}$ which is not observed, due to the equivalence of H-4 and H-6. The low values for '*meta*' coupling across the nitrogen atoms was noted, *cf.* $J_{2,6}$ for pyridine −0.13 c./s.,[3509] but, to the present time, no satisfactory explanation has been presented for the size of such couplings. From the relatively constant internal chemical shifts observed for the neat liquid and for dilute solutions in carbon tetrachloride, association effects must have little influence on the magnetic properties of the molecule. Later,[2411] the spectrum of a deuterochloroform solution was measured

* By T. J. Batterham, Department of Medical Chemistry, Australian National University, Canberra, Australia.

TABLE XIXa. Proton Magnetic Resonance Data[2411, 2607] for Pyrimidine and its *C*-Methyl Derivatives[a] (*New*)

Pyrimidine	Solvent	Chemical shifts (p.p.m.) δ_2	δ_4	δ_5	δ_6	Coupling constants (c./s.) $J_{2,5}$	$J_{4,5}$	$J_{4,6}$	$J_{5,6}$
Parent[b]	$CDCl_3$	9.26	8.78	7.36	8.78	1.5	5.0	2.5	5.0
	Me_2CO	9.17	8.80	7.48	8.80	1.45	5.0	—	5.0
	C_6H_{12}	9.16	8.60	7.08	8.60	1.60	5.0	—	5.0
	$(CD_3)_2SO$	9.26	8.87	7.58	8.87				
2-Methyl	$CDCl_3$	—	8.63	7.12	8.63	(0.6)	4.9	—	4.9
	Me_2CO	(2.62)	8.68	7.23	8.68	(0.55)	4.9	—	4.9
	C_6H_{12}	(2.62)	8.49	6.84	8.49	(0.6)	4.9	—	4.9
4-Methyl	$CDCl_3$	9.09	—	7.21	8.59	1.4	(0.4)	—	5.1
	C_6H_{12}	8.90	—	6.98	8.41	1.45	(0.55)	(0.3)	5.16
5-Methyl	$CDCl_3$	9.04	8.57	—	8.57				
	Me_2CO	8.99	8.65	(2.32)	8.65				
	C_6H_{12}	8.95	8.45	(2.20)	8.45				

[a] Chemical shifts in parentheses are those of the methyl substituents; coupling constants in parentheses are between the methyl protons and the indicated ring-proton.

[b] ^{13}C-couplings obtained on neat liquid: $J_{^{13}C-2,H} = 206.0$, $J_{^{13}C-4,H} = 181.8$, $J_{^{13}C-5,H} = 168.0$ c./s.

with respect to internal tetramethylsilane (TMS), confirming the earlier work and giving chemical shift values which are more meaningful by to-day's standards (Table XIXa). Analysis of the ^{13}C-satellites of each multiplet gave $J_{4,6}$ and the three $^{13}C—H$ couplings, as well as the known coupling constants (Table XIXa). A rough correlation was obtained between the $^{13}C—H$ couplings and the chemical shifts of the respective protons. Long-range couplings between the methyl groups and protons in positions '*ortho*' or '*para*' to them were observed for the three mono-*C*-methylpyrimidines (**2**–**4**). The concept of a total substituent shift, compared with specific methyl shifts, was used to discuss the transmission of the inductive effect throughout the ring. Spectra of these three methyl pyrimidines have been measured more recently at 60 Mc./s. with complete confirmation of the earlier work.[2607] Data are summarized in Table XIXa.

Only one specific study of simple mono-substituted pyrimidines has been made[2607] and due to synthetic difficulties few 4- or 5-substituted derivatives were included. Spectra of 2-substituted pyrimidines (Table XIXb) show a doublet and triplet, typical of an A_2X system, with the H-4,6 doublet considerably broadened by coupling to the adjacent nitrogen atom. Those of 5-substituted derivatives consist of two broad singlets, while 4-substituted pyrimidines give typical ABX patterns with small couplings involving the X-nucleus. The *para* substituent effects were studied using the corresponding 2- and 5-substituted derivatives, e.g., (**5**) and (**6**), to check the reciprocity of transmission of electronic effects between these two non-equivalent *para* positions. Chemical shifts of the $+M$ 2-substituted pyrimidines follow the mesomeric order, $NMe_2 > NH_2 > OMe > Me$, as established for other aromatics.[3510–3512] As in benzenes, the amino group gives a much larger '*para*' shift than does the methoxy group, quite distinct from the almost equal effect of these groups in π-excessive five-membered ring systems such as thiophene. A 5-methoxy substituent causes a *para* shift of H-2 similar to that in the reverse system, but the 5-cyano group does not. More data is required before substituent effects in these compounds can be understood. Representative data on substituted pyrimidines are collected in Table XIXb.

Decoupling experiments have been used to determine the relative signs of couplings in a number of methyl- and amino-pyrimidines.[3517] $J_{2,5}$ and $J_{2,6}$ were found to have the same sign as $J_{5,6}$ and hence were considered to be positive. Coupling constants between *C*-methyl groups and *ortho* ring protons were negative and those with *meta* ring protons, though very small, were considered to be positive.

Spectroscopic methods have long been used to determine the pre-

TABLE XIXb. Proton Magnetic Resonance Data[2607] for Mono-substituted Pyrimidines Uninvolved in Studies of Tautomeric Equilibria[a] (*New*)

2-Substituent	In Me_2CO $\delta_{4,6}$	δ_5	$J_{5,6}$	In C_6H_{12} $\delta_{4,6}$	δ_5	$J_{5,6}$
Cyano	9.04	7.86	5.10			
Acetyl	8.99	7.69	5.00			
Thiocyanato	8.84	7.55	5.00	8.53	7.05	4.80
Chloro	8.78	7.55	4.90	8.47	7.06	4.75
Bromo	8.72	7.57	4.80	8.38	7.06	4.80
Iodo	8.58	7.57	4.90	8.28	7.06	4.90

4-Substituent	In $(CD_3)_2SO$ δ_2	δ_5	δ_6	$J_{2,5}$	$J_{5,6}$
Acetamido	8.83	8.02	8.61	1.30	5.80
Amino	8.39	6.44	8.04	1.25	6.15

5-Substituent	In Me_2CO δ_2	$\delta_{4,6}$	δ_5	In C_6H_{12} δ_2	$\delta_{4,6}$	In $(CD_3)_2SO$ δ_2	$\delta_{4,6}$	δ_5
Carboxy						9.28	9.00	
Cyano	9.44	9.26						
Bromo	9.15	8.96		8.99	8.65			
Hydroxy	8.69	8.39				8.70	8.37	
Methoxy	8.78	8.50	(3.95)[b]	8.72	8.25	8.87	8.60	(3.93)[b]

[a] Chemical shifts in δ (p.p.m.); coupling constants, J (c./s.)
[b] Methoxy group.

dominant tautomer in potentially mobile pyrimidines by comparison of the compound with 'fixed tautomers', usually specifically methylated derivatives. Nuclear magnetic resonance can also be used in this way, in many cases providing greater sensitivity in these comparisons and in others providing new and highly specific criteria for the predominance of certain forms.

With the simple amino- and 'hydroxy'-pyrimidines, p.m.r. spectra have confirmed the predominant forms established by studies of ultra-violet spectra and pK_a values. Thus, with '2-hydroxypyrimidines' the H-5 resonance occurs 0.67 p.p.m. towards higher field than that of 2-methoxypyrimidine (**7**),[2607] confirming the 1,2-dihydro-2-oxo-structure (**8**) established by ultra-violet methods.[3036] On the other hand, 2-amino, 2-methylamino, and 2-dimethylamino-pyrimidines have very

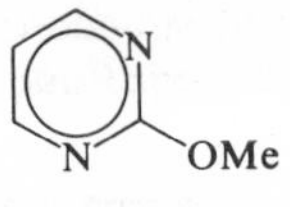

(7) 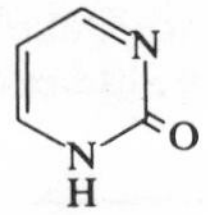 (8) 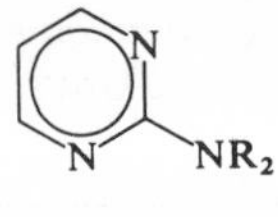 (9) 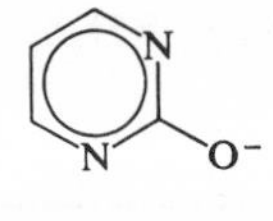 (10)

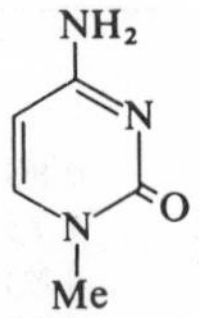

(11) 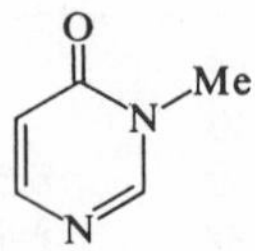 (12) 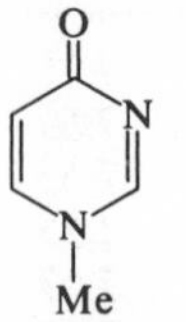 (13) 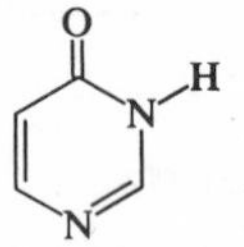 (14)

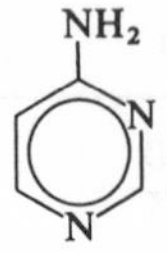

(15) 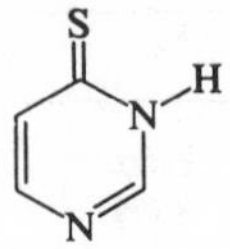 (16) 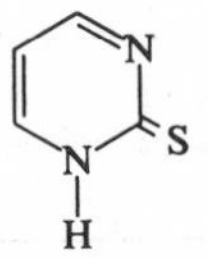 (17) 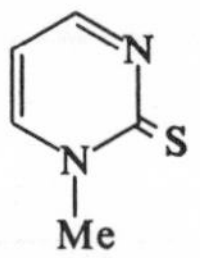 (18)

similar ring-proton shifts,[2607] reflecting the aromatic nature of these compounds (**9**, R = H or Me). Unfortunately, conversion of the fully aromatic structure, as observed in the anion (**10**) of 2-hydroxypyrimidine, into the dihydro-oxo form of the free base (**8**) causes only a small rise in $J_{4,5}(J_{5,6})$ and this value is similar to that observed in other non-tautomeric pyrimidines (see Table XIXb). The size of this coupling constant (5.3 c./s.) is unusual compared with those observed for more highly substituted 2-oxopyrimidines such as 1-methylcytosine (**11**) where $J_{5,6}$ = 7.2 c./s.[2987] In the case of '4-hydroxypyrimidine' the predominant oxo-structure in aqueous solution, previously established by Raman and infra-red spectroscopy,[1458] was confirmed.[3038] Comparison of the spectrum of the dihydro-oxo-compound with those of its 1- and 3-methyl derivatives showed the chemical shift of H-5 (δ 6.51 p.p.m.) in the parent to be closer to that from the 3-methyl derivative (**12**; 6.55 p.p.m.) than that from the 1-methyl derivative (**13**; 6.33 p.p.m.). This evidence was considered sufficient to confirm the 3,4-dihydro-4-oxo- structure (**14**) of the parent pyrimidine. Apparent pK_a values (pK_1 = 1.8, pK_2 = 8.5) obtained from the plot of chemical shift *versus* pH and H_0 were in rough agreement with those obtained by the more accurate potentiometric and spectroscopic methods (pK_1 = 1.69, pK_2 = 8.60).[726] Spectra of 4-aminopyrimidine (**15**) and 3,4-dihydro-4-

TABLE XIXc. Proton Magnetic Resonance Data[2607, 3038] for Potentially Tautomeric Mono-substituted Pyrimidines and for the Model Compounds Used to Determine the Predominant Tautomer (*New*)

		Chemical shifts[a]				Coupling constants (c./s.)		
Pyrimidine	Solvent	δ_2	δ_4	δ_5	δ_6	$J_{2,5}$	$J_{2,6}$	$J_{5,6}$
1,2-Dihydro-2-oxo-	Me_2CO	—	8.33	6.40	8.33			5.3
	$(CD_3)_2SO$	—	8.24	6.34	8.24			—
2-Methoxy-	Me_2CO	(3.93)	8.60	7.07	8.60			4.8
	C_6H_{12}	(3.88)	8.36	6.70	8.36			4.7
1,2-Dihydro-2-thio-	Me_2CO	—	8.25	6.84	8.25			—
1,2-Dihydro-1-methyl-2-thio-	$(CD_3)_2SO$	—	8.50	6.89	8.50			5.35
2-Methylthio-	Me_2CO	(2.50)	8.61	7.14	8.61			4.95
	C_6H_{12}	(2.43)	8.33	6.72	8.33			4.75
2-Amino-	Me_2CO	—	8.27	6.56	8.27			—
	$(CD_3)_2SO$	—	8.22	6.54	8.22			—
2-Methyl-amino-	Me_2CO	—	8.28	6.52	8.28			4.85
	C_6H_{12}	—	8.15	6.26	8.15			4.70
2-Dimethyl-amino-	Me_2CO	—	8.31	6.52	8.31			4.80
	C_6H_{12}	—	8.14	6.22	8.14			4.65
3,4-Dihydro-4-oxo-[b]	D_2O	8.37	—	6.51	7.98	1.0	1.2	7.15
	D_2SO_4[c]	9.26	—	6.85	8.13	0.8	1.7	7.7
	NaOD[d]	8.22	—	6.30	7.91	1.1	1.0	6.3
	$(CD_3)_2SO$	8.17	—	6.29	7.86	1.05	0.7	6.75
3,4-Dihydro-3-methyl-4-oxo-[b]	D_2O	8.42	—	6.55	7.99	0.8	0.6	6.8
	D_2SO_4	9.46	—	6.92	8.19	0.6	1.7	7.7
1,4-Dihydro-1-methyl-4-oxo-	D_2O	8.36	—	6.33	7.79	0.6	2.5	7.5
	D_2SO_4	9.18	—	6.88	8.09	0.7	2.3	7.9
4-Methoxy-	D_2O	8.68	—	6.95	8.45	1.2	0.8	6.2
	D_2SO_4	9.23	—	7.44	8.74	1.0	1.6	7.2
3,4-Dihydro-4-thio-[b]	$(CD_3)_2SO$	8.26	—	7.13	7.80	—	—	6.05

[a] Values in parentheses are for substituent methyl groups.
[b] Named as a 3,4- instead of a 1,6-dihydropyrimidine for convenience in this Table.
[c] 2.13 *N*-D_2SO_4.
[d] 1.44 *N*-NaOD.

thiopyrimidine (**16**) have been measured in a number of solvents[2607] but no discussion of tautomerism is given. Comparison of the chemical shifts of H-5 in 1,2-dihydro-2-thiopyrimidine (**17**) and its 1-methyl derivative (**18**) shows them to be almost identical,[2607] confirming the established structure of the thio-compound. Data for these mono-substituted tautomeric pyrimidines are summarized in Table XIXc.

A large amount of effort, much of it overlapping, has gone into the study of cytosine (**19**) and its derivatives. Data are collected in Table XIXd. While interest in a compound of such biological importance is to be expected, much of the work was prompted by a misinterpretation of the spectrum of deoxycytidine (**20**; R = deoxyribosyl) by L. Gatlin and J. C. Davis.[3513] J. P. Kokko *et al.*[3039] first reported the spectrum of cytosine and introduced the use of deuterated dimethyl sulphoxide as a standard solvent for these substances. Quite a lot has been said about the validity of comparisons between aqueous solutions and those in dimethyl sulphoxide, but with these tautomeric pyrimidines the similarity of spectra in the two solvents, in cases where they can both be measured, leaves little doubt that the same species are present in both solutions. The zwitterionic structure (**21**) was suggested for cytosine and structure (**22**) for its cation. These structures were criticised by A. R. Katritzky and A. J. Waring[2987] who compared the spectra of cytosine with that of its 1-methyl derivative (**23**) and of the cytosine cation with that of the cation of 1,N^4-dimethylcytosine (**24**). The similarities between the spectra show conclusively that the normally accepted structures for cytosine (**19**) and its cation (**25**) are indeed correct. In the spectrum of the 1,N^4-dimethylcytosine cation, H-5 and H-6 gave rise to *two* AB patterns of different intensities and this was assigned to the presence of the *cis-trans* isomers (**26**) and (**27**). With solutions of this

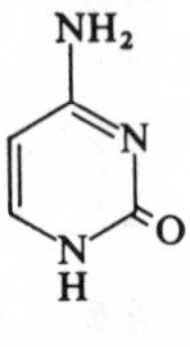

NH_2, N, O, N, H **(19)**; NH_2, N, O, N, R **(20)**; NH_3^+, N, N, O^- **(21)**; NH_3^+, N, O, N, H **(22)**; NH_2, N, O, N, Me **(23)**

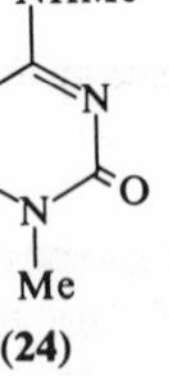

NHMe, N, O, N, Me **(24)**; NH_2, NH^+, O, N, H **(25)**; H, Me, N, N, O, N, Me **(26)**; Me, H, N, N, O, N, Me **(27)**; Me, Me, N, N, O, N, Me **(28)**

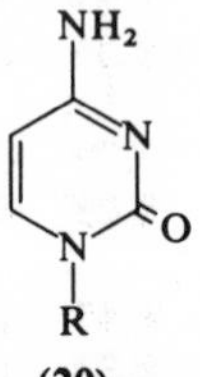

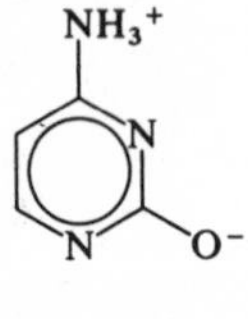

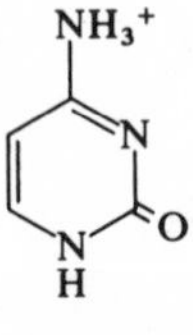

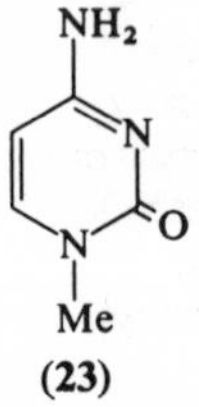

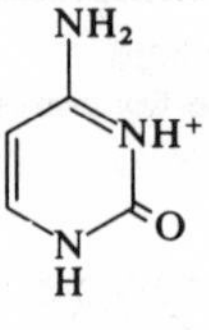

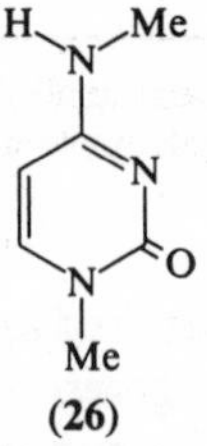

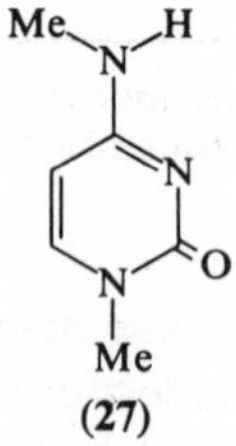

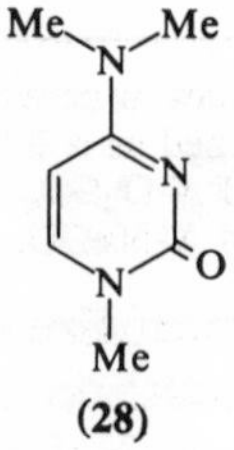

cation in liquid sulphur dioxide, the N^4-methyl groups of the two isomers had different chemical shifts, and coupling with the adjacent N—H group was observed. However, in deuterium sulphate, the chemical shift difference is extremely small. The cation of 1,N^4,N^4-trimethylcytosine (**28**) similarly shows a chemical shift difference between the two N^4-methyl groups in liquid sulphur dioxide but not in deuterium sulphate. In 4N-, but not in concentrated sulphuric acid, 1,N^4-dimethylcytosine shows two peaks for the N^4-methyl group, due to coupling with the adjacent N—H group. Structure (**25**) for the cytosine cation was also deduced from spectra of solutions in anhydrous trifluoroacetic acid.[3269]

Undoubtedly the best work on the structure of cytosine derivatives comes from E. D. Becker, H. T. Miles, and others.[3050, 3514, 3746] These workers, doubtful of the validity of arguments based on the relative chemical shifts expected for different tautomers or on the position and area of peaks from N—H protons, have relied primarily on the synthesis of ^{15}N-analogues, proton spin-decoupling, and temperature variable effects in their extensive study of the nucleoside analogue, 1-methylcytosine (**20**, R = Me). The normal spectrum of this compound in deuterated dimethyl sulphoxide contains two broad peaks (1 proton each) on the low-field side of the doublet from H-6, while that of the exocyclic ^{15}N-analogue shows each of these signals as doublets, split by the ^{15}N-nucleus (J = 94 c./s.). This information unequivocally proves the amino nature of the 4-substituent. At higher temperatures, these peaks broaden, collapse, and finally reappear as a singlet, consistent with increased rotation around the C(4)—NH_2 bond. Liquid sulphur dioxide was used as solvent for low-temperature studies ($-60°$) where three N—H signals were observed, two of which were coupled to the ^{15}N-nucleus in the labelled analogue (J = 94 c./s.). Decoupling experiments showed the third downfield signal to be coupled to H-5 ($J_{3,5}$ = 2.5 c./s.), eliminating the possibility that protonation had occurred on the oxygen atom. Interconversion of the geometrical isomers of the N^4-methyl compounds was studied and rough thermodynamic data obtained.[3514]

Complete ^{15}N-labelling has since been used[3049] to confirm the above results and to provide direct evidence of protonation at N-3. The amino protons from the completely labelled cation, under normal conditions, gave a pair of doublets (J = 94.2 c./s.) as expected. However, one doublet was further split by 4.4 c./s., assigned to coupling between ^{15}N-3 and one of the amino protons. At low temperature (liquid SO_2, $-30°$) a clean doublet, absent from spectra measured in dimethyl sulphoxide at room temperature, was present. The value of this

TABLE XIXd. Proton Magnetic Resonance Data[2987, 3514] for Cytosine and Its Derivatives[a] (*New*)

Cytosine	Salt	Solvent[b]	Chemical shifts (p.p.m.)[c] δ_5	δ_6	δ_1	δ_3	δ_7	Coupling constants (c./s.) $J_{5,6}$	$J_{3,5}$	$J_{6,7}$	$J_{7,Me}$	$J_{^{15}N,7}$[d]	$J_{^{15}N,Me}$[d]	$J_{^{15}N,5}$[d]
Parent	—	$(CD_3)_2SO$	5.62	7.36				7.2						
	HCl	$(CD_3)_2SO$	6.10	7.80				8.0						
	HCl	D_2O	6.15	7.74				8.0						
1-Methyl-	—	$(CD_3)_2SO$	5.67	7.62	(3.23)		7.00	7.0				90		
	HCl	$(CD_3)_2SO$	6.18	8.12	(3.35)		8.85, 10.00	8.0				92.6, 91.4		0.7
	HCl	SO_2	6.28	7.82	(3.55)	11.44	7.45, 7.83	7.0	2.5			93.8, 96.2		0.7
	HI	SO_2	6.33	7.87	(3.58)	11.02	7.28, 7.45	7.5	2.1					
1,3-Dimethyl-	—	$(CD_3)_2SO$	5.70	7.02	(3.20)	(3.18)	7.35	7.0						
	HCl	SO_2	6.42	7.80	(3.58)	(3.65)	7.25, 7.80	7.5						
	HI	$(CD_3)_2SO$	6.13	8.05	(3.40)	(3.43)	9.34	8.0						
		SO_2	6.37	7.82	(3.58)	(3.67)	7.00, 7.82	7.5						
1,7-Dimethyl-	—	$(CD_3)_2SO$	5.75	7.62	(3.27)		(2.80), 7.50	7.0			5.0		94.0	1.5
	HCl	$(CD_3)_2SO$	6.40	8.32	(3.40)		(3.05), 9.54	7.7						
			6.35	8.02	(3.37)		(3.02), 11.02	7.7			5.0		94.0	1.5
	HCl	SO_2	6.38	7.70	(3.54)	11.09	(3.23), 8.65	7.8	2.6	0.7	5.0		93.5	1.5
		[e]	6.52	7.90	(3.52)	11.54	(3.18), 8.03	7.8	2.3					
	D_2SO_4	4*N*-D_2SO_4	6.22	7.93	(3.44)			8.0						
		[e]	6.46	7.71	(3.46)		(3.10)	8.0						
	HI	$(CD_3)_2SO$	6.03	7.97	(3.37)	10.89	(3.02), 9.87	7.5			5.0		94.0	
		[e]	6.33	8.25	(3.37)		(2.97), 8.39	7.5			5.0		94.0	
	HI	SO_2	6.35	7.77	(3.58)	11.50	(3.33), 7.50	8.0	2.6					
		[e]	6.35	7.97	(3.58)	10.92	(3.25), 7.23	8.0						
1,7,7-Trimethyl-	HCl	$(CD_3)_2SO$	6.48	8.28	(3.5)		(3.5)	8.0						
	HCl	SO_2	6.39	7.90	(3.56)		(3.50, 3.45)	8.0						
	D_2SO_4	4*N*-D_2SO_4	6.31	7.84	(3.44)		(3.32)	8.0						

[a] In this Table, the nitrogen atom attached to C-4 is regarded as position 7.
[b] SO_2 = liquid sulphur dioxide at $-60°$.
[c] Values in parentheses are for methyl substituents.
[d] From the 7-^{15}N-derivative.
[e] Alternative geometric isomer; see text.

splitting, 94.0 c./s., indicates $^{15}N—H$ coupling and is compatible only with protonation at the N-3-position.

Spectra of 1-methylcytosine are almost identical with those obtained from the pyrimidine ring of cytosine riboside and deoxyriboside and, by analogy, the tautomeric structures of the pyrimidine ring system in both nucleosides is thought to be identical with that of the methylcytosine.

The structures of the stable tautomer of uracil (**29**) and its simple derivatives have always been supported by these p.m.r. studies. Details, often incomplete, of their spectra have been published a number of times, e.g., refs. 728, 3039, 3269, and the fine structure is clearly visible in a reported spectrum[2607] (40 Mc./s., deuterodimethylsulphoxide, room temperature). The doublets from H-5 and H-6 are further split by H-1, and in the former case, by H-3 as well. Thus, H-6 gives a quartet, $J_{5,6} = 7.8$, $J_{1,6} = 5.7$ c./s., and H-5 a double triplet, $J_{5,6} = 7.8$, $J_{1,5} = J_{3,5} = 1.4$ c./s. Elevation of the temperature increases the rate of exchange of the amide protons and causes complete decoupling of H-1 and H-3 from H-5 and H-6.

The p.m.r. spectrum of the doubly ^{15}N-labelled uracil was extremely complex[3049] and the dideutero derivative (**30**) was prepared to simplify the peak patterns from H-5 and H-6. Thus, the pattern from H-6 reduced to the expected quartet and that from H-5 to an octet from which the H-5–^{15}N splittings of 4.4 and 2.5 c./s. were extracted. Assignment of these couplings to specific ^{15}N-atoms was not possible. In the conversion of this uracil to similarly labelled cytosine, the corre-

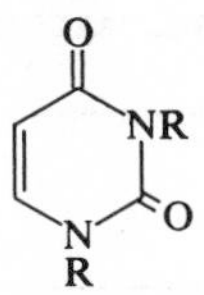

(**29**) R = H
(**30**) R = D

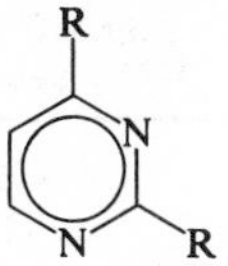

(**31**) R = Cl
(**32**) R = OMe

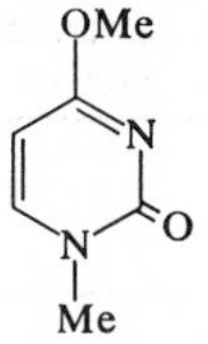

(**33**)

spondingly labelled 2,4-dichloro- (**31**) and dimethoxy-pyrimidine (**32**) and 1,2-dihydro-4-methoxy-1-methyl-2-oxopyrimidine (**33**) were prepared. P.m.r. spectra of the three intermediates and the ^{15}N n.m.r. spectrum of the pyrimidine were measured and briefly discussed.[3049]

P.m.r. spectra of fifteen substituted uracils were used in an attempt to measure the proton mobility within the series and to relate this to biological activity.[3041] It was shown that the chemical shift of the amide proton correlates linearly with both the Hammett σ-constant and the group dipole moment of substituents in the 5-position and that the chemical shift of H-6 also correlates linearly with this dipole moment, provided values for a 5-carboxy substituent are ignored. It must be realised, however, that chemical shifts were measured from *external* TMS without compensation for bulk diamagnetic susceptibility and this casts doubts on the numerical validity of this work.

More recently, the structure of the predominant form of 4,6-dihydroxypyrimidine has been the cause of some contention. The first p.m.r. study[3038] disproved the suggestion based on ultra-violet spectral evidence[2997] that the dioxo form (**34**) predominated since the spectrum contained only two singlets (each one proton) which could be attributed to ring-protons. It is important to note that the ultra-violet spectra were obtained of aqueous solutions while dimethyl sulphoxide was used as solvent for the p.m.r. work. However, addition of 50% deuterium oxide to the dimethyl sulphoxide solutions caused no change in the spectrum of the ring protons, leaving little doubt that both solutions contain the same species in comparable proportions. From a comparison of the spectrum of the dihydroxy compound with those of 4-hydroxy-6-methoxypyrimidine (**35**) and 4,6-dimethoxypyrimidine (**36**) (Table XIXe), either of the dihydro-oxo tautomers (**37**) or (**38**) may predominate. Further comparison with spectra of the 1- and 3-methyl derivatives (**39**) and (**40**) led to the conclusion that 3,4-dihydro-6-hydroxy-4-oxopyrimidine (**37**) was the main species present, presumably in equilibrium with its tautomeric equivalent (**41**). On standing with deuterium oxide, particularly with a trace of acid, all protons except

TABLE XIXe. Chemical Shifts (p.p.m.) for 4,6-Dihydroxypyrimidine and Its Methyl Derivatives[3038] (*New*)

Pyrimidine	Solvent	δNH or OH	δ_2	δ_5	δ_{OMe}	δ_{NMe}
4,6-Dihydroxy-	$(CD_3)_2SO$	11.8	8.09	5.32		
4,6-Dimethoxy-	D_2O		8.32	6.22	3.96	
4-Hydroxy-6-methoxy-	$(CD_3)_2SO$	12.0	8.12	5.57	3.80	
	D_2O		8.19	5.83	3.90	
1,4-Dihydro-6-methoxy-1-methyl-4-oxo-	D_2O		9.15	6.04	3.53	2.86
1,6-Dihydro-4-methoxy-1-methyl-6-oxo-	D_2O		8.31	5.88	3.91	3.53

H-2 exchange. This suggests that the predominant tautomers are in equilibrium with a small amount of the dioxo-form (**34**). These conclusions have been criticised by A. R. Katritzky *et al.*[3052] who, without much practical evidence, suggest the zwitterion (**42**) as the main species, and by a Russian group[3053–3056] who have proposed a number of structures including the di-zwitterion (**43**) as well as both (**37**) and (**42**). Personally, I feel that the original p.m.r. interpretation is completely sound.

The spectrum of 5-nitropyrimidine (**44**) in deuterium oxide was unexceptional, showing two broad singlets[2688] but, after acidification, three singlets were observed, δ 6.49, 8.31, 8.66 p.p.m., indicative of the addition of water across one of the C═N bonds (see ref. 3516). From the spectrum of the 2-methyl derivative, the site of hydration was located as C-4, N-3 (or C-6, N-1) giving the adduct structure (**45**). These conclusions were confirmed by examination of the ultra-violet spectra and pK_a values for the different species involved.

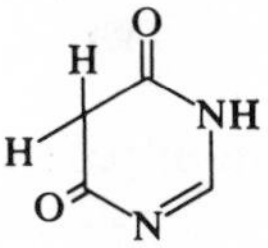

(**34**) 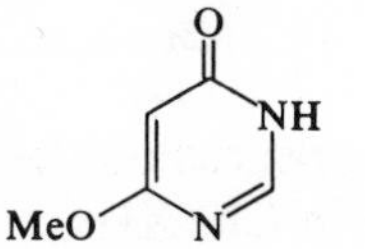 (**35**) 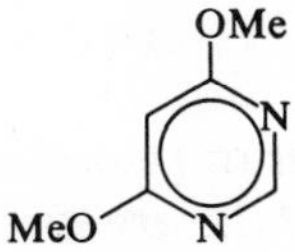 (**36**)

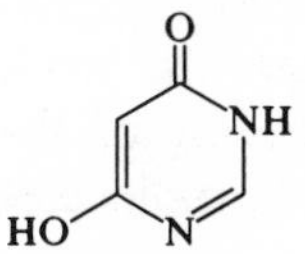

(**37**) 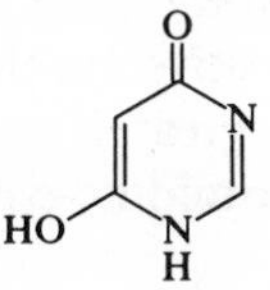 (**38**) 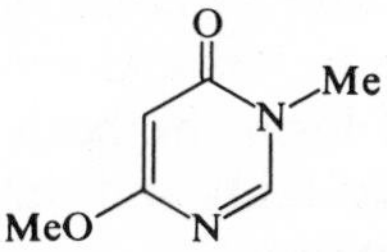 (**39**) 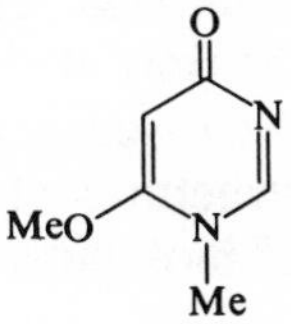 (**40**)

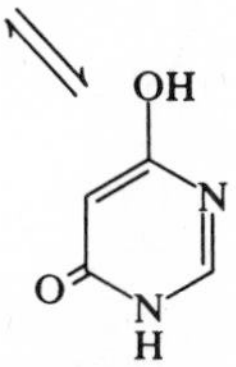

(**41**) 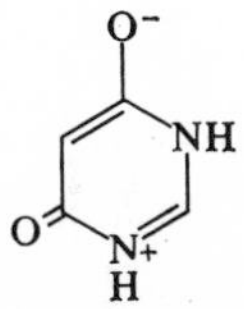 (**42**) 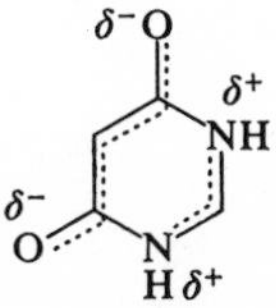 (**43**)

Rapid deuterium exchange of the *C*-methyl protons of a number of 4- and/or 6-methyl-1,2-dihydro-2-oxo(or imino)pyrimidines in deuterium oxide caused trouble in the measurement of their spectra and led to a more detailed investigation of the phenomenon.[2861] Small but consistent shifts allowed the assignment of the downfield *C*-methyl peak in the spectrum of 1,2-dihydro-1,4,6-trimethyl-2-oxo(or imino)pyrimidine (**46**; X = O or NH) to the 6-methyl group (*cf.* refs. 2911, 3515). Deuteration of these compounds was found to be acid-base catalysed and a concerted mechanism was proposed for the acid-catalysed reaction.

Recently, in publications centred on the Dimroth Rearrangement, many data for dihydro-oxo(or imino)pyrimidines and their derivatives have been presented (e.g., refs. 2562, 2700). Similarly, information on all types of substituted pyrimidine has been included in the more modern papers on pyrimidine synthesis. The Tables presented here do not cover all this data but rather give typical examples which may be used for comparison.

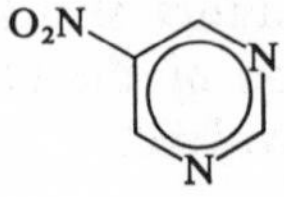

O_2N N N (44) O_2N H OD ND $\overset{+}{N}$ D (45) Me N Me N X (46)

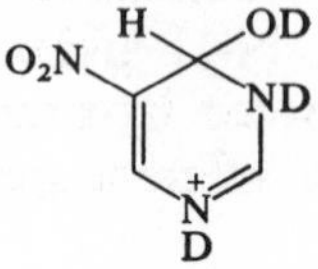

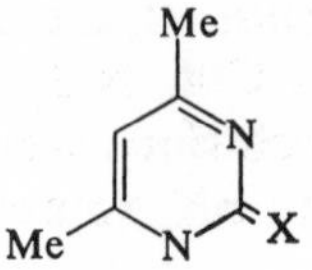

A number of papers have appeared on the spectra of reduced pyrimidines and these are discussed in Ch. XII.

In conclusion, p.m.r. spectroscopy has made a considerable impact on modern pyrimidine chemistry, particularly in the field of tautomeric equilibria. This is sure to continue for some time, and with the production of more powerful and more sensitive spectrometers the work discussed here will form the basis for a greater understanding of the structures of nucleic acids and other pyrimidine-containing molecules of biological importance.

APPENDIX

Systematic Tables of Simple Pyrimidines

Introduction (*H* 501)

As explained in the Preface, the following Tables supplement those in the original volume. Together, the two sets of Tables now cover those simple pyrimidines (as previously defined, *H* 501) revealed in Beilstein's *Handbuch* and in the Indices of *Chemical Abstracts* up to and including volume 66, June 1967.

Conventions used in the supplementary Tables are similar to those (*H* 502) in the original Tables, but some additional terms have been introduced: H in the reference column indicates that earlier data on the compound will be found in the corresponding Table of the original volume; DNP stands for 2,4-dinitrophenylhydrazone, PH for phenylhydrazone, SC for semicarbazone, and TSC for thiosemicarbazone; the trivial name [*pseudo*]uracil, is used to avoid confusion in naming mono-*N*-alkyl derivatives of 4,6-dihydroxypyrimidine; and query marks are used to indicate some doubt in structure, interpretation, or other aspect of the data recorded.

Finally, attention is again drawn to the paragraphs (*H* 503) on how to use the Tables, and to the fact that references below 2170 are listed only in the original volume (*H* 624 et seq.).

TABLE XX. Alkyl and Aryl Pyrimidines (*H* 503)

Pyrimidine	M.P.(°C)	References
5-benzyl-	47	2241
5-benzyl-4,6-dimethyl-2-phenyl-	96-97	3786
4-t-butyl-	67/14mm.; pic.142-143	2163
4-butyl-6-methyl-	99-100/17mm.	301
2,4-dimethyl-	pic.126-130	H, 2554
2,5-dimethyl-	64-65/15mm.	H, 3787, 3788
4,5-dimethyl-	169-175/at.; pic.162	H, 2108, 2163
4,6-dimethyl-	24-26; 154 to 162/at.	H, 301, 431
4,6-dimethyl- (*N*-oxide)	112	428
2,5-dimethyl-4-phenyl-	160-163/15mm.	2329
4,6-dimethyl-2-phenyl-	276/at.	H, 3789
2,4-diphenyl-	71-72; 197-198/5mm.	H, 3790
4,6-diphenyl-	pic.168-169	H, 301, 792, 3544
4-ethyl-	89-90/75mm.; pic.85-87	301
5-ethyl-	-13; 96/60mm.; pic.136	2108
2-ethyl-4-methyl-	160/at.	H, 3789
4-ethyl-5-methyl-	72/11mm.; pic.111	2163
5-ethyl-2-methyl-	65-66/13mm.; pic.143	3788
5-ethyl-4-methyl-	194/at.	H, 3789
2-ethyl-5-methyl-4-phenyl-	168/15mm.	2329
5-ethyl-2-methyl-4-phenyl-	163/15mm.	2329
2-ethyl-4-phenyl-	61-62	2329
4-ethyl-2-phenyl-	135-140/5mm.	3790

TABLE XX

Pyrimidine	M.P.(^{o}C)	References
4-ethyl-6-phenyl-	pic.185-186	H, 792
5-ethyl-2-phenyl-	88-89/0.02mm.; pic.109	3788
5-ethyl-4-phenyl-	162/13mm.; pic.119	2163
5-ethyl-4-phenyl-2-propyl-	175/15mm.; pic.160	2329
5-ethynyl-4-methyl-2-phenyl-	64-65	458
4-isobutyl-	75-76/12mm.; pic.78-79	301, 2163
5-isopropyl-	97/40mm.; pic.116 or 124	2108
2-methyl-	pic.112-114	H, 2451, 3787
4-methyl-	137-138/at.	H, 301, 606, 3789, 3791
4-methyl- (1-*N*-oxide)	45-47; pic.105	2418
4-methyl- (3-*N*-oxide)	82-83; pic.137	2418
5-methyl-	32; pic.142	H, 606, 2108, 2163, 3788
5-methyl- (*N*-oxide)	-	2418
2-methyl-4-phenyl-	53	2329
4-methyl-2-phenyl-	25	H, 3790
4-methyl-6-phenyl-	pic.203-204	H, 301, 3789
5-methyl-2-phenyl-	69-70; pic.154	2241, 3788
5-methyl-4-phenyl-	31-32; pic.139	2108, 2163, 2241
5-methyl-4-phenyl-2-propyl-	173/12mm.	2329
4-methyl-6-phenyl-(*N*-oxide)	138	428
5-methyl-2,4,6-triphenyl-	-	H, 3792
2-phenyl-	100/5mm.	H, 606, 3793
4-phenyl-	66; pic.162 to 165	H, 301, 606, 1610, 2163
2-phenyl-4-propyl-	153-155/10mm.	3790

Pyrimidine	M.P.(°C)	References
4-phenyl-2-propyl-	58	2329
4-phenyl-6-propyl-	175-177/20mm.; pic.164-166	301
4-propyl-	94/50mm.; pic.88-89	301
pyrimidine- (*N*-oxide)	95-96	618, 2415
2,4,5-trimethyl-	107/12mm.; pic.198	3794
2,4,6-triphenyl-	185-186	3786
4-vinyl-	-	3795

TABLE XXI. Aminopyrimidines (*H* 505)

5-acetamido-	EtI 165	H, 3625
4-acetamido-6-amino-	325	3286, 3796
5-acetamido-4-amino-6-azido-	-	2466
4-acetamido-2-amino-6-methyl-	261-262	2791
4-acetamido-6-*N*-amylacetamido-	81	3286
4-acetamido-6-anilino-	231	3286
4-acetamido-6-benzylamino-	181	3286
5-acetamido-4,6-diazido-	-	2466
4-acetamido-2,6-dimethyl-	-	H, 2529, 3869
5-acetamido-2,4-dimethyl-	114-115	887
4-acetamido-6-ethyl-2-methyl-	163-164	2539
4-acetamido-6-isoamylamino-	169	3286
4-acetamido-2-isopropyl-6-methyl-	122-124	2451
2-acetamido-4-methyl-	149 to 154	2955, 3797

TABLE XXI

Pyrimidine	M.P.(°C)	References
4-acetamido-2-methyl-	141-142	245, 2541
4-acetamido-6-methyl-	124-125	2451
4-acetamido-2-methyl-6-phenyl-	169-170	2539, 2540
4-acetamido-2-methyl-6-propyl-	132-133	2539
4-acetamido-6-piperidino-	203	3286
5-acetamido-2,4,6-trimethyl-	171-172	887
5-acetimidoyl-	180-181	2806
2-allylamino-	pic.156-157	2334, 2986
4-allylamino-2-amino-	117-119	2555
4-allylamino-6-amino-	146	623
4-allylamino-2-methyl-	109/1.6mm.	2555
1-allyl-1,2-dihydro-2-imino-	HBr 214; pic.149-150	2986
1-allyl-1,2-dihydro-2-propylimino-	HBr 115; pic.93	3416
5-allyl-4-hydrazino-6-methyl-2-propyl-	dipic.143	2224, 3798
2-allylimino-1,2-dihydro-1-propyl-	HI 92; pic.104	3416
4-*N*-allylthioureido-6-amino-	211	3286
2-amidino	125-126 (?)	889, 2806
5-amidino-	HCl 213	2806
2-amino-	-	H, 1029, 2171, 2765, 3789, 3799-3801
4-amino-	-	H, 2607
5-amino-	-	H, 2158, 2323
2-*N*-aminoamidino-	109-110	3802
4-amino-5-α-aminoethyl-2-methyl-	103	1595
4-amino-5-β-aminoethyl-2-methyl-	101	1595

Pyrimidine	M.P.(^{o}C)	References
4-amino-5-aminomethyl-	>290	H, 2190, 3803
4-amino-5-aminomethyl-2-amyl-	HCl 188-189	3804
4-amino-5-aminomethyl-2-anilino-	HCl 220-223	3805
4-amino-5-aminomethyl-2-benzyl-	-	H, 3806
4-amino-5-aminomethyl-2-benzylamino-	HCl 196-198	3805
4-amino-5-aminomethyl-2-butyl-	HCl 212 to 223	H, 3804, 3806
4-amino-5-aminomethyl-2-cyclohexylamino-	HCl 201-203	3805
4-amino-5-aminomethyl-2,6-dimethyl-	HCl 298; pic.218	H, 886, 3807
4-amino-5-aminomethyl-2-ethyl-	HCl 229 (?)	H, 3806
4-amino-5-aminomethyl-2-hexylamino-	HCl 202-204	3805
4-amino-5-aminomethyl-2-isopropyl-	HCl 257-260	3804
4-amino-5-aminomethyl-2-methyl-	HCl 260-265	H, 3808
4-amino-5-aminomethyl-2-phenyl-	-	H, 3806
4-amino-5-aminomethyl-2-piperidino-	HCl 232-235	3805
4-amino-5-aminomethyl-2-propyl-	HCl 210 to 260	2190, 3804, 3806
2-amino-4-amyl-	-	H, 3790
4-amino-6-amylamino-	115	623, 3809
4-amino-5-anilino-	216-217	2356
2-amino-4-anilino-6-methyl-	-	H, 2623, 2629
4-amino-6-anilino-2-methyl-	HCl 101 (?)	H, 916
2-amino-4-azido-6-phenyl-5-prop-2'-ynyl-	191	3810, 3811

Pyrimidine	M.P.(^{o}C)	References
4-amino-6-benzamido-	217	3286
4-amino-5-benzamido-6-methyl-	195-196	3186
4-amino-6-benzylamino-	211; HCl 165	623, 3809
5-amino-4-benzylamino-	136-137	2673, 2675
5-amino-4-α-benzylhydrazino-	136	2673
5-amino-2,4-bisbenzylamino-	128-130	2463
5-amino-4,6-bisbenzylamino-	182-184	2463
5-amino-2,4-bisbenzylamino-6-methyl-	sul.210-212	2463, 3812
5-amino-2,4-bisbutylamino-	sul.208-210	2463
5-amino-4,6-bisbutylamino-	156	2463
5-amino-2,4-bisbutylamino-6-methyl-	sul.195	2463
5-amino-2,4-biscyclohexylamino-	sul.235	2463
5-amino-4,6-biscyclohexylamino-	296-297 or 300	784, 2463
5-amino-2,4-biscyclohexylamino-6-methyl-	sul.347	2463
2-amino-4,6-bisdimethylamino-	200	2627
4-amino-2,6-bisdimethylamino-	116-117	2837
5-amino-2,4-bisethylamino-	sul.210 to 217	2461, 2463
5-amino-4,6-bisethylamino-	210-211	2463
5-amino-2,4-bisethylamino-6-methyl-	sul.185	2461, 2463
5-amino-2,4-bisisobutylamino-	sul.211	2463
5-amino-4,6-bisisobutylamino-	189	2463

Pyrimidine	M.P.(^oC)	References
5-amino-2,4-bisisobutylamino-6-methyl-	sul.203-205	2463, 3812
5-amino-2,4-bisisopropylamino-	sul.208-209	2463
5-amino-4,6-bisisopropylamino-	236	2463
5-amino-2,4-bisisopropylamino-6-methyl-	sul.189-190	2463
5-amino-4,6-bismethylamino-	-	H, 3813
5-amino-2,4-bismethylamino-	sul.220	2462
5-amino-4,6-bis-α-methylhydrazino-	180	2673
5-amino-2,4-bispropylamino-	sul.210-212	H, 2463
5-amino-4,6-bispropylamino-	208-209	2463
4-amino-6-t-butyl-	171	2574
5-amino-4-t-butyl-	117-118	2574
4-amino-6-butylamino-	118	623
4-amino-2-butyl-5,6-dimethyl-	123-124	2349
4-amino-2-s-butyl-5,6-dimethyl-	109; pic.170	3814
2-amino-4-butyl-6-methyl-	77-79	3815
4-amino-6-*N*-butyl-(thioureido)-	231	3286
4-amino-5-cyclohexenyl-	156-158	2531, 2898
4-amino-6-cyclohexylamino-	203	623, 3809
4-amino-6-cyclohexylamino-2-methyl-	190	916
4-amino-6-cyclopropyl-	151-153	3816
5-amino-2,4-dianilino-	164-166	H, 2463

TABLE XXI

Pyrimidine	M.P.(^{o}C)	References
5-amino-4,6-dianilino-	227-228	2463
5-amino-2,4-dianilino-6-methyl-	sul.>300	2463
5-amino-2,4-diaziridino-6-methyl-	impure	3817
4-amino-2,6-dibenzyl-5-phenyl-	-	H, 2895
4-amino-2,5-diethyl-	-	2270
4-amino-6-diethylamino-	132	623
4-amino-6-diethylamino-2-methyl-	127	916
5-amino-4-diethylamino-6-methylamino-	117-120	2460
4-amino-2,6-diethyl-5-methyl-	193	H, 2895
4-amino-2,6-dihydrazino-	143	3818
2-amino-1,4-dihydro-4-imino-1-methyl-(or tautomer)	HCl 274; pic.255	2288, 2624
4-amino-1,6-dihydro-6-imino-1-methyl-(or tautomer)	HCl 268-269	753
2-amino-4,5-dimethyl-	216-217	H, 2298
2-amino-4,6-dimethyl-	-	H, 427, 2861
2-amino-2,5-dimethyl-	pic.225	H, 1479, 3819, 3820
4-amino-2,6-dimethyl-	-	H, 2334, 3259, 3821
5-amino-2,4-dimethyl-	107-108	887
2-amino-4-dimethylamino-	155-156	2624, 2627
4-amino-2-dimethylamino-	151 to 155	2624, 2837
4-amino-6-dimethylamino-	202 to 209	623, 3822
5-amino-4-dimethylamino-	103-109	3823
5-amino-2-dimethylamino-4-ethylamino-	sul.195-198	2461
4-amino-6-dimethylamino-2-methyl-	185	914

Pyrimidine	M.P.(^oC)	References
5-amino-2-dimethylamino-4-methylamino-	194-196	2927
5-amino-4-dimethylamino-6-methylamino-	151-154	2433, 2454
4-amino-6-dimethylamino-2-phenyl-	-	3824
2-amino-4,6-diphenyl-	HI 212	H, 2627
4-amino-6-dipropylamino-	99	623
4-amino-6-dipropylamino-2-methyl-	130	916
2-amino-4-ethyl-	136 to 140	H, 2298, 3790
2-amino-5-ethyl-	140 to 142	H, 2627, 3801, 3825, 3826
4-amino-2-ethyl-	184	2554, 3827
4-amino-5-ethylamino-	213	2356
4-amino-6-ethylamino-	193	623
5-amino-4-ethylamino-	195-198	717
4-amino-6-ethylamino-2-methyl-	185	916
4-amino-2-ethyl-5,6-dimethyl-	197-198	H, 2539
4-amino-5-ethyl-2,6-dipropyl-	109-111	H, 2791
4-amino-2-ethyl-5-methyl-	175-177	2270
4-amino-6-ethyl-2-methyl-	148-149	2539
5-amino-2-ethyl-4-methyl-	71; 133/12mm.	887
5-β-aminoethyl-4-methyl-	165/2mm.	3828
4-amino-6-ethylureido-	chars	3286
4-amino-6-formamido-	276	H, 3286
4-amino-5-formamido-1,6-dihydro-6-imino-1-(γ-methyl-β-butenyl)-	-	3829
4-amino-5-formamidomethyl-	172; pic.205 or 265	2190, 3830
4-amino-5-formamido-2-methylamino-	202; HCO_2H 195	2454

TABLE XXI

Pyrimidine	M.P.(^{o}C)	References
4-amino-2-β-formamido-α-methylpropyl-5,6-dimethyl-	209-210	3814
4-amino-6-hexylamino-	115	623, 3809
4-amino-2-hexylamino-6-methyl-	95	2573
5-amino-4-hydrazino-	160; HCl 234	2673
2-amino-4-hydrazino-6-methyl-	-	H, 813
4-amino-2-hydrazino-6-methyl-	184-186	813, 3831
5-amino-4-hydrazino-6-methyl-	215	2227
4-amino-6-isoamylamino-	145	623, 3809
2-amino-4-isobutyl-	-	H, 3790
4-amino-6-isobutylamino-	125	623, 3809
4-amino-6-isobutylamino-2-methyl-	93	916
2-amino-4-isopropyl-	117-119	3832-3834
2-amino-5-isopropyl-	73	3826
4-amino-5-isopropyl-	141	2531, 2898
4-amino-6-isopropylamino-	177	623
4-amino-6-isopropylamino-2-methyl-	110	916
4-amino-5-isopropylaminomethyl-2-methyl-	HCl 285	3835
4-amino-5-isopropylidene-aminomethyl-2-methyl-	152	3835
4-amino-2-isopropyl-6-methyl-	212	2451
2-amino-4-methyl-	-	H, 577, 1029, 2299, 2861, 3790
2-amino-5-methyl-	192-193	3793, 3826
4-amino-2-methyl-	pic.167	H, 2451, 2541, 3836

Pyrimidine	M.P.(^{o}C)	References
4-amino-6-methyl-	193 to 197	H, 2574, 2631, 3241
5-amino-2-methyl-	159	887
5-amino-4-methyl-	-	H, 2574
2-amino-4-methylamino-	163	1372, 2624
4-amino-2-methylamino-	131-133	2288
4-amino-5-methylamino-	197	H, 2356
4-amino-6-methylamino-	205 to 211; HCl 214	623, 753
5-amino-4-methylamino-	211-213	H, 3823
4-amino-6-methylamino-2-phenyl-	-	3824
5-amino-2-methyl-4,6-bismethylamino-	146; HI 217	2454
5-amino-4-methyl-2,6-bismethylamino-	sul.218-220	2463
5-amino-4-methyl-2,6-bispropylamino-	sul.195	2463
5-aminomethyl-4-dimethylamino-2-methyl-	HCl 260	H, 2681
4-amino-5-methyl-2,6-diphenyl-	168-170	H, 3837
5-amino-4-α-methylhydrazino-	136	2673
5-aminomethyl-2-methyl-	81	3838, cf.2681
4-amino-2-methyl-6-methylamino-	250	H, 916
5-amino-4-methyl-6-methylamino-	165-167	2563, 2675
5-aminomethyl-2-methyl-4,6-bismethylamino-	HCl 250	2681
2-amino-4-methyl-6-phenyl-	175	H, 3839
2-amino-5-methyl-4-phenyl-	186; pic.225	2329
4-amino-2-methyl-6-phenyl-	-	H, 2539

TABLE XXI

Pyrimidine	M.P.(^{o}C)	References
4-amino-2-methyl-6-piperidino-	-	H, 916
4-amino-6-methyl-2-piperidino-	137-138	524
4-amino-2-methyl-5-piperidinomethyl-	80-84	3840
4-amino-2-methyl-6-propyl-	134-135	2539
4-amino-2-methyl-6-propylamino-	122	916
4-amino-2-methyl-5-thioformamidomethyl-	186	H, 3803
2-amino-4-phenyl-	pic.242	H, 1029, 2299, 2329, 3790
2-amino-5-phenyl-	161-163	H, 3826
4-amino-5-phenyl-	157	H, 2366, 3841
4-amino-6-phenyl-	226-228	301, 2574, 2595
5-amino-4-phenyl-	113-115	2574
4-amino-6-piperidino-	185-186	623, 3809
2-amino-4-propyl-	123	H, 3790
4-amino-6-propylamino-	140	623
4-amino-5-thioformamidomethyl-	-	H, 3803
5-amino-2,4,6-trimethyl-	116-118(not 175)	887, 2180, cf. 2179
2-amylamino-	72-74/0.4mm.; pic.117	2619, 2627
4-amylamino-	115/0.4mm.	H, 2619
2-amylamino-4,6-dimethyl-	145/8mm.; pic.93	2573
4-amylamino-2,6-dimethyl-	158/10mm.; HCl 155	2573
4-amyl-2-anilino-	-	3790
2-amylguanidino-4,6-dimethyl-	208-209	2234
2-amyl-4-hydrazino-6-methyl-	68-72	2224, 3798

Pyrimidine	M.P.(^{o}C)	References
2-anilino-	116	H, 3789
4-anilino-2,6-dimethyl-	102	H, 2153
2-anilino-4-ethyl-	55-56	3790
2-anilino-4-isobutyl-	49-50	3790
2-anilino-4-methyl-	-	H, 3790
2-anilino-4-phenyl-	137-138	3790
4-anilino-5-phenyl-	112	3841
2-anilino-4-propyl-	177/7mm.; 55	3790
2-azido- (or tautomer)	123-125	2619, 3340, 3342, 3842
4-azido- (or tautomer)	77-79	2466
2-azido-4,6-dimethyl- (or tautomer)	153	3340, 3842, 3843
2-benzamido-	141	H, 3844
4-*N*-benzoylthioureido-2,6-dimethyl-	155	3351
2-benzylamidino-	80	2806
4-benzylamidino-	72	2806
2-benzylamino-	pic.157	H, 2626, 2986
4-benzylamino-	98-100 or 105-107	2674, 3185
2-benzylamino-4,6-dimethyl-	107 to 111	H, 2148, 2153
4-benzylamino-2,6-dimethyl-	78-81; MeI 192	2153
4-benzylamino-5-methyl-	105-107	2674
1-benzyl-2-benzylimino-1,2-dihydro-	HCl 164-165	2626
4-benzyl-2,6-dihydrazino-	154-155	3845, 3846
1-benzyl-1,2-dihydro-2-imino-	HCl 194; HBr 154-159; pic.165	2633, 2986
1-benzyl-1,2-dihydro-2-isopropylimino-	HCl 168-169	2626
1-benzyl-1,2-dihydro-2-methylimino-	HCl 208-209; pic.134-135	2626

TABLE XXI

Pyrimidine	M.P.(^{o}C)	References
2-benzylideneamino-	76	3847
2-benzylidenehydrazino-	177	2985
2-benzylidenehydrazino-4,6-dimethyl-	160	3867
2-benzylimino-1,2-dihydro-1-isopropyl-	HI 211; HCl 138	2626
2-benzylimino-1,2-dihydro-1-methyl-	HI 185; pic.168	2626
4-benzylimino-1,4(or 3,4)-dihydro-1,2,6(or 2,3,6)-trimethyl-	HI 190-192	2153
4,6-bisallylamino-	163	623
4,6-bisbenzylamino-	234 to 237	623, 2153
4,6-bisbenzylamino-2-methyl-	192	2747
4,6-bisbutylamino-	154	623
2,4-bisdiethylamino-6-hydrazino-	-	2708
2,4-bisdimethylamino-	98/0.9mm.; 47-49	2624
4,6-bisdimethylamino-	108	827
2,4-bisdimethylamino-6-methyl-	132/11mm.; pic.185	487
4,6-bisdimethylamino-2-methyl-	88-89	487
4,6-bisethylamino-	187	623
4,6-bismethylamino-5-methyliminomethyl-	162-164	2594
2-butylamino-	120/24mm.; pic.130	H, 2152, 2697, 2986
2-t-butylamino-	92/12mm.; 71	2334, 2694
4-butylamino-	63; pic.128	2697
4-butylamino-6-butylimino-1,4-dihydro-1-methyl- (or tautomer)	HI 121	623
2-butylamino-4,6-dimethyl-	82/0.02mm.; pic.112	2573

Pyrimidine	M.P.(°C)	References
2-s-butylamino-4,6-dimethyl-	125/12mm.; pic.145	2694
2-t-butylamino-4,6-dimethyl-	104/12mm.; 53	2694
4-butylamino-2,6-dimethyl-	108 or 158/1mm.; HCl 172	2153, 2573
4-s-butylamino-2,6-dimethyl-	66; HCl 224	2694
4-t-butylamino-2,6-dimethyl-	157	2694
1-butyl-1,2-dihydro-2-imino-	HI 115 or 153; pic.165-167; HCl 151	2627, 2986
1-butyl-1,2-dihydro-2-methylimino-	pic.139	3416
2-butyl-4-hydrazino-6-methyl-	70-72	2224, 3798
2-t-butyl-4-hydrazino-6-methyl-	73	2224
4-butyl-6-hydrazino-2-methyl-	83-85	2224
4-butyl-6-hydrazino-2-propyl-	142/0.4mm.	2224
2-butylimino-1,2-dihydro-1-methyl-	HI 134; pic.87	3416
2-cyanoamino-	261	3793
4-cyanoamino-2,6-dimethyl-	251-252	2791
4-cyanoamino-6-methyl-2-propyl-	196-198	2791
2-cyclohexylamino-	92-94	2619
4-cyclohexylamino-2,6-dimethyl-	135/0.5mm.	H, 2153
4-cyclohexyl-6-hydrazino-2-propyl-	-	3798
4-cyclohexylimino-1,4(or 3,4)-dihydro-1,2,6(or 2,3,6)-trimethyl-	-	2153
4,6-diacetamido-	276	3286

TABLE XXI

Pyrimidine	M.P.(^{o}C)	References
2,4-diamino-	145-146	H, 2203, 2624, 2745
2,5-diamino-	206	H, 3848
4,5-diamino-	205; HCl 244,	H, 2686, 2835, 3189
4,6-diamino-	270-271	H, 2296, 3191
2,4-diamino-5-aminomethyl-	-	H, 2169, 3803
4,6-diamino-5-aminomethyl-2-methyl-	201-202	2681
2,4-diamino-5-amyl-6-methyl-	142-144	3849
2,4-diamino-5-anilino-	200-202	2567, 3850
2,4-diamino-6-anilino-	178 to 184	H, 1272, 2690, 2739
2,5-diamino-4-anilino-	sul.225	H, 2472
4,5-diamino-6-anilino-	171-172	H, 2933
2,5-diamino-4-anilino-6-methyl-	212	H, 2472
4,6-diamino-5-benzamido-	286	3186
2,4-diamino-5-benzyl-	196	H, 3851
2,4-diamino-6-benzylamino-	131-134	3852
4,5-diamino-6-benzylamino-	153	H, 783
4,6-diamino-5-*N*-benzylformamido-	273	3853
4,6-diamino-5-benzylideneamino-2-phenyl-	211-212	3332
2,4-diamino-5-benzyl-6-methyl-	188-190	H, 2343
2,4-diamino-6-benzyl-5-phenyl-	218	2283
2,5-diamino-4-butylamino-	sul.204-206	2462
4,5-diamino-6-butylamino-	143	2365

Pyrimidine	M.P.(^{o}C)	References
2,4-diamino-5-butyl-6-methyl-	150-152	3849
2,4-diamino-5-cyclohex-1'-en-1'-yl-6-methyl-	249-251	2343
2,4-diamino-5-cyclohexylamino-	HCl 250-251	2462
2,4-diamino-6-cyclohexylamino-	HCl 222	3854
2,5-diamino-4-cyclohexylamino-	sul.177-178	2462
4,5-diamino-2-cyclohexylamino-	sul.186	2462
4,5-diamino-6-cyclohexylamino-	215; sul.200	783
4,5-diamino-6-diethylamino-	crude	3856
4,5-diamino-6-diethylamino-2-methyl-	117-118	971
2,4-diamino-3,6-dihydro-6-imino-3-methyl-(or tautomer	HI 310	2288
4,5-diamino-1,2-dihydro-2-imino-1-methyl-(or tautomer)	HI 264	2454
4,5-diamino-3,6-dihydro-6-imino-1-methyl-(or tautomer)	HI 260	2454
4,5-diamino-2,6-dimethyl-	252-254	H, 874
2,4-diamino-5-dimethylamino-	HCl 242	2462
2,5-diamino-4-dimethylamino-	sul.198-199	2462
4,5-diamino-2-dimethylamino-	sul.216	H, 2462
4,5-diamino-6-dimethylamino-	sul.165	H, 783, 3822, 3855
4,5-diamino-6-dipropylamino-	108-110	3856

TABLE XXI

Pyrimidine	M.P.(^{o}C)	References
4,5-diamino-2-ethylamino-	178	2461
2,4-diamino-6-ethyl-5-methyl-	168-170	2343
2,4-diamino-6-ethyl-5-phenyl-	241 to 245	H, 2343, 3857, 3858
4,5-diamino-6-hydrazino-	HCl >200	2466
2,4-diamino-5-isoamyl-6-methyl-	163-164	3849
4,5-diamino-2-methyl-	-	H, 2835
4,5-diamino-6-methyl-	-	H, 874, 2835
2,4-diamino-5-methylamino-	HCl 226	2462
2,5-diamino-4-methylamino-	sul.213-215	2462
4,5-diamino-2-methylamino-	sul.202-204; pic.216	361, 2454, 2462, 3859
4,5-diamino-6-methylamino-	187-189	H, 753
4,6-diamino-5-*N*-γ-methyl-β-butenylformamido-	285	3829, 3853
4,6-diamino-5-*N*-methylformamido-	285	3853
4,5-diamino-6-methyl-2-methylamino-	HCl 250	874
2,4-diamino-5-methyl-6-phenyl-	-	H, 3860
2,4-diamino-6-methyl-5-phenyl-	249 to 256	H, 2343, 3857
2,4-diamino-6-phenyl-	164-165	H, 2343
4,5-diamino-2-phenyl-	146	3861
4,6-diamino-2-phenyl-	197	H, 2297
4,6-diazido-	106-107	2754
2,4-diazido-5-azidomethyl-	liquid	2596
2,4-diazido-6-methyl-	127-128	H, 3862

Pyrimidine	M.P.(°C)	References
2,4-diaziridino-	87-88	3817
4,6-diaziridino-	72	2728
2,4-diaziridino-5-methyl-	108; HCl 208	2728
2,4-diaziridino-6-methyl-	106-107	3817
2-diethylamino-4,6-dimethyl-	97/12mm.; 41; pic.126	2334, 2694
4-diethylamino-2,6-dimethyl-	118/12mm.; HCl 190	H, 2694
2,4-diethyl-6-hydrazino-	86-87	2224, 3798
4,6-dihydrazino-	223-224	2757
2,4-dihydrazino-6-methyl-	215-216; HCl 200	H, 813, 2452, 3863
4,6-dihydrazino-5-*N*-methylacetamido-	240-245	2752
2,4-dihydrazino-5-methyl-6-phenyl-	203-205	3845, 3846
2,4-dihydrazino-6-phenyl-	164-165	3845, 3846
1,2-dihydro-2-imino-1,4-dimethyl-	HI 253; HCl 227	2602, 2861
1,2-dihydro-2-imino-1-methyl-	HCl 265	H, 2151, 2627
1,4-dihydro-4-imino-1-methyl-	pic.175	H, 2781
1,2-dihydro-2-imino-1-methyl-4-methylamino-	HI 217; HCl 301	2624
1,4-dihydro-4-imino-1-methyl-2-methylamino-	HI 228-231; HCl 289	2288
1,2-dihydro-2-imino-1-propyl-	HBr 152; pic.124	2986
1,2-dihydro-2-imino-1-prop-2'-ynyl-	HBr 195-196	3415
1,2-dihydro-2-imino-1,4,6-trimethyl-	HI 275; HCl 244; pic.169	2627
1,2-dihydro-1-methyl-2-methylimino-	HI 216-218; pic.147	2627

TABLE XXI

Pyrimidine	M.P.(^{o}C)	References
2-dimethylamino-1,4-dihydro-4-imino-1-methyl-	HI 281; HCl 287	2624
4-dimethylamino-1,2-dihydro-2-imino-1-methyl-	HI 286; HCl 298; pic.187	2624, 2627
2-dimethylamino-1,6-dihydro-1-methyl-6-methylimino-	HI 146-148; HCl 197; pic.167-169	2624
4-dimethylamino-1,2-dihydro-1-methyl-2-methylimino-	HI 221; HCl 267	2624
4-dimethylamino-1,6-dihydro-1-methyl-6-methylimino-	HI 268	2776
4-dimethylamino-2,6-dimethyl-	130/25mm.; pic.177	H, 2334
2-dimethylamino-5-ethynyl-4-methyl-	80-84/0.1mm.	458
4-dimethylamino-2-methyl-	103-105/23mm.; pic.211 or 217	2451, 2698
4-dimethylamino-6-methyl-	120/19mm.; 47	487
2-dimethylamino-4-methylamino-	77-78	2624
4-dimethylamino-2-methylamino-	98	2627
4-dimethylamino-6-methylamino-	138-141 or 146-148	827, 2433
2-dimethylamino-methyleneamino-	HCl 212	3864
2-dimethylamino-4-vinyl-	-	H, 3795
4-β-dimethylaminovinyl-	-	2426
2,4-dimethyl-6-methylamino-	126	2334
4,6-dimethyl-2-methylamino-	99-100; pic.201	H, 2627
4,6-dimethyl-2-β-phenylhydrazino-	166-167	2148

Pyrimidine	M.P.(^{o}C)	References
4,6-dimethyl-2-piperidino-	60-63; HCl 155	H, 2699
2,4-dimethyl-6-propylamino-	154/10mm.; HCl 205	2573
4,6-dimethyl-2-propylamino-	130/11mm.; pic.125	2573
2,4-dimethyl-6-thioureido-	223-224	3351
2-dipropylamino-4,6-dimethyl-	119/12mm.; HCl 159; pic.84	2694
4-dipropylamino-2,6-dimethyl-	136/12mm.; HCl 164	2694
2-*C*-ethoxy-*C*-iminomethyl-	HCl crude	2806
4-*C*-ethoxy-*C*-iminomethyl-	29-30	2806, 3865
5-*C*-ethoxy-*C*-iminomethyl-	87	2806, 3865
2-ethylamino-	58-60; HI 119; pic.160 to 168	2152, 2627, 2986
2-ethylamino-4,6-dimethyl-	85/8mm.; pic.156	2573
4-ethylamino-2,6-dimethyl-	136/10mm.; HCl 198	2573
1-ethyl-1,2-dihydro-2-imino-	HI 189; HCl 233; pic.178-179	2627, 2986
5-ethyl-1,2-dihydro-2-imino-1-methyl-	HI 206; HCl 255; pic.148	2627
1-ethyl-1,2-dihydro-2-methylimino-	HI 156-157	2627, 3416
2-ethyl-4-hydrazino-6-methyl-	150	2224, 3798, 3866
4-ethyl-6-hydrazino-2-methyl-	127-128	2224, 3798
5-ethyl-4-hydrazino-6-methyl-2-propyl-	62-65	2224, 3798
2-ethyl-4-hydrazino-6-propyl-	188/20mm.	2224, 3798

Pyrimidine	M.P.(^{o}C)	References
4-ethyl-6-hydrazino-2-propyl-	186/25 mm.	2224, 3798
2-ethylimino-1,2-dihydro-1-methyl-	HI 160-161	2627, 3416
5-ethyl-2-methylamino-	51; pic.168	2627
2-hexylamino-	144/12mm.; pic.101	2573
4-hexylamino-	61-63	3185
2-hexylamino-4,6-dimethyl-	100/0.02mm.; pic.102	2573
4-hexylamino-2,6-dimethyl-	134/0.02mm.; HCl 149	2573
2-hydrazino-	110 to 113	H, 813, 2619
2-hydrazino-4,6-dimethyl-	-	H, 427, 2188, 3867
4-hydrazino-2,6-dimethyl-	181 or 187	2554, 3259
4-hydrazino-5,6-dimethyl-	215-217	813
4-hydrazino-5,6-dimethyl-2-phenyl-	180; HCl 259	813
4-hydrazino-5,6-dimethyl-2-propyl-	78-80	2224, 3798
4-hydrazino-2,6-dipropyl-	48-50	2224, 3798
4-hydrazino-2-isopropyl-6-methyl-	86	2224, 3798
2-hydrazino-4-methyl-	89-91; HCl 194-198	H, 813
4-hydrazino-2-methyl-	122-124; HCl 214	813, 2224
4-hydrazino-5-methyl-	205-206	813
4-hydrazino-6-methyl-	140 to 144	H, 813, 2224
4-hydrazino-6-methyl-2,5-dipropyl-	109-111	2224
4-hydrazino-6-methyl-2-phenyl-	96	3798

Pyrimidine	M.P.(°C)	References
4-hydrazino-2-methyl-6-propyl-	100-102	2224, 3798
4-hydrazino-6-methyl-2-propyl-	93-95	2224
4-hydrazino-2-phenyl-	83-84; HCl 250	813
4-hydrazino-5-phenyl-	145	H, 3841
4-hydrazino-6-phenyl-	139-140	813
4-hydrazino-6-piperidino-	122-123	3868
4-hydrazino-2,5,6-trimethyl-	164-166	2224
2-hydroxyamino-4,6-dimethyl-	209-211	2148
4-hydroxyiminomethyl-	153-154	2301, 2426
5-hydroxyiminomethyl-	155-156; MeI 196-198	3242
4-hydroxyiminomethyl-2,6-dimethyl-	213	2301
4-hydroxyiminomethyl-6-methyl-	150	2301
4-hydroxyiminomethyl-2-phenyl-	138	2301
4-*N*-iodoacetamido-2,6-dimethyl-	156-158	3869
2-iodoamino-4,6-dimethyl-	crude	3869
2-isoamylamino-4,6-dimethyl-	146/10mm.; pic.110	2694
4-isoamylamino-2,6-dimethyl-	166/10mm.; HCl 190	2694
2-isobutylamino-4,6-dimethyl-	133/20mm.; 30; pic.140	2694
4-isobutylamino-2,6-dimethyl-	170/20mm.; HCl 207	2694
2-isopropylamino-	28; pic.174	2626
2-isopropylamino-4,6-dimethyl-	112/12mm.; pic.173	2694

TABLE XXI

Pyrimidine	M.P.(^{o}C)	References
4-isopropylamino-2,6-dimethyl-	100; HCl 249	2694
2-methylamino-	-	2151, 2986, 3421
4-methylamino-	pic.162-163	2781
4-methylamino-5-phenyl-	105	3841
4-methyl-2-methylamino-	55-56; pic.183-184	H, 2602
4-methyl-6-piperidino-	178/2mm.	H, 1313
2-phenylamidino-	117-118	2806
4-phenylamidino-	126	2806
5-phenylamidino-	172	2806
2-propylamino-	115/20mm.; 19.5; pic.151-152	2627, 2986
2-prop-2'-ynylamino-	108-109	3415
2,4,5,6-tetra-amino-	sul.>360	H, 979, 2933, 3332
2,4,5-triamino-	-	H, 3870
2,4,6-triamino-	pic.$\nless$300	H, 2610
4,5,6-triamino-	-	H, 1245, 2296, 2933, 3244, 3871
2,4,5-triamino-6-anilino-	232-234	2260
2,4,6-triamino-5-anilino-	250-252	2260
2,4,5-triamino-6-benzylamino-	185-186	3852
2,4,5-triamino-5-benzylideneamino-	159-161 or 163-164	3332, 3872
2,4,6-triamino-5-butyl-	205-206	H, 3849
2,4,6-triamino-5-isoamyl-	144-145	3849
2,4,5-triamino-6-methyl-	242-244	H, 1275
4,5,6-triamino-2-methyl-	-	H, 2933
2,4,6-triamino-5-phenyl-	-	H, 3873

Pyrimidine	M.P.(^{o}C)	References
4,5,6-triamino-2-phenyl-	-	H, 1169
4,5,6-triamino-2-piperidino-	crude	2473
2,4,6-triamino-5-propyl-	185-186	3849
2,4,6-triaziridino-	127-129	2727, 2730
4,5,6-trismethylamino-	149; HI 192	2454

TABLE XXII. Carboxypyrimidines (*H* 516)

4-acetyl-	67	2806
5-acetyl-2-carbamoylmethyl-4-methyl-	125-127	3875
4-azidocarbonyl-	80-90	3346
5-benzoyl-4,6-dicarboxy-2-phenyl-	184	3786
2-β-butoxycarbonylvinyl-	-	3009
2-carbamoyl-	167	889, 2806
4-carbamoyl-	191-192 or 197	2806, 3346
5-carbamoyl-	215	H, 2323, 2680, 2806
4-carbamoyl-2,6-dimethyl-	-	H, 1147, 1313, 2184
5-carbamoyl-2,4-dimethyl-	188-189 or 192	887, 3007, 3874
2-carbamoyl-4,6-diphenyl-	-	3354
5-carbamoyl-2-ethyl-4-methyl-	167-169	887
5-carbamoyl-2-methyl-	218-219	2681
2-carbamoylmethyl-5-carboxy-4-methyl-	222-224	3875

Pyrimidine	M.P.(^{o}C)	References
2-carbamoylmethyl-4-ethoxycarbonyl-6-methyl-	202	3875
2-carbamoylmethyl-5-ethoxycarbonyl-4-methyl-	143	3875
5-carbamoylmethyl-2-methyl-	186	3838
5-carbamoylmethyl-	170-173	2577, 3876
2-carbamoylmethyl-4,5-dimethyl-	182	3794
2-carbamoylmethyl-4,6-dimethyl- (?)	-	3877
5-carbamoylmethyl-2-methyl-	183-185	3876
2-carboxy-	-	H, 2607, 3787
4-carboxy-	-	H, 606, 2156
5-carboxy-	268-270	H, 606, 2680, 2924
4-carboxy-2-carboxymethyl-6-methyl-	218-220	2873
4-carboxy-2,6-dimethyl-	192	H, 2184
5-carboxy-2,4-dimethyl-	188-189	3874
2-carboxy-4,6-diphenyl-	-	3354
2-carboxy-5-methyl-	-	H, 3787
4-carboxy-2-methyl-	204-206	3878
5-carboxymethyl-	168 or 172-173	2577, 3876
5-carboxy-2-methyl-	196	2681
2-carboxymethyl-4,5-dimethyl-	258	3794
5-carboxymethyl-2-methyl-	183-185	3876
5-carboxy-4-methyl-2-thiocyanato-	127-129	2783
5-carboxy-2,4,6-trimethyl-	205	887
5-chlorocarbonyl-	80-85/3mm.	2680

Pyrimidine	M.P.(^{o}C)	References
5-chlorocarbonyl-4-methyl-2-phenyl-	HCl 176-178	3879
2-cyano-	-	H, 2607, 2619, 2806, 3445
4-cyano-	31	2806
5-cyano-	85-86	2607, 2680, 2806
2-cyano-4,6-dimethyl-	-	H, 1313
4-cyano-2,6-dimethyl-	-	H, 1313
5-cyano-2,4-dimethyl-	51-53	3007
2-cyano-4,6-diphenyl-	195-196	3354
5-cyano-4-methoxycarbonyl-2-methyl-	74-76	2175
5-cyano-2-methyl-	72	2681
5-cyano-4-phenyl-	95-96	2367
4,5-dicarbamoyl-2-methyl-	234-236	2175
4,6-dicarboxy-	218; H_2O 211	431
4,5-dicyano-2-methyl-	36-38	2175
4,5-diethoxycarbonyl-2-methyl-	118/0.09mm.	2175
4,5-diethoxycarbonyl-2-α-oxocyclohexyl-	59-60	3880
4-diethoxymethyl-	65-67/0.001mm.	2301
4-diethoxymethyl-2,6-dimethyl-	50-59/0.001mm.	2301
4-diethoxymethyl-2-methyl-	104/7mm.	2301
4-diethoxymethyl-6-methyl-	61-65/0.001mm.	2301
4-diethoxymethyl-6-methyl-2-phenyl-	110-119/0.001mm.	2301
4-diethoxymethyl-2-phenyl-	123/0.002mm.	2301
2-diethylcarbamoyl-	35-36	2806
4-diethylcarbamoyl-	36-37	2806

Pyrimidine	M.P.(^{o}C)	References
5-diethylcarbamoyl-	108-113/0.2mm.	2680
4,6-diformyl-	bis Ph hydrazone 184	3881
4-diformylmethyl-	-	2426
2-ethoxycarbonyl-	64-65	2806
5-ethoxycarbonyl-	103/12mm.; 16 or 39	2323, 2680
5-ethoxycarbonyl-2,4-dimethyl-	113-116/12mm.	887, 3874
5-ethoxycarbonyl-2-ethyl-4-methyl-	118-120/10mm.	887
5-ethoxy-carbonylmethyl-	87/0.05mm.; 33-34	2577, 3876
5-ethoxycarbonyl-2-methyl-	70-71/0.45mm.	2654, 2681
5-ethoxycarbonyl-methyl-2-methyl-	104/3mm.	3838, 3876, 3882
5-ethoxycarbonyl-methyl-4-methyl-	132-134/10mm.	2576
5-ethoxycarbonyl-methyl-2-phenyl- (?)	-	2576
5-ethoxycarbonyl-4-phenyl-	159-161	2369
4-formyl-	42-44/0.7mm.; DNP 279; H_2O 93	2301, 3009
5-formyl-	38-41	2732
4-formyl-2,6-dimethyl-	oxime 213	2301
4-formyl-2-methyl-	H_2O 66	2301
4-formyl-6-methyl-	53	2301
4-formyl-6-methyl-2-phenyl-	oxime 166	2301
4-formyl-2-phenyl-	118	2301
5-hydrazinocarbonyl-	124	2323
4-hydrazinocarbonyl-2,6-dimethyl-	194-195	2184

Pyrimidine	M.P.(°C)	References
5-hydrazinocarbonyl-methyl-	141-142	2577
5-hydrazinocarbonyl-methyl-2-methyl-	158 or 160	3838, 3882
5-hydrazinocarbonyl-4-methyl-2-phenyl-	185-187	1128
2-methoxycarbonyl-	104-105	2607, 3787
4-methoxycarbonyl-	70-71	2806, 3009
5-methoxycarbonyl-	-	H, 2806
4-methoxycarbonyl-2,6-dimethyl-	-	2184
4-β-methoxycarbonyl-ethyl-	-	H, 3009
2-methoxycarbonyl-5-methyl-	-	H, 3787
5-methoxycarbonyl-2-methyl-	78	2681
4-β-methoxycarbonyl-vinyl-	86-88	3009, 3883
2-thiocarbamoyl-	225	2806
4-thiocarbamoyl-	236	2806
5-thiocarbamoyl-	170	2680, 2806
2-thiocyanato-	107-109	2607

TABLE XXIII. Halogenopyrimidines (*H* 518)

5-allyl-4-chloro-6-methyl-2-propyl-	145-146/30mm.	2224
5-allyl-2,4,6-trichloro-	96-98/3mm.; 39	2559
2-amyl-4-chloro-6-methyl-	134-138/25mm.	2224
5-amyl-2,4,6-trichloro-	95/0.25mm.	2559
2-benzyl-5-bromo-4.6-dichloro-	84-87	2575

Pyrimidine	M.P.(^{o}C)	References
2-benzyl-4,6-dichloro-	58-60	3884
4,6-bistribromomethyl-	125-126	431
2-bromo-	56-57	2604
5-bromo-	-	H, 791, 2323
5-bromo-4-butyl-	92-96/8mm.	2924
5-bromo-4-t-butyl-	104-106/16mm.	2574
5-bromo-4-t-butyl-6-chloro-	61-62	2574
5-bromo-2-chloro-	79	2573
5-bromo-2-chloro-4,6-dimethyl-	81	2519
2-bromo-4-chloro-6-methyl-	33-34	2605
5-bromo-4-chloro-6-methyl-	55	2574
5-bromo-4-chloro-6-phenyl-	83-84	2519, 2574, 2595
2-bromo-4-chloro-5-propyl-	130/18mm.	2606
5-bromo-2-dibromomethyl-	66-68	2575
5-bromo-2,4-dichloro-6-methyl-	114/0.3mm.	H, 2147, 2714
5-bromo-4,6-dichloro-2-phenyl-	118-119	2575
5-bromo-2,4-difluoro-6-methyl-	74/20mm.	2613
2-bromo-4,6-dimethyl-	71	2605
5-bromomethyl-	HBr 108-113	2579
5-bromo-4-methyl-	72-73/14mm.	2574
5-bromomethyl-4,6-dichloro-2-chloromethyl-	127-129	2238
5-bromomethyl-2-methyl-	HBr >200	2651
4-bromomethyl-2,5,6-trichloro-	56-57	2236

Pyrimidine	M.P.(^{o}C)	References
5-bromomethyl-2,4,6-trichloro-	134	H, 2236
5-bromo-4-phenyl-	89-90	2574, 2595
5-bromo-2-tribromomethyl-	-	H, 2806
4-t-butyl-6-chloro-	36-38	2574
2-butyl-4-chloro-5,6-dimethyl-	82/0.05mm.	2349
2-butyl-4,6-dichloro-	108/62mm.	2242
5-butyl-4,6-dichloro-	128/11mm.	2242, 2522
5-butyl-2,4,6-trichloro-	83/0.3mm.	2559, 3885
2-chloro-	-	H, 2182, 2604, 3886
4-chloro-2,6-bistrifluoromethyl-	44/15mm.	2193
4-chloro-2-chloromethyl-	95/10mm.	2192, 2819, 3887
4-chloro-2-chloromethyl-6-methyl-	29; HCl 126	3888, 3889
4-chloro-6-cyclopropyl-	98/12mm.	3816
4-chloro-2-cyclopropyl-6-methyl-	104-107/12mm.	3816
4-chloro-2,6-difluoro-	-	2157
2-chloro-4,6-dimethyl-	38-39	H, 2182, 2695, 3886
4-chloro-2,6-dimethyl-	-	H, 813, 2695
4-chloro-5,6-dimethyl-	51-52	H, 2349
4-chloro-5,6-dimethyl-2-phenyl-	125-127	813
4-chloro-5,6-dimethyl-2-propyl-	122-126/22mm.	2224
2-chloro-4,6-diphenyl-	115-116	2556
4-chloro-2,6-dipropyl-	122-125/18mm.	2224
4-chloro-2-ethyl-	75/18mm.; -5	2554
4-chloro-2-ethyl-6-methyl-	93/20mm.	2224

TABLE XXIII

Pyrimidine	M.P.(^{o}C)	References
4-chloro-5-ethyl-6-methyl-2-propyl-	130-132/22mm.	2224
4-chloro-2-isopropyl-6-methyl-	85/12mm. or 92/10mm.	1026, 2451
2-chloro-4-methyl-	-	H, 3886
4-chloro-2-methyl-	-	H, 2541
4-chloro-5-methyl-	25-27	606, 3412; cf.805
4-chloro-6-methyl-	60/22mm.	H, 606, 2574
2-chloromethyl-4,6-dimethyl-	33	3890
4-chloro-6-methyl-2,5-dipropyl-	131-137/10mm.	2224
4-chloromethyl-6-methyl-	pic.115	431
4-chloro-2-methyl-6-phenyl-	57-58	2556
4-chloro-5-methyl-6-phenyl-	54-55	2241
4-chloro-6-methyl-2-propyl-	95-100/30mm.	2224
4-chloro-2-methyl-6-trifluoromethyl-	-	2516
4-chloro-6-methyl-2-trifluoromethyl-	48/2mm.	2193
4-chloro-6-phenyl-	96-97	2574, 2595
2-chloro-4-trichloromethyl-	44-46; 112/12mm.	2419, 2597, 3891
4-chloro-2-trichloromethyl-	105-110/12mm.	2420
5-chloro-2,4,6-trifluoro-	115/at.	2618
4-chloro-2-trifluoromethyl-	45/4mm.	2193
5-α-chlorovinyl-4-methyl-2-phenyl-	145/0.1mm.	458
2,4-dibromo-	65-67	2784
4-dibromomethyl-5-methyl-	98-100; HCl 140	2156

Pyrimidine	M.P.(^{o}C)	References
2,4-dichloro-	-	H, 2182
4,6-dichloro-	-	H, 1258
2,4-dichloro-5-β-chloroethyl-6-methyl-	-	H, 3892
2,4-dichloro-5-chloromethyl-	62-63	2596
4,6-dichloro-2-chloromethyl-	45	2237
4,6-dichloro-2-cyclopropyl-	105-107/12mm.	3816
2,4-dichloro-6-ethyl-	-	H, 2226
4,6-dichloro-2-ethyl-	-	H, 2238
2,4-dichloro-5-fluoro-	38 to 43	2209, 2570, 3893
2,4-dichloro-6-fluoro-	-	2157
4,6-dichloro-2-fluoro-	-	2157
2,4-dichloro-5-iodo-	71-72	3894
2,4-dichloro-5-iodomethyl-	52-53	2596
2,4-dichloro-6-isopropyl-	83/5mm.	2226
4,6-dichloro-2-isopropyl-	47-48/0.6mm.	2238
4,6-dichloro-5-isopropyl-	112/71mm.	2242
4-dichloromethyl-	100/15mm.; 53	2156
2,4-dichloro-5-methyl-	-	H, 2703
4,6-dichloro-5-methyl-	-	H, 2240
2-dichloromethyl-4,6-dimethyl-	75	3890
4,6-dichloro-5-methyl-2-phenyl-	110	2241
4,6-dichloro-2-phenyl-	93-95	H, 3884
4,6-dichloro-5-phenyl-	109 or 112-114	2522, 3895
2,4-dichloro-6-propyl-	91/5mm.	2226
4,6-dichloro-2-propyl-	-	H, 2238

TABLE XXIII

Pyrimidine	M.P.(^{o}C)	References
4,6-dichloro- 2-trichloromethyl-	75-77	2421
2,4-dichloro- 5-trifluoromethyl-	126/650mm.	3896
4,6-dichloro- 2-trifluoromethyl-	38/1mm.	2193
2,4-difluoro-	-	2157
4,6-difluoro-	-	2157
2,4-difluoro-6-methyl-	135-137/at.	2613
4,6-diiodo-	107-108	2609
4,6-dimethyl- 2-trichloromethyl-	126	3890
4,6-dimethyl- 2-trifluoromethyl-	38	3897
2-iodo-	ca 29	2607
4-methyl-6-phenyl- 2-trifluoromethyl-	90	3897
5-methyl- 4-trichloromethyl-	57/0.001mm.; HCl 164-167	2156
4-phenyl- 6-tribromomethyl-	92; HCl 179	2156
2,4,5,6-tetrabromo-	167	H, 2600
2,4,5,6-tetrachloro-	-	H, 3591, 3898
2,4,5,6-tetrafluoro-	89/at.	2612, 2616, 2618
4-tribromomethyl-	83-84; HCl 153	2156
2,4,5-trichloro-	94-95/12mm.	H, 2571
2,4,6-trichloro-	95/10mm.	H, 2558, 2561
4,5,6-trichloro- 2-chloromethyl-	29	2236
2,4,5-trichloro- 6-ethyl-	88/5mm.	2226
2,4,6-trichloro- 5-fluoro-	81/15mm.	2600
2,4,6-trichloro- 5-hexyl-	107/0.3mm.	2559

Pyrimidine	M.P.(^{o}C)	References
2,4,6-trichloro-5-isoamyl-	89/0.3mm.	2559
2,4,5-trichloro-6-isopropyl-	89/5mm.	2226
2,4,6-trichloro-5-isopropyl-	69-71	2238
4,5,6-trichloro-2-isopropyl-	39	2238
4-trichloromethyl-	24; 96/10mm.; HCl 135	2156
2,4,5-trichloro-6-methyl-	115-120/12mm.; 20-21	H, 2236, 2714, 3899
2,4,6-trichloro-5-methyl-	136/25mm.	H, 487
4,5,6-trichloro-2-methyl-	68	2236
2,4,6-trichloro-5-1'-methylbutyl-	95/0.25mm.	2559
2,4,5-trichloro-6-phenyl-	87-88	2226
4,5,6-trichloro-2-phenyl-	125-126	2238
2,4,5-trichloro-6-propyl-	98/5mm.	2226
2,4,6-trichloro-5-propyl-	30-32	2238
4,5,6-trichloro-2-propyl-	65/0.75mm.	2238
2,4,6-trifluoro-	98-99/at.; 59-60/180mm.	2612-2615, 2618

TABLE XXIV. Nitropyrimidines (*H* 521)

Pyrimidine	M.P.(^{o}C)	References
2-benzyl-5-nitro-	108-109	H, 2688
2-methyl-5-nitro-	59-60	H, 2688
5-nitro-	57-58	2688

TABLE XXIV

Pyrimidine	M.P.(°C)	References
2,4,6-trimethyl-5-phenylazo-	crude	2180

TABLE XXV. Oxypyrimidines without *C*- or *N*-Alkyl Groups (*H* 521)

Pyrimidine	M.P.(°C)	References
2-acetoxy-1,4-dihydro-1-hydroxy-4-oxo-	175-177	2200
5-β-acetoxyethyl-2,4-dihydroxy-	204-205	2202
5-acetoxy-2-hydroxy-	179-180	2585
Alloxan	-	H, 2635, 3023, 3156
Alloxan/1,3-bis-hydroxymethyl-	-	3158
Alloxan/1-hydroxymethyl-	-	3158
5-amyloxy-2,4-dihydroxy-	-	3900
5-benzoyloxy-2,4-dihydroxy-	287	3901
5-benzoyloxy-2-hydroxy-	195	2585
2-benzyloxy-	115/0.4mm.	2626
5-benzyloxy-	pic.125-127	2161
5-benzyloxy-2,4-dihydroxy-	283 to 296	79, 2210
5-benzyloxy-4,6-dihydroxy-	223-224	135
4-benzyloxy-2,6-dimethyl- (*N*-oxide)	-	2415
4-benzyloxy-6-hydroxy-	196-197	3001
5-benzyloxy-2-hydroxy-	196	2585
5-benzyloxy-4-hydroxy-	90-93	2161, 2584
2,4-bistrimethylsiloxy-	31-33; 116/12mm.	3070, 3902

Pyrimidine	M.P.(^{o}C)	References
5-boro-2,4-dihydroxy-	330	2923
5-boro-2,4-dimethoxy-	115-117	2923
2-butoxy-	96-98/18mm.	2511
2-s-butoxy-	89-90/18mm.	2511
4-butoxy-	pic.85-86	2619
5-butoxy-4-butoxymethyl-2,6-dihydroxy-	106-107	112
5-s-butoxy-4-s-butoxymethyl-2,6-dihydroxy-	163-164	112
5-butoxy-2,4-dihydroxy-	-	3900
5-butoxymethyl-2,4-dihydroxy-	215	1168
2,4-diallyloxy-	85/0.5mm.; 122/7mm.	2555, 3479
2,4-dibenzyloxy-	235/2mm.; 41-44	2829
2,4-dibenzyloxy-5-boro-	crude	2923
2,4-dibutoxy-	crude	2903
2,4-di-t-butoxy-	74-78	2829
4,6-di-t-butoxy-	84	3001
2,4-diethoxy-5-ethoxymethyl-	90-100/0.001mm.	2596
2,5-dihydroxy-	$\nless$300	2585
4,5-dihydroxy-	300; HBr >270	H, 2584
4,6-dihydroxy-	>300	H, 1258, 2332, 2333, 3244, 3871, 3904, 3905
2,4-dihydroxy-5-αβ-dihydroxyethyl-	-	3906
2,4-dihydroxy-6-β-hydroxyethyl-	208-210	2674
2,4-dihydroxy-5-hydroxymethyl-	-	H, 2202, 3907
4,6-dihydroxy-2-hydroxymethyl-	207 or $\nless$305	2236, 2237

Pyrimidine	M.P.(^{o}C)	References
2,4-dihydroxy-5-isobutoxy-6-isobutoxymethyl-	131	112
2,4-dihydroxy-5-isopropoxy-	-	3900
2,4-dihydroxy-5-isopropoxy-6-isopropoxymethyl-	205-206	112
2,4-dihydroxy-5-methoxy-	341-345	2161, 2586
4,6-dihydroxy-2-methoxy-	193-195	H, 2578, 3908, 3909
4,6-dihydroxy-5-methoxy-	275-280	3910
2,4-dihydroxy-5-methoxy-6-methoxymethyl-	188	112
4,6-dihydroxy-5-methoxy-2-methoxymethyl-	290	3911
2,4-dihydroxy-5-methoxymethyl-	203	1168
4,6-dihydroxy-2-methoxymethyl-	210	3912, 3913
4,6-dihydroxy-5-phenoxy-	297-299	2240
2,4-dihydroxy-5-propoxy-	-	3900
2,4-dihydroxy-5-propoxy-6-propoxymethyl-	123-124	112
2,4-di-isopropoxy-	111/11mm.	2829
2,4-dimethoxy-	pic.129-130	H, 1223
4,6-dimethoxy-	85/16mm.; 30-31	827, 2682
2,4-diphenoxy-	112-113	H, 3914
2-ethoxy-	77-78/20mm.	2511, 2619
5-ethoxy-4,6-dihydroxy-2-hydroxymethyl-	209-210	2236

Pyrimidine	M.P.(^{o}C)	References
5-ethoxy-4-ethoxymethyl-2,6-dihydroxy-	168-170	H, 112
2-ethoxymethyl-4,6-dihydroxy-	219	3913
5-ethoxymethyl-2,4-dihydroxy-	212 or 218	1168, 2160
5-ethoxymethyl-4-hydroxy-	98	2190
4-ethoxy-2-trimethylsiloxy-	116/17mm.	3902
2-hydroxy-	HCl 210	H, 2154, 2604, 3800(cf. 2597)
4-hydroxy-	-	H, 3038
5-hydroxy-	211-212	H, 2161, 2323
4-hydroxy-2,6-bismethoxymethyl-	98-99	3913
2-hydroxy-4,6-dimethoxy-	193-194	2601
4-β-hydroxyethyl-	130-135/10mm.	2425
4-β-hydroxyethyl-2,6-dipropoxy-	128-130/0.4mm.	2425
4-hydroxy-2-hydroxymethyl-	196-197	2192, 2819
4-hydroxy-6-hydroxymethyl-	179-181; HCl 200	957
2-hydroxy-4-methoxy-	206-208	H, 3399
4-hydroxy-5-methoxy-	208 to 220	H, 2161, 3915, 3916
4-hydroxy-6-methoxy-	211-212	2997, 3054
4-hydroxy-5-methoxy-2-methoxymethyl-	112-113	3911
4-hydroxy-5-methoxy-6-methoxymethyl-	147-151	3911
4-hydroxy-6-methoxymethyl-	155	3913
2-isobutoxy-	103-104/17mm.	2630, 2697

TABLE XXV

Pyrimidine	M.P.(^{o}C)	References
2-isopropoxy-	90-91/18mm.	2511
2-methoxy-	69-70/22mm.	H, 2511, 2759
5-methoxy-	76/10mm.; pic.127	H, 792, 2161
2-propoxy-	92-93/22mm.; pic.92-93	2511, 2619
tetrahydroxy-	214	H, 135
2,4,6-triallyloxy-	-	2768
2,4,5-triethoxy-	33-34; HCl 104	2606
2,4,6-triethoxy-	-	H, 1114
2,4,5-trihydroxy-	-	H, 2210, 3907
2,4,6-trihydroxy-	245-247	H, 975, 1001, 1243, 3917-3921
2,4,6-trihydroxy-5-methoxy-	$\nless$360	2586
2,4,6-trimethoxy-	-	H, 2843
4,5,6-trimethoxy-	56-58	3910
2,4,6-tripropoxy-	178-180/20mm.	2575
Uracil	-	H, 2196, 2201, 3563, 3922, 3923

TABLE XXVI. Oxypyrimidines with *C*-Alkyl but without *N*-Substituents (*H* 522)

Pyrimidine	M.P.(^{o}C)	References
5-β-acetoxyethyl-4-methyl-	137-138/7mm.	2328
4-acetoxymethyl-6-methyl-	pic.136	431
5-allyl-2,4-bistrimethylsiloxy-	140-142/12mm.	3924

Pyrimidine	M.P.(^{o}C)	References
5-allyl-2,4-dihydroxy-	280-281	2194, 3407
5-allyl-4-hydroxy-2-methyl-	153-154	2194
5-allyl-4-hydroxy-6-methyl-2-propyl-	112-113	2224
5-allyl-4-hydroxy-2-phenyl-	195-196	2555
4-allyloxy-2-methyl-	47/0.5mm.	2194
4-allyloxy-2-phenyl-	105/0.1mm.	2555
5-amyl-2,4-dihydroxy-6-methyl-	234	1146
2-amyl-4-hydroxy-5,6-dimethyl-	94	2349
2-amyl-4-hydroxy-6-methyl-	80-81	2224
Barbituric acid/5-ethyl-5-hydroxy-	-	H, 2258
Barbituric acid/5-hydroxy-5-methyl-	228-230	3925
2-benzyl-4,6-dihydroxy-	313-314	2688, 3884
5-benzyl-4,6-dihydroxy-2-methyl-	330	H, 2750
4-benzyl-2,6-dihydroxy-5-phenyl-	218-220	3926
5-benzyl-4,6-dihydroxy-2-phenyl-	341 to 345	H, 2750, 3927
2-benzyl-5-ethoxymethyl-4-hydroxy-	148	2190
4-benzyl-6-hydroxy-	180	3928
5-benzyl-2-hydroxy-	HCl 218	3825
2-benzyl-4-hydroxy-5,6-diphenyl-	225-232	3926
4-benzyl-6-hydroxy-2,5-diphenyl-	236-241	3926
5-benzyl-4-hydroxy-2-methoxymethyl-6-methyl-	114-115	2877
2-benzyl-4-hydroxy-6-methyl-5-phenyl-	175-177	3926

TABLE XXVI

Pyrimidine	M.P.(^{o}C)	References
4-benzyl-6-hydroxy-2-methyl-5-phenyl-	220-222	3926
5-benzyloxy-4,6-dihydroxy-2-methyl-	263-264	135
2-benzyloxy-4,6-dimethyl-	160-165/2.5mm.	431
4-benzyloxy-2,6-dimethyl-	141-143/3mm.; pic.143	1149
4-benzyloxy-6-hydroxy-2-methyl-	177-179	3001
4-benzyloxy-6-hydroxy-5-methyl-	221-222	3001
5-benzyloxy-4-hydroxy-2-methyl-	186	79
4-benzyloxy-6-hydroxy-2-phenyl-	231-233	3001
4-benzyloxy-6-hydroxy-5-phenyl-	191-192	3001
5-benzyloxy-4-hydroxy-2-phenyl-	218-220	3476
4-benzyloxy-6-methoxy-5-methyl-	27	3001
4-benzyloxy-6-methyl-	141/4mm.; pic.142-144	1148
4-butoxy-2-t-butoxy-6-methyl-	131-132/7mm.	1313
4-butoxy-2-methoxy-6-methyl-	pic.98-99	1313
4-butoxy-6-methyl-	pic.96-97	1148
5-butyl-2,4-dihydroxy-	298-300	H, 2347
5-butyl-4,6-dihydroxy-	284 to 293	2242, 2522
5-butyl-2,4-dihydroxy-6-methyl-	243	H, 1146
5-butyl-2,4-dihydroxy-6-propyl-	186	1146
5-butyl-4-ethyl-2,6-dihydroxy-	213	1146
4-t-butyl-6-hydroxy-	211	H, 2574

Pyrimidine	M.P.(°C)	References
2-butyl-4-hydroxy-5,6-dimethyl-	145-150/0.02mm.; 120	2349, 3929
2-butyl-4-hydroxy-6-methyl-	120-121	H, 2224
2-t-butyl-4-hydroxy-6-methyl-	174-175	2224
4-butyl-6-hydroxy-2-methyl-	65-67	2224
4-butyl-6-hydroxy-5-methyl-2-propyl-	118	3930
4-butyl-6-hydroxy-2-propyl-	63-65	2224
4-cyclohexyl-6-hydroxy-2-propyl-	110-111	2224
5-cyclopent-1'-en-1'-yl-2,4-dihydroxy-	335	2952
2-cyclopropyl-4,6-dihydroxy-	313	3816
5-cyclopropyl-4,6-dihydroxy-	246	3816
4-cyclopropyl-6-hydroxy-	163-165	3816
4-cyclopropyl-6-hydroxy-2-methoxy-	161-163	3816
2-cyclopropyl-4-hydroxy-6-methyl-	191-193	3816
4-cyclopropyl-6-hydroxy-2-methyl-	204-206	3816
2,4-diallyloxy-5-methyl-	42-43	3479
2,4-dibenzyl-6-hydroxy-5-phenyl-	184-186	H, 3926
4,6-dibenzyloxy-2-methyl-	27-29	2423, 3001
4,6-dibenzyloxy-5-methyl-	66	3001
4,6-dibenzyloxy-2-phenyl-	91-92	3001
4,6-dibenzyloxy-5-phenyl-	87	3001
2,4-dibutoxy-6-methyl-	145-147/8mm.	H, 1313
2,4-di-s-butoxy-6-methyl-	126-127/7mm.; pic.98-99	1313

TABLE XXVI

Pyrimidine	M.P.(°C)	References
2,4-diethoxy-5-ethyl-6-methyl-	130/8mm.	H, 937
2,4-diethoxy-5-methyl-	-	H, 3931
2,4-diethoxy-6-methyl-	118-119/15mm.	H, 2699
5,5-diethyl-4,5-dihydro-6-hydroxy-2-methyl-4-oxo-	EtOH 132-135; HCl 251	2242
2,4-diethyl-6-hydroxy-	89-90	2224
2,4-dihydroxy-5,6-dimethyl-	298	H, 140
2,5-dihydroxy-4,6-dimethyl-	-	H, 2161
2,4-dihydroxy-5-β-hydroxyethyl-6-methyl-	265 or 270	587, 3892
4,6-dihydroxy-2-hydroxymethyl-5-methyl-	289	2238
2,4-dihydroxy-5-isoamyl-6-methyl-	247	1146
2,4-dihydroxy-5-isobutoxy-6-methyl-	238-240	112
2,4-dihydroxy-5-isobutyl-6-methyl-	244	1146
4,6-dihydroxy-2-isopropyl-	296-297	2238
4,6-dihydroxy-5-isopropyl-	>310	2242
2,4-dihydroxy-5-methoxymethyl-6-methyl-	239	2652, 2824
4,6-dihydroxy-2-methoxymethyl-5-methyl-	290	3911
4,6-dihydroxy-2-methoxymethyl-5-phenyl-	309	2877
2,4-dihydroxy-5-methyl-	320-325	H, 805, 1465, 2196, 2306, 2347, 2650, 3104, 3932-3935
2,4-dihydroxy-6-methyl-	332-334	H, 976, 1276, 2863, 2999, 3936

Pyrimidine	M.P.(^{o}C)	References
4,5-dihydroxy-2-methyl-	312-313 or 317	H, 79, 2191, 2210
4,6-dihydroxy-2-methyl-	-	H, 2997, 3905
4,6-dihydroxy-5-methyl-	313	H, 2240
4,6-dihydroxy-2-methyl-5-phenoxy-	330-332	2240
4,6-dihydroxy-2-methyl-5-phenyl-	340	H, 3905
4,6-dihydroxy-5-methyl-2-phenyl-	340	H, 2241
2,4-dihydroxy-6-methyl-5-phosphonomethyl-	283	3937
2,4-dihydroxy-6-phenyl-	262	H, 2226, 3938
4,5-dihydroxy-2-phenyl-	212-215	3476
4,6-dihydroxy-2-phenyl-	-	H, 3001, 3905
4,6-dihydroxy-5-phenyl-	-	H, 2523
2,4-dihydroxy-6-propyl-	-	H, 2226
4,6-dihydroxy-2-propyl-	-	H, 2523
2,4-dimethoxy-6-methyl-	66-68	H, 1148, 1313
4,5-dimethoxy-2-methyl-	54	2698
4,6-dimethoxy-2-methyl-	52-53	487, 3001
4,6-dimethoxy-5-methyl-	87	3001
4,6-dimethoxy-2-phenyl-	60-61	2575, 3001
2-ethoxy-4,6-dimethyl-	-	H, 2629
4-ethoxy-2,6-dimethyl-	92-94/35mm.; pic.125	2451
4-ethoxy-5-ethyl-2-hydroxy-6-methyl-	178-180	937
4-ethoxy-2-hydroxy-5-methyl-	-	H, 805, 936
4-ethoxy-2-hydroxy-6-methyl-	196	1114
4-ethoxy-2-hydroxy-6-methyl-5-propyl-	-	H, 927
4-ethoxy-2-isopropyl-6-methyl-	89-90/12mm.	1026

TABLE XXVI

Pyrimidine	M.P.(^{o}C)	References
4-ethoxy-6-methyl-	85/3mm.; pic.106	1148, 2451
4-ethoxy-6-methyl-(*N*-oxide)	121	1148
4-ethoxymethyl-6-hydroxy-2,5-dimethyl-	-	H, 3930
5-ethoxymethyl-4-hydroxy-2-methyl-	-	H, 1475, 1476
5-ethoxymethyl-4-hydroxy-2-phenyl-	157	2190
5-ethoxymethyl-4-hydroxy-2-propyl-	131	2190
4-ethoxy-5-methyl-2-trimethylsiloxy-	53-54	3902
2-ethyl-4,6-dihydroxy-	-	H, 2523
4-ethyl-2,6-dihydroxy-	-	H, 2226
5-ethyl-2,4-dihydroxy-	-	H, 1465, 2347, 3933
5-ethyl-4,6-dihydroxy-	-	H, 2523
5-ethyl-4,6-dihydroxy-2-methoxymethyl-	281-282	3911
5-ethyl-2,4-dimethoxy-6-methyl-	113-115/11mm.	H, 937
2-ethyl-4-hydroxy-	116	2554
5-ethyl-2-hydroxy-	HCl 217-220	3825
5-ethyl-4-hydroxy-2,6-bismethoxymethyl-	81-82	3911
5-ethyl-2-hydroxy-4-methoxy-6-methyl-	194-195	937
2-ethyl-4-hydroxy-6-methyl-	162-164	H, 2224, 3939
4-ethyl-6-hydroxy-2-methyl-	122-123	2224
4-ethyl-6-hydroxy-2-methyl-5-phenyl-	212-216	3926, 3940
5-ethyl-4-hydroxy-6-methyl-2-propyl-	121-122	H, 2224
2-ethyl-4-hydroxy-6-propyl-	67-68	2224

Pyrimidine	M.P.(^oC)	References
4-ethyl-6-hydroxy-2-propyl-	82-84	2224
5-ethyl-2,4,6-trihydroxy-	197	H, 1001, 1243, 3941
5-hexyl-4-hydroxy-6-methyl-2-propyl-	-	3942 (?)
4-hydroxy-2,6-bismethoxymethyl-5-methyl-	105-106	3911
2-hydroxy-4,5-dimethyl-	196; HCl 264	H, 1028
2-hydroxy-4,6-dimethyl-	-	H, 2182
4-hydroxy-2,5-dimethyl-	175-176	H, 1475, 3003, 3820
4-hydroxy-2,6-dimethyl-	192 to 206	H, 813, 3930, 3939
4-hydroxy-5,6-dimethyl-	-	H, 2349
5-hydroxy-4,6-dimethyl-	138; H_2O 90; pic.176-177	2533
4-hydroxy-2,5-dimethyl-6-phenyl-	180-181	3930
5-hydroxy-4,6-dimethyl-2-phenyl-	152-153; pic.177	2533
4-hydroxy-2,6-dimethyl-5-propyl-	157-159	2307
4-hydroxy-5,6-dimethyl-2-propyl-	132-133	H, 2224
4-hydroxy-2,6-dimethyl-5-prop-2'-ynyl-	223-224	2307
2-hydroxy-4,6-diphenyl-	237-239	2556, 3544
4-hydroxy-2,5-diphenyl-	298	3943
4-hydroxy-2,6-dipropyl-	71-73	2224
4-β-hydroxyethyl-6-methyl-2-propoxy-	148-183(?)/12mm.	2425
4-hydroxy-5-β-hydroxyethyl-2,6-dimethyl-	192-195	3944
4-hydroxy-5-β-hydroxyethyl-2-methyl-	172-173	2514

TABLE XXVI

Pyrimidine	M.P.(^{o}C)	References
4-hydroxy-5-β-hydroxyethyl-6-methyl-	155-156	H, 587, 3892
4-hydroxy-2-hydroxymethyl-6-methyl-	200-203	3889
4-hydroxy-5-hydroxymethyl-2-methyl-	-	H, 3003
4-hydroxy-2-isobutyl-5,6-dimethyl-	112	2349
4-hydroxy-2-isopropyl-5-methyl-	160-162	3626
4-hydroxy-2-isopropyl-6-methyl-	172-173	H, 2224, 3939
2-hydroxy-4-methoxy-5,6-dimethyl-	231	140
4-hydroxy-5-methoxy-6-methoxymethyl-2-methyl-	165-167	3911
2-hydroxy-4-methoxy-5-methyl-	182-183	805
2-hydroxy-4-methoxy-6-methyl-	-	H, 3945
4-hydroxy-5-methoxy-2-methyl-	214-215	2586
4-hydroxy-5-methoxy-6-methyl-	158-159	3243
4-hydroxy-6-methoxy-5-methyl-	234-235	3054
4-hydroxy-6-methoxymethyl-2,5-dimethyl-	-	H, 3911, 3930
4-hydroxy-2-methoxymethyl-5-methyl-	140-141	3911
4-hydroxy-2-methoxymethyl-6-methyl-	103	3913
4-hydroxy-5-methoxymethyl-2-methyl-	-	H, 3946
4-hydroxy-6-methoxymethyl-2-methyl-	173-174	3913
4-hydroxy-6-methoxymethyl-5-methyl-	140-141	3911

Pyrimidine	M.P.(^{o}C)	References
4-hydroxy-5-methoxymethyl-2-propyl-	-	3947
4-hydroxy-5-methoxy-2-phenyl-	206-208	2161
2-hydroxy-4-methyl-	-	H, 2861
2-hydroxy-5-methyl-	210	3065
4-hydroxy-2-methyl-	-	H, 2541
4-hydroxy-5-methyl-	-	H, 606
4-hydroxy-6-methyl-	pic.182	H, 399, 606, 1149, 2451, 2574, 3928, 3948
5-hydroxy-4-methyl-	190	3949
5-hydroxymethyl-2,4-dimethyl-	60-61	3006
4-hydroxy-5-methyl-2,6-diphenyl-	260	H, 3950
4-hydroxy-6-methyl-2,5-diphenyl-	242-244	3926
4-hydroxy-6-methyl-2,5-dipropyl-	101-102	2224
5-hydroxymethyl-4-methoxy-2-methyl-	101-102	2651
5-hydroxymethyl-2-methyl-	105	2651, 3882
4-hydroxy-2-methyl-6-phenyl-	-	H, 2539
4-hydroxy-5-methyl-6-phenyl-	193-194	2241
4-hydroxy-6-methyl-2-phenyl-	215-216	H, 2892, 3951
4-hydroxy-6-methyl-2-phenyl-5-propyl-	147-148	2307
4-hydroxy-6-methyl-2-phenyl-5-prop-2'-ynyl-	218-220	2307
4-hydroxy-2-methyl-5-propoxymethyl-	-	H, 3946
4-hydroxy-2-methyl-6-propyl-	89-91	2224

TABLE XXVI

Pyrimidine	M.P.(^{o}C)	References
4-hydroxy-6-methyl-2-propyl-	-	H, 3939
4-hydroxy-5-phenyl-	176-178	H, 2367
4-hydroxy-6-phenyl-	271 or 272-274	H, 301, 2519, 2574
5-hydroxy-2-phenyl-	148-151	2161
4-hydroxy-2-phenyl-6-propyl-	141	2892
4-hydroxy-6-propyl-	110-112	3928
4-hydroxy-2,5,6-trimethyl-	178	H, 2224, 3952
5-hydroxy-2,4,6-trimethyl-	152-154; H_2O 94; pic.160-161	2533
4-hydroxy-2,5,6-triphenyl-	290-294	H, 3926, 3953
4-isopropoxy-2-methyl-	pic.146-148	2698
2-methoxy-4,6-dimethyl-	34-37; pic.137-138	H, 1313
4-methoxy-2,5-dimethyl-	35/0.5mm.	2818
4-methoxy-2,6-dimethyl-	70/14mm.; pic.126-127	H, 1313, 2746
2-methoxy-4-methyl-	pic.119-120	H, 2746
4-methoxy-2-methyl-	pic.159 or 167-168	487, 2698
4-methoxy-6-methyl-	-	H, 1148, 1149, 2451
4-methoxy-6-methyl-(*N*-1-oxide)	135-136	1373
5-methoxy-2-phenyl-	57-58	2161
5-methyl-2,4-bistrimethylsiloxy-	63-65	3070, 3071, 3902, 3954
4-methyl-2,6-diphenoxy-	-	H, 2699, 3914
4-methyl-2,6-dipropoxy-	151/22mm.	H, 2699
2-methyl-4-phenoxy-	pic.155-158	2698
4-methyl-6-phenoxy-	125-127/7mm.; pic.172-174	1148
2,4,5-trihydroxy-6-methyl-	>220	H, 112

Pyrimidine	M.P.(°C)	References
2,4,6-trihydroxy-5-methyl-	-	H, 3955
4,5,6-trihydroxy-2-methyl-	>290	H, 135
2,4,5-trihydroxy-6-phenyl-	-	3956

TABLE XXVII. Oxypyrimidines with *N*-Substituents (*H* 530)

Alloxan/1,3-dimethyl-	TSC 254-255	H, 3160, 3164
Alloxan/1-methyl-	TSC 260-263	H, 3160, 3164
1-allyl-4-allyloxy-1,2-dihydro-2-oxo-	-	2555
4-allyloxy-1,2-dihydro-1-methyl-2-oxo-	110-112	2555
1-allyl-1,2,3,4-tetrahydro-3-methyl-2,4-dioxo-	-	2555
Barbituric acid/1,5-diethyl-	115-116	1434
Barbituric acid/5-ethyl-1-methyl-	102-103	H, 1434
Barbituric acid/1-methoxy-	185-187	2199
Barbituric acid/1,3,5-trimethyl-	-	H, 740
1-benzyl-1,2-dihydro-2-oxo-	138-139	2626, 2762
1-benzyl-1,4-dihydro-4-oxo-	142-144	3187
1-benzyl-1,6-dihydro-6-oxo-	102-105	3187
1-benzyloxy-1,2-dihydro-4,6-dimethyl-2-oxo-	132	H, 2869
4-benzyloxy-1,6-dihydro-1,5-dimethyl-6-oxo-	86-88	3001
4-benzyloxy-1,2-dihydro-1-methyl-2-oxo-	156-157	2829

TABLE XXVII

Pyrimidine	M.P.(^{o}C)	References
4-benzyloxy-1,6-dihydro-1-methyl-6-oxo-	117	3001
4-benzyloxy-1,6-dihydro-1-methyl-6-oxo-5-phenyl-	127-129	3001
1-benzyl-1,2,3,4-tetrahydro-6-methyl-2,4-dioxo-3-phenyl- (?)	223-224	3947 (cf. 1314, 2966)
1-benzyl-1,2,3,6-tetrahydro-4-methyl-2,6-dioxo-3-phenyl-	-	H, 2966
4-butoxy-1-butyl-1,2-dihydro-2-oxo-	54	3903
1-butyl-1,2-dihydro-2-oxo-	40-41	2511
1-s-butyl-1,2-dihydro-2-oxo-	48-50	2511
1-t-butyl-1,2-dihydro-2-oxo-	132; pic.150	2511
1-butyl-1,2,3,4-tetrahydro-3,6-dimethyl-2,4-dioxo-	41	1145
1-butyl-1,2,3,6-tetrahydro-3,4-dimethyl-2,6-dioxo-	56	1145
5-butyl-1,2,3,4-tetrahydro-1,6-dimethyl-2,4-dioxo-3-phenyl-	145	3958
5-s-butyl-1,2,3,4-tetrahydro-1,6-dimethyl-2,4-dioxo-3-phenyl-	98	3958
1-butyl-1,2,3,4-tetrahydro-5-isopropyl-6-methyl-2,4-dioxo-3-phenyl-	108	3958
1-cyclohexyl-3-ethyl-1,2,3,6-tetrahydro-4-methyl-2,6-dioxo-	109-112	1144
1-cyclohexyl-1,2,3,6-tetrahydro-3,4-dimethyl-2,6-dioxo-	178	1144
5-cyclopent-1'-en-1'-yl-1,2,3,4-tetrahydro-1,3-dimethyl-2,4-dioxo-	149-150	2952
1,3-diallyl-1,2,3,4-tetrahydro-2,4-dioxo-	-	2555

Pyrimidine	M.P.(^{o}C)	References
1,3-dicyclohexyl-1,2,3,4-tetrahydro-6-methyl-2,4-dioxo-	137	3957
5,5-diethyl-1,4,5,6-tetrahydro-1,2-dimethyl-4,6-dioxo-	102-103; H_2O 129; HI 192	2242, 2997
1,3-diethyl-1,2,3,4-tetrahydro-6-hydroxy-5-methyl-2,4-dioxo-	35	2305
1,2-dihydro-1,4-dimethyl-2-oxo-	156-157; pic.159	2630, 2746, 2861
1,2-dihydro-1,6-dimethyl-2-oxo-	89-90; HCl 229; pic.213	2602, 2630
1,6-dihydro-1,2-dimethyl-6-oxo-	63-65	H, 487
1,6-dihydro-1,2-dimethyl-6-oxo-4-phenyl-	108	3959
1,6-dihydro-2,4-dimethyl-6-oxo-1-phenyl-	92-93	3960
1,2-dihydro-1-hydroxy-4,6-dimethyl-2-oxo-	HBr 235-240	2869
1,6-dihydro-4-hydroxy-2-methyl-6-oxo-1-phenyl-(or tautomer)	173-176	3961
1,2-dihydro-1-isopropyl-2-oxo-	90; pic.167	2511, 2762
1,2-dihydro-1-methoxy-4,6-dimethyl-2-oxo-	-5	2869
1,6-dihydro-4-methoxy-1,2-dimethyl-6-oxo-	84-86	3001
1,6-dihydro-4-methoxy-1,5-dimethyl-6-oxo-	89	3001, 3054
1,6-dihydro-4-methoxy-1-methyl-6-oxo-	146-147	2760, 3053, 3054
1,6-dihydro-4-methoxy-1-methyl-6-oxo-5-phenyl-	105-106	3001
1,2-dihydro-1-methyl-2-oxo-	126 to 131-132	H, 1108, 2511

TABLE XXVII

Pyrimidine	M.P.(°C)	References
1,4-dihydro-1-methyl-4-oxo-	-	H, 2676
1,6-dihydro-1-methyl-6-oxo-	-	H, 2512(!), 3187
1,2-dihydro-1-methyl-2-oxo-5,6-diphenyl-	251	3962
1,6-dihydro-1-methyl-6-oxo-4-phenyl-	201	3928
1,2-dihydro-2-oxo-1-propyl-	36	2511
1,6-dihydro-1,2,4-trimethyl-6-oxo-	58-61	H, 3959
1,6-dihydro-1,2,5-trimethyl-6-oxo-	28-30; HI 180	2818
4-ethoxy-1,2-dihydro-1,5-dimethyl-2-oxo-	108-110	H, 3954
4-ethoxy-1,2,3,6-tetrahydro-1,3-dimethyl-2,6-dioxo-	133	739
4-ethoxy-1,2,3,6-tetrahydro-1,3,5-trimethyl-2,6-dioxo-	124-125	740
2-ethyl-1,6-dihydro-4-methyl-6-oxo-1-phenyl-	107-108	3963
1-ethyl-1,4-dihydro-4-oxo-	HCl 232-234	3187
1-ethyl-1,6-dihydro-6-oxo-	60-64	3187
1-ethyl-1,2,3,4-tetrahydro-3,6-dimethyl-2,4-dioxo-	-	H, 1276
1-ethyl-1,2,3,6-tetrahydro-3,4-dimethyl-2,6-dioxo-	106-108	1276
5-ethyl-1,2,3,4-tetrahydro-2,4-dioxo-1,3-diphenyl-6-propyl-	180	2339
1-ethyl-1,2,3,4-tetrahydro-5-isopropyl-6-methyl-2,4-dioxo-3-phenyl-	138	3958

Pyrimidine	M.P.(^{o}C)	References
4-ethyl-1,2,3,6-tetrahydro-5-methyl-2,6-dioxo-1,3-diphenyl-	168-169	2339
hexahydro-5-hydroxymethylene-1,3-dimethyl-2,4,6-trioxo-	Na ≯350	3265
hexahydro-5-hydroxy-1,3,5-trimethyl-2,4,6-trioxo-	107	2120
hexahydro-1,3,5,5-tetramethyl-2,4,6-trioxo-	108-109	H, 3001
1,2,3,4-tetrahydro-1,3-bis-β-hydroxyethyl-2,4-dioxo-	153	3771, 3964
1,2,3,4-tetrahydro-1,3-diisopropyl-6-methyl-2,4-dioxo-	93-94	3957
1,2,3,4-tetrahydro-1,3-dimethyl-2,4-dioxo-	124	H, 739, 2776, 3381, 3931
1,2,3,4-tetrahydro-1,6-dimethyl-2,4-dioxo-3,5-diphenyl-	122	3958
1,2,3,4-tetrahydro-1,6-dimethyl-2,4-dioxo-3-phenyl-	205	H, 1144
1,2,3,4-tetrahydro-2,4-dioxo-1,3-bistrimethylsilyl-	116/12mm.	3965
1,2,3,4-tetrahydro-2,4-dioxo-1,3,6-triphenyl-	188	2339
1,2,3,4-tetrahydro-5-2'-hydroxycyclopentyl-1,3-dimethyl-2,4-dioxo-	132-133	2945
1,2,3,4-tetrahydro-5-hydroxy-1,3-dimethyl-2,4-dioxo-	-	H, 890, 1365
1,2,3,4-tetrahydro-6-hydroxy-1,3-dimethyl-2,4-dioxo-	120-123	H, 739, 978, 996
1,2,3,4-tetrahydro-3-β-hydroxyethyl-1-methyl-2,4-dioxo-	136-138	3012, 3966

TABLE XXVII

Pyrimidine	M.P.(°C)	References
1,2,3,4-tetrahydro-6-hydroxy-1,3,5-trimethyl-2,4-dioxo-	-	H, 2120
1,2,3,4-tetrahydro-5-isobutyl-1,6-dimethyl-2,4-dioxo-3-phenyl-	183	3958
1,2,3,4-tetrahydro-1-isopropyl-3,6-dimethyl-2,4-dioxo-	162	3967
1,2,3,4-tetrahydro-5-isopropyl-1,6-dimethyl-2,4-dioxo-3-phenyl-	125	3958
1,2,3,4-tetrahydro-6-isopropyl-2,4-dioxo-1,3-diphenyl-	193-194	2339
1,2,3,4-tetrahydro-5-isopropyl-6-methyl-2,4-dioxo-3-phenyl-1-propyl-	124-126	3958
1,2,3,4-tetrahydro-6-methoxy-1,3,5-trimethyl-2,4-dioxo-	107-108	740
1,2,3,4-tetrahydro-5-methyl-2,4-dioxo-1,3-diphenyl-6-propyl-	146	2339
1,2,3,4-tetrahydro-5-methyl-2,4-dioxo-1,3-bistrimethylsilyl-	121-123/12mm.	3965
1,2,3,4-tetrahydro-1,3,5,6-tetramethyl-2,4-dioxo-	131-132	H, 487
1,2,3,4-tetrahydro-1,3,6-trimethyl-2,4-dioxo-	-	H, 3968, 3969
Uracil/5-allyl-3,6-dimethyl-	214	1146
Uracil/5-amyl-3-butyl-6-methyl-	61-63	1146
Uracil/5-amyl-3,6-dimethyl-	139	1146
Uracil/1-benzyl-	-	H, 1002, 2337, 2963

Pyrimidine	M.P.(^{o}C)	References
Uracil/3-benzyl-5,6-diethyl-	-	3970
Uracil/3-benzyl-5,6-dimethyl-	-	3970
Uracil/1-benzyl-5-methyl-	161-163	H, 1003
Uracil/3-benzyl-6-methyl-	194	H, 2985
Uracil/1-benzyloxy-	185	2200
Uracil/3-benzyloxy-	157-158	2200
Uracil/1-benzyloxy-5-methyl-	153-154	3972
Uracil/1-butyl-	100-102	2963, 3903
Uracil/3-butyl-	152-153	1464
Uracil/3-butyl-5,6-dimethyl-	117	3973
Uracil/5-butyl-3,6-dimethyl-	134	1146
Uracil/5-butyl-3-ethyl-6-methyl-	95-96	1146
Uracil/5-butyl-6-ethyl-3-methyl-	103-105	3974
Uracil/3-s-butyl-5-β-hydroxyethyl-6-methyl-	-	3975
Uracil/3-butyl-5-isoamyl-6-methyl-	107-108	1146
Uracil/3-butyl-5-isobutyl-6-methyl-	80-81	1146
Uracil/3-butyl-5-isopropyl-6-methyl-	211-212	3976
Uracil/3-butyl-5-methoxy-6-methyl-	-	3973
Uracil/1-butyl-5-methyl-	140	2344
Uracil/1-butyl-6-methyl-	128 to 136	2863, 3977, 3978
Uracil/3-butyl-5-methyl-	-	3979
Uracil/3-s-butyl-5-methyl-	-	3979
Uracil/3-butyl-6-methyl-	176 to 183	1351, 2864, 3977

Pyrimidine	M.P.(°C)	References
Uracil/3-s-butyl-6-methyl-	118-120	3975, 3980
Uracil/5-butyl-3-methyl-	65	2146, 2347
Uracil/5-butyl-6-methyl-3-phenyl-	195	3958
Uracil/5-butyl-3-methyl-6-propyl-	94-96	3974
Uracil/5-butyl-6-methyl-3-propyl-	115	1146
Uracil/5-butyl-3-phenyl-	147	2347, 3933
Uracil/1-cyclohexyl-	217-218	2963
Uracil/3-cyclohexyl-	-	3982
Uracil/3-cyclohexyl-5-β-hydroxyethyl-6-methyl-	-	3979
Uracil/3-cyclohexyl-5-hydroxymethyl-6-methyl-	175-176	3983
Uracil/3-cyclohexyl-5-methoxymethyl-6-methyl-	-	3983
Uracil/1-cyclohexyl-6-methyl-	233-235	888
Uracil/3-cyclohexyl-5-methyl-	-	3982
Uracil/3-cyclohexyl-6-methyl-	238-239	3957
Uracil/1-γ-cyclopentenyl-5-methyl-	168	2344
Uracil/3-γ-cyclopentenyl-6-methyl-	-	3982
Uracil/3-cyclopentyl-5,6-dimethyl-	-	3982
Uracil/3-cyclopentyl-6-methyl-	-	3983
Uracil/3,5-dibutyl-6-methyl-	77	1146, 3976
Uracil/1-3',4'-dihydroxy-cyclopentyl-5-methyl-	215	3984
Uracil/1,5-dimethyl-	281 to 288	H, 373, 1003, 3971

Pyrimidine	M.P.(°C)	References
Uracil/1,6-dimethyl-	-	H, 888, 1351, 3968
Uracil/3,5-dimethyl-	204 to 216	H, 2146, 2347, 3954, 3971
Uracil/3,6-dimethyl-	-	H, 1276, 1351, 3968
Uracil/3,6-dimethyl-5-propyl-	193 or 198	H, 1146, 3974
Uracil/3,6-diphenyl-	286-288	3938
Uracil/5-ethoxymethyl-1-methyl-	126-127	2847
Uracil/5-ethyl-3,6-dimethyl-	229	1146, 3974
Uracil/1-ethyl-6-methyl-	195-196	H, 888, 1114, 3977
Uracil/3-ethyl-6-methyl-	197-198	H, 1276, 1351, 3977
Uracil/5-ethyl-3-methyl-	147 or 153	2146, 2347
Uracil/5-ethyl-6-methyl-3-phenyl-	273	3958
Uracil/5-ethyl-3-phenyl-	217	2347
Uracil/1-hexyl-6-methyl-	108-109	3977
Uracil/3-hexyl-6-methyl-	162-163	3977
Uracil/1-hydroxy-	280-286	2200, 2964
Uracil/3-hydroxy-	290-293	2200
Uracil/1-ε-hydroxyamyl-	78-80	3985
Uracil/1-γ-hydroxy-cyclopentyl-5-methyl-	219-220(*trans*); 189-190(*cis*)	3984
Uracil/1-β-hydroxyethyl-	136-137	3771, 3964
Uracil/3-β-hydroxyethyl-	172-173	3964
Uracil/5-β-hydroxyethyl-1(or 3)methyl-	255-256	2514
Uracil/1-3'-hydroxy-4'-hydroxymethyl-cyclopentyl-5-methyl-	212-213	3064

Pyrimidine	M.P.(^{o}C)	References
Uracil/1-3'-hydroxy-4'-methoxycyclopentyl-5-methyl-	184-186	3984
Uracil/1-hydroxy-5-methyl-	231-235	3972
Uracil/3-hydroxy-5-methyl-	226-228	3972
Uracil/1-hydroxy-6-methyl-	270	3936
Uracil/5-hydroxy-1-methyl-	299	H, 2198
Uracil/5-hydroxy-3-methyl-	250-252	H, 2198
Uracil/5-hydroxymethyl-3-isopropyl-6-methyl-	-	3982
Uracil/3-isoamyl-5,6-dimethyl-	-	3973
Uracil/5-isoamyl-3,6-dimethyl-	138	1146
Uracil/3-isobutyl-5,6-dimethyl-	-	3973
Uracil/3-isobutyl-6-methyl-	-	3986
Uracil/3-isopropyl-5,6-dimethyl-	-	3973
Uracil/3-isopropyl-5-methoxymethyl-6-methyl-	117-118	3981
Uracil/1-isopropyl-5-methyl-	213-216	1003
Uracil/3-isopropyl-6-methyl-	193	3957
Uracil/5-isopropyl-3-phenyl-	228	3958
Uracil/1-methoxy-	200-202	2200
Uracil/3-methoxy-	205-207	2200
Uracil/5-methoxy-1-methyl-	244-245	2198
Uracil/5-methoxy-3-methyl-	229	2198
Uracil/1-methyl-	232 to 235	H, 373, 888, 1002, 2311, 2643, 2963, 3261, 3381
Uracil/3-methyl-	180 to 184	H, 2347, 2643

Pyrimidine	M.P.(^{o}C)	References
Uracil/5-methyl-1-phenyl-	199	373, 3987
Uracil/5-methyl-3-phenyl-	256	2347
Uracil/6-methyl-1-phenyl-	272-274 or 276	888, 3977, 3987
Uracil/6-methyl-3-phenyl-	142(?)	H, 3980
Uracil/6-methyl-3-phenyl-5-propyl-	272	3958
Uracil/5-methyl-1-propyl-	138	743
Uracil/6-methyl-1-propyl-	173	H, 888
Uracil/6-methyl-3-propyl-	183-184	H, 1144
Uracil/1-phenyl-	247	H, 373, 1498
Uracil/3,5,6-trimethyl-	220	H, 1146
[*pseudo*]Uracil/2-benzyl-1-phenyl-	212-213	3988
[*pseudo*]Uracil/1,5-dimethyl-	298-300	3051, 3054
[*pseudo*]Uracil/1,2-diphenyl-	212-213	3988
[*pseudo*]Uracil/1-ethyl-2-phenyl-	PNB 128-129	3989
[*pseudo*]Uracil/1-methyl-	155-156	2995, 3051, 3053
[*pseudo*]Uracil/2-methyl-1-phenyl-	260-261	3988, 3989

TABLE XXVIII. Sulphonylpyrimidines (*H* 535)

4-cyclopropyl-6-methylsulphonyl-	100/0.006	3816
4,6-dimethyl-2-phenylsulphonyl-	158-159	2771
4,6-dimethyl-2-sulphamoyl-	200-201	H, 3990, 3991

TABLE XXVIII

Pyrimidine	M.P.(°C)	References
4-methyl-2,6-bisphenylsulphonyl-	126	2742
2-methyl-4,6-disulpho-	>320; Na >320	3992
2-methylsulphinyl-	152-154/0.5mm.	2619
4-methylsulphinyl-	47-49	2619
5-methylsulphinyl-	84-86	2619
2-methylsulphonyl-	73-74	2619
4-methylsulphonyl-	53-54	2619
5-methylsulphonyl-	135-136	2619
2-phenylsulphinyl-	118-119	2619
2-phenylsulphonyl-	99-100	2619

TABLE XXIX. Thiopyrimidines (*H* 536)

4-allylthio-2-benzylthio-	173/0.9mm.	2555
4-allylthio-2-methyl-	67/0.1mm.	2555
4-allylthio-2-methylthio-	126/1.3mm.	2555
4-allylthio-2-phenyl-	129/0.1mm.	2555
5-amyl-2-mercapto-	195-196	3825
4-benzyl-2,6-dimercapto-	250-251	3846
5-benzyl-2-mercapto-	213	3825
2-benzylthio-4,6-dimethyl-	-	H, 2873, 3993
2,4-bisbenzoylthio-	103	3994
2,4-bisbenzylthio-6-methyl-	37-39	2165
2,4-bismethylthio-	98/0.25mm.; 9-10	827, 2936
4,6-bismethylthio-	52-54(cf.119)	2165, 2609 (cf.3995)
2,4-bismethylthio-5-methylthiomethyl-	40-41	2596

Pyrimidine	M.P.(^{o}C)	References
5-butyl-2-mercapto-	203-205	3825
2-butylthio-4-methyl-	112-114/4mm.	2174
4-cyclopropyl-6-methylthio-	125-128/13mm.	3816
1,2-dihydro-1-methyl-2-thio-	189-191	2173
1,4-dihydro-1-methyl-4-thio- (& 3-isomer)	246 (& 98)	2173
2,4-dimercapto-	300	H, 2165
4,6-dimercapto-	245-246 or 252	831, 2165, 3995
2,4-dimercapto-5-methyl-	284-285	H, 3104, 3997
2,4-dimercapto-6-methyl-	>290 or >360	H, 281, 2165
2,4-dimercapto-5-methyl-6-phenyl-	280-281	3846
4,6-dimercapto-2-methylthio-	>360	2165
2,4-dimercapto-6-phenyl-	268-270	H, 3846
4,6-dimercapto-5-phenyl-	238-242	H, 2522
2,4-dimethyl-6-methylthio-	28; pic.150	2746
4,5-dimethyl-2-methylthio-	66-70/20mm.	2174
4,6-dimethyl-2-methylthio-	pic.109	H, 2746
4,6-dimethyl-2-phenylthio-	68-69	2771
2,4-diselenyl-	-	H, 3224
Dithiouracil/1-benzyl-	169	2337
Dithiouracil/1-methyl-	261	827, 2337
Dithiouracil/3-methyl-	213	281
5-ethyl-2-ethylthio-	82-84	3825
5-ethyl-2-mercapto-	210-212	3825
4-ethyl-6-mercapto-2-methylthio-	189	2753
2-ethylthio-	-	H, 2697
4-ethylthio-	89-90/12mm.; pic.105-106	2697

TABLE XXIX

Pyrimidine	M.P.(^{o}C)	References
5-ethylthio-	96-98/10mm.; pic.98	791
4-ethylthio-2-isopropyl-6-methyl-	115-116/11mm.	1026
2-ethylthio-4-methyl-	-	H, 2174
2-ethylthio-5-methyl-	78-79/0.02mm.	3788
5-isopropyl-2-mercapto-	239-242	3825
2-isopropyl-4-mercapto-6-methyl-	160-161	1026
5-isopropyl-4-methyl-2-methylthio-	100-107/4.5mm.	2174
2-isopropylthio-	106-107/12mm.	2697
2-mercapto-	229-230	H, 431, 2165, 2173, 2512, 3800
4-mercapto-	188 or 190-192	H, 2165, 2173, 2642
2-mercapto-4,6-dimethyl-	-	H, 431
2-mercapto-4-methyl-	216-220	H, 431, 1029
2-mercapto-5-methyl-	233-235	3788, 3825
4-mercapto-6-methyl-2-methylthio-	239	H, 2165
4-mercapto-6-methyl-5-phenyl-	180	3996
2-mercapto-4-methylthio-	190-192	2581
4-mercapto-2-methylthio-	201 or 203	H, 2165, 3999
4-mercapto-2-methylthio-6-phenyl-	234-237	2753
4-mercapto-2-methylthio-6-propyl-	193	2753
2-mercapto-5-phenyl-	225-228	3825
4-mercapto-5-phenyl-	167-169	3996
2-mercapto-5-propyl-	208	3825
4-methyl-2,6-bismethylthio-	43-44	2165
4-methyl-2,6-bisphenylthio-	56	2742
5-methyl-2,4-diselenyl-	186	2779

Pyrimidine	M.P.(°C)	References
4-methyl-2-methylthio-	101-102/20mm.	H, 2746
4-methyl-6-methylthio-	-	H, 3998
5-methyl-2-methylthio-	118-120/16mm.	3788
4-methyl-6-phenylthio- (3-*N*-oxide)	107-108	H, 1373
4-methyl-2-propylthio-	125-128/17mm.	2174
2-methylthio-	-	H, 2173
4-methylthio-	68/0.9mm.; 86-87/12mm.	2173, 2642
5-methylthio-	42-43; 71/2mm.	2619
5-methylthio-2-phenyl-	47-48	2304
2-phenylthio-	45; 145/0.8mm.	2619
1,2,3,4-tetrahydro-1,3-dimethyl-2,4-dithio-	121	H, 827
2,4,6-trimercapto-	>360	H, 2165
2,4,6-trimercapto-5-phenyl-	242-244	2522
2,4,6-trismethylthio-	114-116	2165

TABLE XXX. Amino-Carboxypyrimidines (*H* 537)

4-acetamido-6-carboxy-2-methyl-	247-249	2451
4-acetamido-5-cyano-	124-125	2367
4-acetamido-5-ethoxycarbonyl-	100-101	4000
4-acetamido-5-ethoxycarbonyl-2-methyl-	94-95	2698
2-allylamino-4-amino-5-carbamoyl-	222-223	4001
2-allylamino-4-amino-5-cyano-	171-172	2273
4-amino-2-amyl-5-cyano-	149-150	3804

Pyrimidine	M.P.(^{o}C)	References
4-amino-2-anilino-5-carbamoyl-	246-247	2273
4-amino-2-anilino-5-cyano-	234-235	2273
4-amino-5-azidocarbonyl-2-methyl-	140	1386
4-amino-2-aziridino-5-ethoxycarbonyl-	112-113	2205
4-amino-2-benzylamino-5-carbamoyl-	180-181	2273
4-amino-2-benzylamino-5-cyano-	177-179	2273
4-amino-2-benzyl-5-carbamoylmethyl-	238	H, 3806
4-amino-2-butylamino-5-cyano-	161; HCl 225	4002, 4003
4-amino-2-butyl-5-carbamoylmethyl-	242	3806
4-amino-2-butyl-5-cyano-	143-147	H, 3804
2-amino-5-carbamoyl-	317	2376
4-amino-5-carbamoyl-	258 to 261	H, 2367, 2544, 4000, 4004
4-amino-5-carbamoyl-2-dimethylamino-	290; HCl 293	2273, 4003, 4005
4-amino-5-α-carbamoylethyl-2-methyl-	231	1595
4-amino-5-carbamoyl-2-hexylamino-	155	2273
4-amino-5-carbamoyl-2-methyl-	-	H, 1386
4-amino-5-carbamoyl-2-methylamino-	268-270	2273
5-amino-1-carbamoylmethyl-1,4-dihydro-4-imino-	HI -	2722
4-amino-5-carbamoylmethyl-2-ethyl-	233-234	H, 3806
4-amino-5-carbamoylmethyl-2-phenyl- (& 2-propyl analogue)	177 (& 215)	3806

Pyrimidine	M.P.(°C)	References
4-amino-5-carbamoyl-2-piperidino-	202-204; HCl 280-282	4003
2-amino-4-carboxy-	-	H, 2954, 2955, 3004
2-amino-5-carboxy-	>290	H, 2278
4-amino-5-carboxy-	275 to 284	H, 2366, 4006
4-amino-6-carboxy-	256-257	2954
4-amino-5-β-carboxyethyl-2-methyl-	233	1595
4-amino-2-carboxy-6-methyl-	245-246	2451
4-amino-5-carboxy-2-methyl-	-	H, 3836
5-amino-1-carboxymethyl-1,4-dihydro-4-imino- (internal salt)	ca.300	2722
2-amino-5-carboxy-4-phenyl-	271-274	4007
2-amino-5-cyano-	>260	H, 2276, 4008
4-amino-5-cyano-	255-256; pic.189	H, 2367, 2369, 4004, 4009
4-amino-5-cyano-2,6-biscyanomethyl-	220-224	3335, 4010
4-amino-5-cyano-2-cyclohexylamino-	182-183	2273
4-amino-5-cyano-2-diethylamino-	237	4002, 4003
4-amino-5-cyano-2,6-dimethyl-	227; pic.195	H, 886, 3807
4-amino-5-cyano-2-dimethylamino-	235 to 239; HCl 255	3803, 4002, 4005
4-amino-5-cyano-2-dimethylamino-6-methyl-	202	4003
4-amino-5-cyano-2-hexylamino-	134-135	2273
4-amino-5-cyano-2-*N*-hydroxyanilino-	184-186	2330

Pyrimidine	M.P.(^{o}C)	References
4-amino-5-cyano-2-isopropyl-	150-151	3804
4-amino-5-cyano-2-methyl-	-	H, 459, 1274
4-amino-5-cyano-6-methyl-	217-219	2273
4-amino-5-cyano-2-methylamino-	225	2273, 3803
4-amino-5-cyanomethyl-2-methyl-	255	4011
4-amino-5-cyano-2-methyl-6-phenyl-	253-254	4012
4-amino-5-cyano-2-piperidino-	212-213	2273
4-amino-5-cyano-2-propyl-	165	3803
4-amino-5-cyano-2-propylamino-	167-169	2273
2-amino-4-diethoxymethyl-	137-138	H, 2301
4-amino-5-diethoxymethyl-2-methyl-	66-68	3835
4-amino-5-dimethoxymethyl-2-methyl-	108-109	3835
4-amino-2-dimethylamino-5-hydrazinocarbonyl-	deriv.	4013
5-amino-4,6-dithiocyanato- (?)	>250	4014
2-amino-5-ethoxycarbonyl-	141	H, 2278
4-amino-5-ethoxycarbonyl-	104 or 105	2366, 4000
4-amino-5-ethoxycarbonyl-2-hydrazino-	172-174	2148
4-amino-5-ethoxycarbonyl-2-*N*-hydroxyanilino-	150-152	2330
2-amino-5-ethoxycarbonyl-4-methyl-	220-222	H, 1128
4-amino-5-ethoxycarbonyl-2-methyl-	-	H, 3425
2-amino-5-ethoxycarbonyl-4-phenyl-	156	4007
4-amino-5-ethoxycarbonyl-2-thiocyanato-	152-154	2783

Pyrimidine	M.P.(^{o}C)	References
4-amino-5-formyl-2,3-dihydro-2-imino-3-methyl- (or tautomer)-	202-203; pic.195	2376
4-amino-5-formyl-2,3-dihydro-3-methyl-2-methylimino- (or tautomer)	ca 205; HI 241	2376
2-amino-5-β-formylethyl-4-methyl-	derivs	3190
4-amino-5-formyl-2-methyl-	-	H, 2698
4-amino-5-hydrazinocarbonyl-	197	4000
4-amino-5-α-hydrazinocarbonylethyl-2-methyl-	263	1595
4-amino-5-β-hydrazinocarbonylethyl-2-methyl-	201	1595
4-amino-2-hydrazino-5-hydrazinocarbonyl-	247-248	4015
2-amino-4-methoxycarbonyl-	230	3004
4-amino-2-methyl-5-thiocyanatomethyl-	150-198	1620
4-amino-5-*N*-phenylcarbamoyl-	258	4006
5-anilino-4-carboxy-2-methyl-	218	994
4-anilino-5-cyano-	168	2907
4-anilino-5-ethoxycarbonyl-	103-104	4000
2-anilino-5-ethoxycarbonyl-4-methyl-	102-104	2148
4-anilino-5-ethoxycarbonyl-2-methyl-	85-86	4015
4-benzylamino-5-ethoxycarbonyl-	47-48	4000
2-benzylamino-5-ethoxycarbonyl-4-methyl-	105-106	1128
4-benzylamino-5-ethoxycarbonyl-2-methyl-	70-71	4015

TABLE XXX

Pyrimidine	M.P.(^{o}C)	References
2-butylamino-5-ethoxycarbonyl-4-methyl-	71-73	1128
5-carbamoyl-1,2-dihydro-2-imino-1-methyl-	HI 240; pic.257	2376
5-carbamoyl-4-dimethylamino-2-methyl-	244-245	2698
5-carbamoyl-2-methylamino-	250; pic.229	2376
4-carboxy-2,6-bismethylamino-	314	2214
5-carboxy-2-diethylamino-4,6-dimethyl-	146	2519
5-carboxy-4-dimethylamino-2-methyl-	182-184	2698
5-carboxy-4-hydrazino-2-methyl-	340	2537
5-carboxy-2-methylamino-	308-309	2376
5-carboxy-2-methyl-4-methylamino-	HCl 127-130	2653
5-cyano-2,4-bisdimethylamino-	92-93; HCl 195	4003
5-cyano-4,6-bisdimethylamino-2-methyl-	123	2681
1-cyano-1,2-dihydro-4,6-dimethyl-2-methylimino-	120-122	427
1-cyano-1,2-dihydro-2-imino-4,6-dimethyl-	-	427
5-cyano-1,2-dihydro-2-imino-1-methyl-	HI 231; HCl 228	2376
5-cyano-1,2-dihydro-1-methyl-2-methylimino-	HI 209; HCl 242	2376
5-cyano-4-dimethylamino-	114	2907
5-cyano-4-dimethylamino-2-methyl-	132-134	2681, 2698
5-cyano-4-hydrazino-2,6-diphenyl-	209-210	2370
5-cyano-2-methylamino-	203; pic.172	2376

Pyrimidine	M.P.(^{o}C)	References
5-cyano-2-methyl-4,6-bismethylamino-	196	2681
2-cyclohexylamino-5-ethoxycarbonyl-4-methyl-	111-112	1128
2,4-diamino-5-azidocarbonyl-	155	1386
2,4-diamino-5-carbamoyl-	>320	459
2,4-diamino-5-carbamoyl-6-methyl-	240-241	2273
2,4-diamino-6-carboxy-	345	2214
2,4-diamino-5-cyano-	>320	H, 459
4,6-diamino-5-cyano-2-methyl-	397-398 (!)	H, 4016
2,4-diamino-5-cyano-6-phenyl-	253	4012
4,6-diamino-5-cyano-2-phenyl-	240(?) or 148-149	269, 4016
4,6-diamino-5-ββ-diethoxyethyl-	167-169	2011
4,5-diamino-2-dimethylamino-6-ethoxycarbonyl-	165	2467
2,4-diamino-5-ethoxycarbonyl-	-	H, 4015, 4017
4,5-diamino-6-ethoxycarbonylmethyl-	156	2785
2,4-diamino-5-formyl-	263-264; oxime 290-291	2169
2,4-diamino-5-hydrazinocarbonyl-	266-268	1386
2,4-dianilino-5-ethoxycarbonyl-	187-188	4015
1-ββ-dimethoxyethyl-1,2-dihydro-2-imino- (?)	pic.135	4018
4-dimethylamino-5-ethoxycarbonyl-	90/0.65mm.	4000
4-dimethylamino-5-ethoxycarbonyl-2-methyl-	124-126/6mm.	2698

TABLE XXX

Pyrimidine	M.P.(°C)	References
5-ethoxycarbonyl-2-ethylamino-4-methyl-	102-104	1128
5-ethoxycarbonyl-2-hydrazino-4-methyl-	173-174	1128, 2148
5-ethoxycarbonyl-4-isopropylamino-	90/0.07	4000
5-ethoxycarbonyl-4-methylamino-	63	4000
5-ethoxycarbonyl-2-methyl-4-methylamino-	67-68	2653
5-ethoxycarbonyl-4-methyl-2-methylamino-	95-96	1128
5-ethoxycarbonyl-2-methyl-4-α-methylhydrazino-	crude	2537
5-ethoxycarbonyl-2-methyl-4-β-phenylhydrazino-	182-184	2537
5-ethoxycarbonyl-4-methyl-2-piperidino-	49-51	2148
5-formyl-4-methyl-2,6-dipiperidino-	117	524
5-formyl-4-methyl-2-piperidino-6-thiocyanato-	172	524
4-hydrazino-5-hydrazinocarbonyl-	183	4000
4-hydrazino-5-β-hydrazinocarbonylethyl-2-methyl-	182	1595
5-methoxycarbonyl-2-methyl-4-methylamino-	93-94	2653
2-methyl-4-methylamino-5-methylcarbamoyl-	189	2653
2,4,6-triamino-5-ββ-diethoxyethyl-	88-100 or 111-112	2011, 4019
2,4,5-triamino-6-ethoxycarbonyl-	H_2O 205-206	2467

TABLE XXXI. Amino-Halogenopyrimidines (*H* 539)

Pyrimidine	M.P.(°C)	References
4-acetamido-6-amino-5-bromo-	181	3286
5-acetamido-4-amino-2-chloro-	209-211	2144
5-acetamido-4-amino-6-chloro-	232-233	2466
5-acetamido-4-azido-6-chloro-	176-177	2466
5-acetamido-4-benzylamino-6-chloro-2-trifluoromethyl-	187-188	2735
4-acetamido-6-chloro-	156	3286
4-acetamido-6-chloro-2-diallylamino-	91-93	2837
2-acetamido-4-chloro-6-methyl-	137	4020
5-acetamido-4-chloro-6-methylamino-	232-234	2435
4-acetamido-2,6-dichloro-	184-185	4021
5-acetamido-4,6-dichloro-	149-150	2752
4-allylamino-2-amino-6-chloro-	120	2738
5-allyl-2-amino-4-chloro-6-methyl-	145-146	2566
4-amino-5-aminomethyl-2-heptafluoropropyl-	91-92	2268
4-amino-5-aminomethyl-2-pentafluoroethyl-	134-135	2268
4-amino-5-aminomethyl-2-trifluoromethyl-	147-148	353, 1000, 2268
4-amino-2-amyl-5-bromomethyl-	-	3804
2-amino-4-anilino-5-bromo-6-methyl-	151-153	H, 2631
2-amino-4-anilino-6-chloro-	-	4022

Pyrimidine	M.P.(^{o}C)	References
5-amino-4-anilino-6-chloro-	175-176	1469
5-amino-4-azido-6-chloro- (or tautomer)	171-173	2466, 2934
4-amino-6-benzylamino-5-bromo-	145	3286
2-amino-4-benzylamino-6-chloro-	129	2738
5-amino-4-benzylamino-6-chloro-	207-209	1469
5-amino-4-benzylamino-2-chloro-6-trifluoromethyl-	179-180	2735
5-amino-4-benzylamino-6-chloro-2-trifluoromethyl-	184-185	2735
5-amino-4-benzylamino-6-fluoro-	148-152	4023
4-amino-2-benzyl-5-bromomethyl-	HBr 196	3806
5-amino-4-α-benzylhydrazino-6-chloro-	105-110	2673
2-amino-4,6-bistrichloromethyl-	-	4022
4-amino-2,6-bistrifluoromethyl-	149-150	2193
2-amino-5-bromo-	237-238	H, 3826
4-amino-5-bromo-	-	H, 1674
2-amino-5-bromo-4-chloro-6-methyl-	-	H, 2631
2-amino-5-bromo-4-dibromomethyl-	163; pic.172	2156
2-amino-5-bromo-4,6-dichloro-	235-236	3842
4-amino-5-bromo-2,6-dichloro-	155-157	3842
2-amino-5-bromo-1,4-dihydro-4-imino-1-methyl-	HI 236; HCl 289	2624
2-amino-5-bromo-4,6-dimethyl-	183-184	H, 2842, 4024

Pyrimidine	M.P.(°C)	References
4-amino-5-bromo-2,6-dimethyl-	142-143	H, 2631
2-amino-5-bromo-4-dimethylamino-	114-115	2624
4-amino-5-bromo-2-dimethylamino-	139-140	2624
4-amino-5-bromo-6-isopropylamino-	138	3286
2-amino-4-bromo-6-methyl-	152-154	H, 2564
4-amino-5-bromo-6-methyl-	194	H, 2631
2-amino-5-bromo-4-methylamino-	145-146	2624
4-amino-5-bromo-2-methylamino-	120-121	2624
4-amino-5-bromomethyl-2-butyl-	-	H, 3806
4-amino-5-bromomethyl-2-ethyl-	183-185(?); HBr 180	H, 2270, 3806, 4025
4-amino-5-bromomethyl-2-heptafluoropropyl-	131-133	2268
4-amino-5-bromomethyl-2-methyl-	-	H, 3356, 4026, 4027
4-amino-5-bromomethyl-2-pentafluoroethyl-	145-146	2268
2-amino-5-bromo-4-methyl-6-phenyl-	125-128	2842
4-amino-5-bromomethyl-2-phenyl-	198-200; HBr 165	H, 3806, 4025
2-amino-5-bromo-4-methyl-6-piperidino-	97-100	2631
4-amino-5-bromomethyl-2-propyl-	HBr 294-295	3806
2-amino-5-bromo-4-methyl-6-propyl-	95-97	2165
4-amino-5-bromomethyl-2-trifluoromethyl-	186-188; HBr 132	2268, 2516
4-amino-5-bromomethyl-2-γγγ-trifluoropropyl-	-	4028

TABLE XXXI

Pyrimidine	M.P.(^{o}C)	References
2-amino-4-butylamino-6-chloro-	97-98	2505, 2738
5-amino-4-butylamino-6-chloro-	78-79	4029
5-amino-4-butylamino-2-chloro-6-trifluoromethyl-	172/2mm.	2735
5-amino-4-butylamino-6-chloro-2-trifluoromethyl-	98-100	2735
4-amino-5-butylformamido-6-chloro-	161	3288
2-amino-5-chloro-	233	H, 3826, 4030
4-amino-2-chloro-	-	H, 2674
4-amino-6-chloro-	215-217	H, 2593, 2777, 3809, 4031
5-amino-4-chloro-	110 or 123	751, 2158
4-amino-6-chloro-2-chloromethyl-	112	2290
5-amino-4-chloro-6-cyclohexylamino-	138	4029
5-amino-4-chloro-6-cyclopentylamino-	140-142	4029
4-amino-6-chloro-2,5-diethyl-	-	2270
4-amino-6-chloro-2-diallylamino-	91-93	4033
4-amino-6-chloro-2-diethylamino-	124-125	2837, 4033
2-amino-4-chloro-1,6-dihydro-6-imino-1-methyl- (or tautomer)	ca.192	2288, 2624
2-amino-4-chloro-3,6-dihydro-6-imino-3-methyl- (or tautomer)	HI 270; HCl 273	2624
2-amino-5-chloro-4,6-dimethyl-	192-193	H, 2623
4-amino-5-chloro-2,6-dimethyl-	164-168	H, 2623, 4034

Pyrimidine	M.P.(^oC)	References
2-amino-4-chloro-6-dimethylamino-	-	H, 2624
4-amino-6-chloro-2-dimethylamino-	152-153; HCl 226	H, 2624, 2837
5-amino-4-chloro-6-αβ-dimethylhydrazino-	79-81	2752
2-amino-4-chloro-6-ethyl-	121-122	H, 2564
4-amino-6-chloro-2-ethyl-	133	915
2-amino-4-chloro-6-ethylamino-	154	H, 2505, 2738
5-amino-4-chloro-6-ethylamino-	148-149	717
5-amino-2-chloro-4-ethylamino-6-trifluoromethyl-	159-161	2735
5-amino-4-chloro-6-ethylamino-2-trifluoromethyl-	91-92	2735
5-amino-4-chloro-2-ethyl-6-ethylamino-	94-95	4035
2-amino-4-chloro-5-fluoro-	110	2209
4-amino-2-chloro-5-fluoro-	191-193 or 197	2209, 2748, 3893
4-amino-6-chloro-5-formamido-	254	3288
2-amino-4-chloro-6-hydrazino-	196-197	4036
5-amino-4-chloro-6-hydrazino-	184	2673
2-amino-4-chloro-5-iodo-	-	2588
2-amino-4-chloro-6-methyl-	-	H, 524, 4032
4-amino-5-chloro-6-methyl-	196-197	H, 2623
5-amino-4-chloro-6-methyl-	98-99	2227
2-amino-4-chloro-6-methylamino-	164	H, 2482, 2624, 2738
4-amino-6-chloro-2-methylamino-	198-200	2288, 2443

Pyrimidine	M.P.(^{o}C)	References
5-amino-4-chloro-6-methylamino-	167	H, 2673
5-amino-2-chloro-4-methylamino-6-trifluoromethyl-	185-187	2735
5-amino-4-chloro-6-methylamino-2-trifluoromethyl-	140-142	2735
4-amino-5-chloromethyl-2,6-dimethyl-	HCl 253	886
4-amino-5-chloromethyl-2-ethyl-	HCl 176-178	4037
4-amino-5-chloromethyl-2-heptafluoropropyl-	110-112	2268
5-amino-4-chloro-6-α-methylhydrazino-	203-204	2673
4-amino-5-chloromethyl-2-methyl-	-	H, 2231, 4038
5-amino-2-chloro-4-methyl-6-methylamino-	133	2334
4-amino-5-chloromethyl-2-pentafluoroethyl-	134-136	2268
2-amino-4-chloro-6-methyl-5-propyl-	175-176	4039
4-amino-5-chloromethyl-2-trifluoromethyl-	191-192	353, 2268
4-amino-6-chloro-2-phenyl-	-	4040
4-amino-6-chloro-5-phenyl-	195-197	2522, 4041
2-amino-4-chloro-6-β-phenylhydrazino-	224-225	4042
2-amino-4-chloro-6-phenyl-5-prop-2'-ynyl-	175-177	3810, 3811
4-amino-6-chloro-2-propyl-	126	H, 915
2-amino-4-chloro-6-propylamino-	105	2738
4-amino-6-chloro-2-trifluoromethyl-	148-151	2193

Pyrimidine	M.P.(^{o}C)	References
5-amino-4-cyclopentylamino-6-fluoro-	125-127	2455
2-amino-4,6-dichloro-	-	H, 2503, 2623
4-amino-2,5-dichloro-	191-193	2571
4-amino-2,6-dichloro-	265 to 271	H, 2280, 2920, 4043
5-amino-2,4-dichloro-	122	H, 750, 2462
5-amino-4,6-dichloro-	142 to 147	H, 751, 2227, 4014
2-amino-5,6-dichloro-1,4-dihydro-4-imino-1-methyl-	173; HCl 270	2286
5-amino-4,6-dichloro-2-ethyl-	86/0.3mm.	4035
4-amino-2,6-dichloro-5-ethylformamido-	221	3288
4-amino-2,6-dichloro-5-formamido-	224	H, 3288
2-amino-4,5-dichloro-6-methyl-	211-212	2623
5-amino-2,4-dichloro-6-trifluoromethyl-	72	2193
5-amino-4,6-dichloro-2-trifluoromethyl-	56-59	2193
2-amino-4,6-difluoro-	sublimes <215	2610
4-amino-2,6-difluoro-	215-216	2610
5-amino-4,6-difluoro-	157-159	2455
2-amino-5-fluoro-	192-193	2209
5-amino-4-fluoro-6-methylamino-	142-144	2455
4-amino-2-fluoromethyl-5-methyl-	158-160	2116
4-amino-5-formamidomethyl-2-trifluoromethyl-	204-205	353
2-amino-5-iodo-4,6-dimethyl-	183 or 192	H, 2623, 3869
4-amino-5-iodo-2,6-dimethyl-	141-142	H, 3869

TABLE XXXI

Pyrimidine	M.P.(^{o}C)	References
4-amino-5-iodomethyl-2-methyl-	HI 212	2836
2-amino-4-methyl-6-trifluoromethyl-	124 or 128	2516, 2564
4-amino-6-methyl-2-trifluoromethyl-	173-174	2193
4-amino-5-thioformamidomethyl-2-trifluoromethyl-	184-185	353
4-amino-2,5,6-trichloro-	168	H, 2690
4-amino-2,5,6-trifluoro-	158	2618
4-amino-2-trifluoromethyl-	180-181	2193
4-amino-6-trifluoromethyl-	165-170	443
5-anilino-2,4-dichloro-	95-97	2567, 3850
4-anilino-2,5,6-trichloro-	83-84	2690
4-anilino-2-trifluoromethyl-	126-128	2217
4-azido-5-azidomethyl-2-chloro-	-	2596
5-azidomethyl-2,4-dichloro-	-	2596
4-aziridino-5-bromo-2-chloro-6-methyl-	127-128	2714
4-aziridino-6-chloro-	53-55	2729
4-aziridino-2-chloro-5-fluoro-	74-76	2570
2-aziridino-4-chloro-6-methyl-	-	3817
4-aziridino-2-chloro-5-methyl-	77-79	2255
4-aziridino-2-chloro-6-methyl-	-	3817
2-aziridino-4,6-dichloro-	120-121	H, 2255
4-aziridino-2,6-dichloro-	110-111	H, 2255
4-aziridino-2,5-dichloro-6-methyl-	115-116	2714
4-benzylamino-2-chloro-	132-133	2674

Pyrimidine	M.P.(^{o}C)	References
4-benzylamino-6-chloro-	121	623
4-benzylamino-2-chloro-5-methyl-	129-130	2674
2,4-bisdiethylamino-6-fluoro-	23	2618
2,4-bisdimethylamino-5,6-difluoro-	96-97	2618
5-bromo-2-butylamino-	97-98	2746
5-bromo-2-t-butylamino-	79	2746
5-bromo-4-butylamino-	147/12mm.; pic.144-145	2746
5-bromo-4-t-butylamino-	132/12mm.; pic.186-187	2746
5-bromo-4-t-butyl-6-hydrazino-	116-117	2574
5-bromo-2-diethylamino-4,6-dimethyl-	109-110/0.6mm.	2519
5-bromo-4,6-dihydrazino-	150	2756
5-bromo-1,2-dihydro-2-imino-1-methyl-	HCl 261; pic.184-185	2627
5-bromo-1,2-dihydro-2-imino-1-methyl-4-methylamino-(or tautomer)	HI 255; HCl 281	2624
5-bromo-1,4-dihydro-4-imino-1-methyl-2-methylamino-(or tautomer)	HI 250; HCl 209-210	2624
5-bromo-1,2-dihydro-2-imino-1,4,6-trimethyl-	HCl 250; pic.191-192	2626
5-bromo-1,2-dihydro-1-methyl-2-methylimino-	HI 238; pic.210-211	2627
5-bromo-4-dimethylamino-1,2-dihydro-2-imino-1-methyl-	HI 250; HCl 227-230	2624
5-bromo-4-dimethylamino-1,2-dihydro-1-methyl-2-methylimino-	HI 186; HCl 208-209	2624
5-bromo-2-dimethylamino-4-methylamino-	70-71	2624

TABLE XXXI

Pyrimidine	M.P.(^{o}C)	References
5-bromo-4-dimethylamino-2-methylamino-	121-122	2624
5-bromo-4,6-dimethyl-2-methylamino-	130; pic.164	2626
5-bromo-4,6-dimethyl-2-piperidino-	52-53	2623
5-bromo-2-ethylamino-	122; pic.158	2626
5-bromo-1-ethyl-1,2-dihydro-2-imino-	HCl 258; pic.195	2626
5-bromo-2-hexylamino-	70	2573
5-bromo-2-hydrazino-	205-206	4044
5-bromo-4-hydrazino-6-methyl-	194-195	2574
5-bromo-4-hydrazino-6-methyl-2-propyl-	89-90	3798
5-bromo-4-hydrazino-6-phenyl-	182-183	2574, 2595
5-bromo-2-methylamino-	121; pic.181	2627
5-bromomethyl-4-dimethylamino-2-methyl-	HBr 221	2651
5-bromo-4-methyl-2-methylamino-	87-88; pic.153	2450, 2602
5-bromomethyl-2-methyl-4-methylamino-	HBr 215-220	2653
2-butylamino-5-chloro-4,6-dimethyl-	40-42; pic.100	2623
4-butylamino-2,5,6-trichloro-	62-65	2690
4-chloro-2,6-bisdiethylamino-	133-138/0.25mm.	2708, 4045
5-chloro-2,4-bisdiethylamino-6-fluoro-	117-119/2mm.	2618
4-chloro-2,6-bisdimethylamino-	-	H, 2624
4-chloro-2,6-bisethylamino-	66-68	3992, 4045

Pyrimidine	M.P.(^{o}C)	References
4-chloro-2,6-bisisopropylamino-	78-83	4045
2-chloro-4,6-bismethylamino-	258-259	2624
4-chloro-2,6-bismethylamino-	133	H, 2624
4-chloro-2-β-cyclohexylhydrazino-6-diethylamino-	139-141	2708
4-chloro-2-cyclopropylamino-6-methyl-	HCl 74-76	2564
5-chloro-2-diethylamino-4,6-difluoro-	55-57/0.6mm.	2706, 2707
4-chloro-2-diethylamino-6-hydrazino-	HCl 225	2708
4-chloro-6-diethylamino-2-hydrazino-	98-99	2708
4-chloro-2,6-dihydrazino-	>280	4036
4-chloro-1,6-dihydro-6-imino-1,2-dimethyl-	HI 232-234	719
4-chloro-3,6-dihydro-6-imino-2,3-dimethyl-	HI 244-246	719, 2781
4-chloro-1,6-dihydro-6-imino-1-methyl-	HI 207	2781
5-chloro-1,2-dihydro-2-imino-1-methyl-	HI 260; pic.181-182	2633
4-chloro-2,3-dihydro-2-imino-3-methyl-6-methylamino- (or tautomer)	HI 222; HCl 239	2624
4-chloro-3,6-dihydro-6-imino-3-methylamino- (or tautomer)	HI 245-247; HCl 273	2624
4-chloro-3,6-dihydro-3-methyl-6-methylimino-	HI 223; HCl 238	2781
4-chloro-6-dimethylamino-	102-103	827
4-chloro-6-dimethylamino-2,3-dihydro-2-imino-3-methyl-	HI 204; HCl 233	2624

Pyrimidine	M.P.(^{o}C)	References
4-chloro-6-dimethylamino-2,3-dihydro-3-methyl-2-methylimino-	HI 195-197; HCl 201	2624
2-chloro-4-dimethylamino-5-methyl-	180-182	2703
4-chloro-2-dimethylamino-6-methyl-	-	H, 2564
2-chloro-4-dimethylamino-6-methylamino-	206-207	2624
4-chloro-2-dimethylamino-6-methylamino-	HCl 219-221	H, 2624
4-chloro-6-dimethylamino-2-methylamino-	181-182	2624
5-chloro-4,6-dimethyl-2-piperidino-	57-58	2623
4-chloro-2,6-dipiperidino-	-	H, 2922
4-chloro-6-ethylamino-5-formamido-2-trifluoromethyl-	110-111	2735
4-chloro-2-ethylamino-6-methyl-	92-94	2564
4-chloro-6-ethylamino-2-methyl-	73-75	3992
4-chloro-6-ethylamino-2-methylamino-	85-88	4045
4-chloro-1-ethyl-1,2-dihydro-2-imino-6-methyl-	HI 234-235	2707
4-chloro-2-ethylmethylamino-6-methyl-	58-59/0.3mm.	2564
2-chloro-4-hydrazino-	290-291	2753
4-chloro-6-hydrazino-	164-165	H, 3868
5-chloro-2-hydrazino-	181-182	4044
4-chloro-6-hydrazino-2-methyl-	170-172	3798
4-chloro-6-hydrazino-5-methylamino-	159	2752

Pyrimidine	M.P.(^{o}C)	References
4-chloro-6-hydrazino-2-propyl-	90-92	3798
4-chloro-2-isopropylamino-6-methyl-	HCl 100-102	2564
2-chloro-4-methylamino-	128-129	2288
4-chloro-2-methylamino-	123-124	2288
4-chloro-5-methylamino-6-α-methylhydrazino-	119-121	2752
2-chloro-5-methyl-4-methylamino-	180-182	2703
4-chloro-6-methyl-2-methylamino-	136	H, 2564
4-chloro-6-methyl-2-piperidino-5-αβββ-tetrachloroethyl-	115	524
4-chloro-6-methyl-2-propylamino-	49-50	2564
4-chloro-6-β-phenylhydrazino-	160-163	4042
4-chloro-6-piperidino-	78	623
5-α-chlorovinyl-2-dimethylamino-4-methyl-	100/0.08mm.	458
2,4-diamino-6-anilino-5-bromo-	208-209	2621
2,4-diamino-6-anilino-5-chloro-	218	2690
2,4-diamino-5-bromo-	-	H, 2624
4,6-diamino-5-bromo-	213	H, 827
2,4-diamino-5-bromo-6-chloro-	218	2621, 4046
2,4-diamino-5-bromo-6-piperidino-	155-156	4047
2,5-diamino-4-butylamino-6-chloro-	125-126	2505, 2761
2,4-diamino-6-chloro-	200-201	H, 2285
4,5-diamino-6-chloro-	253-254	H, 2568, 2733, 4048

TABLE XXXI

Pyrimidine	M.P.(°C)	References
4,5-diamino-6-chloro-2-ethyl-	262	4035
2,5-diamino-4-chloro-6-ethylamino-	208-209	2505
4,5-diamino-2-chloro-6-methyl-	265	H, 874
4,5-diamino-6-chloro-2-methyl-	239-243	2835
4,6-diamino-5-chloro-2-phenyl-	218	2297
4,5-diamino-2-chloro-6-trifluoromethyl-	226-227	2193
4,5-diamino-6-chloro-2-trifluoromethyl-	243-244	2193
2,4-diamino-5,6-dichloro-	-	H, 2621, 4046
2,4-diamino-5-fluoro-	166-167	2748
2,4-diamino-6-fluoro-	198-199	2610
4,5-diamino-6-fluoro-	218-220	2455
4,6-diamino-2-trifluoromethyl-	243	2193
2,4-dianilino-5-bromo-6-methyl-	111-112	2714
2,4-dianilino-5-chloro-6-methyl-	HCl 284-285	2714
2,4-diaziridino-5-bromo-6-chloro-	135-136	2255
4,6-diaziridino-5-bromo-2-chloro-	126-128	2728
2,4-diaziridino-6-chloro-	-	H, 2255 (neg.)
2,4-diaziridino-6-chloro-5-fluoro-	128	2728
4,6-diaziridino-2,5-dibromo-	135-150 (?)	2728
4,6-diaziridino-2,5-dichloro-	125-126	2728
2,4-diaziridino-5-fluoro-	94-95	2570
2,4-diaziridino-6-fluoro-	61	2618

Pyrimidine	M.P.(°C)	References
4,5-dichloro-2,6-bisdiethylamino-	160-162	2618
2,4-dichloro-6-diethylamino-	62-64	2708
4,6-dichloro-2-diethylamino-	132-134/15mm.	2708
4,5-dichloro-2,6-dihydrazino-	>280	4036
2,4-dichloro-5-dimethylamino-	90-91	2462
4,6-dichloro-2-dimethylamino-	55-56	H, 2564
2,4-dichloro-6-ethylamino-	98-100	3992
4,6-dichloro-2-ethylamino-	61-64	4045
4,6-dichloro-5-hydroxyiminoethyl-	ca 100	2758
4,6-dichloro-2-isopropylamino-	liquid	4045
4,6-dichloro-5-*N*-methylacetamido-	149-150	2752
2,4-dichloro-5-methylamino-	148-149	2462
4,6-dichloro-2-methylamino-	158	H, 4045
4,6-dichloro-5-methylamino-	78-79	2752
2,4-dichloro-6-β-phenylhydrazino-	172	4042
4,6-dichloro-2-piperidino-	81-82 (?)	2707
1,2-dihydro-2-imino-5-iodo-1-methyl-	HI 239; pic.205	2633
4-dimethylamino-2,6-difluoro-	85-86	2618
4-hydrazino-2-propyl-6-trifluoromethyl-	66-67	2224, 3798
5-iodo-2-methylamino-	136-137; pic.209	2633
2,4,5-triamino-6-chloro-	227	H, 2503, 4049

TABLE XXXI

Pyrimidine	M.P.(^{o}C)	References
2,4,5-triamino-6-trifluoromethyl-	198; sul.175	443, 999
4,5,6-triamino-2-trifluoromethyl-	263-264	2193
2,4,6-triaziridino-5-bromo-	157-159	2730
2,4,6-triaziridino-5-chloro-	168-170	2730
2,4,6-triaziridino-5-fluoro-	120-121	2730
2,4,5-trichloro-6-diethylamino-	50-51	2690
4,5,6-trichloro-2-diethylamino-	76-78	2690
4,5,6-trichloro-2-piperidino-	73-74	2707

TABLE XXXII. Amino-Nitropyrimidines (*H* 545)

4-acetamido-5-methylnitrosoamino-	137	3494
2-amino-4-anilino-6-methyl-5-nitro-	179	H, 2472
4-amino-2-azido-5-nitro-	181-182	2712
2-amino-4-benzylamino-6-methyl-5-nitro-	203	2798, 4050
4-amino-2-benzylamino-6-methyl-5-nitro-	142-143	2413
4-amino-2-benzylamino-5-nitro-	-	4051
4-amino-6-benzylamino-5-nitro-	202	H, 783
2-amino-4,6-bismethylamino-5-nitroso-	-	3813

Pyrimidine	M.P.(°C)	References
2-amino-4-butylamino-5-nitro-	135-136	2462
4-amino-2-butylamino-5-nitro-	118-120	2462
4-amino-6-butylamino-5-nitro-	140-170(!)	2365
2-amino-4-cyclohexylamino-6-methyl-5-nitro-	185	2798, 4050
4-amino-2-cyclohexylamino-6-methyl-5-nitro-	141-142	2413
2-amino-4-cyclohexylamino-5-nitro-	199-200	2462
4-amino-2-cyclohexylamino-5-nitro-	131-132	2462
4-amino-6-cyclohexylamino-5-nitro-	203-204	783
4-amino-6-diethylamino-2-methyl-5-nitro-	109-110	971
4-amino-1,6-dihydro-6-imino-1-methyl-5-nitro-(or tautomer)	HI 223; pic.214	2626
2-amino-4-dimethylamino-5-nitro-	142-143	2626, 2462
4-amino-2-dimethylamino-5-nitro-	213-215	H, 2462
4-amino-6-dimethylamino-5-nitro-	162-163 or 135(?)	H, 783, 3855
4-amino-6-dimethylamino-5-nitroso-2-phenyl-	-	3824, 4040
2-amino-4-dimethylamino-6-phenylazo-	192-193	4052
2-amino-4,6-dimethyl-5-nitro-	220	H, 2787
4-amino-2,6-dimethyl-5-nitro-	159-160	874
4-amino-6-dipropylamino-5-nitro-	115-117	3856
4-amino-2-ethylamino-5-nitro-	179	2461

TABLE XXXII

Pyrimidine	M.P.(^{o}C)	References
4-amino-2-hydrazino-5-nitro-	>225	2712
4-amino-6-hydrazino-5-nitro-	>264	2466
2-amino-4-methylamino-5-nitro-	249-250	2462, 2711
4-amino-2-methylamino-5-nitro-	-	H, 2462
4-amino-6-methylamino-5-nitro-	248; pic.265	H, 753, 2626
4-amino-6-methylamino-5-nitroso-2-phenyl-	-	3824
2-amino-4-methylamino-6-phenylazo-	185-187	4052
4-amino-6-methyl-2-methylamino-5-nitro-	198-199	874
4-amino-6-methylnitroamino-	269	2433
2-amino-4-methyl-5-nitro-6-piperidino-	96	2798, 4050
4-amino-5-methylnitrosoamino-	139	3494
4-amino-5-nitro-2-piperidino-	153 or 156	2712, 4053
4-amino-5-nitro-2-propylamino-	-	4051
4-amino-5-nitro-6-ureido-	>400	2164
2-benzyl-4,6-dihydrazino-5-nitro-	170-171	2688
2,4-bisbenzylamino-6-methyl-5-nitro-	118-119	2463
2,4-bisbenzylamino-5-nitro-	176 to 180	2463, 4054
4,6-bisbenzylamino-5-nitro-	115-116	2463
2,4-bisbutylamino-6-methyl-5-nitro-	94	2463
2,4-bisbutylamino-5-nitro-	129-131	2463

Pyrimidine	M.P.(°C)	References
4,6-bisbutylamino-5-nitro-	49-50	2463
2,4-biscyclohexylamino-6-methyl-5-nitro-	118-119	2463
2,4-biscyclohexylamino-5-nitro-	166	2463
4,6-biscyclohexylamino-5-nitro-	136-137	784, 2463
2,4-bisdiethylamino-5-nitro-	52	2712
2,4-bisdimethylamino-5-nitro-	83-84	H, 2712
2,4-bisethylamino-6-methyl-5-nitro-	125 or 128	2461, 2463
2,4-bisethylamino-5-nitro-	172 or 175	2461, 2463
4,6-bisethylamino-5-nitro-	83-84	2463
2,4-bisisobutylamino-6-methyl-5-nitro-	86-88	2463
2,4-bisisobutylamino-5-nitro-	137-138	2463
4,6-bisisobutylamino-5-nitro-	77-78	2463
2,4-bisisopropylamino-6-methyl-5-nitro-	75-76	2463
2,4-bisisopropylamino-5-nitro-	160-161	2463
4,6-bisisopropylamino-5-nitro-	127-128	2463
2,4-bismethylamino-5-nitro-	251 to 263	H, 2462, 2463, 2711
4,6-bismethylamino-5-nitro-	193-194	H, 2463
2-butylamino-5-nitro-	124	H, 2746
2-t-butylamino-5-nitro-	128	2746
4-cyclohexylamino-2-dimethylamino-5-nitro-	150-152	2927
4,6-diacetamido-2-amino-5-nitroso-(?)	199-200	3333, 4055

Pyrimidine	M.P.(^{o}C)	References
4,6-diacetamido-5-nitroso-2-piperidino-	186	3333
2,4-diamino-6-anilino-5-nitroso-	255	2260
2,4-diamino-6-benzylamino-5-nitroso-	-	3852
4,6-diamino-2-benzylamino-5-nitroso-	224-226	4056
4,6-diamino-2-benzyl-5-nitroso-	247-248	1169
2,4-diamino-6-cyclohexylamino-5-nitroso-	225-226	3854
4,6-diamino-2-dimethylamino-5-nitroso-	282	H, 2297, 2473
4,6-diamino-2-ethyl-5-nitroso-	254	1169
4,6-diamino-2-hydroxyamino-5-nitroso-	>300	2473
2,4-diamino-6-methyl-5-nitro-	235-236	H, 1275, 2790
4,6-diamino-2-methyl-5-nitro-	>360	H, 2445
4,6-diamino-2-methyl-5-nitroso-	306	H, 1169
2,4-diamino-6-methyl-5-phenylazo-	224-226	979
2,4-diamino-5-nitro-	>350	H, 748, 2711
4,6-diamino-5-nitro-	>400	H, 2164, 2332, 2333, 3244, 3871
4,6-diamino-5-nitro-2-phenyl-	269-270	2445
4,6-diamino-5-nitroso-	-	H, 1169 (neg.)
4,6-diamino-5-nitroso-2-phenyl-	243-244	H, 1169
4,6-diamino-5-nitroso-2-piperidino-	195-197	2473

Pyrimidine	M.P.(°C)	References
4,6-diamino-5-nitroso-2-propyl-	228 or 231	2294, 2868
4,6-diamino-5-phenylazo-	285-287	H, 2296
2,4-dianilino-6-methyl-5-nitro-	144-145	2463
2,4-dianilino-5-nitro-	198 to 203	H, 1372, 2463
4,6-dianilino-5-nitro-	168-169	2463
2,4-diazetidino-6-methyl-5-nitro-	141-142	2714
2,4-diaziridino-6-diethylamino-5-nitro-	64	2731
4,6-diaziridino-2-diethylamino-5-nitro-	107	2731
2,4-diaziridino-6-methyl-5-nitro-	117-118 or 120-122	2714, 3817
2,4-diaziridino-5-nitro-6-piperidino-	97-98	2731
4,6-diaziridino-5-nitro-2-piperidino-	155-160	2731
4-diethylamino-6-methylamino-5-nitro-	165-166	2460
4,6-dihydrazino-2-methyl-5-nitro-	198	2688
2,4-dihydrazino-5-nitro-	270	2712
4,6-dihydrazino-5-nitro-	203 or 206	2712, 4057
4-dimethylamino-1,2-dihydro-2-imino-1-methyl-5-nitro-	HI 236	2626
2-dimethylamino-4-ethylamino-5-nitro-	98	2461
2-dimethylamino-4-hydrazino-5-nitro-	221	2712
4-dimethylamino-2-hydrazino-5-nitro-	182-183	2712
2-dimethylamino-4-methylamino-5-nitro-	179-181	2927
4-dimethylamino-6-methylamino-5-nitro-	96-97	2433, 2454

Pyrimidine	M.P.(°C)	References
4-dimethylamino-6-nitroamino-	241	2433
2-hydrazino-5-phenylazo-	210-211	1124
2-methyl-4,6-bismethylamino-5-nitro-	200-201	2454
4-methyl-2,6-bismethylamino-5-nitro-	228 or 235	2463, 2713
4-methyl-5-nitro-2,6-bispropylamino-	98-99	2463
4-methyl-5-nitro-2,6-dipiperidino-	99-100	2714
5-nitro-2,4-bispropylamino-	-	H, 2463
5-nitro-4,6-bispropylamino-	62-63	2463
5-nitro-2,4-dipiperidino-	113-114	2712
2,4,6-triacetamido-5-nitroso-(?)	214	3333, 4055
2,4,6-triamino-5-nitro-	>350	H, 2164
2,4,6-triamino-5-nitroso-	345-346	H, 1169, 4058
2,4,6-triamino-5-phenylazo-	-	H, 979, 2853
2,4,6-triaziridino-5-nitro-	156-160	2730

TABLE XXXIII. Amino-Oxypyrimidines Without *N*-1 or *N*-3-Substituents (*H* 547)

Pyrimidine	M.P.(°C)	References
4-acetamido-5-β-acetoxyethyl-2,6-dihydroxy-	268-269	2202
2-acetamido-5-acetoxy-4-hydroxy-	ca 200	2210
5-acetamido-4-amino-2,6-dihydroxy-	-	H, 4059-4062

Pyrimidine	M.P.(°C)	References
2-acetamido-5-amino-4-hydroxy-	230-234	2464
5-acetamido-4-amino-6-hydroxy-	-	1259
5-acetamido-2-amino-4-hydroxy-6-methyl-	295-298 or 305-308	2223, 2303
5-acetamido-4-amino-6-hydroxy-2-methyl-	-	4059
4-acetamido-6-butoxy-	95-96	2837
5-acetamido-2,4-diamino-6-hydroxy-	-	H, 4059
5-acetamido-2,4-diethoxy-	136	750
5-acetamido-2,4-dihydroxy-6-methylamino-	>280	1005
4-acetamido-2,6-dimethoxy-	183-185	2920
4-acetamido-2-dimethylamino-6-ethoxy-	166-167	2837
4-acetamido-2-dimethylamino-6-isopropoxy-	156-157	2837, 2841
4-acetamido-2-dimethylamino-6-methoxy-	187-188	2837
4-acetamido-2-dimethylamino-6-propoxy-	165-166	2837
4-acetamido-6-ethoxy-	130-131	2837
4-acetamido-2-ethoxy-6-hydroxy-	258-259	4043, 4063
4-acetamido-2-hydroxy-	326-328	H, 3227, 4064
4-acetamido-5-hydroxy-	179-180	2584
4-acetamido-6-hydroxy-	288-289	1106, 3286
5-acetamido-4-hydroxy-2,6-dimethyl-	275-277	H, 2223
4-acetamido-4-hydroxy-2-methoxy-	275-280	4043
2-acetamido-4-hydroxy-6-methyl-	220-221	H, 4020
4-acetamido-6-hydroxy-5-methyl-	303	1106

TABLE XXXIII

Pyrimidine	M.P.(°C)	References
2-acetamido-4-hydroxy-6-phenyl-	254-255	4065
4-acetamido-6-isopropoxy-	105-106	2837
2-acetamido-5-methoxy-	160	4066
4-acetamido-6-methoxy-	138-139	2837
2-acetamido-6-propoxy-	135-136	2837
5-allyl-2-amino-4-hydroxy-	202-206	2194
4-allyloxy-2-amino-	85-87	2194
4-allyloxy-6-amino-5-methoxy-	41-42	2524
4-allyloxy-6-amino-5-methyl-	98-103	3479
4-amino-5-aminomethyl-2-hydroxy-	-	H, 3803
2-amino-5-γ-aminopropyl-4-hydroxy-6-methyl-	HCl 300	3262
2-amino-4-amylamino-5-formamido-6-hydroxy-	233	2738
2-amino-5-amyl-4-hydroxy-6-methyl-	260-262	3849
2-amino-4-anilino-6-hydroxy-	145	2430
2-amino-5-anilino-4-hydroxy-	294-295	2567
4-amino-2-anilino-6-hydroxy-	274-275	H, 2430
4-amino-6-anilino-2-hydroxy-	344-345	2475
4-amino-5-benzamido-6-hydroxy-2-methyl-	>350	3186
5-amino-4-benzylamino-2,6-dihydroxy-	213-214	4067
2-amino-4-benzylamino-5-formamido-6-hydroxy-	250	2738
2-amino-4-benzylamino-6-hydroxy-	217 or 223	2286, 2738, 3228
4-amino-6-benzylamino-2-hydroxy-	306-308	2475

Pyrimidine	M.P.(°C)	References
5-amino-4-benzylamino-2-hydroxy-	218-223	4054
2-amino-4-benzylamino-6-hydroxy-5-*N*-methylformamido-	218-224	4068
4-amino-2-benzyl-5-diethylaminoethyl-6-hydroxy-	174	4069
2-amino-4-benzyl-6-hydroxy-	278-280	4070
2-amino-4-benzyl-6-hydroxy-5-β-hydroxyethyl-	216-217	2229
4-amino-2-benzyl-5-hydroxymethyl-	HCl 214	3806
4-amino-5-benzyloxy-	142-143	2584
2-amino-5-benzyloxy-4-hydroxy-	253-254 or 243	H, 79, 2210
4-amino-2-benzyloxy-5-hydroxymethyl-	129-130	4071
4-amino-2,5-bismethoxymethyl-	86-88	4072
2-amino-5-butoxy-	71 to 75	2212, 4073, 4074
2-amino-5-s-butoxy-	62-64	2212, 4074
4-amino-6-butoxy-	126-127	2837
4-amino-6-t-butoxy-	66-67	2837
4-amino-6-butoxy-2-ethoxy-	139/1mm.	2840
4-amino-5-butoxy-6-β-ethoxyethoxy-	98-99	2524
2-amino-5-s-butoxy-4-hydroxy-	227-230	2212
2-amino-5-t-butoxy-4-hydroxy-	246-248	2212
4-amino-6-butoxy-2-isobutoxy-	82-83	2840
2-amino-4-butoxy-6-methyl-	120/4mm.	2580

Pyrimidine	M.P.(^{o}C)	References
5-amino-4-butylamino-2,6-dihydroxy-	228	4075
2-amino-4-butylamino-5-formamido-6-hydroxy-	238	2738
2-amino-4-butylamino-6-hydroxy-	230	2738
5-amino-4-butylamino-2-hydroxy-	crude	281
2-amino-5-butyl-4-hydroxy-6-methyl-	266-268	3849
4-amino-2-butyl-5-hydroxymethyl-	HCl 196	3806
2-amino-5-butyl-4-hydroxy-6-phenyl-	309-314	H, 2941
2-amino-4-cyclohexylamino-5-formamido-6-hydroxy-	254-255	2738
2-amino-4-cyclohexylamino-6-hydroxy-	280	2738
4-amino-6-cyclohexylamino-2-hydroxy-	323 or 329	2475, 3854
2-amino-5-cyclohexyloxy-	72-73	4076
2-amino-4-cyclopentylamino-5-formamido-6-hydroxy-	256	2738
2-amino-4-cyclopentylamino-6-hydroxy-	273	2738
4-amino-2,6-dibenzyloxy-	108	4077
4-amino-2,6-dibutoxy-	160-165/2mm.	2840
4-amino-2,6-di-s-butoxy-	147-152/2mm.	2840
4-amino-2,6-diethoxy-	108 or 112	2840, 4043
4-amino-5,6-diethoxy-	83-84	2524
5-amino-2,4-diethoxy-	64-66	749, 750, 887
4-amino-5-diethylaminoethyl-6-hydroxy-2-methyl-	204	4069
4-amino-5-diethylaminoethyl-6-hydroxy-2-phenyl-	174	4069

Pyrimidine	M.P.(°C)	References
4-amino-6-diethylamino-2-hydroxy-	311-313	2475
4-amino-5-diethylaminomethyl-6-hydroxy-2-methyl-	175	H, 4078
4-amino-2,5-diethyl-6-hydroxy-	202-204	2270
2-amino-4,5-dihydro-5,6-dihydroxy-4-oxo-5-methyl-	-	2878
2-amino-4,5-dihydroxy-	$\nless$340	H, 79, 2210
2-amino-4,6-dihydroxy-	-	H, 2286
4-amino-2,5-dihydroxy-	-	H, 3907
4-amino-2,6-dihydroxy-	-	H, 752, 2503, 3348, 4079
4-amino-5,6-dihydroxy-	>290	2210
5-amino-2,4-dihydroxy-	-	H, 994
5-amino-4,6-dihydroxy-	HCl 249	H, 2158, 4080
4-amino-2,6-dihydroxy-5-β-hydroxyethyl-	308-310	2202
4-amino-2,6-dihydroxy-5-β-hydroxypropyl-	252-253	2202
2-amino-4,6-dihydroxy-5-methoxy-	$\nless$350	2212, 4081
4-amino-2,6-dihydroxy-5-methyl-	353-355	H, 2306
5-amino-2,4-dihydroxy-6-methyl-	270 to 274	H, 4082, 4083
2-amino-4,6-dihydroxy-5-methylamino-	HCl -	4084
4-amino-2,6-dihydroxy-5-methylamino-	HCl 268-270	4068
5-amino-2,4-dihydroxy-6-methylamino-	260 or >280; HCl >350	1005, 4067, 4085
4-amino-2,6-dihydroxy-5-*N*-methylformamido-	ca.310	4068
2-amino-4,6-dihydroxy-5-α-methylureido-	crude	4084

Pyrimidine	M.P.(^{o}C)	References
5-amino-2,4-dihydroxy-6-phenyl-	235-240	H, 3956
2-amino-4,6-dihydroxy-5-propoxy-	>300	4086
4-amino-2,6-diisopropoxy-	65	2840
2-amino-4,5-dimethoxy-	88	2212
2-amino-4,6-dimethoxy-	pic.199	H, 2765, 2843, 2844
4-amino-2,5-dimethoxy-	180-181	2205, 2586
4-amino-2,6-dimethoxy-	147 to 152	H, 2280, 2593, 2840, 2920, 3259, 4043, 4087-4089
4-amino-5,6-dimethoxy-	88-89	2524, 4090
5-amino-2,4-dimethoxy-	86 to 89	749, 887, 994, 2897
5-amino-4,6-dimethoxy-	95-96	4057
2-amino-4,5-dimethoxy-6-methyl-	117	2764
5-amino-4,6-dimethoxy-2-methyl-	97-98	H, 887
4-amino-2-dimethylamino-6-ethoxy-	68 or 86-87	2837, 2841
4-amino-6-dimethylamino-2-ethoxy-	136-137	2837, 4091
4-amino-5-dimethylaminoethyl-6-hydroxy-2-methyl-	217-218	4069
2-amino-4-dimethylamino-6-hydroxy-	>350	2286
4-amino-2-dimethylamino-6-hydroxy-	286-291	H, 3228, 4092
4-amino-6-dimethylamino-2-hydroxy-	324-326	2475, 2837
5-amino-4-dimethylamino-6-hydroxy-	162-163	2130
5-amino-2-dimethylamino-4-hydroxy-6-methylamino-	HCl 265	2476

Pyrimidine	M.P.(°C)	References
4-amino-2-dimethylamino-6-isopropoxy-	104	2841
4-amino-6-dimethylamino-2-isopropoxy-	105-106	2837, 4091
4-amino-2-dimethylamino-6-methoxy-	93-94	2837, 2841
4-amino-6-dimethylamino-2-methoxy-	158-159	2837
4-amino-5-dimethylaminomethyl-6-hydroxy-2-methyl-	286	H, 4078
4-amino-2-dimethylamino-6-propoxy-	64	2841
4-amino-6-dimethylamino-2-propoxy-	105-106	2837
5-amino-2,4-diphenoxy-	130-132	994
4-amino-2,6-dipropoxy-	42	2840
2-amino-5-ethoxy-	109-111 or 114-115	2212, 4073, 4074
4-amino-2-ethoxy-	82-83 (cf.152) HCl 168	H, 2677, 2837 (cf.35)
4-amino-6-ethoxy-	146 or 152	2837, 3241
2-amino-5-ethoxy-4,6-dihydroxy-	>300	H, 4081
4-amino-6-β-ethoxyethoxy-5-β-methoxyethyl-	66-67	4041
2-amino-5-ethoxy-4-ethoxymethyl-6-hydroxy-	224-225	H, 2980
4-amino-6-β-ethoxyethoxy-5-phenyl-	103-104	4041
2-amino-4-ethoxy-6-hydroxy-	292-294	2610
4-amino-2-ethoxy-5-hydroxymethyl-	148-150	2205, 4071
4-amino-2-ethoxy-6-isobutoxy-	129/1mm.	2840
4-amino-2-ethoxy-6-isopropoxy-	130/1mm.	2840

Pyrimidine	M.P.(^{o}C)	References
2-amino-5-ethoxy-4-methoxy-	110 or 122	2212, 4093
4-amino-2-ethoxy-6-methoxy-	144-145 (cf.108 or 170)	2837, 4094, 4095 (cf. 4043, 4087)
4-amino-6-ethoxy-2-methoxy-	112-113 (cf.142-143)	2837, 4094, 4095 (cf.4043)
4-amino-6-ethoxy-5-methoxy-	64-65	2524, 4096
4-amino-6-ethoxy-2-methoxymethyl-	99	3913
2-amino-4-ethoxy-6-methyl-	99-100; H_2O 36-38	H, 2580, 2623, 2629
4-amino-5-ethoxymethyl-	79; pic.186	2190
4-amino-5-ethoxy-2-methyl-	144	2521
4-amino-6-ethoxy-2-methyl-	122; pic.177	914
2-amino-4-ethoxy-6-methylamino-	123-126	1372
4-amino-5-ethoxymethyl-2-ethyl-	pic.158-160	H, 4025
4-amino-2-ethoxymethyl-6-methoxy-	- (?)	3913
4-amino-5-ethoxymethyl-2-methoxymethyl-	63-65	4072
4-amino-5-ethoxymethyl-2-methyl-	90	H, 920, 922, 923, 1476, 4025, 4026
4-amino-6-ethoxymethyl-2-methyl-	90	H, 2231
4-amino-5-ethoxymethyl-2-phenyl-	132-134	2190, 4025
4-amino-5-ethoxymethyl-2-propyl-	42	2190
4-amino-2-ethoxy-6-phenyl-	108-109	4063
4-amino-2-ethoxy-6-propoxy-	136/1mm.	2840
4-amino-6-ethylamino-2-hydroxy-	336-337	2475

Pyrimidine	M.P.(^{o}C)	References
4-amino-2-ethylamino-5-hydroxymethyl-	133-136	3062
2-amino-5-ethyl-4,5-dihydro-5,6-dihydroxy-4-oxo-	-	2878
4-β-aminoethyl-2,6-dihydroxy-	256-257	2674
5-β-aminoethyl-2,4-dihydroxy-	HBr 309	2514
5-amino-4-ethyl-2,6-dihydroxy-	206-207	1464
5-β-aminoethyl-2,4-dihydroxy-6-methyl-	HBr 312	2514
4-amino-5-*N*-ethylformamido-2,6-dihydroxy-	340	H, 3288
2-amino-4-ethyl-6-hydroxy-	249-251	H, 2564
4-amino-2-ethyl-6-hydroxy-	-	H, 915
5-β-aminoethyl-4-hydroxy-2,6-dimethyl-	HCl 240	3944
2-amino-5-ethyl-4-hydroxy-6-methyl-	286	H, 3849
4-amino-2-ethyl-5-hydroxymethyl-	114; HCl 174	2270, 3806
4-amino-5-ethyl-2-hydroxy-6-methyl-	296-298	H, 937
4-amino-6-ethyl-2-methoxy-	82	4063
4-amino-5-ethyl-6-methoxy-2-methoxymethyl-	83-84	3911
4-amino-5-formamido-2,6-dihydroxy-	>330	4098
4-amino-5-formamido-2-hydroxy-	-	2360, 4099
4-amino-5-formamido-6-hydroxy-	287-288	H, 4100
2-amino-5-formamido-4-hydroxy-6-isoamylamino-	230	2738
2-amino-5-formamido-4-hydroxy-6-isobutylamino-	250-252	2738

Pyrimidine	M.P.(^{o}C)	References
4-amino-5-formamido-2-hydroxy-6-methyl-	-	H, 2360
2-amino-5-formamido-4-hydroxy-6-methylamino-	>350	2483, 2738
2-amino-5-formamido-4-hydroxy-6-(β-methylbutyl)amino-	243	2738
2-amino-5-formamido-4-hydroxy-6-propylamino-	277-279	2738
2-amino-5-hexyl-4-hydroxy-6-methyl-	255; AcOH 157	3849, 4101
2-amino-4-hydrazino-6-hydroxy-	>300	4102
4-amino-2-hydrazino-6-hydroxy-	255 or 260	4103, 4104
5-amino-4-hydrazino-6-hydroxy-	245-248	2466
2-amino-4-hydroxy-	276-278	H, 1379, 2231, 2766
4-amino-2-hydroxy-	HCl 267-268	H, 2272, 2745, 2766, 2915, 4064
4-amino-5-hydroxy-	>250	2584
4-amino-6-hydroxy-	-	H, 357
5-amino-4-hydroxy-	208	H, 2158
4-amino-2-hydroxy-5,6-dimethyl-	332-334	H, 140
4-amino-6-hydroxy-2,5-dimethyl-	-	H, 4097
4-amino-2-hydroxy-5-2'-hydroxycyclopentyl-	244-245	2952
2-amino-4-hydroxy-5-β-hydroxyethyl-	262 or 268-270	2202, 2514
4-amino-2-hydroxy-5-β-hydroxyethyl-	275	2202
2-amino-4-hydroxy-5-hydroxyethyl-6-methyl-	270	H, 2229
2-amino-4-hydroxy-5-hydroxyethyl-6-phenyl-	299	H, 2229

Pyrimidine	M.P.(^{o}C)	References
4-amino-2-hydroxy-5-hydroxymethyl-	-	H, 2670, 4105
4-amino-6-hydroxy-2-hydroxymethyl-	265	2290
2-amino-4-hydroxy-5-γ-hydroxypropyl-6-phenyl-	265-268	4070
2-amino-4-hydroxy-6-isoamylamino-	258	2738
2-amino-4-hydroxy-5-methoxy-	sul.264	H, 2212, 3228, 4074
4-amino-6-hydroxy-5-methoxy-	230-231	3910
2-amino-4-hydroxy-5-methoxy-6-methyl-	223-224	2228
4-amino-6-hydroxy-2-methoxy-5-methyl-	237-238	883
2-amino-4-hydroxymethyl-	145	3004
2-amino-4-hydroxy-6-methyl-	-	H, 3867
4-amino-2-hydroxy-5-methyl-	287-292	H, 805, 3972
4-amino-2-hydroxy-6-methyl-	-	H, 3090
4-amino-6-hydroxy-2-methyl-	-	H, 745, 4097
4-amino-6-hydroxy-5-methyl-	243	1106
5-amino-4-hydroxy-6-methyl-	221-222	2227, 2563
2-amino-4-hydroxy-6-methylamino-	255-257 or 265	1372, 2483, 3228
4-amino-2-hydroxy-6-methylamino-	>360	2475
4-amino-6-hydroxy-2-methylamino-	228-230	3228, 4092
5-amino-2-hydroxy-4-methylamino-	220	H, 4106
5-amino-4-hydroxy-6-methylamino-	sul. -	H, 2182

Pyrimidine	M.P.(^{o}C)	References
4-amino-5-hydroxymethyl-2,6-dimethyl-	185	886
4-amino-5-hydroxymethyl-2-methoxy-	171 to 174	2205, 3369, 3372, 4071
4-amino-5-hydroxymethyl-2-methoxymethyl-	128-130	4072
4-amino-5-hydroxymethyl-2-methyl-	194 or 196	H, 1478, 4107
4-amino-5-hydroxymethyl-2-methylamino-	142-144	3062
5-amino-2-hydroxy-4-methyl-6-methylamino-	ca.295	H, 2334
4-amino-6-hydroxy-2-methyl-5-methylaminomethyl-	HCl 278	H, 4078
4-amino-2-hydroxy-6-methyl-5-phenyl-	214	4108
4-amino-5-hydroxymethyl-2-phenyl-	HCl 199	H, 3806
2-amino-4-hydroxy-6-methyl-5-propyl-	274 or 293-294	3849, 4039
4-amino-2-hydroxy-6-methyl-5-propyl-	218-219	H, 927
4-amino-5-hydroxymethyl-2-propyl-	116; HCl 168	3804, 3806
4-amino-5-hydroxymethyl-2-propylamino-	114-117	3062
2-amino-4-hydroxy-6-phenyl-	304	H, 3839, 4109
4-amino-2-hydroxy-5-phenyl-	335	H, 4108
4-amino-6-hydroxy-2-phenyl-	264-265	H, 3861
2-amino-4-hydroxy-6-phenylhydrazino-	250-252	4052
2-amino-4-hydroxy-6-piperidino-	334-335	3854
4-amino-2-hydroxy-6-piperidino-	330	3854

Pyrimidine	M.P.(°C)	References
2-amino-4-hydroxy-5-propoxy-	238-239	2212
4-amino-6-hydroxy-2-propyl-	290	H, 915
2-amino-4-hydroxy-6-propylamino-	257	2738
2-amino-5-isobutoxy-	66-67	4073
4-amino-6-isobutoxy-	126-127	2837
2-amino-5-isopropoxy-	74-75	4066, 4073
4-amino-2-isopropoxy-	75-76	2837
4-amino-6-isopropoxy-	89 or 93-94	2837, 3241
4-amino-2-isopropoxy-6-methoxy-	98-99	2837, 4095
4-amino-6-isopropoxy-2-methoxy-	154-156/3mm.	2840
4-amino-6-isopropoxy-5-methoxy-	111-112	2524, 4090, 4096
4-amino-6-isopropoxy-2-methoxymethyl-	84	3913
2-amino-4-isopropoxy-6-methyl-	67-68	2580
2-amino-4-isopropoxy-6-methylamino-	105-107	2482
4-amino-5-isopropoxymethyl-2-methoxymethyl-	HCl 69-70	4072
2-amino-4-methoxy-	118-120	H, 2765, 2766, 4110
2-amino-5-methoxy-	80-83 or 86-88	2212, 4073, 4074, 4081
4-amino-2-methoxy-	167 to 170	2677, 2678, 2765, 2766, 2837
4-amino-5-methoxy-	118	2584
4-amino-6-methoxy-	151 or 156-157; pic.202	827, 2765, 2837, 3294
5-amino-4-methoxy-	63 or 77	2158, 4111

Pyrimidine	M.P.(^{o}C)	References
4-amino-5-methoxy-2,6-bismethoxymethyl-	80-81	3911
4-amino-2-methoxy-5,6-dimethyl-	180-181	2280
4-amino-6-methoxy-2,5-dimethyl-	110	4112
5-amino-4-methoxy-2,6-dimethyl-	86-87	887
2-amino-5-β-methoxyethoxy-	80-81	4066
4-amino-2-β-methoxyethyl-6-methyl-	254-255	3913
4-amino-5-β-methoxyethyl-2-methyl-	108-113; pic.209	923
4-amino-6-methoxy-2-methoxymethyl-	93-94	3912, 3913
2-amino-4-methoxy-6-methyl-	156-157	H, 2580
2-amino-5-methoxymethyl-	124-126	4076
4-amino-2-methoxymethyl-	111-112	3913
4-amino-5-methoxy-2-methyl-	93-99	4113
4-amino-5-methoxy-6-methyl-	122	3243
5-amino-4-methoxy-2-methyl-	55-57	887
5-amino-4-methoxy-6-methyl-	70-72	2763
2-amino-4-methoxy-6-methylamino-	135-137	2482
5-amino-4-methoxy-6-methylamino-	139-140	2725
4-amino-2-methoxymethyl-5-methyl-	124-125	4072
4-amino-2-methoxymethyl-5-propoxymethyl-	HCl 47-48	4072
4-amino-5-methoxymethyl-2-propyl-	-	4114, 4115

Pyrimidine	M.P.(°C)	References
4-amino-2-methoxy-6-phenoxy-	137-138	4043, 4087, 4094
4-amino-2-methoxy-6-propoxy-	161-165/3mm.	2840
4-amino-5-methoxy-6-propoxy-	70-71	2524, 4096
5-aminomethyl-2,4-dihydroxy-	HCl 254; pic.223	H, 2071, 4116
5-aminomethyl-4,6-dimethoxy-2-methyl-	54	2681
2-aminomethyl-4-hydroxy-6-methyl-	HCl 274; HBr 266	2875, 4117
5-aminomethyl-4-hydroxy-2-methyl-	HCl 283	H, 1475, 2175
5-aminomethyl-4-methoxy-2-methyl-	46	H, 2681
2-amino-4-methyl-6-propoxy-	61-62	H, 2580
2-amino-5-phenoxy-	120	2212
2-amino-5-propoxy-	75 or 84	2212, 4066, 4073, 4074
4-amino-2-propoxy-	77-78	2837
4-amino-6-propoxy-	132-133	2837
2-amino-4,5,6-trihydroxy-	$\nless$320	H, 135
5-amino-2,4,6-trihydroxy-	-	H, 609
2-amino-4,5,6-trimethoxy-	110 or 113-115	2212, 4118
4-amylamino-2-hydroxy-	105-107	3185
2-amylguanidino-4-hydroxy-6-methyl-	274	2234
4-amyloxy-4-methyl-2-piperidino-	142/0.5mm.; HCl 68	2699
4-anilino-2,6-dihydroxy-	331-332 or 338-340	2802, 2803, 3320
5-anilino-2,4-dihydroxy-	>300	H, 2567, 3850
5-anilino-2,4-dihydroxy-6-methyl-	-	2567, 3850

TABLE XXXIII

Pyrimidine	M.P.(^{o}C)	References
4-anilino-2-ethoxy-	121-122	2690
4-anilino-6-ethoxy-	122-123	2690
4-anilino-2-hydroxy-	266	H, 2153
4-anilino-6-hydroxy-	245-250	2153
2-anilino-4-hydroxy-6-methyl-	244	H, 2148
4-anilino-6-hydroxy-2-methyl-	270	H, 2153
4-anilino-5-hydroxymethyl-2-methyl-	132-133	4015
2-anilino-4-hydroxy-6-phenyl-	281	4109
4-azido-2,6-dimethoxy-	41-42	2611
5-azidomethyl-2,4-dihydroxy-	197-201	2640
4-aziridino-6-methoxy-	91-93/5mm.	2729
Barbituric acid/5-hydrazono-	>300	4119
4-benzamido-5-benzoyloxy-	188-191	2584
4-benzamido-5-benzyloxy-	176-178	2584
4-benzamido-2,6-diethoxy-	76	2663
4-benzamido-2,6-dimethoxy-	139	2663
4-benzamido-5-hydroxy-	156-158	2584
2-benzylamino-4,6-dihydroxy-	>250; pic.204	H, 4120
4-benzylamino-2,6-dihydroxy-	299 to 310	2485, 2802, 3228, 4067
4-benzylamino-2,6-dihydroxy-5-*N*-methylformamido-	238-240	4068
4-benzylamino-5-formamido-2,6-dihydroxy-	271-272	4067
4-benzylamino-2-hydroxy-	217 to 221	H, 2153, 2674, 4054
4-benzylamino-6-hydroxy-	230-234	2153
2-benzylamino-4-hydroxy-6-methyl-	145 or >250	H, 2148, 4120

Pyrimidine	M.P.(°C)	References
4-benzylamino-2-hydroxy-5-methyl-	264-265	2674
4-benzylamino-6-hydroxy-2-methyl-	225-227	2153
4-benzyl-2-hydrazino-6-hydroxy-	191	H, 4121
5-benzyl-2-hydrazino-4-hydroxy-6-methyl-	215-216	H, 4121
2,4-bisdimethylamino-5-ethoxymethyl-	161; pic.205	4122
4-butoxy-2-hydrazino-6-methyl-	44-45	2580
4-butoxy-6-methyl-2-piperidino-	156-158/0.6mm. HCl 158	2699
4-butylamino-2,6-dihydroxy-	265	4075
4-butylamino-2-hydroxy-	170-171	281
2-butylamino-4-hydroxy-6-methyl-	91	H, 2871
5-butyl-2-dimethylamino-4-hydroxy-6-methyl-	103	4123
5-butyl-2-hydrazino-4-hydroxy-6-methyl-	201-202	3867
4-cyclohexylamino-2,6-dihydroxy-	321-322 or 336-338	2803, 2838
2-cyclopropylamino-4-hydroxy-6-methyl-	220-223	2564
4,5-diacetamido-2,6-dihydroxy-	-	H, 2112, 4060, 4061
5,5-diallyl-4,5-diamino-2,5-dihydro-2-oxo-	284-286	3920
2,4-diallyloxy-6-amino-	48-53	2678
2,4-diamino-5-amyl-6-hydroxy-	238-240	3857
2,5-diamino-4-anilino-6-hydroxy-	crude	1272
4,5-diamino-6-anilino-2-hydroxy-	sul.236-239	2481

Pyrimidine	M.P.(^{o}C)	References
4,5-diamino-6-benzylamino-2-hydroxy-	sul.207	2481
2,4-diamino-5-benzylideneamino-6-methoxy-	153-154	4084
2,4-diamino-6-benzyloxy-	108-109	H, 2285
2,5-diamino-4-benzyloxy-6-methylamino-	148-150	2925
2,5-diamino-4-butylamino-6-hydroxy-	208-210	2761
2,5-diamino-4-butylamino-6-methoxy-	112-113	2761
2,4-diamino-5-butyl-6-hydroxy-	231 or 258	3849, 3857
4,6-diamino-5,5-diethyl-2,5-dihydro-2-oxo-	283-285	3920
4,6-diamino-2,5-dihydro-2-oxo-5,5-dipropyl-	303-304	3920
2,4-diamino-5,6-dihydroxy-	>300	H, 2210
2,5-diamino-4,6-dihydroxy-	-	H, 609
4,5-diamino-2,6-dihydroxy-	-	H, 4124, 4125
2,4-diamino-5-β-dimethylaminoethyl-6-hydroxy-	HCl 238	4069
4,5-diamino-2-dimethylamino-6-hydroxy-	HCl 287	H, 2476, 2493, 4092
4,5-diamino-6-dimethylamino-2-hydroxy-	HCl 258 (mono); 211-213 (di)	2481
2,5-diamino-4-ethoxy-	128-129	748
2,4-diamino-5-ethoxymethyl-	162; pic.204	4122
2,4-diamino-5-ethyl-6-hydroxy-	267 or 289-291	3849, 3857
4,6-diamino-5-formamido-2-hydroxy-	>350	H, 909
2,4-diamino-5-hexyl-6-hydroxy-	216-217	3849
2,4-diamino-6-hydroxy-	279	H, 2285, 2286, 2503

Pyrimidine	M.P.(°C)	References
4,5-diamino-2-hydroxy-	HCl 205-210; sul.>300	H, 2472, 4106, 4126
4,5-diamino-6-hydroxy-	-	H, 1245, 2478
4,6-diamino-2-hydroxy-	>360	H, 2481, 2610, 2614
2,4-diamino-6-hydroxy-5-hydroxymethyl-	>290	2286
2,4-diamino-6-hydroxy-5-isoamyl-	260-261	3849
2,4-diamino-5-hydroxymethyl-	231-234	2169, 3062
2,4-diamino-6-hydroxy-5-methyl-	280 or 308-310	2288, 3857
2,5-diamino-4-hydroxy-6-methyl-	275; HCl 255-260	H, 2303
4,5-diamino-2-hydroxy-6-methyl-	-	H, 2472
4,5-diamino-6-hydroxy-2-methyl-	255	H, 745
2,4-diamino-6-hydroxy-5-methylamino-	HCl 259-262	4068, 4167
2,5-diamino-4-hydroxy-6-methylamino-	204-210; HCl 238	1372, 2483, 2851
4,5-diamino-2-hydroxy-6-methylamino-	206-210; HCl 303	2481
4,5-diamino-6-hydroxy-2-methylamino-	HCl 275-277	H, 2851
2,4-diamino-6-hydroxy-5-α-methylbutyl-	-	4128
2,4-diamino-6-hydroxy-5-*N*-methylformamido-	>360	2358, 4068
2,4-diamino-6-hydroxy-5-neopentyl-	-	4129
2,4-diamino-6-hydroxy-5-phenyl-	283-285	3849
4,5-diamino-6-hydroxy-2-phenyl-	228-230	H, 3861
4,5-diamino-6-hydroxy-2-piperidino-	crude	2473

Pyrimidine	M.P.(^{o}C)	References
2,4-diamino-6-hydroxy-5-propyl-	238-240	3849
2,4-diamino-6-isopropoxy-	105-107	2285
2,5-diamino-4-isopropoxy-6-methylamino-	96-98	2482
2,4-diamino-6-methoxy-	162-163; pic.242	H, 2285, 2843
4,6-diamino-2-methoxy-	pic.212	2843
2,5-diamino-4-methoxy-6-methylamino-	171-173	2482
2,4-dianilino-6-hydroxy-	214-216	2430
5-diazo-2,4-dihydroxy-("diazouracil")	H_2O 210	4119
2,4-dibenzyloxy-6-hydrazino-	150	2558
5-diethylamino-2,4-dihydroxy-	258-262	2643
5-β-diethylaminoethyl-4-hydroxy-2,6-dimethyl-	98; HCl 209	3944
4-diethylamino-2-hydroxy-5-methyl-	232	805
2,4-dihydroxy-6-hydroxyamino-	280	2757
2,4-dihydroxy-5-hydroxyiminomethyl-	-	2820
2,4-dihydroxy-5-*N*-methylacetamido-6-methylamino-	AcOH 270-280	1005
2,4-dihydroxy-5-methylamino-	290 or 313	H, 827, 2462
2,4-dihydroxy-6-methylamino-	290 to 302	H, 553, 1005, 2737, 4067
4,6-dihydroxy-5-methylamino-	240	H, 2523, 3905
2,4-dihydroxy-6-α-methylhydrazino-	284-285	4130
2,4-dihydroxy-6-methylhydrazonomethyl-	274-275	2947

Pyrimidine	M.P.(^{o}C)	References
4,6-dihydroxy-2-methyl-5-methylamino-	242	2523
2,4-dihydroxy-5-methyl-6-methylhydrazonomethyl-	261-262	2947
2,4-dihydroxy-6-methyl-5-piperidinomethyl-	355-357	4131
4,6-dihydroxy-2-methyl-5-piperidinomethyl-	>300	4078
2,4-dihydroxy-6-β-phenylhydrazino-	283-284	4130
2,4-dihydroxy-6-piperidino-	315-316	2803
4,6-dihydroxy-5-piperidino-	285-287	2523, 3905
2,4-dihydroxy-6-piperidinoamino-	303-304	3254
2,4-dihydroxy-5-piperidinomethyl-	>320	2071, 2160
2,4-dimethoxy-6-methylamino-	136-138 or 144-145	1005, 3259
2,4-dimethoxy-6-α-methylhydrazino-	105-106	2769, 3259
2-dimethylamino-4,6-dihydroxy-	356	H, 2564
4-dimethylamino-2,6-dihydroxy-	317-320	H, 2438, 2614, 2926
5-dimethylamino-2,4-dihydroxy-	306-308	H, 2462
5-dimethylamino-2,4-dihydroxy-6-methyl-	295-298	4132
2-dimethylamino-5-ethoxy-	MeI 120	4066
4-dimethylamino-2-hydroxy-	249 to 259	2676, 2745, 3086, 4133
4-dimethylamino-6-hydroxy-	286	H, 2676
2-dimethylamino-4-hydroxy-6-methylamino-	198-200 or 201-203	2476, 3228
4-dimethylamino-5-hydroxymethyl-2-methyl-	125-126	3005
2-dimethylamino-4-methoxy-	49/0.5mm.	2676

TABLE XXXIII

Pyrimidine	M.P.(^{o}C)	References
4-dimethylamino-2-methoxy-	154/18mm.	2745
4-dimethylamino-6-methoxy-	136/20mm.	827
5-dimethylaminomethyl-2,4-dihydroxy-	pic.247-248	4116
5-dimethylaminomethylene-amino-2,4-dihydroxy-	HCl 212	4134
4-dimethylaminomethyl-2,5,6-trihydroxy-	180-182	3476
2-β-ethoxyethyl-4-hydrazino-6-methyl-	-	3798
4-ethoxy-2-hydrazino-6-methyl-	70-73 or 82-83	813, 2580
4-ethoxy-6-methyl-2-piperidino-	166-168/10mm. HCl 146	2699
4-ethylamino-2,6-dihydroxy-	288 or 304	2485, 4067
5-ethylamino-2,4-dihydroxy-	268-274; pic.193	H, 2643
4-ethylamino-5-formamido-2,6-dihydroxy-	326	4067
4-ethylamino-2-hydroxy-	214	281
2-ethylamino-4-hydroxy-6-methyl-	170-174	2564
4-ethylamino-5-hydroxymethyl-2-methyl-	161-162	3005
4-ethyl-2-hydrazino-6-hydroxy-	202	2452
4-ethyl-6-hydrazino-2-hydroxy-	ca 300	2753
5-ethyl-2-hydrazino-4-hydroxy-6-methyl-	232 then 320	3867
2-ethylmethylamino-4-hydroxy-6-methyl-	152-155	2564
5-formamido-2,4-dihydroxy-	312	H, 4135
5-formamido-2,4-dihydroxy-6-methyl-	290	4135

Pyrimidine	M.P.(°C)	References
5-formamido-2,4-dihydroxy-6-methylamino-	>350	2470, 4067
4-formamido-2,6-dimethoxy-	205	2593
5-hexyl-2-hydrazino-4-hydroxy-6-methyl-	202-203	2225
4-hydrazino-2,6-dihydroxy-	290	4102, 4130
4-hydrazino-2,6-dihydroxy-5-methyl-	275-276	4130
2-hydrazino-4,6-dimethoxy-	97-98	2601
4-hydrazino-2,6-dimethoxy-	120-122; HCl 143	2769, 3259
2-hydrazino-4-hydroxy-	194-195	1128
4-hydrazino-2-hydroxy-	305-310	1174, 3206
2-hydrazino-4-hydroxy-5,6-dimethyl-	333	3867
2-hydrazino-4-hydroxy-6-hydroxymethyl-	230	813
2-hydrazino-4-hydroxy-5-methyl-	224 or 225	H, 813, 4121
2-hydrazino-4-hydroxy-6-methyl-	ca 245-246	H, 813, 2580, 2999, 3867, 4036
4-hydrazino-2-hydroxy-6-methyl-	H_2O 251; 285	H, 813
2-hydrazino-4-hydroxy-5-methyl-6-phenyl-	245-246	4136
2-hydrazino-4-hydroxy-6-methyl-5-propyl-	215-216	3867
2-hydrazino-4-hydroxy-6-phenyl-	220 or 224-225	H, 3867, 4121
4-hydrazino-2-hydroxy-6-phenyl-	>300	2753
2-hydrazino-4-hydroxy-6-propyl-	186	H, 4121
4-hydrazino-2-hydroxy-6-propyl-	ca 245	2753
2-hydrazino-4-isopropoxy-6-methyl-	50-51 or 135-136	813, 2580

TABLE XXXIII

Pyrimidine	M.P.(°C)	References
2-hydrazino-4-methoxy-	120-121	813
4-hydrazino-5-methoxy-	184-185	2161
2-hydrazino-4-methoxy-6-methyl-	114-116 or 201	H, 813, 2580, 4036
2-hydrazino-4-methyl-6-propoxy-	70-71	2580
4-hydroxy-2,6-bismethylamino-	241-244	3228
2-hydroxy-4-hydroxyamino-	HCl 215 or 221	3090, 3258
2-hydroxy-4-hydroxyamino-5-hydroxymethyl-	HCl 260	3258
2-hydroxy-4-hydroxyamino-5-methyl-	242, 262, or 270	3089, 3090, 3258
4-hydroxy-2-hydroxyamino-6-methyl-	233	H, 2148
2-hydroxy-4-hydroxyiminomethyl-	226	2954
2-hydroxy-4-hydroxyiminomethyl-6-methyl-	286	2954
4-hydroxy-2-isopropylamino-6-methyl-	-	2564
2-hydroxy-4-methoxyamino-	HCl 195-196	3258
2-hydroxy-4-*N*-methylacetamido-	193-194	2984
2-hydroxy-4-methylamino-	ca 272	H, 2998, 3087, 4106
4-hydroxy-2-methylamino-	214-215	2288
4-hydroxy-6-methylamino-	251	H, 827
2-hydroxy-4-α-methylhydrazino-	240	3206
5-hydroxymethyl-4-isopropylamino-2-methyl-	161-162	3005
2-hydroxy-5-methyl-4-methylamino-	235	805
4-hydroxy-6-methyl-5-methyliminomethyl-2-piperidino-	197-198	2490

Pyrimidine	M.P.(^{o}C)	References
4-hydroxy-6-methyl-2-α-methylhydrazino-	187-189	813
5-hydroxymethyl-2-methyl-4-methylamino-	171-172	2653, 3005
5-hydroxymethyl-2-methyl-4-propylamino-	111-112	3005
4-hydroxy-6-methyl-2-phenylhydrazino-	217 or 221	813, 2148
4-hydroxy-6-methyl-2-piperidino-	185-186	H, 2148, 2490
4-hydroxy-2-piperidino-	156-157	2811, 4137
4-isoamyloxy-6-methyl-2-piperidino-	HCl 122-124	2699
4-isobutoxy-6-methyl-2-piperidino-	HCl 108-109	2699
4-isopropoxy-6-methyl-2-piperidino-	HCl 123-125	2699
4-isopropoxy-2,5,6-tristrimethylsilylamino-	105/0.001mm.	4138
4-methoxy-2-methylamino-	55-57	2288
4-methoxy-6-methylamino-	-	2776
4-methoxy-6-methyl-2-piperidino-	HCl 148-149	2699
4-methoxy-2,5,6-tristrimethylsilylamino-	111/0.001mm.	4138
4-methyl-2-piperidino-6-propoxy-	HCl 91; pic.143	2699
2,4,5-triamino-6-benzyloxy-	145-147	H, 2285
2,4,5-triamino-6-hydroxy-	HCl >340	H, 559, 909, 4139-4141
4,5,6-triamino-2-hydroxy-	>360	H, 909, 2481
2,4,5-triamino-6-isopropoxy-	149-150	2285
2,4,5-triamino-6-methoxy-	176-178	H, 2493
2,4,6-trihydroxy-5-methylamino-	-	H, 1277
2-trimethylsiloxy-4-trimethylsilylamino-	122-123	3070, 3071

TABLE XXXIV. Amino-Oxypyrimidines with *N*-1- or *N*-3-Substituents (*H* 561)

Pyrimidine	M.P.(°C)	References
5-acetamido-4-amino-1,2,3,6-tetrahydro-1,3-dimethyl-2,6-dioxo-	278-279	H, 4060
4-acetamido-1,6-dihydro-1,5-dimethyl-6-oxo-	189-190	1106
4-acetamido-3,6-dihydro-3,5-dimethyl-6-oxo-	273-274	1106
4-acetamido-1,2-dihydro-1-methyl-2-oxo-	274	H, 3363
4-acetamido-1,6-dihydro-1-methyl-6-oxo-	303-304	1106
5-acetamido-1,2,3,4-tetrahydro-1,3-dimethyl--6-*N*-methylacetamido-2,4-dioxo-	186-189	H, 2112
5-acetamido-1,2,3,4-tetrahydro-1,3-dimethyl-6-methylamino-2,4-dioxo-	247-249	739
4-acetamido-1,2,3,6-tetrahydro-1,3,5-trimethyl-2,6-dioxo-	221-223	883
1-allyl-6-amino-3,5-diethyl-1,2,3,4-tetrahydro-2,4-dioxo-	166-167	4142
5-allyl-2-amino-1,6-dihydro-1-β-hydroxyethyl-6-oxo-4-phenyl-	193-194	4143
1-allyl-2-amino-1,6-dihydro-5-methyl-6-oxo-4-phenyl-	177-179	4143
1-allyl-6-amino-1,2-dihydro-2-oxo-	240	3479
1-allyl-6-amino-3-βγ-dihydroxypropyl-1,2,3,4-tetrahydro-2,4-dioxo-	163-165	4144

Pyrimidine	M.P.(^{o}C)	References
1-allyl-6-amino-3-ethyl-1,2,3,4-tetrahydro-5-methyl-2,4-dioxo-	83-86	4142
1-allyl-6-amino-5-ethyl-1,2,3,4-tetrahydro-3-methyl-2,4-dioxo-	133-134	4142
4-allylamino-1,2,3,6-tetrahydro-1,3-dimethyl-2,6-dioxo-	163-165	2305
1-allyl-6-amino-1,2,3,4-tetrahydro-3,5-dimethyl-2,4-dioxo-	171-172	4142
1-allyl-6-amino-1,2,3,4-tetrahydro-3-β-hydroxypropyl-2,4-dioxo-	170-172	4144
1-allyl-5,6-diamino-3-ethyl-1,2,3,4-tetrahydro-2,4-dioxo-	138-141 or 145	4144-4146
5-allylmethylamino-1-cyclohexyl-1,2,3,6-tetrahydro-3,4-dimethyl-2,6-dioxo-	188-190/3mm.	1352
4-amino-5-benzamido-1,2,3,6-tetrahydro-1,3-dimethyl-2,6-dioxo-	287-289	2498
4-amino-6-benzylamino-1,2-dihydro-1-methyl-2-oxo-	304-306	2475
4-amino-1-benzyl-1,2-dihydro-2-oxo-	286	2337
4-amino-5-benzylideneamino-2,3-dihydro-2-imino-6-methoxy-3-methyl-	HI 207-209	4084
4-amino-1-benzyloxy-1,2-dihydro-2-oxo-	215-217	2262
2-amino-5-butyl-1,6-dihydro-1-β-hydroxyethyl-6-oxo-4-phenyl-	183-184	4147
5-amino-1-butyl-1,2,3,4-tetrahydro-3,6-dimethyl-2,4-dioxo-	96-97	4148

Pyrimidine	M.P.(^{o}C)	References
5-amino-1-butyl-1,2,3,6-tetrahydro-3,4-dimethyl-2,6-dioxo-	49	1145
1-amino-3-butyl-1,2,3,4-tetrahydro-2,4-dioxo-	69-71	3260
5-amino-1-cyclohexyl-3-ethyl-1,2,3,6-tetrahydro-4-methyl-2,6-dioxo-	122-124	1145, 4148
5-amino-1-cyclohexyl-1,2,3,6-tetrahydro-3,4-dimethyl-2,6-dioxo-	158-159	1145, 4148, 4149
4-amino-1,3-diethyl-1,2,3,6-tetrahydro-5-methyl-2,6-dioxo-	155; H_2O 100	4142
5-amino-1,3-diethyl-1,2,3,4-tetrahydro-6-methyl-2,4-dioxo-	95-96	4150
4-amino-1,6-dihydro-1,5-dimethyl-6-oxo-	190-191	1106
4-amino-3,6-dihydro-3,5-dimethyl-6-oxo-	277-278	1106
2-amino-1,6-dihydro-1,5-dimethyl-6-oxo-4-phenyl-	216-217	4143, 4147
2-amino-1,4-dihydro-1-δ-hydroxybutyl-6-methyl-4-oxo-	190	2871
2-amino-1,6-dihydro-1-β-hydroxyethyl-5-hydroxymethyl-6-oxo-4-phenyl-	240-241	4147
2-amino-1,6-dihydro-1-β-hydroxyethyl-5-methyl-6-oxo-4-phenyl-	217-218	4147
2-amino-1,6-dihydro-1-β-hydroxyethyl-6-oxo-4-phenyl-5-propyl-	183-184	4147
2-amino-1,6-dihydro-1-β-hydroxyethyl-6-oxo-4-phenyl-5-prop-2'-ynyl-	206-208	4147
4-amino-2,3-dihydro-3-hydroxy-5-methyl-2-oxo-	262-266	2972

Pyrimidine	M.P.(^{o}C)	References
4-amino-1,2-dihydro-1-hydroxy-2-oxo-	265-270	2262
4-amino-2,3-dihydro-3-hydroxy-2-oxo-	265-275	2262
2-amino-1,6-dihydro-1-β-hydroxypropyl-5-methyl-6-oxo-4-phenyl-	ca 170	4147
2-amino-1,6-dihydro-1-γ-hydroxypropyl-5-methyl-6-oxo-4-phenyl-	ca 202	4147
4-amino-1,6-dihydro-2-methoxy-1,5-dimethyl-6-oxo-	208-210	883
4-amino-1,2-dihydro 6-methoxy-1-methyl-2-oxo-	273-283	2475
4-amino-1,6-dihydro-2-methoxy-1-methyl-6-oxo-	233-235	H, 909
4-amino-1,2-dihydro-1-methyl-6-methylamino-2-oxo-	273-275	2475
2-amino-1,4-dihydro-1-methyl-4-oxo-	275-280 or 283-285	2215, 2288
2-amino-1,6-dihydro-1-methyl-6-oxo-	257-260 or 262-266	2215, 2288
4-amino-1,2-dihydro-1-methyl-2-oxo-	HCl 285-300	H, 1174, 2288, 2987
4-amino-1,6-dihydro-1-methyl-6-oxo-	184-185	1106
4-amino-2,3-dihydro-3-methyl-2-oxo-	213; HCl 242-245; pic.246 or 240	2168, 2984, 2987
2-amino-1,6-dihydro-5-methyl-6-oxo-4-phenyl-1-propyl-	240-242	4147
2-amino-1,6-dihydro-5-methyl-6-oxo-4-phenyl-1-prop-2'-ynyl-	ca 215	4147

Pyrimidine	M.P.(^{o}C)	References
2-amino-1,6-dihydro-6-oxo-4-phenyl-1,5-diprop-2'-ynyl-	220-222	4147
2-amino-1,6-dihydro-6-oxo-4-phenyl-1-prop-2'-enyl-	212-214	4151
2-amino-4-dimethylamino-1,6-dihydro-1-methyl-5-nitro-6-oxo-	208-211	2444
4-amino-6-dimethylamino-1,2-dihydro-1-methyl-2-oxo-	265-268	2475
2-amino-1-β-dimethylaminoethyl-1,6-dihydro-5-methyl-6-oxo-4-phenyl-	195-197	4147
4-amino-3-β-dimethylaminoethyl-1-ethyl-1,2,3,6-tetrahydro-2,6-dioxo-	177-179	4144
4-amino-3-dimethylamino-5-formamido-1,2,3,6-tetrahydro-1-methyl-2,6-dioxo-	198-199	2291
4-amino-5-dimethylamino-1,2,3,6-tetrahydro-1,3-dimethyl-2,6-dioxo-	194	2109
4-amino-3-dimethylamino-1,2,3,6-tetrahydro-1-β-hydroxyethyl-2,6-dioxo-	192-193	2291
4-amino-3-dimethylamino-1,2,3,6-tetrahydro-1-methyl-2,6-dioxo-	219-220	2291
2-amino-5-β-ethoxyethyl-1,6-dihydro-1-β-hydroxyethyl-6-oxo-4-phenyl-	135-136	4147
2-amino-1-β-ethoxyethyl-1,6-dihydro-5-methyl-6-oxo-4-phenyl-	172-173	4147
4-amino-5-ethylamino-1,2,3,6-tetrahydro-1,3-dimethyl-2,6-dioxo-	131; pic.194; HCl 251	3289

Pyrimidine	M.P.(^{o}C)	References
2-amino-5-ethyl-1,6-dihydro-1-β-hydroxyethyl-6-oxo-4-phenyl-	210-212	4147
2-amino-1-ethyl-1,6-dihydro-5-methyl-6-oxo-4-phenyl-	185-186	4147
5-α-aminoethylidenehexahydro-1,3-dimethyl-2,4,6-trioxo-	248-250	3265
4-amino-5-ethyl-1,2,3,6-tetrahydro-1,3-dimethyl-2,6-dioxo-	196-198 or 208-209	883, 4142
2-amino-5-formamido-1,6-dihydro-1-methyl-6-oxo-	264-266	2483
4-amino-5-formamido-1,6-dihydro-1-methyl-6-oxo-	266-267 or 245-247	3188, 4100
4-amino-5-formamido-3,6-dihydro-3-methyl-6-oxo-	275-280	4100
4-amino-5-formamido-1,2,3,6-tetrahydro-1,3-dimethyl-2,6-dioxo-	252	H, 4098
5-aminohexahydro-1,3,5-trimethyl-2,4,6-trioxo-	105-106; HCl 352	4152
5-aminomethylenehexahydro-1,3-dimethyl-2,4,6-trioxo-	229 to 233	2120, 3265, 4153
5-amino-1,2,3,4-tetrahydro-1,3-dimethyl-6-methylamino-2,4-dioxo-	143	739
4-amino-1,2,3,6-tetrahydro-1,3-dimethyl-2,6-dioxo-	HCl 246-250	H, 883, 1437
4-amino-1,2,3,6-tetrahydro-1,3-dimethyl-2,6-dioxo-5-propyl-	164-165	883
4-amino-1,2,3,6-tetrahydro-1,3-dimethyl-5-methylamino-2,6-dioxo-	177-178; HCl 261; pic.194	3289
4-amino-1,2,3,6-tetrahydro-5-isopropyl-1,3-dimethyl-2,6-dioxo-	174-176	883

Pyrimidine	M.P.(°C)	References
1-amino-1,2,3,4-tetrahydro-3-methyl-2,4-dioxo-	158-159	3260
1-amino-1,2,3,6-tetrahydro-3-methyl-2,6-dioxo-	164-165	3261
4-amino-1,2,3,6-tetrahydro-1,3,5-trimethyl-2,6-dioxo-	246 or 252	883, 4142
5-amino-1,2,3,4-tetrahydro-1,3,6-trimethyl-2,4-dioxo-	167 to 171	H, 1145, 2121, 4150
4-amino-1,3,5-triethyl-1,2,3,6-tetrahydro-2,6-dioxo-	169-170	H, 4142
4-anilino-3-dimethylamino-1,2,3,6-tetrahydro-1-methyl-2,6-dioxo-	170	2246
4-anilino-1,2,3,6-tetrahydro-1,3-dimethyl-2,6-dioxo-	182 or 187	740, 3320
4-anilino-1,2,3,6-tetrahydro-1-methyl-2,6-dioxo-3-piperidino-	198-199	2246
4-azido-1,2,3,6-tetrahydro-1,3-dimethyl-2,6-dioxo-	149-151	739
Barbituric acid/1-dimethylamino-	183-184	2246
4-benzylamino-1,3-dibutyl-1,2,3,6-tetrahydro-2,6-dioxo-	94-95	2485
4-benzylamino-5-dimethylamino-1,2,3,6-tetrahydro-1,3-dimethyl-2,6-dioxo-	89	2109
4-benzylamino-3-dimethylamino-1,2,3,6-tetrahydro-1-methyl-2,6-dioxo-	154-155	2246

Pyrimidine	M.P.(^{o}C)	References
4-benzylamino-1,2,3,6-tetrahydro-1,3-dimethyl-2,6-dioxo-	149-150	2109, 2153, 2485
5-benzylamino-1,2,3,4-tetrahydro-1,3-dimethyl-6-methylamino-2,4-dioxo-	137	1005
4-benzylamino-1,2,3,6-tetrahydro-2,6-dioxo-1,3-diphenyl-	232-234	2485
4-benzylamino-1,2,3,6-tetrahydro-1-methyl-2,6-dioxo-3-piperidino-	178-179	2246
4-benzylamino-1,2,3,6-tetrahydro-1,3,5-trimethyl-2,6-dioxo-	101-102	2305
1-benzyl-4-benzylamino-1,6-dihydro-6-oxo- (?)	127-130	2153
1-benzyl-2-benzylidenehydrazino-1,6-dihydro-4-methyl-6-oxo-	196-198	2985
1-benzyl-4-benzyloxy-1,2-dihydro-2-oxo-	134-135	2829
1-benzyl-2-hydrazino-1,6-dihydro-4-methyl-6-oxo-	177-179	2985
4-β-benzylhydrazino-3-dimethylamino-1,2,3,6-tetrahydro-1-methyl-2,6-dioxo-	186-187	3254
4-β-benzylhydrazino-1,2,3,6-tetrahydro-1,3-dimethyl-2,6-dioxo-	229-231	3254
4-β-benzylhydrazino-1,2,3,6-tetrahydro-1-methyl-2,6-dioxo-3-piperidino-	167-168	3254
1-benzylideneamino-1,2,3,4-tetrahydro-3-methyl-2,4-dioxo-	119-121	3260

Pyrimidine	M.P.(^oC)	References
5-benzylideneamino-1,2,3,4-tetrahydro-1,3,6-trimethyl-2,4-dioxo-	158-160	2121
1,3-bisbenzylideneamino-1,2,3,4-tetrahydro-2,4-dioxo-	158	3260
4-butylamino-1,2,3,6-tetrahydro-1,3-dimethyl-2,6-dioxo-	142-143	739, 2485
4-butylamino-1,2,3,6-tetrahydro-1,3-dimethyl-2,6-dioxo-5-phenyl-	145-146	2305
4-butylamino-1,2,3,6-tetrahydro-1-methyl-2,6-dioxo-3-piperidino-	116-117	2246
4-butylamino-1,2,3,6-tetrahydro-1,3,5-trimethyl-2,6-dioxo-	70-71	740
1-butyl-5-dimethylamino-1,2,3,4-tetrahydro-3,6-dimethyl-2,4-dioxo-	151-155/5mm.	1352
1-butyl-5-dimethylamino-1,2,3,6-tetrahydro-3,4-dimethyl-2,6-dioxo-	145-148/4mm.; HCl 131-134	1352
4-cyclohexylamino-1,2,3,6-tetrahydro-1,3-dimethyl-2,6-dioxo-	165-166	2838, 2929
1-cyclohexyl-5-diallylamino-1,2,3,6-tetrahydro-3,4-dimethyl-2,6-dioxo-	65	4149
1-cyclohexyl-5-diethylamino-1,2,3,6-tetrahydro-3,4-dimethyl-2,6-dioxo-	95	1352
1-cyclohexyl-5-dimethylamino-3-ethyl-1,2,3,6-tetrahydro-4-methyl-2,6-dioxo-	113	1352

Pyrimidine	M.P.(°C)	References
1-cyclohexyl-5-dimethylamino-1,2,3,6-tetrahydro-3,4-dimethyl-2,6-dioxo-	103	1352
1-cyclohexyl-5-dipropylamino-1,2,3,6-tetrahydro-3,4-dimethyl-2,6-dioxo-	65	1352
1-cyclohexyl-5-ethylamino-1,2,3,6-tetrahydro-3,4-dimethyl-2,6-dioxo-	181-187/3.5mm.	1352
1-cyclohexyl-5-ethylmethylamino-1,2,3,6-tetrahydro-3,4-dimethyl-2,6-dioxo-	84-85	1352
1-cyclohexyl-1,2,3,6-tetrahydro-3,4-dimethyl-5-methylamino-2,6-dioxo-	211	1352, 4149
5-cyclohexyl-1,2,3,4-tetrahydro-1,3-dimethyl-6-methylamino-2,4-dioxo-	126-128	2305
4,5-diamino-2,3-dihydro-2-imino-6-methoxy-3-methyl-	HCl -	4084
2,5-diamino-1,6-dihydro-4-methoxy-1-methyl-6-oxo-	195-200	2444
4,5-diamino-1,6-dihydro-2-methoxy-1-methyl-6-oxo-	-	H, 909
2,5-diamino-1,6-dihydro-1-methyl-4-methylamino-6-oxo-	HCl -	2483
2,4-diamino-1,6-dihydro-1-methyl-6-oxo-	275-277	4092
4,5-diamino-1,2-dihydro-1-methyl-2-oxo-	ca 250; pic.239	H, 2288, 4106
4,5-diamino-1,6-dihydro-1-methyl-6-oxo-	195; sul.249	827, 2130, 3398
4,5-diamino-2,3-dihydro-3-methyl-2-oxo-	crude	281
4,5-diamino-3,6-dihydro-3-methyl-6-oxo-	222-223	2492

Pyrimidine	M.P.(°C)	References
4,6-diamino-1,2-dihydro-1-methyl-2-oxo-	sul.327	2481
4,5-diamino-3-β-dimethylaminoethyl-1-ethyl-1,2,3,6-tetrahydro-2,6-dioxo-	83-86	4144
4,5-diamino-3-dimethylamino-1,2,3,6-tetrahydro-1-β-hydroxyethyl-2,4-dioxo-	-	2291
4,5-diamino-3-dimethylamino-1,2,3,6-tetrahydro-1-methyl-2,6-dioxo-	157-159	2291
4,5-diamino-1-ethyl-1,2,3,6-tetrahydro-3-β-methylallyl-2,6-dioxo-	-	4145
2,4-diamino-5-formamido-1,6-dihydro-1-methyl-6-oxo-	>350	H, 2493
4,5-diamino-1,2,3,6-tetrahydro-1,3-dimethyl-2,6-dioxo-	208	H, 2838, 3289, 4098
1,3-diamino-1,2,3,4-tetrahydro-2,4-dioxo-	185-186	3260
4,5-diamino-1,2,3,6-tetrahydro-1-β-hydroxyethyl-2,6-dioxo-3-piperidino-	184-186	2291
4,5-diamino-1,2,3,6-tetrahydro-1-methyl-2,6-dioxo-3-piperidino-	177-179	2291
4,5-diamino-1,2,3,6-tetrahydro-1-methyl-3-β-methylallyl-2,6-dioxo-	149-153	4145, 4146
1,3-dibutyl-1,2,3,4-tetrahydro-6-methylamino-2,4-dioxo-	109-111	2485
1,3-diethyl-4-hydrazino-1,2,3,6-tetrahydro-2,6-dioxo-	183-185	2305

Pyrimidine	M.P.(°C)	References
1,3-diethyl-1,2,3,4-tetrahydro-6-methylamino-2,4-dioxo-	170-171	2305
1,3-diethyl-1,2,3,4-tetrahydro-5-methyl-6-methylamino-2,4-dioxo-	103-104	2305
1,2-dihydro-4-hydroxyamino-1,5-dimethyl-2-oxo-	214-215; HCl 208-210	3090
1,2-dihydro-4-hydroxyamino-1,6-dimethyl-2-oxo-	260-261	3090
1,2-dihydro-4-hydroxyamino-1-methyl-2-oxo-	224-225; HCl 218-219	3090, 3258
1,4-dihydro-4-imino-2-methoxy-1-methyl-	128	668
1,6-dihydro-6-imino-4-methoxy-1-methyl-	HI 143-144	2781
1,4-dihydro-6-methoxy-1-methyl-4-methylimino-	HI 215-217	2781
1,2-dihydro-1-methyl-4-methylamino-2-oxo-	HCl 215-225	H, 2987
1,2-dihydro-1-methyl-6-methylamino-2-oxo-	pic.212	3374
1,6-dihydro-1-methyl-4-methylamino-6-oxo-	188 or 194	827, 2776, 2781
2-dimethylamino-1,6-dihydro-1-methyl-6-oxo-	95-96	2676, 2776
4-dimethylamino-1,2-dihydro-1-methyl-2-oxo-	182-183; HCl 191-199	H, 2676, 2987, 3087
4-dimethylamino-1,6-dihydro-1-methyl-6-oxo-	156-157 or 160	827, 2676, 2776
4-dimethylamino-2,3-dihydro-3-methyl-2-oxo-	116-117	2676
4-dimethylamino-1,2,3,6-tetrahydro-1,3-dimethyl-2,6-dioxo-	79-80	739

Pyrimidine	M.P.(°C)	References
4-dimethylamino-1,2,3,6-tetrahydro-1,3-dimethyl-2-oxo-6-phenylimino-	90-92	2569
1-dimethylamino-1,2,3,4-tetrahydro-6-hydroxy-3-methyl-2,4-dioxo-	92-93	2246
1-dimethylamino-1,2,3,4-tetrahydro-6-isopropylamino-3-methyl-2,4-dioxo-	126-127	2246
1-dimethylamino-1,2,3,4-tetrahydro-3-methyl-2,4-dioxo-6-β-phenylhydrazino-	205-207	3254
4-dimethylamino-1,2,3,6-tetrahydro-1-methyl-2,6-dioxo-3-piperidino-	109-110	2246
1-dimethylamino-1,2,3,4-tetrahydro-3-methyl-2,4-dioxo-6-piperidinoamino-	161-162	3254
4-dimethylamino-1,2,3,6-tetrahydro-1,3,5-trimethyl-2,6-dioxo-	78-79	2305
4-ββ-dimethylhydrazino-1,2,3,6-tetrahydro-1,3-dimethyl-2,6-dioxo-	218-219	3254
4-ethylamino-1,2-dihydro-1-methyl-2-oxo-	-	4154
4-ethylamino-1,2,3,6-tetrahydro-1,3-dimethyl-2,6-dioxo-	165	739
4-ethylamino-1,2,3,6-tetrahydro-1,3-dimethyl-2,6-dioxo-5-phenyl-	166-167	2305
4-ethylamino-1,2,3,6-tetrahydro-1,3-dimethyl-2,6-dioxo-5-propyl-	96	2305
4-ethylamino-1,2,3,6-tetrahydro-1,3,5-trimethyl-2,6-dioxo-	102-103	2305

Pyrimidine	M.P.(^{o}C)	References
5-ethyl-4-ethylamino-1,2,3,6-tetrahydro-1,3-dimethyl-2,6-dioxo-	117-118	2305
5-ethyl-1,2,3,4-tetrahydro-1,3-dimethyl-2,4-dioxo-6-piperidino-	122	2305
5-ethyl-1,2,3,4-tetrahydro-1,3-dimethyl-2,4-dioxo-6-propylamino-	95	2305
5-ethyl-1,2,3,4-tetrahydro-1,3-dimethyl-6-methylamino-2,4-dioxo-	164-165	2305
5-ethyl-1,2,3,4-tetrahydro-6-isopropylamino-1,3-dimethyl-2,4-dioxo-	137-139	2305
5-formamido-1,6-dihydro-2-hydroxy-1-methyl-4-methylamino-6-oxo-	>360	2440
5-formamido-1,2,3,4-tetrahydro-1,3-dimethyl-2,4-dioxo-	-	H, 4062
4-formamido-1,2,3,6-tetrahydro-1,3,5-trimethyl-2,6-dioxo-	174-176	883
hexahydro-1,3-dimethyl-5-methylaminomethylene-2,4,6-trioxo-	230-231	2120
hexahydro-5-hydroxyimino-1,3-dimethyl-2,6-dioxo- (or tautomer)	-	1437
2-hydrazino-1,6-dihydro-1,4-dimethyl-6-oxo-	201 to 207; pic.196-197	813, 4155, 4165
4-hydrazino-1,2-dihydro-1-methyl-2-oxo-	173	813
4-hydrazino-1,2,3,6-tetrahydro-1,3-dimethyl-2,6-dioxo-	212 or 216-218	739, 4156
4-hydrazino-1,2,3,6-tetrahydro-1-methyl-2,6-dioxo-3-piperidino-	212-213	3254

TABLE XXXIV

Pyrimidine	M.P.(^{o}C)	References
4-hydrazino-1,2,3,6-tetrahydro-1,3,5-trimethyl-2,6-dioxo-	163 or 174-176	740, 2305
5-hydrazonohexahydro-1,3-dimethyl-2,4,6-trioxo-(or tautomer)	-	4119
1,2,3,4-tetrahydro-1,3-diisopropyl-6-methylamino-2,4-dioxo-5-phenyl-	175	2305
1,2,3,4-tetrahydro-1,3-dimethyl-5,6-bis(*N*-methylacetamido)-2,4-dioxo-	223-224	H, 2112
1,2,3,4-tetrahydro-1,3-dimethyl-5,6-bismethylamino-2,4-dioxo-	144; pic.165	H, 1005
1,2,3,4-tetrahydro-1,3-dimethyl-2,4-dioxo-6-piperidinoamino-	213-214	3254
1,2,3,4-tetrahydro-1,3-dimethyl-2,4-dioxo-6-propylamino-	170	2485
1,2,3,4-tetrahydro-1,3-dimethyl-2,4-dioxo-5-ureido-	-	H, 4062
1,2,3,4-tetrahydro-1,3-dimethyl-5-*N*-methylacetamido-6-methylamino-2,4-dioxo-	153-154	H, 1005, 2109
1,2,3,4-tetrahydro-1,3-dimethyl-6-methylamino-2,4-dioxo-	244-245	739, 2485
1,2,3,4-tetrahydro-1,3-dimethyl-6-methylamino-2,4-dioxo-5-phenyl-	255	2305
1,2,3,4-tetrahydro-1,3-dimethyl-6-methylamino-2,4-dioxo-5-piperidino-	142	1005

Pyrimidine	M.P.(^{o}C)	References
1,2,3,4-tetrahydro-1,3-dimethyl-6-methylamino-2,4-dioxo-5-propyl-	110-111	2305
1,2,3,4-tetrahydro-1,3-dimethyl-6-α-methylhydrazino-2,4-dioxo-	122	4157
1,2,3,4-tetrahydro-6-hydroxyamino-1,3-dimethyl-2,4-dioxo-	146-148	740
1,2,3,4-tetrahydro-3-β-hydroxyethyl-4-β-hydroxyethylimino-1-methyl-2-oxo-	103-104	3012
1,2,3,4-tetrahydro-3-β-hydroxyethyl-4-imino-1-methyl-2-oxo-	194-196	3012
1,2,3,4-tetrahydro-4-imino-1,3-dimethyl-2-oxo-	-	H, 2745
1,2,3,4-tetrahydro-6-isoamylamino-1,3-dimethyl-2,4-dioxo-	141-143	2485
1,2,3,4-tetrahydro-6-isobutylamino-1,3-dimethyl-2,4-dioxo-	160-161	2485
1,2,3,4-tetrahydro-6-isopropylamino-1,3-dimethyl-2,4-dioxo-	121-122	2929
1,2,3,4-tetrahydro-6-isopropylamino-1,3,5-trimethyl-2,4-dioxo-	122-123	2305
1,2,3,4-tetrahydro-5-isopropyl-1,3-dimethyl-6-methylamino-2,4-dioxo-	86-88	2305
1,2,3,4-tetrahydro-6-isopropylidenehydrazino-1,3-dimethyl-2,4-dioxo-	144-145	2558
1,2,3,4-tetrahydro-3-methyl-2,4-dioxo-1,6-dipiperidino-	129-130	2246

Pyrimidine	M.P.(°C)	References
1,2,3,4-tetrahydro-3-methyl-2,4-dioxo-1-piperidino-6-piperidinoamino-	180-181	3254
1,2,3,4-tetrahydro-3-methyl-6-methylamino-2,4-dioxo-1-piperidino-	198-199	2246
1,2,3,4-tetrahydro-1,3,5-trimethyl-2,4-dioxo-6-propylamino-	91	2305
1,2,3,4-tetrahydro-1,3,5-trimethyl-6-methylamino-2,4-dioxo-	162-163	740, 2305
2,4,5-triamino-1,6-dihydro-1-methyl-6-oxo-	HCl >300	494, 2493
2,4,5-triamino-3,6-dihydro-3-methyl-6-oxo-	sul. -	4084
4,5,6-triamino-1,2-dihydro-1-methyl-2-oxo-	sul. >360	2481
Uracil/5-acetamido-6-amino-1-dimethylamino-	ca 260	2291
Uracil/5-acetamido-6-amino-1-methyl-	295	H, 4060
Uracil/5-acetamido-6-amino-3-methyl-	>350	4158
Uracil/6-acetamido-3,5-dimethyl-	230-232	883
Uracil/6-acetamido-1-methyl-	254-255	2831
Uracil/1-allyl-5-amino-	225-230	3479
Uracil/1-amino-	244-245	3260
Uracil/3-amino-	190-200; HBr 230-235	3261
Uracil/5-amino-6-benzylamino-3-ethyl-	HCl >330	2928
Uracil/5-amino-6-benzylamino-3-methyl-	HCl >330	2560
Uracil/6-amino-5-benzylamino-3-methyl-	211	2489

Pyrimidine	M.P.(^{o}C)	References
Uracil/6-amino-1-benzyloxy-	235-237	2199
Uracil/6-amino-1-benzyloxy-5-methyl-	314-318	3972
Uracil/6-amino-1,5-bisdimethylamino-	248-250	2291
Uracil/5-amino-3-butyl-	145	1464
Uracil/6-amino-3-butyl-	242-247	4159
Uracil/6-amino-1,5-dimethyl-	327	883
Uracil/6-amino-1-dimethylamino-	291-293	2291
Uracil/6-amino-1-dimethylamino-5-formamido-	275	2291, 4160
Uracil/6-amino-5-dimethylamino-1-methyl-	240-241	4161
Uracil/6-amino-5-ethyl-1-methyl-	284-285	883
Uracil/6-amnno-6-formamido-1-hydroxy-	295-300	2316
Uracil/6-amino-5-formamido-3-methyl-	>350	H, 909
Uracil/6-amino-3-hexyl-	223-232	4159
Uracil/6-amino-1-hydroxy-	315	2199, 2316
Uracil/6-amino-1-hydroxyethyl-	253-254	H, 4077
Uracil/6-amino-1-hydroxy-5-methyl-	252-258	3972
Uracil/6-amino-5-isopropyl-1-methyl-	248-250	883
Uracil/6-amino-3-methyl-	327 or 330	909, 2560
Uracil/5-amino-1-methyl-6-methylamino-	HCl 238	2480
Uracil/5-amino-3-methyl-6-methylamino-	>350; HCl >350	2489
Uracil/6-amino-1-methyl-5-methylamino-	183-185 or 210	4162, 4163

TABLE XXXIV

Pyrimidine	M.P.(^{o}C)	References
Uracil/6-amino-1-methyl-5-propyl-	278-282	883
Uracil/6-amino-1-piperidino-	211-212	4164
Uracil/6-anilino-1-dimethylamino-	260-261	2246
Uracil/6-anilino-1-methyl-	308-310	3320
Uracil/6-anilino-3-methyl-	336-338	3320
Uracil/6-anilino-3-phenyl-	308-310	3320
Uracil/6-benzylamino-1-dimethylamino-	202-203	2246
Uracil/6-benzylamino-3-ethyl-	285	2928
Uracil/6-benzylamino-5-formamido-3-methyl-	238	2560
Uracil/6-benzylamino-1-methyl-	292-294	2485
Uracil/6-benzylamino-3-methyl-	300-302 or 282	2485, 2560
Uracil/6-benzylamino-3-phenyl-	267-269	2485
Uracil/6-benzylamino-1-piperidino-	242-243	2246
Uracil/1-benzyl-6-benzylamino-	255-257	2485
Uracil/3-benzyl-6-benzylamino-	286-289	2485
Uracil/1-benzylideneamino-	224-225	3260
Uracil/3-benzylideneamino-	188-189	3261
Uracil/6-butylamino-3-methyl-	242-244	2485
Uracil/6-butylamino-3-phenyl-	306-308	2485
Uracil/6-cyclohexylamino-3-methyl-	310-312	2838
Uracil/6-cyclohexylamino-3-phenyl-	311-313	2838

Pyrimidine	M.P.(^{o}C)	References
Uracil/5,6-diamino-1-dimethylamino-	220	2291, 4160
Uracil/5,6-diamino-1-hydroxy-	>300	2316
Uracil/5,6-diamino-1-methyl-	>330	H, 4098
Uracil/5,6-diamino-3-methyl-	>340; HCl >350	H, 909
Uracil/5,6-diamino-1-piperidino-	-	2291
Uracil/1-β-diethylamino-ethyl-6-methyl-	186-189	3191
Uracil/1-β-dimethylamino-ethyl-5-methyl-	liquid	3191
Uracil/1-β-dimethylamino-ethyl-6-methyl-	HCl 270-272	3191
Uracil/6-ethylamino-1-methyl-5-*N*-methylformamido-	>225	4166
Uracil/5-formamido-3-methyl-6-methylamino-	350-352	2489
Uracil/6-β-formylhydrazino-3-methyl-	222-223	2755
Uracil/6-hydrazino-3-methyl-	223 or 236-238	2755, 4157
Uracil/1-methyl-6-methylamino-	276	2480
Uracil/3-methyl-6-methylamino-	290 or 300-302	2485, 2560
Uracil/3-methyl-6-α-methylhydrazino-	207-209	2440

TABLE XXXV. Amino-Sulphonylpyrimidines (*H* 566)

Pyrimidine	M.P.(^{o}C)	References
2-allylamino-5-allylsulphamoyl-	165-167	1474
5-allylsulphamoyl-2-amino-	225-226	3232
2-amino-5-benzylsulphamoyl-	255-257	3232
2-amino-5-butylsulphamoyl-	239-241	3232
2-amino-5-t-butylsulphamoyl-	186-187	3232
2-amino-5-cyclohexylsulphamoyl-	205-206	3232
2-amino-5-diethylsulphamoyl-	174-175	3232
2-amino-5-dimethylsulphamoyl-	240-241	3232
4-amino-2,6-dimethyl-5-sulphomethyl-	>310	873
4-amino-5-ethyl-2-ethylsulphonyl-6-methyl-	108-109	937
2-amino-5-ethylsulphamoyl-	226-228	3232
4-amino-2-ethylsulphonyl-5,6-dimethyl-	134-135	140
4-amino-2-ethylsulphonyl-5-methyl-	135-136	H, 805, 936
4-amino-2-ethylsulphonyl-6-methyl-5-propyl-	-	H, 927
4-amino-2-methyl-5-sulphomethyl-	-	H, 873, 4167, 4168
2-amino-5-phenylsulphamoyl-	244	3232
2-amino-5-propylsulphamoyl-	235-236	3232
2-amino-5-sulphamoyl-	283-285	1474, 3232
4-amino-6-sulphino-	≮360	2777
2-amino-5-sulpho-	305-307 or 326-328	1474, 3232

TABLE XXXV

Pyrimidine	M.P.(°C)	References
4-aziridino- 6-methylsulphonyl-	112-113	2749
2-butylamino- 5-butylsulphamoyl-	165-167	1474
2-cyclohexylamino- 5-cyclohexylsulphamoyl-	156-158	1474
4,6-diamino- 2-methylsulphonyl- 5-sulphamoyl-	238-239	2364
4,6-diamino-5-sulphamoyl-	217-218	2364
2,4-diamino-6-sulphino-	$\not<$360	2503
2-diethylamino- 5-diethylsulphamoyl-	72-73	1474
4-diethylamino- 2-ethylsulphonyl- 5-methyl-	139-140	805, 936
2-dimethylamino- 5-dimethylsulphamoyl-	200-201	1474
4-dimethylamino- 6-methylsulphonyl-	142-143	2749
2-ethylamino- 5-ethylsulphamoyl-	191-192	1474
2-ethylsulphonyl-5-methyl- 4-methylamino-	138-140	805
2-methylamino- 5-methylsulphamoyl-	194-195	1474
2-propylamino- 5-propylsulphamoyl-	162-163	1474
2,4,6-triamino-5-sulpho-	>310	2364

TABLE XXXVI. Amino-Thiopyrimidines (*H* 567)

Pyrimidine	M.P.(^{o}C)	References
4-acetamido-5-acetylthiomethyl-2-methyl-	116	873
5-acetamido-4-amino-6-mercapto-	280	4014
5-acetamido-4-amino-6-mercapto-2-methyl-	-	3060
4-acetamido-6-benzylthio-	115	3286
5-acetamido-2,4-diacetylthio-	$\nless$300	747
4-acetamido-2-dimethylamino-6-ethylthio-	155-156	2837, 2841
4-acetamido-2-dimethylamino-6-isopropylthio-	186-187	2837
4-acetamido-2-dimethylamino-6-propylthio-	165-167	2837
4-allylamino-2-benzylthio-	65-67	2555
5-allyl-4-amino-2-benzylthio-	66-67	2555
4-allylamino-2-methylthio-	83-84	2555
4-allylthio-2-amino-	133-135	2555
4-allylthio-2,5,6-triamino-	149-151	2778
2-amino-5-anilino-4-mercapto-	223-224	2567
5-amino-2-anilino-4-mercapto-	218	748
4-amino-5-benzamido-2,6-dimercapto-	288-290	3186
4-amino-5-benzamido-6-mercapto-	283-286	3186

Pyrimidine	M.P.(^{o}C)	References
2-amino-4-benzylamino-6-mercapto-	212	2738
2-amino-4-benzylamino-6-methylthio-	115	2738
4-amino-1-benzyl-1,2-dihydro-2-thio-	252	2337
2-amino-4-benzylthio-	178-180	2165
4-amino-6-benzylthio-	140 or 134	2165, 3286
2-amino-4-benzylthio-6-ethylamino-	113	2738
4-amino-6-benzylthio-5-ethylformamido-	180	3288
4-amino-6-benzylthio-5-formamido-	227	H, 3288
5-amino-4-benzylthio-6-mercapto-	183	751
2-amino-4-benzylthio-6-methyl-	118-120	2165
2-amino-4-benzylthio-6-methylamino-	180	2738
2-amino-4,6-bisbenzylthio-	134-136	2165
2-amino-4,6-bisbenzylthio-5-phenyl-	207-209	2165
2-amino-4,6-bisethylthio-	54	H, 2165
4-amino-2,6-bisethylthio-	78-80	4043, 4169
2-amino-4,6-bismethylthio-	116-118	2165, 2601
4-amino-2,6-bismethylthio-	121-123	2165, 4170
5-amino-4,6-bismethylthio-	79	2165
2-amino-4,6-bismethylthio-5-phenyl-	128-129	2165
2-amino-4,6-bispropylthio-	85-87	2165
2-amino-4-butylamino-6-mercapto-	231	2738
2-amino-4-butylthio-6-methyl-	70-72	2165, 2580

TABLE XXXVI

Pyrimidine	M.P.(^{o}C)	References
2-amino-1,4-dihydro-4-imino-1-methyl-6-methylthio- (or tautomer)	HI 282; HCl 288	2288
2-amino-1,6-dihydro-6-imino-1-methyl-4-methylthio- (or tautomer)	192; HI 250; HCl 282	2288
5-amino-1,4-dihydro-4-imino-1-methyl-2-methylthio-	HI 235; HCl 277	2288
2-amino-1,6-dihydro-1-methyl-6-thio-	233-234	2776
4-amino-1,2-dihydro-1-methyl-2-thio-	268	2337
2-amino-4,6-dimercapto-	267	H, 2165
4-amino-2,6-dimercapto-	309 or >360	2165, 3060
5-amino-2,4-dimercapto-	>270	H, 747
5-amino-4,6-dimercapto-	>330	H, 751, 2158
2-amino-4,6-dimercapto-5-phenyl-	266-268	2165
4-amino-2-dimethylamino-6-isopropylthio-	100	2841
2-amino-4-ethylamino-6-ethylthio-	120	2738
2-amino-4-ethylamino-6-mercapto-	224	2738
2-amino-4-ethylamino-6-methylthio-	168	2738
2-amino-4-ethylthio-	155	H, 2165
4-amino-6-ethylthio-	145 or 147-149	2165, 3241
4-amino-2-ethylthio-5,6-dimethyl-	91	H, 140
2-amino-4-ethylthio-6-methyl-	123-125	2165, 2580
2-amino-4-ethylthio-6-methylamino-	118	2738
4-amino-6-ethylthio-2-methylthio-	crude	3216

Pyrimidine	M.P.(^{o}C)	References
4-amino-5-formamido-2,6-bismethylthio-	260	3288
4-amino-5-formamido-6-methylthio-	222	3853
5-amino-4-hydrazino-6-mercapto-	222-224	2227
4-amino-5-hydroxyiminomethyl-2-methylthio-	201-202	1620
2-amino-4-isopropylthio-6-methyl-	108-109	2580
2-amino-4-mercapto-	231-233	2165, 2776
4-amino-2-mercapto-	-	H, 2275
4-amino-6-mercapto-	300 or 306	2165, 2777, 3241
5-amino-4-mercapto-	207	751
2-amino-4-mercapto-6-methyl-	321	H, 2165, 4032
4-amino-6-mercapto-2-methyl-	298	3060
5-amino-4-mercapto-6-methyl-	305-306	2227
2-amino-4-mercapto-6-methylamino-	295	2738
5-amino-2-mercapto-4-methylamino-	-	H, 2458
4-amino-5-mercaptomethyl-2-methyl-	161-163; pic.210; HCl 212	H, 873, 1620
5-amino-2-mercapto-4-methyl-6-methylamino-	244 or 280-285	2563, 2675
4-amino-5-mercaptomethyl-2-methylthio-	138-139	1620
4-amino-2-mercapto-6-methyl-5-phenyl-	300	4171
4-amino-2-mercapto-5-phenyl-	-	H, 4171
4-amino-6-mercapto-5-phenyl-	232-235	2522

TABLE XXXVI

Pyrimidine	M.P.(^{o}C)	References
2-amino-4-mercapto-6-propylamino-	268	2738
2-amino-4-methylamino-6-methylthio-	134	2738
4-amino-5-*N*-γ-methyl-β-butenylformamido-6-methylthio-	167	3853
2-amino-4-methyl-6-methylthio-	147 or 152	H, 2165, 2580
5-amino-4-methyl-6-methylthio-	85-86	2227
4-amino-2-methyl-5-methylthiomethyl-	176-178	1620
2-amino-4-methyl-6-phenylthio-	228	H, 2742
2-amino-4-methyl-6-propylthio-	107	2580
2-amino-4-methylthio-	152-154	2165, 2776
2-amino-5-methylthio-	156 or 161	2304, 4172
4-amino-2-methylthio-	123-125	2288
4-amino-6-methylthio-	168-170; pic.213-215	2165, 2781, 3241
5-amino-2-methylthio-	105	2586
4-amino-2-methylthio-5-methylthiomethyl-	139-140	1620
4-amino-1,2,3,6-tetrahydro-1,3-dimethyl-2,6-dithio-	273-275	4173
5-amino-2,4,6-trimercapto-	$\nless$300	4174
5-benzamido-4,6-dimercapto-	146-147	4014
4-benzylamino-2-mercapto-	249-253	H, 3185, 4054
4-benzylamino-2-methylthio-	93-95	2674
4-benzyloxyamino-2-methylthio-	97-99	2262
2-benzylthio-4-diethylamino-	79-81	2703

Pyrimidine	M.P.(^{o}C)	References
2-benzylthio-4-dimethylamino-5-methyl-	86-87	2703
2-benzylthio-4-dimethylamino-6-methyl-	190/2mm.; HCl 194	4175
2-benzylthio-5-methyl-4-methylamino-	105-106	2703
4-butylamino-2-mercapto-	214 or 220-223	281, 3185
2-butylthio-4-hydrazino-6-methyl-	76-78	2580
4-butylthio-2-hydrazino-6-methyl-	52-54	2580
2,4-diamino-6-benzylthio-	146-148	H, 2778
2,5-diamino-4-butylamino-6-mercapto-	174-175	2761
2,5-diamino-4,6-dimercapto-	-	4174 (?)
4,5-diamino-2,6-dimercapto-	-	H, 2159
4,6-diamino-5-dimethylamino-2-mercapto-	277-279	4176
2,5-diamino-4-ethylamino-6-methylthio-	crude (?)	2738
2,5-diamino-4-ethylthio-	87	748
2,4-diamino-6-mercapto-	301 or >360	H, 2165, 2288, 2503, 2778
2,5-diamino-4-mercapto-	235	748
4,5-diamino-6-mercapto-	-	H, 2933, 3058
4,6-diamino-2-mercapto-	>360	H, 2165
4,5-diamino-2-mercapto-6-methyl-	H_2O 250	H, 2472
4,5-diamino-6-mercapto-2-methyl-	285	H, 3060
4,5-diamino-6-mercapto-2-phenyl-	ca 270	3861
2,5-diamino-4-methylamino-6-methylthio-	crude	2738

Pyrimidine	M.P.(^{o}C)	References
4,5-diamino-6-methylamino-2-methylthio-	pic.195-208	H, 2743
4,6-diamino-5-methylamino-2-methylthio-	172	2743
4,5-diamino-6-methyl-2-methylthio-	194-195	874
2,4-diamino-6-methyl-5-phenylthio-	173 or 177	2773, 4177
2,4-diamino-6-methylthio-	202-204	2288, 2487, 2778
4,5-diamino-6-methylthio-	-	H, 3853
4,6-diamino-2-methylthio-	-	H, 3192
4,5-diamino-6-methylthio-2-phenyl-	97	3861
4,6-diamino-2-methylthio-5-thioformamido-	-	H, 2747
2,4-diamino-5-phenylthio-	217-220	2773
2,4-diamino-6-propylthio-	107-109	2778
4,5-diamino-1,2,3,6-tetrahydro-1,3-dimethyl-2,6-dithio-	-	4173
4,6-dianilino-2-methylthio-	132	2589
4-diethylamino-2-ethylthio-5-methyl-	75-76; HCl 224	805
1,4-dihydro-4-imino-1-methyl-2-methylthio-	HI 225; HCl 237-238	2288
1,4-dihydro-4-imino-1-methyl-6-methylthio-	HCl 251; HI 234-235	2781
1,6-dihydro-6-imino-1-methyl-4-methylthio-	HI 186-187	2781
1,6-dihydro-1-methyl-4-methylamino-6-thio-	167-168	2776
1,4-dihydro-1-methyl-4-methylimino-6-methylthio-	HI 200	2781
2-dimethylamino-4-mercapto-	161-162	2776

Pyrimidine	M.P.(°C)	References
4-dimethylamino-2-mercapto-	265-270	H, 4133
4-dimethylamino-6-mercapto-	276-280	2676
4-dimethylamino-2-mercapto-5-methyl-	242-245	2703
4-dimethylamino-2-mercapto-6-methyl-	292-296	H, 2703
2-dimethylamino-4-methylthio-	29-30; pic.185	2776
4-dimethylamino-2-methylthio-	168/18mm.; 40	2745, 2776
4-dimethylamino-6-methylthio-	54-56	2676
Dithiouracil/5,6-diamino-1-methyl-	>300	2159
4-ethylamino-2-mercapto-	226	281
2-ethylthio-4-hydrazino-6-methyl-	92-94 or 95-96	H, 813, 2580
4-ethylthio-2-hydrazino-6-methyl-	96-97	2580
2-ethylthio-5-methyl-4-methylamino-	58-60; HCl 229	805, 928
4-hexylamino-2-mercapto-	200-204	3185
4-hexylamino-2-methylthio-	218/12mm.; HCl 173	2573
2-hydrazino-4,6-bismethylthio-	106	2601
2-hydrazino-4-isopropylthio-6-methyl-	62-65	2580
4-hydrazino-2-isopropylthio-6-methyl-	103-104	2580
4-hydrazino-2-mercapto-	276-279	H, 2753
4-hydrazino-2-mercapto-6-methyl-	272-275	3218
2-hydrazino-4-methyl-6-methylthio-	122 or 123-124	2452, 2580
4-hydrazino-6-methyl-2-methylthio-	139 to 144	813, 2452, 2580, 2938, 4036

TABLE XXXVI

Pyrimidine	M.P.(°C)	References
2-hydrazino-4-methyl-6-propylthio-	64-65	2580
4-hydrazino-6-methyl-2-propylthio-	104-105	2580
4-hydroxyiminomethyl-6-methyl-2-methylthio-	218-219	2301
4-hydroxyiminomethyl-2-methylthio-	165	2301
4-isoamylamino-2-mercapto-	198-199	3185
4-mercapto-6-methylamino-	258-265	2781
2-mercapto-5-methyl-4-methylamino-	237-241	H, 2703
4-methylamino-6-methylthio-	117-118; pic.199-201	2781
2,4,5-triamino-6-benzylthio-	177-178	2778
4,5,6-triamino-2-benzylthio-	sul. 195-198	2296
2,4,5-triamino-6-butylthio-	89-90	2778
2,4,5-triamino-6-ethylthio-	150-151	2778
2,4,5-triamino-6-mercapto-	-	H, 2503, 3061
4,5,6-triamino-2-mercapto-	-	H, 4178
2,4,5-triamino-6-methylthio-	191-192	2487, 2778
4,5,6-triamino-2-methylthio-	185-186	H, 2933, 4178
2,4,5-triamino-6-propylthio-	145-146	2778

TABLE XXXVII. Amino-Oxypyrimidines with Other Functional Groups (*H* 573)

Pyrimidine	M.P.(°C)	References
4-acetamido-2-chloro-6-ethoxy-	215-216	2837
4-acetamido-6-chloro-2-ethoxy-	194-196	2837, 4043
5-acetamido-4-chloro-6-hydroxy-	212-213	4080
4-acetamido-2-chloro-6-methoxy-	216-217	2837
4-acetamido-6-chloro-2-methoxy-	196 to 198	2837, 4043
2-acetamido-5-β-cyanoethyl-4-hydroxy-6-methyl-	202-203	4179
2-acetamido-5-β-cyanoethyl-4-hydroxy-6-phenyl-	230-232	4179
2-acetamido-5-cyano-4-hydroxy-6-methyl-	285-286	524
4-acetamido-5-fluoro-2-hydroxy-	235-237	4180
2-acetamido-5-β-formylethyl-4-hydroxy-6-methyl-	158-160	2220, 4007, 4181
2-acetamido-5-formylethyl-4-hydroxy-6-phenyl-	149-152	4007
2-acetamido-5-formyl-4-hydroxy-6-methyl-	240	524
2-acetamido-4-hydroxy-5-nitro-	294-296	2464
4-acetamido-6-hydroxy-2-sulphamoyl-	-	4182
5-acetyl-4-amino-1,6-dihydro-1-methyl-6-oxo-	214-215	2389

Pyrimidine	M.P.(^{o}C)	References
5-acetyl-4-amino-1,2,3,6-tetrahydro-1,3-dimethyl-2,6-dioxo-	206	883
5-acetyl-4-dimethylamino-1,2,3,6-tetrahydro-1,3-dimethyl-2,6-dioxo-	121-122	883
5-acetyl-1,2,3,4-tetrahydro-1,3-dimethyl-6-*N*-methylacetamido-2,4-dioxo-	113-114	883
2-allylamino-5-carbamoyl-4-hydroxy-	278	4183
2-allylamino-5-carboxy-4-hydroxy-	197	4183
1-allyl-6-amino-3-βγ-dihydroxypropyl-1,2,3,4-tetrahydro-5-nitroso-2,4-dioxo-	189-191	4144
2-allylamino-5-ethoxycarbonyl-4-hydroxy-	218-219	4183
1-allyl-6-amino-1,2,3,4-tetrahydro-3-β-hydroxypropyl-5-nitroso-2,4-dioxo-	216-218	4144
2-amino-4-anilino-1,6-dihydro-1-methyl-5-nitro-6-oxo-	292-294	2718
2-amino-4-anilino-6-hydroxy-5-nitro-	345-348	4184
4-amino-6-anilino-2-hydroxy-5-nitroso-	272	2475
2-amino-4-anilino-6-hydroxy-5-phenylazo-	>300	1272
4-amino-5-azidocarbonyl-2-hydroxy-	165	1386
4-amino-5-benzoyl-1,6-dihydro-1-methyl-6-oxo-	225-227	2389
2-amino-4-benzylamino-1,6-dihydro-1-methyl-5-nitro-6-oxo-	190-192	2444

Pyrimidine	M.P.(°C)	References
2-amino-4-benzylamino-6-hydroxy-5-nitro-	317 or 335	2444, 2465
2-amino-4-benzylamino-6-hydroxy-5-nitroso-	265-270	2286
4-amino-6-benzylamino-2-hydroxy-5-nitroso-	>360	2475
4-amino-5-benzyl-6-chloro-2-methoxymethyl-	-	2877 (?)
4-amino-5-benzylnitrosoamino-2,4-dihydroxy-	211	2489
4-amino-3-benzyloxy-6-chloro-2,3-dihydro-5-methyl-2-oxo-	218-221	3972
4-amino-3-benzyloxy-6-chloro-2,3-dihydro-2-oxo-	185-195	2262
2-amino-4-benzyloxy-6-fluoro-	100-102	2610
4-amino-2-benzyloxy-6-fluoro-	94-95	2610
4-amino-5-bromo-2,6-dihydroxy-	306	2628
4-amino-5-bromo-2,6-dimethoxy-	170-172	2920, 4024
4-amino-5-bromo-2-dimethylamino-6-hydroxy-	232-234	4137
2-amino-5-bromo-4-ethoxy-6-methyl-	108-109	2623
2-amino-5-bromo-4-hydroxy-	275	H, 2842
4-amino-5-bromo-2-hydroxy-	240-242	H, 827
4-amino-5-bromo-6-hydroxy-	268	827
2-amino-5-bromo-4-hydroxy-6-methyl-	249	H, 2842
2-amino-5-bromo-4-hydroxy-6-trifluoromethyl-	303	443

TABLE XXXVII

Pyrimidine	M.P.(^{o}C)	References
2-amino-5-bromo-4-methoxy-	118	2606
2-amino-5-bromo-4-methoxy-6-methyl-	153-155	2623
2-amino-5-bromo-4-methyl-6-phenoxy-	171	2631
4-amino-5-bromo-1,2,3,6-tetrahydro-1,3-dimethyl-2,6-dioxo-	210	H, 2628
4-amino-5-butoxy-6-chloro-	103-105	2524
2-amino-4-t-butylamino-1,6-dihydro-1-methyl-5-nitro-6-oxo-	217	2716
2-amino-4-t-butylamino-6-hydroxy-5-nitro-	291-292	2716
4-amino-3-butyl-2,3-dihydro-5-nitro-2-oxo-	unisolated	281
2-amino-4-carbamoyl-6-hydroxy-	320	2216
4-amino-5-carbamoyl-2-hydroxy-	>320	H, 459
2-amino-4-carboxy-1-β-carboxyethyl-1,6-dihydro-6-oxo-	299-302	2215
2-amino-4-carboxy-5-diethylaminomethyl-6-hydroxy-	312	4185
2-amino-4-carboxy-1,6-dihydro-1-methyl-6-oxo-	290-292	2215
5-amino-4-carboxy-2,6-dihydroxy-	-	H, 4186
2-amino-1-β-carboxyethyl-1,4-dihydro-4-oxo-	-	2215
2-amino-1-β-carboxyethyl-1,6-dihydro-6-oxo-	unisolated	2215
4-amino-3-β-carboxyethyl-2,3-dihydro-2-oxo-	>290	3363
2-amino-5-β-carboxyethyl-4-hydroxy-6-methyl-	301-303	H, 2232

Pyrimidine	M.P.(^{o}C)	References
2-amino-5-β-carboxyethyl-4-hydroxy-6-phenyl-	259-261	4179
2-amino-4-carboxy-6-hydroxy-	-	H, 2215
4-amino-2-carboxy-6-hydroxy-	$\nless$350	2334
4-amino-5-carboxy-2-hydroxy-	263	H, 459
4-amino-6-carboxy-2-hydroxy-	293-294	2214
2-amino-4-carboxy-6-hydroxy-5-methyl-	311-313	2391
2-amino-5-carboxy-4-hydroxy-6-methyl-	245-246	524
4-amino-3-β-carboxypropyl-2,3-dihydro-2-oxo-	>290	3363
5-amino-4-carboxy-1,2,3,6-tetrahydro-1,3-dimethyl-2,6-dioxo-	-	3968
2-amino-5-β-chlorocarbonylethyl-4-hydroxy-6-methyl-	150-200 (crude)	2232
2-amino-5-β-chlorocarbonylethyl-4-hydroxy-6-phenyl-	130-165 (crude)	4179
4-amino-6-chloro-2,3-dihydro-3-hydroxy-5-methyl-2-oxo-	232-235	3972
2-amino-4-chloro-1,6-dihydro-1-methyl-5-nitro-6-oxo-	275-276	2444
2-amino-4-chloro-1,6-dihydro-1-methyl-6-oxo-	305	2444
4-amino-6-chloro-1,2-dihydro-1-methyl-2-oxo-	220-225	2475
4-amino-5-chloro-2,6-dihydroxy-	325	4187
4-amino-5-chloro-2,6-dimethoxy-	163-165	4188

TABLE XXXVII

Pyrimidine	M.P.(°C)	References
5-amino-4-chloro-2-dimethylamino-6-hydroxy-	233-235	2444
4-amino-2-chloro-6-ethoxy-	133-134	2837
4-amino-6-chloro-2-ethoxy-	128-129	2677, 2837, 2839, 4043
4-amino-6-chloro-5-ethoxy-	119-120	2524
2-amino-5-chloro-4-ethoxy-6-hydroxy-	216-218	2239
4-amino-5-chloro-6-ethoxy-2-methyl-	161	2239
4-amino-6-chloro-2-ethoxymethyl-	-	3913 (?)
4-amino-5-chloro-6-ethoxy-2-phenyl-	92	2239
2-amino-5-β-chloroethyl-4-hydroxy-	HCl 214-215	2514
4-amino-6-chloro-5-ethyl-2-methoxymethyl-	168-169	3911
2-amino-4-chloro-6-hydroxy-	-	H, 4189
2-amino-5-chloro-4-hydroxy-	306-307	2610
4-amino-5-chloro-2-hydroxy-	291-292	2610
4-amino-6-chloro-2-hydroxy-	HCl >360	H, 2475, 2610, 2837
5-amino-4-chloro-6-hydroxy-	HCl 240-242	2466
2-amino-4-chloro-6-hydroxy-5-methoxy-	280-281	2212
2-amino-5-chloro-4-hydroxy-6-methyl-	ca 260	2623
2-amino-4-chloro-6-hydroxy-5-nitro-	275 to >350	2442, 2444, 2457, 2465, 4184
2-amino-4-chloro-6-isopropoxy-	85-86	2482
4-amino-6-chloro-2-isopropoxy-	134-135	2837
4-amino-6-chloro-5-isopropoxy-	139-141	2524

Pyrimidine	M.P.(^{o}C)	References
2-amino-4-chloro-5-methoxy-	146-148	2212
2-amino-4-chloro-6-methoxy-	170 to 174	H, 2543, 2843
2-amino-5-chloro-4-methoxy-	118-120	2571
4-amino-2-chloro-5-methoxy-	182 or 189-190	2161, 2586
4-amino-2-chloro-6-methoxy-	187-188	2837
4-amino-6-chloro-2-methoxy-	128 to 130	2678, 2837, 4043
4-amino-6-chloro-5-methoxy-	179-181	3910, 4096
5-amino-4-chloro-6-methoxy-	79-80	2158, 4111
4-amino-6-chloro-5-β-methoxyethyl-	113-114	4041
2-amino-4-chloro-5-methoxy-6-methyl-	124-125	2228
2-amino-5-chloro-4-methoxy-6-methyl-	138-141	2623
4-amino-6-chloro-2-methoxymethyl-	102	3912
5-amino-2-chloro-4-methoxy-6-methyl-	72-77	2763
5-amino-4-chloro-6-methoxy-2-methyl-	60-61	887
4-amino-6-chloro-2-methoxymethyl-5-methyl-	150-151	3911
2-amino-4-chloro-6-methoxy-5-nitro-	177-179	2718
4-amino-5-chloromethyl-2-hydroxy-	HCl >300	2639
4-amino-5-chloromethyl-2-methoxy-	HCl -	4071
4-amino-5-chloromethyl-2-methoxymethyl-	HCl 202	4072

TABLE XXXVII

Pyrimidine	M.P.(°C)	References
4-amino-6-chloro-2-propoxy-	114-115	2837
4-amino-5-chlorosulphonyl-2,6-dihydroxy-	265	4190
5-amino-4-chloro-1,2,3,6-tetrahydro-1,3-dimethyl-2,6-dioxo-	120-121	2441
2-amino-5-cyano-1,4-dihydro-1-β-hydroxyethyl-4-oxo-	248	743
4-amino-5-cyano-6-ethoxy-2-methyl-	233-235	H, 4191
4-amino-1-β-cyanoethyl-1,2-dihydro-2-oxo-	249-250	3363
2-amino-5-β-cyanoethyl-4-hydroxy-6-methyl-	266-267	2232
2-amino-5-β-cyanoethyl-4-hydroxy-6-phenyl-	315-317	4179
4-amino-5-cyano-2-hydroxy-	340	H, 459, 2897
4-amino-5-cyano-2-methoxy-	221-222	H, 2273
4-amino-6-cyclohexylamino-2-hydroxy-5-nitroso-	290-291	3854
4-amino-3-βγ-dibromopropyl-1,2-dihydro-2-oxo-	207-210	3479
2-amino-4,6-dichloro-5-ethoxy-	188-189	H, 4081
2-amino-4,6-dichloro-5-methoxy-	216-217	2212, 4081
4-amino-5-ββ-diethoxyethyl-2,6-dihydroxy-	>335	2011, 4019
4-amino-5-ββ-diethoxyethyl-2-ethyl-6-hydroxy-	233-235	4192
4-amino-5-ββ-diethoxyethyl-6-hydroxy-	186 to 198-199	2011, 4019, 4193
4-amino-5-ββ-diethoxyethyl-6-hydroxy-2-methyl-	249-250	4192

Pyrimidine	M.P.(^{o}C)	References
4-amino-5-ββ-diethoxyethyl-6-hydroxy-2-phenyl-	174-176	4192
4-amino-5-ββ-diethoxyethyl-6-hydroxy-2-propyl-	207-209	4192
4-amino-5-ββ-diethoxy-α-methylethyl-6-hydroxy-	-	4194
2-amino-5-γγ-diethoxypropyl-4-hydroxy-6-methyl-	177-180	2220
2-amino-1,6-dihydro-4-methoxy-1-methyl-5-nitro-6-oxo-	215-218	2444
2-amino-1,6-dihydro-1-methyl-4-methylamino-5-nitro-6-oxo-	270-272	2444, 2716
4-amino-1,2-dihydro-1-methyl-5-nitro-2-oxo-	271-273	H, 4106
4-amino-1,6-dihydro-1-methyl-5-nitro-6-oxo-	284 (not 184)	827, 2626
2-amino-4,6-dihydroxy-5-nitro-	>400	H, 2164
4-amino-2,6-dihydroxy-5-nitro-	>360	H, 2445
2-amino-4,5-dihydroxy-6-nitroso-	>300	2210
4-amino-2,6-dihydroxy-5-nitroso-	-	H, 752, 2881
5-amino-2,4-dihydroxy-6-sulpho-	-	H, 4195
5-amino-4,6-dihydroxy-2-trifluoromethyl-	257-259	2193
2-amino-4-dimethylamino-6-hydroxy-5-nitro-	250-251	2444
4-amino-2-dimethylamino-6-hydroxy-5-nitro-	287-290 or 305-306	2444, 2445
4-amino-2-dimethylamino-6-hydroxy-5-nitroso-	263	H, 4092

Pyrimidine	M.P.(^{o}C)	References
4-amino-6-dimethylamino-2-hydroxy-5-nitroso-	>360	2475
4-amino-3-dimethylamino-1-methyl-1,2,3,6-tetrahydro-5-nitroso-2,6-dioxo-	-	4160 (?)
4-amino-3-dimethylamino-1,2,3,6-tetrahydro-1-hydroxyethyl-5-nitroso-2,6-dioxo-	201-203	2291
4-amino-3-dimethylamino-1,2,3,6-tetrahydro-1-methyl-5-nitroso-2,6-dioxo-	288-289	2291
4-amino-3-β-ethoxycarbonylethyl-2,3-dihydro-2-oxo-	277	3363
4-amino-6-ethoxycarbonyl-2-hydroxy-5-nitro-	240	2467
4-amino-5-ethoxycarbonyl-2-methoxy-	151-153	2205, 3369
4-amino-2-ethoxy-6-ethoxycarbonyl-5-nitro-	120	2467
2-amino-4-ethoxy-6-fluoro-	120-123	2610
4-amino-2-ethoxy-6-methyl-5-nitro-	168	H, 874
4-amino-5-ethoxymethyl-2-trifluoromethyl-	126-127	353
2-amino-4-ethoxy-5-nitro-	227	748
4-amino-6-ethoxy-5-nitro-	180	2164
2-amino-4-ethoxy-6-phenylazo-	136-138	4052
2-amino-4-ethylamino-1,6-dihydro-1-methyl-5-nitro-6-oxo-	257-259	2444
2-amino-4-ethylamino-6-hydroxy-5-nitro-	342-344	2445
2-amino-4-ethylamino-6-hydroxy-5-phenylazo-	284	1170

Pyrimidine	M.P.(^{o}C)	References
4-amino-3-ethyl-2,3-dihydro-5-nitro-2-oxo-	unisolated	281
2-amino-4-ethyl-5-fluoro-6-hydroxy-	-	4196
4-amino-5-fluoro-1,2-dihydro-1-methyl-2-oxo-	297-299	2748
2-amino-5-fluoro-4,6-dihydroxy-	-	4196
2-amino-4-fluoro-6-hydroxy-	>280	2610
2-amino-5-fluoro-4-hydroxy-	271-274	2209, 2610
4-amino-5-fluoro-2-hydroxy-	297	2610, 4197
4-amino-5-fluoro-6-hydroxy-	>290	2191
4-amino-6-fluoro-2-hydroxy-	$\nless$300	2610
2-amino-5-fluoro-4-hydroxy-6-methyl-	>300	4196
4-amino-5-fluoro-6-hydroxy-2-methyl-	322-325	2191
2-amino-5-fluoro-4-hydroxy-6-pentafluoroethyl-	265-268	4196
2-amino-5-fluoro-4-methoxy-	139-141	2209
4-amino-5-fluoro-2-methoxy-	190-191	2209
2-amino-4-fluoromethyl-6-hydroxy-	250-260	4196
2-amino-5-formyl-4,6-dimethoxy-	224-228; TSC >335	2593
4-amino-5-formyl-2,6-dimethoxy-	189-190; TSC >335	2593
2-amino-5-β-formylethyl-4-hydroxy-6-methyl-	acetal 180	2566, 4181
2-amino-5-β-formylethyl-4-hydroxy-6-phenyl-	acetal 228	4007
4-amino-5-formyl-2-hydroxy-	unisolated	1168
2-amino-5-formyl-4-hydroxy-6-methyl-	269-270	H, 524

TABLE XXXVII

Pyrimidine	M.p.(^{o}C)	References
4-amino-5-formyl-1,2,3,6-tetrahydro-1,3-dimethyl-2,6-dioxo-	194-196	883
2-amino-4-hydrazino-1,6-dihydro-1-methyl-5-nitro-6-oxo-	189	2444
4-amino-6-hydroxy-2-hydroxyamino-5-nitroso-	262	2473
2-amino-4-hydroxy-5-iodo-	-	2588
4-amino-2-hydroxy-5-iodo-	-	H, 2634
2-amino-4-hydroxy-6-methoxycarbonyl-	293-294	2214, 2216
2-amino-4-hydroxy-6-methoxy-5-nitro-	>274	2718
4-amino-6-hydroxy-2-methoxy-5-nitroso-	-	909
2-amino-4-hydroxy-6-methylamino-5-nitro-	360 or >360	2444, 2716, 4184
4-amino-6-hydroxy-2-methylamino-5-nitro-	350-352	2445, 2851
2-amino-4-hydroxy-6-methylamino-5-nitroso-	>300	1372
4-amino-2-hydroxy-6-methylamino-5-nitroso-	>360	2475
4-amino-6-hydroxy-2-methylamino-5-nitroso-	>320	4092
4-amino-5-hydroxymethyl-2-methylsulphonyl-	172-173	3062
2-amino-4-hydroxy-6-methyl-5-nitro-	-	H, 2787
4-amino-6-hydroxy-2-methyl-5-nitro-	260 or $\nless$333	2445, 4198, 4199
4-amino-6-hydroxy-2-methyl-5-nitroso-	-	H, 2881
4-amino-6-hydroxy-2-methyl-5-phenylazo-	320	745, 4200
4-amino-2-hydroxy-6-methyl-5-sulpho-	290-291	1136

Pyrimidine	M.P. (^{o}C)	References
4-amino-6-hydroxy-2-methyl-5-sulpho-	305-306	2364
4-amino-5-hydroxymethyl-2-trifluoromethyl-	180-181	1000, 2268
2-amino-4-hydroxy-5-nitro-	-	824
4-amino-2-hydroxy-5-nitro-	>360	824, 2711
4-amino-6-hydroxy-5-nitroso-2-phenyl-	255	H, 3861
4-amino-2-hydroxy-5-nitroso-6-piperidino-	270	3854
4-amino-6-hydroxy-5-nitroso-2-piperidino-	243	2473
4-amino-6-hydroxy-5-nitroso-2-trifluoromethyl-	162 (crude)	2217
4-amino-6-hydroxy-5-nitro-2-trifluoromethyl-	238-240	2217
2-amino-4-hydroxy-5-phenylazo-	239-240	2211
2-amino-4-hydroxy-6-phenylazo-	223-224	4052
4-amino-6-hydroxy-5-phenylazo-	-	H, 4201
2-amino-4-hydroxy-5-phenylazo-6-trifluoromethyl-	280-282	443, 1242
2-amino-4-hydroxy-6-phenyl-5-β-(phenylcarbamoyl)ethyl-	305-306	4179
4-amino-2-hydroxy-5-sulpho-	-	1136
4-amino-2-hydroxy-6-sulpho-	>360	2475
4-amino-2-hydroxy-5-trifluoroacetamido-	$\nless$300	2360
2-amino-4-hydroxy-6-trifluoromethyl-	>170 or 282	443, 4196
4-amino-6-hydroxy-2-trifluoromethyl-	252	2193
5-amino-4-hydroxy-6-trifluoromethyl-	222	443

TABLE XXXVII

Pyrimidine	M.P.(^oC)	References
2-amino-5-iodo-4-methoxy-	155-157	2571
2-amino-4-isopropoxy-6-methylamino-5-nitroso-	180-181	2482
4-amino-2-methoxy-5-methoxycarbonyl-	208-210	2205
2-amino-4-methoxy-6-methylamino-5-nitroso-	210-211	2482
4-amino-5-methoxy-2-methylsulphonyl-	168-169	2205
2-amino-4-methoxy-5-nitro-	227	748
2-amino-4-methoxy-6-phenylazo-	145-146	4052
4-amino-1,2,3,6-tetrahydro-1,3-dimethyl-2,6-dioxo-5-sulphamoyl-	218-220; H_2O 119	2364
4-amino-1,2,3,6-tetrahydro-1,3-dimethyl-5-*N*-methylsulphamoyl-2,6-dioxo-	203-204	2364
4-amino-1,2,3,6-tetrahydro-1,3-dimethyl-5-nitro-2,6-dioxo-	232-235	2445
4-amino-1,2,3,6-tetrahydro-1,3-dimethyl-5-nitroso-2,6-dioxo-	-	H, 4202
4-amino-1,2,3,4-tetrahydro-5-methoxysulphonyl-1,3-dimethyl-2,6-dioxo-	206-208	2364
2-amylamino-5-carbamoyl-4-hydroxy-	238	4183
2-amylamino-5-carboxy-4-hydroxy-	177	4183
2-amylamino-5-ethoxycarbonyl-4-hydroxy-	180-182	4183
4-anilino-5-chloro-2,6-diethoxy-	131	2690
4-anilino-2,5-dichloro-6-ethoxy-	124	2690
4-anilino-5,6-dichloro-2-ethoxy-	66	2690

Pyrimidine	M.P.(^{o}C)	References
4-anilino-2-dimethylamino-6-hydroxy-5-nitro-	294-296	2444
4-anilino-3-dimethylamino-1,2,3,6-tetrahydro-1-methyl-5-nitroso-2,6-dioxo-	203-204	2246
2-anilino-4-ethoxy-5-nitro-	150	748
2-anilino-4-methoxy-5-nitro-	183	748
4-aziridino-5-fluoro-2-methoxy-	66	2729
4-benzamido-6-chloro-2-methoxy-	109	2663
4-benzamido-6-chloro-2-propoxy-	86	2663
4-benzamido-5-fluoro-2-hydroxy-	256-257	4180
4-benzylamino-5-bromo-2-hydroxy-	192-195	4047
4-benzylamino-5-bromo-1,2,3,6-tetrahydro-1,3-dimethyl-2,6-dioxo-	137	2109
4-benzylamino-2,6-dihydroxy-5-nitroso-	225-360	2485, 4067
4-benzylamino-2-dimethylamino-6-hydroxy-5-nitro-	261-262	2444
4-benzylamino-3-dimethylamino-1,2,3,6-tetrahydro-1-methyl-5-nitroso-2,6-dioxo-	180 (crude)	2246
2-benzylamino-5-ethoxycarbonyl-4-hydroxy-	241-242	2148
4-benzylamino-2-hydroxy-5-nitro-	225-228	4054
4-benzylamino-1,2,3,6-tetrahydro-1-methyl-5-nitroso-2,6-dioxo-3-piperidino-	170 (crude)	2246

TABLE XXXVII

Pyrimidine	M.P. (^{o}C)	References
1-benzyl-2-benzylidenehydrazino-5-ethoxycarbonyl-1,6-dihydro-6-oxo-	204-206	4203
5-benzyl-4-chloro-2-methoxymethyl-6-methylamino-	130-131	2877
5-benzyl-4-chloro-2-methoxymethyl-6-piperidino-	176/0.25mm.	2877
2-benzylidenehydrazino-5-cyano-4-hydroxy-	313	4204
2-benzylidenehydrazino-5-ethoxycarbonyl-1-ethyl-1,6-dihydro-6-oxo-	217-219	4203
2,4-bisdimethylamino-6-hydroxy-5-nitro-	265-267	2444
5-bromo-2,4-dihydroxy-6-methylamino-	$\nless$320	1005
5-bromo-2,4-dimethoxy-6-methylamino-	104-106	1005
5-bromo-2-dimethylamino-4-hydroxy-6-methyl-	232-233	2811, 4137
5-bromo-4-dimethylamino-1,2,3,6-tetrahydro-1,3-dimethyl-2,6-dioxo-	98	1005
5-bromo-4-hydroxy-6-methyl-2-piperidino-	242-243	2811, 4137
5-bromo-4-hydroxy-2-piperidino-	201-202	2811, 4137
5-bromo-1,2,3,4-tetrahydro-1,3-dimethyl-6-methylamino-2,4-dioxo-	152	1005
5-butoxycarbonyl-2-cyclopentylamino-4-hydroxy-	150-152	4205
5-butoxycarbonyl-4-hydroxy-2-neopentylamino-	175-176	4205
2-butylamino-5-butylcarbamoyl-4-hydroxy-	206	4183

Pyrimidine	M.P.(^{o}C)	References
2-butylamino-5-carbamoyl-4-hydroxy-	205	4183
2-s-butylamino-5-carbamoyl-4-hydroxy-	218-220	4183
2-butylamino-5-carboxy-4-hydroxy-	-	4183 (?)
2-s-butylamino-5-carboxy-4-hydroxy-	177	4183
4-butylamino-2,6-dihydroxy-5-nitroso-	198	4075
4-butylamino-3-dimethylamino-1,2,3,6-tetrahydro-1-methyl-5-nitroso-2,6-dioxo-	151-152	2246
2-butylamino-5-ethoxycarbonyl-4-hydroxy-	198-200	4183
2-s-butylamino-5-ethoxycarbonyl-4-hydroxy-	163-165	4183
2-butylamino-5-*N*-ethylcarbamoyl-4-hydroxy-	-	4183
2-butylamino-4-hydroxy-5-*N*-methylcarbamoyl-	-	4183
4-butylamino-2-hydroxy-5-nitro-	205	281
2-butylamino-4-hydroxy-5-*N*-propylcarbamoyl-	-	4183
4-butylamino-1,2,3,6-tetrahydro-1,3-dimethyl-5-nitroso-2,6-dioxo-	261	2485
4-s-butylamino-1,2,3,6-tetrahydro-1,3-dimethyl-5-nitroso-2,6-dioxo-	108-110	2838
4-butylamino-1,2,3,6-tetrahydro-1-methyl-5-nitroso-2,6-dioxo-3-piperidino-	182-183	2246

Pyrimidine	M.P.(^{o}C)	References
5-*N*-butylcarbamoyl-2-cyclopentylamino-4-hydroxy-	243-244	4205
5-carbamoyl-2-cyclopentylamino-4-hydroxy-	281	4205
5-carbamoyl-2-diethylamino-4-hydroxy-	258-260	4206
5-carbamoyl-2-dimethylamino-4-hydroxy-	>300	4206
5-carbamoyl-2-ethylamino-4-hydroxy-	308	4206
5-carbamoyl-2-ethylmethylamino-4-hydroxy-	265	4206
5-carbamoyl-2-hexylamino-4-hydroxy-	195-198	4206
5-carbamoyl-4-hydroxy-2-isoamylamino-	263	4206
5-carbamoyl-4-hydroxy-2-isobutylamino-	278	4206
5-carbamoyl-4-hydroxy-2-isopropylamino-	215-218	4206
5-carbamoyl-4-hydroxy-2-methylamino-	156	4206
5-carbamoyl-4-hydroxy-2-β-methylbutylamino-	281	4205
5-carbamoyl-4-hydroxy-6-methyl-2-piperidino-	290	524
5-carbamoyl-4-hydroxy-2-neopentylamino-	314	4205
5-carbamoyl-4-hydroxy-2-propylamino-	296	4206
5-carboxy-2-cyclopentylamino-4-hydroxy-	205	4205
4-carboxy-5-diethylaminomethyl-2,6-dihydroxy-	246	4185

Pyrimidine	M.P.(^{o}C)	References
4-carboxy-2,6-dihydroxy-5-piperidinomethyl-	231-232	4185
5-carboxy-2-ethylamino-4-hydroxy-	205	4183
5-carboxy-2-hexylamino-4-hydroxy-	178	4183
5-carboxy-4-hydroxy-2-isoamylamino-	186	4183
5-carboxy-4-hydroxy-2-isobutylamino-	197	4183
5-carboxy-4-hydroxy-2-isopropylamino-	188	4183
5-carboxy-4-hydroxy-2-methylamino-	217	4183
5-carboxy-4-hydroxy-2-β-methylbutylamino-	183	4205
5-carboxy-4-hydroxy-2-neopentylamino-	314	4205
5-carboxy-4-hydroxy-2-propylamino-	20 (?)	4183
5-carboxymethyl-2-cyanoamino-4-hydroxy-	247-250	2208
5-carboxyvinyl-4-hydroxy-6-methyl-2-piperidino-	231	524
4-chloro-2-dimethylamino-6-hydroxy-	-	H, 2476
4-chloro-2-dimethylamino-6-hydroxy-5-nitro-	252-253	2717
4-chloro-6-dimethylamino-2-methoxy-	88-89	2676
4-chloro-2-dimethylamino-6-methoxy-5-nitro-	114-116	2718
4-chloro-3-dimethylamino-1,2,3,6-tetrahydro-1-methyl-2,6-dioxo-	127-128	2246
4-chloro-2-ethylamino-6-methoxy-	70-73	4045
4-chloro-5-formamido-1,6-dihydro-2-hydroxy-1-methyl-6-oxo-	225-226	2439, 2440

Pyrimidine	M.P.(^oC)	References
2-chloro-4-hydrazino-5-methoxy-	148-150	2161
4-chloro-6-hydroxy-5-methylamino-	HCl 247-248	2752
4-chloro-2-isopropylamino-6-methoxy-	-	4045
5-chloro-1,2,3,4-tetrahydro-1,3-dimethyl-6-methylamino-2,4-dioxo-	181	1005
4-chloro-1,2,3,6-tetrahydro-1-methyl-2,6-dioxo-3-piperidino-	177-178	2246
1-cyano-3,5-diethyl-1,2,3,4-tetrahydro-6-hydroxy-2-imino-4-oxo-	168-171	1434
5-cyano-2-dimethylamino-4-hydroxy-	294-296	1441, 4206
5-cyano-2-dimethylamino-4-methoxy-	154-155	4003
1-cyano-5-ethyl-1,2,3,4-tetrahydro-6-hydroxy-2-imino-3-methyl-4-oxo-	177	1434
5-cyano-2-hydrazino-4-hydroxy-	>320	4204
5-cyano-6-hydroxy-6-methyl-2-piperidino-	292-294	524
5-cyano-4-hydroxy-2-piperidino-	266	4207
4-cyclohexylamino-2,6-dihydroxy-5-nitro-	288-290	2927
4-cyclohexylamino-2,6-dihydroxy-5-nitroso-	247-249	2838
4-cyclohexylamino-3-dimethylamino-1,2,3,6-tetrahydro-1-methyl-5-nitroso-2,6-dioxo-	217-218	2246
4-cyclohexylamino-1,2,3,6-tetrahydro-1,3-dimethyl-5-nitroso-2,6-dioxo-	140 or 150	2838, 2929

Pyrimidine	M.P.(°C)	References
4-cyclohexylamino-1,2,3,6-tetrahydro-1-methyl-5-nitroso-2,6-dioxo-3-piperidino-	210-211	2246
2-cyclohexylguanidino-5-ethoxycarbonylmethyl-4-hydroxy-	189-190	2208
2-cyclohexylguanidino-5-hydrazinocarbonylmethyl-4-hydroxy-	245-247	2208
2-cyclopentylamino-5-ethoxycarbonyl-4-hydroxy-	186-187	4205
2-cyclopentylamino-5-ethylcarbamoyl-4-hydroxy-	278	4205
2-cyclopentylamino-4-hydroxy-5-methylcarbamoyl-	295	4205
2-cyclopentylamino-4-hydroxy-5-propylcarbamoyl-	274	4205
2,4-diamino-6-benzyloxy-5-nitroso-	205-212	H, 2285
2,4-diamino-5-bromo-6-hydroxy-	244 or 290	H, 1224, 2628
2,5-diamino-4-chloro-1,6-dihydro-1-methyl-6-oxo-	244	2444
2,4-diamino-5-chloro-6-ethoxy-	132	2239
2,4-diamino-5-chloro-6-hydroxy-	341-342	H, 2239
2,4-diamino-5-cyano-6-ethoxy-	220-221	269
2,4-diamino-5-ββ-diethoxyethyl-6-hydroxy-	180-181; H_2O 156-158	2011, 4019
4,6-diamino-5-ββ-diethoxyethyl-2-hydroxy-	>320	4019

TABLE XXXVII

Pyrimidine	M.P.(°C)	References
4,6-diamino-5-ββ-diethoxyethyl-2-methoxy-	149-151	2011
2,4-diamino-1,6-dihydro-1-methyl-5-nitro-6-oxo-	343 to 347	2444, 2716
2,4-diamino-3,6-dihydro-3-methyl-5-nitro-6-oxo-	262	2909
2,4-diamino-1,6-dihydro-1-methyl-5-nitroso-6-oxo-	-	4092
4,6-diamino-1,2-dihydro-1-methyl-5-nitroso-2-oxo-	320	2481
4,5-diamino-6-ethoxycarbonyl-2-hydroxy-	234	2467
4,5-diamino-2-ethoxy-6-ethoxycarbonyl-	201	2467
2,4-diamino-6-hydroxy-5-nitro-	>360	2164, 2445, 2716
4,6-diamino-2-hydroxy-5-nitro-	>350	2164
2,4-diamino-6-hydroxy-5-nitroso-	-	H, 2473, 4140
4,6-diamino-2-hydroxy-5-nitroso-	300 or >360	H, 1169, 2481
2,4-diamino-6-hydroxy-5-phenylazo-	-	H, 2853
4,6-diamino-2-hydroxy-5-phenylazo-	>300	2853
4,5-diamino-6-hydroxy-2-sulphino-	188-190	2478
2,4-diamino-6-hydroxy-5-sulpho-	>330	2364
4,6-diamino-2-hydroxy-5-sulpho-	>330	2364
4,5-diamino-6-hydroxy-2-trifluoromethyl-	284-285	2217
2,4-diamino-6-isopropoxy-5-nitroso-	244-246	2285
2,4-diamino-6-methoxy-5-nitroso-	217-224	H, 2285

Pyrimidine	M.P.(°C)	References
2,4-diaziridino-5-bromo-6-methoxy-	115-116	2728
4,6-diaziridino-5-bromo-2-methoxy-	135-136	2728
2,4-diaziridino-5-chloro-6-methoxy-	111-113	2728
4,6-diaziridino-5-chloro-2-methoxy-	138-140	2728
2,4-diaziridino-6-methoxy-5-nitro-	140-145	2731
4,6-diaziridino-2-methoxy-5-nitro-	165-175	2731
4-diethylamino-2-dimethylamino-6-hydroxy-5-nitro-	238-240	2444
2-diethylamino-5-ethoxycarbonyl-4-hydroxy-	92-93	4206
2,4-dihydroxy-6-methylamino-5-nitro-	>300	2438
2,4-dihydroxy-6-methylamino-5-nitroso-	>280 or >350	553, 1005, 4067
4-dimethylamino-2,6-dihydroxy-5-nitro-	247	2909
2-dimethylamino-5-ethoxycarbonyl-4-hydroxy-	165-167	4206
2-dimethylamino-4-ethoxy-5-nitro-	104-106	2718
2-dimethylamino-4-ethylamino-6-hydroxy-5-nitro-	282	2444
2-dimethylamino-5-formyl-4-hydroxy-6-methyl-	-	H, 524
2-dimethylamino-4-hydrazino-6-hydroxy-5-nitro-	215-216	2444
2-dimethylamino-4-hydroxy-6-methylamino-5-nitro-	310-312	2444

Pyrimidine	M.P.(^{o}C)	References
2-dimethylamino-4-hydroxy-6-methylamino-5-nitroso-	255	2476
4-dimethylamino-6-hydroxy-5-nitro-	216-217	2130
2-dimethylamino-4-hydroxy-5-phenylazo-	194-195	2211
5-dimethylamino-1,2,3,4-tetrahydro-1,3-dimethyl-2,4-dioxo-6-sulpho-	264	4208
1-dimethylamino-1,2,3,4-tetrahydro-6-isopropylamino-3-methyl-5-nitroso-2,4-dioxo-	167-168	2246
1-dimethylamino-1,2,3,4-tetrahydro-3-methyl-6-methylamino-5-nitroso-2,4-dioxo-	228-229	2246
1-dimethylamino-1,2,3,4-tetrahydro-3-methyl-5-nitroso-2,4-dioxo-6-piperidinoamino-	115	3254
2-*NN*-dimethylguanidino-5-ethoxycarbonylmethyl-4-hydroxy-	167	2208
2-*NN*-dimethylguanidino-5-hydrazinocarbonylmethyl-4-hydroxy-	230-231	2208
5-ethoxycarbonyl-2-ethylamino-4-hydroxy-	205-206	4206
5-ethoxycarbonyl-2-ethylmethylamino-4-hydroxy-	137-138	4206
5-ethoxycarbonyl-2-hexylamino-4-hydroxy-	176-177	4206
5-ethoxycarbonyl-2-hydrazino-4-hydroxy-	235-238	4015, 4036
5-ethoxycarbonyl-4-hydroxy-2-isoamylamino-	210-211	4206
5-ethoxycarbonyl-4-hydroxy-2-isobutylamino-	208-209	4206

Pyrimidine	M.P.(^{o}C)	References
5-ethoxycarbonyl-4-hydroxy-2-isopropylamino-	193-195	4206
5-ethoxycarbonyl-4-hydroxy-2-methylamino-	242	4206
5-ethoxycarbonyl-4-hydroxy-2-β-methylbutylamino-	182-183	4205
5-ethoxycarbonyl-4-hydroxy-2-neopentylamino-	218-219	4205
5-ethoxycarbonyl-4-hydroxy-2-propylamino-	212	4206
5-ethoxycarbonylmethyl-4-hydroxy-2-phenylguanidino-	211	2208
4-ethylamino-2,6-dihydroxy-5-nitroso-	253 or 360	2485, 4067
4-ethylamino-2-hydroxy-5-nitro-	273	H, 281
4-ethylamino-1,2,3,6-tetrahydro-1,3-dimethyl-5-nitroso-2,6-dioxo-	<315	2485
5-fluoro-2-hydroxy-4-methylamino-	267-268	3086
5-hydrazinocarbonylmethyl-4-hydroxy-2-phenylguanidino-	266-267	2208
4-hydrazino-6-hydroxy-5-nitro-	>250	2466
4-hydrazino-1,2,3,6-tetrahydro-1,3-dimethyl-5-nitroso-2,6-dioxo-	168	4209
4-hydroxy-2-isoamylamino-5-*N*-methylcarbamoyl-	282-283	4210
2-hydroxy-4-methylamino-5-nitro-	325	H, 2711
4-hydroxy-2-methylamino-5-nitro-	326	2711
4-hydroxy-6-methylamino-5-nitroso-	229-231	3182
4-hydroxy-2-methylamino-6-phenylazo-	200	4052

Pyrimidine	M.P.(°C)	References
4-hydroxy-2-β-methylbutylamino-5-*N*-methylcarbamoyl-	250-252	4205
4-hydroxy-5-*N*-methylcarbamoyl-2-neopentylamino-	292-293	4205
2-hydroxy-4-methyl-6-methylamino-5-nitro-	280	H, 2334
4-hydroxy-2-methyl-6-methylamino-5-nitro-	288-289	2458
4-hydroxy-6-methyl-5-β-nitrovinyl-2-piperidino-	282	524
4-methoxy-2-methylamino-5-nitro-	207-208	2711
4-methoxy-2-methylamino-6-phenylazo-	136-138	4052
1,2,3,4-tetrahydro-1,3-dimethyl-6-methylamino-5-nitroso-2,4-dioxo-	148-150 or 153-260	739, 2485
1,2,3,4-tetrahydro-1,3-dimethyl-5-nitroso-2,4-dioxo-6-piperidinoamino-	120	3254
1,2,3,4-tetrahydro-1,3-dimethyl-5-nitroso-2,4-dioxo-6-propylamino-	224-226	2485
1,2,3,4-tetrahydro-4-imino-1,3-dimethyl-5-nitroso-2-oxo-	-	4211
1,2,3,4-tetrahydro-6-isobutylamino-1,3-dimethyl-5-nitroso-2,4-dioxo-	273	2485
1,2,3,4-tetrahydro-6-isopropylamino-1,3-dimethyl-5-nitroso-2,4-dioxo-	120 or 130	2838, 2929
1,2,3,4-tetrahydro-3-methyl-6-methylamino-5-nitroso-2,4-dioxo-1-piperidino-	230-231	2246

Pyrimidine	M.P.(^{o}C)	References
1,2,3,4-tetrahydro-3-methyl-5-nitroso-2,4-dioxo-1-piperidino-6-piperidinoamino-	125	3254
Uracil/5-acetamido-6-chloro-3-methyl-	217-218	2440
Uracil/5-amino-1-amyloxycarbonyl-(?)	99-100	3346
Uracil/6-amino-5-benzylnitrosoamino-3-methyl-	181	2489
Uracil/6-amino-5-bromo-1-methyl-	265	H, 4163
Uracil/5-amino-6-carboxy-1-methyl-	250-300	3968
Uracil/5-amino-1-γ-carboxypropyl-	127	2336
Uracil/5-amino-6-chloro-3-methyl-	243	2441
Uracil/1-amino-5-cyano-6-ethyl-	202-203	2546
Uracil/1-amino-5-cyano-6-methyl-	244-245	2546
Uracil/5-amino-1-βγ-dibromopropyl-	92-95	3479
Uracil/6-amino-1-dimethylamino-5-nitroso-	221-222	2291
Uracil/5-amino-1-β-ethoxycarbonylethyl-	140-141	2336
Uracil/6-amino-1-hydroxyethyl-5-nitroso-	>250	H, 4077
Uracil/6-amino-1-hydroxy-5-nitroso-	224	2316
Uracil/6-amino-1-methyl-5-nitroso-	-	H, 4212
Uracil/6-amino-3-methyl-5-nitroso-	>350	909
Uracil/6-amino-5-nitroso-1-piperidino-	173	4213

Pyrimidine	M.P.(^{o}C)	References
Uracil/6-anilino-1-dimethylamino-5-nitroso-	147-148	2246
Uracil/6-anilino-3-methyl-5-nitroso-	195-197	3320
Uracil/6-benzylamino-1-dimethylamino-5-nitroso-	150	2246
Uracil/6-benzylamino-3-methyl-5-nitroso-	188 or 195-360	2560
Uracil/6-benzylamino-5-nitroso-3-phenyl-	360	2485
Uracil/6-benzylamino-5-nitroso-1-piperidino-	242-243	2246
Uracil/3-benzyl-6-benzylamino-5-nitroso-	360	2485
Uracil/6-butylamino-3-methyl-5-nitroso-	238-240	2485
Uracil/6-butylamino-5-nitroso-3-phenyl-	230-232	2485
Uracil/6-carboxy-3-methyl-5-piperidinomethyl-	206	4185
Uracil/6-cyclohexylamino-3-methyl-5-nitro-	240-242	2927
Uracil/6-cyclohexylamino-3-methyl-5-nitroso-	233-235	2838
Uracil/6-cyclohexylamino-5-nitroso-3-phenyl-	239-240	2838
Uracil/1-β-diethylaminoethyl-5-nitro-	171	3191
Uracil/6-dimethylamino-3-methyl-5-nitro-	218-220	2444
Uracil/6-β-formylhydrazino-3-methyl-5-nitro-	210 then 251	2755
Uracil/6-hydrazino-3-methyl-5-nitro-	212 or 217	2444, 2755
Uracil/1-methyl-6-methylamino-5-nitro-	260	2909
Uracil/3-methyl-6-methylamino-5-nitro-	329-330	2444

Pyrimidine	M.P.(^{o}C)	References
Uracil/1-methyl-6-methylamino-5-nitroso-	>320	2480
Uracil/3-methyl-6-methylamino-5-nitroso-	263, 267 or 294	2440, 2485, 2560
Uracil/3-methyl-6-α-methylhydrazino-5-nitro-	180	2444
Uracil/3-methyl-6-α-methylhydrazino-5-phenylazo-	215-216	2440
[*pseudo*]Uracil/2-amino-1-methyl-5-nitro-	300-302	2718

TABLE XXXVIII. Amino-Thiopyrimidines with Other Functional Groups (*H* 581)

5-acetamido-2-ethylthio-4-thiocyanato-	185-186	750
4-allylamino-5-ethoxycarbonyl-2-methylthio-	44-45	4015
2-allylthio-4-amino-5-ethoxycarbonyl-	92-94	4214
4-amino-5-amyl-6-chloro-2-methylthio-	214	2204
4-amino-5-benzoyl-4-mercapto-2-phenyl-	240	2320
2-amino-4-benzylthio-6-carboxy-	234-235	2214
4-amino-2-benzylthio-5-cyano-	174-176	H, 4215
4-amino-5-bromo-6-chloro-2-methylthio-	164-165	2842

TABLE XXXVIII

Pyrimidine	M.P.(°C)	References
4-amino-5-bromo-2-ethylthio-6-methyl-	113-114	2631
2-amino-5-bromo-4-mercapto-6-methyl-	207	2165
4-amino-5-bromomethyl-2-methylthio-	HBr ca 300	H, 1620, 4216
2-amino-5-bromo-4-methyl-6-methylthio-	140-142	2165
4-amino-5-butyl-6-chloro-2-methylthio-	215-216	2204
4-amino-2-butylthio-6-chloro-	81-82	3216
4-amino-5-carbamoyl-2-ethylthio-	220	H, 459
4-amino-5-carbamoyl-2-mercapto-	>320	459
4-amino-5-carbamoyl-2-methylthio-	280-281	H, 3062, 4217
4-amino-6-carbamoyl-2-methylthio-	294	2214
4-amino-5-carbamoyl-2-propylthio-	-	3803 (?)
2-amino-5-β-carboxyethyl-4-mercapto-6-methyl-	222-225	4179
2-amino-5-β-carboxyethyl-4-mercapto-6-phenyl-	253-254	4179
2-amino-4-carboxy-6-mercapto-	291-293	2214
4-amino-6-carboxy-2-mercapto-	272-274	2214
2-amino-4-carboxy-6-methylthio-	255	2214
4-amino-6-chloro-5-ethyl-2-methylthio-	215-216	2204
2-amino-4-chloro-6-ethylthio-	109-110	2165
4-amino-6-chloro-2-ethylthio-	78-79	2678

Pyrimidine	M.P.(°C)	References
5-amino-2-chloro-4-ethylthio-	94	750
4-amino-6-chloro-5-formyl-2-methylthio-	189-190	2593
4-amino-6-chloro-5-hexyl-2-methylthio-	194-195	2204
4-amino-6-chloro-2-isopropylthio-	98	2589
2-amino-4-chloro-6-mercapto-	>360	2165
5-amino-2-chloro-4-mercapto-	>300	750
5-amino-4-chloro-6-mercapto-	ca 200 (or >300 ?)	751, 2158, 2227, 4218
5-amino-2-chloro-4-mercapto-6-trifluoromethyl-	132	2193
5-amino-4-chloro-6-mercapto-2-trifluoromethyl-	175-176	2193
4-amino-6-chloro-5-methyl-2-methylthio-	242	2204
4-amino-5-chloromethyl-2-methylthio-	HCl >300	4219
2-amino-4-chloro-5-methylthio-	199-200	4172
2-amino-4-chloro-6-methylthio-	106-108	2165
4-amino-6-chloro-2-methylthio-	-	H, 2475, 2587
5-amino-4-chloro-6-methylthio-	95-96	2158, 4218
4-amino-6-chloro-2-methylthio-5-propyl-	204-205	2204
2-amino-4-chloro-5-phenylthio-	189-192	2203, 4220
2-amino-4-chloro-6-propylthio-	105-106	2165

TABLE XXXVIII

Pyrimidine	M.P.(°C)	References
4-amino-6-chloro-2-propylthio-	97	2589
4-amino-5-cyano-2,6-bismethylthio-	231-232	269
2-amino-5-β-cyanoethyl-4-mercapto-6-methyl-	242-244	4179
4-amino-5-cyano-2-ethylthio-	140 to 145	H, 459, 4215
4-amino-5-cyano-2-mercapto-	-	H, 459, 4215
4-amino-5-cyano-6-methyl-2-methylthio-	237 or 238-240	2273, 4003
4-amino-5-cyano-2-methylthio-	235-237	4005, 4183
4-amino-5-cyano-6-methylthio-2-phenyl-	163-165	269
4-amino-6-diethoxymethyl-2-mercapto-	195-196	4221
5-amino-2,4-dimercapto-6-trifluoromethyl-	>240	2193
4-amino-6-dimethylamino-2-methylthio-5-phenylazo-	198-199	2822
4-amino-5-ethoxycarbonyl-2-ethylthio-	102-103	H, 1620, 4214
4-amino-5-ethoxycarbonyl-2-mercapto-	-	H, 1609
4-amino-6-ethoxycarbonyl-2-mercapto-5-nitro-	202-204	2467
4-amino-5-ethoxycarbonyl-6-mercapto-2-phenyl-	202	2320
4-amino-5-ethoxycarbonyl-2-propylthio-	97-99	4214
4-amino-2-ethylthio-5-fluoro-	94-95	77, 4197, 4222
4-amino-2-ethylthio-5-thiocarbamoyl-	184	3803
5-amino-2-ethylthio-4-thiocyanato-	234	750
4-amino-5-formyl-2-methylthio-	183-184	1620

Pyrimidine	M.P.(^{o}C)	References
4-amino-2-mercapto-6-methyl-5-nitro-	220-221	H, 874
2-amino-4-mercapto-5-nitro-	-	748
4-amino-6-mercapto-5-nitro-	223	3823
4-amino-2-mercapto-6-trifluoromethyl-	203-205	443
4-amino-6-methyl-2-methylthio-5-nitro-	155	874
4-anilino-2-benzylthio-5-ethoxycarbonyl-	76-77	4015
4-anilino-5-carboxy-2-methylthio-	-	2598 (?)
5-anilino-4,6-dichloro-2-methylthio-	130-131	2260
4-anilino-5-ethoxycarbonylmethyl-2-methylthio-	100-101	4015
2-anilino-4-mercapto-5-nitro-	-	748
4-aziridino-5-bromo-6-chloro-2-methylthio-	115-116	2255
4-aziridino-5-bromo-6-methyl-2-methylthio-	75-77	2255
4-aziridino-5-bromo-2-methylthio-	93-94	2255
4-aziridino-6-chloro-5-fluoro-2-methylthio-	80	2729
4-aziridino-6-chloro-2-methylthio-	88-90	2255
4-aziridino-6-chloro-2-methylthio-5-nitro-	114-116	2731
4-aziridino-6-chloro-2-methylthio-5-phenyl-	108-110	2255
4-aziridino-5,6-dichloro-2-methylthio-	106	2729
4-aziridino-5-ethoxycarbonyl-2-methylthio-	85-86	2255

Pyrimidine	M.P.(^{o}C)	References
4-aziridino-5-fluoro-2-methylthio-	64-65	2729
4-aziridino-6-methoxycarbonyl-2-methylthio-	96-99	2255
4-benzylamino-5-ethoxycarbonyl-2-methylthio-	68-69	4015
4-benzylidenehydrazino-5-ethoxycarbonyl-2-methylthio-	148-149	4223
4,6-bismethylamino-2-methylthio-5-nitro-	190	2454
5-bromo-4-hydrazino-2-methylthio-	148-150	2756
4-t-butylamino-5-ethoxycarbonyl-2-methylthio-	63-64	4015
4-chloro-2-dimethylamino-6-mercapto-	304	2776
4-chloro-2-dimethylamino-6-methylthio-	51-52	2776
4-chloro-6-hydrazino-2-methylthio-	157	4036
4-chloro-6-mercapto-5-methylamino-	215-217	4224
4-chloro-6-α-methylaziridino-2-methylthio-	82-84	2255
4,6-diamino-2-benzylthio-5-nitroso-	199-200	2296
2,4-diamino-5-bromo-6-phenylthio-	183-185	2621
4,6-diamino-5-ββ-diethoxyethyl-2-mercapto-	221 or 227-228	2011, 4019
4,5-diamino-6-ethoxycarbonyl-2-mercapto-	202	2467

Pyrimidine	M.P.(^{o}C)	References
2,4-diamino-6-mercapto-5-phenylazo-	257-259	2503
2,4-diamino-6-methylthio-5-nitroso-	>260	2487
4,6-diamino-2-methylthio-5-nitroso-	261-262	1169, 2433
2,4-diamino-6-methylthio-5-phenylazo-	180-182	2487, 2853
4,6-diamino-2-methylthio-5-phenylazo-	235	2853
4,6-diamino-2-methylthio-5-sulphamoyl-	198-199	2364
4,6-diamino-2-methylthio-5-sulpho-	265-267	2364
4,6-diaziridino-5-fluoro-2-methylthio-	107-108	2728
4,6-diaziridino-2-methylthio-5-nitro-	138-139	2728
2-dimethylamino-1,6-dihydro-1-methyl-6-thio-	75-76	2776
4-dimethylamino-1,2-dihydro-1-methyl-2-thio-	203-204	2676, 2776
4-dimethylamino-1,6-dihydro-1-methyl-6-thio-	182-183	2676, 2776
4-dimethylamino-6-mercapto-5-nitro-	182-185	3823
5-ethoxycarbonyl-4-hydrazino-2-methylthio-	100-101	4223
5-ethoxycarbonyl-4-isopropylidenehydrazino-2-methylthio-	164-166	4223
5-ethoxycarbonyl-4-methylamino-2-methylthio-	93-94	4015
5-ethoxycarbonyl-4-α-methylhydrazino-2-methylthio-	107-109	4223

TABLE XXXVIII

Pyrimidine	M.P.(°C)	References
5-ethoxycarbonyl-4-α-methyl-β-propylidenehydrazino-2-methylthio-	66-67	4223
5-ethoxycarbonyl-2-methylthio-4-piperidino-	64-65	4015
5-ethoxycarbonyl-2-methylthio-4-trimethylhydrazino-	92-93	4015
5-formamido-2,4-dimercapto-6-trifluoromethyl-	255-256	2193
5-formyl-4-methyl-6-methylthio-2-piperidino-	284-285	524
4-mercapto-6-methylamino-5-nitro-	202	3823
4-methylamino-6-methylcarbamoyl-2-methylthio-	191-192	2214

TABLE XXXIX. Amino - Oxy - Thiopyrimidines (*H* 585)

Pyrimidine	M.P.(°C)	References
4-acetamido-5-acetoxymethyl-2-methylthio-	141-142	3062
5-acetamido-4-amino-6-hydroxy-2-mercapto-	-	4225
4-acetamido-1,6-dihydro-1-methyl-2-methylthio-6-oxo-	251	1106
5-acetamido-4,6-dihydroxy-2-mercapto-	284	2257
5-acetamido-4-hydroxy-6-methyl-2-methylthio-	248-252	2223

Pyrimidine	M.P.(^{o}C)	References
4-acetamido-6-methoxy-2-methylthio-	164-165	1106
4-acetamido-1,2,3,6-tetrahydro-1,3-dimethyl-6-oxo-2-thio-	-	4173(?)
5-acetoxymethyl-4-amino-2-methylthio-	138; HCl 159	3062
4-allyliminomethyl-6-hydroxy-2-mercapto-	171-172	2219
2-allylthio-4-amino-5-hydroxymethyl-	98	4214
4-amino-5-amyl-6-hydroxy-2-mercapto-	257	2204
4-amino-5-amyl-6-hydroxy-2-methylthio-	167-168	2204
4-amino-5-benzamidomethyl-2-methylthio-	186-187	3062
4-amino-3-benzyl-1,2,3,6-tetrahydro-1-methyl-6-oxo-2-thio-	236-237	4226
4-amino-2-benzylthio-5-formamido-1,6-dihydro-1-methyl-6-oxo-	263-264	4100
4-amino-2-benzylthio-5-formamido-6-hydroxy-	236-237	4100
4-amino-5-bromo-6-hydroxy-2-methylthio-	-	H, 2842
4-amino-5-bromo-6-methoxy-2-methylthio-	138-140	2631
4-amino-6-butoxy-2-methylthio-	crude	3216
4-amino-5-butyl-6-hydroxy-2-mercapto-	251-252	2204
4-amino-5-butyl-6-hydroxy-2-methylthio-	170-171	2204
2-amino-4-butylthio-6-hydroxy-	240-242	2165
4-amino-2-butylthio-6-hydroxy-	190-191	3216

Pyrimidine	M.P.(^{o}C)	References
4-amino-2-butylthio-6-methoxy-	63-65	3216
4-amino-5-butyryloxymethyl-2-methylthio-	110-111	3062
2-amino-4-cyclohexylthio-6-hydroxy-	185	2165
4-amino-5-ββ-diethoxyethyl-5,6-dihydro-2-mercapto-5-methyl-6-oxo-	216	2011
4-amino-5-ββ-diethoxyethyl-6-hydroxy-2-mercapto-	>315 or >360	2011, 4019, 4227
4-amino-5-ββ-diethoxy-α-methylethyl-6-hydroxy-2-mercapto-	-	4194
4-amino-1,3-diethyl-1,2,3,6-tetrahydro-6-oxo-2-thio-	178-182	4226
4-amino-1,6-dihydro-1,5-dimethyl-2-methylthio-6-oxo-	173	1106
4-amino-1,6-dihydro-5-methoxy-1-methyl-2-methylthio-6-oxo-(?)	206-207	3910
4-amino-1,6-dihydro-1-methyl-2-methylthio-5-nitroso-6-oxo-	234	H, 3398
4-amino-1,6-dihydro-1-methyl-2-methylthio-6-oxo-	257	H, 1106
4-amino-3,6-dihydro-3-methyl-2-methylthio-6-oxo-	272	827
4-amino-2,6-dihydroxy-5-γ-mercaptopropyl-	320-325	2908
4-amino-5-βγ-dihydroxypropyl-6-hydroxy-2-mercapto-	246-247	2011
4-amino-5,6-dimethoxy-2-methylthio-	115-117	3910

Pyrimidine	M.P.(°C)	References
4-amino-2-ethoxy-6-ethylthio- (?)	165-169/1.5mm.	4043
4-amino-6-ethoxy-2-isopropylthio-	74-75	2837, 4095
5-amino-2-ethoxy-4-mercapto-	127	749
4-amino-6-ethoxy-2-methylthio-	93 to 96	2586, 2587, 2837, 3216
4-amino-2-ethoxy-6-phenylthio- (?)	108-109	4043
4-amino-5-ethyl-6-hydroxy-2-mercapto-	283-284	2204
4-amino-5-ethyl-6-hydroxy-2-methylthio-	215-217	2204
2-amino-4-ethylthio-6-hydroxy-	248	2165
4-amino-2-ethylthio-6-hydroxy-	218-219	H, 3219, 4104
4-amino-2-ethylthio-5-hydroxymethyl-	156-157	H, 1620, 4214
4-amino-2-ethylthio-5-isopropoxymethyl-	72-73; HBr 173	4219
4-amino-2-ethylthio-6-methoxy-	116 to 118, (107)	2678, 2837, (3216), 4095
4-amino-6-ethylthio-2-methoxy-	83-84	2678, 4043(?)
4-amino-5-formamido-1,6-dihydro-1-methyl-2-methylthio-6-oxo-	256-257	4100
4-amino-5-formamido-1,2,3,6-tetrahydro-1,3-dimethyl-4-oxo-2-thio-	304-305	4173
4-amino-5-hexyl-6-hydroxy-2-mercapto-	245-246	2204
4-amino-5-hexyl-6-hydroxy-2-methylthio-	165-166	2204
4-amino-6-hydroxy-5-β-hydroxyethyl-2-mercapto-	299-300	2202

TABLE XXXIX

Pyrimidine	M.P.(°C)	References
4-amino-6-hydroxy-5-β-hydroxypropyl-2-mercapto-	262	2202
4-amino-6-hydroxy-2-isopropylthio-	219 (or 300)	2587, (3216)
2-amino-4-hydroxy-6-mercapto-	>360	1378, 2165
4-amino-2-hydroxy-6-mercapto-	355	2165, 2503
4-amino-6-hydroxy-2-mercapto-	>360	H, 2165, 2478
2-amino-4-hydroxy-5-β-mercaptoethyl-	272-273	2908
2-amino-4-hydroxy-5-β-mercaptoethyl-6-methyl-	288-292	2908
4-amino-6-hydroxy-2-mercapto-5-γ-mercaptopropyl-	-	2908
4-amino-6-hydroxy-2-mercapto-5-methoxy-	275-280	3910
5-amino-4-hydroxy-2-mercapto-6-methyl-	320-322	H, 2227
4-amino-6-hydroxy-2-mercapto-5-nitroso-	-	H,2478, 2881
5-amino-4-hydroxy-2-mercapto-6-phenyl-	243	3956
4-amino-6-hydroxy-2-mercapto-5-propyl-	247-248	2204
2-amino-4-hydroxy-5-γ-mercaptopropyl-6-methyl-	256-258	2908
2-amino-4-hydroxy-5-γ-mercaptopropyl-6-phenyl-	300	2908
4-amino-6-hydroxy-5-methoxy-2-methylthio-	203	3910
4-amino-5-hydroxymethyl-2-mercapto-	229-232	1620
4-amino-5-hydroxymethyl-2-methylthio-	126-127	1608, 3062, 4228
4-amino-5-hydroxymethyl-2-propylthio-	138-139	4214

Pyrimidine	M.P.(^{o}C)	References
2-amino-4-hydroxy-5-methylthio-	251-252	4172
2-amino-4-hydroxy-6-methylthio-	274-276	2165
4-amino-2-hydroxy-6-methylthio-	294	2165, 2777
4-amino-6-hydroxy-2-methylthio-	267 or 272	H, 2475, 3216
4-amino-6-hydroxy-2-methylthio-5-propyl-	193-194	2204
2-amino-4-hydroxy-5-phenylthio	258-259	2203, 4220
2-amino-4-hydroxy-6-propylthio-	228-232	2165
4-amino-6-hydroxy-2-propylthio-	210 or 213	2587, 3216
4-amino-5-isopropoxymethyl-2-methylthio-	105-108	4219
4-amino-6-isopropoxy-2-methylthio-	126/0.15 mm.; pic.178	2586
4-amino-2-isopropylthio-6-methoxy-	116-117	2837, 4095
2-amino-4-mercapto-6-methoxy-	263	2765
4-amino-2-mercapto-5-ethoxy-	244	2521
5-amino-4-mercapto-6-methoxy-	214-215	2227
4-amino-5-methoxymethyl-2-methylthio-	104-106	1620, 4219
2-amino-4-methoxy-6-methylthio-	89	2765
4-amino-2-methoxy-6-methylthio-	94-95	4043(?), 4229
4-amino-5-methoxy-2-methylthio-	134-136	2205, 2586
4-amino-6-methoxy-2-methylthio-	143-144	2589, 2837, 3216, 4095

Pyrimidine	M.P.(^{o}C)	References
4-amino-6-methoxy-2-propylthio-	99-100	2837, 4095
4-amino-2-methylthio-5-propionyloxymethyl-	159-161	3062
4-amino-2-methylthio-6-propoxy-	crude	3216
4-amino-1,2,3,6-tetrahydro-1,3-dimethyl-5-nitroso-6-oxo-2-thio-	216 or 220	3935, 4173
4-amino-1,2,3,6-tetrahydro-1,3-dimethyl-6-oxo-2-thio-	288 to 293	2935, 4173, 4226
4-amino-1,2,3,6-tetrahydro-1,3-dimethyl-2-oxo-6-thio-	283-286	4173
4-amino-1,2,3,6-tetrahydro-3-β-hydroxyethyl-1-methyl-6-oxo-2-thio-	ca 240	4226
4-amino-1,2,3,6-tetrahydro-1-methyl-3-β-methylallyl-6-oxo-2-thio-	212-213	4146
4-amino-1,2,3,6-tetrahydro-1-methyl-6-oxo-3-phenyl-2-thio-	288-289	4226
5-anilino-4,6-dihydro-2-mercapto-	265-266	2260
5-anilino-4,6-dihydroxy-2-methylthio-	268-269	2260
4-anilino-6-hydroxy-5-mercapto-	182-183	4230
5-anilino-4-hydroxy-2-mercapto-	280-282	2567
4-aziridino-5-hydroxymethyl-2-methylthio-	156-157	2255
5-aminomethylene-1,3-diethylhexahydro-4,6-dioxo-2-thio-	250-251	4231
4-butyramido-5-butyryloxymethyl-2-methylthio-	88-89	3062

Pyrimidine	M.P.(^{o}C)	References
4-carboxy-5-diethylaminomethyl-6-hydroxy-2-mercapto-	274	4185
4-cyclopentyliminomethyl-5-ethyl-6-hydroxy-2-mercapto-	172-173	2219
4-cyclopropyliminomethyl-5-ethyl-6-hydroxy-2-mercapto-	211-212	2219
4,5-diamino-1,6-dihydro-1-methyl-2-methylthio-6-oxo-	212	H, 3398
4,5-diamino-2-hydroxy-4-mercapto-	sul. >270	2159
4,5-diamino-6-hydroxy-2-mercapto-	-	H, 2478
2,4-diamino-6-hydroxy-5-γ-mercaptopropyl-	256-258	2908
4,5-diamino-2-hydroxy-6-methylthio-	ca 260	4126
4,5-diamino-1,2,3,6-tetrahydro-1,3-dimethyl-2-oxo-6-thio-	-	4173
4,5-diamino-1,2,3,6-tetrahydro-1,3-dimethyl-6-oxo-2-thio-	240-243	2935, 4173
1,6-dihydro-1-methyl-4-methylamino-2-methylthio-5-nitroso-6-oxo-	247-248	2440
1,6-dihydro-1-methyl-4-α-methylhydrazino-2-methylthio-6-oxo-	190-191	2440
5-dimethylaminomethyl-4-hydroxy-2-mercapto-	194	4232
5-dimethylaminomethyl-4-hydroxy-2-mercapto-6-methyl-	185	4232
4-dimethylamino-1,2,3,6-tetrahydro-1,3-dimethyl-2-oxo-4-thio-	122-124	2569

Pyrimidine	M.P.(^{o}C)	References
5-formamido-1,6-dihydro-1-methyl-4-methylamino-2-methylthio-6-oxo-	273-275	2440
4-hydrazino-5-methoxy-2-methylthio-	72-73	2584
4-hydroxy-6-hydroxyiminomethyl-2-mercapto-5-methyl-	245	H, 2219
4-hydroxy-6-isobutyliminomethyl-2-mercapto-	149-150	2219
4-hydroxy-6-isopropyliminomethyl-2-mercapto-	141-142	2219
4-hydroxy-2-mercapto-6-propyliminomethyl-	147-148	2219
2-methylthio-4-propionamido-5-propionyloxymethyl-	93-94	3062
2-Thiouracil/6-amino-1-benzyl-	262	2484
2-Thiouracil/6-amino-3-butyl-	212	4159
2-Thiouracil/6-amino-1,5-dimethyl-	302	1106
2-Thiouracil/6-amino-1-dimethylamino-	203 or 216-218	2291, 4164, 4233
2-Thiouracil/6-amino-1-dimethylamino-5-nitroso-	198-200	2291
2-Thiouracil/6-amino-1-ethyl-	249-250	4234
2-Thiouracil/6-amino-1-isoamyl-	251-252	4235
2-Thiouracil/6-amino-1-isoamyl-5-nitroso-	ca 200	4235
2-Thiouracil/6-amino-5-nitroso-1-piperidino-	-	4213 (?)
2-Thiouracil/6-amino-1-piperidino-	211-212	4164
2-Thiouracil/1-anilino-	257	2352

Pyrimidine	M.P.(°C)	References
2-Thiouracil/5,6-diamino-1-benzyl-	242	2484
2-Thiouracil/5,6-diamino-1-dimethylamino-	-	2291
2-Thiouracil/5,6-diamino-1-isoamyl-	234-237	4235
2-Thiouracil/5,6-diamino-1-methyl-	258-259	H, 2159, 2492
2-Thiouracil/5,6-diamino-1-piperidino-	-	2291
4-Thiouracil/5,6-diamino-1-methyl-	>300	2159, 3314

TABLE XL. Aminopyrimidines with Two Minor Functional Groups (*H* 588)

2-acetamido-4-chloro-5-β-cyanoethyl-6-methyl-	140-145	4179
4-amino-6-carbamoyl-2-chloro-	333-335	2214
4-amino-5-carbamoyl-2-methylsulphonyl-	216-218	3062
4-amino-5-carbamoyl-2-trifluoromethyl-	291-293	1000
2-amino-4-carboxy-5-chloro-	-	4236
4-amino-5-carboxy-2-trifluoromethyl-	312-314	1000
4-amino-6-chloro-5-cyano-2-dimethylamino-	-	4217 (?)
2-amino-4-chloro-5-β-cyanoethyl-6-methyl-	215-216	4179
2-amino-4-chloro-5-β-cyanoethyl-6-phenyl-	238-239	4179

Pyrimidine	M.P.(^{o}C)	References
2-amino-4-chloro-5-cyano-6-methyl-	258-259	524
4-amino-2-chloro-6-ethoxycarbonyl-5-nitro-	171-172	2467
4-amino-6-chloro-5-formyl-	163-165	2594
2-amino-4-chloro-6-methoxycarbonyl-	135-137	2214
2-amino-4-chloro-6-methyl-5-nitro-	ca 160	H, 4237
4-amino-2-chloro-6-methyl-5-nitro-	171	H, 4238
4-amino-2-chloro-5-nitro-	220-221	H, 4051, 4239
4-amino-6-chloro-5-nitro-	156-157	H, 1469, 2721 4240
2-amino-4-chloro-6-nitroamino-	227-228	2432, 2718
2-amino-4-chloro-6-phenylazo-	153-154	4042
5-amino-2-chloro-4-thiocyanato-	>300	750
4-amino-5-cyano-2,6-bistrifluoromethyl-	176-178	2217
4-amino-5-cyano-2-heptafluoropropyl-	148-149	2268
4-amino-5-cyano-2-methylsulphonyl-	211-214	3062
4-amino-5-cyano-2-pentafluoroethyl-	177-179	2268
2-amino-4-cyano-6-phenylazo-	208	4052
4-amino-5-cyano-2-trifluoromethyl-	245-246	1000
2-amino-4,6-dichloro-5-formyl-	240-250	2593
4-amino-2,6-dichloro-5-formyl-	185-186	2593
2-amino-4-diethoxycarbonylmethyl-6-methyl-5-nitro-	129	2787, 2788

Pyrimidine	M.P.(^{o}C)	References
2-amino-4-dimethoxycarbonylmethyl-6-methyl-5-nitro-	145	2788
4-amino-2-dimethylamino-6-ethoxycarbonyl-5-nitro-	183	2467
2-amino-4-ethoxycarbonylmethyl-6-methyl-5-nitro-	118	2788
4-amino-6-ethoxycarbonylmethyl-5-nitro-	226	2785
4-amino-5-ethoxycarbonyl-2-methylsulphonyl-	163-164	4015
4-amino-5-ethoxycarbonyl-2-trifluoromethyl-	150-151	1000
4-amino-6-fluoro-5-nitro-	161-163	2455
2-amino-5-nitro-4-thiocyanato-	209-210	748
4-anilino-6-chloro-5-formyl-	99-101	2594
4-anilino-2-chloro-6-methyl-5-nitro-	120-122	2472
2-anilino-4-chloro-6-phenylazo-	153-154	4042
2-anilino-5-nitro-4-thiocyanato-	199-200	748
4-aziridino-2-chloro-6-methoxycarbonyl-	146-147	2255
4-benzylamino-6-chloro-5-nitro-	127-131	4198
4-chloro-5-cyano-2-dimethylamino-	149-150	1441, 4003
4-chloro-5-cyano-6-dimethylamino-2-methyl-	104-105	2681
4-chloro-5-cyano-6-methyl-2-piperidino-	133-134	524
4-chloro-5-cyano-2-piperidino-	115	4207

Pyrimidine	M.P.(^oC)	References
4-chloro-6-diethylamino-5-nitro-	33-34	2460
4-chloro-6-dimethylamino-5-formyl-	140-141	2594
2-chloro-4-dimethylamino-5-nitro-	116-117	2626, 2712
4-chloro-2-dimethylamino-5-nitro-	140	H, 2461
4-chloro-2-dimethylamino-6-phenylazo-	94-95	4042
4-chloro-5-formyl-6-methylamino-	158-159	2594
4-chloro-5-formyl-6-methyl-2-piperidino-	91-92	524
2-chloro-4-methylamino-5-nitro-	86 or 90-91	H, 2711, 3783
4-chloro-2-methylamino-5-nitro-	(122) or 172-173	(2711), 3783
4-chloro-2-methylamino-6-phenylazo-	172-173	4042
2-chloro-4-methyl-6-methylamino-5-nitro-	104 or 111	2334, 2563
4,6-diamino-2-carboxy-5-phenylazo-	275	2334
4,5-diamino-2-chloro-6-ethoxycarbonyl-	270	2467
2,4-diamino-6-chloro-5-nitro-	220-222	2432, cf.2443
2,4-diamino-6-chloro-5-phenylazo-	242-243	2503, 2853
2,4-diamino-6-chloro-5-sulpho-	>330	2364
2,4-diamino-6-ethoxycarbonyl-5-nitro-	186	2467
4,6-diamino-5-nitroso-2-trifluoromethyl-	>360	2294
2,4-diamino-5-nitro-6-trifluoromethyl-	188-190	999

Pyrimidine	M.P.(°C)	References
4,6-diamino-5-nitro-2-trifluoromethyl-	293	2193
2,4-diamino-5-phenylazo-6-trifluoromethyl-	235-236	443
4,6-diamino-5-phenylazo-2-trifluoromethyl-	326-329	2217
4,6-diaziridino-2-chloro-5-nitro-	130-135	2728
4,6-diaziridino-5-fluoro-2-methylsulphonyl-	151-152	2728
2,4-dichloro-6-diethylamino-5-nitro-	56	2731
4,6-dichloro-2-diethylamino-5-nitro-	95-96	2731
2,4-dichloro-5-nitro-6-piperidino-	72-73	2731
5-ethoxycarbonyl-4-hydrazino-2-trifluoromethyl-	80-82	2217

TABLE XLI. Carboxy-Halogenopyrimidines (*H* 589)

2-benzyl-4-chloro-5-ethoxycarbonylmethyl-	195-198/5mm.	3806
5-bromo-2-carbamoyl-	209	889, 2806
5-bromo-2-carboxy-	191-192	H, 2806
5-bromo-4-carboxy-2-methyl-	172	H, 2182
5-bromo-2-formyl-	TSC 230; DNP 227	2575
5-bromo-2-methoxycarbonyl-	148-149	889, 2806
5-butoxycarbonyl-4-chloro-	98-101/0.7mm.	2323
2-butyl-4-chloro-5-ethoxycarbonylmethyl-	-	3806

TABLE XLI

Pyrimidine	M.P.(^{o}C)	References
2-carbamoyl-4-chloro-6-methyl-	177-178	1313
4-carbamoyl-2-chloro-6-methyl-	181-182	2184
4-carbamoyl-2,6-dichloro-	170	2214
5-carbamoylmethyl-4-chloro-2-methyl-	228-230	3876
5-carboxy-2-chloro-4,6-dimethyl-	152	2519
4-carboxy-2-chloro-6-methyl-	182	2184
5-carboxy-4-chloro-6-phenyl-	136	2519
4-carboxy-2,6-dichloro-	115-117	2214
5-carboxymethyl-4-chloro-2-methyl-	122-125	3876
4-carboxy-2,5,6-trichloro-	120	2591
4-chloro-5-β-chloroethyl-6-methyl-	100-110/2mm.	3892
2-chloro-4-cyano-	50-51	2954
2-chloro-5-cyano-	130-132	2276
4-chloro-5-cyano-2,6-diphenyl-	181	2370
4-chloro-2-cyano-6-methyl-	65	1313
2-chloro-5-ethoxycarbonyl-	45	H, 2278
4-chloro-5-ethoxycarbonyl-	74-75/0.05mm.	2323
4-chloro-5-α-ethoxycarbonylethyl-2-methyl-	99-100/0.05mm.	1595
4-chloro-6-ethoxycarbonyl-5-fluoro-2-methyl-	45-46	2191
4-chloro-5-ethoxycarbonylmethyl-	96/0.04mm. or 104/0.5mm.	2577, 3876
4-chloro-5-ethoxycarbonyl-2-methyl-	100/1mm.	H, 4015
4-chloro-5-ethoxycarbonylmethyl-2-ethyl-	146/6mm.; 42	3806

Pyrimidine	M.P.(^{o}C)	References
4-chloro-5-ethoxycarbonylmethyl-2-methyl-	120/4mm.; 41	H, 3838
4-chloro-5-ethoxycarbonylmethyl-2-phenyl-	182/6mm.; 80	2576, 3806
4-chloro-5-ethoxycarbonylmethyl-2-propyl-	165/25mm.	3806
4-chloro-5-ethoxycarbonyl-2-trifluoromethyl-	41-42	1000
2-chloro-4-hydrazinocarbonyl-6-methyl-	156-157	2184
4-chloro-6-methoxycarbonyl-	60-61	2954
2-chloro-4-methoxycarbonyl-6-methyl-	111-113	2184
2,4-dichloro-6-chlorocarbonyl-	109/5mm.	2591
2,4-dichloro-5-cyano-	110-112/2mm.; 62-63	2680
4,6-dichloro-5-cyano-	145	2758
2,4-dichloro-5-cyano-6-methyl-	93-94	2155
4,6-dichloro-5-dimethoxymethyl-	92-93/0.05mm.	2758
2,4-dichloro-5-ethoxycarbonylmethyl-6-methyl-	52-53	2576
4,6-dichloro-5-formyl-	62 or 70	2593, 2594
4,6-dichloro-5-formyl-2-methyl-	103-105	2594
4,6-dichloro-5-formyl-2-phenyl-	151-153	2594
4,5-diethoxycarbonyl-2-chloro-	31-32	3880
2,4,6-tribromo-5-cyano-	212-214	4241

TABLE XLI

Pyrimidine	M.P.(^{o}C)	References
2,4,6-trichloro-5-cyano-	122-123	4242, 4243
2,4,5-trichloro- 6-methoxycarbonyl-	56-58	2591

TABLE XLII. Carboxy-Oxypyrimidines (*H* 590)

1-acetoxy-3-acetyl- 1,2,3,6-tetrahydro-2,6- dioxo- (or isomer ?)	142-143	2200
1-acetyl-3-benzoyl- 1,2,3,4-tetrahydro- 2,4-dioxo-	119-120	3170
1-acetyl-3-benzoyl- 1,2,3,4-tetrahydro- 5-methoxy-2,4-dioxo-	145-150	4244
1-acetyl-1,6-dihydro- 6-oxo- (?)	117-120	791
4-acetyl-2,6-dihydroxy-	265-266	H, 2218
5-acetyl-2,4-dihydroxy-	285 or 292	H, 2317, 2637
5-acetyl-4-hydroxy- 6-methyl-2-phenyl-	263	2314
5-acetyl-1,2,3,4- tetrahydro-6-hydroxy- 1,3-dimethyl-2,4-dioxo-	95-97	H, 883
5-acetyl-2,4,6-trihydroxy-	295-297	4245
5-alllycarbamoyl-1,2,3,4- tetrahydro- 1,3-dimethyl-2,4-dioxo-	133	4246
5-allylcarbamoyl- 2,4,6-trihydroxy-	258-259	4247
1-allyl-5-cyano- 1,2-dihydro-2-oxo-	147-148	2277
1-allyl-5-ethoxycarbonyl- 1,2-dihydro-2-oxo-	115-118	2277

Pyrimidine	M.P.(°C)	References
1-allyl-5-ethoxycarbonyl-1,2,3,6-tetrahydro-3-methyl-2,6-dioxo-	90	4246
5-*N*-allyl(thiocarbamoyl)-2,4,6-trihydroxy-	>360	4247
4-amylcarbamoyl-2,6-dihydroxy-	260-270	3444
4-amyloxycarbonyl-2,6-dihydroxy-	174-175	3432
1-benzoyl-1,2,3,6-tetrahydro-3-methyl-2,6-dioxo-	150-152	3170
2-benzyl-5-carboxy-4,6-dimethoxy-	154-155	2575
1-benzyl-5-ethoxycarbonyl-1,2-dihydro-2-oxo-	131-132	4248
2-benzyl-5-ethoxycarbonylmethyl-4-hydroxy-	176	H, 3806
4-benzyloxy-2-carbamoyl-6-methyl-	83-85	1148
4-benzyloxy-2-cyano-6-methyl-	66-68	1148
1,3-bis-β-cyanoethyl-1,2,3,4-tetrahydro-2,4-dioxo-	78-79	4249
1,3-bis-ββ-diethoxyethyl-1,2,3,4-tetrahydro-2,4-dioxo-	liquid	3154
5-bromoacetyl-2,4-dihydroxy-	240-250	2218
4-butoxy-2-carbamoyl-6-methyl-	83-84	1148
4-s-butoxy-2-carbamoyl-6-methyl-	78-79	1313
4-butoxycarbonyl-2,6-dihydroxy-	184	3436
5-butoxycarbonyl-2,4-dihydroxy-	235-236	3437
5-butoxycarbonyl-4-hydroxy-	191-192	2367

TABLE XLII

Pyrimidine	M.P.(^{o}C)	References
4-butoxycarbonylmethyl-2,6-dihydroxy-	165 or 176	2917, 4250, 4251
5-butoxycarbonylmethyl-2,4-dihydroxy-	187	4250
4-butoxy-2-cyano-6-methyl-	127-128/5mm.	1148
4-butoxy-5-ethoxycarbonyl-	127/0.23mm.	4000
4-butylcarbamoyl-2,6-dihydroxy-	274-275	3441
5-butylcarbamoyl-2,4-dihydroxy-	290-291	3431, 3440
4-butyl-6-carbamoyl-2-hydroxy-	271-273	2184
5-t-butylcarbamoylmethyl-4-hydroxy-2,6-dimethyl-	206-207	2876
5-butyl-4-carboxy-2,6-dihydroxy-	281	2345, 4252
2-butyl-5-ethoxycarbonylmethyl-4-hydroxy-	141	3806
4-butyl-6-hydrazinocarbonyl-2-hydroxy-	173-175	2184
5-carbamoyl-4,6-dimethoxy-	242-244	2519, 2575
5-carbamoyl-4-ethoxy-	138-140	2366
2-carbamoyl-4-ethoxy-6-methyl-	122-123	1148
4-carbamoyl-6-ethyl-2-hydroxy-	310	2184
4-carbamoyl-6-hydroxy-	$\nless$400	957
5-carbamoyl-4-hydroxy-	266-268 or 276-278	2367, 2544, 4000
2-carbamoyl-4-hydroxy-6-methyl-	285-286	1149
4-carbamoyl-2-hydroxy-6-methyl-	$\nless$360	2184
4-carbamoyl-6-hydroxy-2-methyl-	>300	2216

Pyrimidine	M.P.(°C)	References
4-carbamoyl-2-hydroxy-6-propyl-	286-287	2184
5-carbamoyl-2-methoxy-4,6-dimethyl-	237-238	2519
5-carbamoyl-4-methoxy-6-phenyl-	230-231	2519
5-carbamoyl-1,2,3,4-tetrahydro-6-hydroxy-1,3-dimethyl-2,4-dioxo-	218-221	3266, 4253, 4254
2-carbamoyl-4-methoxy-6-methyl-	-	H, 1148
4-carbamoylmethyl-2,6-dihydroxy-	283 or 295	2674, 4251
5-carbamoylmethyl-4-hydroxy-	257-259	3876
5-carbamoylmethyl-4-hydroxy-2,6-dimethyl-	258-260	3876
2-carbamoyl-4-methyl-6-phenoxy-	150-152	1148
4-carbamoyl-1,2,3,6-tetrahydro-1,3-dimethyl-2,6-dioxo-	239	4255
5-carbamoyl-2,4,6-trihydroxy-	>360	H, 4242, 4254
5-carboxy-4,6-diethoxy-	107-109	2575
2-carboxy-4,6-dihydroxy-	220	2334
4-carboxy-2,6-dihydroxy-(orotic acid)	340	H, 752, 3432, 3436, 4256-4259
5-carboxy-2,4-dihydroxy-	297	H, 2071, 4264
4-carboxy-2,6-dihydroxy-5-hydroxymethyl-	lactone 312	1168, 4185
4-carboxy-2,6-dihydroxy-5-methyl-	327	H, 2345
4-carboxy-5,6-dihydroxy-2-methyl-	>250	2191
4-carboxy-2,6-dihydroxy-5-phenyl-	330-340	4252
5-carboxy-4,6-dimethoxy-	197 or 201	2519, 2575

TABLE XLII

Pyrimidine	M.P.(°C)	References
5-carboxy-4,6-dimethoxy-2-methyl-	209-210	2575
5-carboxy-4,6-dimethoxy-2-phenyl-	183-185	2575
4-carboxy-5-ethyl-2,6-dihydroxy-	304	2345
5-carboxy-2-ethyl-4,6-dimethoxy-	148-151	2575
4-carboxy-5-formyl-2,6-dihydroxy-	unisolated	1168
4-carboxy-2-hydroxy-	>180	2954
4-carboxy-6-hydroxy-	264	H, 957, 2214
5-carboxy-4-hydroxy-	236-237	H, 2367
4-carboxy-2-hydroxy-6-methyl-	214-215	2184
4-carboxy-6-hydroxy-2-methyl-	284-285	H, 2216
5-carboxy-2-methoxy-4,6-dimethyl-	152-153	2519
5-carboxy-4-methoxy-6-phenyl-	189-190	2519
4-carboxymethyl-2,6-dihydroxy-	-	H, 4257
5-carboxymethyl-4,6-dimethoxy-	177-179	2575
5-carboxymethyl-4-hydroxy-	207-210	3876
5-carboxymethyl-4-hydroxy-2,6-dimethyl-	233 or 241	2876, 3876
5-carboxymethyl-2,4,6-trimethoxy-	161-163	2575
4-carboxy-1,2,3,6-tetrahydro-1,3-dimethyl-2,6-dioxo-	150-151	H, 4255, 4261
4-carboxy-1,2,3,6-tetrahydro-5-hydroxymethyl-1,3-dimethyl-2,6-dioxo-	lactone 155-156	4255

Pyrimidine	M.P.(^{o}C)	References
4-carboxy-1,2,3,6-tetrahydro-1,3,5-trimethyl-2,6-dioxo-	lactone 212	4255
5-carboxy-2,4,6-triethoxy-	138-140	2575
5-carboxy-2,4,6-trimethoxy-	207-209	2575
4-chlorocarbonyl-2,6-dihydroxy-	-	3436
5-chlorocarbonyl-2,4-dihydroxy-	-	3437
5-carboxy-1,2-dihydro-1,4-dihydroxy-2-oxo-	205-207	2199
5-cyano-1,2-dihydro-1-methyl-2-oxo-	233-234	2276, 4248
5-cyano-2,4-dihydroxy-	294 or 320	H, 2897, 4262
5-cyano-2,4-dihydroxy-6-methyl-	352	2155, 2546
5-cyano-2,4-dimethoxy-	143	2897
5-cyano-4,6-dimethoxy-	208-210	2575
5-cyano-4,6-dimethoxy-2-methyl-	135	2681
5-cyano-4-ethoxy-2,6-diphenyl-	128	2370
2-cyano-4-ethoxy-6-methyl-	59-61	1148
5-cyano-4-ethyl-2,6-dihydroxy-	280-281	2546
5-β-cyanoethyl-2,4,6-trihydroxy-	183-186	4263
5-cyano-2-hydroxy-	262 or 266-268	2276, 4008
5-cyano-4-hydroxy-	244-245	2544
5-cyano-4-hydroxy-2,6-dimethoxy-	217-218	4242
5-cyano-4-hydroxy-2,6-diphenyl-	350-356	2370
5-cyano-4-hydroxy-2-methoxy-	199-201	2205

TABLE XLII

Pyrimidine	M.P.(°C)	References
5-cyano-4-hydroxy-6-methyl-2-phenyl-	290-291	4264
2-cyano-4-methoxy-	53	1223
2-cyano-4-methoxy-6-methyl-	-	H, 1148
5-cyano-4-methoxy-2-methyl-	73-76	2175
2-cyanomethyl-4-hydroxy-6-methyl-	320	2892
2-cyano-4-methyl-6-phenoxy-	104-106	1148
5-cyano-2,4,6-trihydroxy-	>360	H, 4242
4-cyclohexylcarbamoyl-2,6-dihydroxy-	327-328	3432
5-cyclopentylcarbamoylmethyl-4-hydroxy-2,6-dimethyl-	300-304	2876
1,3-dibenzoyl-1,2,3,4-tetrahydro-2,4-dioxo-	154-156	3170
1,3-dibenzoyl-1,2,3,4-tetrahydro-5-methoxy-2,4-dioxo-	176-177	4244
1,3-dibenzoyl-1,2,3,4-tetrahydro-5-methyl-2,4-dioxo-	149-151	3170
1,3-dibenzyl-4-carboxy-1,2,3,6-tetrahydro-2,6-dioxo-	194-196	4261
5-dibromoacetyl-2,4-dihydroxy-	219	2637
4,5-dicarbamoyl-2-hydroxy-	>300	4265
2-diethoxymethyl-2-hydroxy-	143-144	2301
4-diethoxymethyl-6-hydroxy-	122-124	H, 957
4-diethoxymethyl-1,2,3,6-tetrahydro-1,3-dimethyl-2,6-dioxo-	82	487
4-diethylcarbamoyl-2,6-dihydroxy-	234-236	3432

Pyrimidine	M.P.(°C)	References
1,2-dihydro-4-methyl-1-methylcarbamoylmethyl-2-oxo-5-phenyl- (?)	206	4266
2,4-dihydroxy-6-isoamylcarbamoyl-	260-268	3444
2,4-dihydroxy-6-isoamyloxycarbonyl-	176-177	3432
2,4-dihydroxy-6-isobutoxycarbonyl-	220-221	3432
2,4-dihydroxy-6-isobutoxycarbonylmethyl-	190	2917, 4251
2,4-dihydroxy-6-isobutylcarbamoyl-	290	3444
2,4-dihydroxy-6-isopropoxycarbonyl-	208-209	3432
2,4-dihydroxy-6-isopropoxycarbonylmethyl-	217-220	2917, 4251
2,4-dihydroxy-6-isopropylcarbamoyl-	285-300	3444
2,4-dihydroxy-6-methoxycarbonyl-	238 or 248-250	H, 2216, 2591
2,4-dihydroxy-6-methoxycarbonylmethyl-	220	H, 2917
2,4-dihydroxy-5-methylcarbamoyl-	>300	3440
2,4-dihydroxy-5-propionyloxy-	242-245	3901
2,4-dihydroxy-6-propoxycarbonyl-	172-173	3432
2,4-dihydroxy-6-propoxycarbonylmethyl-	183-184	2917
2,4-dihydroxy-6-propylcarbamoyl-	275 or 278	3441, 3444
2,4-dihydroxy-5-thiocarbamoyl-	310-312	4116
2,4-dimethoxy-6-methoxycarbonyl-	108-109	2591
5-ethoxycarbonyl-1,2-dihydro-1-methyl-2-oxo-	HBr 189	4248

TABLE XLII

Pyrimidine	M.P.(^{o}C)	References
5-ethoxycarbonyl-1,4-dihydro-6-methyl-4-oxo-1,2-diphenyl-	155-157	2313
5-ethoxycarbonyl-1,2-dihydro-2-oxo-1-phenyl-	193-194	4248
5-ethoxycarbonyl-2,4-dihydroxy-	-	H, 2537
4-ethoxycarbonyl-5,6-dihydroxy-2-methyl-	218	2191
5-ethoxycarbonyl-2-hydroxy-	HBr 186	2278
5-ethoxycarbonyl-4-hydroxy-	186 to 192	H, 2323, 2367, 2544
5-ethoxycarbonyl-2-hydroxy-4-methyl-	250-252	H, 2637
5-ethoxycarbonyl-4-hydroxy-2-methyl-	190	H, 920
5-ethoxycarbonyl-4-hydroxy-6-methyl-	-	2321
5-ethoxycarbonyl-4-hydroxy-2-phenyl-	214-215	H, 4264
5-ethoxycarbonyl-4-methoxy-	86/0.45mm.	4000
5-ethoxycarbonyl-4-methoxy-2-methyl-	47-48	2651
4-ethoxycarbonylmethyl-2,6-dihydroxy-	191-192	H, 2917
5-ethoxycarbonylmethyl-2-ethyl-4-hydroxy-	166	H, 3806
5-ethoxycarbonylmethyl-4-hydroxy-	156 or 158-160	2577, 3876
5-ethoxycarbonylmethyl-4-hydroxy-2,6-dimethyl-	178-180	3876
5-ethoxycarbonylmethyl-4-hydroxy-2-methyl-	181	H, 3838
5-ethoxycarbonylmethyl-4-hydroxy-2-phenyl-	169-175 or 179	2576, 3806
5-ethoxycarbonylmethyl-4-hydroxy-2-propyl-	148-149	3806

Pyrimidine	M.P.(^{o}C)	References
5-ethoxycarbonyl-4-phenoxy-	65	4000
4-ethoxy-5-ethoxycarbonyl-	82-85/0.1mm.; 28-30	2366, 4000, 4267
4-ethoxy-5-ethoxycarbonyl-2-methyl-	52	2698
4-ethylcarbamoyl-2,6-dihydroxy-	290 or 295-300	3440, 3444
5-ethylcarbamoyl-2,4-dihydroxy-	280-285	3440
4-ethyl-6-hydrazinocarbonyl-2-hydroxy-	237	2184
1-ethylthiocarbonyl-1,2,3,4-tetrahydro-3-methyl-2,4-dioxo-	103	3347
4-formyl-2,6-dihydroxy-	TSC 320; PH 346	H, 957, 2947
5-formyl-2,4-dihydroxy-	302-303; TSC 320	1168, 2155, 2665, 3063
5-formyl-4,6-dihydroxy-	245-250; PH 257	2594
4-formyl-2,6-dihydroxy-5-methyl-	209-211	H, 2947
5-formyl-2,4-dihydroxy-6-methyl-	DNP >300; oxime 260	524, 2155
5-formyl-4,6-dihydroxy-2-methyl-	240 or 300; PH 220 or >240	2155, 2594
5-formyl-4,6-dihydroxy-2-phenyl-	287 or 345; PH 247	524, 2594
5-formyl-4,6-dimethoxy-	137 or 143; TSC 223	2575, 2758
5-formyl-4,6-dimethoxy-2-phenyl-	TSC 212	2575
5-β-formylethyl-4-hydroxy-6-methyl-		3190
4-formyl-6-hydroxy-	acetal 124	H, 957

TABLE XLII

Pyrimidine	M.P.(°C)	References
5-formyl-2-hydroxy-4,6-dimethyl-	PH 231	2155
5-formyl-4-hydroxy-2,6-dimethyl-	DNP 305; oxime 240	2155
5-formyl-4-hydroxy-6-methoxy-	218-222	2758
5-formyl-1,2,3,4-tetrahydro-1,3-dimethyl-2,4-dioxo-	DNP 320	3117
5-formyl-2,4,6-trihydroxy-	330; DNP 302	H, 2155
5-formyl-2,4,6-trimethoxy-	130-133; TSC 223	2575
5-hydrazinocarbonyl-2,4-dihydroxy-	245	2756
5-hydrazinocarbonyl-4-hydroxy-	208	4000
4-hydrazinocarbonyl-2-hydroxy-6-methyl-	235-236	2184
4-hydrazinocarbonyl-2-hydroxy-6-propyl-	187-189	2184
4-hydrazinocarbonylmethyl-2,6-dihydroxy-	>290	H, 4251
4-hydroxy-2,6-dimethyl-5-methylcarbamoylmethyl-	272-275	2876
4-hydroxy-2,6-dimethyl-5-prop-2'-ynylcarbamoylmethyl-	257-258	2876
4-hydroxy-5-isopropylcarbamoylmethyl-2,6-dimethyl-	227-228	2876
4-hydroxy-5-isopropylidene-hydrazinocarbonyl-	262	4000
4-hydroxy-5-methoxycarbonyl-	209-210	2323, 2544
4-hydroxy-6-methoxycarbonyl-	226-228	2954
2-hydroxy-4-methoxycarbonyl-6-methyl-	176-178	2184

Pyrimidine	M.P.(^{o}C)	References
4-hydroxy-6-methoxycarbonyl-2-methyl-	242-243	2216
4-hydroxy-2-methoxy-5-methoxycarbonyl-	199-201	2205
4-hydroxy-5-methylcarbamoylmethyl-2-phenyl-	310-315	2876
4-hydroxy-2-methyl-5-methylcarbamoylmethyl-	215-217	2876
1,2,3,4-tetrahydro-5,6-dimethoxycarbonyl-2,4-dioxo-1,3-diphenyl-	259-260	4268
1,2,3,4-tetrahydro-6-hydroxy-1,3-dimethyl-2,4-dioxo-5-phenylcarbamoyl-	159-161	3266
1,2,3,4-tetrahydro-6-hydroxy-1,3-dimethyl-5-methylcarbamoyl-2,4-dioxo-	174	4253
1,2,3,4-tetrahydro-6-hydroxy-5-methoxycarbonyl-1,3-dimethyl-2,4-dioxo-	152-155	3266
1,2,3,4-tetrahydro-6-methoxycarbonyl-1,3-dimethyl-2,4-dioxo-	76-77	4255
1,2,3,4-tetrahydro-5(or 6)-methoxycarbonyl-2,4-dioxo-1,3-diphenyl-	193-195	4268
Uracil/5-acetoacetyl-6-methyl-1-phenyl-	157-158; DNP 219	3987
Uracil/1-acetyl-	189 or 190-191	985, 4269
Uracil/5-acetyl-1-α-carboxyethyl-	174	2317
Uracil/5-acetyl-1-carboxymethyl-	200	2317
Uracil/5-acetyl-1-α-carboxy-β-methylbutyl-	187	2317

TABLE XLII

Pyrimidine	M.P.(^{o}C)	References
Uracil/5-acetyl-1-α-carboxy-γ-methylbutyl-	153	2317
Uracil/5-acetyl-1-α-carboxy-β-methylpropyl-	196	2317
Uracil/5-acetyl-1-αβ-dicarboxyethyl-	209	2317
Uracil/1-acetyl-5-methoxy-	205-206	4244
Uracil/1-acetyl-5-methyl-	195 to 197	985, 4271
Uracil/5-acetyl-1-methyl-	234	2317
Uracil/5-acetyl-1-phenyl-	269	2317
Uracil/1-amyloxycarbonyl-	127-128	3346
Uracil/3-benzoyl-	200 or 206	3170, 4249
Uracil/3-benzoyl-5-methyl-	215	3170
Uracil/3-benzyl-6-carboxy-	226-227	H, 4261
Uracil/3-benzyl-6-ethoxycarbonyl-	139-141	4261
Uracil/1-benzyloxy-5-carboxy-	235	2199
Uracil/1-benzyloxy-5-ethoxycarbonyl-	165-168 and 170	2199
Uracil/3-butyl-6-carboxy- (?)	>260	4272
Uracil/5-butyl-6-carboxy-3-methyl-	245	2347, 4252
Uracil/5-butyl-6-carboxy-3-phenyl-	255	2345
Uracil/3-s-butyl-6-methyl-5-thiocyanato-	157-158	3981
Uracil/1-δ-carbamoylbutyl-	163-164	3382
Uracil/1-β-carbamoylethyl-	222	3428
Uracil/1-carbamoylmethyl-	298	3427, 3435
Uracil/1-δ-carboxybutyl-	134-135	3382
Uracil/6-carboxy-1,5-dimethyl-	288	2347, 4252
Uracil/1-β-carboxyethyl-	179 or 183-185	3363, 3428

Pyrimidine	M.P.(^{o}C)	References
Uracil/3-β-carboxyethyl-	169-174	3363
Uracil/6-carboxy-3-ethyl- (?)	228-231	H, 4272
Uracil/1-β-carboxyethyl-	180	4273
Uracil/6-carboxy-5-ethyl-3-methyl-	287	2347
Uracil/6-carboxy-5-ethyl-3-phenyl-	288	2345
Uracil/6-carboxy-3-hexyl- (?)	159-164	4272
Uracil/6-carboxy-5-hydroxymethyl-3-methyl-	lactone 267	4185
Uracil/1-carboxymethyl-	285 or 295	H, 3382, 3427
Uracil/6-carboxy-1-methyl-	277-278	H, 4261
Uracil/6-carboxy-3-methyl-	316-323	H, 2215, 4252, 4261
Uracil/1-carboxymethyl-5-cyano-6-methyl-	277-279	2546
Uracil/6-carboxy-3-methyl-5-phenyl-	267-270	4252
Uracil/6-carboxy-5-methyl-3-phenyl-	316	2345
Uracil/6-carboxy-3-phenyl-	278-280	H, 888, 4252
Uracil/1-δ-cyanobutyl-	115	3382
Uracil/5-cyano-1,6-dimethyl-	297-298	2546
Uracil/5-cyano-1-ethoxycarbonylmethyl-	137-141	2318
Uracil/1-β-cyanoethyl-	219	3428, 4249
Uracil/1-β-cyanoethyl-5-methyl-	195-196	4273
Uracil/5-cyano-6-ethyl-1-methyl-	239-240	2546
Uracil/5-cyano-6-ethyl-1-phenyl-	-	2546
Uracil/5-cyano-1-β-hydroxyethyl-	194	743

Pyrimidine	M.P.(°C)	References
Uracil/5-cyano-1-β-hydroxypropyl-	223	743
Uracil/5-cyano-6-methyl-1-phenyl-	332-334	2546
Uracil/1-ββ-diethoxyethyl-	91 or 93	3154, 4270
Uracil/6-dimethoxymethyl-3-methyl-	137	2219
Uracil/1-ethoxycarbonyl-	130-132	3346
Uracil/3-ethoxycarbonylethyl-	101-103	3363
Uracil/1-ethoxycarbonylmethyl-	141-142	3382, 3427
Uracil/1-ethylthiocarbonyl-	194	3347
Uracil/1-formylmethyl-	H_2O 213-214 or 207	3153, 3154, 4270
Uracil/1-hydrazinocarbonylmethyl-	241	3427
Uracil/1-methoxycarbonyl-	192-195	3346
Uracil/1-δ-methoxycarbonylbutyl-	115-116	3985
Uracil/1-β-methoxycarbonylethyl-	110	3428

TABLE XLIII. Carboxy-Sulphonylpyrimidines (*H* 598)

(no entries)

TABLE XLIV. Carboxy-Thiopyrimidines (*H* 599)

Pyrimidine	M.P.(°C)	References
5-acetyl-2-benzyl-4-mercapto-6-methyl-	194	4274
5-acetyl-2-butyl-4-mercapto-6-methyl-	141	4274

Pyrimidine	M.P.(^{o}C)	References
5-acetyl-2-t-butyl-4-mercapto-6-methyl-	156	4274
5-acetyl-2,4-dimethyl-6-methylthio-	47	4274
5-acetyl-2-ethoxycarbonyl-4-mercapto-6-methyl-	173	4274
5-acetyl-2-ethyl-4-mercapto-6-methyl-	154	4274
5-acetyl-4-mercapto-2,6-dimethyl-	149	4274
5-acetyl-4-mercapto-6-methyl-2-phenyl-	216-218	2314
5-acetyl-4-mercapto-6-methyl-2-propyl-	152	4274
5-acetyl-4-methyl-6-methylthio-2-phenyl-	51	4274
4-amyl-5-carbamoyl-2-methylthio-	83-85	2184
4-amyl-6-hydrazinocarbonyl-2-methylthio-	103	2184
4-amyl-6-methoxycarbonyl-2-methylthio-	57-58	2184
5-benzoyl-4-mercapto-6-methyl-2-phenyl-	219	2314
2-benzyl-5-ethoxycarbonyl-1,4-dihydro-6-methyl-1-phenyl-4-thio-	182-183	2313
2-benzyl-5-ethoxycarbonyl-4-mercapto-6-methyl-	162-164	2313
4-butyl-2-butylthio-4-hydrazinocarbonyl-	82-83	2184
4-butyl-6-carbamoyl-2-ethylthio-	56-57	2184
4-butyl-6-carbamoyl-2-methylthio-	87-88	2184
4-butyl-2-ethylthio-6-hydrazinocarbonyl-	64-65	2184
4-butyl-6-hydrazinocarbonyl-2-methylthio-	83-84	2184

TABLE XLIV

Pyrimidine	M.P.(°C)	References
4-butyl-6-methoxycarbonyl-2-methylthio-	37	2184
4-butylthio-5-ethoxycarbonyl-2-methylthio-	177/1.0mm.	4015
4-t-butylthio-5-ethoxycarbonyl-2-methylthio-	153/0.4mm.	4015
5-carbamoyl-4,6-bismethylthio-	214-215	2519
4-carbamoyl-6-ethyl-2-methylthio-	126	2184
4-carbamoyl-6-hexyl-2-methylthio-	79-80	2184
4-carbamoyl-6-methyl-2-methylthio-	183-184	2184
4-carbamoyl-2-methylthio-6-propyl-	138-139	2184
4-carboxy-2,6-bismethylthio-	145-146	2214
5-carboxy-2,4-bismethylthio-	201-203	2165
5-carboxy-4,6-bismethylthio-	225 or 230	2519, 2575
4-carboxy-2,6-dimercapto-	281-282	2214
5-carboxy-2,4-dimercapto-	261-263	2165
4-carboxy-2-mercapto-	-	2954
4-carboxy-6-mercapto-	247-248	2954
4-carboxy-6-mercapto-2-methylthio-	235	2214
5-carboxy-4-methyl-2-methylthio-	169-171	1128
4-carboxy-2-methylthio-	208-210	H, 2608
5-cyano-1,4-dihydro-6-methyl-1,2-diphenyl-4-thio-	288-290	2319
5-cyano-4-mercapto-	>300	2367

Pyrimidine	M.P.(°C)	References
5-cyano-4-mercapto-2,6-dimethyl-	212	4274
5-cyano-4-mercapto-2,6-diphenyl-	253	4274
5-cyano-4-mercapto-2-methyl-6-phenyl-	272	4274
5-cyano-4-mercapto-6-methyl-2-phenyl-	230	2319
4-cyano-2-methylthio-	82-84	2608, 2954
5-cyano-4-methylthio-	94-96	2366, 2367
4-diethoxymethyl-5-ethoxycarbonyl-6-mercapto-2-phenyl-	147	4274
4-diethoxymethyl-6-methyl-2-methylthio-	103/0.001mm.	2301
4-diethoxymethyl-2-methylthio-	116/0.001mm.	2301
4,6-dimercapto-2-trifluoromethyl-	150-151	2193
5-ethoxycarbonyl-2,4-bismethylthio-	86-88	2936, 4015
5-ethoxycarbonyl-1,4-dihydro-1,6-dimethyl-2-phenyl-4-thio-	152 or 156	2313, 2319
5-ethoxycarbonyl-1,4-dihydro-2,6-dimethyl-1-phenyl-4-thio-	209 or 215	2313, 2319
5-ethoxycarbonyl-1,4-dihydro-6-methyl-1,2-diphenyl-4-thio-	215 or 220	2313, 2319
5-ethoxycarbonyl-1,4-dihydro-1,2,6-trimethyl-4-thio-	165 or 167	2313, 2319
5-ethoxycarbonyl-4-ethylthio-	25; 108/0.4mm.	4000
5-ethoxycarbonyl-4-ethylthio-2-methylthio-	168/0.8mm.	4015
5-ethoxycarbonyl-4-isopropylthio-2-methylthio-	161/0.8mm.	4015

TABLE XLIV

Pyrimidine	M.P.(^{o}C)	References
5-ethoxycarbonyl-4-mercapto-	149-150	4000
5-ethoxycarbonyl-4-mercapto-2,6-dimethyl-	142	4274
5-ethoxycarbonyl-4-mercapto-2-methyl-	157-158	H, 2654
5-ethoxycarbonyl-4-mercapto-6-methyl-2-phenyl-	140 or 150	2313, 2314
5-ethoxycarbonylmethyl-4-mercapto-2-methyl-	185	2654, 3882
5-ethoxycarbonyl-2-methyl-4-methylthio-	57-58	2651
5-ethoxycarbonyl-4-methyl-2-methylthio-	53-54	1128
5-ethoxycarbonyl-4-methyl-6-methylthio-2-phenyl-	51	4274
5-ethoxycarbonyl-2-methylthio-4-propylthio-	187/2.5mm.	4015
5-ethoxycarbonyl-2-methylthio-4-thiocyanato-	161	2936
4-ethyl-6-methoxycarbonyl-2-methylthio-	74	2184
4-formyl-6-methyl-2-methylthio-	87; oxime 219	2301
4-formyl-2-methylthio-	68; oxime 165	2301
4-hexyl-6-hydrazinocarbonyl-2-methylthio-	103-105	2184
4-hexyl-6-methoxycarbonyl-2-methylthio-	63-64	2184
4-hydrazinocarbonyl-6-methyl-2-methylthio-	176-177	2184
4-hydrazinocarbonyl-2-methylthio-6-propyl-	114-115	2184
4-methoxycarbonyl-6-methyl-2-methylthio-	111-112	2184
4-methoxycarbonyl-2-methylthio-6-propyl-	61-62	2184

TABLE XLV. Halogeno-Nitropyrimidines (*H* 600)

Pyrimidine	M.P.(^{o}C)	References
2-benzyl-4,6-dichloro-5-nitro-	66-69	2688
5-bromo-2-nitroamino-	189	4044
4-chloro-2,6-dimethyl-5-nitro-	106-108/12mm.	887
2-chloro-5-nitro-	110-111	H, 4275
5-chloro-2-nitroamino-	173	4044
4-chloro-6-phenylazo-	72-73	4042
4,6-dichloro-2-ethyl-5-nitro-	78/0.1mm.	4035
2,4-dichloro-6-isopropyl-5-nitro-	37; 134-136/12mm.	2435
2,4-dichloro-6-methyl-5-nitro-	-	H, 2461
2,4-dichloro-5-nitro-	-	H, 748, 2461, 4237
4,6-dichloro-5-nitro-	102-104	H, 796, 3244, 3871
2,4-dichloro-5-nitro-6-trifluoromethyl-	48/0.3mm.	999
4,6-dichloro-5-nitro-2-trifluoromethyl-	49-50; 72/6mm.	2193
2,4-dichloro-6-phenylazo-	86-88	4042
4,6-difluoro-5-nitro-	crude 185-195/at.	2455
2,4,6-trichloro-5-nitro-	-	H, 4276

TABLE XLVI. Halogeno-Oxypyrimidines (*H* 600)

Pyrimidine	M.P.(^{o}C)	References
5-allyl-4-hydroxy-2-trifluoromethyl-	153-154	2194
4-allyloxy-5-allyloxymethyl-2-chloro-	110-120/0.001mm.	2596
5-allyloxymethyl-2,4-dichloro-	85-95/0.001mm.	2596

TABLE XLVI

Pyrimidine	M.P.(^{o}C)	References
4-allyloxy-2-trifluoromethyl-	42/0.4mm.	2194, 2555
4-amyloxy-2-chloro-6-methyl-	159/18mm.; HCl 89-91	2699
Barbituric acid/5-bromo-5-ethyl-	202-204	H, 4277
Barbituric acid/5-chloro-5-ethyl-	193	H, 4277
Barbituric acid/5,5-dichloro-	211-215	H, 2622
Barbituric acid/5-ethyl-5-fluoro-	204-205	4278
2-benzyl-5-bromo-4,6-dihydroxy-	>200	2575
2-benzyl-5-bromo-4,6-dimethoxy-	114-115	2575
5-benzyl-4-chloro-6-hydroxy-2-methoxymethyl-	159	2877
5-benzyl-4-chloro-6-methoxy-2-methoxymethyl-	133/0.18mm.	2877
5-benzyl-4-chloro-2-methoxymethyl-6-methyl-	145/0.05mm.	2877
5-benzyl-4,6-dichloro-2-ethoxymethyl-	147/0.3mm.	2877
5-benzyl-4,6-dichloro-2-methoxymethyl-	42-43; 125/0.03mm.	2877
5-benzyloxy-4-chloro-	126-128	2347, 2584
1-benzyloxy-4-chloro-1,2-dihydro-2-oxo-	175-177	2262
5-benzyloxy-2,4-dichloro-	88-90	2210
5-bromo-4-t-butyl-6-hydroxy-	163-164	2574
5-bromo-4-chloro-2,6-dimethoxy-	96-98	1005
2-bromo-4-chloro-5-ethoxy-	43-46	2606
5-bromo-4-chloro-6-hydroxy-	249	2240
5-bromo-4-chloro-6-methoxy-	70 or 77	2519, 2575

Pyrimidine	M.P.(^{o}C)	References
5-bromo-4-chloro-6-methoxy-2-phenyl-	91-93	2575
5-bromo-1-cyclohexyl-1,2,3,6-tetrahydro-3,4-dimethyl-2,6-dioxo-	175 or 178-181	1352, 3967
2-bromo-4,5-diethoxy-	49; HCl 135	2606
5-bromo-4,6-diethoxy-	34-35	2575
5-bromo-1,2-dihydro-4-methoxy-1-methyl-2-oxo-	148-149	3046
5-bromo-1,2-dihydro-1-methyl-2-oxo-	210-211	2746
5-bromo-1,6-dihydro-1-methyl-6-oxo-	158-159	2746
4-bromo-2,6-dihydroxy-	$\nless$340	2611
5-bromo-2,4-dihydroxy-	296 to 312	H, 266, 998, 1463, 2196, 4279, 4280
5-bromo-4,6-dihydroxy-	264	H, 2842
5-bromo-2,4-dihydroxy-6-methyl-	242 to 248	H, 2625, 2714, 2842, 2863, 4132
5-bromo-4,6-dihydroxy-2-methyl-	ca 240	2575
4-bromo-2,6-dimethoxy-	90-91	2611
5-bromo-2,4-dimethoxy-	51-52; 125/17mm.	H, 2606
5-bromo-4,6-dimethoxy-	148-150	2519, 2575
5-bromo-2,4-dimethoxy-6-methyl-	76-77	2623, 2714
5-bromo-4,6-dimethoxy-2-methyl-	116-119	2575
5-bromo-4,6-dimethoxy-2-phenyl-	103	2575
5-bromo-2-ethyl-4,6-dihydroxy-	223-224	2575
5-β-bromoethyl-2,4-dihydroxy-	262-263	2514

TABLE XLVI

Pyrimidine	M.P.(°C)	References
5-β-bromoethyl-2,4-dihydroxy-6-methyl-	268-270	2514
5-bromo-2-ethyl-4,6-dimethoxy-	67-69	2575
5-bromo-2-hydroxy-	234-235	H, 2154
5-bromo-4-hydroxy-	252-255	H, 2191
5-bromo-2-hydroxy-4,6-dimethyl-	223; HBr >300	H, 2623
5-bromo-4-hydroxy-6-methoxy-	234-235	3054
5-bromo-4-hydroxy-6-methyl-	214 to 233	2574, 2575, 2623
5-bromo-4-hydroxy-6-phenyl-	242 or 246-247	2519, 2574, 2595
5-bromo-4-methoxy-	66-67 or 72-74	2575, 2595
5-bromo-2-methoxy-4,6-dimethyl-	61-62	2519
5-bromo-4-methoxy-6-methyl-	79-80	2575
5-bromo-4-methoxy-6-phenyl-	88-89	2519
5-bromomethyl-2,4-dihydroxy-	>330	2160, 2650
5-bromomethyl-2,4-dihydroxy-6-methyl-	>350	2652, 2824
5-bromo-1,2,3,4-tetrahydro-1,3-diisopropyl-6-methyl-2,4-dioxo-	-	4281
5-bromo-1,2,3,4-tetrahydro-1,3-dimethyl-2,4-dioxo-	184-185	H, 666, 890, 998, 1112
5-bromo-2,4,6-triethoxy-	52-54	2575
5-bromo-2,4,6-trihydroxy-	200	H, 2628
5-bromo-2,4,6-trimethoxy-	144-145	2575
5-bromo-2,4,6-tripropoxy-	130-140/0.1mm.	2575
4-butoxy-2-chloro-6-methyl-	133/12mm.; HCl 102-105	2580, 2699

Pyrimidine	M.P.(°C)	References
5-butoxy-4,6-dichloro-	128-132/12mm.	2524
4-butoxy-2,5,6-trichloro-	120/3.5 mm.	2618
5-butyl-4-chloro-1,3-diethyl-1,2,3,6-tetrahydro-2,6-dioxo-	120/2mm.	2305
2-butyl-4-chloro-6-hydroxy-	121-122	2242
5-butyl-4-chloro-6-hydroxy-	140	2242
1-s-butyl-5-chloro-1,2,3,6-tetrahydro-3,4-dimethyl-2,6-dioxo-	-	3967 (?)
1-t-butyl-5-chloro-1,2,3,6-tetrahydro-3,4-dimethyl-2,6-dioxo-	-	3967 (?)
5-butyl-4-chloro-1,2,3,6-tetrahydro-2,6-dioxo-1,3-diphenyl-	135-136	2305
4-butyl-5-fluoro-6-hydroxy-	88 or 107	4282, 4283
4-chloro-2,6-bismethoxymethyl-	77-78/0.15mm.	3913
4-chloro-2,6-bismethoxymethyl-5-methyl-	83/0.025mm.	3911
4-chloro-5-cyclohexyl-1,2,3,6-tetrahydro-1,3-dimethyl-2,6-dioxo-	137-138	2305
5-chloro-1-cyclohexyl-1,2,3,6-tetrahydro-3,4-dimethyl-2,6-dioxo-	-	4281 (?)
1-3'-chlorocyclopentyl-1,2,3,4-tetrahydro-3,5-dimethyl-2,4-dioxo-(trans)	169-171	3984
4-chloro-6-cyclopropyl-2-propoxy-	82-83/0.13mm.	3816
5-chloro-2,4-diethoxy-6-fluoro-	71-73	2618
4-chloro-1,3-diethyl-1,2,3,6-tetrahydro-2,6-dioxo-	86-88	2305

Pyrimidine	M.P. (^{o}C)	References
4-chloro-1,3-diethyl-1,2,3,6-tetrahydro-5-methyl-2,6-dioxo-	25	2305
4-chloro-1,6-dihydro-1-methyl-6-oxo-	87-88	827
4-chloro-2,6-dihydroxy-	296 to 307	H, 553, 2438, 2442, 2611, 3270, 4284
5-chloro-2,4-dihydroxy-	323-325	H, 4280
5-chloro-4,6-dihydroxy-	300	H, 3181
5-chloro-4,6-dihydroxy-2-hydroxymethyl-	>360	2236
5-chloro-2,4-dihydroxy-6-isopropyl-	260	2226
5-chloro-4,6-dihydroxy-2-isopropyl-	>365	2238
4-chloro-2,6-dihydroxy-5-methyl-	266-267	4130
5-chloro-2,4-dihydroxy-6-methyl-	>300	H, 2160, 3899
5-chloro-4,6-dihydroxy-2-methyl-	>305	2236
5-chloro-2,4-dihydroxy-6-phenyl-	270-272	H, 2226
5-chloro-4,6-dihydroxy-2-phenyl-	331-332	H, 2238, 2575
5-chloro-2,4-dihydroxy-6-propyl-	242-244	2226
5-chloro-4,6-dihydroxy-2-propyl-	294	2238
2-chloro-4,6-dimethoxy-	101	2601
4-chloro-2,6-dimethoxy-	-	H, 2843, 4169
4-chloro-5,6-dimethoxy-	53-55	3910, 4285
5-chloro-2,4-dimethoxy-6-methyl-	67-69	2623
2-chloro-4-ethoxy-5-ethoxymethyl-	32-33	2596
2-chloro-4-ethoxy-5-fluoro-	35-36	2748

Pyrimidine	M.P. (^{o}C)	References
5-chloro-4-ethoxy-6-hydroxy-2-methyl-	216-218	2239
5-chloro-4-ethoxy-6-hydroxy-2-phenyl-	274	2239
2-chloro-4-ethoxy-6-methyl-	81/4mm.;39-41; HCl 148-149	813, 2580, 2699
4-chloro-2-ethoxymethyl-	HCl 118	2819
4-chloro-5-ethoxymethyl-2-phenyl-	91	2190
4-chloro-5-ethyl-2,6-bismethoxymethyl-	87/0.04mm.	3911
5-chloro-2-ethyl-4,6-dihydroxy-	318-319	2238
5-chloro-4-ethyl-2,6-dihydroxy-	260-261	2226
4-chloro-2-ethyl-6-hydroxy-	161	2554
5-β-chloroethyl-4-hydroxy-2,6-dimethyl-	HCl 176	3944
4-chloro-5-ethyl-1,2,3,6-tetrahydro-1,3-dimethyl-2,6-dioxo-	71-72	2305
4-chloro-5-fluoro-2,6-dihydroxy-	232	4286
2-chloro-5-fluoro-4-hydroxy-	177 then 228-243	2748
4-chloro-5-fluoro-2-methoxy-	86-87/21mm.	2209
2-chloro-5-hydroxy-	195-196	2585
4-chloro-6-hydroxy-	-	H, 2240, 2777
5-chloro-2-hydroxy-	236	H, 2154
5-chloro-4-hydroxy-	180-182	H, 2191, 4283
4-chloro-6-hydroxy-5-isopropyl-	174-175	2242
5-chloro-4-hydroxy-6-methoxy-	244-245	3054
5-chloro-4-hydroxy-2-methoxymethyl-	153	3911

TABLE XLVI

Pyrimidine	M.P.(^{o}C)	References
4-chloro-6-hydroxy-2-methoxymethyl-5-phenyl-	185-186	2877
4-chloro-2-hydroxy-5-methyl-	-	805 (?)
4-chloro-6-hydroxy-2-methyl-	231-233	H, 2191
4-chloro-6-hydroxy-5-methyl-	200 or 203	2240, 2242
5-chloro-4-hydroxy-2-methyl-	227-228	4282
5-chloro-4-hydroxy-6-methyl-	209-211	2623
4-chloro-5-hydroxymethyl-2-methyl-	82-83	2651
4-chloro-6-hydroxy-5-phenoxy-	194-195	2240
2-chloro-4-isoamyloxy-6-methyl-	147/15mm.; HCl 95-98	2699
2-chloro-4-isobutoxy-6-methyl-	131/18mm.; HCl 125-128	2699
2-chloro-4-isopropoxy-6-methyl-	112/14mm.; HCl 150-160	813, 2580, 2699
2-chloro-4-methoxy-	55	H, 1223
2-chloro-5-methoxy-	75-77	H, 2161
4-chloro-5-methoxy-	64 to 71	2161, 2586, 3915
4-chloro-6-methoxy-	80/18mm.	H, 827
2-chloro-4-methoxy-5,6-dimethyl- (?)	65-66	140
4-chloro-5-methoxy-2-methoxymethyl-	73-74	3911
4-chloro-5-methoxy-6-methoxymethyl-	116/11mm.	3911
4-chloro-5-methoxy-6-methoxymethyl-2-methyl-	70-72/0.1mm.	3911
2-chloro-4-methoxy-6-methyl-	32-35; HCl 164	H, 813, 2580, 2699

Pyrimidine	M.P.(^{o}C)	References
4-chloro-5-methoxy-2-methyl-	86 or 98-99	2586, 4113
4-chloro-5-methoxy-6-methyl-	55-56/7mm.	3243
4-chloro-6-methoxymethyl-	90-93/12mm.	3913
4-chloro-6-methoxy-2-methyl-	77-78/8mm.	4287
4-chloro-6-methoxymethyl-2,5-dimethyl-	56-57/0.2mm.	3911
4-chloro-2-methoxymethyl-5-methyl-	59/0.025mm.	3911
4-chloro-2-methoxymethyl-6-methyl-	73/0.6mm.	3913
4-chloro-6-methoxymethyl-2-methyl-	90-100/11mm.	3913
4-chloro-6-methoxymethyl-5-methyl-	36-37	3911
4-chloro-5-methoxymethyl-2-propyl-	-	3947
4-chloro-6-methoxy-2-phenyl-	70-71	2575
5-chloromethyl-2,4-dihydroxy-	270 to 355	H, 2071, 2115, 2160, 2640, 2650
5-chloromethyl-2,4-dihydroxy-6-methyl-	225	H, 3937
4-chloromethyl-5-fluoro-2,6-dihydroxy-	240-241	77
2-chloro-4-methyl-6-phenoxy-	77-79	2699
4-chloro-6-methyl-2-phenoxy-	65	3914
2-chloro-4-methyl-6-propoxy-	123/13mm.; HCl 132	2580, 2699
4-chloro-1,2,3,6-tetrahydro-1,3-diisopropyl-2,6-dioxo-5-phenyl-	156-157	2305

TABLE XLVI

Pyrimidine	M.P.(°C)	References
4-chloro-1,2,3,6-tetrahydro-1,3-dimethyl-2,6-dioxo-	110 or 112	H, 739, 4156
4-chloro-1,2,3,6-tetrahydro-1,3-dimethyl-2,6-dioxo-5-phenyl-	134-136	2305
4-chloro-1,2,3,6-tetrahydro-1,3-dimethyl-2,6-dioxo-5-propyl-	48-50	2305
4-chloro-1,2,3,6-tetrahydro-2,6-dioxo-1,3-diphenyl-	141-143	2485
4-chloro-1,2,3,6-tetrahydro-5-isopropyl-1,3-dimethyl-2,6-dioxo-	52-53	2305
4-chloro-1,2,3,6-tetrahydro-5-methyl-2,6-dioxo-1,3-diphenyl-	172-173	2305
4-chloro-1,2,3,6-tetrahydro-1,3,5-trimethyl-2,6-dioxo-	135-136	740, 4288
5-chloro-2,4,6-triethoxy-	44-48	2575
5-chloro-2,4,6-trihydroxy-	280-290	2622
4-chloro-2,5,6-trimethoxy-	73	2586
5-chloro-2,4,6-trimethoxy-	130-132	2575
1-cyclohexyl-1,2,3,6-tetrahydro-5-iodo-3,4-dimethyl-2,6-dioxo-	187-188	1352
2,4-diamyloxy-5,6-dichloro-	112-115/3mm.	2618
2,4-dibenzyloxy-5-bromo-	87-89	2923
2,4-dibenzyloxy-6-chloro-	55	2558
2,5-dibromo-4,6-dihydroxy-	-	3054 (?)
5,5-dibromohexahydro-1,3-dimethyl-2,4,6-trioxo-	172-173	H, 3265
2,5-dibromo-4-methoxy-	85	2606
1,3-dibromo-1,2,3,4-tetrahydro-2,4-dioxo-	-	4289 (?)
1,3-dibromo-1,2,3,4-tetrahydro-6-methyl-2,4-dioxo-	-	4289 (?)

Pyrimidine	M.P.(^{o}C)	References
2,4-dibutoxy-5-chloro-6-fluoro-	114/0.5mm.	2618
4,6-dibutoxy-5-chloro-2-fluoro-	70/1mm.	2618
1,3-dibutyl-6-chloro-1,2,3,4-tetrahydro-2,4-dioxo-	184/10mm.	2485
4,5-dichloro-2,6-dimethoxy-	74-76	H, 4188
4,6-dichloro-5-ethoxy-	102-107/12mm.	2524
2,4-dichloro-5-ethoxymethyl-	80-90/0.001mm.	2596
4,6-dichloro-2-ethoxymethyl-	49/0.03mm.	3913
4,6-dichloro-5-ethyl-2-methoxymethyl-	69/0.07mm.	3911
4,5-dichloro-6-hydroxy-	212-213	2240
4,6-dichloro-2-hydroxy-	262	2165
4,6-dichloro-5-isopropoxy-	108-113	2524
2,4-dichloro-5-methoxy-	68 or 74	2161, 2586
2,4-dichloro-6-methoxy-	-	H, 3992
2,5-dichloro-4-methoxy-	51-53	2571
4,6-dichloro-2-methoxy-	59-60	2578, 3908, 3909
4,6-dichloro-5-methoxy-	57-58	2524, 3910
4,6-dichloro-5-β-methoxyethoxy-	98-100/0.1mm.	4041
2,4-dichloro-5-β-methoxyethoxymethyl-	80-90/0.001mm.	2596
4,5-dichloro-6-methoxy-2-methoxymethyl-	59-60	3911
4,6-dichloro-5-methoxy-2-methoxymethyl-	40	3911
2,4-dichloro-5-methoxymethyl-	95-97/1.5mm.	2579
4,5-dichloro-2-methoxymethyl-	40	3911

TABLE XLVI

Pyrimidine	M.P.(^{o}C)	References
4,6-dichloro-2-methoxymethyl-	41	3912, 3913
4,6-dichloro-2-methoxymethyl-5-methyl-	31	3911
4,6-dichloro-2-methoxymethyl-5-phenyl-	128-131/0.04mm.	2877
4-dichloromethyl-6-hydroxy-2-methyl-	155-156	2518
4,6-dichloro-5-phenoxy-	91-92	2240
2,4-diethoxy-5,6-difluoro-	43	2618
2,4-diethoxy-5-fluoro-	18-19	2767
2,4-diethoxy-6-fluoro-	59/0.25mm.	2618
4,6-difluoro-2-hydroxy-	K $\nless$280	2614
5-difluoromethyl-2,4-dihydroxy-	285-300	2664, 2665
2,4-dihydroxy-5-iodo-	270-273	H, 2623, 4280
2,4-dihydroxy-6-iodo-	279-280	2611
2,4-dihydroxy-5-iodomethyl-	252	2650
2,4-dihydroxy-5-iodo-6-methyl-	240 or 283	2623, 4290
2,4-dihydroxy-5-trifluoromethyl-	239-243	2264, 2664, 2666, 2667, 4260
2,4-dihydroxy-6-trifluoromethyl-	220 to 232	443, 999, 4196
4,6-dihydroxy-2-trifluoromethyl-	265	2193
2,4-dimethoxy-5-trifluoromethyl-	55-56	3896
4-ethoxy-5-fluoro-1,2-dihydro-1-methyl-2-oxo-	135-136	2748
5-ethoxymethyl-2-fluoromethyl-4-hydroxy-	180	2116
5-fluoro-1,2-dihydro-4-methoxy-1-methyl-2-oxo-	193-194	2572
4-fluoro-2,6-dihydroxy-	240 or 245	2614, 2737

Pyrimidine	M.P.(^{o}C)	References
5-fluoro-2,4-dihydroxy-	282-284	77, 2191, 4197, 4222, 4291
5-fluoro-2,4-dihydroxy-6-methyl-	>300	4196
5-fluoro-4,6-dihydroxy-2-methyl-	>300	4292
5-fluoro-2,4-dihydroxy-6-pentafluoroethyl-	221-222	4196
5-fluoro-2,4-dihydroxy-6-phenyl-	290-295	4196
5-fluoro-2,4-dihydroxy-6-trifluoromethyl-	224-227	4196
4-fluoro-2,6-dimethoxy-	51 or 54-55	2610, 2611, 2614
5-fluoro-2,4-dimethoxy-	50-51	2572
5-fluoro-2-hydroxy-	172-173	2679
5-fluoro-4-hydroxy-	204-205	2191, 4282, 4283
5-fluoro-4-hydroxy-2,6-dimethyl-	177-178	4196
5-fluoro-4-hydroxy-6-isobutyl-2-methyl-	114-116	4282
5-fluoro-4-hydroxy-2-methoxy-	206-207	77, 4197, 4222
5-fluoro-2-hydroxy-4-methyl-	198-200	2679
5-fluoro-4-hydroxy-2-methyl-	218-220	2191, 4282
5-fluoro-4-hydroxy-6-methyl-	143-144; HCl 177-178	4283
5-fluoro-4-hydroxy-2-methyl-6-pentafluoroethyl-	105-106	4196
5-fluoromethyl-2,4-dihydroxy-	270-272	4196
2-fluoromethyl-4,6-dihydroxy-	>250	2116
2-fluoromethyl-4,6-dihydroxy-5-methyl-	≮300	2116

TABLE XLVI

Pyrimidine	M.P.(°C)	References
4-fluoromethyl-6-hydroxy-	-	2518
2-fluoromethyl-4-hydroxy-5-methyl-	-	2116
2-fluoromethyl-4-hydroxy-6-methyl-	166-168	2116
4-fluoromethyl-6-hydroxy-2-methyl-	204-207	4196
5-fluoro-1,2,3,4-tetrahydro-1,3-dimethyl-2,4-dioxo-	128-130	4293
4-hydroxy-2,6-bistrifluoromethyl-	117-118	2193
4-hydroxy-5-hydroxymethyl-2-trifluoromethyl-	166-167	2268
4-hydroxy-5-iodo-	254-265	4283
4-hydroxy-5-iodo-6-methyl-	238-239	2623, 2629
4-hydroxy-5-methyl-4-phenyl-6-trifluoromethyl-	>300	3239
4-hydroxy-2-methyl-6-trifluoromethyl-	136 or 142-144	2516, 4196
4-hydroxy-6-methyl-2-trifluoromethyl-	140-141	2193
4-hydroxy-2-propyl-6-trifluoromethyl-	83-84	2224
4-hydroxy-2-trifluoromethyl-	167-168	2193
4-hydroxy-6-trifluoromethyl-	162-163	443
4-iodo-2,6-dimethoxy-	173-175	2611
5-iodo-2,4-dimethoxy-	71	3894
1,2,3,4-tetrahydro-1-3'-iodocyclopentyl-3,5-dimethyl-2,4-dioxo (trans)	170-171	3984
2,4,5-trichloro-6-ethoxy-	66	2618
2,4,6-trichloro-5-methoxy-	67-68	2586
4,5,6-trichloro-2-methoxymethyl-	41-42	3911

Pyrimidine	M.P.(^{o}C)	References
2,4,5-trifluoro-6-hydroxy-	121	2618
5-trifluoromethyl-2,4-bistrimethylsiloxy-	crude	4294
Uracil/1-allyl-5-fluoro-	126-127	3903
Uracil/3-amyl-5-fluoro-	126-127	3903
Uracil/1-benzyl-5-bromo-	203-205	H, 1002
Uracil/3-benzyl-6-chloro-	198 or 209-210	2485, 4295
Uracil/3-benzyl-6-chloro-5-fluoro-	163-166	4286
Uracil/3-benzyl-5-chloro-6-methyl-	-	3970
Uracil/1-benzyl-5-fluoro-	170-171	3903
Uracil/3-benzyloxy-6-chloro-5-methyl-	187-189	3972
Uracil/5-bromo-3-t-butyl-	-	3979
Uracil/5-bromo-3-butyl-6-methyl-	158-160	3973
Uracil/5-bromo-3-s-butyl-6-methyl-	155 or 157-160	3981, 4296
Uracil/5-bromo-3-t-butyl-6-methyl-	-	3973, 4296
Uracil/5-bromo-3-cyclohexyl-	-	3979
Uracil/5-bromo-3-cyclohexyl-6-methyl-	-	3983
Uracil/5-bromo-3-cyclopentyl-6-propyl-	-	3982
Uracil/5-bromo-1-isopropyl-	203-205	1002
Uracil/5-bromo-3-isopropyl-6-methyl-	158-159	2864, 4296
Uracil/5-bromo-1-methyl-	267-272	H, 2643
Uracil/5-bromo-3-methyl-	238-241	H, 2643
Uracil/5-bromo-6-methyl-3-phenyl-	-	H, 4296
Uracil/5-bromo-3-phenyl-	-	3979

Pyrimidine	M.P.(^{o}C)	References
Uracil/3-s-butyl-5-chloro-	-	3979 (?)
Uracil/3-t-butyl-5-chloro-	-	3979 (?)
Uracil/3-butyl-5-chloro-6-methyl-	Na -	3973
Uracil/3-s-butyl-5-chloro-6-methyl-	153-154	3980, 3981
Uracil/3-t-butyl-5-chloro-6-methyl-	-	3973 (?)
Uracil/5-chloro-6-chloromethyl-3-isopropyl-	-	3981 (?)
Uracil/5-chloro-3-cyclohexyl-6-methyl-	-	3983 (?)
Uracil/1-3'-chlorocyclopentyl-5-methyl- (trans)	169-171	3984
Uracil/1-β-chloroethyl-	163-165	3771
Uracil/6-chloro-3-ethyl-	215-217	4295
Uracil/6-chloro-3-ethyl-5-fluoro-	171-172	4286
Uracil/5-chloro-6-ethyl-3-phenyl-	-	3970
Uracil/6-chloro-5-fluoro-3-methyl-	215-217	4286
Uracil/6-chloro-5-fluoro-3-phenyl-	222-223	4286
Uracil/6-chloro-3-hydroxy-5-methyl-	215-220	3972
Uracil/5-chloro-3-isopropyl-6-methyl-	-	4296
Uracil/6-chloro-3-methyl-	277 to 282	2439, 2485, 2560, 3270, 4295, 4297
Uracil/5-chloromethyl-3-isopropyl-6-methyl-	-	3982
Uracil/5-chloro-6-methyl-3-phenyl-	-	3970
Uracil/6-chloro-3-phenyl-	214 or 270	2485, 4295

Pyrimidine	M.P.(^{o}C)	References
Uracil/1-3',4'-dibromocyclopentyl-5-methyl- (a single geometric isomer)	229-231	3984
Uracil/5-fluoro-3-ε-hydroxyamyl-	93-94	3903
Uracil/5-fluoro-1-methyl-	264-265	2748, 3045
Uracil/5-fluoro-3-methyl-	-	3045
Uracil/5-iodo-3-isopropyl-6-methyl-	-	3981 (?)

TABLE XLVII. Halogeno-Sulphonylpyrimidines (*H* 604)

5-bromo-4-chloro-6-methyl-2-methylsulphonyl-	140-142	2255
5-bromo-4,6-dichloro-2-methylsulphonyl-	169-171	2255
2-chloro-5-chlorosulphonyl-	66-67	1474
4-chloro-5-ethyl-2-ethylsulphonyl-6-methyl-	74-75	937
4-chloro-2-ethylsulphonyl-5,6-dimethyl-	78	140
4-chloro-2-ethylsulphonyl-5-fluoro-	96	2600
4-chloro-2-ethylsulphonyl-5-methyl-	65-66	H, 936
4-chloro-5-fluoro-2-methylsulphonyl-	103	2600
2-chloro-4-methyl-6-phenylsulphonyl-	129	2742
4-chloro-6-methylsulphonyl-	125-126	2749
2,4-dichloro-5-chlorosulphonyl-	97-99	4298

TABLE XLVII

Pyrimidine	M.P.(°C)	References
4,6-dichloro-2-ethylsulphonyl-5-fluoro-	80	2600
4,6-dichloro-5-fluoro-2-methylsulphonyl-	106-107	2600
4,6-dichloro-2-methylsulphonyl-	119	2165
2,4-dichloro-5-sulpho-	Na -	4298
2-ethylsulphonyl-5-fluoro-	145/0.3mm.	2679
5-fluoro-4-methyl-2-methylsulphonyl-	71	2679
5-fluoro-2-methylsulphonyl-	80	2679
4-iodo-6-methylsulphonyl-	124-127	2609
2-methylsulphonyl-4-phenyl-6-trifluoromethyl-	151-152	3239
4,5,6-trichloro-2-methylsulphonyl-	141	2600

TABLE XLVIII. Halogeno-Thiopyrimidines (*H* 605)

Pyrimidine	M.P.(°C)	References
5-allyl-2-benzylthio-4-chloro-	196/1.3mm.	2555
2-benzylthio-4-chloro-	49-50	H, 2703
2-benzylthio-4-chloro-5-methyl-	49-50	2703
4,6-bisbenzylthio-5-bromo-	95-97	2165
4,6-bisbenzylthio-5-chloro-	86-88	2165
5-bromo-4,6-bismethylthio-	155	2165, 2519
5-bromo-4,6-bispropylthio-	44-46	2165
5-bromo-4-chloro-6-methyl-2-methylthio-	70	H, 2255
5-bromo-4,6-dichloro-2-methylthio-	83-85	2255
5-bromo-4,6-dimercapto-	213	2165

Pyrimidine	M.P.(^{o}C)	References
5-bromo-2-mercapto-	180	2746
5-bromo-4-mercapto-	185	2746
5-bromomethyl-4-mercapto-2-methyl-	HBr >260	2651
5-bromomethyl-2-methyl-4-methylthio-	HBr 275	2651
5-bromo-2-methylthio-	67	H, 2746
5-bromo-4-methylthio-	78	2746
5-butyl-4-chloro-6-mercapto-	176-178	2522
2-butylthio-4-chloro-6-methyl-	102-105/0.3mm.	2580
4-butylthio-2-chloro-6-methyl-	124-127/4mm.	2580
5-chloro-4,6-bisethylthio-	58-59	2165
5-chloro-4,6-bismethylthio-	118-120	2165
2-chloro-4,6-bismethylthio-	67	2601
5-chloro-4,6-dimercapto-	215-217	2165
4-chloro-5-ethyl-2-ethylthio-6-methyl-	-	H, 937
4-chloro-6-ethyl-2-methylthio-	136-137/14mm.	2753
4-chloro-2-ethylthio-5-fluoro-	108/12mm.	77, 2515, 2600, 2679, 4222
2-chloro-4-ethylthio-6-methyl-	96-97/3mm.	2580
4-chloro-2-ethylthio-5-methyl-	136-137/12mm.	H, 2582
4-chloro-2-ethylthio-6-methyl-	89-90/0.5mm.	H, 2580
4-chloro-2-ethylthio-6-methyl-5-propyl-	142/4mm.	H, 927
4-chloro-5-fluoro-6-fluoromethyl-2-methylthio-	125-128/16mm.	2679

TABLE XLVIII

Pyrimidine	M.P.(°C)	References
4-chloro-5-fluoro-6-methyl-2-methylthio-	73	2679
4-chloro-5-fluoro-2-methylthio-	108-112/15mm.; 221/at.	2600, 2679
2-chloro-4-isopropylthio-6-methyl-	77-79/3mm.	2580
4-chloro-2-isopropylthio-6-methyl-	102-104/0.5mm.	2580
4-chloro-6-mercapto-	crude	2158
4-chloro-6-mercapto-5-phenyl-	220-225	2522
2-chloro-4-methyl-6-methylthio-	107-110/5mm.	H, 2580
4-chloro-6-methyl-2-methylthio-	147/32mm.	H, 1351, 2580
2-chloro-4-methyl-6-phenylthio-	94	2742
2-chloro-4-methyl-6-propylthio-	103-105/2mm.	2580
4-chloro-6-methyl-2-propylthio-	106-107/0.5mm.	2580
2-chloro-4-methylthio-	66-68	H, 2581
4-chloro-6-methylthio-	51 or 52-54	2158, 2749
4-chloro-2-methylthio-6-phenyl-	62	2753
4-chloro-2-methylthio-6-propyl-	153-154/17mm.	2753
4,6-dichloro-2-ethylthio-5-fluoro-	103-106/4mm.	2600
4,6-dichloro-5-fluoro-2-methylthio-	56	2600
2,4-dichloro-5-isopropylthiomethyl-	100-110/0.001mm.	2596
4,6-dichloro-2-methylthio-	40 to 43	H, 2165, 2586, 2589
2,4-dichloro-5-methylthiomethyl-	80-100/0.001mm.	2596

Pyrimidine	M.P.(^{o}C)	References
4,6-dichloro-2-methylthio-5-phenyl-	108-110	2255
2-ethylthio-5-fluoro-	92/11mm.	2679
5-fluoro-4-methyl-2-methylthio-	97-100/17mm.	2679
5-fluoro-2-methylthio-	87/13mm.	2679
4-iodo-2-methylthio-	46-47; HI 142	2608
4-iodo-6-methylthio-	50-51	2609
2-mercapto-5-methyl-4-phenyl-6-trifluoromethyl-	140-141	3239
4-mercapto-6-methyl-2-trifluoromethyl-	86-87	2193
4-mercapto-2-trifluoromethyl-	87-90	2193
2-methylthio-4-phenyl-6-trifluoromethyl-	75	3239
4,5,6-trichloro-2-methylthio-	60	2600

TABLE XLIX. Nitro-Oxypyrimidines (*H* 606)

4-allyloxy-2-hydroxy-5-nitro-	230	3479
2-benzyl-4,6-dihydroxy-5-nitro-	254	2688
1-butyl-1,2,3,6-tetrahydro-3,4-dimethyl-5-nitro-2,6-dioxo-	78	1145
1-cyclohexyl-3-ethyl-1,2,3,6-tetrahydro-4-methyl-5-nitro-2,6-dioxo-	133-135	1145
1-cyclohexyl-1,2,3,6-tetrahdyro-3,4-dimethyl-5-nitro-2,6-dioxo-	138	1145

TABLE XLIX

Pyrimidine	M.P.(^{o}C)	References
2,4-diallyloxy-5-nitro-	108-110/8mm.	3479
2,4-diethoxy-5-nitro-	42-44 or 45	H, 749, 887
4,6-diethoxy-5-nitro-	61-62	2164
1,3-diethyl-1,2,3,4-tetrahydro-6-methyl-5-nitro-2,4-dioxo-	85-86	4150
1,2-dihydro-1-isopropyl-5-nitro-2-oxo-	109-110	2762
1,2-dihydro-4-methoxy-1-methyl-5-nitro-2-oxo-	151-152	2897
1,2-dihydro-1-methyl-5-nitro-2-oxo-	170-171	3483
2,4-dihydroxy-6-isopropyl-5-nitro-	>230	2438
4,6-dihydroxy-2-methoxy-5-nitro-	185	3909
2,4-dihydroxy-6-methyl-5-nitro-	290	4082, 4083, 4299
2,4-dihydroxy-5-nitro-	295 to 311	H, 749, 824, 2263, 2572, 4239
4,6-dihydroxy-5-nitro-	>300	H, 2332, 2333, 3244, 3871
2,4-dihydroxy-6-phenylazo-	243-244	4300
2,4-dihydroxy-6-phenyl-5-phenylazo-	218-225	3956
2,4-dimethoxy-6-methyl-5-nitro-	83	H, 4299
4,6-dimethoxy-2-methyl-5-nitro-	116-117	H, 887
2,4-dimethoxy-5-nitro-	94-96	H, 994
2-ethoxy-4-hydroxy-5-nitro-	165-166	824
4-ethoxy-2-hydroxy-5-nitro-	198-199	824
2-ethoxy-5-nitro-	52-54	2762
1-ethyl-1,2-dihydro-5-nitro-2-oxo-	111-113	2762
2-ethyl-4,6-dihydroxy-5-nitro-	>260	4035

Pyrimidine	M.P.(°C)	References
4-hydroxy-6-methoxy-5-nitro-	242-243	3054
2-hydroxy-5-nitro-	201-202	H, 3483
4-hydroxy-5-nitro-	192	2322
4-hydroxy-5-nitro-2-phenyl-	278	2322
2-isopropoxy-5-nitro-	45-47	2762
4-methoxy-2,6-dimethyl-5-nitro-	42-43	887
2-methoxy-5-nitro-	69-70	2431, 2746
4-methoxy-5-nitro-	39-40	2562
5-nitro-2,4-diphenoxy-	105 or 108-110	748, 994
1,2,3,4-tetrahydro-1,3-dimethyl-5-nitro-2,4-dioxo-	159-161	H, 824, 2263
1,2,3,4-tetrahydro-6-hydroxy-1,3-dimethyl-5-nitro-2,4-dioxo-	147-148	H, 2445, 4301
1,2,3,4-tetrahydro-6-hydroxy-1,3-dimethyl-5-nitroso-2,4-dioxo-	145-146	H, 4301
1,2,3,4-tetrahydro-1,3,6-trimethyl-5-nitro-2,4-dioxo-	150 or 153-154	H, 1145, 4150
2,4,6-trihydroxy-5-phenylazo-	288-289	2853
Uracil/3-benzyl-5-nitro-	235	2263
Uracil/3-butyl-5-nitro-	162	1464
Uracil/1-methyl-5-nitro-	-	H, 3497
Uracil/3-methyl-5-nitro-	-	H, 2263, 3497
Uracil/5-nitro-3-phenyl-	299-301	2263

TABLE L. Nitro-Sulphonylpyrimidines (*H* 608)

No entries

TABLE LI. Nitro-Thiopyrimidines (*H* 608)

Pyrimidines	M.P.(^{o}C)	References
2,4-dimercapto-5-nitro-	213	H, 748
2-mercapto-5-nitro-	217	2746
2-methylthio-5-nitro-	84-85	H, 2746

TABLE LII. Oxy-Sulphonylpyrimidines (*H* 608)

Pyrimidines	M.P.(^{o}C)	References
5-amylsulphinyl-2,4-dihydroxy-	207	3002
5-amylsulphonyl-2,4-dihydroxy-	320	3002
5-azidosulphonyl-2,4-dihydroxy-6-methyl-	193	4302
4-benzylsulphonyl-2,6-dimethoxy-	98	2774
2-benzylsulphonyl-5-methoxy-	110-112	4303
5-butoxysulphonyl-2,4-dihydroxy-6-methyl-	183-185	4302
5-chlorosulphonyl-2,4-dihydroxy-	>300	H, 1136, 4190
5-chlorosulphonyl-2,4-dihydroxy-6-methyl-	255-260	4302
2,4-dihydroxy-5-methoxysulphonyl-6-methyl-	204-205	4302
2,4-dihydroxy-6-methyl-5-propoxysulphonyl-	188-190	4302
2,4-dihydroxy-5-methylsulphinyl-	228	3002
2,4-dihydroxy-6-methyl-5-sulpho-	150-151; Na 260	3863

TABLE LII.

Pyrimidine	M.P.(^{o}C)	References
2,4-dihydroxy-5-methylsulphonyl-	342	3002
2,4-dihydroxy-6-methylsulphonyl-	-	H, 3497
2,4-dihydroxy-6-phenylsulphonyl-	278-280	4138
2,4-dihydroxy-5-sulphamoyl-	≮300	H, 1136
2,4-dihydroxy-5-sulpho-	Na -	1136
2,4-dimethoxy-6-phenylsulphonyl-	158-159	2771, 4169
4,5-dimethoxy-6-phenylsulphonyl-	96-100	3910, 4285
5-dimethylsulphamoyl-2,4-dihydroxy-	-	H, 4190
4-ethoxy-5-ethyl-2-ethylsulphonyl-6-methyl-	212-215/10mm.	937
4-ethoxy-2-ethylsulphonyl-5-methyl-	-	H, 805, 936
5-ethoxysulphonyl-2,4-dihydroxy-6-methyl-	176	4302
5-ethyl-2-ethylsulphonyl-4-methoxy-6-methyl-	45-47	937
5-ethylsulphinyl-2,4-dihydroxy-	222	3002
5-ethylsulphonyl-2,4-dihydroxy-	284	H, 3002
4-ethylsulphonyl-2,6-dimethoxy-	108-111	2771, 4169
4-ethylsulphonyl-5,6-dimethoxy-	87-88	3910, 4285
2-ethylsulphonyl-5-methoxy-	75-76	4303
2-ethylsulphonyl-4-methoxy-5,6-dimethyl-	65-66	140
2-ethylsulphonyl-4-methoxy-5-methyl-	67-68	805, 936
2-ethylsulphonyl-4-methoxy-6-methyl-5-propyl-	232/7mm.	H, 927

TABLE LII

Pyrimidine	M.P.(^oC)	References
5-hydrazinosulphonyl-2,4-dihydroxy-6-methyl-	221-222	4302
4-hydroxy-5-methoxy-2-methylsulphonyl-	220-222	4244
4-hydroxy-5-methoxy-6-methylsulphonyl-	232-233	2205
4-hydroxy-6-methyl-2-sulpho-	K 300	2999
4-hydroxy-2-methyl-5-sulphomethyl-	325	H, 4167
4-hydroxy-6-methylsulphonyl-	223-224	2749
2-hydroxy-5-sulpho-	>300	1474
4-methoxy-2-methyl-6-sulpho-	Na 279-280	3992
4-methoxy-6-methylsulphonyl-	98-99	2682
5-methoxy-2-methylsulphonyl-	115 or 118-119	2198, 4303

TABLE LIII. Oxy-Thiopyrimidines (*H* 609)

5-β-acetoxyethyl-2-hydroxy-4-mercapto-	216-219	2202
5-acetylthio-2,4-bistrimethylsilyloxy-	53-55	4304
5-acetylthio-2,4-dihydroxy-	254-255	4304
5-acetylthiomethyl-2,4-dihydroxy-	244-246	2650
5-acetylthio-1,2,3,4-tetrahydro-6-hydroxy-1,3-dimethyl-2,4-dioxo-	157-161	2304
1-allyl-2-benzylthio-1,4-dihydro-4-oxo-	64-66	2555, 4305

Pyrimidine	M.P.(^{o}C)	References
1-allyl-2-benzylthio-1,6-dihydro-6-oxo-	42-43	2555, 4305
5-allyl-2-benzylthio-4-hydroxy-	158-159	2194, 4305
5-allyl-1,6-dihydro-1,4-dimethyl-2-methylthio-6-oxo-	86	1146, 3974, 4306
1-allyl-1,4-dihydro-2-methylthio-4-oxo-	95-97	2555
1-allyl-1,6-dihydro-2-methylthio-6-oxo-	52-55	2555
5-allyl-4-hydroxy-2-methylthio-	152-154	2194, 4305
4-allyloxy-2-benzylthio-	154/0.8mm.	2194, 2555
4-allyloxy-2-methylthio-	88/0.5mm.	2194, 2555
5-amyl-1,6-dihydro-1,4-dimethyl-2-methylthio-6-oxo-	56	1146,3974
5-amyl-4-hydroxy-6-methyl-2-methylthio-	154	1146
5-amyloxy-4-hydroxy-2-mercapto-	-	3900 (?)
5-amylthio-2,4-dihydroxy-	255	3002
1-benzyl-1,6-dihydro-4-methyl-2-methylthio-	107-108	2985
1-benzyl-2-ethylthio-1,4-dihydro-5-methyl-4-oxo-	123	H, 4307
5-benzyloxy-1,6-dihydro-1-methyl-2-methylthio-6-oxo-	93-94	2585
1-benzyloxy-1,4-dihydro-2-methylthio-4-oxo-	149	2964
5-benzyloxy-2-ethylthio-4,6-dihydroxy-	164-165	135
5-benzyloxy-2-ethylthio-4-hydroxy-	189-190	79
5-benzyloxy-4-hydroxy-2-mercapto-	226 or 230-232	79, 2161

TABLE LIII

Pyrimidine	M.P.(^{o}C)	References
5-benzyloxy-4-hydroxy-2-methylthio-	180-181	2585
5-benzyloxy-4-mercapto-	156-158	2161
5-benzyloxy-2-methylthio-	69-70	2585
2-benzylthio-4-β-butenyloxy-	140/0.04mm.	4305
4-benzylthio-2,6-dihydroxy-	242	2165
4-benzylthio-2,6-dimethoxy-	141/0.1mm.	2774
2-benzylthio-4-hydroxy-	194 or 196-199	H, 981, 2703
4-benzylthio-6-hydroxy-	238-239	2165
2-benzylthio-4-hydroxy-5-methyl-	204-205	H, 2703
2-benzylthio-4-hydroxy-6-methyl-	183-184	H, 2147
2-benzylthio-4-hydroxy-5-α-methylallyl-	110-111	4305
4-β-butenyloxy-2-methylthio-	85/1.3mm.	4305
5-butoxy-4-butoxymethyl-6-hydroxy-2-mercapto-	110-111	112
5-s-butoxy-4-s-butoxymethyl-6-hydroxy-2-mercapto-	142-143	112
5-butoxy-4-hydroxy-2-mercapto-	(?)	3900
4-butoxy-6-methyl-2-methylthio-	130-136/9mm.	1351
5-butyl-1,6-dihydro-1,4-dimethyl-2-methylthio-6-oxo-	71	1146, 3974
5-butyl-1,6-dihydro-1-methyl-2-methylthio-6-oxo-4-propyl-	liquid	3974
5-butyl-4-ethyl-1,6-dihydro-1-methyl-2-methylthio-6-oxo-	liquid	3974
5-butyl-4-ethyl-6-hydroxy-2-mercapto-	185	1146

Pyrimidine	M.P.(^{o}C)	References
2-butyl-4-hydroxy-6-mercapto-	230-240	2242
5-butyl-4-hydroxy-6-mercapto-	207-209	2242
5-butyl-4-hydroxy-2-mercapto-6-propyl-	147-153	1146
5-butyl-4-hydroxy-6-methyl-2-methylthio-	159-160	H, 1146, 3867
2-butyl-4-hydroxy-6-methylthio-	126	2242
5-butyl-4-hydroxy-6-methylthio-	219-220	2242
2-butylthio-4-hydroxy-6-methyl-	-	4308 (?)
4-cyclopropyl-6-hydroxy-2-mercapto-	236-238	3816
4-cyclopropyl-6-hydroxy-2-methylthio-	196-198	3816
5,5-diethyl-1,4,5,6-tetrahydro-1,2-dimethyl-4-oxo-6-thio-	H_2O 142-143	2242
5,5-diethyl-1,4,5,6-tetrahydro-1,2-dimethyl-6-oxo-4-thio-	151-152	2242
1,3-diethyl-1,2,3,4-tetrahydro-4-oxo-2-thio-	66	H, 2352
1,4-dihydro-1,6-dimethyl-2-methylthio-4-oxo-	207	1351
1,6-dihydro-1,4-dimethyl-2-methylthio-6-oxo-	94	H, 813, 1351
1,6-dihydro-1,4-dimethyl-2-methylthio-6-oxo-5-propyl-	85-86	1146, 3974
1,6-dihydro-4-hydroxy-1-methyl-2-methylthio-6-oxo- (or tautomer)	195-197	2439
1,6-dihydro-5-isoamyl-1,4-dimethyl-2-methylthio-6-oxo-	164-166/9mm.	1146, 4306

TABLE LIII

Pyrimidine	M.P.(^{o}C)	References
1,2-dihydro-1-isopropyl-4-methylthio-2-oxo-	143-145	2762
1,6-dihydro-5-methoxy-4-methoxymethyl-1-methyl-2-methylthio-6-oxo-	46-47	2198
1,4-dihydro-5-methoxy-1-methyl-2-methylthio-4-oxo-	180-181	2198
1,6-dihydro-5-methoxy-1-methyl-2-methylthio-6-oxo-	140-141	2198
1,6-dihydro-4-methoxy-1-methyl-6-thio-	138-139	2760
1,2-dihydro-1-methyl-4-methylthio-2-oxo-	123-124	H, 4309
1,4-dihydro-1-methyl-6-methylthio-4-oxo-	196-197; pic.191	2760
1,6-dihydro-1-methyl-4-methylthio-6-oxo-	174-175	827, 2760
1,6-dihydro-4-methyl-2-methylthio-6-oxo-1-propyl-	crude	1351
1,2-dihydro-4-methylthio-2-oxo-1-propyl-	101-103	2762
1,6-dihydro-1,4,5-trimethyl-2-methylthio-6-oxo-	85-86	1146, 3974
4,5-dihydroxy-6-hydroxymethyl-2-mercapto-	268-269	3476
4,6-dihydroxy-5-isobutyl-2-isopropylthio-	390-392	4310
4,6-dihydroxy-5-isopropyl-2-isopropylthio-	271-272	4310
2,4-dihydroxy-5-mercapto-	-	H, 4190
2,4-dihydroxy-6-mercapto-	245	2165
4,6-dihydroxy-2-mercapto-	ca 245 or >360	2165, 2519
2,4-dihydroxy-5-β-mercaptoethyl-6-methyl-	310	2908

Pyrimidine	M.P.(^{o}C)	References
2,4-dihydroxy-5-mercaptomethyl-	272-274	2650
4,5-dihydroxy-2-mercapto-6-methyl-	>310	112
2,4-dihydroxy-5-γ-mercaptopropyl-6-methyl-	283-285	2908
2,4-dihydroxy-5-methylthio-	300	3002
4,6-dihydroxy-2-methylthio-	>360	H, 2165, 2454
2,4-dihydroxy-5-methylthiomethyl-	258-260	2650
4,6-dihydroxy-2-methylthio-5-phenyl-	299-301	2255
2,4-dihydroxy-5-propylthio-	253	3002
4,6-dihydroxy-2-selenyl-	193-210	2131
2,4-dimethoxy-6-phenylthio-	65-69	2771, 4169
4,5-dimethoxy-6-phenylthio-	160-170/0.001mm.	3910, 4285
5-ethoxy-4-ethoxymethyl-6-hydroxy-2-mercapto-	174-176	H, 112
4-ethoxy-5-ethyl-2-ethylthio-6-methyl-	139-140/5mm.	937
4-ethoxy-2-ethylthio-5-methyl-	-	H, 928
4-ethoxy-2-ethylthio-6-methyl-	105-106/1mm.	H, 1276
4-ethoxy-2-ethylthio-6-methyl-5-propyl-	147/4mm.	H, 927
5-ethoxy-4-hydroxy-2-mercapto-	-	3900
4-ethoxy-6-methyl-2-methylthio-	40	H, 1351, 2623
2-ethoxy-4-methyl-6-phenylthio-	190/1.5mm.	2742
2-ethoxy-4-methylthio-	liquid	2762
4-ethoxy-1,2,3,6-tetrahydro-1,3-diphenyl-2,6-dithio-	250	4311

TABLE LIII

Pyrimidine	M.P.(°C)	References
5-ethyl-1,6-dihydro-1,4-dimethyl-2-methylthio-6-oxo-	114-116	1146, 3974
1-ethyl-1,6-dihydro-4-methyl-2-methylthio-6-oxo-	54-55	1351
1-ethyl-1,2-dihydro-4-methylthio-2-oxo-	227	2762
1-ethyl-2-ethylthio-1,6-dihydro-4-methyl-6-oxo-	106/1mm.	1276
5-ethyl-2-ethylthio-4-hydroxy-6-methyl-	141	H, 937
5-ethyl-2-ethylthio-4-methoxy-6-methyl-	162/22mm.	937
4-ethyl-2-hydroxy-6-mercapto-	223	2753
4-ethyl-6-hydroxy-2-mercapto-	-	H, 2226 (?)
5-ethyl-4-hydroxy-6-methyl-2-methylthio-	201-202	3867
4-ethyl-6-hydroxy-2-methylthio-	157	H, 2753
4-ethyl-1,2,3,6-tetrahydro-5-methyl-2-oxo-1,3-diphenyl-6-thio-	207	2339
1-ethyl-1,2,3,4-tetrahydro-3-methyl-4-oxo-2-thio-	56	2352
1-ethyl-1,2,3,6-tetrahydro-3-methyl-6-oxo-2-thio-	73	2352
5-ethyl-1,2,3,4-tetrahydro-2-oxo-1,3-diphenyl-6-propyl-4-thio-	194-196	2339
2-ethylthio-1,4-dihydro-1,5-dimethyl-4-oxo-	154-155	3954
2-ethylthio-1,6-dihydro-1,5-dimethyl-6-oxo-	60-62	3954
5-ethylthio-2,4-dihydroxy-	258	3002
4-ethylthio-2,6-dimethoxy-	145/12mm.	2771

Pyrimidine	M.P.(^{o}C)	References
4-ethylthio-5,6-dimethoxy-	84/0.05mm.	3910, 4285
2-ethylthio-4-hydroxy-5,6-dimethyl-	-	H, 140
2-ethylthio-4-hydroxy-5-β-hydroxyethyl-6-methyl-	156-157	587
2-ethylthio-4-hydroxy-5-methyl-	159-161	H, 2582
2-ethylthio-4-hydroxy-6-methyl-	143-145	H, 1114, 4312
2-ethylthio-4-hydroxy-6-methyl-5-propyl-	93	H, 927
2-ethylthio-4-methoxy-5,6-dimethyl-	138/5mm.	140
2-ethylthio-4-methoxy-5-methyl-	145-147/17mm.	805, 3954
2-ethylthio-4-methoxy-6-methyl-5-propyl-	147/4mm.	H, 927
5-ethylthiomethyl-2,4-dihydroxy-	244-246	2650
5-hexyl-4-hydroxy-2-mercapto-	171-172	H, 2204
5-hexyl-4-hydroxy-2-mercapto-6-methyl-	183 or 190-192	2204, 2225
2-hydroxy-4,6-bismethylthio-	212-213	2601
4-hydroxy-2,6-bismethylthio-	197	2165
2-hydroxy-4,6-dimercapto-	266-267	2165
4-hydroxy-2,6-dimercapto-	262-264	2165
4-hydroxy-5,6-dimethyl-2-methylthio-	216-217	H, 3867
4-hydroxy-5-2'-hydroxycyclopentyl-2-mercapto-	303-304	2952
4-hydroxy-5-2'-hydroxycyclopentyl-2-methylthio-	180-181	2952
4-hydroxy-5-β-hydroxyethyl-2-mercapto-	248-249	2514

Pyrimidine	M.P.(°)	References
4-hydroxy-5-β-hydroxyethyl-2-mercapto-6-methyl-	258 or 260-265	587, 3892
4-hydroxy-5-β-hydroxyethyl-6-methyl-2-methylthio-	198-199	587
4-hydroxy-5-β-hydroxyethyl-2-methylthio-	184-185	2202
4-hydroxy-6-hydroxymethyl-2-methylthio-	219-220	957
4-hydroxy-5-isoamyl-6-methyl-2-methylthio-	126	1146
4-hydroxy-5-isobutoxy-6-isobutoxymethyl-2-mercapto-	158-160	112
4-hydroxy-5-isobutoxy-2-mercapto-6-methyl-	261-262	112
4-hydroxy-5-isobutyl-6-isopropoxy-2-isopropylthio- (?)	130-131	4310
4-hydroxy-5-isobutyl-6-isopropoxy-2-mercapto-	160-161	4310
4-hydroxy-5-isobutyl-3-isopropyl-2-isopropylthio-6-oxo- (?)	111-112	4310
4-hydroxy-5-isobutyl-6-methyl-2-methylthio-	135-136	1146
4-hydroxy-5-isopropoxy-6-isopropoxymethyl-2-mercapto-	205-206	112
4-hydroxy-6-isopropoxy-5-isopropyl-2-isopropylthio- (?)	177-178	4310
4-hydroxy-6-isopropoxy-5-isopropyl-2-mercapto- (?)	196-197	4310
4-hydroxy-5-isopropoxy-2-mercapto-	-	3900

Pyrimidine	M.P.(^{o}C)	References
4-hydroxy-5-isopropyl-6-mercapto-	272-273	2242
4-hydroxy-6-isopropyl-2-mercapto-	-	H, 2226 (?)
2-hydroxy-4-mercapto-	290 or 300	H, 1108, 2165
4-hydroxy-2-mercapto-	300 or 310-312	H, 981, 2165, 2352, 4313
4-hydroxy-6-mercapto-	242 or 247	1379, 2165, 2242, 2760
4-hydroxy-2-mercapto-5,6-dimethyl-	279-280	H, 140
5-hydroxy-2-mercapto-4,6-dimethyl-	>230	2161
4-hydroxy-5-β-mercaptoethyl-2,6-dimethyl-	276	2908
4-hydroxy-2-mercapto-5-β-mercaptoethyl-6-methyl-	224-227	2908
4-hydroxy-2-mercapto-5-γ-mercaptopropyl-6-methyl-	262	2908
4-hydroxy-2-mercapto-5-methoxy-	281 or 288-290	H, 2161, 3915
4-hydroxy-2-mercapto-5-methoxy-6-methoxymethyl-	188-190	112, 2198
4-hydroxy-2-mercapto-5-methoxy-6-methyl-	-	3243
4-hydroxy-2-mercaptomethyl-	-	3083
2-hydroxy-4-mercapto-5-methyl-	255-257	H, 2674
4-hydroxy-2-mercapto-5-methyl-	279-280	H, 3935
4-hydroxy-2-mercapto-6-methyl-	330	H, 1436, 2519, 2874
4-hydroxy-6-mercapto-5-methyl-	ca 200	2242
4-hydroxy-2-mercapto-5-methyl-6-phenyl-	-	H, 2241
4-hydroxy-2-mercapto-6-methyl-5-propyl-	210	H, 927

Pyrimidine	M.P.(^{o}C)	References
4-hydroxy-2-mercapto-5-phenoxy-	-	H, 3900
2-hydroxy-4-mercapto-6-phenyl-	310-313	2753
4-hydroxy-2-mercapto-6-phenyl-	270-273	H, 4232
4-hydroxy-2-mercapto-5-propoxy-	-	3900
4-hydroxy-2-mercapto-5-propoxy-6-propoxymethyl-	136-137	112
2-hydroxy-4-mercapto-6-propyl-	193	2753
4-hydroxy-2-mercapto-6-propyl-	-	H, 3220
4-hydroxy-5-γ-mercaptopropyl-2,6-dimethyl-	215-218	2908
4-hydroxy-5-methoxy-6-methoxymethyl-2-methylthio-	153-154	2198
5-hydroxy-4-methoxymethyl-2-methylthio-	190-191	4314
4-hydroxy-2-methoxy-5-methylthio-	170-171	3002
4-hydroxy-5-methoxy-2-methylthio-	196-198	2198, 2586, 4315
4-hydroxy-6-methoxy-2-methylthio-	191	2589
4-hydroxy-5-α-methylallyl-2-methylthio-	131-132	4305
5-hydroxymethyl-4-mercapto-2-methyl-	193-195	2651
2-hydroxy-5-methyl-4-methylthio-	214-215	H, 2581
4-hydroxy-5-methyl-2-methylthio-	232-234	H, 813
4-hydroxy-5-methyl-6-methylthio-	236-237	2242

Pyrimidine	M.P.(°C)	References
4-hydroxy-6-methyl-2-methylthio-	218-222	H, 399, 3867
4-hydroxy-6-methyl-2-methylthio-5-propyl-	181-182	H, 3867
2-hydroxy-4-methylthio-	Na 290	H, 2762
4-hydroxy-6-methylthio-	230 or 233-234	1379, 2165, 2760
5-hydroxy-2-methylthio-	169 or 177	2585, 2586
4-hydroxy-2-methylthio-6-phenyl-	238 or 254	H, 2753, 3867
4-hydroxy-2-methylthio-6-propyl-	158	H, 2753
4-hydroxy-2-selenyl-	-	H, 2131
2-isopropoxy-4-methylthio-	liquid	2762
4-mercapto-5-methoxy-	-	H, 2161
4-mercapto-6-methoxy-	193-194	2760
4-mercapto-6-methoxy-5-methyl-	220-222	2242
4-mercapto-5-methoxy-2-phenyl-	198-199	2161
4-methoxy-6-methyl-2-methylthio-	35	H, 813, 1351
2-methoxy-4-methylthio-	liquid	2762
4-methoxy-6-methylthio-	60/0.3mm.; pic.117-118	2609, 2682, 2760
5-methoxy-2-methylthio-	69	2586
5-methoxy-4-methylthio-	75	2161
4-methyl-2-methylthio-6-propoxy-	130-137/9mm.	1351
4-methylthio-2-propoxy-	liquid	2762
1,2,3,4-tetrahydro-1,3-dimethyl-2-oxo-4-thio-	131-132	H, 827
1,2,3,4-tetrahydro-1,3-dimethyl-4-oxo-2-thio-	109	2352
1,2,3,4-tetrahydro-6-hydroxy-1,3-dimethyl-4-oxo-2-thio-	183	4316

Pyrimidine	M.P.(°C)	References
1,2,3,4-tetrahydro-6-isopropyl-2-oxo-1,3-diphenyl-4-thio-	206-207	2339
1,2,3,4-tetrahydro-5-methyl-2-oxo-1,3-diphenyl-6-propyl-4-thio-	158	2339
1,2,3,4-tetrahydro-2-oxo-1,3,6-triphenyl-4-thio-	190	2339
2-Thiobarbituric acid/5-benzylidene-	-	4317
2-Thiobarbituric acid/1-methyl-	202	H, 4318
2-Thiouracil/1-allyl-	197-198	4305
2-Thiouracil/1-benzyl-	233	H, 2352
2-Thiouracil/1-benzyloxy-	226	2964
2-Thiouracil/1,5-dimethyl-	227 or 230-232	H, 372, 3971
2-Thiouracil/3,5-dimethyl-	207-208	3971
2-Thiouracil/3,6-dimethyl-	265	H, 1276
2-Thiouracil/1-ethyl-	239	H, 2352
2-Thiouracil/3-ethyl-	177	H, 2352
2-Thiouracil/3-ethyl-6-methyl-	202-203	1276
2-Thiouracil/1-hydroxy-	203 and 277	2964
2-Thiouracil/5-β-hydroxyethyl-3-isopropyl-6-methyl-	-	3975
2-Thiouracil/1-β-hydroxyethyl-5-methyl-	204	743
2-Thiouracil/1-β-hydroxypropyl-5-methyl-	191	743
2-Thiouracil/3-isopropyl-5,6-dimethyl-	-	3982
2-Thiouracil/1-methyl-	226-228	372, 2311, 2352
2-Thiouracil/3-methyl-	207 or 283	2195, 2352
2-Thiouracil/5-methyl-1-phenyl-	202-203	372

Pyrimidine	M.P.(°C)	References
2-Thiouracil/5-methyl-1-propyl-	200	743
2-Thiouracil/1-phenyl-	236	372
4-Thiouracil/1-methyl-	193-194	1174
4-Thiouracil/3-methyl-	183-184	2641
4-Thio[*pseudo*]uracil/5,5-diethyl-2-methyl-	192-195	2242
4-Thio[*pseudo*]uracil/1-methyl-	209-211	2760
4-Thio[*pseudo*]uracil/3-methyl-	209-210	2760
Uracil/3-isopropyl-6-methyl-5-methylthiomethyl-	-	3982
Uracil/1-β-mercaptopropyl-5-methyl-	158	743
[*pseudo*]Uracil/1-methyl-2-methylthio-	195-197	2440

TABLE LIV. Oxy-Thiopyrimidines with Other Functional Groups (*H* 616)

Pyrimidine	M.P.(°C)	References
5-acetyl-4,6-dihydroxy-2-mercapto-	285	2245
4-acetyl-6-hydroxy-2-mercapto-	274-276	2218
5-acetyl-2-hydroxy-4-mercapto-6-methyl-	262	2314
2-amylthio-4-ethoxycarbonyl-5-fluoro-6-hydroxy-	93-94	2517
1-benzoyl-2-benzoylthio-1,4(or 1,6)-dihydro-5-methoxy-4(or 6)-oxo-	101-102	4244

Pyrimidine	M.P.(^{o}C)	References
1-benzoyl-1,4(or 1,6)-dihydro-5-methoxy-2-methylthio-4(or 6)-oxo-	106-107	4244
5-benzoyl-4-ethoxy-6-mercapto-2-phenyl-	221	2320
5-benzyloxy-4-chloro-2-methylthio-	80-81	2585
2-benzylthio-4-carboxy-6-hydroxy-	273-274	H, 2214
2-benzylthio-4-chloro-6-hydroxy-	207-208	2589
2-benzylthio-4-ethoxycarbonyl-5-fluoro-6-hydroxy-	155-158	2517
2-benzylthio-5-fluoro-4-hydroxy-	216-218	4197, 4222
5-bromo-1,2-dihydro-6-methoxy-1,4-dimethyl-2-thio- (or isomer)	237-238	2575
5-bromo-4,6-dihydroxy-2-methylthio-	>360	2255
5-bromo-4-ethoxy-6-methyl-2-methylthio-	122-125/3mm.	2623
5-bromo-2-ethylthio-4-hydroxy-	-	2623
5-bromo-4-hydroxy-2-mercapto-6-methyl-	256	H, 2147
5-bromo-4-hydroxy-6-methyl-2-methylthio-	246	H, 2842
5-bromo-4-hydroxy-2-methylthio-	252	H, 2842
5-bromo-4-methoxy-6-methyl-2-methylthio-	70-71	2575
4-butyl-5-fluoro-6-hydroxy-2-mercapto-	258-259	4283
2-butylthio-4-carboxy-6-hydroxy-	227-228	2214
2-butylthio-4-ethoxycarbonyl-5-fluoro-6-hydroxy-	113-115	2517

Pyrimidine	M.P.(^{o}C)	References
5-carbamoylmethyl-4-hydroxy-2-mercapto-	277-280	3876
5-β-carboxyethyl-4-hydroxy-2-mercapto-6-methyl-	-	4179
4-carboxy-2-ethylthio-5-fluoro-6-hydroxy-	168-169	H, 4197
4-carboxy-2-ethylthio-6-hydroxy-	247-248	2214
5-carboxy-2-ethylthio-4-hydroxy-	-	H, 3456
4-carboxy-5-fluoro-6-hydroxy-2-mercapto-	-	4197
4-carboxy-5-fluoro-6-hydroxy-2-methylthio-	199-201	4197, 4222
4-carboxy-2-hydroxy-6-mercapto-	307-308	2214
4-carboxy-6-hydroxy-2-mercapto-	320-321	H, 2216, 2348
4-carboxy-2-hydroxy-6-methylthio-	228-230	2214
4-carboxy-6-hydroxy-2-methylthio-	253-254	H, 957, 2214
4-carboxy-6-hydroxy-2-propylthio-	236-237	2214
4-carboxy-6-methoxy-2-methylthio-	233-234	2214
4-carboxymethyl-6-hydroxy-2-mercapto-	-	4232
5-β-chlorocarbonylethyl-4-hydroxy-2-mercapto-6-methyl-	-	4179
5-chlorocarbonyl-4-hydroxy-2-mercapto-	crude	3437
4-chloro-1,6-dihydro-1-methyl-2-methylthio-6-oxo-	111-112	2440
5-chloro-4,6-dihydroxy-2-methylthio-	255-259	4319

Pyrimidine	M.P.(^{o}C)	References
5-chloro-2,4-dimethoxy-6-methylthio-	92-94	4188
4-chloro-6-hydroxy-2-isopropylthio-	159	2587
5-chloro-4-hydroxy-6-methyl-2-methylthio-	269-270	H, 2160
4-chloro-6-hydroxy-2-methylthio-	208 or 215-216	2165, 2589, 2777
4-chloro-5-methoxy-2-methylthio-	73 to 82	2205, 2584, 2586, 4315
4-chloro-6-methoxy-2-methylthio-	39-40	4320
4-chloromethyl-6-hydroxy-2-methylthio-	180-181	H, 2867
1-β-cyanoethyl-1,6-dihydro-2-methylthio-6-oxo-	-	3363
5-β-cyanoethyl-4-hydroxy-2-mercapto-6-methyl-	-	4179
4-cyano-2-ethylthio-6-hydroxy-	210-211	2611
5-cyano-4-hydroxy-2-mercapto-	265-272	H, 1609
5-cyano-2-hydroxy-4-mercapto-6-methyl-	245-250	2319, 4274
5-cyano-4-hydroxy-2-methylthio-	220-222	1441, 4003
4-dichloromethyl-6-hydroxy-2-mercapto-	198-199	2518
4-dichloromethyl-6-hydroxy-2-methylthio-	155-156	2518
5-ββ-diethoxyethyl-4,5-dihydro-6-hydroxy-2-mercapto-5-methyl-4-oxo-	94-96	2011
4-αα-diethoxyethyl-6-hydroxy-2-mercapto-	144-147	2218
4-diethoxymethyl-6-hydroxy-2-mercapto-	-	H, 752
4-diethoxymethyl-6-hydroxy-2-methylthio-	-	H, 957

Pyrimidine	M.P.(^{o}C)	References
4,6-dihydroxy-2-methylthio-5-nitro-	223-224 or 245	2454, 4319
4-dimethoxymethyl-5-ethyl-6-hydroxy-2-mercapto-	135-136	2219
4-dimethoxymethyl-6-hydroxy-2-mercapto-	180-181	2219
4-dimethoxymethyl-6-hydroxy-2-mercapto-5-methyl-	155-156	2219
4-ethoxycarbonyl-2-ethylthio-5-fluoro-6-hydroxy-	164 or 168-169	2517, 4222
4-ethoxycarbonyl-5-fluoro-2-hexylthio-6-hydroxy-	78-80	2517
4-ethoxycarbonyl-5-fluoro-6-hydroxy-2-isobutylthio-	113-115	2517
4-ethoxycarbonyl-5-fluoro-6-hydroxy-2-isopropylthio-	155-158	2517
4-ethoxycarbonyl-5-fluoro-6-hydroxy-2-methylthio-	183-184	4321
4-ethoxycarbonyl-5-fluoro-6-hydroxy-2-propylthio-	154-156	2517
5-ethoxycarbonyl-4-hydroxy-2-mercapto-	243-244	H, 3437
5-ethoxycarbonyl-2-hydroxy-4-mercapto-6-methyl-	251	4274
5-ethoxycarbonyl-4-hydroxy-2-methylthio-	-	H, 2598, 4217
5-ethoxycarbonylmethyl-4-hydroxy-2-mercapto-	179-180	H, 2577
5-ethoxycarbonylmethyl-4-hydroxy-2-methylthio-	188-189	4015
4-ethoxy-5-ethoxycarbonyl-2-hydroxy-6-mercapto-	240	2320
4-ethoxy-5-ethoxycarbonyl-6-mercapto-2-phenyl-	163	2320
2-ethoxy-4-mercapto-5-nitro-	133	749
5-ethyl-4-formyl-6-hydroxy-2-mercapto-	148-150	2219

Pyrimidine	M.P.(^{o}C)	References
2-ethylthio-5-fluoro-4-hydroxy-	190 or 192-193	4197, 4222
2-ethylthio-5-fluoro-4-hydroxy-6-methyl-	190-192	4196
2-ethylthio-5-fluoro-4-hydroxy-6-pentafluoroethyl-	159-160	4196
2-ethylthio-5-fluoro-4-hydroxy-6-phenyl-	235-240	4196
2-ethylthio-5-fluoro-4-hydroxy-6-trifluoromethyl-	136-138	4196
2-ethylthio-4-formyl-6-hydroxy-	oxime 256	H, 2611
4-ethylthio-6-hydroxy-5-nitro-	214	2240
2-ethylthio-4-hydroxy-6-trifluoromethyl-	176-178	4196
5-fluoro-4,6-dihydroxy-2-mercapto-	Na >300	4319
5-fluoro-4,6-dihydroxy-2-methylthio-	219-220	4319
5-fluoro-4-fluoromethyl-6-hydroxy-2-methylthio-	221-222	77
5-fluoro-4-hydroxy-2-mercapto-	225-226	77, 2191, 4222
5-fluoro-4-hydroxy-2-mercapto-6-methyl-	294-297	4283
5-fluoro-4-hydroxy-6-methyl-2-methylthio-	269-270	2679
5-fluoro-2-hydroxy-4-methylthio-	222-224	2581
5-fluoro-4-hydroxy-2-methylthio-	241-243	2679, 4197, 4222, 4291
4-fluoromethyl-6-hydroxy-2-methylthio-	230-231	2518
5-formyl-4,6-dihydroxy-2-mercapto-	DNP 295	H, 3446

Pyrimidine	M.P.(^{o}C)	References
4-formyl-6-hydroxy-2-mercapto-	251	H, 957, 2219
5-formyl-4-hydroxy-2-mercapto-	DNP 305	3446
4-formyl-6-hydroxy-2-mercapto-5-methyl-	233-234	H, 2219, 3267
5-formyl-4-hydroxy-2-mercapto-6-methyl-	300; *p*-NPH 322	2155, 3446
5-formyl-4-hydroxy-2-mercapto-6-propyl-	DNP 319	3446
5-formyl-4-hydroxy-6-methyl-2-methylthio-	300	2155
4-formyl-6-hydroxy-2-methylthio-	133	957
4-hydroxy-2-mercapto-6-methoxycarbonyl-	255	2216
4-hydroxy-2-mercapto-5-methylcarbamoylmethyl-	310-315	2876
4-hydroxy-2-mercapto-6-methyl-5-methylcarbamoylmethyl-	299-303	2876
4-hydroxy-2-mercapto-6-methyl-5-phenylazo-	-	2227, 2563
4-hydroxy-2-mercapto-5-nitro-	255-257	2263
4-hydroxy-2-mercapto-5-phenylazo-6-trifluoromethyl-	168-172	443, 1242
4-hydroxy-2-mercapto-6-phenyl-5-phenylazo-	215-217	3956
4-hydroxy-6-methoxycarbonyl-2-methylthio-	253-254	2214
4-hydroxy-5-methylcarbamoylmethyl-2-methylthio-	213-216	2876
4-hydroxy-6-methyl-5-methylcarbamoylmethyl-2-methylthio-	266-269	2876

Pyrimidine	M.P.(^{o}C)	References
4-mercapto-6-methoxy-5-nitro-	165-166	2227
4-methoxy-6-methoxycarbonyl-2-methylthio-	100-102	2214
2-Thiobarbituric acid/5-acetyl-1-methyl-	210	2245
2-Thiouracil/3-amyl-6-methyl-5-thiocyanato-	-	3982
2-Thiouracil/5-bromo-6-methyl-3-phenyl-	-	3981 (?)
2-Thiouracil/1-β-carboxyethyl-	192-193	4322
2-Thiouracil/1-carboxymethyl-	265	2352
2-Thiouracil/1-carboxymethyl-5-methyl-	246-247	372
2-Thiouracil/6-carboxy-3-phenyl-	H_2O 226	888
2-Thiouracil/5-chloro-3-cyclohexyl-6-methyl-	-	3982
2-Thiouracil/1-β-cyanoethyl-	231-232	4322
2-Thiouracil/6-dimethoxymethyl-3,5-dimethyl-	93-94	2219
2-Thiouracil/6-dimethoxymethyl-3-ethyl-	103-104	2219
2-Thiouracil/6-dimethoxymethyl-3-ethyl-5-methyl-	98-100	2219
2-Thiouracil/6-dimethoxymethyl-3-methyl-	148-149	2219
2-Thiouracil/6-dimethoxymethyl-5-methyl-3-phenyl-	156-157	2219
2-Thiouracil/4-dimethoxymethyl-5-methyl-3-propyl-	88-90	2219

Pyrimidine	M.P.(°C)	References
2-Thiouracil/6-dimethoxymethyl-3-phenyl-	161-162	2219
2-Thiouracil/6-dimethoxymethyl-3-propyl-	106-107	2219
2-Thiouracil/3-ethyl-6-formyl-	155-157	2219
2-Thiouracil/3-ethyl-6-formyl-5-methyl-	180-182	2219
2-Thiouracil/6-formyl-3,5-dimethyl-	213-215	2219
2-Thiouracil/6-formyl-3-methyl-	215-216	2219
2-Thiouracil/6-formyl-5-methyl-3-phenyl-	225-227	2219
2-Thiouracil/6-formyl-5-methyl-3-propyl-	152-153	2219
2-Thiouracil/6-formyl-3-phenyl-	259-260	2219
2-Thiouracil/6-formyl-3-propyl-	134-135	2219
4-Thiouracil/5-acetyl-1-methyl-	215-216	2319
4-Thiouracil/5-acetyl-6-methyl-1-phenyl-	262	2319
4-Thiouracil/5-cyano-6-methyl-1-phenyl-	265	2319
4-Thiouracil/5-ethoxycarbonyl-1,6-dimethyl-	215	2319
4-Thiouracil/5-ethoxycarbonyl-6-methyl-1-phenyl-	197-199	2319

TABLE LV. Oxypyrimidines with Two Minor Functional Groups (*H* 621)

Pyrimidine	M.P.(^{o}C)	References
1-acetyl-5-bromo-3-butyl-1,2,3,4-tetrahydro-6-methyl-2,4-dioxo-	54-55	3967, 4281
1-acetyl-5-bromo-3-s-butyl-1,2,3,4-tetrahydro-6-methyl-2,4-dioxo-	-	3967 (?)
1-acetyl-5-bromo-3-t-butyl-1,2,3,4-tetrahydro-6-methyl-2,4-dioxo-	-	3967 (?)
Barbituric acid/5-chloro-5-nitro-	-	2635
1-benzoyl-1,4(or 1,6)-dihydro-5-methoxy-2-methylsulphonyl-4(or 6)-oxo-	145-147	4244
5-bromo-4-carboxy-2,6-dihydroxy-	$2H_2O$ 288	H, 1463, 4279
5-bromo-4-carboxy-6-hydroxy-	206-207	447
5-bromo-4-carboxy-1,2,3,6-tetrahydro-1,3-dimethyl-2,6-dioxo-	211	4255
4-butoxycarbonyl-5-fluoro-2,6-dihydroxy-	184	3436
5-carbamoyl-4-chloro-6-methoxy-	145	2519
4-carbamoyl-5-fluoro-6-hydroxy-	$\nless$360	2191
4-carbamoyl-5-fluoro-6-hydroxy-2-methyl-	308-309	2191
4-carboxy-5-chloro-2,6-dihydroxy-	286-288	H, 2591
4-carboxy-5-chloro-6-hydroxy-2-methyl-	185	2191

Pyrimidines	M.P.(^{o}C)	References
5-carboxy-4-chloro-6-methoxy-	141 or 145	2519, 2575
5-carboxy-4-chloro-6-methoxy-2-phenyl-	177-178	2575
4-carboxy-2,6-dihydroxy-5-iodo-	-	2454
4-carboxy-2,6-dihydroxy-5-nitro-	-	H, 2454
4-carboxy-5-fluoro-2,6-dihydroxy-	H_2O 250 to 255	77, 2517, 3436, 4197, 4321, 4323
4-carboxy-5-fluoro-6-hydroxy-	215 or 221	2191, 4282
4-carboxy-5-fluoro-6-hydroxy-2-methyl-	183 or 215	2191, 4282
4-carboxy-2-hydroxy-6-trifluoromethyl-	impure	4196
4-carboxy-6-hydroxy-2-trifluoromethyl-	232-234	2217
2-chloro-5-cyano-4-ethoxy-6-methyl-	134-136	2155
4-chloro-5-cyano-6-hydroxy-	210-215	2758
5-chloro-2,4-dihydroxy-6-methoxycarbonyl-	255-256	2591
5-chloro-4,6-dihydroxy-2-methoxymethyl-	300	3911
4-chloro-2,6-dihydroxy-5-nitro-	220-222	2438
4-chloro-2,6-dihydroxy-5-phenylazo-	230-240	2506
5-chloro-2,4-dimethoxy-6-methylsulphonyl-	180-182	4188
5-chloro-4-ethoxycarbonyl-6-hydroxy-2-methyl-	156	2191
4-chloro-6-hydroxy-5-nitro-	198-199	2240, 2466
2-chloro-4-methoxy-6-methyl-5-nitro-	60-63	2763

TABLE LV

Pyrimidine	M.P.(^{o}C)	References
4-chloro-6-methoxy-2-methyl-5-nitro-	51-52	887
4-chloro-6-methoxy-2-methylsulphonyl-	90-91	4320
4-chloro-6-methoxy-5-nitro-	65-66	2227, 4111
4-chloro-6-methoxy-2-nitroamino-	120-122	2718
4-chloro-1,2,3,6-tetrahydro-1,3-dimethyl-5-nitro-2,6-dioxo-	65-68 or 80-83	2441, 2718
2,4-dichloro-6-methoxy-5-nitro-	48-50	2731
4,6-dichloro-2-methoxy-5-nitro-	73-75	3909
2,4-dihydroxy-6-methoxycarbonyl-5-nitro-	199-200	697
4,6-dihydroxy-5-nitroso-2-trifluoromethyl-	157-159	2217
2,4-dihydroxy-5-nitro-6-trifluoromethyl-	39-41	999
4,6-dihydroxy-5-nitro-2-trifluoromethyl-	148-150	2193
4-dimethoxymethyl-6-hydroxy-2-trifluoromethyl-	93-94	2217
4-ethoxycarbonyl-2,6-dihydroxy-5-nitro-	-	H, 697
4-ethoxycarbonyl-5-fluoro-6-hydroxy-	235	4282
4-ethoxycarbonyl-5-fluoro-6-hydroxy-2-methyl-	289	2191
5-ethoxycarbonyl-4-phenoxy-2-trifluoromethyl-	65-66	1000
5-fluoro-4-hydroxy-6-methoxycarbonyl-2-methyl-	199-201	4282

Pyrimidine	M.P.(^{o}C)	References
Uracil/1-acetyl-5-bromomethyl-	168	4324
Uracil/1-acetyl-5-fluoro-	128-129	4325
Uracil/1-acetyl-5-iodo-	167-168	3894, 4326
Uracil/3-bromo-6-methyl-5-nitro-	170-171	3982
Uracil/1-β-carboxyethyl-5-fluoro-	185	4273
Uracil/1-β-carboxyethyl-5-nitro-	268-270	2336
Uracil/6-carboxy-1-methyl-5-nitro-	-	H, 3968
Uracil/1-γ-carboxypropyl-5-nitro-	178-180	2336
Uracil/6-chloro-3-methyl-5-nitro-	195-197	2439, 2440, 2444
Uracil/1-β-cyanoethyl-5-fluoro-	233-234	4273
Uracil/1-β-ethoxycarbonylethyl-5-nitro-	170-171	2336
Uracil/1-γ-ethoxycarbonylpropyl-5-nitro-	133-134	2336

TABLE LVI. Thiopyrimidines with Two Minor Functional Groups (*H* 623)

2-benzylthio-4-chloro-5-ethoxycarbonyl-	206/1mm.	H, 4015
5-bromo-4-carboxy-2-methylthio-	HCl 136-137	H, 4327
4-carbamoyl-6-chloro-2-methylthio-	204-205	2214
5-carbamoylmethyl-4-chloro-2-methylthio-	168-170	4015

Pyrimidine	M.P.(°C)	References
4-carboxy-5-chloro-2-methylthio-	-	H, 4327
4-chloro-5-cyano-2-methylthio-	67-68	4328
4-chloro-5-ethoxycarbonylmethyl-2-methylthio-	148-149/1mm.	4015
4-chloro-5-ethoxycarbonyl-2-methylthio-	59-61	H, 2598, 2936
2-β-chloroethyl-5-ethoxycarbonyl-4-mercapto-6-methyl-	128	4274
4-chloro-6-methoxycarbonyl-2-methylthio-	118-119	2214
2-chloromethyl-5-ethoxycarbonyl-1,4-dihydro-6-methyl-1-phenyl-4-thio-	130	2313
4,6-dichloro-2-methylthio-5-nitro-	61-63	2454, 2600
5-ethoxycarbonyl-4-iodo-2-methylthio-	72-74	2608
2-ethylthio-5-nitro-4-thiocyanato-	131	749

TABLE LVII. Pyrimidines with Three Minor Functional Groups (*New*)

Pyrimidine	M.P.(°C)	References
4-chloro-6-ββ-diethoxyvinyl-5-nitro-	100	2785
4-chloro-5-ethoxycarbonyl-2-methylsulphonyl-	129-130	3062
2-chloro-5-nitro-4-thiocyanato-	141	749
2,4-dichloro-6-ethoxycarbonyl-5-nitro-	37-38	697
2,4-dichloro-6-methoxycarbonyl-5-nitro-	90-91	697

2170. Taylor and Ehrhart, *J. Amer. Chem. Soc.*, 82, 3138 (1960).

2171. Kobayashi, *J. Pharm. Soc. Japan*, 82, 445 (1962).

2172. Kobayashi and Sakata, *J. Pharm. Soc. Japan*, 82, 447 (1962).

2173. Albert and Barlin, *J. Chem. Soc.*, 1962, 3129.

2174. Saikawa, *J. Pharm. Soc. Japan*, 84, 207 (1964).

2175. Schwan and Tieckelmann, *J. Het. Chem.*, 2, 202 (1965).

2176. Baker and Shapiro, *J. Med. Chem.*, 6, 664 (1963).

2177. Morimoto, *J. Pharm. Soc. Japan*, 82, 462 (1962).

2178. Kurzer and Godfrey, *Angew. Chem.*, 75, 1157 (1963).

2179. Bell and Caldwell, *J. Org. Chem.*, 26, 3534 (1961).

2180. Urban, Chopard-Dit-Jean, and Schnider, *Gazzetta*, 93, 163 (1963).

2181. Sherman and Von Esch, *J. Med. Chem.*, 8, 25 (1965).

2182. Kosolapoff and Roy, *J. Org. Chem.*, 26, 1895 (1961).

2183. Budesinsky and Roubinek, *Coll. Czech. Chem. Comm.*, 29, 2341 (1964).

2184. Budesinsky and Roubinek, *Coll. Czech. Chem. Comm.*, 26, 2871 (1961).

2185. Longo and Mugnaioni, *Boll. chim. farm.*, 100, 430 (1961).

2186. Budesinsky and Musil, *Coll. Czech. Chem. Comm.*, 26, 2865 (1961).

2187. Mariella and Zelko, *J. Org. Chem.*, 25, 647 (1960).

2188. Giuliano and Leonardi, *Farmaco* (Ed. Sci.), 11, 389 (1956); through *Chem. Zentr.*, 1957, 12756.

2189. Takamizawa, *Vitamin*, 30, 195 (1964).

2190. Takamizawa and Hirai, *Chem. Pharm. Bull.*, 12, 393 (1964).

2191. Budesinsky, Jelinek, and Prikryl, *Coll. Czech. Chem. Comm.*, 27, 2550 (1962).

2192. Shvachkin, Syrtsova, and Filatova, *J. Gen. Chem.*, 33, 2487 (1963).

2193. Inoue, Saggiomo, and Nodiff, *J. Org. Chem.*, 26, 4504 (1961).

2194. Minnemeyer, Egger, Holland, and Tieckelmann, *J. Org. Chem.*, 26, 4425 (1961).

2195. Sano, *Chem. Pharm. Bull.*, 10, 313 (1962).

2196. Parkanyi and Sorm, *Coll. Czech. Chem. Comm.*, 28, 2491 (1963).

2197. Shapira, *J. Org. Chem.*, 27, 1918 (1962).

2198. Budesinsky, Prikryl, and Svatek, *Coll. Czech. Chem. Comm.*, 29, 2980 (1964).

2199. Klötzer, *Monatsh.*, 95, 265 (1964).

2200. Klötzer, *Monatsh.*, 95, 1729 (1964).

2201. Parkanyi, *Chem. Listy*, 56, 652 (1962).

2202. Fissekis, Myles, and Brown, *J. Org. Chem.*, 29, 2670 (1964).

2203. Roth and Hitchings, *J. Org. Chem.*, 26, 2770 (1961).

2204. Watanabe, Tsuji, and Toyoshima, *Chem. Pharm. Bull.*, 11, 495 (1963).

2205. Koppel, Springer, Robins, and Cheng, *J. Org. Chem.*, 27, 3614 (1962).

2206. Dinan, Minnemeyer, and Tieckelmann, *J. Org. Chem.*, 28, 1015 (1963).

2207. Morrison and Jeffrey, *Canad. Pharm. J.*, (Sc. Sect.), 95, 276 (1962).

2208. Takagi and Ueda, *Chem. Pharm. Bull.*, 12, 607 (1964).

2209. Biressi, Carissimi, and Ravenna, *Gazzetta*, 93, 1268 (1963).

2210. Chesterfield, Hurst, McOmie, and Tute, *J. Chem. Soc.*, 1964, 1001.

2211. Alexander and Jeffrey, *Canad. Pharm. J.*, (Sc. Sect.), 94, 45 (1961).

2212. Budesinsky, Bydzovsky, Kopecky, Prikryl, and Svab, *Czechoslov. Farm.*, 10, 14 (1961); *Chem. Abstr.*, 55, 25972 (1961).

2213. Roth, Falco, Hitchings, and Bushby, *J. Med. Pharm. Chem.*, 5, 1103 (1962).

2214. Daves, Baiocchi, Robins, and Cheng, *J. Org. Chem.*, 26, 2755 (1961).

2215. Angier and Curran, *J. Org. Chem.*, 26, 1891 (1961).

2216. Chkhikvadze, Britikova, and Magidson, *J. Gen. Chem.*, 34, 161 (1964).

2217. Barone, *J. Med. Chem.*, 6, 39 (1963).

2218. Ross, Acton, Skinner, Goodman, and Baker, *J. Org. Chem.*, 26, 3395 (1961).

2219. Piantadosi, Skulason, Irvin, Powell, and Hall, *J. Med. Chem.*, 7, 337 (1964).

2220. Baker and Morreal, *J. Pharm. Sci.*, 52, 840 (1963).

2221. Stachel, *Chem. Ber.*, 95, 2172 (1962).

2222. Huffman and Schaefer, *J. Org. Chem.*, 28, 1816 (1963).

2223. Tanaka, Sugawa, Nakimori, Sanno, Ando, and Imai, *Chem. Pharm. Bull.*, 12, 1024 (1964).

2224. Miller and Rose, *J. Chem. Soc.*, 1963, 5642.

2225. Levin, Gul'kina, and Kukhtin, *J. Gen. Chem.*, 33, 2673 (1963).

2226. Gershon, Braun, and Scala, *J. Med. Chem.*, 6, 87 (1963).

2227. Taylor, Barton, and Paudler, *J. Org. Chem.*, 26, 4961 (1961).

2228. Naito, Hirata, and Kobayashi, Jap. Pat., 7766 (1965); through *Chem. Abstr.*, 63, 1801 (1965).

2229. Wagle and Kulkarni, *Indian J. Chem.*, 1, 449 (1963).

2230. Biniecki, Gutkowska, and Koslowska, *Acta Polon. Pharm.*, 19, 443 (1962).

2231. Taborska and Biniecki, *Dissertationes Pharmaceuticae*, 15, 389 (1963).

2232. Baker, Santi, Almaula, and Werkheiser, *J. Med. Chem.*, 7, 24 (1964).

2233. Furukawa, *Chem. Pharm. Bull.*, 10, 1215 (1962).

2234. Mamalis, Green, Outred, and Rix, *J. Chem. Soc.*, 1962, 3915.

2235. Wagner, U. S. Pat., 3,177,216 (1965); through *Chem. Abstr.*, 63, 1800 (1965).

2236. Gershon, Dittmer, and Braun, *J. Org. Chem.*, 26, 1874 (1961).

2237. Shvachkin and Sirtsova, *J. Gen. Chem.*, 33, 2848 (1963).

2238. Gershon, Braun, Scala, and Rodin, *J. Med. Chem.*, 7, 808 (1964).

2239. Taylor, Loeffler, and Mencke, *J. Org. Chem.*, 28, 509 (1963).

2240. Khromov-Borisov and Kheifets, *J. Gen. Chem.*, 34, 1321 (1964).

2241. Robba and Moreau, *Bull. Soc. chim. France*, 1960, 1648.

2242. Brown and Teitei, *J. Chem. Soc.*, 1964, 3204.

2243. Pfleiderer and Nubel, *Annalen*, 631, 168 (1960).

2244. Vazakas and Bennetts, *J. Med. Chem.*, 7, 342 (1964).

2245. Ziegler and Steiner, *Monatsh.*, 96, 548 (1965).

2246. Kazmirowski and Carstens, *J. prakt. Chem.*, [4], 26, 101 (1964).

2247. Hesse, Goldhahn, and Fürst, *Arch. Pharm.*, 295, 598 (1962).

2248. Crossley, Miller, Hartung, and Moore, *J. Org. Chem.*, 5, 238 (1940).

2249. Dolze, Goldhahn, and Fürst, *Arch. Pharm.*, 294, 190 (1961).

2250. Bose and Garrett, *J. Amer. Chem. Soc.*, 84, 1310 (1962).

2251. Bose and Garrett, *Tetrahedron*, 19, 85 (1963).

2252. Kleineberg and Ziegler, *Monatsh.*, 94, 502 (1963).

2253. Moye, *Austral. J. Chem.*, 17, 1309 (1964).

2254. Bretschneider, Dehler, and Klötzer, *Monatsh.*, 95, 207 (1964).

2255. Koppel, Springer, and Cheng, *J. Org. Chem.*, 26, 1884 (1961).

2256. Carissimi, Grasso, Grumelli, Milla, and Ravenna, *Farmaco* (Ed. Sci.), 17, 390 (1964).

2257. Singh and Berlinguet, *Canad. J. Chem.*, 42, 605 (1964).

2258. Horsch and Probst, *Arch. Pharm.*, 296, 249 (1963).

2259. Buzas, Egnell, and Moczar, *Bull. Soc. chim. France*, 1962, 267.

2260. O'Brien, Baiocchi, Robins, and Cheng, *J. Med. Pharm. Chem.*, 5, 1085 (1962).

2261. Budesinsky and Protiva, "Synthetische Arzneimittel", Akademie-Verlag, Berlin, 1961, pp. 238-257.

2262. Klötzer, *Monatsh.*, 96, 169 (1965).

2263. Prystas and Gut, *Coll. Czech. Chem. Comm.*, 28, 2501 (1963).

2264. Mertes and Saheb, *J. Pharm. Sci.*, 52, 508 (1963).

2265. Piskala and Gut, *Coll. Czech. Chem. Comm.*, 28, 2376 (1963).

2266. Fanshawe, Bauer, and Safir, *J. Org. Chem.*, 30, 1278 (1965).

2267. Stenbuck, Baltzly, and Hood, *J. Org. Chem.*, 28, 1983 (1963).

2268. Barone and Tieckelmann, *J. Org. Chem.*, 26, 598 (1961).

2269. Takamizawa and Hirai, Jap. Pat., 6236 (1965); through *Chem. Abstr.*, 63, 1803 (1965).

2270. Ogawa and Kasahara, *Vitamin*, 28, 238 (1963); *Chem. Abstr.*, 60, 5494 (1964).

2271. Cottis and Tieckelmann, *J. Org. Chem.*, 26, 79 (1961).

2272. Hill and Le Quesne, *J. Chem. Soc.*, 1965, 1515.

2273. Taylor, Knopf, Meyer, Holmes, and Hoefle, *J. Amer. Chem. Soc.*, 82, 5711 (1960).

2274. Budesinsky, Perina, and Sluka, *Czechoslov. Farm.*, 11, 345 (1962).

2275. Dymicky and Caldwell, *J. Org. Chem.*, 27, 4211 (1962).

2276. Takamizawa, Hirai, Sato, and Tori, *J. Org. Chem.*, 29, 1740 (1964).

2277. Takamizawa and Hirai, *Chem. Pharm. Bull.*, 12, 1418 (1964).

2278. Takamizawa and Hirai, *Chem. Pharm. Bull.* 12, 804 (1964).

2279. Takamizawa and Hirai, *J. Org. Chem.*, 30, 2290 (1965).

2280. Bretschneider, Klötzer, and Spiteller, *Monatsh.*, 92, 128 (1961).

2281. Baltzly, Sheehan, and Stone, *J. Org. Chem.*, 26, 2352 (1961).

2282. Julia and Chastrette, *Bull. Soc. chim. France*, 1962, 2247.

2283. Rosenkrantz, Citarel, Heinsohn, and Becker, *J. Chem. Eng. Data*, 8, 237 (1963).

2284. Bergmann, Kalmus, Ungar-Waron, and Kwietny-Gourin, *J. Chem. Soc.*, 1963, 3729.

2285. Pfleiderer and Lohrmann, *Chem. Ber.*, 94, 12 (1961).

2286. Rembold and Schramm, *Chem. Ber.*, 96, 2786 (1963).

2287. Taylor and Cheng, *J. Org. Chem.*, 25, 148 (1960).

2288. Brown and Jacobsen, *J. Chem. Soc.*, 1962, 3172.

2289. Nakata and Ueda, *J. Pharm. Soc. Japan*, 80, 1065 (1960).

2290. Shvachkin and Syrtsova, *J. Gen. Chem.*, 33, 3805 (1963).

2291. Kazmirowski, Dietz, and Carstens, *J. prakt. Chem.*, [4], 19, 150 (1963).

2292. Kazmirowski, Dietz, and Carstens, *J. prakt. Chem.*, [4], 19, 162 (1963).

2293. Gagnon, Boivin, and Giguere, *Canad. J. Chem.*, 29, 1028 (1951).

2294. Osdene, U. S. Pat. 3,180,869 (1965); through *Chem. Abstr.*, 63, 1807 (1965).

2295. Gold and Bayer, *Chem. Ber.*, 94, 2594 (1961).

2296. Berezovskii and Yurkevich, *J. Gen. Chem.*, 32, 1655 (1962).

2297. Taylor and Jefford, *J. Amer. Chem. Soc.* 84, 3744 (1962).

2298. Bredereck, Effenberger, and Botsch, *Chem. Ber.*, 97, 3397 (1964).

2299. Bredereck, Effenberger, Botsch, and Rehn, *Chem. Ber.*, 98, 1081 (1965).

2300. Bredereck, Effenberger, and Simchen, *Chem. Ber.*, 98, 1078 (1965).

2301. Bredereck, Sell, and Effenberger, *Chem. Ber.*, 97, 3407 (1964).

2302. Baker and Jordaan, *J. Het. Chem.*, 2, 21 (1965).

2303. Baker and Sachdev, *J. Pharm. Sci.*, 53, 1020 (1964).

2304. Gompper, Euchner, and Kast, *Annalen*, 675, 151 (1964).

2305. Strauss, *Annalen*, 638, 205 (1960).

2306. Jezdic, *Bull. Inst. Nucl. Sci. "Boris Kidrich"*, 12, 121 (1961); *Chem. Abstr.*, 57, 16605 (1963).

2307. Schulte, Reisch, Mock, and Kauder, *Arch. Pharm.*, 296, 235 (1963).

2308. Hyde, Acton, Baker, and Goodman, *J. Org. Chem.*, 27, 1717 (1962).

2309. Kalm, *J. Org. Chem.*, 26, 2925 (1961).

2310. Shvachkin and Azarova, *J. Gen. Chem.*, 34, 407 (1964).

2311. Sano, *Chem. Pharm. Bull.*, 10, 308 (1962).

2312. Eiden and Nagar, *Arch. Pharm.*, 297, 367 (1964).

2313. de Stevens, Smolinsky, and Dorfman, *J. Org. Chem.*, 29, 1115 (1964).

2314. Goerdeler and Pohland, *Chem. Ber.*, 96, 526 (1963).

2315. Peel, *Chem. and Ind.*, 1960, 1112.

2316. Creswell, Maurer, Strauss, and Brown, *J. Org. Chem.*, 30, 408 (1965).

2317. Dewar and Shaw, *J. Chem. Soc.*, 1961, 3254.

2318. Dewar and Shaw, *J. Chem. Soc.*, 1965, 1642.

2319. Goerdeler and Gnad, *Chem. Ber.*, 98, 1531 (1965).

2320. Goerdeler and Keuser, *Chem. Ber.*, 97, 3106 (1964).

2321. Gompper, Noppel, and Schaefer, *Angew. Chem.*, 75, 918 (1963).

2322. Simchen, *Angew. Chem.*, 76, 860 (1964).

2323. Bredereck, Effenberger, and Schweizer, *Chem. Ber.*, 95, 803 (1962).

2324. Bredereck, Gompper, Rempfer, Klemm, and Keck, *Chem. Ber.*, 92, 329 (1959).

2325. Bredereck, Gompper, Effenberger, Keck, and Heise, *Chem. Ber.*, 93, 1398 (1960).

2326. Drefahl and Hanschmann, *J. prakt. Chem.*, [4], 23, 324 (1964).

2327. Tsatsaronis and Effenberger, *Chem. Ber.*, 94, 2876 (1961).

2328. Volodina, Terent'ev, Roshchupkina, and Mishina, *J. Gen. Chem.*, 34, 469 (1964).

2329. Bredereck, Effenberger, and Treiber, *Chem. Ber.*, 96, 1505 (1963).

2330. Hull and Farrand, *J. Chem. Soc.*, 1963, 6028.

2331. Brown and Harper, Unpublished work.

2332. Fujimoto and Ono, *J. Pharm. Soc. Japan*, 85, 364 (1965).

2333. Fujimoto and Ono, *J. Pharm. Soc. Japan*, 85, 367 (1965).

2334. Brown, England, and Lyall, *J. Chem. Soc.*, 1966(C), 226.

2335. Fanshawe, Bauer, Ullman, and Safir, *J. Org. Chem.*, 29, 308 (1964).

2336. Martinez, Lee, and Goodman, *J. Med. Chem.*, 8, 187 (1965).

2337. Chatamra and Jones, *J. Chem. Soc.*, 1963, 811.

2338. Dewar and Shaw, *J. Chem. Soc.*, 1962, 583.

2339. Bianchetti, Dalla Croce, and Pocar, *Gazzetta*, 94, 606 (1964).

2340. Fusco, Bianchetti, and Pocar, *Gazzetta*, 91, 849 (1961).

2341. Appelquest, U. S. Pat. 2,517,824 (1950); *Chem. Zentr.*, 1951, II, 2393.

2342. Modest, Chatterjee, and Protopapa, *J. Org. Chem.*, 30, 1837 (1965).

2343. Modest, Chatterjee, and Kangur, *J. Org. Chem.*, 27, 2708 (1962).

2344. Murdock and Angier, *J. Org. Chem.*, 27, 3317 (1962).

2345. Clerc-Bory and Mentzner, *Bull. Soc. chim. France*, 1958, 436.

2346. Lumieux and Puskas, *Canad. J. Chem.*, 42, 2909 (1964).

2347. Guyot, Chopin, and Mentzner, *Bull. Soc. chim. France*, 1960, 1596.

2348. Moriyama, *J. Pharm. Soc. Japan*, 83, 169 (1963).

2349. Brossi, *Arch. Pharm.*, 296, 299 (1963).

2350. Eiden and Nagar, *Naturwiss.*, 50, 403 (1963).

2351. Sprio, *Ricerca sci.* (Rend. B), [2], 4, 585 (1964).

2352. Warrener and Cain, *Chem. and Ind.*, 1964, 1989.

2353. Capuano and Giammanco, *Gazzetta*, 86, 126 (1956).

2354. Capuano and Giammanco, *Gazzetta*, 86, 109 (1956).

2355. Khmelevskii, *J. Gen. Chem.*, 31, 3123 (1961).

2356. Bredereck, Effenberger, and Rainer, *Annalen*, 673, 82 (1964).

2357. Pfleiderer and Sagi, *Annalen*, 673, 78 (1964).

2358. Haines, Reese, and Todd, *J. Chem. Soc.*, 1962, 5281.

2359. Bredereck, Effenberger, and Resemann, *Chem. Ber.*, 95, 2796 (1962).

2360. Albert, *J. Chem. Soc.*, 1966 (B), 438.

2361. Scarborough, *J. Org. Chem.*, 29, 219 (1964).

2362. Schulte, *Chem. Ber.*, 87, 820 (1954).

2363. Schroeder and Dodson, *J. Amer. Chem. Soc.*, 84, 1904 (1962).

2364. Gilow and Jacobus, *J. Org. Chem.*, 28, 1994 (1963).

2365. Shealy and O'Dell, *J. Org. Chem.*, 29, 2135 (1964).

2366. Schaefer, Huffman, and Peters, *J. Org. Chem.*, 27, 548 (1962).

2367. Huffman, Schaefer, and Peters, *J. Org. Chem.*, 27, 551 (1962).

2368. Schaefer and Peters, *J. Amer. Chem. Soc.*, 81, 1470 (1959).

2369. Kreutzberger and Grundmann, *J. Amer. Chem. Soc.*, 26, 1121 (1961).

2370. Schmidt, *Chem. Ber.*, 98, 346 (1965).

2371. Titherley and Worrall, *J. Chem. Soc.*, 97, 839 (1910).

2372. Kröhnke, Schmidt, and Zecher, *Chem. Ber.*, 97, 1163 (1964).

2373. E. Schmidt, Dissertation, University of Giessen (1962); as quoted in ref. 2372.

2374. Brown and Harper, in "Pteridine Chemistry", Pergamon Press, Oxford, 1964, p. 217 *et seq.*

2375. Perrin and Pitman, *J. Chem. Soc.*, 1965, 7071.

2376. Brown and Paddon-Row, *J. Chem. Soc.*, 1966(C), 164.

2377. Taylor and Ravindranathan, *J. Org. Chem.*, 27, 2622 (1962).

2378. Lister, *Rev. Pure Appl. Chem.*, 13, 30 (1963).

2379. Naylor, Shaw, Wilson, and Butler, *J. Chem. Soc.*, 1961, 4845.

2380. Cresswell and Brown, *J. Org. Chem.*, 28, 2560 (1963).

2381. Richter, Loeffler, and Taylor, *J. Amer. Chem. Soc.*, 82, 3144 (1960).

2382. Dallacker and Steiner, *Annalen*, 660, 98 (1962).

2383. Curran and Angier, *J. Org. Chem.*, 26, 2364 (1961).

2384. Mulvey, Cottis, and Tieckelmann, *J. Org. Chem.*, 29, 2903 (1964).

2385. Taylor and Loeffler, *J. Amer. Chem. Soc.*, 82, 3147 (1960).

2386. Taylor and Borror, *J. Org. Chem.*, 26, 4967 (1961).

2387. Sutcliffe, Zee-Cheng, Cheng, and Robins, *J. Med. Pharm. Chem.*, 5, 588 (1962).

2388. Taylor and Morrison, *J. Amer. Chem. Soc.*, 87, 1976 (1965).

2389. Taylor and Garcia, *J. Org. Chem.*, 29, 2116 (1964).

2390. Effenberger, *Chem. Ber.*, 98, 2260 (1965).

2391. Baker and Jordaan, *J. Het. Chem.*, 2, 162 (1965).

2392. Takagi, Hubert-Habart, and Royer, *Compt. rend.*, 160, 5302 (1965).

2393. Wheatley, *Acta Cryst.*, 13, 80 (1960).

2394. Kim and Hameka, *J. Amer. Chem. Soc.*, 85, 1398 (1963).

2395. de la Vega and Hameka, *J. Amer. Chem. Soc.*, 85, 3504 (1963).

2396. Coulson, *J. Chem. Soc.*, 1963, 5893.

2397. Adamov and Tupitsyn, *Vestn. Leningrad. Univ.*, 17, No 16, *Ser. Fiz. i Khim.*, No 3, 47 (1962); through *Chem. Abstr.*, 59, 7351 (1963).

2398. Tjebbes, *Acta Chem. Scand.*, 16, 916 (1962).

2399. Amos and Hall, *Mol. Phys.*, 4, 25 (1961).

2400. Hückel and Salinger, *Ber.*, 77, 810 (1944).

2401. Hameka and Liquori, *Mol. Phys.*, 1, 9 (1958).

2402. El-Sayed, *J. Chem. Phys.*, 36, 552 (1962).

2403. Simmons and Innes, *J. Mol. Spectroscopy*, 13, 435 (1964).

2404. Favini and Simonetta, *Atti Accad. naz. Lincei, Rend. Classe Sci. fiz. mat. nat.*, 28, 57 (1960); through *Chem. Abstr.*, 55, 105 (1961).

2405. Bliznyukov and Grin, *Trudy Khar'Kov. Farm. Inst.*, 1957, No 1, 26; through *Chem. Abstr.*, 55, 12031 (1961).

2406. Dodd, Hopton, and Hush, *Proc. Chem. Soc.*, 1962, 61.

2407. Börresen, *Acta Chem. Scand.*, 17, 921 (1963).

2408. Janowski, *Bull. Acad. polon. Sci., Ser. Sci. math. astr., phys.*, 11, 621 (1963).

2409. Zahradnik, *Abhandl. deut. Akad. Wiss. Berlin, Kl. Med.*, 1964, 389.

2410. Gronowitz and Hoffman, *Arkiv Kemi*, 16, 459 (1960).

2411. Reddy, Hobgood, and Goldstein, *J. Amer. Chem. Soc.*, 84, 336 (1962).

2412. Ward, *J. Amer. Chem. Soc.*, 84, 332 (1962).

2413. Lister and Timmis, *J. Chem. Soc.*, 1960, 1113.

2414. Taylor and Garcia, *J. Org. Chem.*, 30, 655 (1965).

2415. Shindo, *Chem. Pharm. Bull.*, 8, 33 (1960).

2416. Kubota, *Bull. Chem. Soc. Japan*, 35, 946 (1962).

2417. Kubota and Watanabe, *Bull. Chem. Soc. Japan*, 36, 1093 (1963).

2418. Ogata, Watanabe, Tori, and Kano, *Tetrahedron Letters*, 1964, 19.

2419. Ueno and Koshi, Jap. Pat. 28,657 (1964); through *Chem. Abstr.*, 62, 11827 (1965).

2420. Ueno and Koshi, Jap. Pat. 28,658 (1964); through *Chem. Abstr.*, 62, 11827 (1965).

2421. Ueno and Koshi, Jap. Pat. 6,235 (1965); through *Chem. Abstr.*, 63, 1803 (1965).

2422. Tanaka, Sugawa, Nakamori, Sanno, and Ando, *J. Pharm. Soc. Japan*, 75, 770 (1955).

2423. Springer, Haggerty, and Cheng, *J. Het. Chem.*, 2, 49 (1965).

2424. Profft and Raddatz, *J. prakt. Chem.*, [4], 12, 206 (1961).

2425. Profft and Raddatz, *Wiss. Z. Tech. Hochsch. Chem. Leuna-Merseburg.*, 2, 439 (1960); through *Chem. Abstr.*, 55, 11425 (1961).

2426. Bredereck and Simchen, *Angew. Chem.*, 75, 1102 (1963).

2427. Bredereck and Jentzsch, *Chem. Ber.*, 93, 2410 (1960).

2428. Rice, Dudek, and Barber, *J. Amer. Chem. Soc.*, 87, 4569 (1965).

2429. Collins, *J. Chem. Soc.*, 1963, 1337.

2430. Roy, Ghosh, and Guha, *J. Org. Chem.*, 25, 1909 (1960).

2431. Stempel, Brown, and Fox, *Abstr. 145th Meeting Amer. Chem. Soc.* (New York, 1963), p. 14-O.

2432. O'Brien, Cheng, and Pfleiderer, *J. Med. Chem.*, 9, 573 (1966).

2433. Söll and Pfleiderer, *Chem. Ber.*, 96, 2977 (1963).

2434. Katritzky, Shepherd, and Waring, *Rec. Trav. chim.*, 81, 443 (1962).

2435. Lister, *J. Chem. Soc.*, 1963, 2228.

2436. Baker and Santi, *J. Pharm. Sci.*, 54, 1252 (1965).

2437. Senda, Izumi, Kano, and Tsubota, *Reports of Gifu College of Pharmacy*, 11, 62 (1961); *Chem. Abstr.*, 57, 11194 (1962).

2438. Cresswell and Wood, *J. Chem. Soc.*, 1960, 4768.

2439. Daves, Robins, and Cheng, *J. Amer. Chem. Soc.*, 83, 3904 (1961).

2440. Daves, Robins, and Cheng, *J. Amer. Chem. Soc.*, 84, 1724 (1962).

2441. Liao and Cheng, *J. Het. Chem.*, 1, 212 (1964).

2442. Davoll and Evans, *J. Chem. Soc.*, 1960, 5041.

2443. O'Brien, Baiocchi, Robins, and Cheng, *J. Med. Chem.*, 6, 467 (1963).

2444. Pfleiderer and Walter, *Annalen*, 677, 113 (1964).

2445. Taylor and McKillop, *J. Org. Chem.*, 30, 3153 (1965).

2446. Okuda, Fukuda, Kuniyoshi, and Shinoda, *J. Pharm. Soc. Japan*, 82, 1043 (1962).

2447. Okuda, Fukuda, and Kuniyoshi, Jap. Pat. 10,051 (1965); through *Chem. Abstr.*, 63, 5659 (1965).

2448. Bojarski and Kahl, *Roczniki Chem.*, 38, 1305 (1964).

2449. Saikawa, *J. Pharm. Soc. Japan*, 84, 121 (1964).

2450. Saikawa and Maeda, *J. Pharm. Soc. Japan*, 84, 131 (1964).

2451. Fujita, Yamamoto, Minami, and Takamatsu, *Chem. Pharm. Bull.*, 13, 1183 (1965).

2452. Sugihara and Ito, *J. Pharm. Soc. Japan*, 85, 418 (1965).

2453. Karlinskaya and Khromov-Borisov, *J. Gen. Chem.*, 32, 1847 (1962).

2454. Brown and Jacobsen, *J. Chem. Soc.*, 1965, 3770.

2455. Beaman and Robins, *J. Med. Pharm. Chem.*, 5, 1067 (1962).

2456. Neilson and Wood, *J. Chem. Soc.*, 1962, 44.

2457. Stuart and Wood, *J. Chem. Soc.*, 1963, 4186.

2458. Jacobsen, *J. Chem. Soc.*, 1966(C), 1065.

2459. Pfleiderer and Fink, *Chem. Ber.*, 96, 2964 (1963).

2460. Marsico and Goldman, *J. Org. Chem.*, 30, 3597 (1965).

2461. Pfleiderer and Taylor, *J. Amer. Chem. Soc.*, 82, 3765 (1960).

2462. O'Brien, Noell, Robins, and Cheng. *J. Med. Chem.*, 9, 121 (1966).

2463. Goldner and Carstens, *J. prakt. Chem.*, [4], 12, 242 (1961).

2464. Tong, Lee, and Goodman, *J. Amer. Chem. Soc.*, 86, 5664 (1964).

2465. Stuart, West, and Wood, *J. Chem. Soc.*, 1964 4769.

2466. Temple, McKee, and Montgomery, *J. Org. Chem.*, 30, 829 (1965).

2467. Clark, Kernick, and Layton, *J. Chem. Soc.*, 1964, 3221.

2468. Cresswell, Neilson, and Wood, *J. Chem. Soc.*, 1961, 476.

2469. Clark, Kernick, and Layton, *J. Chem. Soc.*, 1964, 3215.

2470. Cresswell, Neilson, and Wood, *J. Chem. Soc.*, 1960, 4776.

2471. Albert and Clark, *J. Chem. Soc.*, 1964, 1666.

2472. Karlinskaya and Khromov-Borisov, *J. Gen. Chem.*, 32, 1858 (1962).

2473. Cresswell and Strauss, *J. Org. Chem.*, 28, 2563 (1963).

2474. Kuwada, Masuda, Kishi, and Asai, *Chem. Pharm. Bull.*, 8, 798 (1960).

2475. Pfleiderer and Fink, *Annalen*, 657, 149 (1962).

2476. Pfleiderer and Deckert, *Chem. Ber.*, 95, 1597 (1962).

2477. Roy, Ghosh, and Guha, *J. Indian Chem. Soc.*, 36, 651 (1959).

2478. Taylor and Cheng, *J. Org. Chem.*, 25, 148 (1960).

2479. Alekseeva and Pushkareva, *J. Gen. Chem.*, 33, 1693 (1963).

2480. Pfleiderer and Nübel, *Annalen*, 647, 155 (1961).

2481. Pfleiderer and Fink, *Chem. Ber.*, 96, 2950 (1963).

2482. Pfleiderer and Lohrmann, *Chem. Ber.*, 94, 2708 (1961).

2483. Pfleiderer and Rukwied, *Chem. Ber.*, 94, 1 (1961).

2484. Montgomery and Thomas, *J. Org. Chem.*, 28, 2304 (1963).

2485. Goldner, Dietz, and Carstens, *Annalen*, 691, 142 (1966).

2486. Goldner, Dietz, and Carstens, *Naturwiss.*, 51, 137 (1964).

2487. Israel, Protopapa, Chatterjee, and Modest, *J. Pharm. Sci.*, 54, 1626 (1965).

2488. Kalatzis, *J. Chem. Soc. (B)*, 1967, 277.

2489. Nübel and Pfleiderer, *Chem. Ber.*, 98, 1060 (1965).

2490. Bergel, Brown, Leese, Timmis, and Wade, *J. Chem. Soc.*, 1963, 846.

2491. Albert, *J. Chem. Soc.*, 1966(B), 427.

2492. Brown and Jacobsen, *J. Chem. Soc.*, 1965, 1175.

2493. Pfleiderer, Liedek, Lohrmann, and Rukwied, *Chem. Ber.*, 93, 2015 (1960).

2494. Ruzicka and Lycka, *Coll. Czech. Chem.Comm.*, 27, 1790 (1962).

2495. Taylor, Jefford, and Cheng, *J. Amer. Chem. Soc.*, 83, 1261 (1961).

2496. Osdene, in "Pteridine Chemistry," Ed. Pfleiderer and Taylor, Pergamon Press, Oxford, 1964, p. 65.

2497. Taylor and Garcia, *J. Amer. Chem. Soc.*, 86, 4720 (1964).

2498. Taylor and Garcia, *J. Amer. Chem. Soc.*, 86, 4721 (1964).

2499. Goldner, Dietz, and Carstens, *Z. Chem.*, 4, 454 (1964).

2500. Goldner, Dietz, and Carstens, *Tetrahedron Letters*, 1965, 2701.

2501. Timmis, Cooke, and Spickett, in "Chemistry and Biology of Purines," Ed. Wolstenholme and O'Connor, Churchill, London, 1957, p. 134.

2502. Hampshire, Hebborn, Triggle, Triggle, and Vickers, *J. Med. Chem.*, 8, 745 (1965).

2503. Israel, Protopapa, Schlein, and Modest, *J. Med. Chem.*, 7, 792 (1964).

2504. O'Brien, Westover, Robins, and Cheng, *J. Med. Chem.*, 8, 182 (1965).

2505. Shealy, Struck, Clayton, and Montgomery, *J. Org. Chem.*, 26, 4433 (1961).

2506. Pfleiderer and Nübel, *Chem. Ber.*, 93, 1406 (1960).

2507. Elslager and Worth, *J. Med. Chem.*, 6, 444 (1963).

2508. Elslager, Capps, Kurtz, Werbel, and Worth, *J. Med. Chem.*, 6, 646 (1963).

2509. Elslager, U.S. Pat., 3,083,195 (1963); through *Chem. Abstr.*, 59, 10077 (1963).

2510. Kyowa Fermentation Industry, Fr. Pat., 1,415,224 (1965); through *Chem. Abstr.*, 64, 5115 (1966).

2511. Brown and Foster, *J. Chem. Soc.*, 1965, 4911.

2512. Hünig and Oette, *Annalen*, 640, 98 (1961).

2513. Fel'dman and Boksiner, *Biol. Aktivn. Soedin., Akad. Nauk SSSR*, 1965, 79; through *Chem. Abstr.*, 63, 18080 (1965).

2514. Chkhikvadze and Magidson, *J. Gen. Chem.*, 34, 2577 (1964).

2515. Ivin and Nemets, *J. Gen. Chem.*, 35, 1303 (1965).

2516. Biressi, Carissimi, and Ravenna, *Gazzetta*, 95, 1293 (1965).

2517. Riemschneider and Pehlmann, *Z. Naturforsch.*, 20B, 540 (1965).

2518. Ashani and Cohen, *Israel J. Chem.*, 3, 101 (1965).

2519. Mehta, Miller and Mooney, *J. Chem. Soc.*, 1965, 6695.

2520. Paddon-Row, Ph.D. Thesis, Canberra, 1967.

2521. Baganz, Rüger, and Kohtz, *Chem. Ber.*, 98, 2576 (1965).

2522. Budesinsky, Roubinek, and Svatek, *Coll. Czech. Chem. Comm.*, 30, 3730 (1965).

2523. Soemmer, Ger. Pat. 1,200,308 (1965); through *Chem. Abstr.*, 63, 16364 (1965).

2524. Grüssner, Montavon, and Schnider, *Monatsh.*, 96, 1677 (1965).

2525. Leese and Timmis, *J. Chem. Soc.*, 1961, 3816.

2526. Leese and Timmis, *J. Chem. Soc.*, 1961, 3818.

2527. Gompper, Gäng, and Saygin, *Tetrahedron Letters*, 1966, 1885.

2528. Taylor and Abul-Husn, *J. Org. Chem.*, 31, 342 (1966).

2529. Taylor and Morrison, *Angew. Chem.*, 76, 342 (1964).

2530. Albert and Tratt, *Chem. Comm.*, 1966, 243.

2531. Taylor and Berger, *Angew. Chem.*, 78, 144 (1966).

2532. Sekiya and Osaki, *Chem. Pharm. Bull.*, 13, 1319 (1965).

2533. Dornow and Hell, *Chem. Ber.*, 93, 1998 (1960).

2534. Dornow and Helberg, *Chem. Ber.*, 93, 2001 (1960).

2535. Kleineberg and Ziegler, *Monatsh.*, 96, 1352 (1965).

2536. Hanada, Jap. Pat. 15,197 (1965); through *Chem. Abstr.*, 63, 14880 (1965).

2537. Hauser, Peters, and Tieckelmann, *J. Org. Chem.*, 26, 451 (1961).

2538. Hirata, Iwashita, and Teshima, *Nagoya Sangyo Kagaku Kenkyujho Kenkyu Hokoku*, No.9, 83 (1957); through *Chem. Abstr.*, 51, 12074 (1957).

2539. Osborne, Wieder, and Levine, *J. Het. Chem.*, 1, 145 (1964).

2540. Osborne and Levine, *J. Org. Chem.*, 28, 2933 (1963).

2541. den Hertog, van der Plas, Pieterse, and Streef, *Rec. Trav. chim.*, 84, 1569 (1965).

2542. de Stevens, Blatter, and Carney, *Angew. Chem.*, 78, 125 (1966).

2543. Modest, Chatterjee, Foley, and Farber, *Acta Union Internat. Contre Cancer*, 20, 112 (1964).

2544. Bredereck, Simchen, and Traut, *Chem. Ber.*, 98, 3883 (1965).

2545. Grigat and Pütter, *Angew. Chem.* 77, 913 (1965).

2546. Warrener, *Chem. and Ind.*, 1966, 381 and 556.

2547. Baker and Schwan, *J. Med. Chem.*, 9, 73 (1966).

2548. Du Bois, Cochran, and Thomson, *Proc. Soc. Exp. Biol. Med.*, 67, 169 (1948).

2549. Knudsen, *Acta Pharmacol. Toxicol.*, 20, 295 (1963).

2550. Hilton and Nomura, *Weed Res.*, 4, 216 (1964).

2551. Gershon and Parmegiani, *Trans. N.Y. Acad. Sci.*, 25, 638 (1963).

2552. Albert, "Selective Toxicity," 4th edition, Methuen, London, 1968, p.378 *et seq.*

2553. Cheng, *Progr. Med. Chem.*, 6, 67 (1969); Cheng and Roth, *ibid.* in preparation.

2554. Van der Plas, Haase, Zuurdeeg, and Vollering, *Rec. Trav. chim.*, 85, 1101 (1966).

2555. Minnemeyer, Clarke, and Tieckelmann, *J. Org. Chem.*, 31, 406 (1966).

2556. Ciba, Fr.Pat. 1,396,684 (1965); through *Chem. Abstr.*, 63, 9964 (1966).

2557. Goya, Takahashi, and Okano, *J. Pharm. Soc. Japan*, 86, 952 (1966).

2558. Bredereck, Brauninger, Hayer, and Vollmann, *Chem. Ber.*, 92, 2937 (1959).

2559. Gershon, Parmegiani, and D'Ascoli, *J. Med. Chem.*, 10, 113 (1967).

2560. Nübel and Pfleiderer, *Chem. Ber.*, 95, 1605 (1962).

2561. Robison, *J. Amer. Chem. Soc.*, 80, 5481 (1958).

2562. Biffin, Brown, and Lee, *Aust. J. Chem.*, 20, 1041 (1967).

2563. Vincze and Cohen, *Israel J. Chem.*, 4, 23 (1966).

2564. Tedeschi, Belg.Pat. 657,135 (1965); through *Chem. Abstr.*, 64, 19637 (1966).

2565. Baker and Ho, *J. Pharm. Sci.*,54, 1261 (1965).

2566. Baker, Morreal, and Ho, *J. Med. Chem.*, 6, 658 (1963).

2567. Gerns, Perrotta, and Hitchings, *J. Med. Chem.*, 9, 108 (1966).

2568. Ballweg, *Annalen*, 657, 141 (1962).

2569. Bredereck and Humburger, *Chem. Ber.*, 99, 3227 (1966).

2570. Protsenko and Bogodist, *J. Gen. Chem.*, 33, 537 (1963).

2571. Klötzer and Schantl, *Monatsh.*, 94, 1190 (1963).

2572. Prystas and Sorm, *Coll. Czech. Chem. Comm.*, 30, 1900 (1965).

2573. Brown and Lyall, *Aust. J. Chem.*, 17, 794 (1964).

2574. Van der Plas, *Rec. Trav. chim.*, 84, 1101 (1965).

2575. Caton, Grant, Pain, and Slack, *J. Chem. Soc.*, 1965, 5467.

2576. Partyka, U.S.Pat. 3,225,047 (1965); through *Chem. Abstr.*, 64, 6664 (1966).

2577. Zymalkowski and Reimann, *Arch. Pharm.*, 299, 362 (1966).

2578. Bretschneider, Dehler, and Klötzer, *Monatsh.*, 95, 207 (1964).

2579. Haggerty, Springer, and Cheng, *J. Het. Chem.*, 2, 1 (1965).

2580. Takagi and Ueda, *Chem. Pharm. Bull.*, 11, 1382 (1963).

2581. Ueda and Fox, *J. Med. Chem.*, 6, 697 (1963).

2582. Wittenburg, *J. prakt. Chem.*, 33, 165 (1966).

2583. Koppel, Springer, Robins, and Cheng, *J. Org. Chem.*, 27, 181 (1961).

2584. McOmie and Turner, *J. Chem. Soc.*, 1963, 5590.

2585. Hurst, McOmie, and Searle, *J. Chem. Soc.*, 1965, 7116.

2586. Budesinsky, Bydzovsky, Kopecky, Svab, and Vavrina, *Czechoslov. Farm.*, 10, 241 (1961); through *Chem. Abstr.*, 55, 25973 (1961).

2587. Yamaoka, Tsukamoto, and Aso, *J. Agr. Chem. Soc. Japan*, 37, 626 (1963); *Chem. Abstr.*, 63, 9942 (1965).

2588. Abu-Zeid, Abu-Elela, and Ghoneim, *J. Pharm. Sci. U. A. R.*, 4, 41 (1963); *Chem. Abstr.*, 63, 18078 (1965).

2589. Yamaoka and Aso, *J. Org. Chem.*, 27, 1462 (1962).

2590. Portnyagina and Karp, *Ukrain. khim. Zh.*, 32, 1306 (1966); *Chem. Abstr.*, 67, 3058 (1967).

2591. Gershon, *J. Org. Chem.*, 27, 3507 (1962).

2592. Boulton, Hurst, McOmie, and Tute, *J. Chem. Soc. (C)*, 1967, 1202.

2593. Klötzer and Herberz, *Monatsh.*, 96, 1567 (1965).

2594. Bredereck, Simchen, and Santos, *Chem. Ber.*, 100, 1344 (1967).

2595. Van der Plas and Geurtsen, *Tetrahedron Letters*, 1964, 2093.

2596. Brossmer and Röhm, *Annalen*, 692, 119 (1966).

2597. Heitmeier, Spinner, and Gray, *J. Org. Chem.*, 26, 4419 (1961).

2598. Peters, Holland, Bryant, Minnemeyer, Hohenstein, and Tieckelmann, *Cancer Res.*, 19, 729 (1959).

2599. Shadbolt and Ulbricht, *J. Chem. Soc. (C)*, 1967, 1172.

2600. Protsenko and Bogodist, *Ukrain. khim. Zh.*, 32, 378 (1966); through *Chem. Abstr.*, 65, 3869 (1966).

2601. Bee and Rose, *J. Chem. Soc. (C)*, 1966, 2031.

2602. Brown and Paddon-Row, *J. Chem. Soc. (C)*, 1967, 1928.

2603. Caton, Hurst, McOmie, and Hunt, *J. Chem. Soc. (C)*, 1967, 1204.

2604. Bly and Mellon, *J. Org. Chem.*, 27, 2945 (1962).

2605. Bly, *J. Org. Chem.*, 29, 943 (1964).

2606. Bader and Spiere, *J. Org. Chem.*, 28, 2155 (1963).

2607. Gronowitz, Norrman, Gestblom, Mathiasson, and Hoffman, *Arkiv Kemi*, 22, 65 (1964).

2608. Schwan and Tieckelmann, *J. Het. Chem.*, 1, 201 (1964).

2609. Taft and Shepherd, *J. Med. Pharm. Chem.*, 5, 1335 (1962).

2610. Wempen and Fox, *J. Med. Chem.*, 6, 688 (1963).

2611. Horwitz and Tomson, *J. Org. Chem.*, 26, 3392 (1961).

2612. Schroeder, *J. Amer. Chem. Soc.*, 82, 4115 (1960).

2613. Ivin, Slesarev, and Sochilin, *J. Gen. Chem.*, 34, 4120 (1964).

2614. Nemets, Ivin, and Slesarev, *J. Gen. Chem.*, 35, 1429 (1965).

2615. Boudakain, Kober, and Shipkowski, U.S.Pat. 3,280,124 (1966); through *Chem. Abstr.*, 66, 2582 (1967).

2616. Chambers, MacBride, and Musgrave, *Chem. and Ind.*, 1966, 1721.

2617. Imperial Smelting Corporation, Belg. Pat. 660,907 (1965); through *Chem. Abstr.*, 64, 3568 (1966).

2618. Schroeder, Kober, Ulrich, Rätz, Agahigian, and Grundmann, *J. Org. Chem.*, 27, 2580 (1962).

2619. Brown and Ford, *J. Chem. Soc. (C)*, 1967, 568.

2620. Katsnel'son, Del'nik, and Semikolennykh, *Med. Prom. SSSR*, 20, 17 (1966); through *Chem. Abstr.*, 65, 18579 (1966).

2621. Phillips, Mehta, and Strelitz, *J. Org. Chem.*, 28, 1488 (1963).

2622. Ziegler, Salvador, and Kappe, *Monatsh.*, 93, 1376 (1962).

2623. Nishiwaki, *Tetrahedron*, 22, 2401 (1966).

2624. Brown and Teitei, *J. Chem. Soc.*, 1965, 755.

2625. Dubicki, Zielinski, and Starks, *J. Pharm. Sci.*, 53, 1422 (1964).

2626. Brown and Harper, *J. Chem. Soc.*, 1965, 5542.

2627. Brown and Harper, *J. Chem. Soc.*, 1963, 1276.

2628. Granados, Marquez, and Melgarejo, *Anales real Soc. espan. Fis. Quim.*, *Ser. B.*, 58, 479 (1962); *Chem. Abstr.*, 59, 1631 (1963).

2629. Nishiwaki, *Tetrahedron*, 22, 3117 (1966).

2630. Brown and Lee, *Aust. J. Chem.*, 21, 243 (1968).

2631. Nishiwaki, *Chem. Pharm. Bull.*, 10, 1029 (1962).

2632. Bui, Gillet, and Dumont, *Internat. J. Appl. Radiation Isotopes*, 16, 337 (1965).

2633. Brown and Paddon-Row, *J. Chem. Soc. (C)*, 1967, 903.

2634. Chang and Welch, *J. Med. Chem.*, 6, 428 (1963).

2635. Ziegler and Kappe, *Monatsh.*, 95, 1057 (1964).

2636. Barszcz, Tramer, and Shugar, *Acta Biochim. Polon.*, 10, 9 (1963).

2637. Makarov and Popova, *Sintez Prirodn. Soedin. ikh Analogov i Fragmentov, Akad. Nauk SSSR, Otd. Obsch. i Tekhn. Khim.*, 1965, 273; through *Chem. Abstr.*, 65, 2256 (1966).

2638. Svab, Budesinsky, and Vavrina, *Coll. Czech. Chem. Comm.*, 32, 1582 (1967).

2639. Brossmer and Röhm, *Angew. Chem.*, 77, 260 (1965).

2640. Farkas and Sorm, *Coll. Czech.Chem. Comm.*, 26, 893 (1961).

2641. Ueda and Fox, *J. Amer. Chem. Soc.*, 85, 4024 (1963).

2642. Armarego, *J. Chem. Soc.*, 1965, 2778.

2643. Benitez, Ross, Goodman, and Baker, *J. Amer. Chem. Soc.*, 82, 4585 (1960).

2644. Singh and Lal, *J. Indian Chem. Soc.*, 43, 308 (1966).

2645. Ledochowski, Wolski, Ledochowski, and Trzeciak, *Roczniki Chem.*, 37, 1083 (1963).

2646. Ross, Lee, Schelstraete, Goodman, and Baker, *J. Org. Chem.*, 26, 3021 (1961).

2647. Filip and Farkas, *Coll. Czech. Chem. Comm.*, 32, 462 (1967).

2648. Segal and Skinner, *J. Org. Chem.*, 27, 199 (1962).

2649. Montgomery, Fikes, and Johnston, *J. Org. Chem.*, 27, 4080 (1962).

2650. Giner-Sorolla and Medrek, *J. Med. Chem.*, 9, 97 (1966).

2651. Schellenberger and Winter, *Z. physiol. Chem.*, 344, 16 (1966).

2652. Shvachkin and Syrtsova, *J. Gen. Chem.*, 34, 2159 (1964).

2653. Schellenberger and Winter, *Z. physiol. Chem.*, 322, 164 (1960).

2654. Schellenberger, Rödel, and Rödel, *Z. physiol. Chem.*, 339, 122 (1964).

2655. Baker and Kawazu, *J. Med. Chem.*, 10, 316 (1967).

2656. Sprinzl, Farkas, and Sorm, *Coll. Czech. Chem. Comm.*, 32, 1649 (1967).

2657. Farkas and Sorm, *Coll. Czech. Chem. Comm.*, 31, 1413 (1966).

2658. Elderfield and Wood, *J. Org. Chem.*, 26, 3042 (1961).

2659. Ghosh, *J. Med. Chem.*, 9, 423 (1966).

2660. Albert and Reich, *J. Chem. Soc.*, 1961, 127.

2661. Albert, *J. Chem. Soc. (B)*, 1966, 438.

2662. Nitta, Miyamoto, Nebashi, and Yoneda, *Chem. Pharm. Bull.*, 13, 901 (1965).

2663. Nitta, Miyamoto, and Yoneda, Jap. Pat. 14022 (1965); through *Chem. Abstr.*, 63, 13285 (1965).

2664. Mertes, Saheb, and Miller, *J. Med. Chem.*, 9, 876 (1966).

2665. Mertes and Saheb, *J. Med. Chem.*, 6, 619 (1963).

2666. Heidelberger, Parsons, and Remy, *J. Med. Chem.*, 7, 1 (1964).

2667. Heidelberger, Parsons, and Remy, *J. Amer. Chem. Soc.*, 84, 3597 (1962).

2668. Barlin and Brown, in "Topics in Heterocyclic Chemistry," ed. R.N. Castle, Wiley, New York, 1969.

2669. Barlin and Brown, *J. Chem. Soc. (B)*, 1967, 648.

2670. Barlin and Brown, *J. Chem. Soc. (B)*, 1967, 736.

2671. Barlin and Brown, *J. Chem. Soc. (C)*, 1967, 2473.

2672. Brown and Ford, *J. Chem. Soc. (C)*, 1969, in press.

2673. Montgomery and Temple, *J. Amer. Chem. Soc.*, 82, 4592 (1960).

2674. Ballweg, *Annalen*, 673, 153 (1964).

2675. Brown, Ford, and Tratt, *J. Chem. Soc. (C)*, 1967, 1445.

2676. Brown and Teitei, *Aust. J. Chem.*, 18, 199 (1965).

2677. Klötzer and Schantl, *Monatsh.*, 94, 1178 (1963).

2678. Spiteller and Bretschneider, *Monatsh.*, 92, 183 (1961).

2679. Budesinsky, Prikryl, and Svatek, *Coll. Czech. Chem. Comm.*, 30, 3895 (1965).

2680. Godefroi, *J. Org. Chem.*, 27, 2264 (1962).

2681. May and Sykes, *J. Chem. Soc. (C)*, 1966, 649.

2682. Shepherd, Taft, and Krazinski, *J. Org. Chem.*, 26, 2764 (1961).

2683. Rossi, Piselli, Pirola, Bertani, and Passerini, *Farmaco* (Ed. Sci.), 18, 499 (1963).

2684. Aft, *Diss. Abstr.*, 22, 4178 (1962).

2685. Aft and Christensen, *J. Org. Chem.*, 27, 2170 (1962).

2686. Evans, *J. Chem. Soc.*, 1964, 2450.

2687. Albert and Catterall, *J. Chem. Soc. (C)*, 1967, 1533.

2688. Biffin, Brown, and Lee, *J. Chem. Soc. (C)*, 1967, 573.

2689. Shepherd and Fedrick, *Adv. Het. Chem.*, 4, 145 (1965).

2690. Ackermann and Dussy, *Helv. Chim. Acta*, 45, 1683 (1962).

2691. Ackermann, *Helv. Chim. Acta*, 49, 454 (1966).

2692. Shein, Mamaev, Zagulyaeva, and Shvets, *Reaktsionnaya Sposobnost Organ. Soedin., Tartusk Gos. Univ.*, 2, 65 (1965); through *Chem. Abstr.*, 65, 10461 (1966).

2693. Mamaev, Zagulyaeva, Shein, and Shvets, *Reaktsionnaya Sposobnost Organ. Soedin., Tartusk Gos. Univ.*, 2, 61 (1965); through *Chem. Abstr.*, 65, 10461 (1966).

2694. Brown and Lyall, *Aust. J. Chem.*, 18, 741 (1965).

2695. Brown and Lyall, *Aust. J. Chem.*, 18, 1811 (1965).

2696. Smith and Stevens, *J. Chem. Educ.*, 38, 574 (1961).

2697. Brown and Foster, *Aust. J. Chem.*, 19, 1487 (1966).

2698. Mizukami and Hirai, *J. Org. Chem.*, 31, 1199 (1966).

2699. Profft and Raddatz, *Arch. Pharm.*, 295, 649 (1962).

2700. Brown and England, *J. Chem. Soc. (C)*, 1967, 1922.

2701. Chapman and Taylor, *J. Chem. Soc.*, 1961, 1908.

2702. De Brunner, U.S.Pat. 3,184,461 (1965); through *Chem. Abstr.* 63, 8377 (1965).

2703. Koppel, Springer, Robins, and Cheng, *J. Org. Chem.*, 27, 181 (1961).

2704. Declerck, Degroote, de Lannoy, Nasielski-Hinkens, and Nasielski, *Bull. Soc. chim. belg.*, 74, 119 (1965).

2705. Geigy A.-G., Fr.Pat. 1,413,722 (1965); through *Chem. Abstr.*, 64, 5109 (1966).

2706. Kober and Rätz, *J. Org. Chem.*, 27, 2509 (1962).

2707. Kober and Rätz, U.S.Pat. 3,259,623 (1966); through *Chem. Abstr.*, 65, 8930 (1966).

2708. Ciba, Fr.Pat. 1,332,539 (1963); through *Chem. Abstr.*, 60, 2981 (1964).

2709. Berg and Parnell, *J. Chem. Soc.*, 1961, 5275.

2710. Parnell, *J. Chem. Soc.*, 1962, 2856.

2711. Taylor and Thompson, *J. Org. Chem.*, 26, 5224 (1961).

2712. Wiley, Lanet, and Hussung, *J. Het. Chem.*, 1, 175 (1964).

2713. Cohen and Vincze, *Israel J. Chem.*, 2, 1 (1964).

2714. Elderfield and Prasad, *J. Org. Chem.*, 25, 1583 (1960).

2715. Stuart, Wood, and Duncan, *J. Chem. Soc. (C)*, 1966, 285.

2716. Pfleiderer and Zondler, *Chem. Ber.*, 99, 3008 (1966).

2717. Pfleiderer and Nübel, *Chem. Ber.*, 95, 1615 (1962).

2718. Bühler and Pfleiderer, *Chem. Ber.*, 99, 2997 (1966).

2719. Shvachkin and Berestenko, *J. Gen. Chem.*, 32, 1712 (1962).

2720. Shvachkin and Berestenko, *J. Gen. Chem.*, 33, 3842 (1963).

2721. Segal and Shapiro, *J. Med. Pharm. Chem.*, 1, 371 (1959).

2722. Albert and Matsuura, *J. Chem. Soc.*, 1962, 2162.

2723. Lin and Price, *J. Org. Chem.*, 26, 108 (1961).

2724. Ikehara and Ohtsuka, *Chem. Pharm. Bull.*, 9, 27 (1961).

2725. Lister and Timmis, *J. Chem. Soc. (C)*, 1966, 1242.

2726. Lister, *J. Chem. Soc.*, 1960, 3394.

2727. Bogodist and Protsenko, U.S.S.R.Pat. 162,149 (1964); through *Chem. Abstr.*, 61, 13327 (1964).

2728. Protsenko and Bogodist, *Ukrain. khim. Zhur.*, 32, 867 (1966); through *Chem. Abstr.*, 66, 2531 (1967).

2729. Bogodist, *Ukrain. khim. Zhur.*, 32, 1091 (1966); through *Chem. Abstr.*, 66, 37870 (1967).

2730. Bogodist and Protsenko, *Ukrain. khim. Zhur.*, 32, 1094 (1966); *Chem. Abstr.*, 66, 94988 (1967).

2731. Bogodist, *Ukrain. khim. Zhur.*, 33, 87 (1967); through *Chem. Abstr.*, 66, 94989 (1967).

2732. Bredereck, Simchen, Santos, and Wagner, *Angew. Chem.*, 78, 717 (1966).

2733. Lin and Price, *J. Org. Chem.*, 26, 264 (1961).

2734. Temple, Kussner, and Montgomery, *J. Med. Pharm. Chem.*, 5, 866 (1962).

2735. Nagano, Inoue, Saggiomo, and Nodiff, *J. Med. Chem.*, 7, 215 (1964).

2736. Schaeffer and Weimar, *J. Org. Chem.*, 25, 774 (1960).

2737. Wempen and Fox, *J. Med. Chem.*, 7, 207 (1964).

2738. Noell and Robins, *J. Med. Pharm. Chem.*, 5, 558 (1962).

2739. O'Brien, Baiocchi, Robins, and Cheng, *J. Org. Chem.*, 27, 1104 (1962).

2740. Hoffmann-La Roche, Brit.Pat. 946,487 (1964); through *Chem. Abstr.*, 64, 3568 (1966).

2741. Sen and Madan, *J. Indian Chem. Soc.*, 37, 617 (1960).

2742. Profft and Sitter, *Arch. Pharm.*, 296, 151 (1963).

2743. Blackburn and Johnson, *J. Chem. Soc.*, 1960, 4347.

2744. Peters, Minnemeyer, Spears, and Tieckelmann, *J. Org. Chem.*, 25, 2137 (1960).

2745. Brown and Lyall, *Aust. J. Chem.*, 15, 851 (1962).

2746. Brown and Foster, *Aust. J. Chem.*, 19, 2321 (1966).

2747. Blackburn and Johnson, *J. Chem. Soc.*, 1960, 4358.

2748. Durr, *J. Med. Chem.*, 8, 253 (1965).

2749. Nyberg and Cheng, *J. Het. Chem.*, 1, 1 (1964).

2750. Ziegler, Nölken, and Junek, *Monatsh.*, 93, 708 (1962).

2751. Baker, Santi, Coward, Shapiro, and Jordaan, *J. Het. Chem.*, 3, 425 (1966).

2752. Temple, McKee, and Montgomery, *J. Org. Chem.*, 28, 923 (1963).

2753. Camerino, Palamidessi, and Sciaky, *Gazzetta*, 90, 1830 (1960).

2754. Guither, Clark, and Castle, *J. Het. Chem.*, 2, 67 (1965).

2755. Liao, Baiocchi, and Cheng, *J. Org. Chem.*, 31, 900 (1966).

2756. Koppel, Springer, Daves, and Cheng, *J. Pharm. Sci.*, 52, 81 (1963).

2757. Chang, *J. Med. Chem.*, 8, 884 (1965).

2758. Klötzer and Herberz, *Monatsh.*, 96, 1573 (1965).

2759. Daniels, Grady, and Bauer, *J. Amer. Chem. Soc.*, 87, 1531 (1965).

2760. Brown and Teitei, *J. Chem. Soc.*, 1963, 4333.

2761. Warner and Bardos, *J. Med. Chem.*, 9, 977 (1966).

2762. Hopkins, Jonak, Tieckelmann, and Minnemeyer, *J. Org. Chem.*, 31, 3969 (1966).

2763. Naito, Hirata, and Kotake, Jap.Pat. 7281 (1965); through *Chem. Abstr.*, 63, 617 (1965).

2764. Naito, Hirata, and Kobayashi, Jap.Pat. 7900 (1965); through *Chem. Abstr.*, 63, 1802 (1965).

2765. Okuda and Kuniyoshi, *J. Pharm. Soc. Japan*, 82, 1035 (1962).

2766. Karlinskaya and Khromov-Borisov, *J. Gen. Chem.*, 27, 2113 (1957).

2767. Durr, *J. Med. Chem.*, 8, 140 (1965).

2768. Clampitt and Mueller, *J. Polymer Sci.*, 62, 15 (1962).

2769. Naito, Nagase, and Hirata, Jap.Pat. 10793 (1962); through *Chem. Abstr.*, 59, 3939 (1963).

2770. Naito, Nagasa, Hirata, and Okada, Jap.Pat. 20577 (1963); through *Chem. Abstr.*, 60, 2976 (1964).

2771. Klötzer, *Monatsh.*, 92, 1212 (1961).

2772. van Zwieten and Huisman, *Rec. Trav. chim.*, 81, 554 (1962).

2773. Falco, Roth, and Hitchings, *J. Org. Chem.*, 26, 1143 (1961).

2774. Kukolja and Cvetnic, *Croat. Chem. Acta*, 34, 115 (1962).

2775. Taylor and Garcia, *J. Org. Chem.*, 29, 2121 (1964).

2776. Brown and Teitei, *Aust. J. Chem.*, 18, 559 (1965).

2777. Israel, Protopapa, Schlein, and Modest, *J. Med. Chem.*, 7, 5 (1964).

2778. Daves, Noell, Robins, Koppel, and Beaman, *J. Amer. Chem. Soc.*, 82, 2633 (1960).

2779. Mautner, Chu, Jaffe, and Sartorelli, *J. Med. Chem.*, 6, 36 (1963).

2780. Trompen and Huisman, *Rec. Trav. chim.*, 85, 175 (1966).

2781. Brown and Teitei, *J. Chem. Soc.*, 1963, 3535; 1965, 832.

2782. Haco A.-G., Neth.Pat.Appl. 6,516,326 (1966); through *Chem. Abstr.*, 67, 54160 (1967).

2783. Kinugawa, Ochiai, and Yamamoto, *J. Pharm. Soc. Japan*, 83, 1086 (1963).

2784. Geigy A.-G. Fr.Pat. 1,377,696 (1964).

2785. Montgomery and Hewson, *J. Org. Chem.*, 30, 1528 (1965).

2786. Shvachkin, Cyrtsova, Berestenko, and Prokof'ev, *J. Gen. Chem.*, 32, 2060 (1962).

2787. Shvachkin, Syrtsova, Berestenko, and Prokof'ev, *Vestn. Mosk. Univ., Ser.II, Khim.*, 17, No.5, 75 (1962); through *Chem. Abstr.*, 58, 5669 (1963).

2788. Shvachkin and Berestenko, *J. Gen. Chem.*, 33, 3132 (1963).

2789. Shvachkin and Berestenko, *Vestn. Mosk. Univ., Ser.II, Khim.*, 18, No.1, 74 (1963); through *Chem. Abstr.*, 59, 2811 (1963).

2790. Shvachkin and Berestenko, *Vestn. Mosk. Univ., Ser.II, Khim.*, 20, 91 (1965); through *Chem. Abstr.*, 63, 9943 (1965).

2791. Miller and Rose, *J. Chem. Soc.*, 1965, 3357.

2792. Arbuzov and Lugovkin, *J. Gen. Chem.*, 22, 1199 (1952).

2793. Roy, *Diss. Abstr.*, 21, 295 (1960).

2794. Kajihara, Hirata, and Odaka, *J. Chem. Soc. Japan, Pure Chem. Sect.*, 87, 884 (1966).

2795. Zagulyaeva and Mamaev, *Izvest. Akad. Nauk S.S.S.R., Ser. Khim.*, 1965, 2087.

2796. Shvachkin and Berestenko, *J. Gen. Chem.*, 32, 1712 (1962).

2797. Shvachkin and Berestenko, *J. Gen. Chem.*, 33, 2842 (1963).

2798. Shvachkin, Berestenko, and Lapuk, *Vestn. Mosk. Univ., Ser.II, Khim.*, 17, No.3, p.74; through *Chem. Abstr.*, 58, 4564 (1963).

2799. Van Meeteren and van der Plas, *Rec. Trav. chim.*, 86, 15 (1967).

2800. Van Meeteren and van der Plas, *Rec. Trav. chim.*, 86, 567 (1967).

2801. Van Meeteren and van der Plas, *Tetrahedron Letters*, 1966, 4517.

2802. Paul and Sen, *Indian J. Chem.*, 1, 98 (1963).

2803. Paul and Sen, *Indian J. Chem.*, 2, 212 (1964).

2804. Schwan and Tieckelmann, *J. Org. Chem.*, 29, 941 (1964).

2805. Gunar, Mikhailopulo, and Zav'yalov, *Izvest. Akad. Nauk S.S.S.R., Ser. Khim.*, 1966, 1496.

2806. Robba, *Ann. Chim. (France)*, 5, 351 (1960).

2807. Matzke, Austrian Pat. 233,568 (1964); through *Chem. Abstr.*, 61, 5664 (1964).

2808. Matzke, Austrian Pat. 242,146 (1965); through *Chem. Abstr.*, 63, 13282 (1965).

2809. Matzke, Austrian Pat. 242,148 (1965); through *Chem. Abstr.*, 63, 13282 (1965).

2810. Matzke, Ger.Pat. 1,204,235 (1965); through *Chem. Abstr.*, 64, 2103 (1966).

2811. Roth and Schloemer, *J. Org. Chem.*, 28, 2659 (1963).

2812. Roth and Bunnett, *J. Amer. Chem. Soc.*, 87, 340 (1965).

2813. Wellcome Foundation, Brit.Pat. 1,015,784 (1966); through *Chem. Abstr.*, 64, 9742 (1966).

2814. Rajkumar and Binkley, *J. Med. Chem.*, 6, 550 (1963).

2815. Grundmann and Richter, *J. Org. Chem.*, 32, 2308 (1967).

2816. Baker, Lourens, and Jordaan, *J. Het. Chem.*, 4, 39 (1967).

2817. Schellenberger and Winter, *Z. physiol. Chem.*, 322, 173 (1960).

2818. David and Hirshfeld, *Bull. Soc. chim. France*, 1966, 527.

2819. Shvachkin, Filatova, and Syrtsova, *Vestn. Mosk. Univ., Ser.II, Khim.*, 18, No.2, 55 (1963); through *Chem. Abstr.*, 59, 6404 (1963).

2820. Giner-Sorolla and Bendich, *J. Org. Chem.*, 31, 4239 (1966).

2821. Hauptmann, Kluge, Seidig, and Wilde, *Angew. Chem.*, 77, 678 (1965).

2822. Shvachkin and Syrtsova, *Vestn. Mosk. Univ., Ser.II, Khim.*, 16, No.6, 75 (1961); through *Chem. Abstr.*, 57, 8577 (1962).

2823. Shvachkin, Syrtsova, and Savel'ev, *Vestn. Mosk. Univ., Ser.II, Khim.*, 17, No.1, 73 (1962); through *Chem. Abstr.*, 58, 4565 (1963).

2824. Shvachkin and Syrtsova, *Vestn. Mosk. Univ., Ser.II, Khim.*, 18, No.4, 88 (1963); through *Chem. Abstr.*, 59, 10229 (1963).

2825. Shvachkin, Novikova, Reznikova, and Padyukova, *J. Gen. Chem.*, 33, 4022 (1963).

2826. Shvachkin and Shprunka, *Vestn. Mosk. Univ., Ser.II, Khim.*, 19, No.6, 72 (1964); through *Chem. Abstr.*, 62, 6481 (1965).

2827. Shvachkin and Shprunka, *J. Gen. Chem.*, 35, 2251 (1965).

2828. Chadwick, Hampshire, Hebborn, Triggle, and Triggle, *J. Med. Chem.*, 9, 874 (1966).

2829. Prystas and Sorm, *Coll. Czech. Chem. Comm.*, 31, 1035 (1966).

2830. Lyashenko, Kolesova, Aleksandr, and Sheremet'eva, *J. Gen. Chem.*, 34, 2752 (1964).

2831. Serfontein and Schröder, *J. S. African Chem. Inst.*, 19, 38 (1966).

2832. Shvachkin and Berestenko, *J. Gen. Chem.*, 34, 3506 (1964).

2833. Shvachkin, Berestenko, and Mishin, *Vestn. Mosk. Univ., Ser.II, Khim.*, 20, No.4, 89 (1965); through *Chem. Abstr.*, 63, 16453 (1965).

2834. Schulze and Willitzer, *J. prakt. Chem.*, 33, 50 (1966).

2835. Matsuura and Goto, *J. Chem. Soc.*, 1965, 623.

2836. Takamizawa, Hirai, Hamashima, and Hata, *Chem. Pharm. Bull.*, 12, 558 (1964).

2837. Nitta, Okui, and Ito, *Chem. Pharm. Bull.*, 13, 557 (1965).

2838. Goldner, Dietz, and Carstens, *Annalen*, 692, 134 (1966).

2839. Nitta and Okui, Jap. Pat. 10335 (1962); through *Chem. Abstr.*, 59, 5176 (1963).

2840. Nitta and Okui, Jap. Pat. 10336 (1962); through *Chem. Abstr.*, 59, 5177 (1963).

2841. Nitta, Okui, and Ito, Jap. Pat. 220 (1965); through *Chem. Abstr.*, 62, 11826 (1965).

2842. Nishiwaki, *Chem. Pharm. Bull.*, 9, 38 (1961).

2843. Okuda and Kuniyoshi, *J. Pharm. Soc. Japan*, 82, 1031 (1962).

2844. Okuda and Kuniyoshi, Jap. Pat. 3142 (1963); through *Chem. Abstr.*, 59, 11528 (1963).

2845. Zemlicka, Smrt, and Sorm, *Tetrahedron Letters*, 1962, 397.

2846. Zemlicka and Sorm, *Coll. Czech. Chem. Comm.*, 30, 2052 (1965).

2847. Prystas and Sorm, *Coll. Czech. Chem. Comm.*, 31, 1053 (1966).

2848. Arnold, *Coll. Czech. Chem. Comm.*, 24, 4048 (1959).

2849. Bosshard, Mory, Schmid, and Zollinger, *Helv. Chim. Acta*, 42, 1653 (1959).

2850. Papesch and Dodson, *J. Org. Chem.*, 28, 1329 (1963).

2851. Brown and Jacobsen, *J. Chem. Soc.*, 1961, 4413.

2852. Jones, Mian, and Walker, *J. Chem. Soc. (C)*, 1966, 692.

2853. Israel, Schlein, Maddock, Farber, and Modest, *J. Pharm. Sci.*, 55, 568 (1966).

2854. Lynch and Poon, *Canad. J. Chem.*, 45, 1431 (1967).

2855. Brown, in "Mechanisms of Molecular Migrations," ed. B. S. Thyagarajan, Wiley, New York, 1968, Vol.1, p.209.

2856. Clark, Gelling, and Neath, *Chem. Comm.*, 1967, 859.

2857. Biffin, Brown and Porter, *Tetrahedron Letters*, 1967, 2029.

2858. Biffin, Brown, and Porter, *J. Chem. Soc. (C)*, 1968, 2159.

2859. Taylor and Morrison, *J. Org. Chem.*, 32, 2379 (1967).

2860. Shvachkin and Azarova, *J. Gen. Chem.*, 35, 563 (1965).

2861. Batterham, Brown, and Paddon-Row, *J. Chem. Soc. (B)*, 1967, 171.

2862. Khromov-Borisov and Karlinskaya, *J. Gen. Chem.*, 26, 1728 (1956).

2863. Konrat'eva, Gunar, Kogan, and Zav'yalov, *Izvest. Akad. Nauk S.S.S.R., Ser. Khim.*, 1966, 1219.

2864. Bucha, Cupery, Harrod, Loux, and Ellis, *Science*, 137, 537 (1962).

2865. Sekiya and Osaki, *J. Pharm. Soc. Japan*, 86, 854 (1966).

2866. Frankel, Belsky, Gertner, and Zilkha, *J. Chem. Soc. (C)*, 1966, 493.

2867. Cohen, Thom, and Bendich, *J. Org. Chem.*, 27, 3545 (1962).

2868. Osdene, Santilli, McCardle, and Rosenthale, *J. Med. Chem.*, 9, 697 (1966).

2869. Zvilichovsky, *Tetrahedron*, 23, 353 (1967).

2870. Kazmirowski, Dietz, and Carstens, *J. prakt. Chem.*, 35, 136 (1967).

2871. Hayashi, *Nagoya Shiritsu Yakugakubu Kenkyu Nempo*, 12, 42 (1964); through *Chem. Abstr.*, 64, 19605 (1966).

2872. Arct, Prelicz, and Witek, *Roczniki Chem.*, 41, 683 (1967); *Chem. Abstr.*, 67, 54098 (1967).

2873. Brokke, U.S.Pat.3,250,775 (1966); through *Chem. Abstr.*, 65, 3888 (1966).

2874. Simon and Levshina, *Tr. Ukr. Nauch.—Issled. Inst. Eksp. Endokrinol.*, 20, 193 (1965); through *Chem. Abstr.*, 66, 85752 (1967).

2875. Vargha and Balazs, *Studia Univ. Babes-Bolyai, Ser. Chem.*, 10, 79 (1965); through *Chem. Abstr.*, 64, 19606 (1966).

2876. Partyka, U.S.Pat. 3,296,261 (1967); through *Chem. Abstr.*, 66, 55508 (1967).

2877. Ciba, Neth.Pat.Appl. 6,503,747 (1965); through *Chem. Abstr.*, 64, 11221 (1966).

2878. Huang, Chou, Tao, Chien, Lu, Wang, Chen, Wu, and Chang, *Hua Hsueh Hsueh Pao*, 31, 502 (1965); through *Chem. Abstr.*, 64, 17589 (1966).

2879. Wamhoff and Korte, *Chem. Ber.*, 100, 1324 (1967).

2880. Nantka-Namirski and Wojciechowski, *Roczniki Chem.*, 41, 669 (1967); *Chem. Abstr.*, 67, 54099 (1967).

2881. Turkevich and Kovaliv, *Ukrain. khim. Zhur.*, 31, 607 (1965); through *Chem. Abstr.*, 63, 7006 (1965).

2882. Zondler, Forrest, and Lagowski, *J. Het. Chem.*, 4, 124 (1967).

2883. Heidelberger, *Ann. Rev. Pharmacol.*, 7, 101 (1967).

2884. Britikova, Chkhikvadze, and Magidson, *Khim. Geterotsikl. Soedin.*, 1966, 783; through *Chem. Abstr.*, 67, 64344 (1967).

2885. Wolski, Trzeciak, and Ledochowski, *Roczniki Chem.*, 41, 215 (1967); *Chem. Abstr.*, 67, 64343 (1967).

2886. Yuki and Hayakawa, *J. Pharm. Soc. Japan*, 87, 458 (1967).

2887. Brown and Paddon-Row, *J. Chem. Soc. (C)*, 1967, 1856.

2888. Kato, Yamanaka, and Shibata, *Chem. Pharm. Bull.*, 15, 921 (1967).

2889. Kato, Yamanaka, and Shibata, *Tetrahedron*, 23, 2965 (1967).

2890. Streef and den Hertog, *Rec. Trav. chim.*, 85, 803 (1966).

2891. Lahmani, Ivanoff, and Magat, *Compt. rend.*, 263, 1005 (1966).

2892. Ried and Stock, *Annalen*, 700, 87 (1966).

2893. Hubert-Habart, Takagi, Cheutin, and Royer, *Bull. Soc. chim. Fr.*, 1966, 1587.

2894. Hubert-Habart, Takagi, and Royer, *Bull. Soc. chim. Fr.*, 1967, 356.

2895. Meriwether, U.S. Pat. 3,113,945 (1963); through *Chem. Abstr.*, 60, 5519 (1964) and *Chem. Zentr.*, 137, 3-2472 (1966).

2896. Kabbe, *Annalen*, 704, 144 (1967).

2897. Prystas and Sorm, *Coll. Czech. Chem. Comm.*, 31, 3990 (1966).

2898. Taylor and Berger, *J. Org. Chem.*, 32, 2376 (1967).

2899. Gewald, *J. prakt. Chem.*, 32, 26 (1966).

2900. Montgomery, Hewson, Clayton, and Thomas, *J. Org. Chem.*, 31, 2202 (1966).

2901. Lister, *Adv. Het. Chem.*, 6, 1 (1966).

2902. Grigat, Pütter, and Mühlbauer, *Chem. Ber.*, 98, 3777 (1965).

2903. Grigat and Pütter, *Chem. Ber.*, 100, 1385 (1967).

2904. Gunar, Mikhailopulo, Ovechkina, and Zav'yalov, *Khim. Geterotsikl. Soedin.*, 1967, 48; through *Chem. Abstr.*, 67, 54097 (1967).

2905. Goerdeler and Wieland, *Chem. Ber.*, 100, 47 (1967).

2906. Schulze, Willitzer, and Fritzsche, *Chem. Ber.*, 99, 3492 (1966).

2907. Jutz and Müller, *Angew. Chem.*, 78, 1059 (1966).

2908. Wamhoff and Korte, *Chem. Ber.*, 99, 872 (1966).

2909. Zondler and Pfleiderer, *Chem. Ber.*, 99, 2984 (1966).

2910. Nagpal and Dhar, *Tetrahedron*, 23, 1297 (1967).

2911. Paudler and Kuder, *J. Org. Chem.*, 31, 809 (1966).

2912. Saikawa and Wada, Jap.Pat. 20978 (1965); through *Chem. Abstr.*, 64, 2105 (1966).

2913. Wozniki, Dolewski, Jankowski, Karwowski, and Kwiatkowski, *Bull. Acad. polon. Sci., Ser. Sci. math. astron. phys.*, 12, 655 (1964).

2914. Kwiatkowski and Zurawski, *Bull. Acad. polon. Sci., Ser. Sci. math. astron. phys.*, 13, 487 (1965).

2915. Arantz and Brown, in "Synthetic Procedures in Nucleic Acid Chemistry," ed. W. Zorbach and R. Tipson, Wiley, New York, 1968, Vol.1, p.55.

2916. Simon and Meneghini, *Arquivos Istituto Biologico*, 30, 69 (1963); *Chem. Abstr.*, 61, 8130 (1964).

2917. Shvachkin, Berestenko, and Mishin, *Vestn. Mosk. Univ., Ser. II, Khim.*, 18, No. 3, 82 (1963); through *Chem. Abstr.*, 59, 7525 (1963).

2918. Poddubnaya and Lavrenova, *Vestn. Mosk. Univ., Ser. II, Khim.*, 18, No.5, 65 (1963); through *Chem. Abstr.*, 60, 1750 (1964).

2919. Shvachkin, Berestenko, and Tung-Hai, *Khim. Geterotsikl. Soedin., Akad. Nauk Latv. S.S.R.*, 1965, 626; through *Chem. Abstr.*, 64, 801 (1966).

2920. Wojciechowski, *Acta Polon. Pharm.*, 19, 121 (1962); *Chem. Abstr.*, 59, 1633 (1963).

2921. Wojciechowski, Pol. Pat. 48,707 (1964); through *Chem. Abstr.*, 63, 18113 (1965).

2922. Wojciechowski, Pol. Pat. 48,869 (1964); through *Chem. Abstr.*, 63, 18114 (1965).

2923. Liao, Podrebarac, and Cheng, *J. Amer. Chem. Soc.*, 86, 1869 (1964).

2924. Gronowitz and Röe, *Acta Chem. Scand.*, 19, 1741 (1965).

2925. Pfleiderer and Lohrmann, *Chem. Ber.*, 95, 738 (1962).

2926. Pfleiderer, Sagi, and Grözinger, *Chem. Ber.*, 99, 3530 (1966).

2927. Pfleiderer and Bühler, *Chem. Ber.*, 99, 3022 (1966).

2928. Pfleiderer, Bunting, Perrin, and Nübel, *Chem. Ber.*, 99, 3503 (1966).

2929. Bühler and Pfleiderer, *Chem. Ber.*, 100, 492 (1967).

2930. Bredereck, Herlinger, and Graudums, *Chem. Ber.*, 95, 54 (1962).

2931. Bredereck, Gompper, Schuh, and Teilig, in "Newer Methods of Preparative Organic Chemistry," ed. Foerst, Academic Press, New York, 1964, Vol.3, pp. 241-301.

2932. Schmidt, *Chem. Ber.*, 98, 334 (1965).

2933. Shealy, Clayton, and Montgomery, *J. Org. Chem.*, 27, 2154 (1962).

2934. Temple, McKee, and Montgomery, *J. Org. Chem.*, 27, 1671 (1962).

2935. Woolridge and Slack, *J. Chem. Soc.*, 1962, 1863.

2936. Shadbolt and Ulbricht, *Chem. and Ind.*, 1966, 459.

2937. Perrin, "Dissociation Constants of Organic Bases in Aqueous Solution'. Butterworths, London, 1965.

2938. Ganguly, Sen, and Guha, *Naturwiss.*, 48, 695 (1961).

2939. Ganguly and Sen, *Indian J. Chem.*, 1, 364 (1963).

2940. Baker and Santi, *J. Pharm. Sci.*, 56, 380 (1967).

2941. Baker and Shapiro, *J. Pharm. Sci.*, 55, 308 (1966).

2942. Baker, Ho, Coward, and Santi, *J. Pharm. Sci.*, 55, 302 (1966).

2943. Baker and Meyer, *J. Pharm. Sci.*, 56, 570 (1967).

2944. Baker, Jackson, and Meyer, *J. Pharm. Sci.*, 56, 566 (1967).

2945. Albert and Armarego, *Adv. Het. Chem.*, 4, 1 (1965).

2946. Perrin, *Adv. Het. Chem.*, 4, 43 (1965).

2947. Zee-Cheng and Cheng, *J. Het. Chem.*, 4, 163 (1967).

2948. Albert and Matsuura, *J. Chem. Soc.*, 1961, 5131.

2949. Osdene, Santilli, McCardle, and Rosenthale, *J. Med. Chem.*, 10, 165 (1967).

2950. Pfleiderer and Kempter, *Angew. Chem.*, 79, 234a (1967).

2951. Pfleiderer and Kempter, *Angew. Chem.*, 79, 234b (1967).

2952. Fissekis and Markert, *J. Org. Chem.*, 31, 2945 (1966).

2953. O'Brien, Springer, and Cheng, *J. Het. Chem.*, 3, 115 (1966).

2954. Daves, O'Brien, Lewis, and Cheng, *J. Het. Chem.*, 1, 130 (1964).

2955. Whitlock, Lipton, and Strong, *J. Org. Chem.*, 30, 115 (1965).

2956. Triggle and Triggle, *J. Pharm. Sci.*, 54, 795 (1965).

2957. Brown and Evans, *J. Chem. Soc.*, 1962, 4039.

2958. Taylor, Warrener, and McKillop, *Angew. Chem.*, 78, 333 (1966).

2959. Taylor, McKillop, and Warrener, *Tetrahedron*, 23, 891 (1967).

2960. Taylor, Vromen, Ravindranathan, and McKillop, *Angew. Chem.*, 78, 332 (1966).

2961. Taylor, McKillop, and Vromen, *Tetrahedron*, 23, 885 (1967).

2962. Bauer, Nambury, and Hershenson, *J. Het. Chem.*, 3, 224 (1966).

2963. Cheng and Lewis, *J. Het. Chem.*, 1, 260 (1964).

2964. Warrener and Cain, *Angew. Chem.*, 78, 491 (1966).

2965. Warrener and Cain, *Tetrahedron Letters*, 1966, 3225.

2966. Warrener and Cain, *Tetrahedron Letters*, 1966, 3231.

2967. Varner and Bingeman, reported in *Chem. Eng. News*, Jan. 1, 1962, p. 50.

2968. Hilton, Monaco, Moreland, and Gentner, *Weeds*, 12, 129 (1964).

2969. Hoffmann, McGahen, and Sweetser, *Nature*, 202, 577 (1964).

2970. Roy-Burman, Roy, and Sen, *Naturwiss.*, 47, 515 (1960).

2971. Roy-Burman, Sen, and Guha, *Naturwiss.*, 48, 737 (1961).

2972. Roy-Burman and Sen, *Naturwiss.*, 49, 494 (1962).

2973. Glicksman, *J. Electrochem. Soc.*, 109, 352 (1962).

2974. Ochiai and Morita, *Tetrahedron Letters*, 1967, 2349.

2975. Goodman and Krishna, *Rev. Mod. Phys.*, 35, 541 (1963).

2976. Wright, Bauer, and Bell, *J. Het. Chem.*, 3,440 (1966).

2977. Heine, Weese, Cooper, and Durbetaki, *J. Org. Chem.*, 32, 2708 (1967).

2978. Baker and Jordaan, *J. Het. Chem.*, 3, 315 (1966).

2979. Baker and Jordaan, *J. Het. Chem.*, 3, 319 (1966).

2980. Baker and Jordaan, *J. Het. Chem.*, 3, 324 (1966).

2981. Baker and Jordaan, *J. Het. Chem.*, 4, 31 (1967).

2982. Kishikawa and Yuki, *Chem. Pharm. Bull.*, 14, 1365 (1966).

2983. Trattner, Elion, Hitchings, and Sharefkin, *J. Org. Chem.*, 29, 2674 (1964).

2984. Ueda and Fox, *J. Org. Chem.*, 29, 1770 (1964).

2985. Sirakawa, *J. Pharm. Soc. Japan*, 80, 956 (1960).

2986. Goerdeler and Roth, *Chem. Ber.*, 96, 534 (1963).

2987. Katritzky and Waring, *J. Chem. Soc.*, 1963, 3046.

2988. Skaric, Gaspert, and Skaric, *Croat. Chem. Acta*, 36, 87 (1964).

2989. D. D. Bly, personal communication.

2990. A. R. Katritzky, personal communication.

2991. Hurst and Kuksis, *Canad. J. Biochem. Physiol.*, 36, 931 (1958).

2992. Jones, Mian, and Walker, *J. Chem. Soc. (C)*, 1966, 1784.

2993. Kalatzis, *J. Chem. Soc. (B)*, 1969,

2994. Ulbricht, *J. Chem. Soc.*, 1961, 3345.

2995. Kheifets and Khromov-Borisov, *Zhur. Org. Khim.*, 1, 1173 (1965); through *Chem. Abstr.*, 63, 11555 (1965).

2996. Noell and Cheng, *J. Het. Chem.*, 3, 5 (1966).

2997. Brown and Teitei, *Aust. J. Chem.*, 17, 567 (1964).

2998. Lavrenova and Poddubnaya, *J. Gen. Chem.*, 34, 2864 (1964).

2999. Levin and Kukhtin, *J. Gen. Chem.*, 32, 1709 (1962).

3000. Budesinsky, Prikryl, and Svatek, *Coll. Czech. Chem. Comm.*, 32, 1637 (1967).

3001. Prystas and Sorm, *Coll. Czech. Chem. Comm.*, 32, 1298 (1967).

3002. Carpenter and Shaw, *J. Chem. Soc.*, 1965, 3987.

3003. David, Estramareix, Hirshfeld, and Sinay, *Bull. Soc. chim. Fr.*, 1964, 936.

3004. Schütte and Woltersdorf, *Annalen*, 684, 209 (1965).

3005. Schellenberger and Müller, *Z. physiol. Chem.*, 328, 220 (1962).

3006. Schwan, Holland, and Tieckelmann, *J. Het. Chem.*, 1, 299 (1964).

3007. Schwan, Tieckelmann, Holland, and Bryant, *J. Med. Chem.*, 8, 750 (1965).

3008. Ebetino and Amstutz, *J. Med. Chem.*, 7, 389 (1964).

3009. Wong, Brown, and Rapoport, *J. Org. Chem.*, 30, 2398 (1965).

3010. Reznik and Pashkurov, *Izvest. Akad. Nauk S.S.S.R., Ser.Khim.*, 1966, 1613.

3011. Singh and Lal, *J. Indian Chem. Soc.*, 40, 195 (1963).

3012. Mizuno, Okuyama, Hayatsu, and Ukita, *Chem. Pharm. Bull.*, 12, 1240 (1964).

3013. Tuppy and Küchler, *Monatsh.*, 95, 1698 (1964).

3014. Farkas and Sorm, *Coll. Czech. Chem. Comm.*, 28, 1620 (1963).

3015. Jeffrey, Mootz, and Mootz, *Acta Cryst.*, 19, 691 (1965).

3016. Bryan and Tomita, *Nature*, 192, 812 (1961).

3017. Bryan and Tomita, *Acta Cryst.*, 15, 1174 (1962).

3018. Jeffrey and Kinoshita, *Acta Cryst.*, 16, 20 (1963).

3019. Hoogsteen, *Acta Cryst.*, 16, 28 (1963).

3020. Bolton, *Acta Cryst.*, 16, 166 (1963).

3021. Jeffrey, Ghose, and Warwicker, *Acta Cryst.*, 14, 881 (1961).

3022. Craven, *Acta Cryst.*, 17, 282 (1964).

3023. Bolton, *Acta Cryst.*, 17, 147 (1964).

3024. Singh, *Acta Cryst.*, 13, 1036 (1960).

3025. Alexander and Pitman, *Acta Cryst.*, 9, 501 (1956).

3026. Bolton, *Acta Cryst.*, 19, 1051 (1965).

3027. Bolton, *Acta Cryst.*, 16, 950 (1963).

3028. Craven, Martinez-Carrera, and Jeffrey, *Acta Cryst.*, 17, 891 (1964).

3029. Bertinotti, Bonamico, Braibanti, Coppola, and Giacomello, *Ann. Chim. (Italy)*, 49, 825 (1959); through *Chem. Abstr.*, 53, 21973 (1959).

3030. Berthou, Cavelier, Marek, Rerat, and Rerat, *Compt. rend.*, 255, 1632 (1962).

3031. Craven and Mascarenhas, *Acta Cryst.*, 17, 407 (1964).

3032. Craven and Takei, *Acta Cryst.*, 17, 415 (1964).

3033. Gillier, *Compt. rend.*, 257, 427 (1963).

3034. Silverman and Yannoni, *Acta Cryst.*, 18, 756 (1965).

3035. B. M. Craven, personal communication.

3036. Katritzky and Lagowski, *Adv. Het. Chem.*, 1, 311 (1963); 2, 1-80(1963).

3037. Katritzky and Waring, *Chem. and Ind.*, 1962, 695.

3038. Inoue, Furutachi, and Nakanishi, *J. Org. Chem.*, 31, 175 (1966).

3039. Kokko, Goldstein, and Mandell, *J. Amer. Chem. Soc.*, 83, 2909 (1961).

3040. Dekker, *Ann. Rev. Biochem.*, 29, 453 (1960).

3041. Kokko, Mandell, and Goldstein, *J. Amer. Chem. Soc.*, 84, 1042 (1962).

3042. Nakanishi, Suzuki, and Yamazaki, *Bull. Chem. Soc. Japan*, 34, 53 (1961).

3043. Katritzky and Jones, *Chem. and Ind.*, 1961, 722.

3044. Wittenburg, *Chem. Ber.*, 99, 2391 (1966).

3045. Wierzchowski, Litonska, and Shugar, *J. Amer. Chem. Soc.*, 87, 4621 (1965).

3046. Katritzky and Waring, *J. Chem. Soc.*, 1962, 1540.

3047. Cook, *Canad. J. Chem.*, 44, 335 (1966).

3048. Berens and Shugar, *Acta Biochim. Polon.*, 10, 25 (1963).

3049. Roberts, Lambert, and Roberts, *J. Amer. Chem. Soc.*, 87, 5439 (1965).

3050. Miles, Bradley, and Becker, *Science*, 142, 1569 (1963).

3051. Kheifets and Khromov-Borisov, *J. Gen. Chem.*, 34, 3134 (1964).

3052. Katritzky, Popp, and Waring, *J. Chem. Soc. (B)*, 1966, 565.

3053. Kheifets, Khromov-Borisov, and Kol'tsov, *Doklady Akad. Nauk S.S.S.R.*, 166, 635 (1966); through *Chem. Abstr.*, 64, 11207 (1966).

3054. Kheifets and Khromov-Borisov, *Zh. org. Khim.*, 2, 1511 (1966).

3055. Kheifets, Khromov-Borisov, and Kol'tsov, *Zh. org. Khim.*, 2, 1516 (1966).

3056. Kheifets, Khromov-Borisov, Kol'tsov, and Volkenstein, *Tetrahedron*, 23, 1197 (1967).

3057. Mizuno, Ikehara, and Watanabe, *Chem. Pharm. Bull.*, 10, 647 (1962).

3058. Bergmann and Kalmus, *J. Chem. Soc.*, 1962, 860.

3059. Beaman and Robins, *J. Amer. Chem. Soc.*, 83, 4038 (1961).

3060. Ueda, Tsuji, and Momona, *Chem. Pharm. Bull.*, 11, 912 (1963).

3061. McCormack and Mautner, *J. Org. Chem.*, 29, 3370 (1964).

3062. Nairn and Tieckelmann, *J. Org. Chem.*, 25, 1127 (1960).

3063. Brossmer and Ziegler, *Tetrahedron Letters*, 1966, 5253.

3064. Murdoch and Angier, *J. Amer. Chem. Soc.*, 84, 3758 (1962).

3065. Laland and Serek-Hanssen, *Biochem. J.*, 90, 76 (1964).

3066. Helgeland and Laland, *Biochim. Biophys. Acta*, 87, 353 (1964).

3067. Birkofer, Richter, and Ritter, *Chem. Ber.*, 93, 2804 (1960).

3068. Birkofer, Kühlthau, and Ritter, *Chem. Ber.*, 93, 2810 (1960).

3069. Birkofer, Ritter, and Kühlthau, *Angew. Chem.*, 75, 209 (1963).

3070. Nishimura, Shimizu, and Iwai, *Chem. Pharm. Bull.*, 11, 1470 (1963).

3071. Nishimura and Iwai, *Chem. Pharm. Bull.*, 12, 352 (1964).

3072. Nishimura and Iwai, *Chem. Pharm. Bull.*, 12, 357 (1964).

3073. Nishimura and Shimizu, *Agric. Biol. Chem.*, 28, 224 (1964).

3074. Wittenburg, *Z. Chem.*, 4, 303 (1964).

3075. Wittenburg, *Angew. Chem.*, 77, 1043 (1965).

3076. Wittenburg, *Chem. Ber.*, 99, 2380 (1966).

3077. Bräuniger and Koine, *Arch. Pharm.*, 296, 665 (1963).

3078. Bräuniger and Koine, *Arch. Pharm.*, 296, 668 (1963).

3079. Birkofer and Ritter, *Angew. Chem.*, 77, 414 (1965).

3080. Lister, *J. Chem. Soc.*, 1960, 899.

3081. Bobranski and Wagner, *Roczniki Chem.*, 38, 237 (1964).

3082. Albert and McCormack, *J. Chem. Soc.*, 1965, 6930.

3083. Weiss, *Mikrochim. Acta,* 1961, 11.

3084. Lumbroso, Pigenet, Nasielski-Hinskens, and Promel, *Bull. Soc. chim. Fr.,* 1967, 1833.

3085. Daniels, Grady, and Bauer, *J. Org. Chem.*, 27, 4710 (1962).

3086. Wempen, Duschinsky, Kaplan, and Fox, *J. Amer. Chem. Soc.*, 83, 4755 (1961).

3087. Szer and Shugar, *Acta Biochim. Polon.*, 13, 177 (1966).

3088. Wierzchowski, *Postepy Biochem.*, 13, 127 (1967).

3089. Janion and Shugar, *Biochem. Biophys. Res. Comm.*, 18, 617 (1965).

3090. Janion and Shugar, *Biochim. Polon.*, 12, 337 (1965).

3091. W. Hepworth, personal communication.

3092. Cahn, "Survey of Chemical Publications," The Chemical Society, London, 1965, p. 58 *et seq.*

3093. Ulbricht, "Purines, Pyrimidines, and Nucleotides," Pergamon, Oxford, 1964.

3094. Ulbricht, "Introduction to the Chemistry of Nucleic Acids and Related Natural Products," Oldbourne, London, 1966.

3095. Hutchinson, "The Nucleotides and Coenzymes," Methuen, London, 1964.

3096. Jordan, "The Chemistry of Nucleic Acids," Butterworths, London, 1960.

3097. Chargaff, "Essays on Nucleic Acids," Elsevier, Amsterdam, 1963.

3098. Michelson, "The Chemistry of Nucleosides and Nucleotides," Academic, London, 1963.

3099. Steiner and Beins, "Polynucleotides," Elsevier, Amsterdam, 1961.

3100. Cantoni and Davies, "Procedures in Nucleic Acid Research," Harper and Row, New York, 1966, Vol.1–.

3101. Zorbach and Tipson, "Synthetic Procedures in Nucleic Acid Chemistry," Interscience, New York, 1968, Vol.1–.

3102. Scharffenberg and Beltz, "The Nucleic Acids," Pergamon, Oxford, 1962, Vol.1–.

3103. Davidson and Cohn, "Progress in Nucleic Acid Research," Academic, New York, 1963, Vol.1—.

3104. Zee-Cheng, Robins, and Cheng, *J. Org. Chem.*, 26, 1877 (1961).

3105. Mantescu, Genunche, and Balaban, *J. Labelled Compounds*, 1, 178 (1965).

3106. Lewin, *J. Chem. Soc.*, 1964, 792.

3107. Mautner and Bergson, *Acta Chem. Scand.*, 17, 1694 (1963).

3108. DeVoe and Tinoco, *J. Mol. Biol.*, 4, 500 (1962).

3109. Berthod and Pullman, *Compt. rend.*, 257, 2738 (1963).

3110. Pullman and Pullman, *Biochim. Biophys. Acta*, 64, 403 (1962).

3111. Berthod, Giessner-Prettre, and Pullman, *Theoretica Chim. Acta*, 5, 53 (1966).

3112. Gattner and Fahr, *Annalen*, 670, 84 (1963).

3113. Wang, *Photochem. Photobiol.*, 1, 135 (1962).

3114. Scholes, Ward, and Weiss, *Science*, 133, 2016 (1961).

3115. Fikus and Shugar, *Acta Biochim. Polon.*, 13, 39 (1966).

3116. Alcantara and Wang, *Photochem. Photobiol.*, 4, 473 (1965).

3117. Alcantara and Wang, *Photochem. Photobiol.*, 4, 465 (1965).

3118. Wang and Alcantara, *Photochem. Photobiol.*, 4, 477 (1965).

3119. Barszcz and Shugar, *Acta Biochim. Polon.*, 8, 455 (1961).

3120. Scholes, Weiss, and Wheeler, *Nature*, 178, 157 (1956).

3121. Ekert and Monier, *Nature*, 188, 309 (1960).

3122. Hems, *Nature*, 186, 710 (1960).

3123. Nofre, Bobillier, Chabre, and Cier, *Bull. Soc. chim. Fr.*, 1965, 2820.

3124. Daniels and Grimson, *Nature*, 197, 484 (1963).

3125. Wacker, *Progr. Nucleic Acid Res.*, 1, 369 (1963).

3126. Beukers, Ylstra, and Berends, *Rec. Trav. chim.*, 77, 729 (1958).

3127. Beukers and Berends, *Biochim. Biophys. Acta*, 41, 550 (1960).

3128. Beukers and Berends, *Biochim. Biophys. Acta*, 49, 181 (1961).

3129. Wacker, Dellweg, and Lodemann, *Angew. Chem.*, 73, 64 (1961).

3130. Wang, *Nature*, 188, 844 (1960).

3131. Wang, *Nature*, 190, 690 (1961).

3132. Wang, *Nature*, 200, 879 (1963).

3133. Wulf and Fraenkel, *Biochim. Biophys. Acta*, 51, 332 (1961).

3134. Varghese and Wang, *Nature*, 213, 909 (1967).

3135. Wang, *Photochem. Photobiol.*, 3, 395 (1964).

3136. Blackburn and Davies, *Chem. Comm.*, 1965, 215.

3137. Blackburn and Davies, *J. Chem. Soc. (C)*, 1966, 2239.

3138. Anet, *Tetrahedron Letters*, 1965, 3713.

3139. Hollis and Wang, *J. Org. Chem.*, 32, 1620 (1967).

3140. Ishihara, *Photochem. Photobiol.*, 2, 455 (1963).

3141. Wacker, *J. Chim. phys.*, 58, 1041 (1961).

3142. Wacker, Weinblum, Träger, and Moustafa, *J. Mol. Biol.*, 3, 790 (1961).

3143. Smith, *Photochem. Photobiol.*, 2, 503 (1963).

3144. Fürst, Fahr, and Wieser, *Z. Naturforsch.*, 22b, 354 (1967).

3145. Johns, Rapaport, and Delbrück, *J. Mol. Biol.*, 4, 104 (1962).

3146. Wang, *Fed. Proc.*, 24, S-71 (1965).

3147. Rörsch, Beukers, Ijlstra, and Berends, *Rec. Trav. chim.*, 77, 423 (1958).

3148. Wang, Patrick, Varghese, and Rupert, *Proc. Nat. Acad. Sci. U.S.A.*, 57, 465 (1967).

3149. Varghese and Wang, *Science*, 156, 955 (1967).

3150. Ulbricht, *Progr. Nucleic Acid Res. and Mol. Biol.*, 4, 189 (1965).

3151. Gmelin, *Z. physiol. Chem.*, 316, 164 (1959).

3152. Gmelin, *Acta Chem. Scand.*, 15, 1188 (1961).

3153. Shaw and Dewar, *Proc. Chem. Soc.*, 1961, 216.

3154. Martinez and Lee, *J. Org. Chem.*, 30, 317 (1965).

3155. Kjaer, Knudsen, and Larsen, *Acta Chem. Scand.*, 15, 1193 (1961).

3156. Pochon, Herbert, and Pichat, *Comm. energie at. (France)*, *Rappt.* No 1822 (1961); through *Chem. Abstr.*, 55, 25967 (1961).

3157. Pullman, *Acta Cryst.*, 17, 1074 (1964).

3158. Andrews, U.S. Pat. 3,105,075 (1963); through *Chem. Abstr.*, 60, 1767 (1964).

3159. Kwart and Sarasohn, *J. Amer. Chem. Soc.*, 83, 909 (1961).

3160. Kwart, Spayd, and Collins, *J. Amer. Chem. Soc.*, 83, 2579 (1961).

3161. Patterson, Lazarow, and Levey, *J. Biol. Chem.*, 177, 187 (1949).

3162. Clark-Lewis and Edgar, *J. Chem. Soc.*, 1965, 5556.

3163. Clark-Lewis and Edgar, *J. Chem. Soc.*, 1965, 5551.

3164. Heinisch, Ozegowski, and Mühlstädt, *Chem. Ber.*, 97, 5 (1964).

3165. Heinisch, Ozegowski, and Mühlstädt, *Chem. Ber.*, 98, 3095 (1965).

3166. Sax, Migliore, and Baughman, *Analyt. Biochem.*, 3, 150 (1962).

3167. Freifelder, Geiszler, and Stone, *J. Org. Chem.*, 26, 203 (1961).

3168. Fulton and Lyons, *Aust. J. Chem.*, 21, 419 (1968).

3169. Jezdic and Rajinvajin, *Bull. Inst. Nucl. Sci. "Boris Kidrich,"* 12, 127 (1961).

3170. Novacek, Hesoun, and Gut, *Coll. Czech. Chem. Comm.*, 30, 1890 (1965).

3171. Guschlbauer, Favre, and Michelson, *Z. Naturforsch.*, 20b, 1141 (1965).

3172. Nishiwaki, *Tetrahedron*, 23, 2657 (1967).

3173. Sobell and Tomita, *Acta Cryst.*, 17, 122 (1964).

3174. Ghose, Jeffrey, Craven, and Warwicker, *Acta Cryst.*, 13, 1034 (1960).

3175. Cunliffe-Jones, *Spectrochim. Acta*, 21, 747 (1965).

3176. Momigny, Urbain, and Wankenne, *Bull. Soc. roy. Sci. Liège*, 34, 337 (1965).

3177. Van Duijneveldt, Gil, and Murrell, *Theoret. Chim. Acta*, 4, 85 (1966).

3178. Fink, *Arch. Biochem. Biophys.*, 107, 493 (1964).

3179. Letham, *J. Chromatog.*, 20, 184 (1965).

3180. Heidelberger, *Progr. Nucleic Acid Res. and Mol. Biol.*, 4, 1 (1965).

3181. Kheifets and Khromov-Borisov, *J. Gen. Chem.*, 34, 3851 (1964).

3182. Bergmann, Kleiner, Neimann, and Rashi, *Israel J. Chem.*, 2, 185 (1964).

3183. Borodkin, Jonsson, Cocolas and McKee, *J. Med. Chem.*, 10, 290 (1967).

3184. Triggle and Triggle, *J. Med. Chem.*, 10, 285 (1967).

3185. Segal, Hedgcoth, and Skinner, *J. Med. Pharm. Chem.*, 5, 871 (1962).

3186. Fu, Chinoporos, and Terzian, *J. Org. Chem.*, 30, 1916 (1965).

3187. Bauer, Wright, Mikrut, and Bell, *J. Het. Chem.*, 2, 447 (1965).

3188. Townsend and Robins, *J. Org. Chem.*, 27, 990 (1962).

3189. Beaman, Gerster, and Robins, *J. Org. Chem.*, 27, 986 (1962).

3190. Baker, Ho, and Neilson, *J. Het. Chem.*, 1, 79 (1964).

3191. Nagpal and Dhar, *Indian J. Chem.*, 3, 126 (1965).

3192. Schlein, Israel, Chatterjee, and Modest, *Chem. and Ind.*, 1964, 418.

3193. Gupta and Huennekens, *Biochemistry*, 6, 2168 (1967).

3194. Kleine-Natrop, *Acta Allergologica*, 21, 319 (1966).

3195. Chan and Miller, *Aust. J. Chem.*, 20, 1595 (1967).

3196. Clark and Neath, *J. Chem. Soc. (C)*, 1966, 1112.

3197. Armarego, "Quinazolines", Wiley, New York, 1967.

3198. Robins, in "Heterocyclic Compounds," Ed.Elderfield, Wiley, New York, 1967, Vol.8, p. 244 *et. seq.*

3199. Clark, *J. Chem. Soc. (C)* 1967, 1543.

3200. Ismail and Wibberley, *J. Chem. Soc. (C)*, 1967, 2613.

3201. Loader and Timmons, *J. Chem. Soc. (C)*, 1967, 1343.

3202. Korach and Bergmann, *J. Org. Chem.*, 14, 1118 (1949).

3203. Fosse, Hieulle, and Bass, *Compt. rend.*, 178, 811 (1924).

3204. Hayes and Hayes-Baron, *J. Chem. Soc. (C)*, 1967, 1528.

3205. Lingens and Schneider-Bernlöhr, *Angew. Chem.*, 76, 378 (1964).

3206. Lingens and Schneider-Bernlöhr, *Annalen*, 686, 134 (1965).

3207. Levene and Bass, *J. Biol. Chem.*, 71, 167 (1927).

3208. Caputto, Leloir, Cardini, and Paladini, *J. Biol. Chem.*, 184, 333 (1950).

3209. Baron and Brown, *J. Chem. Soc.*, 1955, 2855.

3210. Witzel, *Annalen*, 620, 126 (1959).

3211. Bhattacharya, *Indian J. Chem.*, 5, 62 (1967).

3212. Shimizu, *Ann. Sankyo Res. Lab.*, 19, 1 (1967).

3213. Acheson, Foxton, and Stubbs, *J. Chem. Soc. (C)*, 1968, 926.

3214. Polya and Shanks, *J. Chem. Soc.*, 1964, 4986.

3215. Delia, Olsen, and Brown, *J. Org. Chem.*, 30, 2766 (1965).

3216. Tsuji, Watanabe, Nakadai, and Toyoshima, *Chem. Pharm. Bull.*, 10, 9 (1962).

3217. Van Eyk and Veldstra, *Phytochemistry*, 5, 457 (1966).

3218. Camerino, Palamidessi, and Sciaky, *Gazzetta*, 90, 1821 (1960).

3219. Watanabe, Friedman, Cushley, and Fox, *J. Org. Chem.*, 31, 2942 (1966).

3220. Stanovnik and Tisler, *Arzneimittel-Forschung*, 14, 1004 (1964).

3221. Stanovnik and Tisler, *Farm. Vestnik.*, 14, 129 (1963); through *Chem. Abstr.*, 61, 6894 (1964).

3222. Stanovnik and Tisler, *Bull. sci., Conseil Acad. R.P.F. Yougoslavie*, 9, 2 (1964); through *Chem. Abstr.*, 62, 420 (1965).

3223. Shefter and Mautner, *J. Amer. Chem. Soc.*, 89, 1249 (1967).

3224. Shefter, James, and Mautner, *J. Pharm. Sci.*, 55, 643 (1966).

3225. Sano, *Chem. Pharm. Bull.*, 10, 320 (1962).

3226. Cushley, Wempen, and Fox, *J. Amer. Chem. Soc.*, 90, 709 (1968).

3227. Wempen, Brown, Ueda, and Fox, *Biochemistry*, 4, 54 (1965).

3228. Curran and Angier, *J. Org. Chem.*, 28, 2672 (1963).

3229. Pashkurov and Reznik, *Doklady Akad. Nauk S.S.S.R.*, 171, 874 (1967); through *Chem. Abstr.*, 67, 54097 (1967).

3230. Jen and Wang, *Acta Pharm.Sinica*, 10, 298 (1963).

3231. Doerr, Wempen, Clarke, and Fox, *J. Org. Chem.*, 26, 3401 (1961).

3232. Caldwell, Fidler, and Santora, *J. Med. Chem.*, 6, 58 (1963).

3233. Fel'dman and Biskupskaya, *Sintez Prirodn. Soedin, ikh Analogov i Fragmentov, Akad. Nauk S.S.R., Otd. Obshch. i. Tekhn. Khim.*, 1965, 266; through *Chem. Abstr.*, 65, 3953 (1966).

3234. Fel'dman and Kheifets, *J. Gen. Chem.*, 31, 755 (1961).

3235. Koppell, Springer, Robins, and Cheng, *J. Med. Pharm. Chem.*, 5, 639 (1962).

3236. Barlin and Brown, *J. Chem.Soc. (B)*, 1968, 1435.

3237. Barlin and Brown, *J. Chem.Soc. (B)*, 1969, in press.

3238. Bogodist, U.S.S.R. Pat. 172,811 (1965); through *Chem. Abstr.*, 64, 740 (1966).

3239. Rorig and Wagner, U.S. Pat. 3,149,109 (1964); through *Chem. Abstr.*, 63, 7022 (1965).

3240. Brown, Ford, and Paddon-Row, *J. Chem. Soc. (C)*, 1968, 1452.

3241. Okuda, Kuniyoshi, Oshima, and Nagasaki, *J. Pharm. Soc. Japan*, 82, 1039 (1962).

3242. Ashani, Edery, Zahavy, Künberg, and Cohen, *Israel J. Chem.*, 3, 133 (1965).

3243. Naito, Hirata, and Kobayashi, Jap.Pat. 7513 (1965); through *Chem. Abstr.*, 63, 1802 (1965).

3244. Fel'dman and Zlobina, *Mechenye Biol. Aktivn. Veschchestva, Sb. Statei,* 1962, 53; through *Chem. Abstr.*, 59, 7527 (1963).

3245. Albert and McCormack, *J. Chem. Soc. (C)*, 1968, 63.

3246. Van der Plas and Jongjan, *Tetrahedron Letters*, 1967, 4385.

3247. Johnston and Gallagher, *J. Org. Chem.*, 28, 1305 (1963).

3248. Acheson, Aplin, Gagan, Harrison, and Miller, *Chem. Comm.*, 1966, 451.

3249. Albert and Tratt, *J. Chem. Soc. (C)*, 1968, 344.

3250. Albert, *Chem. Comm.* 1967, 684.

3251. Albert, *J. Chem. Soc. (C)*, 1968, 2076.

3252. Albert and Tratt, *Angew. Chem.*, 78, 596 (1966).

3253. Sugimoto and Matsuura, *Research Bull.* (Nagoya Univ.), 11, 87 (1967).

3254. Kazmirowski and Carstens, *J. prakt. Chem.*, 32, 47 (1966).

3255. Patel and Brown, *Nature*, 214, 402 (1967).

3256. Schuster, *J. Mol. Biol.*, 3, 447 (1961).

3257. Brown and Schell, *J. Mol. Biol.*, 3, 709 (1961).

3258. Brown and Schell, *J. Chem. Soc.*, 1965, 208.

3259. Nagase, Hirata, and Inaoka, *J. Pharm. Soc. Japan*, 82, 528 (1962).

3260. Klötzer and Herberz, *Monatsh.*, 96, 1731 (1965).

3261. Klötzer, *Monatsh.*, 97, 1117 (1966).

3262. Baker and Coward, *J. Pharm. Sci.* 54, 714 (1965).

3263. Roth, Jäger, and Brandes, *Arch. Pharm.*, 298, 885 (1965).

3264. Roth and Brandes, *Arch. Pharm.*, 299, 612 (1966).

3265. Bredereck, Gompper, Effenberger, Popp, and Simchen, *Chem. Ber.*, 94, 1241 (1961).

3266. Bredereck and Richter, *Chem. Ber.*, 98, 131 (1965).

3267. Skulason, Piantadosi, Zambrana, and Irvin, *J. Med. Chem.*, 8, 292 (1965).

3268. Baker and Novotny, *J. Het. Chem.*, 4, 23 (1967).

3269. Jardetzky, Pappas, and Wade, *J. Amer. Chem. Soc.*, 85, 1657 (1963).

3270. Wempen and Fox, *J. Amer. Chem. Soc.*, 86, 2474 (1964).

3271. Ulbricht, *J. Chem. Soc.*, 1965, 6134.

3272. Nishiwaki, *Tetrahedron*, 23, 1153 (1967).

3273. Lafaix and Josien, *J. Chim. phys.*, 62, 684 (1965).

3274. Kwiatkowski, *Acta Phys. Polon.*, 30, 963 (1966).

3275. WHO Scientific Group, *World Health Organization Tech. Report Ser.*, No. 375 (1967).

3276. Van Dijck, Claesen, Vanderhaeghe, and De Somer, *Antibiotics and Chemotherapy*, 9, 523 (1959).

3277. Roy-Burman and Sen, *Biochemical Pharmacol.*, 13, 1437 (1964).

3278. Irwin and Wibberley, *J. Chem. Soc. (C)*, 1967, 1745.

3279. Temple, McKee, and Montgomery, *J. Org. Chem.*, 28, 2257 (1963).

3280. Pfleiderer and Reisser, *Chem. Ber.*, 95, 1621 (1962).

3281. Angier, *J. Org. Chem.*, 28, 1398 (1963).

3282. McLeod, *J. Res. Nat. Bur. Stand.*, 66A, 65 (1962).

3283. Miura, Ikeda, Oohashi, Igarashi, and Okada, *Chem. Pharm. Bull.*, 13, 529 (1965).

3284. Crooks and Robinson, *Chem. and Ind.*, 1967, 547.

3285. Morimoto, *J. Pharm. Soc. Japan*, 82, 465 (1962).

3286. Traverso, Robbins, and Whitehead, *J. Med. Pharm. Chem.*, 5, 808 (1962).

3287. Watanabe and Fox, *Angew. Chem.*, 78, 589 (1966).

3288. Montgomery and Hewson, *J. Org. Chem.*, 26, 4469 (1961).

3289. Bredereck and Gotsmann, *Chem. Ber.*, 95, 1902 (1962).

3290. Temple and Montgomery, *J. Org. Chem.*, 28, 3038 (1963).

3291. Kempter, Rokos, and Pfleiderer, *Angew. Chem.*, 79, 233 (1967).

3292. Nitta, Yonida, and Miyamoto, U.S. Pat. 3,317,534 (1967); through *Chem. Abstr.*, 67, 54156 (1967).

3293. Bredereck, Siegel, and Föhlisch, *Chem. Ber.*, 95, 403 (1962).

3294. Kato, Yamanaka, and Moriya, *J. Pharm. Soc. Japan*, 84, 1201 (1964).

3295. Fürst and Ebert, *Chem. Ber.*, 93, 99 (1960).

3296. Piper and Johnston, *J. Org. Chem.*, 30, 1247 (1965).

3297. Petyunin and Kalugina, *J. Gen. Chem.*, 34, 1255 (1964).

3298. Bergmann, Tamari, and Ungar-Waron, *J. Chem. Soc.*, 1964, 565.

3299. Gutorov and Golovchinskaya, *Khim. Farm. Zhur.*, 1, 28 (1967); through *Chem. Abstr.*, 67, 82193 (1967).

3300. Kropacheva, Sazonov, and Sergievskaya, *J. Gen. Chem.*, 32, 3796 (1962).

3301. Noell and Cheng, *J. Med. Chem.*, 11, 63 (1968).

3302. Pfleiderer and Rukwied, *Chem. Ber.*, 94, 118 (1961).

3303. Viscontini and Bobst, *Helv. Chim. Acta*, 49, 1815 (1966).

3304. Brown and England, *J. Chem. Soc.*, 1965, 1530.

3305. Uno, Machida, Hanai, Ueda, and Sasaki, *Chem. Pharm. Bull.*, 11, 704 (1963).

3306. Kalm, *J. Org. Chem.*, 26, 3026 (1961).

3307. Nitta, Okui, Ito, and Togo, *Chem. Pharm. Bull.*, 13, 568 (1965).

3308. Craveri and Zoni, *Farmaco* (Ed. Sci.), 17, 573 (1962).

3309. Tanabe Seiyaku Co., Fr.Pat. 1,402,805 (1965); through *Chem. Abstr.*, 63, 13280 (1965).

3310. Horstmann, Woerffel, and Wirtz, Fr.Pat. M2753 (1964); through *Chem. Abstr.*, 62, 16267 (1965).

3311. Horstmann, Woerffel, and Wirtz, Ger.Pat. 1,193,952 (1965); through *Chem. Abstr.*, 63, 13280 (1965).

3312. Biffin and Brown, *Tetrahedron Letters*, 1968, 2503.

3313. Lister, *Rev. Pure Appl. Chem.*, 11, 178 (1961).

3314. Bergmann and Tamari, *J. Chem. Soc.*, 1961, 4468.

3315. Bergmann, Rashi, Kleiner, and Knafo, *J. Chem. Soc. (C)*, 1967, 1254.

3316. Balsiger, Fikes, Johnston, and Montgomery, *J. Org. Chem.*, 26, 3386 (1961).

3317. Barlin and Chapman, *J. Chem. Soc.*, 1965, 3017.

3318. Okano, Goya, Takadate, and Ito, *J. Pharm. Soc. Japan*, 86, 649 (1966).

3319. Goldner, Dietz, and Carstens, *Annalen*, 693, 233 (1966).

3320. Goldner, Dietz, and Carstens, *Annalen*, 694, 142 (1966).

3321. Goldner, Dietz, and Carstens, *Annalen*, 698, 145 (1966).

3322. Pfleiderer and Blank, *Angew. Chem.*, 78, 679 (1966).

3323. Pfleiderer, *Angew. Chem.*, 75, 993 (1963).

3324. Elderfield and Mehta, in "Heterocyclic Compounds," Ed. Elderfield, Wiley, New York, 1967, Vol. 9, p.1 *et. seq.*

3325. Baugh and Shaw, *J. Org. Chem.*, 29, 3610 (1964).

3326. Baranov and Gorizdra, *J. Gen. Chem.*, 32, 1220 (1962).

3327. Taylor and Yoneda, *Angew. Chem.*, 79, 901 (1967).

3328. Lister and Timmis, *Chem. and Ind.*, 1963, 819.

3329. Pyl, Melde, and Beyer, *Annalen*, 663, 108 (1963).

3330. Nyberg, Noell, and Cheng, *J. Het. Chem.*, 2, 110 (1965).

3331. Pachter and Nemeth, *J. Org. Chem.*, 28, 1187 (1963).

3332. Pachter, *J. Org. Chem.*, 28, 1191 (1963).

3333. Pachter, Nemeth, and Villani, *J. Org. Chem.*, 28, 1197 (1963).

3334. Pachter and Nemeth, *J. Org. Chem.*, 28, 1203 (1963).

3335. Polansky and Grassberger, *Monatsh.*, 94, 662 (1963).

3336. Iwasaki, *J. Pharm. Soc. Japan*, 82, 1368 (1962).

3337. Wierzchowski and Shugar, *Photochem. Photobiol.*, 2, 377 (1963).

3338. Wierzchowski, Shugar, and Katritzky, *J. Amer. Chem. Soc.*, 85, 827 (1963).

3339. Bulow, *Ber.*, 42, 4429 (1909).

3340. Temple and Montgomery, *J. Org. Chem.*, 30, 826 (1965).

3341. Temple, Coburn, Thorpe, and Montgomery, *J. Org. Chem.*, 30, 2395 (1965).

3342. Temple, Thorpe, Coburn, and Montgomery, *J. Org. Chem.*, 31, 935 (1966).

3343. Temple, Kussner, and Montgomery, *J. Org. Chem.*, 31, 2210 (1966).

3344. Jack, *J. Med. Pharm. Chem.*, 3, 253 (1961).

3345. Buu-Hoi, Rips, and Derappe, *J. Med. Chem.*, 7, 364 (1964).

3346. Dyer, Gluntz, and Tanck, *J. Org. Chem.*, 27, 982 (1962).

3347. Dyer and Richmond, *J. Med. Chem.*, 8, 195 (1965).

3348. Cingolani, Bellomonte, and Sordi, *Farmaco* (Ed. Sci.), 19, 1050 (1964).

3349. Broadbent, Miller, and Rose, *J. Chem. Soc.*, 1965, 3369.

3350. Taylor and Sowinski, *J. Amer. Chem. Soc.*, 90, 1374 (1968).

3351. Barnikov and Bödeker, *Z. Chem.*, 5, 62 (1965).

3352. W. V. Curran and R. B. Angier, Personal communication.

3353. Clarkson and Martin, *Nature*, 192, 523 (1961).

3354. Mamaev and Krivopalov, *Khim. Geterotsikl. Soedin., Akad. Nauk Latv. S.S.R.*, 1966, 145; through *Chem. Abstr.*, 65, 710 (1966).

3355. Matsukawa and Yurugi, *Rev. Japan Lit. Beriberi Thiamine*, 1965, 81-134.

3356. Fel'dman, Petrova, Drok, and Semicheva, *Mechenye Biol. Aktivn. Veschestva, Sb. Statei*, 1962, 20; through *Chem. Abstr.*, 59, 10041 (1963).

3357. Palecek and Janik, *Arch. Biochem. Biophys.*, 98, 527 (1962).

3358. Smith and Elving, *J. Amer. Chem. Soc.*, 84, 2741 (1962).

3359. Kwiatkowski, *Acta Phys. Polon.*, 29, 573 (1966).

3360. Nakajima and Pullman, *Bull. Soc. chim. France*, 1958, 1502.

3361. Fahr, Kleber, and Boebinger, *Z. Naturforsch.*, 21B, 219 (1966).

3362. Cramer and Seidel, *Biochem. Biophys. Acta*, 72, 157 (1963).

3363. Ueda and Fox, *J. Org. Chem.*, 29, 1762 (1964).

3364. Bell, *Biochem. Biophys. Acta*, 47, 602 (1961).

3365. Nowacki and Przybylska, *Bull. Acad. Polon. Sci., Ser. Sci. Biol.*, 9, 279 (1961); through *Chem. Abstr.*, 56, 1725 (1962).

3366. Bell and Foster, *Nature*, 194, 91 (1962).

3367. Watanabe, *J. Pharm. Soc. Japan*, (Western Language Supplement), 59, 133 (1939).

3368. Abderhalden, *Klin. Wochschr.*, 18, 171 (1939).

3369. Koppel, Springer, Robins, and Cheng, *J. Org. Chem.*, 27, 1492 (1962).

3370. Tanaka, Takeuchi, Tanaka, Yonehara, Umezama, and Sumiki, *J. Antibiotics*, 14A, 161 (1961); through *Chem. Abstr.*, 56, 15958 (1962).

3371. Tanaka, Tanaka, Yonehara, and Umezawa, *J. Antibiotics*, 15A, 191 (1962); through *Chem. Abstr.*, 59, 5162 (1963).

3372. Tanaka, Takeuchi, and Yonehara, *J. Antibiotics*, 15A, 197 (1962); through *Chem. Abstr.*, 59, 5162 (1963).

3373. Stevens, Nagarajan, and Haskell, *J. Org. Chem.*, 27, 2991 (1962).

3374. Iwasaki, *J. Pharm. Soc. Japan*, 82, 1358-1395 (1962) [eleven papers].

3375. Fox, Kuwada, Watanabe, Ueda, and Whipple, *Antimicrobial. Agents Chemotherapy*, 1964, 518; through *Chem. Abstr.*, 63, 3477 (1965).

3376. Otaka, Takeuchi, Endo, and Yonehara, *Tetrahedron Letters*, 1965, 1405.

3377. Otaka, Takeuchi, Endo, and Yonehara, *Tetrahedron Letters*, 1965, 1411.

3378. Onuma, Nawata, and Saito, *Bull. Chem. Soc. Japan*, 39, 1091 (1966).

3379. Brown, *Proc. Roy. Aust. Chem. Inst.*, 33, 57 (1966).

3380. Funakoshi, Irie, and Ukita, *Chem. Pharm. Bull.*, 9, 406 (1961).

3381. Mikhailopulo, Gunar, and Zav'yalov, *Izvest. Akad. Nauk S.S.S.R., Ser. Khim.* 1965, 1715.

3382. Baker and Chheda, *J. Pharm. Sci.*, 54, 25 (1965).

3383. Szer and Shugar, *Acta Biochim. Polon.*, 7, 491 (1960).

3384. Falco and Fox, *J. Med. Chem.*, 11, 148 (1968).

3385. Codington, Cushley, and Fox, *J. Org. Chem.*, 33, 466 (1968).

3386. Yamaoka, Otter, and Fox, *J. Med. Chem.*, 11, 55 (1968).

3387. Gold-Aubert and Gysin, *Helv. Chim. Acta*, 44, 105 (1961).

3388. Bojarski, Kahl, and Melzacka, *Roczniki Chem.*, 36, 1259 (1962); through *Chem. Abstr.*, 59, 5158 (1963).

3389. Bojarski and Kahl, *Roczniki Chem.*, 37, 589 (1963); through *Chem. Abstr.*, 59, 8746 (1963).

3390. Bojarski and Kahl, *Roczniki Chem.*, 38, 1493 (1964); through *Chem. Abstr.*, 62, 9127 (1965).

3391. Kahl and Melzacka, *Roczniki Chem.*, 37, 591 (1963); through *Chem. Abstr.*, 59, 8746 (1963).

3392. Bojarski, Kahl, and Melzacka, *Roczniki Chem.*, 39, 875 (1965); through *Chem. Abstr.*, 64, 5088 (1966).

3393. Bojarski and Kahl, *Roczniki Chem.*, 41, 311 (1967); through *Chem. Abstr.*, 67, 54093 (1967).

3394. Kahl and Chytros-Majchrowicz, *Roczniki Chem.*, 40, 1905 (1966); through *Chem. Abstr.*, 67, 11469 (1967).

3395. Bojarski, Kahl, and Melzacka, *Roczniki Chem.*, 40, 1465 (1966); through *Chem. Abstr.*, 67, 11470 (1967).

3396. Lawley and Brookes, *Biochem. J.*, 89, 127 (1963).

3397. McNaught and Brown, *J. Org. Chem.*, 32, 3689 (1967).

3398. Yamaoka and Aso, *J. Agr. Chem. Soc. Japan*, 35, 280 (1961); *Chem. Abstr.*, 60, 520 (1964).

3399. Noell and Cheng, *J. Het. Chem.*, 5, 25 (1968).

3400. Pliml and Prystas, *Adv. Het. Chem.*, 8, 115 (1967).

3401. Prystas and Sorm, *Coll. Czech. Chem. Comm.*, 29, 131 (1964).

3402. Prystas and Sorm, *Coll. Czech. Chem. Comm.*, 29, 2956 (1964).

3403. Prystas and Sorm, *Coll. Czech. Chem. Comm.*, 30, 2960 (1965).

3404. Ulbricht and Rogers, *J. Chem. Soc.*, 1965, 6125.

3405. Ulbricht and Rogers, *J. Chem. Soc.*, 1965, 6130.

3406. Thacker and Ulbricht, *J. Chem. Soc.(C)*, 1968, 333.

3407. Minnemeyer, Tieckelmann, and Holland, *J. Med. Chem.*, 6, 602 (1963).

3408. Ueda and Nishino, *J. Amer. Chem. Soc.*, 90, 1678 (1968).

3409. Ueda, Iida, Ikeda, and Mizuno, *Chem. Pharm. Bull.*, 14, 666 (1966).

3410. Ulbricht, *Proc. Chem. Soc.*, 1962, 298.

3411. Ulbricht, *Angew. Chem.*, 74, 767 (1962).

3412. Lee, Ph.D. Thesis, Canberra, 1968.

3413. Wiberg, Shryne, and Kintner, *J. Amer. Chem. Soc.*, 79, 3160 (1957).

3414. Spinner and White, *Chem. and Ind.*, 1967, 1784.

3415. Iwai and Hiraoka, *Chem. Pharm. Bull.*, 12, 813 (1964).

3416. Brown, England, and Harper, *J. Chem. Soc. (C)*, 1966, 1165.

3417. Brown and Jacobsen, *Tetrahedron Letters*, 1960, No 25, p.17.

3418. Perrin, *J. Chem. Soc.*, 1963, 1284.

3419. Angier and Curran, *J. Amer. Chem. Soc.*, 81, 5650 (1959).

3420. Angier, in "Pteridine Chemistry," Pergamon Press, Oxford, 1964, 232.

3421. Perrin and Pitman, *Aust. J. Chem.*, 18, 763 (1965).

3422. Perrin and Pitman, *Aust. J. Chem.*, 18, 471 (1965).

3423. Pitman, Ph.D. Thesis, Canberra (1965), p.180 *et seq.*

3424. Otomasu, Takahashi, and Ogata, *Chem. Pharm. Bull.*, 12, 714 (1964).

3425. Hirai, *Chem. Pharm. Bull.*, 14, 861 (1966).

3426. Pfleiderer, Grözinger, and Sagi, *Chem. Ber.*, 99, 3524 (1966).

3427. Shvachkin and Azarova,
J. Gen. Chem., 33, 590 (1963).

3428. Shvachkin, Azarova, and Rapanovich,
Vestn. Mosk. Univ., Ser.II, Khim., 18, 68 (1963);
through *Chem. Abstr.*, 59, 15283 (1963).

3429. Hermann and Black,
Appl. Spectroscopy, 20, 413 (1966).

3430. Shvachkin, Berestenko, and Mishin,
J. Gen. Chem., 34, 1687 (1964).

3431. Ross, Goodman, and Baker,
J. Org. Chem., 25, 1950 (1960).

3432. Nagpal, Agarwal, and Dhar,
Indian J. Chem., 3, 356 (1965).

3433. Bono, Cheng, Frei, and Kelly,
Cancer Res., 24, 513 (1964).

3434. Nakatani, *J. Pharm. Soc. Japan*, 84, 1057 (1964).

3435. Shvachkin and Azarova,
Vestn. Mosk. Univ., Ser.II, Khim., 18, 53 (1963);
through *Chem. Abstr.*, 59, 5159 (1963).

3436. Ivin and Nemets, *J. Gen. Chem.*, 35, 1294 (1965).

3437. Nemets and Ivin, *J. Gen. Chem.*, 35, 1299 (1965).

3438. Crosby and Berthold, *J. Med. Chem.*, 6, 334 (1963).

3439. Safonova and Myshkina,
J. Gen. Chem., 34, 1682 (1964).

3440. Klosa, *J. prakt. Chem.*, 26, 43 (1964).

3441. Pazdro, *Diss. Pharm. Pharmacol.*, 18, 359 (1966);
Chem. Abstr., 66, 18691 (1967).

3442. Bredereck, Effenberger, and Resemann,
Angew. Chem., 74, 253 (1962).

3443. Shadbolt and Ulbricht,
J. Chem. Soc. (C), 1968, 1203.

3444. Cingolani, Bellomonte, and Sordi,
Farmaco (Ed. Sci.), 20, 259 (1965).

3445. Case and Koft, *J. Amer. Chem. Soc.*, 81, 905 (1959).

3446. Wiley, Hussung, Hobbs, and Huh,
J. Med. Chem., 7, 358 (1964).

3447. Wiley, Canon, and Hussung,
J. Med. Chem., 6, 333 (1963).

3448. Brossmer and Ziegler, *Angew Chem.*, 79, 322 (1967).

3449. Cheng and Cheng, *J. Org. Chem.*, 33, 892 (1968).

3450. Baker, Ho, and Chheda, *J. Het. Chem.*, 1, 88 (1964).

3451. Taylor and Morrison, *Angew. Chem.*, 77, 859 (1965).

3452. Grundmann, *Fortschr. chem. Forsch.*, 7, 62 (1966).

3453. Zav'yalov, Mikhailopulo, and Gunar, *Izvest. Akad. Nauk S.S.S.R., Ser. Khim.*, 1965, 1887.

3454. Tucci, Takahashi, Tucci, and Li, *J. Inorg. Nucl. Chem.*, 26, 1263 (1964).

3455. Nakatani, Nishikawa, and Mizuta, *J. Pharm. Soc. Japan*, 84, 1051 (1964).

3456. Tucci, Doody, and Li, *J. Phys. Chem.*, 65, 1570 (1961).

3457. Clark, Grantham, and Lydiate, *J. Chem. Soc. (C)*, 1968, 1122.

3458. Clark and Pendergast, *J. Chem. Soc. (C)*, 1968, 1124.

3459. Katritzky, Kingsland, and Tee, *Chem. Comm.*, 1968, 289.

3460. Morrison, Feeley, and Kleopfer, *Chem. Comm.*, 1968, 358.

3461. O'Brien, Weinstock, and Cheng, *J. Med. Chem.*, 11, 387 (1968).

3462. Kwiatkowski and Zurawski, *Acta Phys. Polon.*, 32, 893 (1967).

3463. Caton and McOmie, *J. Chem. Soc. (C)*, 1968, 836.

3464. Klötzer and Herberz, *Monatsh.*, 99, 847 (1968).

3465. Ballweg, *Tetrahedron Letters*, 1968, 2171.

3466. Pullman, Berthod, and Langlet, *Compt. rend. (D)*, 1968, 1063.

3467. Wempen, Miller, Falco, and Fox, *J. Med. Chem.*, 11, 144 (1968).

3468. Wahren, *Z. Chem.*, 6, 181 (1966).

3469. Wahren, *Tetrahedron*, 24, 441 (1968).

3470. Wahren, *Tetrahedron*, 24, 451 (1968).

3471. Myles and Fox, *J. Med. Chem.*, 11, 143 (1968).

3472. Albert, *Angew. Chem.*, 79, 913 (1967).

3473. Thompson, *Chem. Comm.*, 1968, 532.

3474. Shadbolt and Ulbricht, *J. Chem. Soc. (C)*, 1968, 733.

3475. Sugimoto and Matsuura, *Research Bull.* (Nagoya Univ.), 12, 24 (1968).

3476. O'Brien, Weinstock, Springer, and Cheng, *J. Het. Chem.*, 4, 49 (1967).

3477. Haraoka, Sugihara, and Ito, Jap. Pat. 10,052 (1965); through *Chem. Abstr.*, 63, 5659 (1965).

3478. Albert and Yamamoto, *J. Chem. Soc. (C)*, 1968, 1181.

3479. Portnyagina and Karp, *Ukrain. khim. Zhur.*, 31, 215 (1965); through *Chem. Abstr.*, 63, 1785 (1965).

3480. Portnyagina and Karp, *Ukrain. khim. Zhur.*, 31, 83 (1965); through *Chem. Abstr.*, 62, 14665 (1965).

3481. P. N. Edwards, W. Hepworth, and T. W. Thompson, Personal communication (1968).

3482. Hepworth and Thompson, Belg. Pat. 700,109 (1969).

3483. Stempel, Ph. D. Thesis, Cornell, 1968.

3484. Bushby and Hitchings, *Brit. J. Pharmacol.*, 33, 72 (1968).

3485. Hitchings and Burchall, *Adv. Enzymol.*, 27, 417 (1965).

3486. Pines, *Brit. Med. J.*, 1967, vol.3, 202.

3487. Stewart and Anderson, *Brit. Med. J.*, 1965, vol. 2, 682.

3488. Roth and Bunnett, *J. Amer. Chem. Soc.*, 87, 334 (1965).

3489. G. B. Barlin, Personal communication (1968).

3490. Kalatzis, *J. Chem. Soc. (B)*, 1969,

3491. Kalatzis, *J. Chem. Soc. (B)*, 1967, 273.

3492. Okano and Kojima, *J. Pharm. Soc. Japan*, 86, 547 (1966).

3493. Christensen, Rytting, and Izatt, *J. Phys. Chem.*, 71, 2700 (1967).

3494. Albert, *J. Chem. Soc. (B)*, 1966, 427.

3495. Morley and Simpson, *J. Chem. Soc.*, 1949, 1014.

3496. Pohland, Flynn, Jones, and Shive, *J. Amer. Chem. Soc.*, 73, 3247 (1951).

3497. Jonas and Gut, *Coll. Czech. Chem. Comm.*, 27, 716 (1962).

3498. Gottschling and Heidelberger, *J. Mol. Biol.*, 7, 541 (1963).

3499. Taylor and Robinson, *Talanta*, 8, 518 (1961).

3500. Thyagarajan, *Adv. Het. Chem.*, 8, 143 (1967).

3501. Chinoporos, Papathanasopoulos, and Fu, *Chim. Chron.* (A), 32, 35 (1967); through *Chem. Abstr.*, 68, 73591 (1968).

3502. Einstein, Hosszu, Longworth, Rahn, and Wei, *Chem. Comm.*, 1967, 1063.

3503. Bretschneider and Fitz, Swiss Pat. 423,786 (1967); through *Chem. Abstr.*, 68, 2910 (1968).

3504. Brock, *Chem. Z.*, 8, 143 (1968); through *Chem. Abstr.*, 68, 114,536 (1968).

3505. Habicht and Zubiani, Swiss Pat. 443,341 (1967); through *Chem. Abstr.*, 68, 114627 (1968).

3506. Habicht, Swiss Pat. 443,342 (1967); through *Chem. Abstr.*, 68, 114625 (1968).

3507. Chang, Hahn, Kim, and Oh, *J. Korean Chem. Soc.*, 10, 51 (1966); *Chem. Abstr.*, 67, 100,093 (1967).

3508. Upjohn Co., Neth. Pat. Appl. 6,615,385 (1967); through *Chem. Abstr.*, 68, 21947 (1968).

3509. Castellano, Sun, and Kostelnik, *J. Chem. Phys.*, 46, 327 (1967).

3510. Spiesecke and Schneider, *J. Chem. Phys.*, 35, 731 (1961).

3511. Gronowitz and Hoffman, *Arkiv. Kemi*, 16, 539 (1960).

3512. Gronowitz, Sorlin, Gestblom, and Hoffman, *Arkiv. Kemi*, 19, 483 (1962).

3513. Gatlin and Davies, *J. Amer. Chem. Soc.*, 84, 4464 (1962).

3514. Becker, Miles, and Bradley, *J. Amer. Chem. Soc.*, 87, 5575 (1965).

3515. Paudler and Blewitt, *J. Org. Chem.*, 30, 4081 (1965).

3516. Batterham, *J. Chem. Soc. (C)*, 1966, 999.

3517. Rodmar, Rodmar, Khan, and Gronowitz, *Arkiv. Kemi,* 27, 87 (1967).

3518. Grundmann and Richter, *J. Org. Chem.,* 33, 476 (1968).

3519. Rahamim, Sharvit, Mandelbaum, and Sprecher, *J. Org. Chem.,* 32, 3856 (1967).

3520. Okano, Goya, and Matsumoto, *J. Pharm. Soc. Japan,* 87, 1315 (1967).

3521. Banks, Field, and Haszeldine, *J. Chem. Soc. (C),* 1967, 1822.

3522. D'Alcontres, Lo Vecchio, Crisafulli, and Gattuso, *Gazzetta,* 97, 997 (1967).

3523. Brossmer, *Angew. Chem.,* 79, 691 (1967).

3524. Zavyalov, Mikhailopulo, Gunar, and Ovechina, *Izvest. Akad. Nauk S.S.S.R., Ser. Khim.,* 1967, 859; through *Chem. Abstr.,* 68, 12945 (1968).

3525. Aroyan, Kaldrikyan, and Melik-Ogandzhanyan, *Arm. Khim. Zhur.,* 20, 61 (1967); through *Chem. Abstr.,* 67, 100101 (1967).

3526. Foglizzo and Novak, *J. Chim. phys.,* 64, 1484 (1967); through *Chem. Abstr.,* 68, 82703 (1968).

3527. Aroyan and Melik-Ogandzhanyan, *Arm. Khim. Zhur.,* 20, 314 (1967); through *Chem. Abstr.,* 68, 29669 (1968).

3528. Aroyan and Kaldrikyan, *Sin. Geterotsikl. Soedin., Akad. Nauk Arm. S.S.R., Inst. Tonkoi Org. Khim.,* 1966, No.7, p.30; through *Chem. Abstr.,* 68, 114538 (1968).

3529. Fujimoto, Jap. Pat. 4260 (1967); through *Chem. Abstr.,* 68, 29717 (1968).

3530. Beaman and Duschinsky, U.S. Pat. 3,317,532 (1967); through *Chem. Abstr.,* 68, 29715 (1968).

3531. Watanabe, Kobori, and Takahashi, Jap.Pat. 4261 (1967); through *Chem. Abstr.,* 67, 90831 (1967).

3532. Budesinsky and Letovsky, *Czechoslov. Farm.,* 15, 432 (1966); through *Chem. Abstr.,* 67, 90756 (1967).

3533. Takamizawa, Hirai, and Sato, Jap.Pat. 10415 (1965); through *Chem. Abstr.,* 63, 5659 (1965).

3534. Silversmith, *J. Org. Chem.,* 27, 4090 (1962).

3535. Zimmermann, *Angew. Chem.*, 75, 1124 (1963).

3536. Nofre, Cier, Chapurlat, and Mareschi, *Bull. Soc. chim. France*, 1965, 332.

3537. Gaspert and Skaric, *Croat. Chem. Acta*, 35, 171 (1963).

3538. Kreling and McKay, *Canad. J. Chem.*, 40, 143 (1962).

3539. Kretov and Borodavko, *Zhur. org. Khim.*, 2, 364 (1966); through *Chem. Abstr.*, 65, 2255 (1966).

3540. D'Amico, Tung, Campbell, and Mullins, *J. Chem. Eng. Data*, 8, 446 (1963); through *Chem. Abstr.*, 59, 8742 (1963).

3541. Hartmann and Mayer, *J. prakt. Chem.*, 30, 87 (1965).

3542. Ried and Piechaczek, *Annalen*, 696, 97 (1966).

3543. Konyukhov, Sakovich, Krupnova, and Pushkareva, *Zhur. Org. Khim.*, 1, 1487 (1965); through *Chem. Abstr.*, 64, 6650 (1966).

3544. Mamaev, *Biol. Aktivn, Soedin Akad. Nauk S.S.S.R.*, 1965, 38; through *Chem. Abstr.*, 63, 18081 (1965).

3545. Hoffmann and Schulze, D.D.R.Pat. 48617 (1966); through *Chem. Abstr.*, 66, 28793 (1967).

3546. Nast, Ley, and Eholzer, Ger.Pat. 1,248,056 (1967); through *Chem. Abstr.*, 68, 87305 (1968).

3547. Takamizawa and Hirai, Jap.Pat. 14,950 (1967); through *Chem. Abstr.*, 68, 105219 (1968).

3548. Takamizawa and Hirai, Jap.Pat. 14,951 (1967); through *Chem. Abstr.*, 68, 105223 (1968).

3549. Evans, in "Modern Methods in Organic Chemistry," ed. C. J. Timmons, Van Nostrand, London, 1969. Chapter 1.

3550. Evans, *Rev. Pure Appl. Chem. (Australia)*, 15, 23 (1965).

3551. R. F. Evans, Unpublished results.

3552. Skaric, Gaspert, and Jerkunica, *Croat. Chem. Acta*, 38, 1 (1966).

3553. Büchi, Braunschweiger, Kira, and Lauener, *Helv. Chim. Acta*, 49, 2337 (1966).

3554. Mikolajewska and Kotelko, *Acta Polon. Pharm.*, 22, 219 (1965); through *Chem. Abstr.*, 63, 17891 (1965).

3555. Kondo and Witkop, *J. Amer. Chem. Soc.*, 90, 764 (1968).

3556. R. F. Evans and K. N. Mewett, Unpublished results.

3557. Cerutti, Kondo, Landis, and Witkop, *J. Amer. Chem. Soc.*, 90, 771 (1968).

3558. Nussbaum and Duschinsky, *Chem. and Ind.*, 1966, 1142.

3559. Janik and Palecek, *Arch. Biochem. Biophys.*, 105, 225 (1964).

3560. Janik and Elving, *Chem. Rev.*, 68, 295 (1968).

3561. Seebode and Vasey, Brit.Pat. 919,703 (1963); through *Chem. Abstr.*, 59, 7539 (1963).

3562. Lozeron, Gordon, Gabriel, Tautz, and Duschinsky, *Biochemistry*, 3, 1844 (1964).

3563. Wacker, Kornhauser, and Träger, *Z. Naturforsch.*, 20B, 1043 (1965).

3564. Khattak and Green, *Internat. J. Radiation Biol.*, 11, 131 (1966); through *Chem. Abstr.*, 65, 20476 (1966).

3565. Khattak and Green, *Internat. J. Radiation Biol.*, 11, 137 (1966); through *Chem. Abstr.*, 65, 20476 (1966).

3566. Moravek and Leseticky, *Coll. Czech. Chem. Comm.*, 33, 1352 (1968).

3567. Seefelder and Jentzsch, Ger.Pat. 1,189,998 (1965); through *Chem. Abstr.*, 63, 610 (1965).

3568. Jentzsch and Seefelder, *Chem. Ber.*, 98, 1342 (1965).

3569. Faust, Yee, and Sahyun, *J. Org. Chem.*, 26, 4044 (1961).

3570. Adcock, Lawson, and Miles, *J. Chem. Soc.*, 1961, 5120.

3571. Parke Davis and Co., Brit.Pat. 948,897 (1964); through *Chem. Abstr.*, 60, 10690 (1964).

3572. Pfizer and Co., Brit.Pat. 877,306 (1961); through *Chem. Abstr.*, 58, 2458 (1963).

3573. Faust and Sahyun, U.S.Pat. 3,042,674 (1962); through *Chem. Abstr.*, 57, 16568 (1962).

3574. Johnson and Woodburn, *J. Org. Chem.*, 27, 3958 (1962).

3575. Hurwitz and Aschkenasy, Belg.Pat. 637,380 (1964); through *Chem. Abstr.*, 62, 7892 (1965).

3576. Baganz, Domaschke, Fock, and Rabe, *Chem. Ber.*, 95, 1832 (1962).

3577. Bindler, Model, and Zinkernagel, Swiss Pat. 350,301 (1962); through *Chem. Abstr.*, 56, 14301 (1962).

3578. Geigy A.-G., Brit.Pat. 869,181 (1961); through *Chem. Abstr.*, 62, 13156 (1965).

3579. Bindler, Model, and Zinkernagel, Ger.Pat. 1,111,197 (1961); through *Chem. Abstr.*, 59, 8764 (1963).

3580. Wollweber, Hiltmann, Kroneberg, and Stoepel, Belg.Pat. 613,662 (1962); through *Chem. Abstr.*, 57, 15121 (1962).

3581. Langis and Pilkington, U.S.Pat. 3,126,381 (1964); through *Chem. Abstr.*, 60, 14517 (1964).

3582. Ayerst, McKenna, and Harrison Ltd., Brit.Pat. 952,802 (1964); through *Chem. Abstr.*, 61, 4162 (1964).

3583. Drew Chemical Corp., Brit.Pat. 986,057 (1965); through *Chem. Abstr.*, 63, 1793 (1965).

3584. Baganz and Domaschke, *Chem. Ber.*, 95, 1840 (1962).

3585. Baganz and Domaschke, *Angew. Chem.*, 75, 1025 (1963).

3586. Baganz, Rabe, and Repplinger, *Chem. Ber.*, 98, 2572 (1965).

3587. Baganz and Domaschke, *Arch. Pharm.*, 295, 758 (1962).

3588. Baganz and Rabe, *Chem. Ber.*, 98, 3652 (1965).

3589. Spry and Aaron, *J. Org. Chem.*, 31, 3838 (1967).

3590. Phillips, U.S.Pat. 3,041,338 (1962); through *Chem. Abstr.*, 57, 13773 (1962).

3591. Sahyun and Faust, U.S.Pat. 3,009,915 (1962); through *Chem. Abstr.*, 56, 15521 (1962).

3592. Pfizer Corp., Belg.Pat. 658,987 (1965); through *Chem.Abstr.*, 64, 8192 (1966).

3593. Gallaghan and Evans, U.S.Pat. 2,994,695 (1962); through *Chem. Abstr.*, 56, 2461 (1962).

3594. Boehringer und Sohn, Belg.Pat. 623,305 (1963); through *Chem. Abstr.*, 64, 2096 (1966).

3595. Bindler and Model, Ger.Pat. 1,145,623 (1963); through *Chem. Abstr.*, 60, 2947 (1964).

3596. Suzuki, Inoue, and Goto, *Chem. Pharm. Bull.*, 16, 933 (1968).

3597. Hall, *Diss. Abs.*, 22, 4182 (1962).

3598. Merten, Belg. Pat. 611,643 (1962); through *Chem. Abstr.*, 57, 16573 (1962).

3599. Timmler, Ger. Pat. 1,126,392 (1962); through *Chem. Abstr.*, 57, 9859 (1962).

3600. Wayland, U.S.Pat. 3,158,501 (1964); through *Chem. Abstr.*, 62, 4158 (1965).

3601. Chemische Werke Huels A.-G., Belg.Pat. 631,698 (1963); through *Chem. Abstr.*, 61, 1876 (1964).

3602. Schneider, Swiss Pat. 357,400 (1961); through *Chem. Abstr.*, 57, 839 (1962).

3603. Sun Chemical Corp., Brit.Pat. 931,560 (1963); through *Chem. Abstr.*, 60, 1772 (1964).

3604. Merten and Müller, *Angew. Chem.*, 74, 866 (1962).

3605. McKay and Hatton, *J. Amer. Chem. Soc.*, 78, 1618 (1956).

3606. Warner, *J. Org. Chem.*, 28, 1642 (1963).

3607. Wellcome Foundation,Neth. Pat. Appl., 6,510,117 (1966); through *Chem. Abstr.*, 65, 3791 (1966).

3608. Lloyd, U.S.Pat. 3,016,379 (1962); through *Chem. Abstr.*, 57, 2233 (1962).

3609. Goncalves and Barrans, *Compt. rend.*, 258, 3507 (1964).

3610. Evans, *Aust. J. Chem.*, 20, 1643 (1967).

3611. Nikawitz, U.S.Pat. 3,152,122 (1964); through *Chem. Abstr.*, 62, 1670 (1965).

3612. Chemische Werke Albert, Brit.Pat. 848,399 (1960); through *Chem. Abstr.*, 57, 11215 (1962).

3613. Boswell and Williams, U.S.Pat. 3,137,697 (1964); through *Chem. Abstr.*, 61, 7024 (1964).

3614. Petersen, Brandeis, and Fikentscher, Ger.Pat. 1,229,093 (1966); through *Chem. Abstr.*, 66, 37943 (1967).

3615. Zigeuner and Rauter, *Monatsh.*, 96, 1950 (1965).

3616. Petersen, Brandeis, and Fikentscher, Ger.Pat. 1,230,805 (1966); through *Chem. Abstr.*, 66, 46435 (1967).

3617. Hatt and Triffett, *Chem. Comm.*, 1965, 439.

3618. Zigeuner, Fuchs, Brunetti, and Sterk, *Monatsh.*, 97, 36 (1966).

3619. Inoi, Okamoto, and Koizumi, *J. Org. Chem.*, 31, 2700 (1966).

3620. Mamaev and Sedova, *Khim. Geterosikl. Soedin.*, 1967, 5721; through *Chem. Abstr.*, 68, 49545 (1968).

3621. Mamaev and Mikhaleva, *Khim. Geterosikl. Soedin., Akad. Nauk Latv. S.S.R.*, 1965, 948; through *Chem. Abstr.*, 64, 14187 (1966).

3622. Unkovskii, Ignatova, Donskaya, and Zaitseva, *Probl. Organ. Sinteza, Akad. Nauk S.S.S.R., Otd. Obshch. i Tekhn. Khim.*, 1965, 202; through *Chem. Abstr.*, 64, 9719 (1966).

3623. Unkovskii, Ignatova, Gridunov, and Donskaya, U.S.S.R.Pat. 172,761 (1965); through *Chem. Abstr.*, 64, 740 (1966).

3624. Shell Research, Belg.Pat. 623,714 (1963); through *Chem. Abstr.*, 60, 9299 (1964).

3625. Evans and Shannon, *J. Chem. Soc.*, 1965, 1406.

3626. David and Sinay, *Bull. Soc. chim. France*, 1965, 2301.

3627. Iwasaki, *J. Pharm. Soc. Japan*, 82, 1358 (1962).

3628. Hanze, *J. Amer. Chem. Soc.*, 89, 6720 (1967).

3629. Takamizawa and Hirai, Jap.Pat. 1184 (1967); through *Chem. Abstr.*, 66, 55505 (1967).

3630. Barbieri, Bernardi, Palamidessi, and Venturi, *Tetrahedron Letters*, 1968, 2931.

3631. R. F. Evans and M. N. Nolan, Unpublished results.

3632. Iwakura, Nishiguchi, and Nabeya, *J. Org. Chem.*, 31, 1651 (1966).

3633. R. F. Evans and R. G. Hoare, Unpublished results.

3634. Wilson, *Tetrahedron*, 21, 2561 (1965).

3635. Billman and Dorman, *J. Pharm. Sci.*, 51, 1071 (1962).

3636. Billman and Khan, *J. Med. Chem.*, 9, 347 (1966).

3637. Billman and Meisenheimer, *J. Med. Chem.*, 6, 682 (1963).

3638. Billman and Meisenheimer, *J. Med. Chem.*, 8, 540 (1965).

3639. Billman and Khan, *J. Med. Chem.*, 8, 498 (1965).

3640. Billman and Khan, *J. Med. Chem.*, 11, 312 (1968).

3641. Bell and Dunstan, *J. Chem. Soc. (C)*, 1966, 870.

3642. Bayer A.-G., Neth.Pat. Appl. 6,505,669 (1965); through *Chem. Abstr.*, 64, 11233 (1966).

3643. Krässig, *Makromol. Chem.*, 17, 77 (1956).

3644. Shoolery in "Third Workshop on N.M.R. and E.P.R. Spectroscopy, Palo Alto, 1959," Pergamon, New York, 1960, p.40.

3645. Winstead, Strachan, and Heine, *J. Org. Chem.*, 26, 4116 (1961).

3646. Frankel, *Tetrahedron*, 19, Supplement 1, p.213 (1963).

3647. Frankel, U.S.Pat. 3,000,890 (1961); through *Chem. Abstr.*, 57, 7285 (1962).

3648. Piotrowska and Urbanski, *J. Chem. Soc.*, 1962, 1942.

3649. Eckstein, Gluzinski, and Urbanski, *Bull. Acad. Polon. Sci., Ser. Sci. Chim.*, 12, 623 (1964); through *Chem. Abstr.*, 62, 7748 (1965).

3650. Thompson, U.S.Pat. 3,063,999 (1962); through *Chem. Abstr.*, 59, 2835 (1963).

3651. Hodge, Belg.Pat. 660,155 (1965); through *Chem. Abstr.*, 64, 6666 (1966).

3652. Kawahara, Sato, Fumita, and Miwa, Jap.Pat. 7525 (1961); through *Chem. Abstr.*, 58, 13968 (1963).

3653. Borovik and Mamaev, *Izvest. Sibirsk. Otd. Akad. Nauk S.S.S.R., Ser. Khim. Nauk*, 1966, 95; through *Chem. Abstr.*, 65, 10586 (1966).

3654. Winberg, U.S.Pat. 3,239,518 (1966); through *Chem. Abstr.*, 64, 15898 (1966).

3655. Brown, U.S.Pat. 3,214,428 (1965); through *Chem. Abstr.*, 64, 3501 (1966).

3656. Lyle and Anderson, *Adv. Het. Chem.*, 6, 45 (1966).

3657. Bell and Necker, U.S.Pat. 3,054,797 (1962); through *Chem. Abstr.*, 58, 10216 (1963).

3658. Jentzsch, Belg.Pat. 647,590 (1964); through *Chem. Abstr.*, 63, 13282 (1965).

3659. Gmünder and Lindenmann, *Helv. Chim. Acta*, 47, 66 (1964).

3660. Greenhalgh and Bannard, *Canad. J. Chem.*, 39, 1017 (1961).

3661. Geigy A.-G., Belg.Pat. 627,804 (1963); through *Chem. Abstr.*, 60, 13252 (1964).

3662. de Antoni, Fr.Pat. 1,363,904 (1964); through *Chem. Abstr.*, 62, 573 (1965).

3663. Slezak, Brit.Pat. 888,627 (1962); through *Chem. Abstr.*, 59, 641 (1963).

3664. Gordon, U.S.Pat. 3,219,522 (1965); through *Chem. Abstr.*, 64, 6671 (1966).

3665. Tweit, U.S.Pat. 3,025,295 (1962); through *Chem. Abstr.*, 57, 8589 (1962).

3666. D'Angeli, Di Bello, and Giormani, *Gazzetta*, 94, 1342 (1965).

3667. Lambrech and Henesley, Brit.Pat. 889,002 (1962); through *Chem. Abstr.*, 57, 4635 (1962).

3668. Henesley and Lambrech, U.S.Pat. 3,186,990 (1965); through *Chem. Abstr.*, 63, 7018 (1965).

3669. Gordon, U.S.Pat. 2,988,478 (1961); through *Chem. Abstr.*, 57, 6363 (1962).

3670. Wellcome Foundation, Belg.Pat. 626,015 (1963); through *Chem. Abstr.*, 60, 7920 (1964).

3671. McKay and Kreling, *Canad. J. Chem.*, 40, 205 (1962).

3672. Takamizawa and Hirai, Jap.Pat. 7470 (1967); through *Chem. Abstr.*, 68, 29716 (1968).

3673. Schickh, Ger.Pat. 1,190,944 (1965); through *Chem. Abstr.*, 63, 2985 (1965).

3674. Hurwitz, Aschkenasy, and Kelley, Fr.Pat. 1,386,554 (1965); through *Chem. Abstr.*, 63, 1990 (1965).

3675. Bindler, Model, and Zinkernagel, Swiss Pat. 355,154 (1961); through *Chem. Abstr.*, 58, 13971 (1963).

3676. Snipes and Bernhard, *Radiation Res.*, 33, 162 (1968); through *Chem. Abstr.*, 68, 44735 (1968).

3677. Erner, Green, and Mills, U.S.Pat. 3,050,523 (1962); through *Chem. Abstr.*, 57, 15128 (1962).

3678. Kondo and Witkop, *J. Amer. Chem. Soc.*, 90, 3258 (1968).

3679. Bindler, Model, and Zinkernagel, Swiss Pat. 357,402 (1961); through *Chem. Abstr.*, 61, 5661 (1964).

3680. Zigeuner, Fuchs, and Galatik, *Monatsh.*, 97, 43 (1966).

3681. Zigeuner, Adam, and Galatik, *Monatsh.*, 97, 52 (1966).

3682. Lelean and Morris, *Chem. Comm.*, 1968, 239.

3683. Bellamy, Hayes, and Michels, U.S.Pat. 3,001,992 (1961); through *Chem. Abstr.*, 57, 11201 (1962).

3684. Enders and Pusch, *Textile Res. J.*, 36, 322 (1966); through *Chem. Abstr.*, 65, 4011 (1966).

3685. Eckstein, Gluzinski, and Plenkiewiez, *Bull. Acad. Polon. Sci., Ser. Sci. Chim.*, 11, 325 (1963); through *Chem. Abstr.*, 60, 2928 (1964).

3686. Schellenberg, Kaufmann, and Langenbeck, D.D.R.Pat. 48,605 (1966); through *Chem. Abstr.*, 65, 16868 (1966).

3687. Langenbeck, Kaufmann, and Schellenberger, *Annalen*, 690, 42 (1965).

3688. Berni, *Diss. Abs.*, 27, 3446 (1967).

3689. Jameson and Khan, *J. Chem. Soc. (A)*, 1968, 921.

3690. Di Bello, Giormani, and D'Angeli, *Gazzetta*, 97, 787 (1967).

3691. Mammi, Clemente, Del Pra, D'Angeli, Veronese, Toniolo, Rigatti, Coletta, and Boccalon, *Chem. Comm.*, 1968, 741.

3692. Vincendon, Cier, and Nofre, *Compt. rend.*, 260, 711 (1965); *Chem. Abstr.*, 62, 11637 (1965).

3693. Vincendon, Cier, and Nofre, *Bull. Soc. chim. France*, 1965, 1997.

3694. Janssen, *Rec. Trav. chim.*, 81, 650 (1962).

3695. Unkovskii, Ignatova, Zaitseva, and Donskaya, *Khim. Geterotsikl. Soedin., Akad. Nauk Latv. S.S.R.*, 1965, 586; through *Chem. Abstr.*, 64, 560 (1966).

3696. Sheehan, Cruickshank, and Boshart, *J. Org. Chem.*, 26, 2525 (1961).

3697. Rouillier, Delmau, Duplan, and Nofre, *Tetrahedron Letters*, 1966, 4189.

3698. Nofre, Murat, and Cier, *Bull. Soc. chim. France*, 1965, 1749.

3699. Rouillier, Delmau, and Nofre, *Bull. Soc. chim. France*, 1966, 3515.

3700. Chabre, Gagnaire, and Nofre, *Bull. Soc. chim. France,* 1966, 108.

3701. Dias and Truter, *Acta Cryst.,* 17, 937 (1964).

3702. Riddell and Lehn, *Chem. Comm.,* 1966, 375.

3703. Farmer and Hamer, *Tetrahedron,* 24, 829 (1968).

3704. R. F. Evans and Q. N. Porter, Unpublished results.

3705. Milborrow, *Weed Res.,* 5, 332 (1965); through *Chem. Abstr.,* 64, 13313 (1966).

3706. Bindler and Rumpf, Swiss Pat. 360,843 (1962); through *Chem. Abstr.,* 58, 9576 (1963).

3707. Dow Chemical Co., Neth.Pat. Appl. 300,737 (1965); through *Chem. Abstr.,* 64, 13329 (1966).

3708. Hageman and Riddoll, U.S.Pat. 3,226,221 (1965); through *Chem. Abstr.,* 64, 10340 (1966).

3709. Nickell, Gordon, and Goenaga, *Plant Disease Reporter,* 45, 756 (1961); through *Chem. Abstr.,* 56, 9159 (1962).

3710. Sztanyik, Varteresz, Doklen, and Nador, *Internat. Symp. Radiosensitzers Radioprotection Drugs (Milano; 1964),* 1, 515 (1965); through *Chem. Abstr.,* 64, 20157 (1966).

3711. Geigy A.-G., Fr.Pat. 1,343,933 (1963); through *Chem. Abstr.,* 64, 3289 (1966).

3712. Bindler and Model, Ger.Pat. 1,126,393 (1962); through *Chem. Abstr.,* 61, 667 (1964).

3713. Rohm and Haas Co., Neth.Pat. Appl. 6,406,038 (1964); through *Chem. Abstr.,* 63, 11583 (1965).

3714. Fronek and Klos, *Plant Disease Reporter,* 47, 348 (1963); through *Chem. Abstr.,* 59, 14509 (1963).

3715. Abramitis and Reck, U.S.Pat. 3,135,656 (1964); through *Chem. Abstr.,* 61, 6311 (1964).

3716. Shoeb, Mukerjee, Anand, and Dhar, *Indian J. Chem.,* 3, 507 (1965).

3717. Berenschot, King, Stubbs, and Bobalik, U.S.Pat. 3,140,979 (1964); through *Chem. Abstr.,* 61, 6873 (1964).

3718. Short, Biermacher, Dunnigan, and Leth, *J. Med. Chem.,* 6, 275 (1963).

3719. Toyoshima, Morishita, and Otsuka, Jap.Pat. 17,584 (1965); through *Chem. Abstr.,* 63, 18102 (1965).

3720. Tootal Ltd., Neth.Pat. Appl. 6,512,783 (1966); through *Chem. Abstr.*, 65, 9091 (1966).

3721. Alexander, *Amer. Dyestuff Reporter*, 55, 602 (1966).

3722. Tripp, McCall, and O'Connor, *Amer. Dyestuff Reporter*, 52, 598 (1963).

3723. Yamamoto, Oshima, and Suzuki, *Jap.Pat.* 12,716 (1962); through *Chem. Abstr.*, 59, 8923 (1963).

3724. Dan River Mills Inc., Neth.Pat. Appl. 6,513,329 (1966); through *Chem. Abstr.*, 65, 7360 (1966).

3725. Chemische Fabrik Pfersee G.m.b.H., Neth.Pat. Appl. 6,507,857 (1965); through *Chem. Abstr.*, 64, 19878 (1966).

3726. Diamond Alkali Co., Neth.Pat. Appl. 6,514,887 (1966); through *Chem. Abstr.*, 65, 17153 (1966).

3727. Rintelen, Petersen, and Heilmann, Ger.Pat. 1,124,355 (1962); through *Chem. Abstr.*, 57, 334 (1962).

3728. Pohlmann, Spoor, and Schickh, Belg.Pat. 625,362 (1963); through *Chem. Abstr.*, 60, 14696 (1964).

3729. Phillips, U.S.Pat. 3,025,296 (1962); through *Chem. Abstr.*, 57, 2489 (1962).

3730. Pugin and Bindler, U.S.Pat. 3,082,213 (1963); through *Chem. Abstr.*, 60, 700 (1964).

3731. Geigy A.-G., Brit.Pat. 929,257 (1963); through *Chem. Abstr.*, 61, 7028 (1964).

3732. Andress and Gee, U.S.Pat. 3,205,232 (1965); through *Chem. Abstr.*, 63, 18113 (1965).

3733. Capowski and Wascher, U.S.Pat. 3,014,792 (1961); through *Chem. Abstr.*, 56, 15730 (1962).

3734. Lindsay and Mees, Brit.Pat. 935,671 (1963); through *Chem. Abstr.*, 59, 15870 (1963).

3735. Hodge, U.S.Pat. 3,183,188 (1965); through *Chem. Abstr.*, 63, 2830 (1965).

3736. Hodge, Ger.Pat. 1,134,244 (1962); through *Chem. Abstr.*, 57, 12962 (1962).

3737. Barkley, Turner, Pianotti, Carthage, and Schwartz, *Antimicrobial Agents Ann.*, 1960, 507; through *Chem. Abstr.*, 56, 14402 (1962).

3738. Hodge, U.S.Pat. 3,087,891 (1961); through *Chem. Abstr.*, 59, 4945 (1963).

3739. Hodge and Lafferty, Fr.Pat. 1,364,172 (1964); through *Chem. Abstr.*, 62, 7988 (1965).

3740. Cahn, Alano, Hauser, and Herold, *Proc. Meeting Coll. Internat. Neuro-Psychopharmacol. (München; 1962)*, 3, 490 (1964); through *Chem. Abstr.*, 65, 6142 (1966).

3741. Fareri and Pellegrini, U.S.Pat. 3,048,607 (1962); through *Chem. Abstr.*, 59, 7299 (1963).

3742. Frankel, U.S.Pat. 3,041,337 (1962); through *Chem. Abstr.*, 57, 13757 (1962).

3743. Danielson, Belg.Pat. 624,519 (1963); through *Chem. Abstr.*, 59, 2998 (1963).

3744. Braun, U.S.Pat. 3,207,762 (1965); through *Chem. Abstr.*, 63, 17814 (1965).

3745. Finkelstein, Pan, Niesler, Johnson, and Schneider, *J. Pharmacol.*, 125, 330 (1959).

3746. Shoup, Miles, and Becker, *J. Amer. Chem. Soc.*, 89, 6200 (1967).

3747. McBridge and Miller, Fr.Pat. 1,415,468 (1965); through *Chem. Abstr.*, 64, 5111 (1966).

3748. Clark and Perrin, *Quart. Rev.*, 18, 295 (1964).

3749. Riddell, *J. Chem. Soc. (B)*, 1967, 560.

3750. Chang, Kim, and Hahn, *J. Korean Chem. Soc.*, 9, 75 (1965); *Chem. Abstr.*, 64, 17588 (1966).

3751. Ridi, Mangiavacchi, and Franchi, *Boll. chim. farm.*, 106, 664 (1967); *Chem. Abstr.*, 68, 114537 (1968).

3752. Tanabe Seiyaku Co., Fr.Pat. 1,470,072 (1967); through *Chem. Abstr.*, 68, 12990 (1968).

3753. Tanabe Seiyaku Co., Brit.Pat. 1,087,505 (1967); through *Chem. Abstr.*, 68, 114628 (1968).

3754. Calligaris, Linda, and Marino, *Tetrahedron*, 23, 813 (1967).

3755. Kato and Yamamoto, *Chem. Pharm. Bull.*, 15, 1334 (1967).

3756. Asbun and Binkley, *J. Org. Chem.*, 31, 2215 (1966).

3757. Lynch, Macdonald, and Webb, *Tetrahedron*, 24, 3595 (1968).

3758. Schulze and Willitzer, *Chem. Ber.*, 100, 3460 (1967).

3759. Gronowitz and Hallberg,
Acta Chem. Scand., 21, 2296 (1967).

3760. Blackburn and Davies,
J. Chem. Soc. (C), 1966, 1342.

3761. Davies and Puddephatt,
Tetrahedron Letters, 1967, 2265.

3762. Fissekis and Creegan,
J. Org. Chem., 32, 3595 (1967).

3763. Kato, Yamanaka, and Shibata,
J. Pharm. Soc. Japan, 87, 955 (1967).

3764. Bredereck, Simchen, and Traut,
Chem. Ber., 100, 3664 (1967).

3765. Bretschneider and Egg, *Monatsh.*, 98, 1577 (1967).

3766. Bukac and Sebenda, *Coll. Czech. Chem. Comm.*,
32, 3537 (1967).

3767. Prystas, *Coll. Czech. Chem. Comm.*, 32, 4260 (1967).

3768. Prystas, *Coll. Czech. Chem. Comm.*, 32, 4241 (1967).

3769. Hoffmann, *Z. Chem.*, 8, 147 (1968).

3770. Pitha and Ts'o, *J. Org. Chem.*, 33, 1341 (1968).

3771. Prystas and Gut,
Coll. Czech. Chem. Comm., 27, 1054 (1962).

3772. Temple, Smith, and Montgomery,
J. Org. Chem., 33, 530 (1967).

3773. Elad and Rosenthal, *Chem. Comm.*, 1968, 879.

3774. Albert, "Heterocyclic Chemistry," 2nd Edition,
Athlone Press, London, 1968.

3775. Acheson, "An Introduction to the Chemistry of
Heterocyclic Compounds," 2nd Edition,
Interscience, New York, 1967.

3776. Badger, "The Chemistry of Heterocyclic Compounds,"
Academic Press, New York, 1961.

3777. Katritzky and Lagowski, "The Principles of
Heterocyclic Chemistry," Methuen, London, 1967.

3778. Palmer, "The Structure and Reactions of
Heterocyclic Compounds," Arnold, London, 1967.

3779. Morrison and Kloepfer, *J. Amer. Chem. Soc.*,
90, 5037 (1968).

3780. B. W. Arantz and D. J. Brown, Unpublished results.

3781. Albert, in "Synthetic Procedures in Nucleic Acid Chemistry," ed. W. Zorbach and R. Tipson, Wiley, New York, 1969, Vol.2.

3782. Barlin and Perrin, *Quart. Rev.*, 20, 75 (1966).

3783. Barlin, *J. Chem. Soc. (B)*, 1967, 954.

3784. Adman, Gordon, and Jensen, *Chem. Comm.*, 1968, 1019.

3785. Chang, Lou, Cheng, Chi, and Chen, *Acta Chim.Sinica*, 30, 88 (1964).

3786. Kröhnke, Schmidt, and Zecker, *Chem. Ber.*, 97, 1163 (1964).

3787. Holland, Brit. Pat. 777,465 (1957); through *Chem. Abstr.*, 52, 1284 (1958).

3788. Bredereck, Herlinger, and Schweizer, *Chem. Ber.*, 93, 1208 (1960).

3789. Copenhaver, Ger. Pat. 822,086 (1951); through *Chem. Abstr.*, 52, 2094 (1958).

3790. Klimko, Mikhalev, and Skoldinov, *J. Gen. Chem.*, 1258 (1960).

3791. Bredereck, *Org. Synth.*, 43, 77 (1963).

3792. Scala, Bikales, and Becker, *J. Org. Chem.*, 30, 303 (1965).

3793. Maisack, Peukert, and Schoenleben, Brit. Pat. 860,423 (1961); through *Chem. Abstr.*, 55, 17664 (1961).

3794. Dornow and Siebrecht, *Chem. Ber.*, 93, 1106 (1960).

3795. Overberger and Michelotti, *J. Amer. Chem. Soc.*, 80, 988 (1958).

3796. Traverso and Whitehead, U.S. Pat. 2,874,157 (1959); through *Chem. Abstr.*, 53, 12319 (1959).

3797. Miura, Belg. Pat. 638,220 (1964); through *Chem. Abstr.*, 62, 9106 (1965).

3798. Miller and Rose, Brit. Pat. 864,731 (1961); through *Chem. Abstr.*, 55, 24798 (1961).

3799. Protopopova and Skoldinov, *J. Gen. Chem.*, 28, 240 (1958).

3800. Protopopova and Skoldinov, *J. Gen. Chem.*, 27, 1276 (1957).

3801. Klimko, Protopopova, Smirnova, and Skoldinov, *J. Gen. Chem.*, 32, 2961 (1962).

3802. Case, *J. Org. Chem.*, 30, 931 (1965).

3803. Chatterjee and Anand, *J. Sci. Ind. Research (India)*, 18B, 272 (1959); through *Chem. Abstr.*, 54, 6745 (1960).

3804. Merck, Brit. Pat. 911,551 (1962); through *Chem. Abstr.*, 58, 10214 (1963).

3805. Habicht, Swiss Pat. 361,573 (1962); through *Chem. Abstr.*, 59, 8762 (1963).

3806. Verma and Dey, *J. Indian Chem. Soc.*, 40, 283 (1963).

3807. Biggs and Sykes, *J. Chem. Soc.*, 1961, 2595.

3808. Wenz, Goettmann, and Koop, *Annalen*, 680, 82 (1964).

3809. Whitehead and Traverso, U.S. Pat. 2,845,425 (1958); through *Chem. Abstr.*, 53, 2262 (1959).

3810. Wagner, U.S. Pat. 3,281,408 (1966); through *Chem. Abstr.*, 66, 18718 (1967).

3811. Wagner, U.S. Pat. 3,281,420 (1966); through *Chem. Abstr.*, 66, 18719 (1967).

3812. Carstens and Goldner, D.D.R. Pat. 21,593 (1958); through *Chem. Abstr.*, 56, 8727 (1962).

3813. Pachter and Weinstock, U.S. Pat. 3,127,402 (1964); through *Chem. Abstr.*, 60, 14523 (1964).

3814. Taniguchi, *J. Pharm. Soc. Japan*, 78, 329 (1958).

3815. Musil and Budesinsky, Czech. Pat. 86,863 (1957); through *Chem. Abstr.*, 54, 2376 (1960).

3816. Geigy, Nath. Pat. Appl. 6,513,321 (1966); through *Chem. Abstr.*, 65, 8931 (1966).

3817. Popova and Nemets, *Trudy Leningrad. Tekhnol. Inst. im Lensoveta*, 40, 180 (1957); through *Chem. Abstr.*, 54, 19693 (1960).

3818. Kunze, Ger. Pat. 962,165 (1957); through *Chem. Abstr.*, 53, 10262 (1959).

3819. Takamizawa, Ikawa, and Narisada, *J. Pharm. Soc. Japan*, 78, 632 (1958).

3820. Bonvicino and Hennessy, *J. Org. Chem.*, 24, 451 (1959).

3821. Takatori and Asano, *Gifu Yakka Daigaku Kiyo*, 7, 60 (1957); through *Chem. Abstr.*, 52, 5422 (1958).

3822. Angier and Marsico, *J. Org. Chem.*, 25, 759 (1960).

3823. El'tsov, Muravich-Aleksandr, and Roitshtein, *Zhur. org. Khim.*, 3, 205 (1967); through *Chem. Abstr.*, 66, 104986 (1967).

3824. Weinstock, U.S. Pat. 2,963,478 (1960); through *Chem. Abstr.*, 55, 10482 (1961).

3825. Rylski, Sorm, and Arnold, *Coll. Czech. Chem. Comm.*, 24, 1667 (1959).

3826. Protopopova, Klimko, and Skoldinov, *Khim. Nauka i Prom.*, 4, 805 (1959); through *Chem. Abstr.*, 54, 11037 (1960).

3827. Van Winkle, McClure, and Williams, *J. Org. Chem.*, 31, 3300 (1966).

3828. Volodina, Terent'ev, Kudryashova, and Mishina, *J. Gen. Chem.*, 34, 473 (1964).

3829. Denayer, Belg. Pat. 609,114 (1962); through *Chem. Abstr.*, 58, 536 (1963).

3830. Takamizawa and Hirai, Jap. Pat. 22,009 (1964); through *Chem. Abstr.*, 62, 9150 (1965).

3831. Baumbach, Henning, and Hilgetag, *Z. Chem.*, 4, 67 (1964).

3832. Seefelder, Ger. Pat. 1,112,982 (1959); through *Chem. Abstr.*, 56, 5810 (1962).

3833. Eilingsfeld, Seefelder, and Weidinger, *Chem. Ber.*, 96, 2671 (1963).

3834. Martin, Barton, Gott, and Meen, *J. Org. Chem.*, 31, 943 (1966).

3835. Mizukami and Hirai, *Chem. Pharm. Bull.*, 14, 1321 (1966).

3836. David and Estramareix, *Bull. Soc. chim. France*, 1960, 1731.

3837. Schaefer, U.S. Pat. 3,294,798 (1966); through *Chem. Abstr.*, 66, 10758 (1967).

3838. Nakao, Ito, and Muramatsu, *Sankyo Kenkyusho Nempo*, 18, 33 (1966); through *Chem. Abstr.*, 66, 115678 (1967).

3839. Kulkarni, Sabnis, and Kulkarni, *J. Sci. Ind. Res., India*, 19C, 6 (1960); through *Chem. Abstr.*, 54, 22576 (1960).

3840. Kenten, *Biochem. J.*, 69, 439 (1958).

3841. Tsatsaronis and Effenberger, *Chem. Ber.*, 94, 2876 (1961).

3842. Sirakawa, Jap. Pat. 777 (1957); through *Chem. Abstr.*, 52, 4699 (1958).

3843. Temple and Montgomery, *J. Amer. Chem. Soc.*, 86, 2946 (1964).

3844. Uno, Machida, and Hanai, *Chem. Pharm. Bull.*, 14, 756 (1966).

3845. Libermann, Fr. Pat. 1,170,121 (1959); through *Chem. Abstr.*, 55, 9439 (1961).

3846. Libermann and Rouaix, *Bull. Soc. chim. France*, 1959, 1793.

3847. Goerdeler and Ruppert, *Chem. Ber.*, 96, 1630 (1963).

3848. Hensel, *Chem. Ber.*, 97, 96 (1964).

3849. Baker, Ho, and Santi, *J. Pharm. Sci.*, 54, 1415 (1965).

3850. Wellcome Foundation Ltd., Brit. Pat. 971,307 (1964); through *Chem. Abstr.*, 61, 14689 (1964).

3851. Stenbuck and Hood, U.S. Pat. 3,049,544 (1962); through *Chem. Abstr.*, 58, 1478 (1963).

3852. Leonard, Skinner, and Shive, *Arch. Biochem. Biophys.*, 92, 33 (1961).

3853. Denayer, *Bull. Soc. chim. France*, 1962, 1358.

3854. Ghosh and Sen, *J. Indian Chem. Soc.*, 42, 505 (1965).

3855. Almirante, *Ann. Chim. (Italy)*, 49, 333 (1959).

3856. Weiss, Robins, and Noell, *J. Org. Chem.*, 25, 765 (1960).

3857. Furukawa, Seto, and Toyoshima, *Chem. Pharm. Bull.*, 9, 914 (1961).

3858. DeGraw, Ross, Goodman, and Baker, *J. Org. Chem.*, 26, 1933 (1961).

3859. Bergmann, Levin, Kwietny-Govrin, and Ungar, *Biochim. Biophys. Acta*, 47, 1 (1961).

3860. Iwai, Iwashige, Yura, Nakamura, and Shinozaki, *Chem. Pharm. Bull.*, 12, 1446 (1964).

3861. Bergmann, Kalmus, Ungar-Waron, and Kwietny-Govrin, *J. Chem. Soc.*, 1963, 3729.

3862. Reynolds, Van Allan, and Tinker, *J. Org. Chem.*, 24, 1205 (1959).

3863. Feldman and Chzhun-Tszi, *J. Gen. Chem.*, 30, 3835 (1960).

3864. Steiger, U.S. Pat. 3,133,078 (1961); through *Chem. Abstr.*, 61, 8241 (1964).

3865. Reynaud and Moreau, *Bull. Soc. chim. France*, 1960, 2002.

3866. Miller and Rose, Brit. Pat. 859,287 (1961); through *Chem. Abstr.*, 55, 17665 (1961).

3867. Brady and Herbst, *J. Org. Chem.*, 24, 922 (1959).

3868. Postovskii and Smirnova, *Proc. Acad. Sci. (U.S.S.R.)*, 166, 1136 (1966).

3869. Orazi, Corral, and Bertorello, *J. Org. Chem.*, 30, 1101 (1965).

3870. Lewis, Beaman, and Robins, *Canad. J. Chem.*, 41, 1807 (1963).

3871. Davidenkov, *Med. Prom. S.S.S.R.*, 16, No.1, p.25 (1962); through *Chem. Abstr.*, 58, 4565 (1963).

3872. Hoefle and Holmes, U.S. Pat. 3,111,521 (1963); through *Chem. Abstr.*, 60, 2979 (1964).

3873. Rhone-Poulenc, Fr. Pat. (addition), 63, 159 (1955); through *Chem. Abstr.*, 53, 7214 (1959).

3874. Hoffmann-La Roche, Brit. Pat. 800,776 (1958); through *Chem. Abstr.*, 53, 8178 (1959).

3875. Ridi, Papini, and Checchi, *Gazzetta*, 91, 973 (1961).

3876. Massaroli and Signorelli, *Boll. Chim. Farm.*, 105, 400 (1966); through *Chem. Abstr.*, 65, 8903 (1966).

3877. Ridi, Papini, and Checchi, *Gazzetta*, 92, 209 (1962).

3878. Reiner and Eugster, *Helv. Chim. Acta*, 50, 128 (1967).

3879. Morimoto, *J. Pharm. Soc. Japan*, 82, 386 (1962).

3880. Kuehne, *J. Amer. Chem. Soc.*, 84, 837 (1962).

3881. Ried and Gross, *Chem. Ber.*, 90, 2646 (1957).

3882. Schellenberger and Rödel, *Angew. Chem.*, 76, 226 (1964).

3883. Brown and Rapoport, *J. Org. Chem.*, 28, 3261 (1963).

3884. Weissauer, Belg. Pat. 644,765 (1964); through *Chem. Abstr.*, 63, 11740 (1965).

3885. Deebel and Hamm, U.S. Pat. 2,879,150 (1959); through *Chem. Abstr.*, 53, 15107 (1959).

3886. Seefelder and Leuchs, Brit. Pat. 913,910 (1962); through *Chem. Abstr.*, 59, 2836 (1963).

3887. Shvachkin, Novikova, Reznikova, and Padyukova, *Biol. Aktivn. Soedin; Akad. Nauk S.S.S.R.* 1965, 25; through *Chem. Abstr.*, 63, 16453 (1965).

3888. Shvachkin, Syrtsova, and Prokof'ev, *J. Gen. Chem.*, 32, 2431 (1962).

3889. Shvachkin and Krivtsov, *J. Gen. Chem.*, 34, 2164 (1964).

3890. Schroeder, U.S. Pat. 3,118,889 (1964); through *Chem. Abstr.*, 60, 12027 (1964).

3891. Gray and Heitmeier, U.S. Pat. 3,192,216 (1965); through *Chem. Abstr.*, 63, 16363 (1965).

3892. Kuwayama, *J. Pharm. Soc. Japan*, 82, 1028 (1962).

3893. Hoffmann-La Roche, Brit.Pat. 877,318 (1960); through *Chem. Abstr.*, 56, 8724 (1962).

3894. Prystas and Sorm, *Coll. Czech. Chem. Comm.*, 29, 121 (1964).

3895. Weidinger and Wellenreuther, Brit.Pat. 927,974 (1963); through *Chem. Abstr.*, 60, 2987 (1964).

3896. Shen, Lewis, and Ruyle, *J. Org. Chem.*, 30, 835 (1965).

3897. King, *U.S. Dept. Com., Office Tech. Serv.*, AD 258,217 (1961); *Chem. Abstr.*, 58, 4557 (1963).

3898. Benz, U.S. Pat. 3,075,980 (1963); through *Chem. Abstr.*, 59, 1660 (1963).

3899. Schoenauer, Swiss Pat. 372,679 (1963); through *Chem. Abstr.*, 60, 13257 (1964).

3900. Chang, Yang, Wang, and Hu, *Yao Hsueh Hsueh Pao*, 10, 600 (1963); through *Chem. Abstr.*, 60, 14504 (1964).

3901. Wellcome Foundation, Brit.Pat. 980,854 (1965); through *Chem. Abstr.*, 62, 14695 (1965).

3902. Wittenburg, *Z. Chem.*, 4, 303 (1964).

3903. Baker and Jackson, *J. Pharm. Sci.*, 54, 1758 (1965).

3904. Runti, Sindellari, and Nisi, *Ann. Chim. (Italy)*, 49, 1649 (1959).

3905. Sömmer, D.D.R.Pat. 32,753 (1966); through *Chem. Abstr.*, 65, 727 (1966).

3906. Chambers and Kurkov, *Biochemistry*, 3, 326 (1964).

3907. Cier, Lefier, Ravier, and Nofre, *Compt. rend.*, 254, 504 (1962).

3908. Bretschneider and Klötzer, U.S.Pat. 3,091,610 (1963); through *Chem. Abstr.*, 59, 3934 (1963).

3909. Moye, *Aust. J. Chem.*, 17, 1309 (1964).

3910. Bretschneider, Richter, and Klötzer, *Monatsh.*, 96, 1661 (1965).

3911. Ciba, Belg.Pat. 645,062 (1964); through *Chem.Abstr.*, 66, 37942 (1967).

3912. Ciba, Belg.Pat. 641,252 (1964); through *Chem.Abstr.*, 63, 617 (1965).

3913. Ciba, Belg.Pat. 641,253 (1964); through *Chem.Abstr.*, 63, 9962 (1965).

3914. Blanchard, Brit.Pat. 1,010,997 (1965); through *Chem. Abstr.*, 64, 5109 (1966).

3915. Hoffman-La Roche, Brit.Pat. 914,929 (1963); through *Chem. Abstr.*, 59, 1659 (1963).

3916. Yasushige, Tsuruta, Aida, Hyoto, and Noda, Jap.Pat. 27,259 (1964); through *Chem. Abstr.*, 62, 11828 (1965).

3917. Korte and Ludwig, *Annalen*, 615, 94 (1958).

3918. Dick and Drugarin, *Acad. Rep. Populare Romine, Baza Cercetari Stiint Timisoara, Studii Cercetari Stiinte Chim.*, 8, 225 (1961); through *Chem. Abstr.*, 58, 3431 (1963).

3919. Dashkevich, *J. Gen. Chem.*, 32, 2346 (1962).

3920. Shimo, Kawasaki, and Wakamatsu, *J. Chem. Soc. Japan, Ind. Chem. Sect.*, 65, 1370 (1962).

3921. Fel'dman and Shepshelevich, *Mechenye Biol. Aktivn. Veshchestva, Sb. Statei*, 1962, 85; through *Chem. Abstr.*, 59, 7525 (1963).

3922. Chakraborty and Loring, *J. Biol. Chem.*, 235, 2122 (1960).

3923. Ehrensvard and Liljekvist, *Acta Chem. Scand.*, 13, 2070 (1959).

3924. Montgomery and Hewson, *J. Het. Chem.*, 2, 313 (1965).

3925. Doumas and Biggs, *J. Biol. Chem.*, 237, 2306 (1962).

3926. Wajon and Arens, *Rec. Trav. chim.*, 76, 79 (1957).

3927. Ziegler, Kleineberg, and Meindl, *Monatsh.*, 97, 10 (1966).

3928. Maillard, Morin, Vincent, and Bernard, U.S. Pat. 3,047,462 (1962); through *Chem. Abstr.*, 58, 1474 (1963).

3929. Brossi and Schnider, Swiss Pat. 364,788 (1962); through *Chem. Abstr.*, 60, 5518 (1964).

3930. Kano and Makizumi, Jap. Pat. 5722 (1957); through *Chem. Abstr.*, 52, 11967 (1958).

3931. Zemlicka, *Coll. Czech. Chem. Comm.*, 28, 1060 (1963).

3932. Chi and Wu, *Acta Chimica Sinica*, 22, 188 (1956).

3933. Guyot and Mentzer, *Compt. rend.*, 246, 436 (1958).

3934. Fel'dman and Shepshelevich, *Mechenye Biol. Aktivn. Veshchestva, Sb. Statei*, 1962, 65; through *Chem. Abstr.*, 59, 7528 (1963).

3935. Takamizawa, Hayashi, and Ito, Jap. Pat. 24,383 (1963); through *Chem. Abstr.*, 60, 2965 (1964).

3936. Ajello, Ajello, and Sprio, *Ric. Sci., Rend. Sez. B*, 4, 105 (1964); through *Chem. Abstr.*, 61, 5644 (1964).

3937. Berlin and Vasil'eva, *J. Gen. Chem.*, 28, 1063 (1958).

3938. Golovinskii, *Compt. rend. acad. bulgare sci.*, 10, 49 (1957); through *Chem. Abstr.*, 52, 6364 (1958).

3939. Mel'nikov, Shvetsova-Shilovskaya, and Grapov, *Org. Insektofungitsidy i Gerbitsidy*, 1958, 108; through *Chem. Abstr.*, 55, 3600 (1961).

3940. Wajon and Arens, *Rec. Trav. chim.*, 76, 65 (1957).

3941. Giudicelli, Najer, Chabrier, and Joannic-Voisinet, *Ann. pharm. franc.*, 15, 533 (1957); through *Chem. Abstr.*, 52, 11081 (1958).

3942. Miller and Rose, Brit. Pat.897,870 (1962); through *Chem. Abstr.*, 58, 10211 (1963).

3943. Eiden and Nagar, *Naturwiss.*, 50, 403 (1963).

3944. Schuler and Surayni, Ger. Pat. 1,117,128 (1961); through *Chem. Abstr.*, 56, 11600 (1962).

3945. Ridi, Cheichi, and Papini, *Ann. Chim. (Italy)*, 44, 769 (1954).

3946. Takamizawa, Jap. Pat. 1126 (1958); through *Chem. Abstr.*, 53, 1388 (1959).

3947. Hoffman-La Roche, Brit. Pat. 953,875 (1964); through *Chem. Abstr.*, 61, 8321 (1964).

3948. Khalifa, *Bull. Fac. Pharm.* (Cairo), 1, 149 (1961-1962).

3949. Müller and Plieninger, *Chem. Ber.*, 92, 3009 (1959).

3950. Kagan and Suen, *Bull. Soc. chim. France*, 1966, 1819.

3951. Arbuzov and Zoroastrova, *Izvest. Akad. Nauk S.S.S.R., Ser. Khim.*, 1958, 1331.

3952. Asai, *J. Pharm. Soc. Japan*, 83, 471 (1963).

3953. Sprio, *Ric. Sci., Rend Sez. B*, 4, 585 (1964); through *Chem. Abstr.*, 62, 1657 (1965).

3954. Wittenburg, *Chem. Ber.*, 99, 2380 (1966).

3955. Biggs and Doumas, *J. Biol. Chem.*, 238, 2470 (1963).

3956. Bamberg, Hemmerich, and Erlenmeyer, *Helv. Chim. Acta*, 43, 395 (1960).

3957. Lacey and Ward, *J. Chem. Soc.*, 1958, 2134.

3958. Geigy A.-G., Belg. Pat. 619,275 (1962); through *Chem. Abstr.*, 59, 10078 (1963).

3959. Gompper, *Chem. Ber.*, 93, 198 (1960).

3960. Beiersdorf A.-G., Belg. Pat. 620,379 (1963); through *Chem. Abstr.*, 59, 7537 (1963).

3961. Kodak, Belg. Pat. 560,452 (1957); through *Chem. Abstr.*, 54, 126 (1960).

3962. Eiden and Nagar, *Arch. Pharm.*, 297, 367 (1964).

3963. Beiersdorf A.-G., Belg. Pat. 637,891 (1964); through *Chem. Abstr.*, 62, 10449 (1965).

3964. Tuppy and Küchler, *Monatsh.*, 95, 1698 (1964).

3965. Sanko Co., Belg. Pat. 639,885 (1964); through *Chem. Abstr.*, 62, 16364 (1965).

3966. Ukita, Okuyama, and Hayatsu, *Chem. Pharm. Bull.*, 11, 1399 (1963).

3967. Luckenbaugh and Soboczenski, U.S. Pat. 3,235,363 (1966); through *Chem. Abstr.*, 65, 13733 (1966).

3968. Fischer, Neumann, and Roch, *Annalen*, 633, 158 (1960).

3969. Davydov, *Izvest. Akad. Nauk S.S.S.R. Ser. Khim.*, 1963, 571.

3970. Du Pont, Brit.Pat. 968,665 (1964); through *Chem. Abstr.*, 61, 13327 (1964).

3971. Naito, Hirata, Kawakami, and Sano, *Chem. Pharm. Bull.*, 9, 703 (1961).

3972. Klötzer and Herberz, *Monatsh.*, 96, 1721 (1965).

3973. Du Pont, Brit.Pat. 968,664 (1964); through *Chem. Abstr.*, 61, 13327 (1964).

3974. Senda and Fujimura, Jap.Pat. 11,966 (1961); through *Chem. Abstr.*, 56, 5978 (1962).

3975. Loux and Soboczenski, Belg.Pat. 620,174 (1962); through *Chem. Abstr.*, 59, 639 (1963).

3976. Senda and Fujimura, Jap.Pat. 8176 (1961); through *Chem. Abstr.*, 58, 13969 (1963).

3977. Gunar and Zav'yalov, *Dokl. Akad. Nauk S.S.S.R.*, 158, 1358 (1964).

3978. Gunar, Ovechkina, and Zav'yalov, *Izvest. Akad. Nauk S.S.S.R., Ser. Khim.*, 1965, 1076.

3979. Soboczenski, U.S.Pat. 3,235,358 (1966); through *Chem. Abstr.*, 64, 11223 (1966).

3980. Klopping and Loux, Fr.Pat. 1,394,286 (1965); through *Chem. Abstr.*, 63, 4309 (1965).

3981. Loux, Luckenbaugh, and Soboczenski, Belg.Pat. 625,897 (1963); through *Chem. Abstr.*, 60, 14519 (1964).

3982. Du Pont, Brit.Pat. 968,661 (1964); through *Chem. Abstr.*, 61, 13326 (1964).

3983. Du Pont, Brit.Pat. 968,663 (1964); through *Chem. Abstr.*, 61, 13327 (1964).

3984. Murdock and Angier, *J. Amer. Chem. Soc.*, 84, 3748 (1962).

3985. Baker, Jackson, and Chheda, *J. Pharm. Sci.*, 54, 1617 (1965).

3986. Loux and Soboczenski, U.S.Pat. 3,254,082 (1966); through *Chem. Abstr.*, 65, 7193 (1966).

3987. Gunar and Zav'yalov, *Izvest. Akad. Nauk S.S.S.R., Ser. Khim.*, 1965, 747.

3988. Dashkevich, *Dokl. Akad. Nauk S.S.S.R.*, 145, 323 (1962).

3989. Dashkevich and Siraya, *Tr. Leningr. Khim.–Farmatsevt. Inst.*, 1962, No.16, 202; through *Chem. Abstr.*, 60, 9276 (1964).

3990. Pala, *Farmaco*, (Ed. Sci.), 13, 461 (1958).

3991. Omikron-Gagliardi Soc., Brit.Pat. 833,049 (1960); through *Chem. Abstr.*, 55, 1665 (1961).

3992. Matsumoto, Shioyama, and Murata, *Noyaku Seisan Gijutsu*, 9, 17 (1963); through *Chem. Abstr.*, 61, 6300 (1964).

3993. Cranham, Cummings, Johnston, and Stevenson, *J. Sci. Food Agric.*, 9, 143 (1958).

3994. König, Ger.Pat. 1,189,380 (1963); through *Chem. Abstr.*, 62, 16259 (1965).

3995. Agfa A.–G., Brit.Pat. 823,943 (1959); through *Chem. Abstr.*, 54, 19243 (1960).

3996. Perina, Bydzovsky, Budesinsky, and Sluka, Czech.Pat. 111,782 (1964); through *Chem. Abstr.*, 62, 5285 (1965).

3997. Castle, Kaji, Gerhardt, Guither, Weber, Malm, Shoup, and Rhoads, *J. Het. Chem.*, 3, 79 (1966).

3998. Mason, *J. Chem. Soc.*, 1960, 219.

3999. Newman and Karnes, *J. Org. Chem.*, 31, 3980 (1966).

4000. Bredereck, Effenberger, and Schweizer, *Chem. Ber.*, 95, 956 (1962).

4001. Hoefle and Meyer, U.S.Pat. 2,949,466 (1960); through *Chem. Abstr.*, 55, 589 (1961).

4002. Habicht, Swiss Pat. 358,424 (1962); through *Chem. Abstr.*, 58, 3444 (1963).

4003. Ciba, Brit.Pat. 901,749 (1962); through *Chem. Abstr.*, 59, 1660 (1963).

4004. Mautner, *J. Org. Chem.*, 23, 1450 (1958).

4005. Ciba, Brit.Pat. 893,235 (1962); through *Chem. Abstr.*, 58, 12581 (1963).

4006. Bredereck, Effenberger, and Rosemann, *Chem. Ber.*, 95, 2796 (1962).

4007. Baker and Shapiro, *J. Med. Chem.*, 6, 664 (1963).

4008. Takamizawa, Hirai, and Sato, Jap.Pat. 26,733 (1964); through *Chem. Abstr.*, 62, 10448 (1965).

4009. Taylor, Knopf, Cogliano, Barton, and Pfleiderer, *J. Amer. Chem. Soc.*, 82, 6058 (1960).

4010. Anderson, Bell, and Duncan, *J. Chem. Soc.*, 1961, 4705.

4011. Kaiser, U.S.Pat. 3,036,075 (1962); through *Chem. Abstr.*, 58, 1477 (1963).

4012. Dornow and Schleese, *Chem. Ber.*, 91, 1830 (1958).

4013. Ciba, Brit.Pat. 933,158 (1963); through *Chem. Abstr.*, 61, 5665 (1964).

4014. Ishidate and Yuki, *Chem. Pharm. Bull.*, 8, 131 (1960).

4015. Peters, Minnemeyer, Spears, and Tieckelmann, *J. Org. Chem.*, 25, 2137 (1960).

4016. Trofimenko, Little, and Mower, *J. Org. Chem.*, 27, 433 (1962).

4017. Eiden, *Arch. Pharm.*, 295, 516 (1962).

4018. Almirante, Mugnaini, Fritz, and Provinciali, *Boll. Chim. Farm.*, 105, 32 (1966); through *Chem. Abstr.*, 65, 699 (1966).

4019. Wellcome Foundation, Brit.Pat. 812,366 (1959); through *Chem. Abstr.*, 54, 592 (1960).

4020. Parnell, *J. Chem. Soc.*, 1962, 2856.

4021. Kugita, Takeda, and Oine, Jap.Pat. 22,887 (1963); through *Chem. Abstr.*, 60, 2966 (1964).

4022. Thompson, *J. Chem. Soc.*, 1962, 617.

4023. Beaman and Robins, *J. Org. Chem.*, 28, 2310 (1963).

4024. Esche and Wojahn, *Arch. Pharm.*, 299, 56 (1966).

4025. Leichssenring, D.D.R.Pat. 11,159 (1956); through *Chem. Abstr.*, 52, 17295 (1958).

4026. Tursin, *Trudy Vsesoyuz. Nauch. Issledovatel. Vitamin. Inst.*, 6, 22 (1959); through *Chem. Abstr.*, 55, 11428 and 12411 (1961).

4027. Lisnyanskii and Zhdanovich, *Trudy Vsesoyuz. Nauch. Issledovatel. Vitamin. Inst.*, 7, 48 (1961); through *Chem. Abstr.*, 59, 1637 (1963).

4028. Rogers and Clark, U.S.Pat. 3,042,675 (1962); through *Chem. Abstr.*, 57, 16628 (1962).

4029. Montgomery and Temple, *J. Amer. Chem. Soc.*, 80, 409 (1958).

4030. Sokolova and Magidson, *Materialy po Obmenu Peredov. Opytom i Nauch. Dostizhen., v. Khim.—Farm. Prom.*, 1958, 129; through *Chem. Abstr.*, 58, 25956 (1961).

4031. Goldner, *Chem. Tech.*, 12, 495 (1960); through *Chem. Abstr.*, 55, 5517 (1961).

4032. Fel'dman, *Trudy Leningr. Khim.—Farmatsevt. Inst.*, 1962, 25; through *Chem. Abstr.*, 60, 10680 (1964).

4033. Nitta, Okui, and Ito, Jap.Pat. 19,895 (1966); through *Chem. Abstr.*, 66, 55503 (1967).

4034. Brand, Richter, and Rieckhoff, Brit.Pat. 988,398 (1965); through *Chem. Abstr.*, 62, 16267 (1965).

4035. Temple and Montgomery, *J. Org. Chem.*, 31, 1417 (1966).

4036. Hirao, Fujimoto, Kato, and Okazaki, *Kogyo Kagaku Zasshi*, 66, 1682 (1963); through *Chem. Abstr.*, 60, 12008 (1964).

4037. Merck, Brit.Pat. 902,392 (1962); through *Chem. Abstr.*, 57, 15128 (1962).

4038. Suranyi and Wilk, Ger.Pat. 1,150,987 (1963); through *Chem. Abstr.*, 60, 2971 (1964).

4039. Hurt and Stephen, *Tetrahedron Suppl.*, 7, 227 (1966).

4040. Taylor and Weinstock, Brit.Pat. 951,655 (1964); through *Chem. Abstr.*, 61, 4378 (1964).

4041. Hoffmann-La Roche, Belg.Pat. 665,446 (1965); through *Chem. Abstr.*, 65, 728 (1966).

4042. Langley, Brit.Pat. 854,011 (1960); through *Chem. Abstr.*, 55, 13457 (1961).

4043. Nakazawa and Watatani, *Ann. Rep. Takamme Lab.*, 12, 32 (1960); through *Chem. Abstr.*, 55, 6491 (1961).

4044. Sirakawa, Tsujikawa, and Tsuda, U.S.Pat. 3,041,339 (1962); through *Chem. Abstr.*, 57, 13774 (1962).

4045. Fusco, Losco, and Troiani, Ital.Pat. 662,501 (1964); through *Chem. Abstr.*, 65, 20147 (1966).

4046. Phillips and Mehta, Ger.Pat. 1,148,556 (1963); through *Chem. Abstr.*, 59, 11525 (1963).

4047. Wellcome Foundation, Brit.Pat. 1,104,881 (1965); through *Chem. Abstr.*, 64, 9741 (1966).

4048. Lewis, Schneider, and Robins, *J. Org. Chem.*, 26, 3837 (1961).

4049. Shealey, Clayton, O'Dell, and Montgomery, *J. Org. Chem.*, 27, 4518 (1962).

4050. Shvachkin, Berestenko, and Lapuk, *J. Gen. Chem.*, 32, 3893 (1962).

4051. Okumura, Jap.Pat. 18,048 (1964); through *Chem. Abstr.*, 62, 5286 (1965).

4052. Langley, Brit.Pat. 875,717 (1958); through *Chem. Abstr.*, 56, 4780 (1962).

4053. Levin and Tamari, *J. Chem. Soc.*, 1960, 2782.

4054. Fidler and Wood, *J. Chem. Soc.*, 1957, 3980.

4055. Pachter, U.S.Pat. 3,159,629 (1964); through *Chem. Abstr.*, 62, 6496 (1965).

4056. Wellcome Foundation, Brit.Pat. 1,016,656 (1966); through *Chem. Abstr.*, 64, 9747 (1966).

4057. Krackov and Christensen, *J. Org. Chem.*, 28, 2677 (1963).

4058. Laboratoires Lumiere et Institut Merieux, Fr.Pat. 1,364,734 (1964); through *Chem. Abstr.*, 62, 572 (1965).

4059. Acker and Castle, *J. Org. Chem.*, 23, 2010 (1958).

4060. Golovchinskaya and Kolodkin, *J. Gen.Chem.*, 29, 1650 (1959).

4061. Khmelevskii, *J. Gen. Chem.*, 31, 3123 (1961).

4062. Bredereck, Kolb, Saum, and Schmidt, *Rev. Chim. (Roumaine)*, 7, 705 (1962); through *Chem. Abstr.*, 61, 4353 (1964).

4063. Sunagawa and Watatani, Jap.Pat. 5187 (1962); through *Chem. Abstr.*, 59, 642 (1963).

4064. Codington, Fecher, Maguire, Thompson, and Brown, *J. Amer. Chem. Soc.*, 80, 5164 (1958).

4065. Banfield and McGuinness, *J. Chem. Soc.*, 1963, 2747.

4066. Schering A.-G., Neth.Pat.Appl. 299,218 (1965); through *Chem. Abstr.*, 64, 8202 (1966).

4067. Pfleiderer and Nübel, *Annalen*, 631, 168 (1960).

4068. Pfleiderer and Sagi, *Annalen*, 673, 78 (1964).

4069. Schuler and Suranyi, Ger.Pat. 1,122,533 (1962); through *Chem. Abstr.*, 57, 2231 (1962).

4070. Baker, Santi, and Shapiro, *J. Pharm. Sci.*, 53, 1317 (1964).

4071. Merck, Brit.Pat. 884,772 (1961); through *Chem. Abstr.*, 57, 2231 (1962).

4072. Hoffman-La Roche, Belg.Pat. 620,111 (1963); through *Chem. Abstr.*, 60, 4160 (1964).

4073. Priewe and Gutsche, Ger.Pat. 1,145,622 (1963); through *Chem. Abstr.*, 59, 6421 (1963).

4074. Budesinsky, Kopecky, and Prikryl, Czech. Pat. 104,969 (1962); through *Chem. Abstr.*, 60, 8043 (1964).

4075. Kuwada, Masuda, and Asai, *Chem. Pharm. Bull.*, 8, 792 (1960).

4076. Schering A.-G., Brit.Pat. 997,632 (1965); through *Chem. Abstr.*, 63, 16365 (1965).

4077. Lohrmann, Lagowski, and Forrest, *J. Chem. Soc.*, 1964, 451.

4078. Hirano and Yonemoto, Jap.Pat. 2222 (1957); through *Chem. Abstr.*, 52, 5489 (1958).

4079. Drinkard, U.S. Pat. 2,804,459 (1957); through *Chem. Abstr.*, 52, 2096 (1958).

4080. Ishidate and Yuki, *Chem. Pharm. Bull.*, 8, 137 (1960).

4081. Schering A.-G., Brit.Pat. 943,489 (1963); through *Chem. Abstr.*, 60, 5518 (1964).

4082. Chang and Yuan, *Yao Hsueh Hsueh Pao*, 6, 139 (1958); through *Chem. Abstr.*, 53, 14108 (1959).

4083. Fel'dman, Semicheva, and Zhokhovets, *Mechenye Biol. Aktivn. Veshchestva Sb. Statei*, 1962, 26; through *Chem. Abstr.*, 59, 8745 (1963).

4084. Borowitz, Bloom, Rothschild, and Sprinson, *Biochemistry*, 4, 650 (1965).

4085. Birch and Moye, *J. Chem. Soc.*, 1958, 2622.

4086. Chen, *Hua Hsueh Tung Pao*, 1964, 61; through *Chem. Abstr.*, 61, 5508 (1964).

4087. Sunagawa and Nakazawa, Jap.Pat. 4893 (1962); through *Chem. Abstr.*, 59, 642 (1963).

4088. Sunagawa, Nakazawa, and Watatani, Jap.Pat. 10,697 (1962); through *Chem. Abstr.*, 59, 5177 (1963).

4089. Bretschneider and Klötzer, Swiss Pat. 369,454 (1963); through *Chem. Abstr.*, 60, 14517 (1964).

4090. Hoffmann-La Roche, Brit.Pat. 946,488 (1964); through *Chem. Abstr.*, 64, 3568 (1966).

4091. Nitta, Okui, and Ito, Jap.Pat. 13,878 (1965); through *Chem. Abstr.*, 63, 13285 (1965).

4092. Yamada, Chibata, and Kiguchi, *Tanabe Seiyaku Kenkyu Nempo*, 2, 13 (1957); through *Chem. Abstr.*, 52, 1177 (1958).

4093. Budesinsky, Bydzovsky, Kopecky, and Prikryl, Czech.Pat. 97,788 (1960); through *Chem. Abstr.*, 55, 23568 (1961).

4094. Sunagawa, Nakazawa, and Watatani, Jap.Pat. 2291 (1963); through *Chem. Abstr.*, 59, 11529 (1963).

4095. Nitta, Okui, and Ito, Jap.Pat. 19,808 (1966); through *Chem. Abstr.*, 66, 55500 (1967).

4096. Hoffman-La Roche, Fr.Pat. 1,343,491 (1963); through *Chem. Abstr.*, 64, 3567 (1966).

4097. Korte, Paulus, Vogel, and Weitkamp, *Annalen*, 648, 124 (1961).

4098. Ozaki, *J. Pharm. Soc. Japan*, 80, 1798 (1960).

4099. Johns, *J. Biol. Chem.*, 11, 67 (1912).

4100. Elion, *J. Org. Chem.*, 27, 2478 (1962).

4101. Ueda, Tsuji, Tsutsui, and Kano, Jap.Pat. 11,833 (1962); through *Chem. Abstr.*, 59, 10080 (1963).

4102. Imperial Chemical Industries, Brit.Pat. 876,601 (1959); through *Chem. Abstr.*, 56, 4781 (1962).

4103. Tenor and Kröger, *Chem. Ber.*, 97, 1373 (1964).

4104. Levin, Platonova, and Kukhtin, *Izvest. Akad. Nauk S.S.S.R., Ser. Khim.*, 1964, 1475.

4105. Iida, Shibazaki, and Imaizumi, *Tokyo Yakka Daigaku Kenkyu Nempo*, 11, 38 (1961); through *Chem. Abstr.*, 58, 2450 (1963).

4106. Fox and Van Praag, *J. Org. Chem.*, 26, 526 (1961).

4107. Di Bella and Hennessy, *J. Org. Chem.*, 26, 2017 (1961).

4108. Budesinsky and Perina, Czech.Pat. 88,060 (1958); through *Chem. Abstr.*, 54, 2378 (1960).

4109. Sirakawa, *J. Pharm. Soc. Japan*, 80, 1542 (1960).

4110. Soda, Aka, Nishiide, Watanabe, and Ono, Jap.Pat. 26,409 (1964); through *Chem. Abstr.*, 62, 9148 (1965).

4111. Hoffmann-La Roche, Brit.Pat. 914,417 (1963); through *Chem. Abstr.*, 58, 12579 (1963).

4112. Naito, Hirata, and Kobayashi, Jap.Pat. 7767 (1965); through *Chem. Abstr.*, 63, 1803 (1965).

4113. Yasushige, Tsuruta, Aida, Noda, and Hyoto, Jap.Pat. 28,776 (1964); through *Chem. Abstr.*, 62, 11828 (1965).

4114. Ghozi, Fr.Pat. 1,276,663 (1962); through *Chem. Abstr.*, 57, 13779 (1962).

4115. Hoffmann-La Roche, Brit.Pat. 953,876 (1964); through *Chem. Abstr.*, 61, 8322 (1964).

4116. Yamada and Achiwa, *Chem. Pharm. Bull.*, 9, 119 (1961).

4117. Mengelberg, *Chem. Ber.*, 93, 2230 (1960).

4118. Bretschneider and Richter, Swiss Pat. 391,714 (1965); through *Chem. Abstr.*, 66, 37940 (1967).

4119. Fischer and Fahr, *Annalen*, 651, 64 (1962).

4120. Sen and Gupta, *J. Indian Chem. Soc.*, 39, 129 (1962).

4121. Libermann, Fr.Pat. 1,170,115 (1959); through *Chem. Abstr.*, 54, 21147 (1960).

4122. Takamizawa, Hirai, and Seppa, Jap.Pat. 5711 (1965); through *Chem. Abstr.*, 62, 16268 (1965).

4123. McHattie, Brit.Pat. 1,019,227 (1966); through *Chem. Abstr.*, 64, 14197 (1966).

4124. Sherman and Taylor, *Org. Synth.*, 37, 12 (1957).

4125. Pfleiderer, *Chem. Ber.*, 92, 2468 (1959).

4126. Kalmus and Bergmann, *J. Chem. Soc.*, 1961, 760.

4127. Sprecher and Sprinson, *Biochemistry*, 4, 655 (1965).

4128. Baker and Lourens, *J. Het. Chem.*, 2, 344 (1965).

4129. Brändström, *Acta Chem. Scand.*, 13, 619 (1959).

4130. Langley, Brit.Pat. 845,378 (1960); through *Chem. Abstr.*, 55, 6505 (1961).

4131. Buttini, Galimberti, Gerosa, Kramer, and Melandri, *Boll. Chim. Farm.*, 100, 411 (1961); through *Chem. Abstr.*, 57, 8541.

4132. Chernova and Khokhlov, *J. Gen. Chem.*, 30, 1281 (1960).

4133. Kissman and Weiss, *J. Amer. Chem. Soc.*, 80, 2575 (1958).

4134. Steiger, U.S.Pat. 3,153,033 (1964); through *Chem. Abstr.*, 61, 14584 (1964).

4135. Weng, *Yao Hsueh Hsueh Pao*, 7, 253 (1959); through *Chem. Abstr.*, 54, 14256 (1960).

4136. Libermann, Fr.Pat. addn. 72,202 (1960); through *Chem. Abstr.*, 57, 4680 (1962).

4137. Wellcome Foundation, Brit.Pat. 990,857 (1965); through *Chem. Abstr.*, 63, 4310 (1965).

4138. Birkofer, Ritter, and Kühlthau, *Chem. Ber.*, 97, 934 (1964).

4139. Roy and Kundu, *Nature*, 188, 581 (1960).

4140. Szlompek-Nesteruk, Znajek, and Kazimierczak, *Przemysl. Chem.*, 42, 226 (1963); through *Chem. Abstr.*, 60, 4144 (1964).

4141. Kazimierczak and Kazimierczak, Pol.Pat. 49,065 (1965); through *Chem. Abstr.*, 63, 18111 (1965).

4142. Stoll, Ger.Pat. 938,846 (1956); through *Chem. Abstr.*, 53, 6273 (1959).

4143. Rorig and Wagner, U.S.Pat. 3,214,430 (1965); through *Chem. Abstr.*, 64, 3568 (1966).

4144. Schroeder, U.S.Pat. 3,162,635 (1964); through *Chem. Abstr.*, 62, 13162 (1965).

4145. Kalm, U.S.Pat. 2,931,812 (1960); through *Chem. Abstr.*, 54, 17433 (1960).

4146. Kalm, *J. Org. Chem.*, 26, 2925 (1961).

4147. Searle, Brit.Pat. 1,003,802 (1965); through *Chem. Abstr.*, 64, 5112 (1966).

4148. Ogui, Fujimura, Senda, and Suzuki, Jap.Pat.6282 (1959); through *Chem. Abstr.*, 54, 16471 (1960).

4149. Ogui, Fujimura, Senda, and Suzuki, Jap.Pat.5574 (1959); through *Chem. Abstr.*, 54, 14279 (1960).

4150. Papesch, U.S.Pat. 3,056,781 (1962); through *Chem. Abstr.*, 58, 5702 (1963).

4151. Wagner, U.S.Pat. 3,244,717 (1966); through *Chem. Abstr.*, 65, 10598 (1966).

4152. Goldhahn, *Pharmazie*, 12, 549 (1957); through *Chem. Abstr.*, 52, 10105 (1958).

4153. Bredereck, Effenberger, and Sauter, *Chem. Ber.*, 95, 2049 (1962).

4154. Fikus, Wierzchowski, and Shugar, *Photochem. Photobiol.*, 1, 325 (1962).

4155. Sirakawa, *J. Pharm. Soc. Japan*, 78, 1395 (1958).

4156. Tanabe, Asahi, Nishikawa, Shima, Kuwada, Kanzawa, and Ogata, *Takeda Kenkyusho Nempo*, 22, 133 (1963); through *Chem. Abstr.*, 60, 13242 (1964).

4157. Partenheimer, Ger.Pat. 1,186,466 (1965); through *Chem. Abstr.*, 62, 13159 (1965).

4158. Fink, Adams, and Pfleiderer, *J. Biol. Chem.*, 239, 4250 (1964).

4159. Ueda and Kato, Jap.Pat. 13,826 (1961); through *Chem. Abstr.*, 56, 10170 (1962).

4160. Kazmirowski, Carstens, and Donat, D.D.R.Pat. 25,960 (1963); through *Chem. Abstr.*, 62, 4041 (1965).

4161. Gutorov and Golovchinskaya, *Khim.—Farm. Zhur.*, 1, 23 (1967); through *Chem. Abstr.*, 66, 115689 (1967).

4162. Wojciechowski, Pol.Pat. 42,976 (1960); through *Chem. Abstr.*, 55, 27382 (1961).

4163. Wojciechowski, *Acta Polon. Pharm.*, 18, 409 (1961); through *Chem. Abstr.*, 57, 11193 (1962).

4164. Carstens, Kazmirowski, and Donat, D.D.R.Pat. 23,546 (1962); through *Chem. Abstr.*, 58, 9103 (1963).

4165. Allen, Reynolds, Tinker, and Williams, *J. Org. Chem.*, 25, 361 (1960).

4166. Bredereck, Christmann, Koser, Schellenberg, and Nast, *Chem. Ber.*, 95, 1812 (1962).

4167. Bonvicino and Hennessy, *J. Amer. Chem. Soc.*, 79, 6325 (1957).

4168. Masuda, *J. Pharm. Soc. Japan*, 81, 533 (1961).

4169. Bretschneider, Klötzer, Spiteller, and Dehler, *Monatsh.*, 92, 75 (1961).

4170. Sunagawa, Nakazawa, and Watatani, Jap.Pat.18,745 (1962); through *Chem. Abstr.*, 59, 10075 (1963).

4171. Budesinsky and Perina, Czech.Pat. 88,078 (1958); through *Chem. Abstr.*, 54, 2379 (1960).

4172. Bretschneider and Richter, Swiss Pat. 395,106 (1965); through *Chem. Abstr.*, 64, 15896 (1966).

4173. Merz and Stahl, *Beitr. Biochem. Physiol. Naturstoffen, Festschr.*, 1965, 285; through *Chem. Abstr.*, 65, 3877 (1966).

4174. Ciba, Brit.Pat. 815,977 (1959); through *Chem. Abstr.*, 54, 3464 (1960).

4175. Bayer A.-G., Brit.Pat. 810,846 (1959); through *Chem. Abstr.*, 54, 2378 (1960).

4176. Gold and Bayer, *Chem. Ber.*, 94, 2594 (1961).

4177. Hitchings, Falco, and Roth, U.S.Pat. 2,953,567 (1960); through *Chem. Abstr.*, 55, 4546 (1961).

4178. Sato, Iwashige, and Nishimura, Jap.Pat. 6537 (1962); through *Chem. Abstr.*, 58, 13967 (1963).

4179. Baker and Almaula, *J. Het. Chem.*, 1, 263 (1964).

4180. Duschinsky, Gabriel, Hoffer, Berger, Titsworth, and Grunberg, *J. Med. Chem.*, 9, 566 (1966).

4181. Baker and Morreal, *J. Pharm. Sci.*, 51, 596 (1962).

4182. Korman, U.S.Pat. 2,944,057 (1960); through *Chem. Abstr.*, 55, 587 (1961).

4183. Cilag -Chemie, Brit.Pat. 859,716 (1961); through *Chem. Abstr.*, 55, 17669 (1961).

4184. Lohrmann and Forrest, *J. Chem. Soc.*, 1964, 460.

4185. Chkhivadze, Britikova, and Magidson, *Biol. Aktivn. Soedin; Akad. Nauk S.S.S.R.*, 1965, 16.

4186. Goldner and Trampau, D.D.R.Pat. 39,698 (1965); through *Chem. Abstr.*, 64, 2103 (1966).

4187. Schroeder, U.S.Pat. 3,080,364 (1963); through *Chem. Abstr.*, 59, 6420 (1963).

4188. Geigy A.-G., Belg.Pat. 626,639 (1963); through *Chem. Abstr.*, 60, 8043 (1964).

4189. Ross, Skinner, and Shive, *J. Org. Chem.*, 26, 3582 (1961).

4190. Armour and Co., Brit.Pat. 831,067 (1960); through *Chem. Abstr.*, 54, 19727 (1960).

4191. Mower and Dickinson, *J. Amer. Chem. Soc.*, 81, 4011 (1959).

4192. West and Beauchamp, *J. Org. Chem.*, 26, 3809 (1961).

4193. Mizuno, Ikehara, Watanabe, and Suzaki, *Chem. Pharm. Bull.*, 11, 1091 (1963).

4194. Okuma, *J. Antibiotics*, 14A, 343 (1961); through *Chem. Abstr.*, 58, 4557 (1963).

4195. Fischer, Roch, and Neumann, Ger.Pat. 927,631 (1955); through *Chem. Abstr.*, 52, 3874 (1958).

4196. Bergmann, Cohen, and Shahak, *J. Chem. Soc.*, 1959, 3278.

4197. Hoffmann-La Roche, Brit.Pat. 806,584 (1958); through *Chem. Abstr.*, 53, 16171 (1959).

4198. Dyer, Reitz, and Farris, *J. Med. Chem.*, 6, 289 (1963).

4199. Karlinskaya, Khromov-Borisov, and Ivanova, *J. Gen. Chem.*, 34, 3734 (1964).

4200. Craveri and Zoni, *Chimica*, 33, 473 (1957); through *Chem. Abstr.*, 52, 11082 (1958).

4201. Fujimoto and Teranishi, Jap.Pat. 15,038 (1965); through *Chem. Abstr.*, 63, 13285 (1965).

4202. Abramova and Khmelevskii, *Med. Prom. S.S.S.R.*, 18, 35 (1964); through *Chem. Abstr.*, 61, 1865 (1964).

4203. Spickett and Wright. *J. Chem. Soc.(C)*, 1967, 498.

4204. Sirakawa, *J. Pharm. Soc. Japan*, 80, 952 (1960).

4205. Cilag—Chemie, Belg.Pat. 612,974 (1962); through *Chem. Abstr.*, 58, 1473 (1963).

4206. Habicht, Swiss Pat. 358,426 (1962); through *Chem. Abstr.*, 58, 3443 (1963).

4207. Druey, Schmidt, Eichenberger, and Wilhelm, Ger.Pat. 1,088,503 (1960); through *Chem. Abstr.*, 56, 4782 (1962).

4208. Fischer, Eilingsfeld, and Neumann, *Annalen*, 651, 49 (1962).

4209. Pfleiderer and Schündehütte, *Annalen*, 615, 42 (1958).

4210. Habicht, Swiss Pat. 358,425 (1962); through *Chem. Abstr.*, 58, 3443 (1963).

4211. Nesterov and Chubova, U.S.S.R.Pat. 119,879 (1959); through *Chem. Abstr.*, 54, 2375 (1960).

4212. Nesterov, Semikolennykh, Zhukova, Ivanova, and Katsnel'son, *Med. Prom. S.S.S.R.*, 20, 12 (1966); through *Chem. Abstr.*, 65, 709 (1966).

4213. Carstens, Kazmirowski, and Donat, D.D.R.Pat. 23,487 (1962); through *Chem. Abstr.*, 58, 9103 (1963).

4214. Sakuragi, *Arch. Biochem. Biophys.*, 74, 362 (1958).

4215. Cottis and Tieckelmann, *J. Org. Chem.*, 26, 79 (1961).

4216. Blank and Caldwell, *J. Org. Chem.*, 24, 1137 (1959).

4217. Schmidt, Eichenberger, and Wilhelm, *Angew. Chem.*, 73, 15 (1961).

4218. Marchal and Promel, *Bull. Soc. chim. belges*, 66, 406 (1957).

4219. Okuda and Price, *J. Org. Chem.*, 24, 14 (1959).

4220. Wellcome Foundation, Brit.Pat. 951,431 (1964); through *Chem. Abstr.*, 60, 14520 (1964).

4221. Bloch, *Ann. Chim. (France)*, 10, 583 (1965).

4222. Heidelberger and Duschinsky, U.S.Pat. 2,802,005 (1957); through *Chem. Abstr.*, 52, 2100 (1958).

4223. Hauser, Peters, and Tieckelmann, *J. Org. Chem.*, 25, 1570 (1960).

4224. Nemeryuk and Safonova, *Khim. Geterotsikl. Soedin., Akad. Nauk Latv. S.S.R.*, 1966, 470; through *Chem. Abstr.*, 65, 8903 (1966).

4225. Liau, Yamashita, and Matsui, *Agric. Biol. Chem.*, 26, 624 (1962); through *Chem. Abstr.*, 59, 3921 (1963).

4226. Papesch, U.S.Pat. 3,190,881 (1965); through *Chem. Abstr.*, 63, 9964 (1965).

4227. Kopecky and Michl, Czech.Pat. 97,821 (1960); through *Chem. Abstr.*, 56, 4781 (1962).

4228. Okuda and Price, *J. Org. Chem.*, 22, 1719 (1957).

4229. Sunagawa and Watatani, Jap.Pat. 24,970 (1964); through *Chem. Abstr.*, 62, 11826 (1965).

4230. Gewald, *J. prakt. Chem.*, 32, 26 (1966).

4231. Kodak, Belg.Pat. 587,287 (1960); through *Chem. Abstr.*, 56, 2320 (1961).

4232. Schueler, Grabhoefer, and Ulrich, Ger.Pat. 1,163,142 (1964); through *Chem. Abstr.*, 60, 14518 (1964).

4233. Carstens, Kazmirowski, and Donat, Ger.Pat. 1,130,812 (1962); through *Chem. Abstr.*, 57, 13781 (1962).

4234. Papesch, U.S.Pat. 3,299,066 (1967); through *Chem. Abstr.*, 66, 76028 (1967).

4235. Leonard and Laursen, *J. Org. Chem.*, 27, 1778 (1962).

4236. Hooper and Bray, *J. Chem. Phys.*, 30, 957 (1959).

4237. Karlinskaya and Khromov-Borisov, *J. Gen. Chem.*, 30, 1485 (1960).

4238. Woenckhaus, *Chem. Ber.*, 97, 2439 (1964).

4239. Okumura, Kusaka, and Takematsu, *Bull. Chem. Soc. Japan*, 33, 1471 (1960).

4240. Goldner, *Z. Chem.*, 1, 16 (1961).

4241. Andrew and Poole, U.S.Pat. 3,097,910 (1963); through *Chem. Abstr.*, 62, 1768 (1965).

4242. Nakazawa and Watatani, *Ann. Rep. Takamme Lab.*, 12, 37 (1960); through *Chem. Abstr.*, 55, 6491 (1961).

4243. Hrehorowicz, Pol.Pat. 48,293 (1964); through *Chem. Abstr.*, 62, 7775 (1965).

4244. Novacek and Hedrlin, *Coll. Czech. Chem. Comm.* 32, 1045 (1967).

4245. Vul'fson and Zhurin, *J. Gen. Chem.*, 31, 281 (1961).

4246. Whitehead and Traverso, *J. Amer. Chem. Soc.*, 80, 2182 (1958).

4247. Prelicz, *Diss. Pharm. Pharmacol.*, 18, 31 (1966); through *Chem. Abstr.*, 65, 3868 (1966).

4248. Takamizawa and Hirai, Jap.Pat. 2264 (1967); through *Chem. Abstr.*, 66, 95073 (1967).

4249. Chambers, *Biochemistry*, 4, 219 (1965).

4250. Lyuminarskii, Sochilin, and Ivin, *Zhur. Organ. Khim.*, 1, 1335 (1965); through *Chem. Abstr.*, 63, 13255 (1965).

4251. Shvachkin, Berestenko, and Mishin, *Biol. Aktivn. Soedin., Akad. Nauk S.S.S.R.*, 1965, 72; through *Chem. Abstr.*, 64, 9720 (1966).

4252. Lemieux and Puskas, *Canad. J. Chem.*, 42, 2909 (1964).

4253. Scarborough, *J. Org. Chem.*, 26, 3717 (1961).

4254. Scarborough and Gould, *J. Org. Chem.*, 26, 3720 (1961).

4255. Chkhikvadze, Britikova, and Magidson, *Biol. Aktivn. Soedin., Akad. Nauk S.S.S.R.*, 1965, 22; through *Chem. Abstr.*, 63, 18081 (1965).

4256. Rhone-Poulenc, Brit.Pat. 800,709 (1958); through *Chem. Abstr.*, 53, 16172 (1959).

4257. Drehmann and Born, *J. prakt. Chem.*, 5, 200 (1957).

4258. Pichat, Audinot, and Carbonnier, *Bull. Soc. chim. France*, 1959, 1798.

4259. Chkhikvadze and Magidson, *Med. Prom. S.S.S.R.*, 14, No 10, 24 (1960); through *Chem. Abstr.*, 55, 6487 (1961).

4260. Heidelberger, U.S.Pat. 3,201,387 (1965); through *Chem. Abstr.*, 63, 16447 (1965).

4261. Curran and Angier, *J. Org. Chem.*, 31, 201 (1966).

4262. Gut, Moravek, Parkanyi, Prystas, Skoda, and Sorm, *Coll. Czech. Chem. Comm.*, 24, 3154 (1959).

4263. Valcavi and Gaudenzi, *Farmaco, Ed. Sci.*, 18, 847 (1963); through *Chem. Abstr.*, 60, 4142 (1964).

4264. Santilli, Bruce, and Osdene, *J. Med. Chem.*, 7, 68 (1964).

4265. Jones, *J. Org. Chem.*, 25, 956 (1960).

4266. Engelbrecht and Lenke, D.D.R.Pat. 19,629 (1960); through *Chem. Abstr.*, 55, 22346 (1961).

4267. Juby, U.S.Pat. 3,300,496 (1967); through *Chem. Abstr.*, 66, 115722 (1967).

4268. Huisgen, Morikawa, Herbig, and Brunn, *Chem. Ber.*, 100, 1094 (1967).

4269. Fel'dman, Berlin, and Semicheva, *Mechenye Biol. Veshchestva, Sb. Statei*, No 2, 43 (1966); through *Chem. Abstr.*, 66, 65800 (1967).

4270. Dewar and Shaw, *J. Chem. Soc.*, 1962, 583.

4271. Hoffer, *Chem. Ber.*, 93, 2777 (1960).

4272. Ueda and Kato, Jap.Pat. 14,625 (1961); through *Chem. Abstr.*, 56, 10170 (1962).

4273. Paegle, Lidaks, and Shvachkin, *Khim. Geterotsikl. Soedin., Akad. Nauk Latv. S.S.R.*, 1966, 316; through *Chem. Abstr.*, 65, 2256 (1966).

4274. Goerdeler and Wieland, *Chem. Ber.*, 100, 47 (1967).

4275. Signor, Scoffone, Biondi, and Bezzi, *Gazzetta*, 93, 65 (1963).

4276. Bogodist, U.S.S.R.Pat. 178,383 (1966); through *Chem. Abstr.*, 65, 2278 (1966).

4277. Wunderlich, D.D.R.Pat. 44,121 (1965); through *Chem. Abstr.*, 64, 19633 (1966).

4278. Inman, Oesterling, and Tyczkowski, *J. Amer. Chem. Soc.*, 80, 6533 (1958).

4279. Moravek and Filip, *Coll. Czech. Chem. Comm.*, 25, 2697 (1960).

4280. Chang, Kim, Park, and Hahn, *J. Korean Chem. Soc.*, 9, 29 (1965); through *Chem. Abstr.*, 64, 15876 (1966).

4281. Luckenbaugh and Soboczenski, Brit.Pat. 968,666 (1964); through *Chem. Abstr.*, 61, 13327 (1964).

4282. Budesinsky, Jelinek, and Prikryl, Czech.Pat. 107,166 (1963); through *Chem. Abstr.*, 60, 9291 (1964).

4283. Budesinsky, Prikryl, Svab, and Jelinek, Czech.Pat. 108,806 (1963); through *Chem. Abstr.*, 61, 5664 (1964).

4284. McNutt, *J. Amer. Chem. Soc.*, 82, 217 (1960).

4285. Hoffmann-La Roche, Fr.Pat. 1,341,774 (1963); through *Chem. Abstr.*, 60, 10698 (1964).

4286. Gauri, Ger.Pat. 1,232,153 (1967); through *Chem. Abstr.*, 66, 95080 (1967).

4287. Takeda, Jap.Pat. 19,143 (1963); through *Chem. Abstr.*, 60, 2965 (1964).

4288. Strauss, Ger.Pat. 1,141,287 (1962); through *Chem. Abstr.*, 58, 12578 (1963).

4289. Slezak, U.S.Pat. 3,002,975 (1958); through *Chem. Abstr.*, 56, 1462 (1962).

4290. Karlinskaya and Khromov-Borisov, *J. Gen. Chem.*, 30, 899 (1960).

4291. Fel'dman, Berlin, Nurova, Kuznetsova, Alekseeva, Lutts, Chkhikvadze, and Magidson, *Med. Prom. S.S.S.R.*, 19, 12 (1965); through *Chem. Abstr.*, 62, 16239 (1965).

4292. Bergmann, Cohen, and Shahak, *J. Chem. Soc.*, 1959, 3286.

4293. Fikus, Wierzchowski, and Shugar, *Biochem. Biophys. Res. Comm.*, 16, 478 (1964).

4294. Ryan, Acton, and Goodman, *J. Org. Chem.*, 31, 1181 (1966).

4295. Gauri, Ger.Pat. 1,139,505 (1962); through *Chem. Abstr.*, 58, 7952 (1963).

4296. Thompson and Wallick, U.S.Pat. 3,274,196 (1966); through *Chem. Abstr.*, 65, 16981 (1966).

4297. Gauri, Ger.Pat. 1,185,617 (1965); through *Chem. Abstr.*, 62, 13159 (1965).

4298. Seitz, U.S.Pat. 3,086,020 (1963); through *Chem. Abstr.*, 59, 10084 (1963).

4299. Senda, Izumi, Kano, and Tsubota, *Gifu Yakka Daigaku Kiyo*, 11, 62 (1961); through *Chem. Abstr.*, 57, 11194 (1962).

4300. Langley, Brit.Pat. 890,076 (1962); through *Chem. Abstr.*, 57, 4683 (1962).

4301. Goldner, Dietz, and Carstens, *Annalen*, 699, 145 (1966).

4302. Elderfield and Prasad, *J. Org. Chem.*, 26, 3863 (1961).

4303. Budesinsky, Bydzovsky, Kopecky, and Prikryl, Czech.Pat. 109,263 (1963); through *Chem. Abstr.*, 60, 15889 (1964).

4304. Bardos, Kotick, and Szantay, *Tetrahedron Letters*, 1966, 1759.

4305. Dinan, Minnemeyer, and Tieckelmann, *J. Org. Chem.*, 28, 1015 (1963).

4306. Senda and Fujimura, Jap.Pat. 8177 (1961); through *Chem. Abstr.*, 58, 13968 (1963).

4307. Benn, Chatamra, and Jones, *J. Chem. Soc.*, 1960, 1014.

4308. Martello, Lemetre, and Nobile, Ital.Pat. 548,752 (1956); through *Chem. Abstr.*, 53, 4305 (1959).

4309. Ikehara, Ueda, and Ikeda, *Chem. Pharm. Bull.*, 10, 767 (1962).

4310. Aspelund, *Acta Acad. Aboensis, Math. et Phys.*, 21, No 10, p.20 (1958); through *Chem. Abstr.*, 53, 21991 (1959).

4311. Whitehead and Traverso, *J. Amer. Chem. Soc.*, 80, 962 (1958).

4312. Kodak, Belg.Pat. 561,108 (1957); through *Chem. Abstr.*, 54, 4220 (1960).

4313. Shvachkin, Shprunka, and Kazakara, *J. Gen. Chem.*, 34, 3846 (1964).

4314. Ciba, Neth.Pat. Appl. 6,506,572 (1965); through *Chem. Abstr.*, 64, 19634 (1966).

4315. Fedrick, Shepherd, and Svokos, Fr.Pat. M2325 (1964); through *Chem. Abstr.*, 61, 667 (1964).

4316. Shepherd, *J. Chem. Soc.*, 1964, 4410.

4317. Mittra and Rout, *J. Indian Chem. Soc.*, 40, 993 (1963); through *Chem. Abstr.*, 60, 11501 (1964).

4318. Salvesen, *Medd. Norsk Farm. Selskap*, 25, 1 (1963); through *Chem. Abstr.*, 59, 3923 (1963).

4319. Bogodist and Protsenko, *Ukr. Khim. Zhur.*, 31, 1309 (1965); through *Chem. Abstr.*, 64, 12670 (1966).

4320. Geigy A.-G., Fr.Pat. 1,339,747 (1963); through *Chem. Abstr.*, 60, 5518 (1964).

4321. Duschinsky and Heidelberger, U.S.Pat. 2,948,725 (1960); through *Chem. Abstr.*, 55, 17663 (1961).

4322. Shvachkin, Rapanovich, and Azarova, *Vestn. Mosk. Univ., Ser.II, Khim.*, 20, 73 (1965).

4323. Ivin and Nemets, *J. Gen. Chem.*, 34, 4120 (1964).

4324. Hoffmann-La Roche, Neth.Pat. Appl. 6,514,743 (1966); through *Chem. Abstr.*, 65, 10597 (1966).

4325. Hoffmann-La Roche, Brit.Pat. 875,971 (1961); through *Chem. Abstr.*, 57, 15221 (1962).

4326. Amiard and Torelli, Fr.Pat. 1,336,866 (1963); through *Chem. Abstr.*, 60, 3082 (1964).

4327. Grant, von Seeman, and Winthrop, U.S.Pat. 2,876,224 (1959); through *Chem. Abstr.*, 53, 16170 (1959).

4328. Schmidt and Wilhelm, U.S.Pat. 3,014,035 (1961); through *Chem. Abstr.*, 56, 14304 (1962).

Sincere apologies are proffered to French, Czech, and other authors whose names appear in the above references without proper accentuation: appropriate accent marks were unavailable on the English-language typewriter used to prepare the camera-copy.

Index

This index covers the text and all interspersed tables with the exception of Table XVI (The pK_a Values of Some Pyrimidines in Water). It and the appendix tables, XX–LVII, are excluded entirely as already being indices in themselves.

As before (*H* 677), the page numbers immediately following primary entries refer to syntheses or general information. A number in parentheses indicates that although the subject is treated on that page the actual name of the primary entry appears there only in part, as a synonym, or even simply by inference. The letter *f* after a page number indicates that the subject is treated on that and the following page or pages.